Thermodynamic Models for Industrial Applications

Thermodynamic Models for Industrial Applications

From Classical and Advanced Mixing Rules to Association Theories

GEORGIOS M. KONTOGEORGIS
Technical University of Denmark, Lyngby, Denmark

GEORGIOS K. FOLAS
Shell Global Solutions, The Netherlands

WILEY
A John Wiley and Sons, Ltd, Publication

This edition first published 2010

Registered office
John Wiley & Sons Ltd, The Atrium, Southern Gate, Chichester, West Sussex, PO19 8SQ, United Kingdom

For details of our global editorial offices, for customer services and for information about how to apply for permission to reuse the copyright material in this book please see our website at www.wiley.com.

Library of Congress Cataloging-in-Publication Data

Kontogeorgis, Georgios M.
Thermodynamic models for industrial applications : from classical and advanced mixing rules to association theories / Georgios M. Kontogeorgis, Georgios K. Folas.
p. cm.
Includes bibliographical references and index.
ISBN 978-0-470-69726-9 (cloth)
1. Thermodynamics–Industrial applications. 2. Chemical engineering. I. Kontogeorgis, Georgios M. II. Folas, Georgios K. III. Title.
TP155.2.T45K66 2010
660'.2969–dc22

2009028762

A catalogue record for this book is available from the British Library.

ISBN: 978-0-470-69726-9 (Cloth)

Set in 10/12 pt, Times Roman by Thomson Digital, Noida, India

No man lives alone and no books are written in a vacuum either. Our families especially (in Denmark, The Netherlands and Greece) have deeply felt the consequences of the process of writing this book.

I (Georgios Kontogeorgis) would like to dedicate the book to my wife Olga for her patience, support, love and understanding – especially as, during the period of writing of this book, our daughter, Elena, was born.

I (Georgios Folas) would like to thank Georgios Kontogeorgis for our excellent collaboration in writing this monograph during the past two years. I am grateful to my family and wish to dedicate this book to my wife Athanasia for always inspiring and supporting me.

Contents

Preface

Thermodynamics plays an important role in numerous industries, both in the design of separation equipment and processes as well as for product design and optimizing formulations. Complex polar and associating molecules are present in many applications, for which different types of phase equilibria and other thermodynamic properties need to be known over wide ranges of temperature and pressure. Several applications also include electrolytes, polymers or biomolecules. To some extent, traditional activity coefficient models are being phased out, possibly with the exception of UNIFAC, due to its predictive character, as advances in computers and statistical mechanics favor use of equations of state. However, some of these 'classical' models continue to find applications, especially in the chemical, polymer and pharmaceutical industries. On the other hand, while traditional cubic equations of state are often not adequate for complex phase equilibria, over the past 20–30 years advanced thermodynamic models, especially equations of state, have been developed.

The purpose of this work is to present and discuss in depth both 'classical' and novel thermodynamic models which have found or can potentially be used for industrial applications. Following the first introductory part of two short chapters on the fundamentals of thermodynamics and intermolecular forces, the second part of the book (Chapters 3–6) presents the 'classical' models, such as cubic equations of state, activity coefficient models and their combination in the so-called EoS/G^E mixing rules. The advantages, major applications and reliability are discussed as well as the limitations and points of caution when these models are used for design purposes, typically within a commercial simulation package. Applications in the oil and gas and chemical sectors are emphasized but models suitable for polymers are also presented in Chapters 4–6.

The third part of the book (Chapters 7–14) presents several of the advanced models in the form of association equations of state which have been developed since the early 1990s and are suitable for industrial applications. While many of the principles and applications are common to a large family of these models, we have focused on two of the models (the CPA and PC–SAFT equations of state), largely due to their range of applicability and our familiarity with them. Extensive parameter tables for the two models are available in the two appendices on the companion website at www.wiley.com/go/Kontogeorgis. The final part of the book (Chapters 15–20) illustrates applications of thermodynamics in environmental science and colloid and surface chemistry and discusses models for mixtures containing electrolytes. Finally, brief introductions about the thermodynamic tools available for mixtures with biomolecules as well as the possibility of using quantum chemistry in engineering thermodynamics conclude the book.

The book is based on our extensive experience of working with thermodynamic models, especially the association equations of state, and in close collaboration with industry in the petroleum, energy, chemical and polymer sectors. While we feel that we have included several of the exciting developments in thermodynamic models with an industrial flavor, it has not been possible to include them all. We would like, therefore, to apologize in advance to colleagues and researchers worldwide whose contributions may not have been included or adequately discussed for reasons of economy. However, we are looking forward to receiving comments and suggestions which can lead to improvements in the future.

The book is intended both for engineers wishing to use these models in industrial applications (many of them already available in commercial simulators, as stand-alone or in CAPE-Open compliant format) and for students, researchers and academics in the field of applied thermodynamics. The contents could also be used in graduate courses on applied chemical engineering thermodynamics, provided that a course on the fundamentals of applied thermodynamics has been previously followed. For this reason, problems are provided on the companion website at www.wiley.com/go/Kontogeorgis. Answers to selected problems are available, while a full solution manual is available from the authors.

Georgios M. Kontogeorgis
Copenhagen, Denmark

Georgios K. Folas
Amsterdam, The Netherlands

About the Authors

Georgios M. Kontogeorgis has been a professor at the Technical University of Denmark (DTU), Department of Chemical and Biochemical Engineering, since January 2008. Prior to that he was associate professor at the same university, a position he had held since August 1999. He has an MSc in Chemical Engineering from the Technical University of Athens (1991) and a PhD from DTU (1995). His current research areas are energy (especially thermodynamic models for the oil and gas industry), materials and nanotechnology (especially polymers – paints, product design, and colloid and surface chemistry), environment (design CO_2 capture units, fate of chemicals, migration of plasticizers) and biotechnology. He is the author of over 100 publications in international journals and co-editor of one monograph. He is the recipient of the Empirikion Foundation Award for 'Achievements in Chemistry' (1999, Greece) and of the Dana Lim Price (2002, Denmark).

Georgios K. Folas was appointed as technologist in the distillation and thermal conversion department, Shell Global Solutions (The Netherlands) in January 2009. He previously worked as Senior Engineer (Facilities and Flow Assurance) in Aker Engineering & Technology AS (Oslo, Norway). He has an MSc in Chemical Engineering from the Technical University of Athens (2000) and an industrial PhD from DTU (2006), in collaboration with Statoilhydro (Norway). He is the author of 15 publications in international journals and the recipient of the Director Peter Gorm-Petersens Award for his PhD work.

Acknowledgments

We wish to thank all our students and colleagues and especially the faculty members of IVC-SEP Research Center, at the Department of Chemical and Biochemical Engineering of the Technical University of Denmark (DTU), for the many inspiring discussions during the past 10 years which have largely contributed to the shaping of this book. Our very special thanks go to Professor Michael L. Michelsen for the endless discussions we have enjoyed with him on thermodynamics.

In the preparation of this book we have been assisted by many colleagues, friends, current and former students. Some have read chapters of the book or provided material prior to publication, while we have had extensive discussions with others. We would particularly like to thank Professors J. Coutinho, G. Jackson, I. Marrucho, J. Mollerup, G. Sadowski, L. Vega and N. von Solms, Doctors M. Breil, H. Cheng, Ph. Coutsikos, J.-C. de Hemptinne, I. Economou, J. Gabrielsen, A. Grenner, E. Karakatsani I. Kouskoumvekaki, Th. Lindvig, E. Solbraa, N. Sune, A. Tihic, I. Tsivintzelis and W. Yan, as well as the current PhD and MSc students of IVC-SEP, namely A. Avlund, J. Christensen, L. Faramarzi, F. Leon, B. Maribo-Mogensen and A. Sattar-Dar.

All contributions have been highly valuable and we are deeply grateful for them.

List of Abbreviations

AAD %	percentage average absolute deviation: $AAD\,\% = \frac{1}{NP}\sum_{i=1}^{NP} ABS\left(\frac{x_{\exp,i}-x_{calc,i}}{x_{\exp,i}}\right)\cdot 100$ for a property x
AM	arithmetic mean rule (for the cross co-volume parameter, b_{12})
AMP	2-amino-2-methyl-1-propanol
ATPS	aqueous two-phase systems
BCF	bioconcentration factor
BR	butadiene rubber (polybutadiene)
BTEX	benzene–toluene–ethylbenzene–xylene
CCC	critical coagulation concentration
CDI	chronic daily intake
CK–SAFT	Chen–Kreglewski SAFT
CMC	critical micelle concentration
Comb-FV	combinatorial free volume (effect, term, contributions)
COSMO	conductor-like screening model
CPA	cubic-plus-association
CPP	critical packing parameter
CS	Carnahan–Starling
CSP	corresponding states principle
CTAB	hexadecyl trimethylammonium bromide
DBE	dibutyl ether
DDT	dichlorodiphenyltrichloroethane
DEA	diethanolamine
DEG	diethylene glycol
DFT	density functional theory
DH	Debye–Hückel
DiPE	diisopropyl ether
DIPPR	Design Institute for Physical Property (database)
DLVO	Derjaguin–Landau–Verwey–Overbeek (theory)
DME	dimethyl ether
DPE	dipropyl ether
ECR	Elliott's combining rule
EoS	Equation of state
EPA	Environmental Protection Agency
EPE	ethyl propyl ether
ESD	Elliott–Suresh–Donohue (EoS)
EU	European Union
FCC	Face-centered cubic structure (close packed, $Z = 12$)

FH	Flory–Huggins
FOG	first-order groups
FV	Free volume
GC	group contribution (methods, principle)
GCA	group contribution plus association
GCVM	group contribution of Vidal and Michelsen mixing rules
GERG	Group Européen de Recherche Gazière
GLC	gas–liquid chromatography
GLE	gas–liquid equilibria
GM	geometric mean rule (for the cross-energy parameter, a_{12})
HB	hydrogen bonds/bonding
HCB	hexachlorobenzene
HF	Hartree–Fock
HIC	hydrophobic interaction chromatography
HLB	hydrophilic–lipophilic balance
HSP	Hansen solubility parameters
HV	Huron–Vidal mixing rule
IEC	ion-exchange chromatography
LALS	low-angle light scattering
LC	local composition (models, principle, etc.)
LCST	lower critical solution temperature
LCVM	linear combination of Vidal and Michelsen mixing rules
LGT	linear gradient theory
LJ	Lennard-Jones
LLE	liquid–liquid equilibria
LR	Lewis–Randall; long range
mCR-1	modified CR-1 combining rule (for the CPA EoS), equation (9.10)
MC–SRK	Mathias–Copeman SRK
MDEA	methyl diethanolamine
MEA	monoethanolamine
MEG	(mono)ethylene glycol
MEK	methyl ethyl ketone
MHV1	modified Huron–Vidal first order
MHV2	modified Huron–Vidal second order
MM	McMillan–Mayer
MO	molecular orbital
MSA	mean spherical approximation
MW	molecular weight
NLF–HB	lattice–fluid hydrogen bonding (EoS)
NP	number of experimental points
NRHB	non-random hydrogen bonding (EoS)
NRTL	non-random two liquid
PAHs	polynuclear aromatic hydrocarbons
PBA	poly(butyl acrylate)
PBD	polybutadiene
PBMA	poly(butyl methacrylate)
PCBs	polychlorinated biphenyls

PC–SAFT	perturbed-chain SAFT
PDH	Pitzer–Debye–Hückel
PDMS	poly(dimethyl siloxane)
PEA	poly(ethyl acrylate)
PEG	(poly)ethylene glycol
PIB	polyisobutylene
PIPMA	poly(isopropyl methacrylate)
PM	primitive model
PMA	poly(methyl acrylate)
PMMA	poly(methyl methacrylate)
PP	polypropylene
PPA	poly(propyl acrylate)
PR	Peng–Robinson
PS	polystyrene
PSRK	predictive Soave–Redlich–Kwong
PVAc	poly(vinyl acetate)
PVAL	poly(vinyl alcohol)
PVC	poly(vinyl chloride)
PVT	pressure, volume, temperature
PZ	piperazine
QC	quantum chemistry
QM	quantum mechanics
QSAR	quantitative structure–activity relationships
RDF	radial distribution function
RK	Redlich–Kwong
RP-HPLC	reversed-phase high-pressure liquid chromatography
RPM	restrictive primitive model
RST	regular solution theory
SAFT	statistical associating fluid theory
SCFE	supercritical fluid extraction
SDS	sodium dodecyl sulfate
SGE	solid–gas equilibria
SL	Sanchez–Lacombe
SOG	second-order groups
SLE	solid–liquid equilibria
SR	short range
SRK	Soave–Redlich–Kwong (EoS)
SVC	second virial coefficients
SWP	Sako–Wu–Prausnitz (EoS)
TEG	triethylene glycol
THF	tetrahydrofurane
UCST	upper critical solution temperature
UMR–PR	universal mixing rule (with the PR EoS)
UNIFAC	universal quasi-chemical functional group activity coefficient
UNIQUAC	universal quasi-chemical
vdW	van der Waals (EoS)
vdW1f	vdW one-fluid (mixing rules)

VLE	vapor–liquid equilibria
VLLE	vapor–liquid–liquid equilibria
VOR	volatile organic compound
VR	variable range
VTPR	volume-translated Peng–Robinson (EoS)
WHO	World Health Organization
WS	Wong–Sandler
WWF	World Wide Fund for Nature
$\Delta P\%$	average absolute percentage error: $\Delta P\% = \frac{1}{NP}\sum_{i=1}^{NP} ABS\left(\frac{P_{exp,i}-P_{calc,i}}{P_{exp,i}}\right)\cdot 100$ in bubble point pressure P of component i
Δy	average absolute percentage deviation: $\Delta y = \frac{1}{NP}\sum_{i=1}^{NP} ABS\left(y_{exp,i}-y_{calc,i}\right)$ in the vapor phase mole fraction of component i
$\Delta\rho\%$	average absolute percentage deviation: $\Delta\rho\% = \frac{1}{NP}\sum_{i=1}^{NP} ABS\left(\frac{\rho_{exp,i}-\rho_{calc,i}}{\rho_{exp,i}}\right)\cdot 100$ in the liquid density of component i

List of Symbols

a	energy term in the SRK term (bar l^2/mol^2) *or* activity *or* particle radius
a_0	surfactant head area
a_{ij}	non-randomness parameter of molecules of type i around a molecule of type j
a_{mk}, $a_{mk,1}$, $a_{mk,2}$, $a_{mk,3}$	UNIFAC temperature-dependent parameters, K
A	surface area *or* Helmholtz energy *or* Hamaker constant
A_{eff}	effective Hamaker constant
A_i	site A in molecule i
A_{ii}	Hamaker constant of particle/surface i–i
$A_{m,i}$	parameter in Langmuir constant, K/bar
A_{spec}	specific surface area, typically in m^2/g
A_0	area occupied by a gas molecule
$\tilde{a}$	reduced Helmholtz energy
a_0	parameter in the energy term of CPA (bar L^2/mol^2) *or* area of the head of a surfactant molecule
A_1, A_2, A_3	parameters in GERG model for water
A_{123}	Hamaker constant between particles (or surfaces) 1 and 3 in medium 2
b	co-volume parameter (l/mol) of cubic equations of state
B	second virial coefficient
B_j	site B in molecule j
$B_{m,i}$	parameter in Langmuir constant, K
C	molar concentration (often in mol/l or mol/m^3) *or* concentration (in general) *or* the London coefficient
c_1	parameter in the energy term of CPA
$C_{m,i}$	Langmuir constant for component i in cavity m
d	density (eq. 4.29) *or* temperature-dependent diameter
D	Diffusion coefficient *or* dielectric constant
E	modulus of Elasticity
f	fugacity, bar
f	fugacity, bar
F	Force
G	Gibbs energy
G^E, g^E	excess Gibbs energy
g_{ji}/R	Huron–Vidal energy parameter, characteristic of the $j-i$ interaction, K
g	radial distribution function
h	Planck's constant, 6.626×10^{-34} J s
H	enthalpy
H	interparticle or interface distance *or* (H_i) Henry's law constant
I	first ionization potential, J *or* ionic strength

k	Boltzmann's constant, J/K
K_i	Distribution factor e.g. Table 1.3
K	chemical equilibrium constant
k_{12}, k_{ij}	binary interaction parameter (in equations of state)
K_{OW}	octanol–water partition coefficient
K^{ref}	chemical equilibrium constant at the reference temperature
l	parameter in the Hansen–Beerbower–Skaarup equation (eq. 18.8) *or* distance between charges in a molecule (eq. 2.2a or 2.2b)
l_c	length of a surfactant molecule
m	segment number *or* molality
MW, M	molecular weight (molar mass)
N_A	Avogadro's number $= 6.0225 \times 10^{23}$ mol/mol
N_{agg}	aggregation (or aggregate) number
n	refractive index
n_T	true number of moles
n_o	apparent number of moles
P	pressure, bar
P^{sat}	saturated vapor pressure
q	charge
Q	quadrupole moment, C m^2
Q_k	surface area parameter for group k
Q_w	van der Waals surface area
R	gas constant, bar l/mol/K *or* molecular radius
r	radial distance from the center of the cavity, Å *or* intermolecular distance
R_i	the radius of cage i, Å
R_k	volume parameter for group k
S	Harkins spreading coefficient *or* entropy
T	temperature, K
T_c	critical temperature, K
$T_{m,i}$	melting temperature of the component i, K
T_r	reduced temperature
T^{ref}	reference temperature, K
T_0	arbitrary temperature for linear UNIFAC (in the temperature dependency of the energy parameters), see Table 5.7
U	composition variable *or* internal energy
V_A	(van der Waals) potential energy
$\tilde{V}$	reduced volume
V^*	hard-core volume
V	volume
V_c	critical volume
V_f	free volume
V_g	gas volume at STP conditions (= 22 414 cm^3/mol)
$\bar{V}_i$	partial molar volume
V_m	molar volume (L mol^{-1}) *or* maximum volume occupied by a gas (in adsorption in a solid)
V_W^{ICE}	molar volume of ice, l mol^{-1}
V_w	van der Waals volume
$W(r)$	cell potential function, J

X	monomer fraction
X_{A_i}	fraction of A-sites of molecule *i* that are not bonded
x_i	liquid mole fraction of component *i*
y	reduced density, eq. 2.11 *or* 9.12
y_i	vapor mole fraction of component *i*
Z	compressibility factor *or* co-ordination number
Z_i	ionic valence
ΔCp_i	heat capacity change of the component *i* at the melting temperature, J/mol/K
ΔG	Gibbs free energy change (also of micellization)
ΔH	enthalpy change (also of micellization)
$\Delta h_w^{EH-L_0}$	enthalpy differences between the empty hydrate lattice and liquid water, J/mol
ΔH_i^{fus}	heat of fusion of the component *i* at the melting temperature, J/mol
$\Delta\mu_w^0$	chemical potential difference between the empty hydrate and pure liquid water, J/mol
ΔS	entropy change (also of micellization)
$\Delta V_w^{EH-L_0}$	molar volume differences between the empty hydrate lattice and liquid water, J/mol

Greek letters

α_0	electronic polarizability
α	polarizability *or* Kamlet acid parameter *or* distance of closest approach (Chapter 15)
α	reduced energy $(=\frac{\alpha}{bRT})$, eq. (3.16) & Table 6.3
β	Kamlet base parameter
$\beta^{A_iB_j}$	association volume parameter between site *A* in molecule *i* and site *B* in molecule *j* (dimensionless) [in CPA]
γ	mole-based activity coefficient *or* surface *or* interfacial tension
γ_i^C	combinatorial part of activity coefficient for the component *i*
γ_i^r	residual part of activity coefficient for the component *i*
γ^∞	infinite dilution coefficient
$\Gamma(r)$	potential energy–distance function
Γ_k	activity coefficient of group *k* at mixture composition *or* adsorption of compound (*k*)
Γ_k^i	activity coefficient of group *k* at a group composition of pure component *i*
Γ_{max}	maximum adsorption (often in mol/g)
δ	solubility parameter, $(J/cm^3)^{1/2}$
Δ	association strength, l/mol
ε	dispersion energy parameter, association energy, J
ε_0	permittivity of vacuum (free space), $8.854 \times 10^{-12}\ C^2/J/m$
ε_r	dielectric constant (dimensionless)
$\varepsilon^{A_iB_j}$	association energy parameter between site *A* in molecule *i* and site *B* in molecule *j*, bar l/mol
ζ	partial volume fraction *or* zeta potential
η	the reduced fluid density of CPA *or* volume fraction of PC–SAFT
ϑ	contact angle *or* surface area fraction
θ_i	surface area fraction for component *i* in the mixture
Θ	occupancy of cavity *m* by component *i*
κ	association volume of PC–SAFT *or* Debye screening length, eq. 15.25
μ	dipole moment in Debye *or* (μ_i) chemical potential
ν	main electronic absorption frequency in the UV region (about 3×10^{15} Hz)
ν_i	number of cavities of type *i*

ν_{ki}	number of groups of type *k* in molecule *i*
π	surface pressure ($= \gamma_w - \gamma$)
$\Delta\varphi$	electrical potential difference, eq. 19.35
ψ_0	surface potential
ρ	molar density, mol/l
σ	segment diameter, Å
τ_{ji}	Boltzmann factor (in local composition models), eq. (5.1)
Φ	(volume/segment) fraction
$\hat{\varphi}_i$	fugacity coefficient of component *i* in a mixture
ω	acentric factor
χ_{12}	Flory-Huggins (interaction) parameter
Ω	weight-based activity coefficient
Ω_1^∞	infinite dilution weight-based activity coefficient

Superscripts and subscripts

AB	site A–site B
A_iB_j	site A in molecule *i* with site B in molecule *j*
A, *B*, *C*, *D*	site indicators
A	anion or attractive
AB	acid–base interactions
Adh, *A*	adhesion
attr	attractive
assoc	association
b	boiling point/temperature
c or *crit*	critical
C	cation or combinational
chem	chemical
cal	calculated value
Coh	cohesion
comb	combinatorial
comb-fv	combinatorial free volume
d or *disp*	dispersion
DP	data points
DH	Debye-Hückel
E	excess
EH	empty hydrate
eq	equilibrium
excl	excluded
exp	experimental value
fv, FV	free volume
f, *fus*	fusion
FH	Flory-Huggins
g or *gas*	gas
H	hydrate
h or *hb*, *HB*	hydrogen bonding
hc	hard chain

hs	hard sphere
i	gas, solid *or* liquid in expressions for surface or interfacial tensions *or* component index
id	ideal
i, *j*	component indexes
j	gas, solid or liquid in expressions for surface *or* interfacial tensions *or* component index
L or *l*	liquid
LW	London/van der Waals
m	mixture *or* molar *or* molality
max	maximum
mix	mixing
mol	molecular
o, *O*	oil (in the 'broader' sense used in colloid and surface science)
ow	octanol-water
oc	octanal-organic carbon
p	polar
PDH	Pitzer-Debye-Hückel
phys	physical
r	reduced
ref	reference
res, *R*	residual *or* repulsive
rep	repulsive
s, *S*	solid
sat	saturated/saturation
sdw	sediment-water
seg	segment
sl	solid–liquid interface
subl, *sub*	sublimation
s_1s_2	solid 1–solid 2 interface
spec	specific (non-dispersion) effects, e.g. due to polar, hydrogen bonding, metallic *or* specific (in general)
surf	surfactant
sw	soil-water
tr	transition
t	triple point
V, *v*	vapor
VAP, *vop*	vaporization
w, *W*	water
∞	infinite dilution
$+$	acid contribution (acid–base theory)
$\pm$	mean value (in electrolytes)
$-$	base contribution (acid–base theory)

Part A
Introduction

1

Thermodynamics for Process and Product Design

The design of separation processes, chemical and biochemical product design and certain other fields, e.g. material science and environmental assessment, often require thermodynamic data, especially phase equilibria. Table 1.1 summarizes the type of data needed in the design of various separation processes. The importance of thermodynamics can be appreciated as often more than 40% of the cost in many processes is related to the separation units.[1]

The petroleum and chemical industries have for many years been the traditional users of thermodynamic data, though the polymer, pharmaceutical and other industrial sectors are today making use of thermodynamic tools. Moreover, thermodynamic data are important for product design and certain applications in the environmental field, e.g. estimation of the distribution of chemicals in environmental ecosystems. Already several commercial simulators have a wide spectrum of thermodynamic models to choose from and companies often use the so-called 'decision or selection trees', see Figure 1.1, for selecting models suitable for specific applications, either those developed in-house[2] or those suggested by the simulator providers.[3]

Still, it is often questioned whether sufficient data and/or suitable models are available for a particular process or need. Opinions differ even within the same industrial sector and they should also be seen in relation to the time that the various statements have been made.[4,5] The needs, even within the same industrial sector, are not always the same. Dohrn and Pfohl[6] explain why, in the chemical industry, the answer to the question about the availability of thermophysical data can be almost anything from 'we have enough data', or 'we don't have enough data', to 'we have too much data'. These statements can be respectively justified based on the availability of suitable models in process simulators, the existence of difficult separations or the many databases which may be at hand. Data for multicomponent mixtures especially can be scarce and costly even for well-defined mixtures of industrial importance such as water–hydrocarbon–alcohols or glycols. Moreover, Dohrn and Pfohl[6] illustrate, using examples, how similar models may yield different designs even for rather 'simple' mixtures, e.g. in the case of ethylbenzene/styrene with the SRK equation of state. In an earlier study, Zeck[7] presents thermodynamic difficulties and needs, as seen from the chemical industry's point of view. These are summarized in Table 1.2.

As both Tables 1.1 and 1.2 illustrate, different types of phase equilibria data or calculations are needed depending on the problem, especially the separation type involved. The fundamental phase equilibria

Thermodynamic Models for Industrial Applications: From Classical and Advanced Mixing Rules to Association Theories
Georgios M. Kontogeorgis and Georgios K. Folas

Table 1.1 Phase equilibria data needed in the design of specific unit operations

Unit operation	phase equilibria type
Distillation	Vapor–liquid equilibria (VLE)
Azeotropic distillation	VLE, liquid–liquid equilibria (LLE)
Extractive distillation	LLE
Evaporation, drying	Gas–liquid equilibria
Absorption	VLE
Reboiled absorption	Gas–liquid equilibria
Stripping	Gas–liquid equilibria
Extraction	LLE
Supercritical fluid extraction	Gas–liquid and solid–gas equilibria
Adsorption	Vapor–solid equilibria Liquid–solid equilibria
Crystallization	Liquid–solid (vapor) equilibria
Leaching	Liquid–solid equilibria
Bioseparations	LLE
Extraction with aqueous two-phase systems	
Liquid–liquid extraction with reverse micelles	

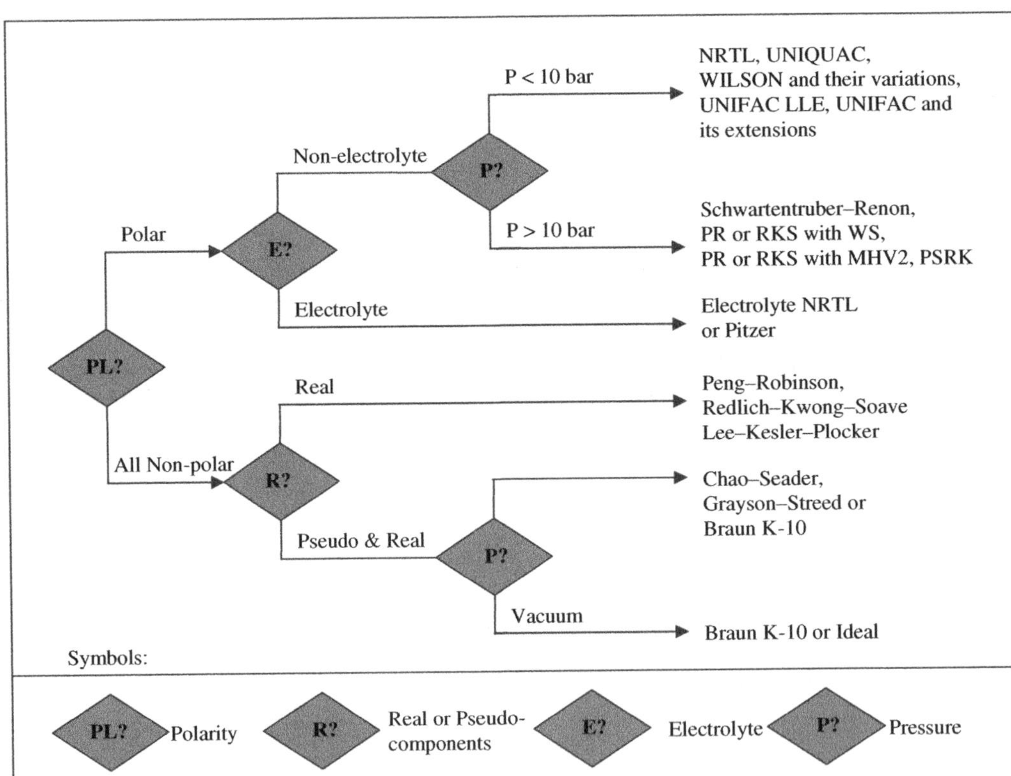

Figure 1.1 Available thermodynamic models in commercial process simulators and an example of a selection tree for choosing appropriate thermodynamic models depending on the type of compounds involved. After Carlson[3]

Table 1.2 Thermodynamic challenges of interest to the process industry. After Zeck[7]

Separation or other process	Deficiencies of existing (1991) models	Goals for the future
Azeotropic distillation	Insufficient precision in the description and estimation of VLE and LLE using one model and one parameter set	Standardized models, one parameter set
Extractive distillation	No available strategy or methodology for selection of solvents	Search strategy for selection of solvents based on molecular parameters
Extraction	Multiple measurements required, insufficient precision in description and prediction	Possibility of basing calculation on binary parameters, estimation based on molecular parameters
Heat exchanger	Insufficient precision and quality of prediction from mixing rules	New mixing rules with improved precision for multicomponent systems
Absorption	Many empirical models, limited extent of application	Efficient models, practical computing time
Adsorption	Estimation of adsorption isotherms, in particular for multicomponent adsorption	Efficient new models, multicomponent adsorption, selection of adsorption medium
Waste water treatment	No available characterization of waste water to enable further treatment	Efficient new models

equation, which is the usual starting point for all phase equilibria problems, is the equality of the fugacities of all components at all phases (α, β, γ, ...):

$$\hat{f}_i^{a} = \hat{f}_i^{\beta} = \hat{f}_i^{\gamma} = \ldots \quad \text{with } i = 1, 2, \ldots, N \tag{1.1}$$

where N is the number of components.

Equation (1.1) holds at equilibria for all compounds in a multicomponent mixture and for all phases (α, β, γ, ...). Using this equation, the 'formal' (mathematical) problem is solved. Fugacity coefficients can be calculated from volumetric data or alternatively from an equation of state (functions of P–V–T). Physically, we can imagine that the fugacity is the 'tendency' of a molecule to leave from one phase to another. Phase equilibrium is a dynamic one, e.g. for VLE the number of liquid molecules going to the vapor phase is, at equilibrium, equal to the number of vapor molecules going to the liquid phase. The basic equation (1.1) may appear in different forms depending on the type of phase equilibria and even the nature of the thermodynamic model used (equation of state, activity coefficient). These forms are sometimes easier to use in practice than the general equation (1.1), although they are naturally all derived from this equation upon well-defined assumptions. The various forms of phase equilibria are summarized in Table 1.3, while Appendix 1.A presents some of the most important fundamental equations in thermodynamics which will find applications in the coming chapters.

The principal thermodynamic models are the equations of state (EoS), which can be expressed as functions of $P(V, T)$ or $V(P, T)$. The fugacity coefficient of a compound in a mixture can be calculated from any of the equivalent equations below:

$$RT \ln \hat{\varphi}_i = RT \ln \frac{\hat{f}_i}{y_i P} = \int_V^{\infty} \left[\left(\frac{\partial P}{\partial n_i} \right)_{T,V,n_j} - \frac{RT}{V} \right] dV - RT \ln \left(\frac{PV}{RT} \right) \tag{1.2}$$

Table 1.3 *Phase equilibrium equations in specific cases including basic equations for equilibrium calculations with equations of state (EoS). The fugacity coefficient of a compound in a mixture is defined as* $\hat{\varphi}_i = \hat{f}/x_iP_i$, *where* x_i *can be the concentration in the liquid, vapor or solid phase. The vapor pressure* P^{sat} *is obtained from correlations based, for example, on the Antoine equation or the DIPPR correlations*

Type of phase equilibrium	Expression
VLE	$y_i\hat{\varphi}_i^V = x_i\hat{\varphi}_i^L \quad i = 1,2,\ldots,N$ $y_i\hat{\varphi}_i^V P = x_i\gamma_i P_i^{sat}\varphi_i^{sat}\exp\left[\frac{V_i^L(P-P_i^{sat})}{RT}\right]$ $y_i\hat{\varphi}_i^V P = x_i\gamma_i P_i^{sat}$
VLE with modified Raoult's law (ideal gas vapor phase)	$y_iP = x_i\gamma_i P_i^{sat}$
Equations for VLE calculations with EoS	$K_i = \frac{\hat{\varphi}_i^L}{\hat{\varphi}_i^V}$ $y_i = x_iK_i$ and $\sum_i y_i = 1$ **(bubble P or T)** $x_i = y_i/K_i$ and $\sum_i x_i = 1$ **(dew P or T)**
GLE – using activity coefficient models for liquid	$f_i^{gas} = y_i\hat{\varphi}_iP = x_i\gamma_i f_i^L$
LLE – EoS (*I* and *II* indicate the two liquid phases)	$\hat{f}_i^{\,I} = \hat{f}_i^{\,II} \Rightarrow (x_i\hat{\varphi}_i^l)^{\mathrm{I}} = (x_i\hat{\varphi}_i^l)^{\mathrm{II}} \quad i = 1,2,\ldots,N$
LLE – activity coefficient model (*I* and *II* indicate the two liquid phases)	$(x_i\gamma_i)^{\mathrm{I}} = (x_i\gamma_i)^{\mathrm{II}} \quad i = 1,2,\ldots,N$
Equations for LLE calculations with EoS – two phase *P–T* flash calculation	$\sum_i z_i\frac{K_i-1}{1-\beta+\beta K_i} = 0$ (Rachford-Rice equation) $K_i = \frac{\hat{\varphi}_i^L}{\hat{\varphi}_i^V}$ $\beta y_i + (1-\beta)x_i = z_i$, where β is the vapor phase fraction, while at equilibrium $K_i = y_i/x_i$
Equations for VLLE calculations with EoS – multiphase *P–T* flash calculation	$\sum_i^{comp} z_i\frac{K_{i,j}-1}{1+\sum_{k=1}^{F-1}\beta_k(K_{i,k}-1)} = 0 \quad j = 1,2,\ldots,F-1$ where F is the number of phases $K_{i,j} = \frac{\hat{\varphi}_{i,F}}{\hat{\varphi}_{i,j}} = \left(\frac{y_{i,j}}{y_{i,F}}\right)$ $j = 1,2,\ldots,F-1$ with F the 'reference' phase $y_{i,F} + \sum_{j=1}^{F-1}\beta_j(y_{i,j}-y_{i,F}) = z_i$
SLE at low pressures Ideal solid phase Solubility of a solid *i* in a liquid *i*, assuming pure solid phase	$x_i\gamma_i = \exp\left[\frac{\Delta H_{fus,i}}{RT_{m,i}}\left(1-\frac{T_{m,i}}{T}\right) + \frac{\Delta C_{p,i}}{R}\left(\frac{T_{m,i}}{T}-1-\ln\frac{T_{m,i}}{T}\right)\right]$
SLE (general equation) Solubility of a solid *i* in a liquid *i* at high pressures, assuming pure solid phase and pressure-independent heat capacity and specific molar volumes	$x_i\gamma_i = \exp\left[-\frac{(v_{0i}^S-v_{0i}^L)(P^+-P)}{RT} + \frac{\Delta H_{fus,i}}{RT_{m,i}}\left(1-\frac{T_{m,i}}{T}\right) + \frac{\Delta C_{p,i}}{RT}(T_{m,i}-T) - \frac{\Delta C_{p,i}}{R}\ln\frac{T_{m,i}}{T}\right]$
SGE (solubility of the pure solid *i* in a supercritical fluid)	$y_i\hat{\varphi}_iP = P_i^{sat}\exp\left[\frac{V_i^s(P-P_i^{sat})}{RT}\right]$

Abbreviations of the phase equilibria types: GLE = gas–liquid equilibria; LLE = liquid–liquid equilibria; SGE = solid–gas equilibria; SLE = solid–liquid equilibria; VLE = vapor–liquid equilibria; VLLE = vapor–liquid–liquid equilibria.

$$RT\ln\hat{\varphi}_i = RT\ln\frac{\hat{f}_i}{y_i P} = \int_0^P\left(\bar{V}_i - \frac{RT}{P}\right)dP \tag{1.3}$$

Equation (1.2) is most suitable for EoS of the type $P(V,T)$, while Equation (1.3) is suitable for EoS of the form $V(P,T)$. In principle, they are suitable for all types of fluid phases, conditions (T, P, *concentration*) and mixtures of any number of components. In practice, however, the situation can be quite different and equilibria types such as SLE, LLE and even complex VLE can often be conveniently handled with activity coefficient models, specifically developed for condensed phases (Table 1.3). Activity coefficients are useful means of representing deviations from ideality. In thermodynamics we picture as ideal the solutions which contain compounds with similar sizes and shapes and where the forces between like and unlike molecules are essentially the same, e.g. methanol/ethanol, pentane/hexane, benzene/toluene or mixtures of isomers. However, the phase equilibrium equations (for VLE and SLE) can take various forms depending on the precise conditions under which the solution is ideal. Ideal solutions do not separate into two liquid phases (i.e. no LLE is present). Table 1.4 summarizes the various definitions of ideality in thermodynamics.

We often use the terminology *gamma–phi* ($\gamma - \varphi$) and *phi–phi* ($\varphi - \varphi$) for the approaches, with the latter implying that an EoS is used for all phases, while in the former case an activity coefficient model is used for the liquid or solids phases. It is apparent from the above discussion that the distinction between the *gamma–phi* ($\gamma - \varphi$) and *phi–phi* ($\varphi - \varphi$) approaches is not of a fundamental character but rather a traditional (and somewhat old-fashioned) one. Such a distinction largely exists due to the fact that classical cubic equations of state EoS, which were the 'first' EoS in the market, in combination with the widely used van der Waals one-fluid mixing rules, are typically suitable 'only' for describing VLE of rather simple systems (e.g. mixtures of hydrocarbons and gases). Thus, numerous activity coefficient models have been developed since the early twentieth century, particularly for complex mixture VLE, LLE and SLE. Moreover, they provided a way for

Table 1.4 *Ideality in thermodynamics. The Dalton, Raoult, Henry and Lewis–Randall 'laws'*

Ideality	Equation
Ideal gas	$\bar{V}_i = \left(\frac{\partial V}{\partial n_i}\right)_{T,P,n_{j\neq i}} = \frac{\partial}{\partial n_i}\left(\frac{nRT}{P}\right) = \frac{RT}{P} \Rightarrow \hat{f}_i = y_i P$
Dalton	$PV = nRT$
Ideal gas or liquid solution at any pressure (Lewis–Randall) Valid in practice at low pressure and mole fractions above 0.9 (it is essentially a representation of 'A molecule feels at home when it is alone with its own kind than with company.')	$V = \sum_i n_i V_i \Rightarrow \bar{V}_i = V_i \Rightarrow \hat{f}_i = y_i f_{pure,i} y_i \quad f_i^v = x_i f_i^l$
Ideal liquid (Raoult) at low pressures (activity coefficient = unity)	$y_i P = x_i P_i^{sat}$ (VLE) $x_i = \exp\left[\frac{\Delta H_i^{fus}}{RT_{m,i}}\left(1 - \frac{T_{m,i}}{T}\right) + \frac{\Delta C_{p,i}}{R}\left(\frac{T_{m,i}}{T} - 1 - \ln\frac{T_{m,i}}{T}\right)\right]$ (SLE)
Henry's law: definition and relationship with infinite dilution activity coefficient Valid at low concentrations, e.g. mole fractions below 0.03	$\hat{f}_i^l = x_i H_i$ $H_i = \lim_{x_i \to 0}\left(\frac{\hat{f}_i^l}{x_i}\right) = \gamma_i^\infty f_i^l$ $H_i = \gamma_i^\infty P_i^{sat} \Rightarrow \gamma_i^\infty = \frac{H_i}{P_i^{sat}}$

Table 1.5 *Typical values of infinite dilution activity coefficients in aqueous systems. Activity coefficient values can be sometimes useful in determining whether phase splitting will occur, as miscible systems have in most cases activity coefficients below 10, while very immiscible systems have activity coefficients above 200, and the activity coefficients of partially miscible systems typically lie in between these two values*

Compound	Activity coefficient of water at infinite dilution in the compound
Methyl ethyl ketone	32
Diethyl ether	160
Chloroform	860
Carbon tetrachloride	10 000
Ethyl acetate	150
Octanol	3700
Benzene	2400
Toluene	12 000
Naphthalene	140 000
Phenanthrene	7 400 000
Hexachlorobenzene	980 000 000
Adenine	7200
Suanine	115 000

fast, simple calculations in the mid twentieth century, when computers were not as powerful as they are today. In addition, activity coefficients help visualize the deviations from ideality, and, as Table 1.5 illustrates, activity coefficients can vary enormously, from far below unity, e.g. for polymer solutions, up to several million for 'complex' pollutants in water. This variation indicates the wide range of intermolecular forces, which are discussed in Chapter 2. In most cases, activity coefficient values are above unity (positive deviations from Raoult's law). Negative deviations from Raoult's law (activity coefficients below one) are present in mixtures exhibiting strong cross-interactions, e.g. chloroform–acetone and nearly athermal hydrocarbon and polymer solutions (mixtures with almost zero heat of mixing). Some common phase diagrams for binary mixtures are presented in Appendix 1.B.

The purpose of this book is not to discuss the 'fundamentals of thermodynamics', i.e. derivations and background of the equations shown in Table 1.3 or the numerical aspects of solving these equations. Excellent textbooks are available[8–13] with the last, by Michelsen and Mollerup, focusing especially on computational aspects of thermodynamic models. It is rather the purpose of this textbook to address how thermodynamics assisted by disciplines like physical chemistry and statistical thermodynamics 'attempt' to identify the 'best' model (EoS, activity coefficient) for specific applications, taking into account the peculiarities of the applications considered:

- For which phases can the model be applied (VLE, LLE, VLLE, SLE, SGE, etc.)? Is there a possibility for the existence of more than two phases at the same or different conditions (e.g. VLE at high temperatures and LLE at low temperatures)?
- Conditions (T, P, concentration).
- Peculiarities (e.g. azeotropic behavior, negative deviations from Raoult's law).
- Type of compounds (hydrocarbons, alcohols, water, polymers, electrolytes, etc.).
- Number and nature of interaction parameters – how can they be obtained?

- Are the models suitable for correlation (*description*) and/or prediction of phase behavior (i.e. calculations when no experimental data are available for determining the model parameters)?
- Simplicity vs. complexity – speed of calculations.
- Performance for multicomponent systems (parameters obtained from binary data).

While specific thermodynamic models often 'come and go', certain general theories, concepts and principles do stay or apply in many models. Examples of such theories and concepts are: group contribution, local composition, corresponding states principle, solubility parameters, free volumes, mixing and combining rules, and association theories (chemical-like, lattice and perturbation theories). It is also the purpose of this book to highlight these concepts and their use in thermodynamic models.

Clearly, a thorough understanding of intermolecular forces is useful both in the interpretation of phase behavior and in the choice and in some cases development of improved models. A short 'practical' introduction on the intermolecular and interparticle forces is presented in the next chapter.

In conclusion, chemical engineering thermodynamics and in particular phase equilibria are important in both process and product design. Different types of phase equilibria (VLE, LLE, SLE, etc.) are important, depending on the application, especially the type of separation method used. The starting point for representing phase equilibria with thermodynamic models is the concept of equality of fugacities in all phases, a criterion which can take more readily used forms depending on the equilibrium type, as shown in Table 1.3. VLE is often easier to represent with thermodynamic models than LLE and VLLE provided that the 'end-points' of a VLE phase diagram (vapor pressures) are well reproduced. Azeotropic mixtures may be more difficult to represent than non-azeotropic ones. LLE phase diagrams for non-polymeric mixtures are typically of the upper critical solution temperature (UCST) type and often rather symmetric with respect to concentration, while LLE for polymer solutions is concentration asymmetric and often both UCST and LCST (Lower Critical Solution Temperature) types of behaviors are present. An auxiliary property typically used for representing phase equilibria of complex mixtures is the activity coefficient, which represents deviations from the ideal behavior as expressed by Raoult's law. Experimental activity coefficient data can be obtained from VLE or SLE data. There are no general thermodynamic models which can describe equally successfully all types of phase equilibria at all conditions. Suitable models for high- and low-pressure phase equilibria for simple as well as complex mixtures including those with solids, polymers, electrolytes and associating fluids will be presented in this book.

Appendix 1.A Important equations from the framework of thermodynamics

1.A.1 Excess and mixing properties

For any property M, e.g. V, H, S, etc., the excess (E) and mixing (mix) values are defined as:

$$\begin{aligned} \Delta M_{mix} &= M - \textstyle\sum_i x_i M_i \\ M^E &= \Delta M_{mix} - \Delta M_{mix}^{ideal} \end{aligned} \tag{1.4}$$

This means that for V and H, where $\Delta V_{mix}^{ideal} = 0, \Delta H_{mix}^{ideal} = 0$, we have:

$$\begin{aligned} \Delta V_{mix} &= V^E \\ \Delta H_{mix} &= H^E \end{aligned} \tag{1.5}$$

Table 1.6 Thermodynamic functions and partial derivatives

Internal energy	$U = TS - PV + \sum \mu_i n_i$
	$dU = TdS - PdV + \sum \mu_i dn_i$
Enthalpy ($H = U + PV$)	$H = U + PV = TS + \sum \mu_i n_i$
	$dH = TdS + VdP + \sum \mu_i dn_i$
Gibbs energy ($G = H - TS$)	$G = U + PV - TS = H - TS = \sum \mu_i n_i$
	$dG = -SdT + VdP + \sum \mu_i dn_i$
Helmholtz energy ($A = G - PV$)	$A = U - TS = -PV + \sum \mu_i n_i$
	$dA = -SdT - PdV + \sum \mu_i dn_i$
Pressure	$-P = \left(\frac{\partial A}{\partial V}\right)_{T,n}$
Volume	$V = \left(\frac{\partial G}{\partial P}\right)_T$
Entropy	$-S = \left(\frac{\partial A}{\partial T}\right)_{V,n}$
Chemical potential	$\mu_i = \left(\frac{\partial A}{\partial n_i}\right)_{T,V,n_j} = \left(\frac{\partial G}{\partial n_i}\right)_{T,P,n_j} = \left(\frac{\partial U}{\partial n_i}\right)_{S,V,n_j} = \left(\frac{\partial H}{\partial n_i}\right)_{S,P,n_j}$
Specific heat capacity at constant V	$C_V = \left(\frac{\partial U}{\partial T}\right)_V$
Specific heat capacity at constant P	$C_P = \left(\frac{\partial H}{\partial T}\right)_P$

This is not the case for the Gibbs energy or Helmholtz energy, because of the entropy term:

$$\Delta S_{mix}^{ideal} = -R \sum_i x_i \ln x_i \tag{1.6}$$

The following equations apply:

$$G^E = H^E - TS^E \tag{1.7}$$

$$\Delta G_{mix} = \Delta H_{mix} - T\Delta S_{mix} \Rightarrow \Delta G_{mix}^{ideal} = RT \sum_i x_i \ln x_i \tag{1.8}$$

1.A.2 Excess Gibbs energy, fugacities and activity coefficients

These are as follows:

$$\frac{g^E}{RT} = \ln \varphi - \sum_i x_i \ln \varphi_i = \sum_i x_i \ln \gamma_i \tag{1.9}$$

$$\ln \gamma_i = \ln \frac{\hat{\varphi}_i}{\varphi_i} \tag{1.10}$$

$$\frac{\Delta G_{mix}}{RT} = \sum_i x_i \ln (x_i \gamma_i) = \sum_i x_i \ln a_i \tag{1.11}$$

1.A.3 Deriving activity coefficients (γ_i) and activities (a_i) from the excess Gibbs energy and the Gibbs energy change of mixing

The equations are:

$$RT \ln \gamma_i = \left(\frac{\partial G^E}{\partial n_i}\right)_{T,P,n_{j\neq i}} = \left(\frac{\partial n g^E}{\partial n_i}\right)_{T,P,n_{j\neq i}} \tag{1.12}$$

$$\ln a_i = \frac{\partial}{\partial n_i}\left(\frac{n\Delta G_{mix}}{RT}\right)_{T,P,n_{j\neq i}} \tag{1.13}$$

Appendix 1.B Common phase diagrams for binary mixtures and phase envelopes

Figure 1.2 (left) presents a *Pxy* diagram for the binary mixture *n*-propanol–water at 363.15 K. The lower curve is the dew point curve; below this curve and for any concentration of *n*-propanol the mixture is vapor. The upper curve is the bubble point curve and for any pressure higher than the bubble point curve the mixture is liquid. At a given composition the pressure along the bubble point curve is the pressure where an infinitesimal bubble of vapor coexists with the liquid, while for pressures between the dew point and bubble point curve the two different phases (vapor and liquid) coexist. Similar observations can be made for the *Txy* diagram presented in Figure 1.2 (right).

The shape of the *Pxy* or *Txy* diagram indicates the deviation from the ideal solution behavior. For mixtures that exhibit moderate deviations from ideal solution behavior such as methanol–ethanol (see Figure 1.7 of Problem 2 on the companion website at www.wiley.com/go/Kontogeorgis) no azeotrope is formed. In the case of larger deviations, and in particular when the mixture components have comparable pure component vapor

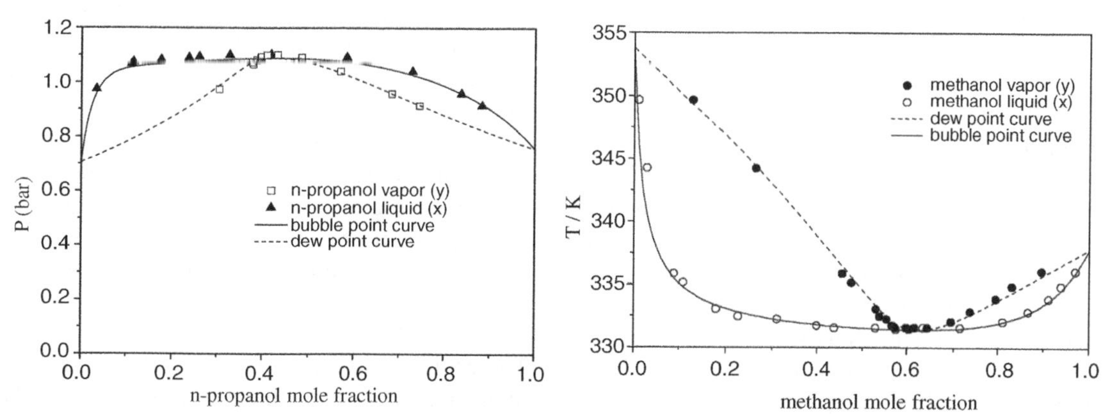

*Figure 1.2 Left: Pxy diagram for the mixture n-propanol–water at 363.15 K. Experimental data are from Ratcliff et al, Can. J. Chem. Eng., **1969**, 47, 148. Right: Txy diagram for the mixture methanol–benzene at 1 bar. Experimental data are from Nagata, J. Chem. Eng. Data, **1969**, 14, 418. The lines are guides to the eye*

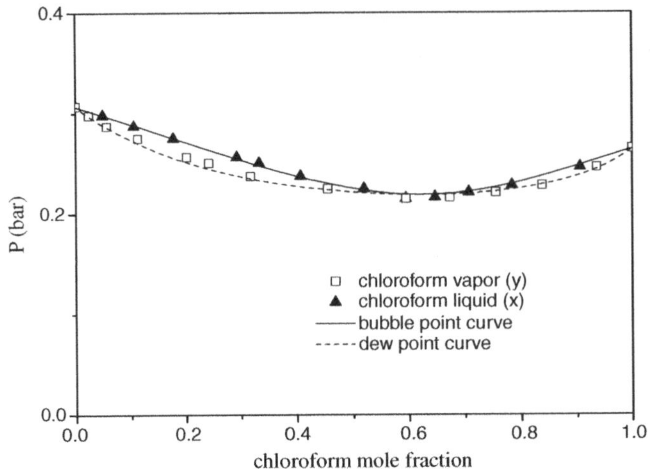

Figure 1.3 *Pxy diagram for the mixture acetone–chloroform at 298.15 K. Experimental data are from Tamir et al., Fluid Phase Equilib.,* ***1981****, 6, 113. The lines are guides to the eye*

pressures, an azeotrope may form. A typical example is shown in Figure 1.2 for mixtures that exhibit positive deviations from Raoult's law; this azeotrope is called minimum boiling since the azeotropic composition at a given pressure has the lowest boiling temperature, as shown in Figure 1.2 (right). Negative deviations from Raoult's law are rather rare; they are found in cases where the components form hydrogen bonds with each other, e.g. when one compound is an electron acceptor and the other an electron donor. Chloroform–acetone is a classical example of a maximum boiling azeotrope, as presented in Figure 1.3.

A more complex phase diagram is presented in Figure 1.4, for the mixture methanol–*n*-heptane at atmospheric pressure. The VLE of the mixture is similar to the one presented in Figure 1.2 exhibiting positive deviations from Raoult's law. As already mentioned, below the bubble point curve a single liquid phase is formed. However, when the temperature is further decreased, and depending on the relative concentration of methanol and *n*-heptane, the mixture becomes partly immiscible and an additional liquid phase is formed. The left part of the LLE curve corresponds to the solubility of methanol in the hydrocarbon phase (i.e. *n*-heptane-rich phase), while the right part represents the solubility of methanol in the polar phase (i.e. methanol-rich phase). The LLE curve is called binodal and ends at the upper critical solution temperature, which is the highest temperature where the mixture is still partly immiscible.

Figure 1.5 presents the phase envelope (*PT* diagram) of the binary mixture ethane–heptane (C_2–*n*-C_7) with the SRK EoS. As can be seen, a typical phase envelope consists of two lines, the dew point line and the bubble point line. The phase envelope separates the single phase region from the two-phase region. At pressures above the bubble point the fluid is in liquid form. At pressures below the bubble point curve, the mixtures separate into two phases, a vapor phase and a liquid phase. The remaining part of the curve is the dew point line. The effect of varying concentration on the phase envelope is also apparent, since different concentrations result in different curves and different critical points. The critical line (or critical locus) of the binary system is also presented on the phase diagram. The critical line represents the *PT* curve through all possible critical points for mixtures of the two components, from pure ethane to pure heptane. The dew point line and the bubble point line are the same curve for a pure component, called the vapor pressure curve.

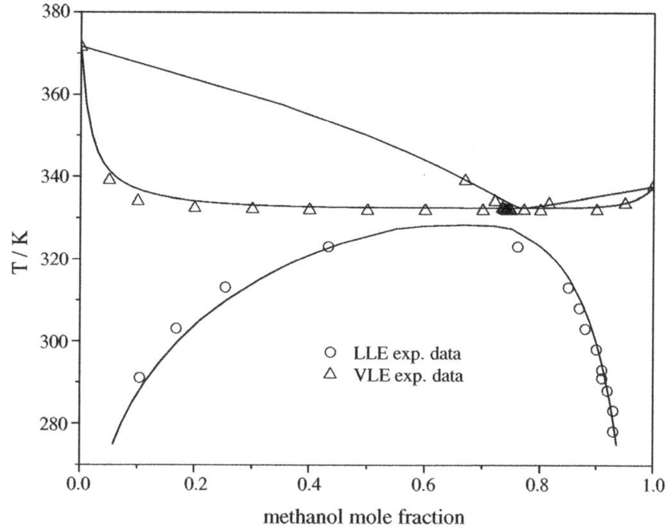

*Figure 1.4 VLE and LLE of methanol–n-heptane at P = 1 bar. Experimental data are from Sørensen and Arlt, Liquid–liquid equilibrium data collection (Binary Systems), DECHEMA Chemistry Data Series, Vol. 5, Part 1, 1980 and Higashiuchi et al., Fluid Phase Equilib., **1987**, 36, 35. The lines are calculations with the CPA equation of state (Chapter 9) using an interaction parameter $k_{12} = 0.005$*

Figure 1.6 presents a classical phase envelope for a seven-component natural gas mixture. At the critical point the liquid and the vapor have identical properties. The point of maximum pressure on the phase diagram (140.3 bar) is called cricondenbar and the point of extreme temperature cricondentherm (336 K). The phase diagram of Figure 1.6 shows an interesting phenomenon, called retrograde condensation. Normally, an

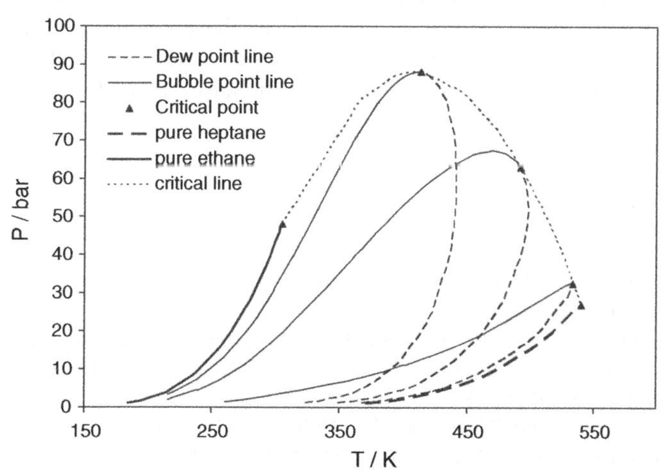

Figure 1.5 Phase envelopes for the ethane–heptane binary mixture with the SRK EoS and $k_{ij} = 0$ at different concentrations. The vapor pressure curves of pure ethane (solid curve) and heptane (dashed curve) are also presented: for a pure component, the bubble point and the dew point lines merge in the vapor pressure curve

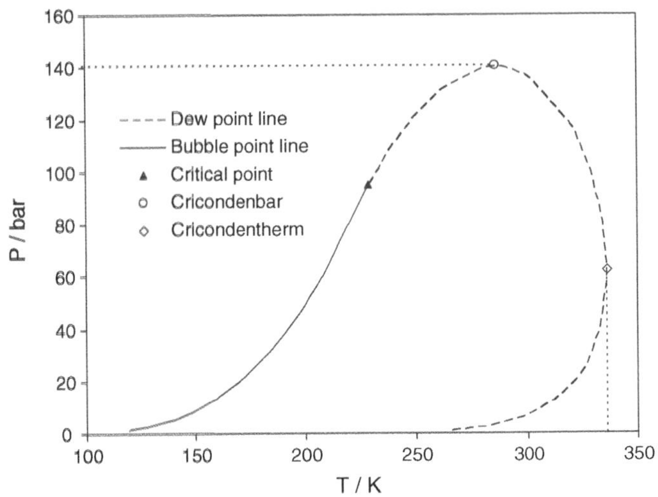

Figure 1.6 *Phase envelope of seven-component mixture*

increase in pressure leads to increased condensation (formation of liquid) and a reduction to reduced liquid formation. Consider now the natural gas mixture in the figure at a temperature of 320 K. As can be seen in Figure 1.6, at a pressure of 130 bar we are in the single phase (vapor) region; a decrease in pressure leads to the formation of a liquid phase, while upon further reduction of the pressure we observe the usual behavior, i.e. the condensed liquid re-evaporates, and below the dew point curve a single vapor phase is again obtained. Retrograde phenomena are common in gas reservoirs and a proper understanding of retrograde behavior is important for efficient production.

This discussion is limited to common phase envelopes. Unusual phase envelopes, however, also exist. Atypical phase envelopes can have two critical points: phase envelopes with an almost vertical increase in pressure at a given temperature (as a result of an LLE) at the phase boundary, or phase envelopes with no critical point location (as a result of a phase split in three phases in the area where the critical point would have been located in case the mixture were a two-phase one). Michelsen and Mollerup[13] discuss such unusual phenomena in more detail.

References

1. J.M. Prausnitz, R.N. Lichtenthaler, E.G. de Azevedo, *Molecular Thermodynamics of Fluid-Phase Equilibria* (3rd edition). Prentice Hall International, 1999.
2. G.M. Kontogeorgis, R. Gani, Introduction to computer aided product design. In: G.M. Kontogeorgis, R. Gani, *Computer-Aided Property Estimation for Process and Product Design*. Elsevier, 2004.
3. E.A. Carlson, *Chem. Eng. Prog.*, **1996**, October, 35–46.
4. C. Tsonopoulos, J.L. Heidman, *Fluid Phase Equilibr.*, **1986**, *29*, 391–414.
5. S. Gupta, J.D. Olson, *Ind. Eng. Chem. Res.*, **2003**, *42(25)*, 6359–6374.
6. R. Dohrn, O. Pfohl, *Fluid Phase Equilib.*, **2002**, *194–197*, 15–29.
7. S. Zeck, *Fluid Phase Equilib.*, **1991**, *70*, 125–140.
8. J.M. Smith, H.C. van Ness, M.M. Abbott, *Introduction to Chemical Engineering Thermodynamics* (7th edition). McGraw-Hill International, 2005.

9. S.I. Sandler, *Chemical and Engineering Thermodynamics* (3rd edition). John Wiley & Sons, Ltd, 1999.
10. J.R. Elliott, C.T. Lira, *Introductory Chemical Engineering Thermodynamics*. Prentice Hall International, 1999.
11. D.P. Tassios, *Applied Chemical Engineering Thermodynamics*. Springer-Verlag, 1993.
12. J. Vidal, *Thermodynamics: Applications in chemical engineering and the petroleum industry*. TECHNIP, IFP Publications, 1997.
13. M.L. Michelsen, J.M. Mollerup, *Thermodynamic Models: Fundamentals & Computational Aspects*. Tie-Line Publications, 2004 and 2007 (2nd edition).

2

Intermolecular Forces and Thermodynamic Models

2.1 General

Consider the following:

- Why is water a liquid at room temperature but a molecule of similar size such as methane a gas at the same conditions?
- Why does ethanol boil at 79 °C but its isomer dimethyl ether at −25 °C?
- Why is methanol miscible with hexane at high temperatures but splits into two liquid phases at lower ones?
- Why can even a simple non-polar molecule like argon exist in the liquid state?
- Why do oil (hydrocarbons) and water not mix?
- Why do polar gases behave non-ideally at low temperatures but much less so at higher ones?
- Why do salts dissociate into ions when they are in water but not in hexane?
- Why are nylon and even more so the aramid fibers so strong?
- Why do glycols 'hate' the aromatic hydrocarbons less compared to the paraffinic ones?
- Why do salts have such high melting points?
- Why are electrolyte solutions often more difficult to describe than non-electrolyte ones?
- Why does acetic acid vapor behave highly non-ideally even at atmospheric pressure?
- Why does ice float on water?

These and many more questions find answers via the understanding of intermolecular forces. The same could be claimed for many questions related to phase behavior and thermodynamic models such as:

- Why do certain solutions like chloroform and acetone exhibit negative deviations from Raoult's law?
- Is there a theoretical explanation for the geometric mean rule typically used for the cross-energy parameter in cubic and other equations of state?
- Why do classical cubic equations of state using the geometric mean rule for the cross-energy parameter typically require negative interaction parameters for mixtures like chloroform and acetone?

Thermodynamic Models for Industrial Applications: From Classical and Advanced Mixing Rules to Association Theories
Georgios M. Kontogeorgis and Georgios K. Folas

- Why is it often stated that cubic (and other) equations of state have problems in satisfactorily representing second virial coefficients, especially at low temperatures?
- Are there any theoretical justifications for the van der Waals one-fluid mixing rules?
- Why do modern molecular theories which explicitly account for dispersion, chain, repulsive and hydrogen bonding effects fail for 'simple' systems like acetone–pentane?

As non-ideality is often due to the presence of intermolecular forces, it is worth looking at the most important of those, including also a short discussion of the forces between surfaces or particles (larger than molecular dimensions). Knowledge of intermolecular forces is useful in many contexts, such as:

- Interpreting and understanding phase behavior.
- Understanding the molecular basis of certain thermodynamic models and principles, e.g. the corresponding states principle (see Chapter 3).
- Choosing suitable thermodynamic models.
- Developing better terms in thermodynamic models like equations of state as well as improved mixing and combining rules.

A useful way to represent intermolecular (and interparticle) forces is via the potential energy–distance function, $\Gamma(r)$, which is related to the intermolecular force:

$$F(r) = -\frac{d\Gamma(r)}{dr} \tag{2.1}$$

The total energy of molecules is the sum of their kinetic energy, depending on the temperature, and the potential energy, depending on their positions and forces.

Table 2.1 summarizes several of the most important intermolecular potential functions of relevance to chemical/biochemical engineering. These are discussed in the next section, while a separate discussion is devoted to hydrogen bonding and other quasi-chemical forces which are often much stronger than the secondary (van der Waals and other) forces presented in Table 2.1. A negative (minus) sign in the potential energy $\Gamma(r)$ indicates attractive forces and a positive (plus) sign indicates repulsive forces.

Table 2.2 presents the expressions for the van der Waals forces (in the form of potential functions) for interactions between particles or surfaces (in colloidal science, the potential energy is typically designated as V and the distance as H, instead of the symbols Γ and r, used for molecules). The expressions shown in Table 2.2 are derived upon integrating the expressions of Table 2.1 (for the van der Waals forces) and considering the different geometries of particles/surfaces.[1] The remaining part of this section presents the expressions for the intermolecular and interparticle potentials when the medium is other than vacuum or air as well as an explanation of the physical properties involved in the equations of Tables 2.1 and 2.2.

The dipole moment (μ) is defined as:

$$\mu = ql \tag{2.2a}$$

where μ is the dipole moment (1 Debye = 3.336×10^{-30} C m), q_i are electric charges (C) and l is the distance between molecules (m). The quadrupole moment (Q) is defined as:

$$Q = \sum_i q_i l_i^2 \tag{2.2b}$$

where the quadrupole moment Q is expressed in C m^2.

Table 2.1 *Intermolecular potential functions for various types of forces (between molecules 1 and 2). Γ is the potential energy and r is the distance between molecules. Γ_{ij} indicates the potential energy between two different molecules i and j, while Γ_{ii} (or Γ_i) is the potential energy between two identical molecules of type i*

Force type	Expression for the potential energy $\Gamma(r)$
Coulomb (electrostatic)	$\Gamma_{12}=\frac{q_1q_2}{(4\pi\varepsilon_0\varepsilon_r)r}=\frac{(z_1z_2)e^2}{(4\pi\varepsilon_0\varepsilon_r)r}$ ε_0 is the dielectric permittivity of vacuum (8.854×10^{-12} C^2 J/m), q_i electric charges (C), z_i ionic valences, ε_r the dielectric constant (dimensionless), r the distance between charged molecules and e the unit charge ($1.602\,18\times10^{-19}$ C)
Van der Waals (general expression)	$\Gamma_{12}=-\frac{C}{r^6}$ C value (J m^6)
Van der Waals – dispersion (London)	$\Gamma_{12}=-\frac{3}{2}\frac{\alpha_{01}\alpha_{02}}{(4\pi\varepsilon_0)^2r^6}\left[\frac{I_1I_2}{I_1+I_2}\right]$ I is the first ionization potential (J) and α_{0i} is the electronic polarizability (C^2 m^2/J)
Van der Waals – polar (Keesom)	$\Gamma_{12}=-\frac{1}{3}\frac{\mu_1^2\mu_2^2}{kT(4\pi\varepsilon_0)^2r^6}$ k is Boltzmann's constant (1.38×10^{-23} J/K), T the temperature and μ the dipole moment
Van der Waals – induction (Debye)	$\Gamma_{12}=-\frac{\alpha_{01}\mu_2^2+\alpha_{02}\mu_1^2}{(4\pi\varepsilon_0)^2r^6}$
Dipole–quadrupole	$\Gamma_{12}=-\frac{\mu_1^2Q_2^2}{kT(4\pi\varepsilon_0)^2r^8}$ Q is the quadrupole moment
Quadrupole–quadrupole	$\Gamma_{12}=-\frac{7}{40}\frac{Q_1^2Q_2^2}{kT(4\pi\varepsilon_0)^2r^{10}}$
Induction quadrupole	$\Gamma_{12}=-\frac{3}{2}\frac{\alpha_{01}Q_2^2+\alpha_{02}Q_1^2}{(4\pi\varepsilon_0)^2r^8}$

Useful constants: $N_A=6.022\times10^{23}$ mol^{-1}, $k=1.38\times10^{-23}$ J/K, dipole moment unit 1 D (Debye) $=3.336\times10^{-30}$ C m, unit or electronic (elementary) charge $e=1.602\times10^{-19}$ J, Planck's constant $h=6.626\times10^{-34}$ J s, ε_0 dielectric permittivity of vacuum $=8.854\times10^{-12}$ C^2/J/m.

The quadrupole moments are due to the concentration of electric charges at four separate points in the molecule, and l in Equation (2.2b) is the distance from some arbitrary origin. Molecules such as benzene, nitrogen, CO and especially CO_2 have appreciable quadrupole moments.

Polarizability is defined as the ease with which the electrons of molecules are displaced by an electric field, e.g. created by an ion or a polar molecule. The total polarizability is the sum of the electronic polarizability and

Table 2.2 *Van der Waals forces between particles/surfaces (according to Israelachvili,[1] 'molecules' with diameters larger than 0.5 nm should be treated as small particles and the van der Waals forces should be estimated with the expressions shown in this table, otherwise the strength of the interaction will be underestimated). V is the potential energy, H is the interparticle/intersurface distance and R is the radius (for spherical particles). A is the so-called Hamaker constant (see Equations (2.8)–(2.10)). C is defined in Table 2.1 and ρ is the number density (molecules/volume)*

Geometry	Expression for the potential energy $V_A(H)$
Surface – molecule	$V_A = -\frac{\pi C \rho}{6H^3}$
Two equal-sized spheres	$V_A = -\frac{A.R}{12H}$
Two infinite-size plates	$V_A = -\frac{A}{12\pi H^2}$
Sphere–plate	$V_A = -\frac{A.R}{6H}$

the orientational polarizability due to polar forces (the Debye–Langevin equation):

$$\alpha = \alpha_0 + \frac{\mu^2}{3kT} \text{ is the total polarizability } (\mathrm{m^2/V \text{ or } C^2\, m^2/J}) \tag{2.3}$$

$$\alpha_0 = 4\pi\varepsilon_0 R^3 \text{ is the electronic polarizability } (\mathrm{m^2/V \text{ or } C^2\, m^2/J}) \tag{2.4}$$

$R =$ molecular radius (m)

All the expressions in Table 2.1 include the electronic polarizabilities. Polarizability is expressed in m^2/V or C^2 m^2/J. In volume units, polarizability is expressed as $\alpha_0/(4\pi\varepsilon_0)$ (reduced polarizability).

Total and electronic polarizabilities can be estimated via dielectric constants, ε, and refractive indexes, n, as shown from the equations below (called the Clausius–Mossotti and Lorentz–Lorentz equations):[1]

$$\frac{\alpha}{4\pi\varepsilon_0} = \left(\frac{\varepsilon - 1}{\varepsilon + 2}\right) R^3 = \left(\frac{\varepsilon - 1}{\varepsilon + 2}\right)\frac{3V}{4\pi} \tag{2.5}$$

$$\frac{\alpha_0}{4\pi\varepsilon_0} = \left(\frac{n^2 - 1}{n^2 + 2}\right) R^3 = \left(\frac{n^2 - 1}{n^2 + 2}\right)\frac{3V}{4\pi} \tag{2.6}$$

The molecular volume V is calculated as $M/(\rho N_A)$ (M is the molecular weight and ρ is the mass density). Notice from the above equations that the electronic polarizability and thus the dispersion forces are related to the refractive index of the compound (n), while the total polarizability and thus also the polar effects are related to the dielectric constants (ε).

When using the Debye and Keesom expressions (induction and dipolar forces) for interactions in a medium other than vacuum or air, the permittivity of vacuum could be simply multiplied by the dielectric constant. However, this cannot be done for the London forces, which are due to fluctuating dipoles. Thus, in the general case, the van der Waals forces between molecules 1 and 2 in a solvent medium 3 can be expressed as:[2]

$$\Gamma = \Gamma_{\nu=0} + \Gamma_{\nu>0} = -\frac{3kT}{r^6} R_1^3 R_2^3 \left[\frac{\varepsilon_1 - \varepsilon_3}{\varepsilon_1 + 2\varepsilon_3}\right]\left[\frac{\varepsilon_2 - \varepsilon_3}{\varepsilon_2 + 2\varepsilon_3}\right] - \frac{\sqrt{3}h\nu R_1^3 R_2^3}{2r^6} \frac{(n_1^2 - n_3^2)(n_2^2 - n_3^2)}{\sqrt{(n_1^2 + 2n_3^2)(n_2^2 + 2n_3^2)}\left[\sqrt{(n_1^2 + 2n_3^2)} + \sqrt{n_2^2 + 2n_3^2}\right]} \tag{2.7}$$

where:

$\Gamma_{\nu=0}$ = zero-frequency contribution due to polar/induction forces (Keesom/Debye)
$\Gamma_{\nu>0}$ = finite frequency contributions due to dispersion (London) forces
R_i = radius of particle i
ε = static dielectric constant
ν = main electronic absorption frequency in the UV region (about 3×10^{15} Hz), assumed to be the same for all three media
n = refractive index in the visible region
k = Boltzmann's constant
h = Planck's constant
T = absolute temperature

Repulsive forces due to overlapping clouds are much less well understood and are typically represented with a potential energy expression ($\Gamma^{rep} = C/r^m$) similar to that of the van der Waals forces but with a higher exponent than 6 (m is typically between 8 and 16, see section 2.4.2). This lack of full understanding of repulsive forces may partially explain why, in several classical equations of state (e.g. SRK and PR), no attempts are made to modify the co-volume parameter in the repulsive term and the same van der Waals repulsive term is used, while many efforts are focused on improving the attractive term.

There are two approaches for estimating the Hamaker constant, the microscopic or London and the macroscopic or Lifshitz ones, which are presented briefly below.

2.1.1 Microscopic (London) approach

Definition of the Hamaker constant:

$$A = \pi^2 C \rho^2 \tag{2.8}$$

where ρ is the number density (molecules/volume) and C is the London coefficient as estimated from Table 2.1.

Hamaker constant of particles (1) in a medium (2):

$$A_{12} = \left(\sqrt{A_{11}} - \sqrt{A_{22}}\right)^2 \tag{2.9a}$$

Hamaker constant in the case of two different types of particles (1 and 3) in a medium 2:

$$A_{123} = \left(\sqrt{A_{11}} - \sqrt{A_{22}}\right)\left(\sqrt{A_{33}} - \sqrt{A_{22}}\right) \tag{2.9b}$$

A_{ii} is the Hamaker constant between particles of type i.

2.1.2 Macroscopic (Lifshitz) approach

Hamaker constant for particles 1 and 2 in a medium 3:[1]

$$A = A_{\nu=0} + A_{\nu>0} = \frac{3kT}{4}\left[\frac{\varepsilon_1 - \varepsilon_3}{\varepsilon_1 + \varepsilon_3}\right]\left[\frac{\varepsilon_2 - \varepsilon_3}{\varepsilon_2 + \varepsilon_3}\right] + \frac{3h\nu}{8\sqrt{2}}\frac{(n_1^2 - n_3^2)(n_2^2 - n_3^2)}{\sqrt{(n_1^2 + n_3^2)(n_2^2 + n_3^2)}\left[\sqrt{(n_1^2 + n_3^2)} + \sqrt{n_2^2 + n_3^2}\right]} \tag{2.10}$$

The symbols were explained in the text following Equation (2.7).

2.2 Coulombic and van der Waals forces

The very strong Coulombic forces partially explain the difficulties associated with constructing suitable theories for electrolyte solutions.[3] These forces are in the range 100–600 kJ/mol, much stronger than the van der Waals forces (often less than 1 kJ/mol) and even the 'quasi-chemicals" hydrogen bonds (10–40 kJ/mol). However, as the distance increases and especially in media with high dielectric constant (or relative permittivity) such as water ($\varepsilon_r = 80$), the Coulombic forces can decrease substantially. For example, for NaCl, the potential energy is (see Problem 4 on the companion website at www.wiley.com/go/Kontogeorgis) $\Gamma = 10.64 \times 10^{-21} = 2.6\, kT$ at contact ($r = 2.76$ Å) in water (as medium) and only 0.0127 kT at a distance of 56 nm. In air and at contact the potential energy is 200 kT. Thus, salts dissociate in ions when they are in water but not in non-polar media.

Some of the most important forces in practical applications involving non-ionic molecules are the van der Waals ones. As seen from Table 2.1:

1. Except for the dispersion forces, which are universal, all the other attractive forces are 'specific' forces; that is, they are only present when the compounds involved have some special characteristics, e.g. electric charges, presence of dipoles or hydrogen bonds. The dispersion forces are the attractive forces that exist even in completely non-polar molecules such as argon and methane, and explain why even these non-polar molecules can, under certain conditions, exist in the liquid state.
2. All van der Waals forces between molecules are rather short range; the potential energy decreases with the inverse of the sixth power of the distance of molecules.
3. The van der Waals forces are always attractive when the molecules are in vacuum or air but they can be repulsive if they are in some medium (if the refractive index of the medium has a value intermediate to that of the two molecules or particles, see Equations (2.7) and (2.10)).
4. Of the three van der Waals forces, only the polar ones depend directly on temperature. Moreover, for polar forces, the potential energy depends on the fourth power of the dipole moment for pure polar fluids and is thus quite important for highly polar molecules (having a dipole moment above 1 Debye).

A comparative evaluation of the van der Waals forces for both like and unlike molecules is shown in Tables 2.3 and 2.4 and in Figure 2.1.

Table 2.3 *Comparison of intermolecular forces between two identical molecules. The C values of the van der Waals forces ($\Gamma = -C/r^6$) for identical molecules are given at 0 °C. C values are expressed in 10^{-79} J m^6; from Prausnitz et al.*[3] *and Tassios.*[4] *Notice that the dispersion contribution of water is 15% of the total van der Waals forces at this temperature, while it is 24% at 298 K*

Molecule	Dipole moment (D)	*C* dipole	*C* induction	*C* dispersion
Argon	0	0	0	50
CCl_4	0	0	0	1460
Cyclohexane	0	0	0	1560
Methane	0	0	0	106
CO_2	0	0	0	116
CO	0.10	0.0018	0.0390	64.3
$CHCl_3$	1.0102	26	20	936
HCl	1.08	24.1	6.14	107
Methanol	1.7	148	18.7	135.4
Ammonia	1.47	82.6	9.8	70.5
Water	1.84	203	10.8	38.1
Acetone	2.87	1200	104	486

CCl_4 = carbon tetrachloride, CO = carbon monoxide, HCl = hydrogen chloride, $CHCl_3$ = chloroform, CO_2 = carbon dioxide.

With respect to the relative significance of the van der Waals forces, we can comment that:

1. Dispersion forces are always quite significant and almost always far from being considered negligible. Actually, except for small, very polar molecules such as water, they will usually exceed the Keesom and Debye contributions and will dominate the van der Waals interactions.
2. Induction forces are generally small, rarely more than 7%, even for polar molecules, and they are normally the weakest of the three van der Waals forces.
3. Dipolar forces are significant for very polar molecules, especially when the dipole moment is above 1 D.
4. Quadrupole effects are much smaller, but can be of some importance in special cases, e.g. for CO_2, N_2 and other molecules with high quadrupole moment.

Table 2.4 *Comparison of intermolecular forces between unlike molecules. The C values of the van der Waals forces ($\Gamma = -C/r^6$) for different molecules are given at 0 °C. C values are expressed in 10^{-79} J m^6; from Prausnitz et al.*[3] *and Tassios*[4]

Molecule	Dipole moment (D)		*C* dipole	*C* induction	*C* dispersion
CCl_4–CC_6	0	0	0	0	1510
CCl_4–ammonia	0	1.47	0	22.7	320
CC_6–ammonia	0	1.47	0	24.5	333
Acetone–CC_6	2.87	0	0	89.5	870
Acetone–methane	2.87	0	0	22	225
CO–HCl	0.1	1.08	0.206	2.30	82.7
Water–HCl	1.84	1.08	69.8	10.8	63.7
Acetone–ammonia	2.87	1.47	315	32.3	185
Acetone–water	2.87	1.84	493	34.5	135

CCl_4 = carbon tetrachloride, CC_6 = cyclohexane, CO = carbon monoxide, HCl = hydrogen chloride.

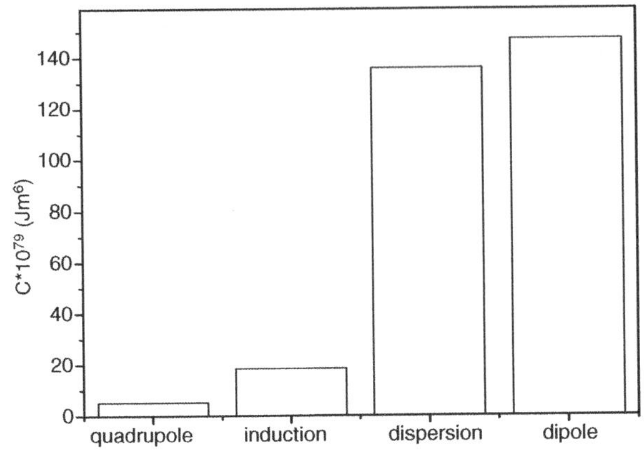

Figure 2.1 Relative magnitudes of intermolecular forces between two molecules of methanol at 0 °C. The C is calculated using the equations of Table 2.1

Despite their short range, van der Waals (vdW) forces can be quite important and in some cases may dominate the behavior of a system:

1. They can affect significantly certain physical properties, as shown in Table 2.5 for the boiling temperature of compounds of varying polarity. The effect of polar forces can be clearly seen, as exemplified by the increasing value of the boiling temperature with increasing dipole moment (polarity).
2. Even though polar vdW forces have the same distance dependency as the other two types of vdW forces, their temperature dependency makes them quite special. Not explicitly accounting for the polar forces may result in problems when molecular theories, e.g. equations of state like SAFT (Statistical Associating Fluid Theory), are used for highly polar mixtures and this is a subject of animated discussions in the literature.[5,6] One example is shown in Figure 2.2, while further discussion of the inclusion of the polarity in advanced equations of state is provided in Chapter 13. As can be seen in Figure 2.2, a model like sPC–SAFT which accounts explicitly for various forces (dispersion, hydrogen bonding) but not for the polar ones cannot describe phase equilibria for highly polar mixtures, even when the polar compounds are (arbitrarily) assumed to be self-associating (i.e. capable of forming hydrogen bonds).
3. The dispersion forces are semi-additive (see Figure 2.3 for *n*-alkanes; also Goodwin[7]) and thus the vdW forces, especially the dispersion ones, often dominate in the case of colloidal particles and surfaces, where the distance dependency is much more pronounced compared to the vdW forces between molecules, as can be seen in Table 2.2.

Table 2.5 *Boiling temperature for compounds of similar molecular weight but varying polarity*

Compound	Molecular weight (g/mol)	Dipole moment (D)	Boiling temperature (K)
Propane	44	0.1	231
Dimethyl-ether	46	1.3	248
Chloromethane	50	1.9	249
Acetaldehyde	44	2.7	294
Acetonitrile	41	3.9	355

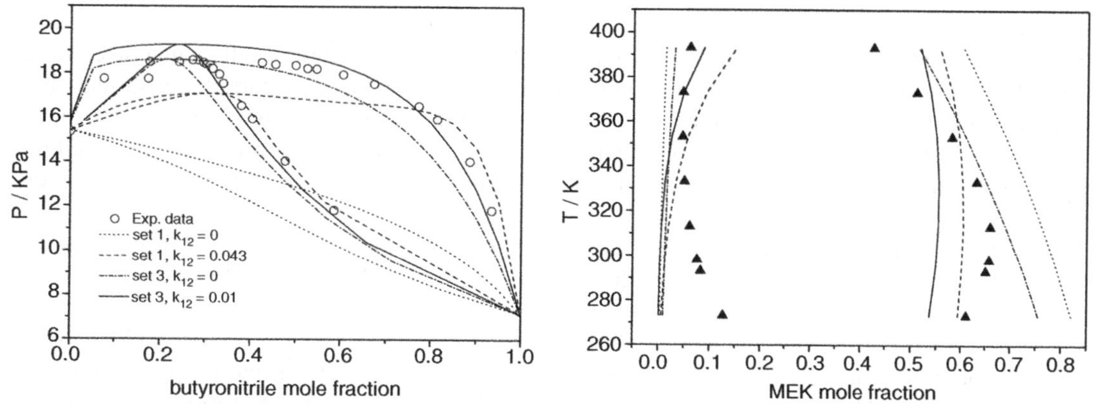

Figure 2.2 *Phase equilibria with the sPC–SAFT equation of state for two polar mixtures without using explicit terms to account for the polarity in sPC–SAFT. The left figure is the VLE of butyronitrile–heptane and the right one shows the methyl ethyl ketone (MEK) and water LLE. Various parameter sets for the nitrile and the ketone are used. For butyronitrile: set 1 = non-associating; set 3 = self-associating. For MEK: solid line ($k_{12} = -0.45$), non-associating; dashed line ($k_{12} = -0.3$), non associating, only solvating with water (cross-association equal to the self-association of water); dotted line ($k_{12} = 0$), self-associating; dashed–dotted line ($k_{12} = -0.03$), self-associating. More information about PC–SAFT and other association models will be provided in the second part of the book, see e.g. Chapter 8*

In particular for the vdW forces between particles, we can state the following – also in connection to the forces between molecules:

1. The expressions in Table 2.2 are derived from the summation of the vdW forces between molecules (shown in Table 2.1). Thus the distance dependency shown in Table 2.2 is derived assuming the exponent

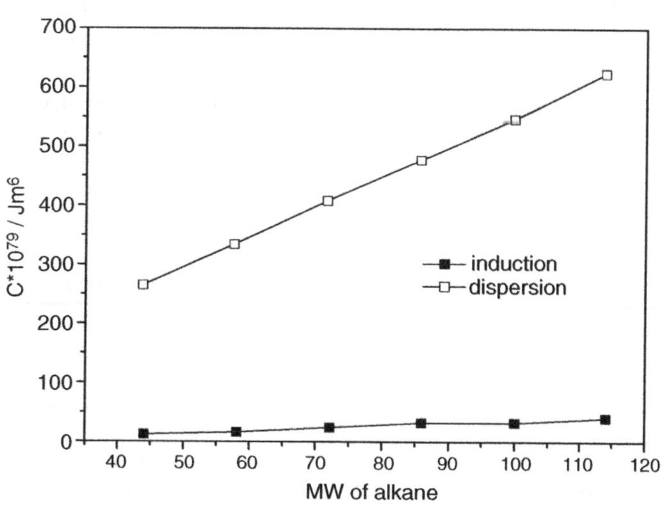

Figure 2.3 *Importance and additivity of dispersion forces. The dispersion and induction parameter for n-alkanes. The C is calculated using the equations of Table 2.1*

$n = 6$ in the intermolecular potential (i.e. validity of the Lennard-Jones potential, eq. 2.15). If another exponent is used (other than $n = 6$), the exponent in the interparticle forces will change accordingly.

2. The vdW forces are always attractive in vacuum, air or between two identical particles or surfaces but can be repulsive between different particles (or surfaces) in a third medium (exactly as was the case for forces between molecules). Repulsive vdW forces have important applications, e.g. predicting immiscibility in polymer blends, engulfing, etc.; they are discussed further in Chapter 16. The vdW forces decrease because of an intervening medium.
3. The most important applications of vdW forces between particles or surfaces are in the understanding of colloid stability via the well-known DLVO theory or in adhesion studies (see e.g. Myers,[8] also Chapter 16 and Israelachvili[1]).
4. There are basically two ways to estimate the Hamaker constants, either via Equation (2.8) combined with the combining rules shown in Equations (2.9a) and (2.9b) (this is the London approach) or via the more accurate and rigorous Lifshitz approach, which in a simplified form is expressed via Equation (2.10).
5. In some cases, particle–particle forces can be measured and experimental values are available for some systems. They agree well with the theoretical calculations of the Hamaker constants via the London or Lifshitz theories.[1]
6. In most cases, the second (non-zero frequency) term in Equation (2.10) (containing the refractive indexes), which is due to the London forces, dominates the Hamaker constant value and thus the value for the forces between particles or surfaces, but for highly polar molecules, e.g. water, the first term in Equation (2.10) (with the dielectric constants) can be significant.

2.3 Quasi-chemical forces with emphasis on hydrogen bonding

2.3.1 Hydrogen bonding and the hydrophobic effect

As can be expected from the results illustrated in Tables 2.6 and 2.7, something other than polarity and the other vdW forces should explain these large differences among the physical properties of certain compounds. These differences are attributed to the strong attractive forces called 'hydrogen bonding forces', typically occurring between H (2.1) and F (4.0), O (3.5) or N (3.0) atoms (the parentheses contain the electronegativities on the Pauling scale). Although Cl (3.0) has an electronegativity equal to that of nitrogen, it is not typically considered to participate in strong hydrogen bonds. Hydrogen bonds are typically much stronger than vdW forces but still less strong compared to the ordinary chemical (covalent) bonds (150–900 kJ/mol, i.e. 100–300 kT).

Table 2.6 *Physical properties for hydrogen bonding and non-hydrogen bonding compounds*

Compound	Boiling temperature T_b (°C)	Enthalpy of vaporization at T_b (kJ/mol)	Water solubility (mass %)	Dipole moment (D)
Dimethyl ether	−25	18.6	7.12	1.3
Acetone	56	29.1	Infinite	2.9
Ethanol	79	42.6	Infinite	1.7
1-Propanol	98	41.4	Infinite	1.7
Water	100	40.7		1.8

Table 2.7 Boiling temperature for compounds of varying polarity and hydrogen bonding degree but of similar molecular weight (in order to have meaningful comparisons). After Israelachvili[1]

Compound	Molecular weight (g/mol)	Dipole moment (D)	Boiling temperature T_b (°C)
Ethane	30	0	−89
Formaldehyde	30	2.3	−19
Methanol	32	1.7	64
n-Butane	58	0	−0.5
Acetone	58	2.9	56
Acetic acid	60	1.7	118
n-Hexane	86	0	69
Ethyl propyl ether	88	1.2	64
1-Pentanol	88	1.7	138

Hydrogen bonds are not the only quasi-chemical interactions of practical importance. Other types of charge-transfer interactions (generally called Lewis acid–Lewis base, LA–LB) also exist. All types of LA–LB interactions including hydrogen bonds are responsible for the creation of weak complexes, for example:

- dimers for organic acids;
- linear or cyclic oligomers for alcohols and phenols;
- hexamers for hydrogen fluoride;
- three-dimensional networks for water.

In particular, the hydrogen bonds of water are very strong and this is demonstrated in water's many special properties, e.g. the maximum density at 4 °C (Figure 2.4), the extensive hydrogen bonds up to high

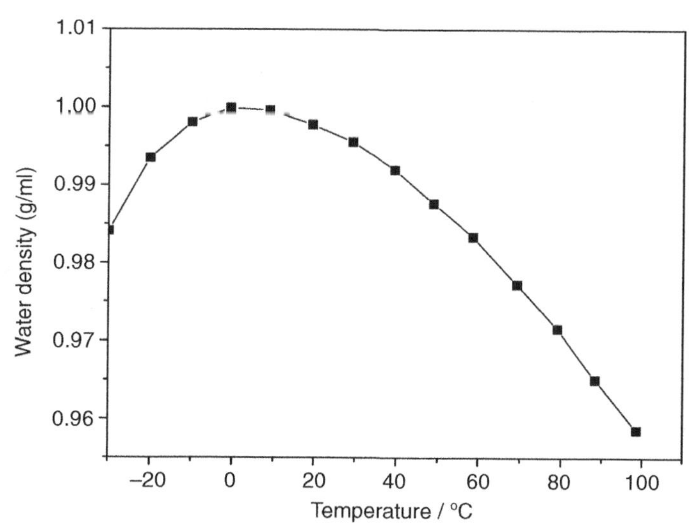

Figure 2.4 *Water density as a function of temperature illustrating the maximum at 4 °C*

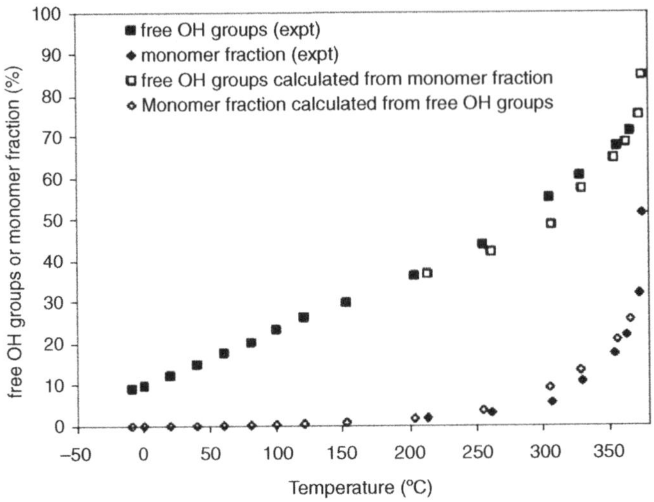

Figure 2.5 *Free OH groups or percentage monomer fraction for pure water as a function of temperature. The experimental data are from Luck, Angew. Chem. Int. Ed. Engl.,* ***1980****, 19, 28. At low to moderate temperatures almost all water molecules are in aggregate form (no monomers are present), while at higher temperatures some hydrogen bonds break and a fraction of water molecules exists in monomeric form. Reprinted with permission from Ind. Eng. Chem. Res., Investigating Models for Associating Fluids Using Spectroscopy by Nicolas von Solms, Georgios M. Kontogeorgis et al., 45, 15, 5368–5374 Copyright (2006) American Chemical Society*

temperatures (Figure 2.5), its high dielectric constant, heat capacity and surface tension, and of course the hydrophobic effect (Table 2.8 and Figure 2.6). According to the hydrophobic effect, water molecules in liquid water are connected with strong extensive hydrogen bonds and they are 'forced' into even more structured cavities (so that hydrogen bonding is restored) if 'foreign' non-polar molecules (alkanes, fluorocarbons, etc.)

Table 2.8 *Change in standard molar Gibbs energy, enthalpy and entropy, all in kJ/mol, for the transfer of hydrocarbons from pure liquids into water at 25 °C.[3,9] Notice the large negative entropy changes due to the hydrophobic effect. In the case of butane, the entropy decrease amounts to 85% of the Gibbs energy of solubili- zation, while for other hydrocarbons the entropic contribution is even larger*

Hydrocarbon	ΔG (kJ/mol)	ΔH (kJ/mol)	$T\Delta S$ (kJ/mol)
Ethane	16.3	−10.5	−26.8
Propane	20.5	−7.1	−27.6
n-Butane	24.7	−3.3	−28.0
n-Pentane	28.6	−2.0	−30.61
n-Hexane	32.4	0	−32.4
Cyclohexane	28.1	−0.1	−28.2
Benzene	19.2	+2.1	−17.1
Toluene	22.6	+1.7	−20.9
Ethylbenzene	26.2	+2.02	−24.2

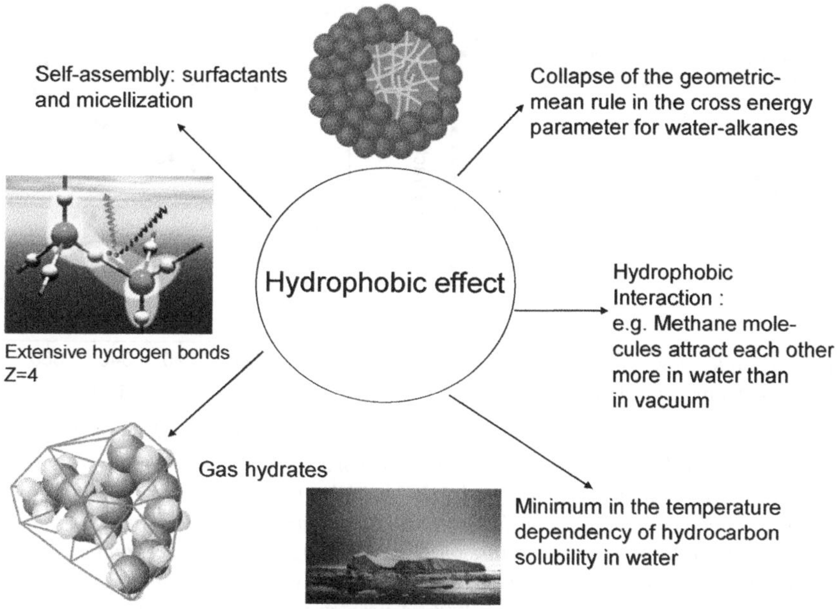

Figure 2.6 *Some implications of the hydrophobic effect, one of the most unique properties of water molecules*

are added in the solution. Thus, water molecules 'like themselves' too much and wish to stick to each other to 'be far away from enemies', especially the very non-polar molecules. This higher degree of local order compared to pure liquid water explains the entropy decrease shown in Table 2.8. It is more this loss of entropy rather than enthalpy changes that leads to the unfavorable positive Gibbs energy change associated with the non-mixing of hydrocarbons and other similar molecules in water.

Related to the hydrophobic effect is also the so-called hydrophobic interaction, a term describing the strong attraction between non-polar (hydrophobic) molecules and surfaces in water. This attraction is often stronger than their interaction in free space and it naturally cannot be explained via the vdW forces which would predict the opposite effect! (That is, lower attraction of the molecules in a medium compared to free space.) Moreover, also associated with the hydrophobic effect is the density–temperature profile (Figure 2.4) which explains why ice floats on water. Without this, life would be impossible in the sea.

2.3.2 Hydrogen bonding and phase behavior

In general, we can differentiate between three categories of hydrogen bonding and other LA–LB interactions:

1. Intermolecular or self-association between different molecules of the same type, e.g. between two like alcohol or acid molecules.
2. Intramolecular association, i.e. between different atoms inside the same molecules such as between H (hydrogen) and O (oxygen) atoms in methoxylethanol.
3. Cross-association or solvation between different types of molecules, e.g. water–methanol or chloroform–acetone.

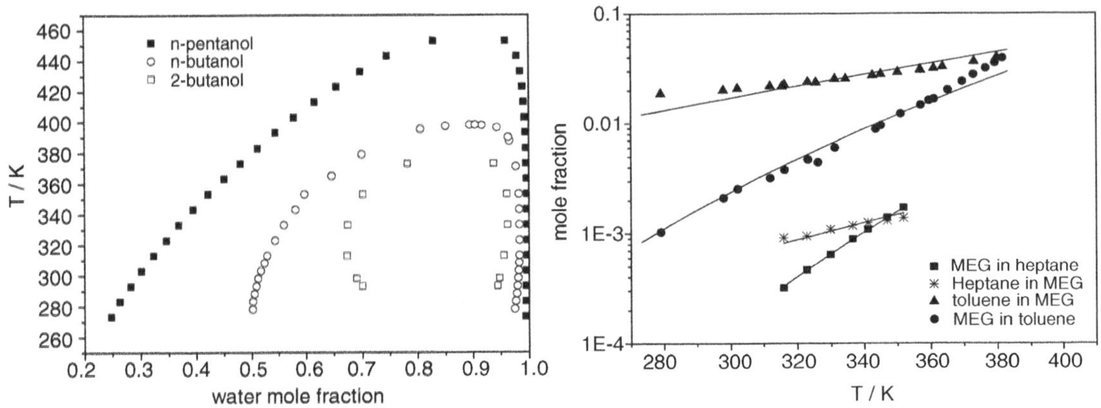

Figure 2.7 *Left: LLE of water with three heavy alcohols. Water is completely miscible with the three low-molecular-weight alcohols (methanol, ethanol, propanol) due to strong solvation effects, but becomes progressively immiscible with the heavier alcohols. Right: LLE for monoethylene glycol (MEG) with heptane and toluene. The effect of solvation is clear, as higher solubilities are observed in the case of the aromatic hydrocarbon. The lines are correlations with the CPA equation of state (Chapter 9). Reprinted with permission from Ind. Eng. Chem. Res., Application of the Cubic-Plus-Association (CPA) Equation of State to Complex Mixtures Georgios K. Folas, Georgios M. Kontogeorgis, M.L. Michelsen, E.H. Stenby 45, 4, 1527–1538 Copyright (2006) American Chemical Society*

In particular for the cross-association/solvation effects which are rather widespread, various possibilities exist depending on the compounds involved:

1. Cross-association between two self-associating compounds, e.g. water–methanol or acetone butanol.
2. Solvation where only one of the compounds is self-associating, e.g. water with ethers or acetone and ethanol with chloroform or acetone.
3. Solvation where none of the compounds is self-associating, e.g. chloroform–acetone.
4. Charge transfer (LA–LB) complexes not due to hydrogen bonding, e.g. nitrobenzene–mesitylene.
5. LA–LB complexes between a polar self-associating compound and an aromatic or olefinic hydrocarbon, e.g. water or alcohols with benzene.

Acid–base interactions can be understood and to some extent quantified with various parameters, e.g. the so-called solvatochromic (or Kamlet–Taft) parameters.[11] Self-associating compounds have both an acid and a base parameter. The associating complexes can be studied with spectroscopic techniques, as discussed in Chapter 7. Examples of these interactions and the dramatic effect they can have on phase behavior can be seen in Figures 2.7[14] and 2.8.

2.4 Some applications of intermolecular forces in model development

There are many examples, besides the understanding of phase diagrams, where knowledge of intermolecular forces can be utilized in model development. A direct use is in molecular simulation,

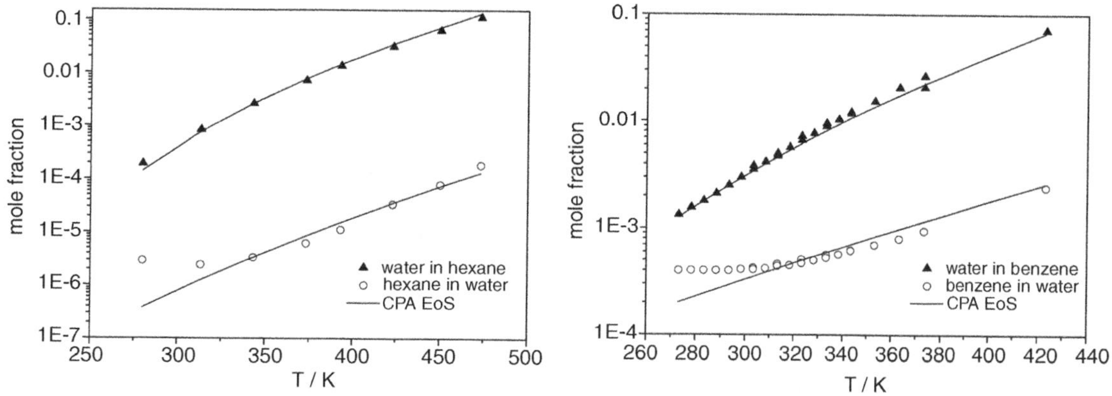

Figure 2.8 *LLE of water–hexane (left) and water–benzene (right), an illustration of charge-transfer complexes and their implications on phase behavior. Hydrocarbons are not miscible in water. However, the solubilities of aromatic hydrocarbons in water are much higher than those of aliphatic hydrocarbons in water. This is due to weak 'charge-transfer complexes' between aromatics and water (due to the so-called π electrons of the aromatic rings). In other words, while both aliphatic and aromatic hydrocarbons are not miscible in water, in the case of aromatics this is 'much less so'. The lines are correlations with the CPA equation of state (Chapter 9) using interaction parameters obtained from generalized correlations. The minimum in the temperature dependency of the hydrocarbon solubility in water is attributed to the hydrophobic effect*

which will not be covered here (for a recent review see Economou[12]). A few examples are mentioned below.

2.4.1 Improved terms in equations of state

In agreement with molecular simulation data (as Figure 2.9 shows), it has been shown that the expression which best describes the 'ideal repulsive fluid' (the hard-sphere one) is, in terms of the compressibility factor, given by the so-called Carnahan–Starling equation[15]:

$$Z = \frac{1+y+y^2-y^3}{(1-y)^3} \quad \text{with} \quad y = \frac{b}{4V} \tag{2.11}$$

The difference from the simple vdW repulsive term:

$$Z = \frac{1}{1-4y} \tag{2.12}$$

is evident.

Thus, novel molecular theories (e.g. SAFT) make use of this improved equation. Despite the success of SAFT theory, it is worth mentioning the statement of Mathias and Klotz[13] about the persistence in using the vdW term in many engineering equations of state: '$RT/(V-b)$ is wrong but somehow will exist in the most popular (reliable) models'.

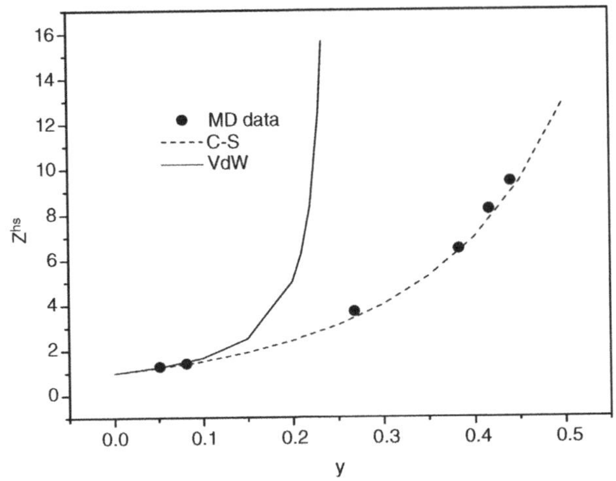

Figure 2.9 *Comparison of the Carnahan–Starling (C–S) and van der Waals (VdW) equations for the repulsive term against the molecular simulation (MD) data*

2.4.2 Combining rules in equations of state

The Mie expression for the intermolecular potential:

$$\Gamma = \Gamma_{repulsive} + \Gamma_{attractive} = \frac{A}{r^m} - \frac{B}{r^n} \tag{2.13}$$

$$\Gamma = \frac{m}{m-n}\left(\frac{m}{n}\right)^{\frac{n}{m-n}} \varepsilon \left[\left(\frac{\sigma}{r}\right)^m - \left(\frac{\sigma}{r}\right)^n\right] \tag{2.14}$$

represents the effect of vdW attractive forces and repulsions (for simplicity we have omitted the subscripts i, j and ij).

The two molecular parameters, energy ε and diameter σ, can be obtained from macroscopic data (vapor phase volumetric data via the second virial coefficients, transport properties such as viscosity and diffusivity). Values obtained from different sources do not always agree well with each other. A popular choice is the Lennard-Jones potential where we make use of the exponent $n = 6$ anticipated from the vdW forces (see Table 2.1), whereas $m = 12$:

$$\Gamma = 4\varepsilon\left[\left(\frac{\sigma}{r}\right)^{12} - \left(\frac{\sigma}{r}\right)^6\right] \tag{2.15}$$

Using these expressions as the starting point, it can be shown that (see Problems 5 and 6 on the companion website at www.wiley.com/go/Kontogeorgis):

$$\Gamma_{12} = \sqrt{\Gamma_1 \Gamma_2} \left(\frac{2\sqrt{I_1 I_2}}{I_1 + I_2} \right) \tag{2.16}$$

where Γ_i is the intermolecular potential between two molecules of the same type i and Γ_{ij} is the intermolecular potential between two different molecules of types i and j.

Equation (2.16) provides some justification for the geometric mean rule for the cross-energy interaction parameter often used in cubic equations of state (and modern molecular theories):

$$a_{12} = \sqrt{a_1 a_2} \tag{2.17}$$

2.4.3 Beyond the Lennard-Jones Potential

Both the Mie and Lennard-Jones (LJ) equations express the potential energy between two molecules only as a function of their distance, not their orientation. This is, strictly speaking, correct only for molecules with spherical force fields such as argon, krypton and xenon, or 'simple' molecules in Pitzer's definition: oxygen, nitrogen and carbon monoxide. For large/polar molecules, their relative orientation does play an important role and their potential function cannot be expected to follow such simple equations. However, the fact that the reduced intermolecular potential is a universal function of a reduced distance gives theoretical validation to the so-called 'two-parameter corresponding states principle' on which many practical engineering models are based, such as the cubic equations of state discussed in Chapter 3. The limitations of the LJ potential become apparent as:

- Different sets of LJ parameters often result from different properties (virial, transport) especially for polar ('non-simple') molecules.
- Even for Ar, one set of LJ parameters cannot fit the second virial coefficients over the extensive temperature range for which they are available (−140 to +150 °C).

Clearly real molecules do not always behave as LJ molecules. Still, the LJ potential is very useful in many practical situations, even for rather complex molecules, but not for very accurate calculations. In cases when even the LJ potential may be difficult to use, its simplified version in the form of the square-well potential can be used. The use of this potential instead of the LJ one largely simplifies the numerical calculations in many cases.

Other useful potential functions include (see Prausnitz *et al.*[3] for an extensive discussion):

- The Sutherland potential (a LJ term for the attraction and a hard sphere for the repulsion).
- The Mie 6 (the exponent of the repulsive term, m, is left to vary, while $n = 6$).
- The Stockmayer potential, which contains an extra term for dipolar effects.
- The three-variable Kihara potential.

These potential functions are summarized in Table 2.9. In the typical cases, the three-parameter potential functions represent properties (e.g. virial coefficients) better than the two-parameter ones. Of special interest is the Kihara potential in which, unlike the LJ one, it is assumed that molecules have a hard core surrounded by soft electron clouds. The LJ potential assumes that molecules are point centers with the possibility of full interpenetration. The physical picture of the molecular interactions represented by the Kihara potential is more realistic than that of LJ but it also has one more adjustable parameter than LJ. With its three parameters fitted to data, the Kihara potential can satisfactorily fit the Ar virial coefficients over the whole temperature range. It generally performs better than the LJ in representing virial coefficient data over extensive temperature ranges and it is especially useful in correlating low-temperature virial coefficients.

2.4.4 Mixing rules

The mixing rule accepted from statistical mechanics for the second virial coefficient:

$$B = \sum_i \sum_j x_i x_j B_{ij} \tag{2.18}$$

provides a limit that mixing rules for the parameters of equations of state should obey. This is indeed the case for the vdW one-fluid mixing rules often used in cubic equations of state:

$$a = \sum_i \sum_j x_i x_j a_{ij} \tag{2.19}$$

$$b = \sum_i \sum_j x_i x_j b_{ij} \tag{2.20}$$

but not for several recent mixing rules (of the EoS/G^E type), discussed in Chapter 6.

The virial equation is given by:

$$Z = 1 + \frac{B}{V} + \frac{C}{V^2} + \frac{D}{V^3} + \dots \tag{2.21}$$

where B, C, D,... are the second, third, fourth,... virial coefficients.

Finally, testing a model's performance against second virial coefficient data is a sensitive test because second virial coefficients directly reflect intermolecular forces. The second virial coefficients are linked to the intermolecular potential Γ via the equation:

$$B = 2\pi N_A \int_0^\infty \left[1 - \exp\left(-\frac{\Gamma}{kT}\right)\right] r^2 dr \tag{2.22}$$

Table 2.9 *Functions of some intermolecular potentials for rather simple molecules. The number of (adjustable) parameters is given in the first column in parentheses*

Name [number of adjustable parameters]	Function	Comments
Ideal gas [0]	$\Gamma = 0$	The simplest of all potentials. Only valid for ideal gases
Hard sphere [1]	$\Gamma = \infty \quad r \leq \sigma$ $\Gamma = 0 \quad r > \sigma$	The 'ideal' repulsive potential. Attractive forces are ignored. Approximation of real potential at high temperatures
Square well [3]	$\Gamma = \infty \quad r < \sigma$ $\Gamma = -\varepsilon \quad \sigma < r < \lambda\sigma$ $\Gamma = 0 \quad r > \lambda\sigma$	A simplified form of the LJ potential. Mathematically simple and useful for practical applications
Triangular well [3]	$\Gamma = \infty \quad r < \sigma$ $\Gamma = \frac{\lambda\varepsilon}{\lambda - 1}\left[\frac{r}{\lambda\sigma} - 1\right] \quad \sigma < r < \lambda\sigma$ $\Gamma = 0 \quad r > \lambda\sigma$	Not widely used. Mainly of academic interest
Sutherland [2]	$\Gamma = \infty \quad r \leq \sigma$ $\Gamma = -\frac{K}{r^6} \quad r > \sigma$	Hard-sphere model combined with London's theory
Mie [4]	$\Gamma = \frac{m}{m-n}\left(\frac{m}{n}\right)^{\frac{n}{m-n}}\varepsilon\left[\left(\frac{\sigma}{r}\right)^m - \left(\frac{\sigma}{r}\right)^n\right]$	Besides the LJ, other known variations are the (7,28) for polyatomic molecules, e.g. cyclohexane and SiF_4, and the (3,9) in connection with lattice theories
Lennard-Jones (Mie 6,12) [2]	$\Gamma = 4\varepsilon\left[\left(\frac{\sigma}{r}\right)^{12} - \left(\frac{\sigma}{r}\right)^6\right]$	One of the most widely used potentials. Rigorously valid for non-polar molecules over a wide range of conditions
Exp-6 (modified Buckingham) [3]	$\Gamma = \frac{\varepsilon}{1-(6/\gamma)}\left\{\frac{6}{\gamma}\exp\left[\gamma\left(1 - \frac{r}{r_{\min}}\right)\right] - \left(\frac{r_{\min}}{r}\right)^6\right\}$	Repulsive term expressed as exponential function of r, in agreement with theoretical calculations
Kihara (rigid spherical core) [3]	$\Gamma = \infty \quad r < 2a$ $\Gamma = 4\varepsilon\left[\left(\frac{\sigma-2a}{r-2a}\right)^{12} - \left(\frac{\sigma-2a}{r-2a}\right)^6\right] \quad r \geq 2a$	Molecules possessing hard cores but surrounded by soft electron clouds. Widely used in gas hydrate studies. Often more successful than the LJ potential for complex fluids
Stockmayer [2]	$\Gamma = 4\varepsilon\left[\left(\frac{\sigma}{r}\right)^{12} - \left(\frac{\sigma}{r}\right)^6\right] + \frac{\mu^2}{r^3}f(\vartheta_i)$	Valid for polar molecules. $f(\vartheta_i)$ is a known function of the angles determining the orientation of the dipoles

2.5 Concluding remarks

Intermolecular forces play a crucial role in understanding phase behavior and thermodynamic model development. Coulombic forces (of importance in electrolyte solutions) are very strong and long range, while most other forces are far less strong and substantially more short range but often equally important. Of the three types of vdW forces, polar and dispersion are the most important ones, with the latter being universal and the former especially strong for highly polar molecules, e.g. those having dipole moments above 1 Debye. Quadrupole forces are less important and more short range than the vdW ones, but can be important at low temperatures for strongly quadrupolar molecules, e.g. CO_2. Quasi-chemical forces, especially hydrogen bonding ones, are very important and often dominating in molecules such as water, alcohols, organic acids, amines, glycols and many biomolecules and polymers. The hydrophobic effect in water and solvating interactions are attributed to the hydrogen bonding or in general to the Lewis acid–Lewis base interactions. For macromolecules and, in general, for particles/droplets in the colloid domain, the vdW forces are much longer range than the forces between molecules. But both intermolecular and interparticle or interfacial forces depend on the intervening medium, which can be quantified via the dielectric constants and refractive indexes or the Hamaker constants. The vdW forces are typically attractive, but exactly because of the presence of an intervening medium, they can be repulsive in some systems consisting of at least two different types of molecules or particles. Intermolecular potential functions can be used to represent intermolecular forces and are often used directly in thermodynamic models. The very simple hard-sphere potential can be mathematically expressed by the Carnahan–Starling equation and is often considered a model term for repulsive forces. The rather simple square-well or the Lennard-Jones potentials include both repulsive and attractive contributions and are realistic model potentials for simple molecules. In the case of the Lennard-Jones one, the distance dependency of the potential energy is the same as that indicated by the vdW forces. Input from the intermolecular forces has found numerous direct or indirect uses in the thermodynamic models. For example, second virial coefficients depend on the intermolecular potential, while the geometric mean rule for the combining rule typically used for the energy parameter in equations of state has its origin in the geometric mean rule for the intermolecular potential, as derived from the dispersion (London) forces. Concepts from intermolecular forces will be used throughout the book in the derivation and explanation of thermodynamic models and phase behavior.

References

1. J. N. Israelachvili, *Intermolecular and Surface Forces*. Academic Press, 1985.
2. A.D. McLachlan, *Proc. R. Soc.*, **1963**, *Ser. A 271, 387*; **1963**, *274*, 80.
3. J.M. Prausnitz, R.N. Lichtenthaler, E.G. de Azevedo, *Molecular Thermodynamics of Fluid-Phase Equilibria* (3rd edition). Prentice Hall International, 1999.
4. D.P. Tassios, *Applied Chemical Engineering Thermodynamics*. Springer-Verlag, 1993.
5. P.K. Jog, W.G. Chapman, *Mol. Phys.*, **1999**, *97(3)*, 307.
6. F. Tumakaka, J. Gross, G. Sadowski, *Fluid Phase Equilib.*, **2005**, *228–229*, 89.
7. J. Goodwin, *Colloids and Interfaces with Surfactants and Polymers: An Introduction*. John Wiley & Sons, Ltd, 2004.
8. D. Myers, *Surfaces, Interfaces, and Colloids: Principles and Applications*. VCH, 1991.
9. S.J. Gilland I. Wadso, *Proc. Natl Acad. Sci. USA*, **1976**, *73*, 2955.
10. N. von Solms, M.L. Michelsen, C.P. Passos, S.O. Derawi, G.M. Kontogeorgis, *Ind. Eng. Chem. Res.*, **2006**, *45*, 5368.

11. M.J. Kamlet, J.M. Abboud, M.H. Abraham, R.W. Taft, *J. Org. Chem.*, **1983**, *48*, 2877.
12. I.G. Economou. Molecular simulation of phase equilibria for industrial applications. In: G.M. Kontogeorgis, R. Gani, *Computer-Aided Property Estimation for Process and Product Design*. Elsevier, 2004.
13. P.M. Mathias, H.C. Klotz, *Chem. Eng. Prog.*, **1994**, *90*, 67.
14. G.K. Folas, G.M Kontogeorgis, M.L. Michelsen, E.H. Stenby, *Ind. Eng. Chem. Res.*, **2006**, *45(4)*, 1527.
15. N.F. Carnahan, K.E. Starling, *J. Chem. Physics*, **1969**, *51*(2), 635.

Part B

The Classical Models

3

Cubic Equations of State: The Classical Mixing Rules

3.1 General

Cubic equations of state are classical *high-pressure* models. From a thermodynamic point of view, the term 'high pressure' refers to pressures high enough so as to have a significant effect on the thermodynamic properties of both phases, typically over 15–20 bar. High-pressure vapor–liquid equilibria (VLE) can be more complex than low-pressure VLE, since at high pressures both phases are non-ideal. At low pressures, the main source of non-ideality is the liquid phase. The non-ideality of the vapor phase at low pressures is typically about 10% and it can be estimated by, for example, corresponding states methods or the virial equation. K-charts and the Chao–Seader are popular methods for high-pressure VLE calculations for systems containing hydrocarbons and gases. By far, though, the most popular method is the equations of state, especially the cubic ones. In most cases, high-pressure VLE as found in nature and the chemical industry apply to mixtures with at least one subcritical and one supercritical component. In many oil- and gas-related systems, hydrocarbons are present together with gases (methane, ethane, CO_2, N_2, etc.) and sometimes water is also present.

The two- and especially the three-parameter cubic equations of state (EoS) represent a family of classical but still very useful and widely applied engineering models. The most well-known EoS are the van der Waals (vdW), Redlich–Kwong (RK) (now mostly of historical value) and especially the Soave–Redlich–Kwong (SRK) and Peng–Robinson (PR) equations; the last two are typically employed in the petroleum and chemical industries. Such cubic EoS are still the primary choice of models today for petrochemicals, gas processing and air separation.[1,2] The most well-known cubic EoS together with the typical expressions often used for estimating their parameters are shown in Table 3.1.

With reference to Table 3.1 and the methods shown for the estimation of the EoS parameters: T_c is the critical temperature, P_c is the critical pressure and T_r is the reduced temperature ($= T/T_c$); and the generalization of the energy parameter as a function of temperature and the acentric factor, ω, was first proposed by Soave.[5] The acentric factor, introduced by Pitzer, represents a measure of the acentricity (non-sphericity) of the molecule:

$$\omega = -\log P_r^{sat}|_{Tr=0.7} - 1.00 \tag{3.1}$$

where P_r^{sat} is the reduced vapor pressure ($= P^{sat}/P_c$).

Thermodynamic Models for Industrial Applications: From Classical and Advanced Mixing Rules to Association Theories
Georgios M. Kontogeorgis and Georgios K. Folas

***Table 3.1** The most important cubic EoS and the 'classical' way of estimating their parameters (based on critical point, T_c and P_c and vapor pressures). EoS such as vdW, SRK and PR equations are called 'cubic' because they are of third degree when solved with respect to volume. Many of these EoS employ the classical vdW repulsive term ($RT/(V-b)$) and different expressions for the attractive term. The repulsive term corrects for the finite volume of molecules (b) and the attractive term accounts for the intermolecular forces (a). If a = b = 0, then the ideal gas equation is obtained (PV = nRT). T_c is the critical temperature, P_c is the critical pressure and ω is the acentric factor. R is the ideal gas constant*

EoS	Equation	Energy and co-volume parameters	Critical Compressibility factor
van der Waals[3] (vdW)	$P=\frac{RT}{V-b}-\frac{a}{V^2}$	$a=\frac{27}{64}\frac{(RT_c)^2}{P_c}$ $b=\frac{1}{8}\frac{RT_c}{P_c}$	$Z_c=0.375$
Redlich–Kwong[4] (RK)	$P=\frac{RT}{V-b}-\frac{a}{V(V+b)\sqrt{T}}$	$a=0.427\,48\frac{(R^2T_c^{2.5})}{P_c}$ $b=0.086\,64\frac{RT_c}{P_c}$	$Z_c=0.333$
Soave–Redlich-Kwong (SRK)	$P=\frac{RT}{V-b}-\frac{a(T)}{V(V+b)}$	$a_c=0.427\,48\frac{(RT_c)^2}{P_c}$ $b=0.086\,64\frac{RT_c}{P_c}$	$Z_c=0.333$
Soave[5]		$a(T)=a_c\left[1+m(1-\sqrt{T_r})\right]^2$ $m=0.48+1.574\omega-0.176\omega^2$	
Peng–Robinson (PR)	$P=\frac{RT}{V-b}-\frac{a(T)}{V(V+b)+b(V-b)}$	$a_c=0.457\,24\frac{(RT_c)^2}{P_c}$ $b=0.077\,80\frac{RT_c}{P_c}$	$Z_c=0.307$
Peng and Robinson[6]		$a(T)=a_c\left[1+m(1-\sqrt{T_r})\right]^2$ $m=0.374\,64+1.542\,26\omega$ $-0.269\,92\omega^2$	

The acentric factor increases with molecular weight, but it is affected by the polar character of molecules as well. The acentric factor is almost linearly related to the critical compressibility factor Z_c. Experimental Z_c values are in the range 0.26–0.28 for about two-thirds of all compounds; they can be as low as 0.22–0.24 for polar compounds (water, methanol, ammonia). It should be noted that several more $m(\omega)$ generalizations in addition to those shown in Table 3.1 have been proposed for both SRK[7,8] and PR[9].

The parameters of cubic (and other) EoS can be estimated in various ways (and not 'just' from critical point/vapor pressures, which is the approach illustrated in Table 3.1). The subject will be discussed later (Sections 3.2 and 3.4.1). Various estimation methods for pure compound parameters are summarized in

Table 3.2 *Methods for estimating the pure compound parameters of cubic EoS*

Method	Equations	Comments
First principles (equations shown here for the vdW EoS but similar ones can be derived for other cubic EoS)	$a = \frac{2\pi C N_A^2}{(n-3)\sigma^{n-3}} = \frac{2\pi C N_A^2}{3\sigma^3}$ $C = 1.05 \times 10^{-76} ab$ J m^6 ($n = 6$ in last equation) a in dm^6 atm/mol^2 b in dm^3/mol $b = \frac{2}{3}\pi N_A \sigma^3 = 4V_{molecule}$	Method rarely used in practice The C parameter includes polar, induction and dispersion contributions (see Tables 2.1 and 3.3)
Critical properties (T_c and P_c)	Table 3.1 (for vdW and RK), based on: $\left(\frac{\partial P}{\partial V}\right)_c = \left(\frac{\partial^2 P}{\partial V^2}\right)_c = 0$	Unsatisfactory vapor pressures away from the critical point
Critical properties and vapor pressure	See Table 3.1 (SRK and PR)	The Soave-type generalization is strictly valid for hydrocarbons and non-polar molecules alone and specific equations such as the so-called Mathias–Copeman[10] method must be used for polar and hydrogen bonding compounds
Vapor pressure and liquid densities	See Tables 3.4 and 3.5 for some values and Section 3.4.1 for a discussion	No accurate representation of the critical point

Table 3.2. Although typically considered purely empirical, these cubic EoS do possess certain characteristics, which justifies the name 'semi-empirical'. These characteristics are summarized in Table 3.3.

The most widely used cubic EoS (Table 3.1) are two-parameter models. Extension to mixtures requires, therefore, mixing rules for the two parameters of the EoS, the energy parameter and the co-volume one. One way to extend the cubic EoS to mixtures, which is widely employed, is via the so-called van der Waals one-fluid (vdW1f) mixing rules (quadratic composition dependency for both parameters) and the classical combining rules, i.e. the geometric mean rule for the cross-energy and the arithmetic mean rule for the cross co-volume parameter:

$$a = \sum_{i=1}^{n}\sum_{j=1}^{n} x_i x_j a_{ij}$$
$$b = \sum_{i=1}^{n}\sum_{j=1}^{n} x_i x_j b_{ij} \tag{3.2}$$

$$a_{ij} = \sqrt{a_i a_j}(1-k_{ij})$$
$$b_{ij} = \frac{b_i + b_j}{2}(1-l_{ij}) \tag{3.3}$$

Table 3.3 *Five reasons why cubic EoS are semi-empirical but not completely empirical models! See Vera and Prausnitz[11] for a discussion of the vdW/cubic EoS theory and their derivation/link to statistical thermodynamics*

Reason	Equation/explanation
They account for free volumes[a]	$V_f = V - V^* = V - b$
They account for the basic vdW forces (see also Chapter 2)	$a = f(C)$ Table 3.2 $\Gamma_{ij} = -\frac{C}{r^6} = -\left(\frac{1}{3}\frac{\mu_i^2\mu_j^2}{kT} + \alpha_i\mu_j^2 + \alpha_j\mu_i^2 + \frac{3}{2}\frac{\alpha_i\alpha_j(I_iI_j)}{I_i+I_j}\right)\frac{1}{(4\pi\varepsilon_0)^2 r^6}$ See Table 2.1 in Chapter 2
Representations of two- or three-parameter corresponding states principle (CSP)	Two-parameter CSP: all fluids have the same compressibility factor Z (= same deviations from ideal gas behavior) at the same reduced temperature and pressure, T_r and P_r (Figure 3.1), i.e. same $Z(T_r, P_r)$ Three-parameter CSP: all fluids have the same compressibility factor Z (= same deviations from ideal gas behavior) at the same reduced temperature and pressure, T_r and P_r, and the same value of the acentric factor, i.e. same $Z(T_r, P_r, \omega)$
VdW1f mixing rules (Equations (3.2)) satisfy the quadratic mixing rule for the second virial coefficient	$B = b - a/RT$ (SRK, PR) $B = \sum_i \sum_j x_i x_j B_{ij}$
Geometric mean rule for cross-energy parameter derived from London theory (Chapter 2)	$\Gamma_{ij} = \sqrt{\Gamma_i\Gamma_j} \rightarrow a_{ij} = \sqrt{a_i a_j}$ (under certain assumptions, see Appendix 3.B)

[a] V^* is the so-called hard-core volume, which in the case of cubic EoS is equal to the co-volume.

[b] However, the concepts of free volume and hard-core volume are rather complex and various possibilities are available. A short discussion is provided in Appendix 3.A.

Notice that mixing rules depend on composition, while the combining rules do not! These mixing and combining rules imply that the mixing of molecules is random.

Of the two interaction parameters, k_{ij} is by far the most important one; it is typically fitted to phase equilibrium data. Two interaction parameters are often needed for complex polar systems and special cases, e.g. solid–gas phase equilibria (SGE), see Figure 3.7 (later). In the often used case where $l_{ij} = 0$, the mixing rule for the co-volume parameter is simplified to:

$$b = \sum_{i=1}^{n} x_i b_i \tag{3.4}$$

The great success of cubic EoS together with the vdW1f mixing rules lies especially in the ability for fast calculations and accurate representation of low- and high-pressure VLE for mixtures of hydrocarbons and hydrocarbons with gases (methane, N_2, CO_2, H_2S, etc.), mixtures which are especially important in the petroleum industry. For more complex mixtures and other types of phase equilibria, cubic EoS can have serious limitations, as explained later.

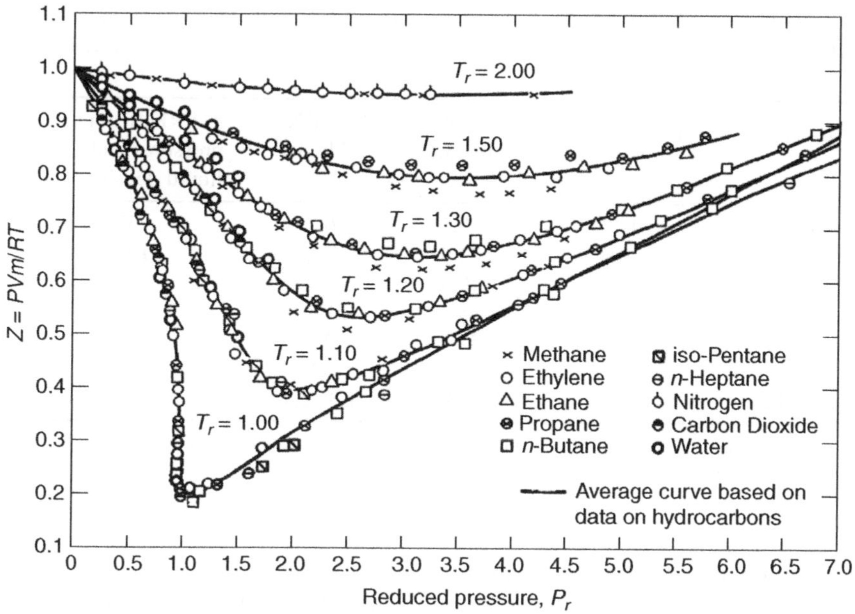

Figure 3.1 *The corresponding states principle: compressibility factor Z versus reduced pressure at various reduced temperatures for 10 gases. Notice that at low reduced pressures, the attractive forces dominate and Z decreases (negative virial coefficients), while Z increases at higher pressures, where repulsive forces dominate (positive virial coefficients). The figure illustrates the two-parameter corresponding states principle. Cubic EoS with a third parameter (acentric factor) are a representation of the three-parameter corresponding states principle, which is more accurate than the two-parameter version. From Laidler and Meisser[12] and G.-J. Su, Ind. Eng. Chem.,* ***1946****, 38(8), 803. Reprinted with permission from American Chemical Society (1946)*

3.2 On parameter estimation

3.2.1 Pure compounds

The equations for estimating the pure EoS parameters shown in Table 3.1 ensure accurate vapor pressure representation (for hydrocarbons and non-polar compounds) and accurate reproduction of the critical point of the compounds. Accurate representation of the vapor pressure of polar compounds and over a wide temperature range requires use of more complex temperature-dependent expressions[10] or a direct fit to vapor pressure data or DIPPR correlations. Densities are not well represented using SRK or PR with the parameters estimated by the method shown in Table 3.1 (critical point and vapor pressures). However, for VLE calculations by far the most important property is the vapor pressure, thus density representation is not always considered equally important to match.

Alternatively, see Table 3.2, cubic EoS parameters can be fitted simultaneously to vapor pressures and liquid densities, an approach widely used, as we will see later, for modern EoS like SAFT. Tables 3.4 and 3.5 compare the EoS parameter values estimated with various approaches for a few compounds. Especially for the co-volume parameter, several estimation methods are included, some of which are based on its physical significance in relation to the vdW volume, V_w, and the free volume, V_f (Table 3.3).

Table 3.4 The co-volume parameter of cubic EoS for some compounds estimated from various methods (all values are in cm^3/mol). σ is the molecular diameter, b_c is the co-volume at the critical point, V_w is the vdW volume and N_A is Avogadro's number

Compound	$b = \frac{2}{3}\pi N_A \sigma^3$	$b = \frac{1}{6}\pi N_A \sigma^3$	b_c (vdW), Table 3.1	b_c (SRK), Table 3.1	b, SRK (from vapor pressures and liquid densities)	$b = 1.41\ V_w$
Propane	169	42.3	90.5	62.7	57.8	52.9
Hexane	265.6	66.4	173.9	120.5	107.9	96.3
Benzene	193	48.3	120.8	83.8	75	68.2
Decane	435	109	305	211	178	154.5
Argon	56	14	32	22		23.3
n-C_{16}			523	362	296	240.3
b/V_w	**3.4–4.5**	**1.0**	**1.9–3.1**	**1.3–2.1**	**1.5–1.7**	**1.41**

For comparison purposes, we can mention that the b/V_w ratio with PR using fitted parameters for vapor pressure and liquid density data is 1.7 for n-C_{16} and similar ratios are also obtained for heavier alkanes.[13,14]

The co-volume parameter of the EoS deserves further discussion. The co-volume or hard-core volume is a measure of the volume of a molecule where other molecules cannot penetrate. Due to molecular packing (which exists in the liquid as well as the solid state), this co-volume is expected to be higher than the molecular (or vdW) volume. Exactly how much higher will depend on the nature of the liquid structure. The reason for including $b = 1.41\ V_w$ in Table 3.4 lies in the well-accepted picture for the liquid state that each molecule has about 10 neighbor molecules, i.e. corresponding to a coordination number $Z=10$. It can be shown that the close-packed (FCC) structure with $Z=12$ (typical in the solid state) corresponds to $V^* = 1.35V_w$,[17] but 1.41 is the coefficient which corresponds to the more realistic picture of $Z=10$ for the liquid state. As will be discussed in Chapter 5, many local composition models (e.g. UNIQUAC and UNIFAC) also employ the assumption $Z=10$. The subject of hard-core volume and free volumes is further discussed in Appendix 3.A.

Table 3.5 The energy parameter of cubic EoS (given as T_c^2/P_c) for a few n-alkanes estimated from various methods (C_x is an n- alkane with x carbon atoms). The methods of Ting et al.[13] and Voutsas et al.[14] are based on PR and pseudo critical parameters (T_c,P_c) which are no longer the experimental ones, but are regressed based on simultaneous fitting to vapor pressure and liquid densities. The last method 'Fitted SRK' is the SRK with parameters also fitted to vapor pressures and liquid densities. The results with the SRK EoS are obtained, however, by using a modified expression for the energy parameter ($a(T) = a_0(1+c_1(1-\sqrt{T_r}))^2$, see Table 3.1) and fitting directly b, c_1 and a_0 to vapor pressures and liquid densities; hence the critical pressure is not used at all, since b is treated as a regressed parameter, while only the experimental T_c is used in the modified expression of the energy parameter. The first method designated as 'Classical' is based on T_c and P_c obtained from experimental data. All values are given in (K^2/bar) × 10^3

Alkane	Classical (exp. T_c and P_c)	Ting *et al.*[13] [PR]	Voutsas *et al.*[14] [PR]	'Fitted SRK', Yakoumis *et al.*[15]	'Fitted SRK' Oliveira *et al.*[16]
C_6	8.56	8.36	8.30	8.014	7.918
C_{16}	36.97	32.61	32.91	32.12	31.14
C_{24}	73.56	53.42	55.67	56.023	56.023
C_{36}	161.78	90.42	92.28	102.93	102.93
C_{40}	188.26	—	102.59	—	—

Table 3.6 *Example of the use of Equation (3.5) for screening GC methods for estimating the critical properties. As illustrated from some sample results (more are available in the publications listed in the text), the Constantinou–Gani[100] GC method usually performs best. For the Joback method it is of paramount importance to employ the experimental boiling point temperature in the estimation of the critical temperature. All the GC methods are described in Poling et al.[18]*

Compound	T_c/P_c (Equation (3.5))	T_c/P_c (Constantinou–Gani)	T_c/P_c (Joback)	T_c/P_c (Ambrose)
D-fructose	22.2	25.7	10.3	13.8
Sucrose	46.7	43.4	21.8	27.6
n-Eicosane	69.3		80.14	
1-Eicosanol	72		77.6	
Hexadecene	48.38	51.64	52.8	
Octadecene	57.38	60.64	62.68	
Decapentanoic acid	48.76	50.94	50.74	

Despite the various possibilities in parameter estimation, by far the most widely used method of parameter estimation with cubic EoS is based on the use of critical properties. Critical properties, if not experimentally available, can be estimated using group contribution (GC) methods, e.g. Joback, Ambrose or Constantinou and Gani methods (Poling *et al.*[18] offer an extensive review of these methods). The following equation can be used for screening estimation methods:[19–21]

$$\frac{T_c}{P_c} = 9.0673 + 0.433\,09\left(Q_w^{1.3} + Q_w^{1.95}\right) \tag{3.5}$$

T_c is the critical temperature in K, P_c is the critical pressure in bar and Q_w is the dimensionless vdW volume surface area, estimated from GC values using the Bondi method, as available in UNIFAC tables.[18] Some examples of screening GC methods using Equation (3.5) are shown in Table 3.6.

3.2.2 Mixtures

The interaction parameter k_{ij} (Equation (3.3)) is typically fitted to phase equilibrium data, although roughly for some systems its physical significance in combination with the vdW/London theory (Chapter 2) can be used to obtain rough estimates, as shown in Table 3.7.

Excellent agreement is obtained between 'experimental' C_{12} and the values calculated by the geometric mean (GM) rule (i.e. k_{ij} is close to zero) for mixtures of non-polar molecules (where dispersion forces dominate), e.g. Ne–methane or HCl–HJ. Serious deviations are seen, however, for polar and especially for aqueous mixtures with alkanes, due to the hydrophobic effect (discussed in Chapter 2). Notice especially the high positive k_{12} value for water–methane. This indicates that water and methane molecules tend to interact with their own kind rather that with each other, which eventually explains the observed immiscibility in water–hydrocarbon mixtures.

Although such calculations are highly simplified, they do give a feeling of some trends that are expected for the k_{ij} parameters. The interaction parameters are low for non-polar mixtures and can be very high for mixtures containing polar and especially hydrogen bonding fluids. These calculations often result in positive k_{ij} values or otherwise indicate that the geometric mean rule is typically not valid for polar mixtures. However, the interaction parameters calculated with this theoretical approach are often much higher than those actually

Table 3.7 Interaction parameters calculated from the vdW forces. C values are from Tassios,[22] Prausnitz et al.[23] and Israelachvili[24]. All C values are in 10^{-79} J m^6 (Chapter 2, Table 2.1). All values are calculated at 273 K except for the last two systems which are at 293 K

Mixture	C_{12} (theory, 'exp.')	C_{12} (GM rule) $[=\sqrt{C_1C_2}]$	Interaction parameter k_{12} $C_{12}=\sqrt{C_1C_2}(1-k_{12})$
CO–HCl	85.21	93.97	0.093
CC_6–CCl_4	1510	1509.2	−0.000 53
Water–HCl	144.3	185.9	0.224
Water–acetone	662.5	671.5	0.0134
Acetone–CC_6	959.5	1671	0.4258
CCl_4–ammonia	342.7	487.7	0.297
HC–HJ	205	214	0.042
Water–methane	67	119.1	0.4374

CO = carbon monoxide, CCl_4 = carbon tetrachloride, CC_6 = cyclohexane, HCl = hydrogen chloride, HJ = hydrogen iodide.

needed to fit the experimental data with cubic EoS. For example, for acetone–cyclohexane $k_{ij}=0.11$ is needed with SRK (compare to the value equal to 0.43 shown in Table 3.7). For acetone–methane and acetone–water the optimum values are negative, but for such complex systems, cubic EoS with vdW1f mixing rules do not provide good results. Thus, we can rarely use such theoretical approaches based on the C values for obtaining interaction parameter values for a wide range of mixtures. k_{ij} values should be obtained from other sources, typically fitted to experimental phase equilibrium (VLE or other) data. For mixtures of hydrocarbons, k_{ij} are close to zero, but for gas–hydrocarbon mixtures non-zero k_{ij} are needed, e.g. in the case of SRK around 0.08 for N_2 with many alkanes, 0.12–0.15 for CO_2/alkanes and 0.05–0.08 for H_2S–alkanes.[25] For more precise calculations, many equations, often in the form of generalized correlations, have been proposed, typically linked to or developed for specific cubic EoS. These correlations permit estimation of the interaction parameter from knowledge of the characteristics of the mixture, e.g. the acentric factor of the components and the reduced temperature. As an example for a translated and modified PR EoS,[9] the following correlations have been proposed:

CO_2/hydrocarbons:[26]

$$k_{ij} = a(\omega_j) + b(\omega_j)T_{ri} + c(\omega_j)T_{ri}^2 \tag{3.6}$$

where T_{ri} is the reduced temperature (T/T_c) of CO_2 and the parameters a, b and c are functions of the acentric factor of the alkane (ω). The equation performs satisfactorily, but a special form was required for the CO_2/methane system.

Methane–alkanes:[2]

$$k_{ij} = -0.134\,09\omega + 2.285\,43\omega^2 - 7.614\,55\omega^3 + 10.465\,65\omega^4 - 5.2351\omega^5 \tag{3.7}$$

where two different correlations are required depending on the chain length of n-alkane for alkane carbon numbers below 20; while for alkane carbon numbers above 20:

$$k_{ij} = -0.046\,33 - 0.043\,67\ln\omega \tag{3.8}$$

Nitrogen–alkanes:[28]

$$k_{ij} = Q(\omega_j) - \frac{T_{ri}^2 + A(\omega_j)}{T_{ri}^3 + C(\omega_j)} \tag{3.9}$$

where Q, A and C are functions of the acentric factor.

H_2S–alkanes:[29]

$$k_{ij} = 0.1029 - 0.1498\omega \tag{3.10}$$

Similar correlations have been developed for these families of mixtures (gas–hydrocarbons) for PR, SRK and other cubic EoS, e.g. Carroll and Mather,[30] for H_2S–alkanes. Stamataki *et al.*[31,32] presented successful applications of cubic EoS using these correlations to high-pressure VLE and volumetric behavior for mixtures of gases with hydrocarbons. It is interesting to note that the chain length dependency of k_{ij} depends on the gas involved. For example, in the case of the PR EoS, k_{ij} increases with increasing carbon number, N_c for nitrogen–alkanes (up to C_{16} for which data are available), but it decreases with N_c for CO_2–alkanes (data available up to C_{44}), while it has a mixed chain length dependency for methane–alkanes (it increases with chain length up to acentric factor values of almost one but k_{ij} decreases with chain length for the heavier alkanes). Temperature has also an effect on k_{ij} (often described by a U-type curve).

Alternative (to the vdW1f) mixing and combining rules have been proposed, see Appendix 3.B and Section 3.4.3 for a discussion. These alternative rules attempt to resolve some of the deficiencies of the classical mixing rules, and indeed they do so in some cases, but due to practical experience and familiarity, the vdW1f mixing and classical combining rules dominate for practical applications, at least for mixtures with gases and hydrocarbons. Considering the results for many mixtures and with different cubic EoS (SRK, PR) we can conclude as follows:

1. For gas–hydrocarbons, the k_{ij} are absolutely necessary and although the correlation of phase behavior is often excellent, the sensitivity to k_{ij} can be high. One example is illustrated in Figure 3.2, using PR with $k_{ij} = 0$ and a fitted interaction parameter. The EoS cannot represent the azeotrope of CO_2–ethane, while an excellent correlation is possible with a large positive value of k_{ij}. At even lower temperatures, even higher k_{ij} values are needed and the performance of the model is worse. This figure illustrates therefore the importance of quadrupole moment effects, especially at low temperatures. At even lower temperatures, e.g. for CO_2–decane, LLE is observed, which cannot be represented well with cubic EoS using the vdW1f mixing rules.
2. When one k_{ij} per binary mixture is used, satisfactory prediction of multicomponent VLE is obtained with SRK or PR for mixtures containing gases and hydrocarbons.[22]
3. Typically, k_{ij} decreases with increasing temperature, since polar and other intermolecular forces become less important at high temperatures.
4. In by far most cases, positive k_{ij} are needed (i.e. the geometric mean rule, Equation (3.3), overestimates the cross-interaction between molecules). Negative k_{ij} are required for several solvating systems, e.g. chloroform–acetone, Figure 3.3. SRK provides good VLE correlation even for chloroform–acetone, which exhibits negative deviations from Raoult's law, as can be seen by the maximum in the *Txy* plot (azeotrope). An explanation for such negative k_{ij} values is based on our understanding of intermolecular forces, as discussed in Chapter 2. The geometric mean rule employed for the cross-energy parameter is derived from the London theory of dispersion forces and cannot be expected to be valid for mixtures with strong interactions. In the case of strongly solvating systems, we may expect that the cross-energy

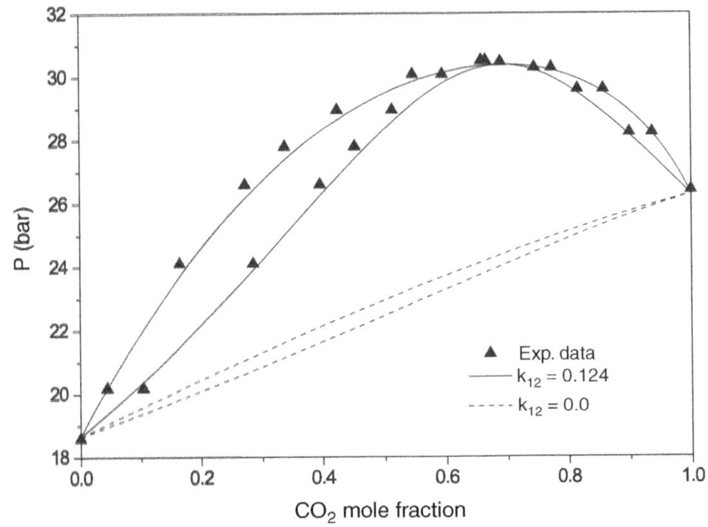

Figure 3.2 *Prediction and correlation of VLE for CO_2–ethane ($T = 263.15$ K) with the PR EoS using the vdW1f mixing rules and the classical combining rules (Equations (3.2)–(3.3)). Experimental data from Fredenslund and Mollerup, J. Chem. Soc., Faraday Trans. 1,* ***1974****, 70, 1653*

term is larger than the value provided by the geometric mean rule and this is why a negative k_{ij} is needed.

5. Good results are obtained with cubic EoS using the vdW1f mixing rules and k_{ij} for several polar systems as shown for two examples in Figure 3.3. This illustrates the great flexibility of cubic EoS and why they are often used beyond their expected range of applicability.

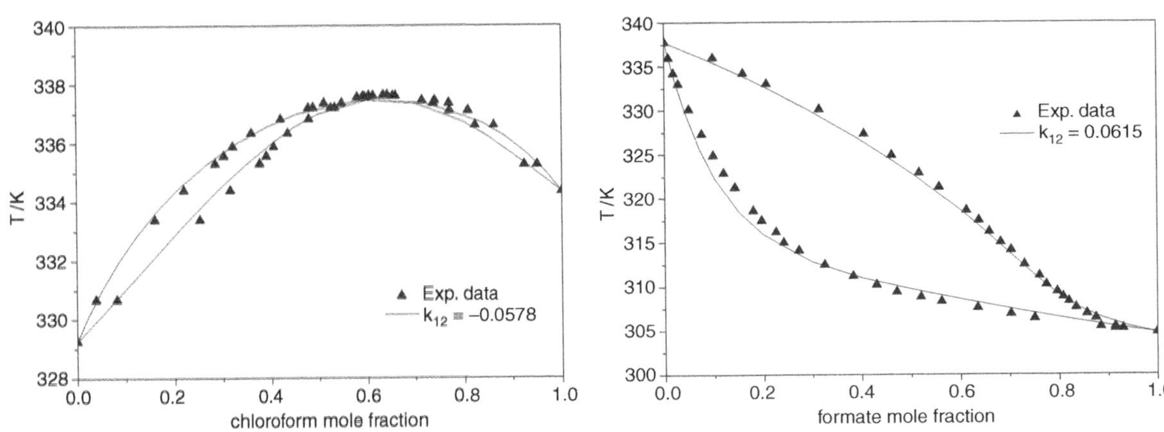

Figure 3.3 *Left: VLE correlation ($P = 1$ bar) for chloroform–acetone with SRK using the vdW1f mixing rules and the classical combining rules (Equations (3.2) and (3.3)) with $k_{12} = -0.0578$. Experimental data from Kojima et al., J. Chem. Eng. Data,* ***1991****, 36, 343. Right: VLE correlation for methyl formate–methanol with SRK at 1 bar using the vdW1f mixing rules and the classical combining rules (Equations (3.2) and (3.3)), using $k_{12} = 0.0615$. Experimental data from Kozub et al., J. Prakt. Chem.,* ***1962****, 17, 282*

3.3 Analysis of the advantages and shortcomings of cubic EoS

Such an analysis should necessarily consider the specific applications and conditions and moreover analyze the behavior against:

- the functional form of cubic EoS;
- the way in which pure compound parameters have been estimated;
- extension to mixtures and the mixing and combining rules chosen.

We can generally conclude that most investigations agree that:

- There are small differences in the VLE correlation among the widely used cubic EoS, e.g. SRK or PR, provided the same way of obtaining the pure parameters and the same mixing/combining rules are used. In some cases, e.g. gas hydrate calculations, absolute fugacity values are needed. It has been shown that SRK results to better fugacities than PR for some gases (methane, nitrogen).
- It is often 'more' important to correct for the combining rule for the energy parameter (rather than the co-volume one), except for asymmetric mixtures with hydrocarbons,[33] where l_{ij} is more important than k_{ij} (see Section 3.4.3 for a discussion of this point).

The discussion of the advantages and shortcomings of cubic EoS below is limited at this stage to the 'more usual' way cubic EoS are used, i.e. with parameters estimated using the critical point and vapor pressure data (approach 2, Table 3.2) and using the vdW1f mixing rules (Equations (3.2)–(3.3)). Additional discussions about the range of applicability of cubic EoS are available in the literature, both from academic[34–36] and industrial points of view.[43,37]

3.3.1 Advantages of cubic EoS

Tsonopoulos and Heidman[38] have nicely summarized these in their 1986 article as follows:

> *Cubic EoS are simple, reliable and allow for direct incorporation of critical conditions. We, in the petroleum industry, continue to find that such simple EoS are very useful high-pressure VLE models, and we found as yet no reason to use complex non-cubic equations of state.*

This statement is indeed largely true even today and the following points highlight the most important advantages of cubic EoS:

1. They are simple models capable of fast calculations.
2. They are applicable over a wide range of pressures and temperatures.
3. They are capable of describing properties of compounds in both liquid and vapor phases, and can therefore be used to predict phase equilibrium properties, such as vapor pressure, heat of vaporization, enthalpies and various other properties (speed of sound, heat capacities, etc.).
4. There is no need, in most cases, for more than one interaction parameter for gas–hydrocarbons and good correlations for such mixtures are obtained.
5. Satisfactory results are obtained both for low- and high-pressure VLE.
6. Often good multicomponent VLE prediction is achieved for mixtures containing hydrocarbons, gases and other non-polar compounds (using interaction parameters from binary data).
7. Many existing databases and correlations are available for k_{ij}.
8. Well-established vdW1f and classical combining rules work well for simple systems and also for correlating VLE of many polar mixtures.

Table 3.8 *Vapor pressure calculation of n-octacosane (n-C_{28}) using PR and various methods for estimating the critical properties, T_c, P_c. Modified from Kontogeorgis and Tassios,[39] where references to the methods below are also provided. % AAD is the average absolute deviation between experimental and calculated vapor pressures*

Estimation method	T_c (K)	P_c (bar)	Acentric factor	% AAD in vapor pressure
Magoulas–Tassios	829.86	7.28	1.193 94	3.1
Ambrose	843.02	8.86	1.153 19	4.4
Constantinou–Gani	823.20	7.19	1.253 19	7.4
Elhassan	854.53	8.16	1.011 77	15.8
Hu	838.89	8.41	1.169 53	4.5
Joback	864.42	6.55	0.827 53	64.6
Teja	842.11	9.69	1.200 80	11.2

Despite the above positive characteristics, there are some negative aspects of cubic EoS:

1. Calculations may, in some cases, be sensitive to the interaction parameter, k_{ij}, especially for gas–hydrocarbons.
2. Interaction parameters, often k_{ij}, depend on temperature.
3. Cubic EoS do not yield liquid volumes in good agreement with experimental values, unless a volume translation is used.[9] PR is better than SRK in that respect, but good liquid volumes do not affect the performance of the EoS for phase equilibria, at least for the VLE calculations.
4. Vapor pressures can be very sensitive to the critical properties, especially the critical temperature, T_c, and at low temperatures (and pressures), as Table 3.8 illustrates. For small compounds, even those with well-known critical properties, problems may occur at very low temperatures and vapor pressures especially close to the triple point. For example, for propane at the triple point (1.6×10^{-9} bar), the deviation between the experimental and predicted vapor pressure is 100% for PR and 70% for a translated form of PR.[22]

However, the above negative aspects are, for practical purposes, often less crucial compared to the serious limitations of cubic EoS discussed in the next section.

3.3.2 Shortcomings and limitations of cubic EoS

The major limitations of cubic EoS are the following:

1. In most cases, predictions (i.e. setting all the interaction parameters to zero) are not possible for binary systems and a k_{ij} fitted to experimental phase equilibrium data is needed even for gases with hydrocarbons. For gas-polar mixtures, often a temperature-dependent interaction parameter is needed, as illustrated in Figure 3.4.
2. Often a poor correlation of complex VLE of polar mixtures is obtained, as shown in Figures 3.5 and 3.6, and as discussed in extensive comparisons in the literature.[40,44] Attempts to correlate polar/strongly self- and cross-associating systems with cubic EoS and vdW1f mixing rules may result in false phase splits (Figure 3.6). Equally poor behavior has been reported for other cross-associating mixtures or systems with polar and non-polar compounds, e.g. acetone–water, butanol–water, amine–alcohol or acetone–hydrocarbons.[41,42]

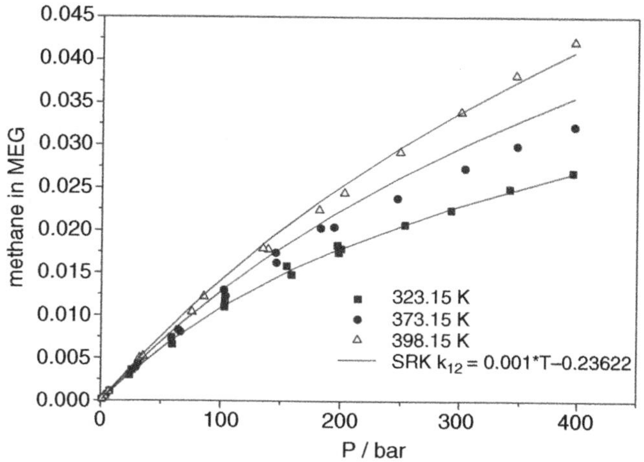

Figure 3.4 *VLE correlation of MEG–methane with SRK using the vdW1f mixing rules (Equations (3.2) and (3.3)) and temperature-dependent interaction parameter: $k_{ij} = 0.001 \times T - 0.236\,22$. Experimental data are from Zheng et al., Fluid Phase Equilib.,* ***1999****, 155, 277; Jou et al., Can. J. Chem. Eng.,* ***1994****, 72*

3. When two interaction parameters are used in cubic EoS (e.g. both k_{ij} and l_j), the models often become highly flexible and can represent complex VLE, e.g. for H_2S–water[23] or solid–gas equilibria (SGE)[45], as shown in Figure 3.7. Unfortunately, in most cases these two interaction parameters cannot be easily generalized as a function of some characteristics of the molecules involved, e.g. molecular weight and polarity.
4. As Tables 3.9–3.10 and Figures 3.8–3.9 show, cubic EoS typically yield positive deviations from Raoult's law (i.e. activity coefficient values above unity) for nearly athermal alkane mixtures, while the experimental data show negative deviations from Raoult's law, which moreover increase with increasing

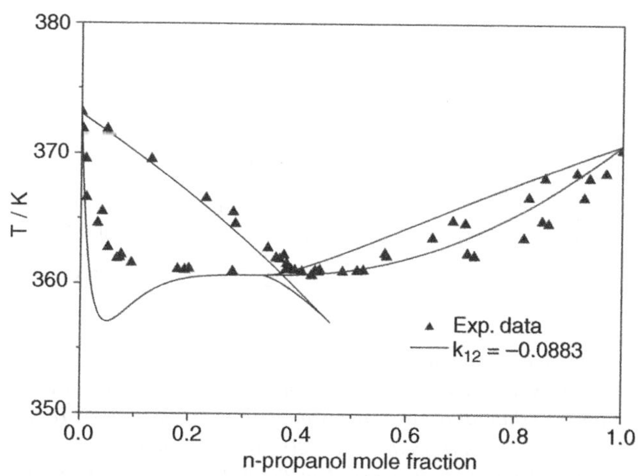

Figure 3.5 *VLE correlation (P = 1 bar) for propanol–water with SRK using the vdW1f mixing rules (Equations (3.2) and (3.3)) and $k_{ij} = -0.0883$. Experimental data are from Ellis and Thwaites, Chem. Process Eng.,* ***1955****, 36, 358 and Murti and VanWinkle, Chem. Eng. Data Ser.,* ***1958****, 3, 72*

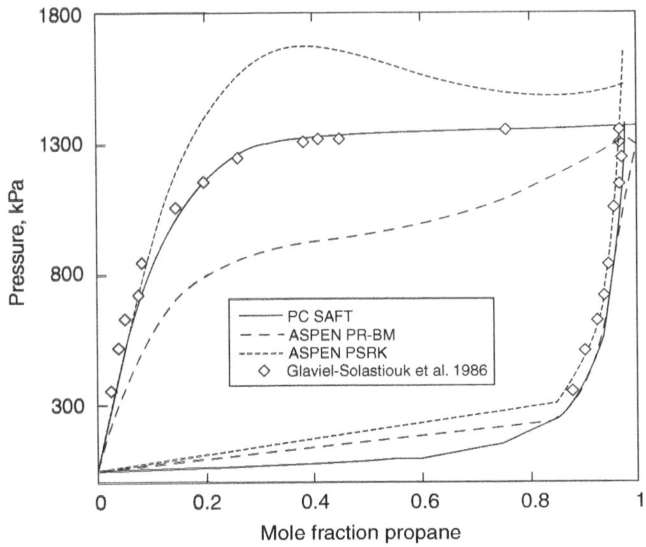

Figure 3.6 *VLE correlation for methanol–propane at 313 K using PC–SAFT and two cubic EoS available in a commercial simulator. The two cubic EoS predict either near ideal solution behavior or erroneous liquid–liquid phase splits. From Yarrison and Chapman.[44] Reprinted with permission from Fluid Phase Equilibria, A systematic study of methanol + n-alkane vapor-liquid and liquid-liquid equilibria using the CK–SAFT and PC–SAFT equations of state by Matt Yarrison and Walter G. Chapman, 226, 195 Copyright (2004) Elsevier*

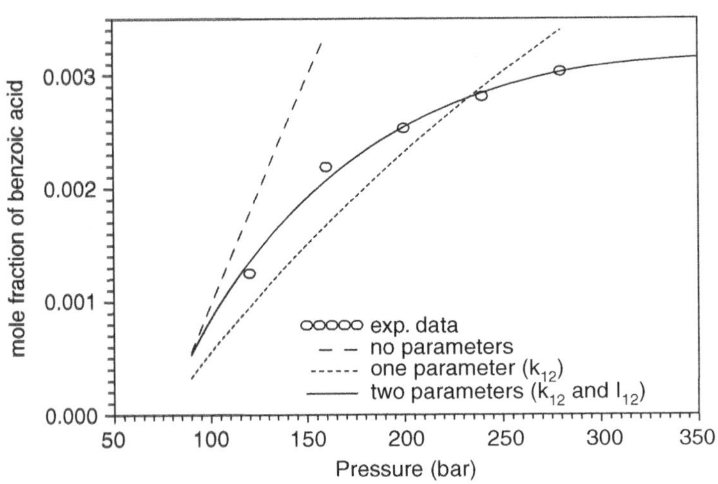

Figure 3.7 *Solid–gas equilibrium (SGE) calculations with PR and the vdW1f mixing rules (Equations (3.2) and (3.3)). The solubility of benzoic acid in CO_2 is shown as a function of pressure, without any interaction parameters, using only k_{ij} (= 0.043; 11% deviation in mole fraction) or using both k_{ij} and l_{ij} (= −0.036 and −0.186, respectively; 3% deviation in mole fraction). A constant k_{ij} = 0.1 gives good predictions (20–30%) in some cases for the solubility of many solids (aromatic hydrocarbons, phenols and ketones) in CO_2. The flexibility of cubic EoS in correlating SGE using two interaction parameters (k_{ij} and l_{ij}) is significant, e.g. the errors in solubility with PR for the solubility of stearic acid (328 K) and cholesterol (323 K) in CO_2 are 92% and 154% with one k_{ij} and they fall to only 13% and 3.5%, respectively, using two interaction parameters. Reprinted with permission from Journal of Supercritical Fluids, Prediction of solid-gas equilibria with the Peng-Robinson equation of state by Coutsikos, Magoulas and Kontogeorgis, 25, 3, 197–212 Copyright (2003) Elsevier*

Table 3.9 *Activity coefficient of heptane at infinite dilution in n-alkanes with Peng–Robinson (using the vdW1f mixing rules and $k_{ij} = 0$) compared to experimental data. Experimental data from Parcher et al., J. Chem. Eng. Data, **1975**, 20, 145. C_x indicates an n-alkane with x carbon atoms*

Heptane +	Experimental	Peng–Robinson ($k_{ij} = 0$)
C_{15}	0.960	1.019
C_{16}	0.912	1.032
C_{18}	0.893	1.037
C_{24}	0.831	1.070
C_{28}	0.768	1.116
C_{32}	0.721	1.173
C_{36}	0.672	1.241

size asymmetry (the activity coefficients become progressively much lower than unity). This is a serious problem which, however, as these figures illustrate, can be resolved by using different rules than the vdW1f mixing and combining rules, e.g. equation (3.11).

5. LLE of highly immiscible systems (Figure 3.10) like water or glycols with alkanes is not correlated satisfactorily. In most cases, a single interaction parameter cannot represent both solubilities in such systems. Even for 'less difficult' mixtures like methanol–alkanes, representation of LLE at low temperatures is not satisfactory using the k_{ij} obtained from VLE (see Figure 3.11 and Gupta and Olson)[47]. In general, LLE is not very well correlated with cubic EoS even for non-polar systems, e.g. CO_2–decane at low temperatures.[48] Other types of phase equilibria such as SLE (Figure 3.11) are often difficult to represent satisfactorily with cubic EoS/vdW1f mixing rules, when associating fluids are involved.
6. Results are poor for complex, multicomponent VLE and LLE, especially in the presence of associating compounds and water.
7. Cubic EoS cannot be easily extended to 'more complex' molecules like electrolytes and biomolecules.

Limitations 1–4 above may largely be attributed to the parameter estimation for the pure compounds and especially their extension to mixtures in the sense that the performance of cubic EoS can be drastically improved with advanced mixing rules, especially the so-called EoS/G^E ones (which are discussed in Chapter 6). Problems 5–7 above (and others) are, however, more serious and can partially be addressed within the framework of cubic EoS with more advanced (than the vdW1f) mixing rules. Entirely different

Table 3.10 *Activity coefficient of n-alkanes at infinite dilution in n-heptane with Peng–Robinson (using the vdW1f mixing rules and $k_{ij} = 0$) compared to experimental data. Experimental data from Kniaź, Fluid Phase Equilib., **1991**, 68, 35 and Parcher et al., J. Chem. Eng. Data, **1975**, 20, 145. C_x indicates an n-alkane with x carbon atoms*

Alkane + heptane	Experimental	Peng–Robinson ($k_{ij} = 0$)
C_{16}	0.912	1.085
C_{18}	0.910	1.202
C_{20}	0.838	1.359
C_{24}	0.729	2.652
C_{26}	0.729	4.090
C_{36}	0.490	219.4

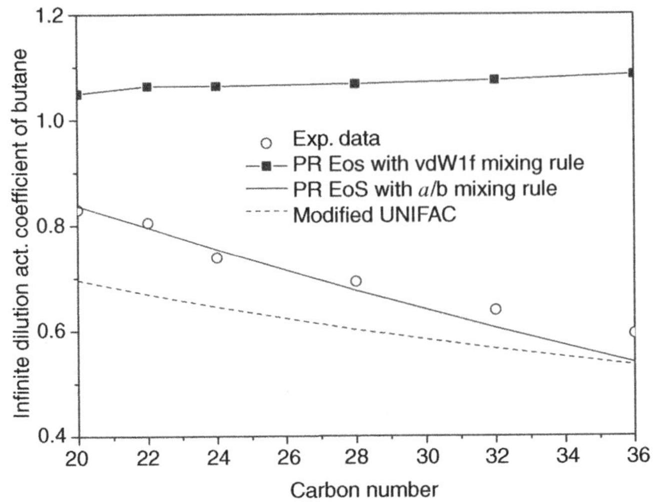

Figure 3.8 *Activity coefficients at infinite dilution of n-butane in alkane solvents at 373 K as a function of the alkane carbon number using the Peng–Robinson EoS. Results are shown using the vdW1f mixing rules (Equations (3.2) and (3.3), with* $k_{ij}=0$*) and the a/b mixing rule:*

$$\frac{a}{b}=\sum_{i=1}^{n}x_i\frac{a_i}{b_i} \tag{3.11}$$

For comparison the results are shown with the modified UNIFAC activity coefficient model[46] *(see also Chapter 5). Experimental data from Parcher et al., J. Chem. Eng. Data,* ***1975****, 20, 145*

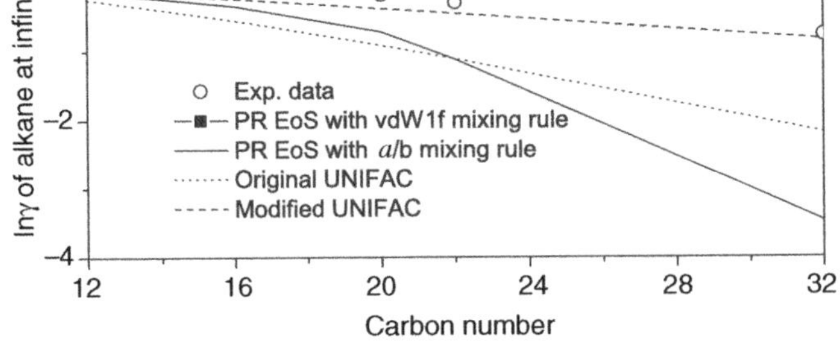

Figure 3.9 *Experimental and predicted alkane activity coefficients at infinite dilution for n-alkane/n-hexane systems using the PR EoS (with the vdW1f and the a/b mixing rules, Equations (3.2)–(3.3) and (3.11)) and with the original and modified UNIFAC activity coefficient models (Chapter 5). Experimental data are from Kniaź, Fluid Phase Equilib.,* ***1991****, 68, 35*

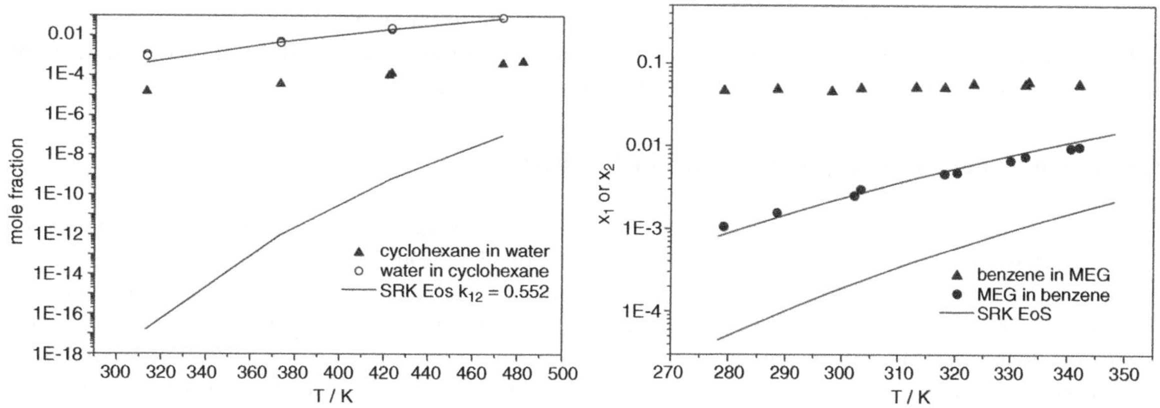

Figure 3.10 *Left: LLE correlation for water–cyclohexane using SRK and the vdW1f mixing rules (Equations (3.2) and (3.3)) and one interaction parameter, $k_{ij} = 0.552$ (notice that the value is close to that reported for water–methane in Table 3.7). The binary interaction parameter is fitted to the water solubility in cyclohexane. It is seen that simultaneous representation of the water and hydrocarbon solubilities is not possible with SRK (and other cubic EoS). Experimental data from Tsonopoulos and Wilson, AIChE J.,* ***1983****, 29(6), 990–999. Right: LLE calculations for MEG–benzene using SRK and the vdW1f mixing rules with $k_{ij} = 0.213$. The binary interaction parameter is fitted to the MEG solubility in benzene phase. It can be seen that SRK using the classical vdW1f mixing rules (Equations (3.2) and (3.3)) erroneously calculates the benzene solubility in the MEG phase to be lower than the MEG solubility in benzene phase. Experimental data from Folas et al., J. Chem. Eng. Data,* ***2006****, 51, 977–983. Reprinted with permission from J. Chem. Eng. Data, Liquid–Liquid Equilibria for Binary and Ternary Systems Containing Glycols, Aromatic Hydrocarbons, and Water: Experimental Measurements and Modeling with the CPA EoS by Georgios K. Folas, Georgios M. Kontogeorgis et al., 51, 3, 977–983 Copyright (2006) American Chemical Society*

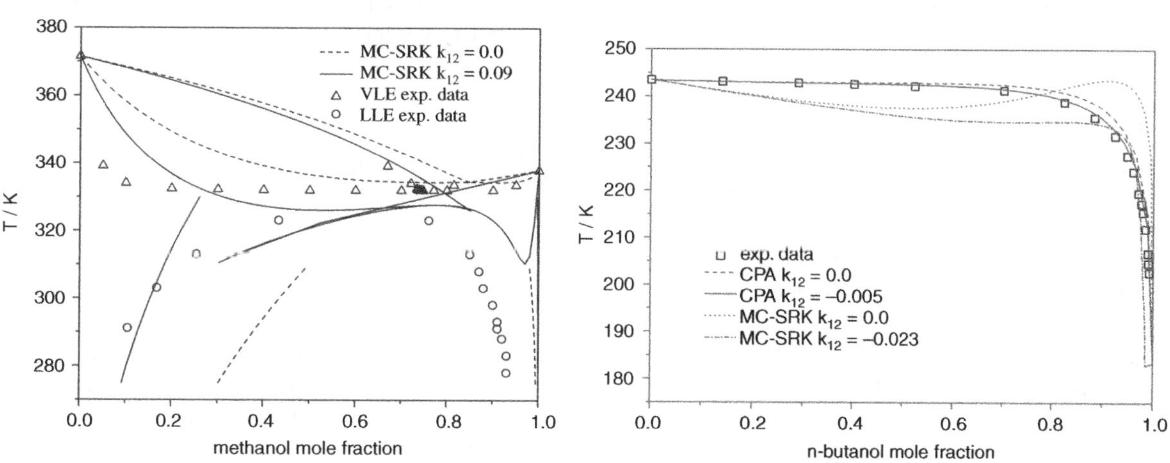

Figure 3.11 *Left: VLE and LLE for methanol–heptane with SRK. The Mathias–Copeman (MC) version of SRK is used, i.e. with methanol parameters which reproduce the vapor pressure of the pure compounds. Experimental data from Sørensen and Arlt, Liquid–liquid equilibrium data collection, DECHEMA Chemistry Data Series, Part 1, Vol. V, 1995 and Budantseva et al., Zh. Fiz. Khim.,* ***1975****, 49, 1844. Right: SLE of butanol–decane with SRK and CPA (see Chapter 9). MC-SRK is used for butanol. Equally poor results are obtained with SRK for the SLE of other mixtures containing polar compounds, e.g. MEG–water. Experimental data are from Plesnar et al., J. Chem. Thermodyn.,* ***1990****, 22, 403. Reprinted with permission from Ind. Eng. Chem. Res., Ten Years with the CPA (Cubic-Plus-Association) Equation of State. Part 1. Pure Compounds and Self-Associating Systems by Georgios M. Kontogeorgis, Georgios K. Folas et al., 45, 14, 4855–4868 Copyright (2006) American Chemical Society*

approaches may be needed for multiphase, multicomponent equilibria in the presence of condensed phases and/or polar and associating compounds (e.g. water and alcohols or glycols), as will be discussed in the third part of the book (Chapters 7–14), which is devoted to association theories.

3.4 Some recent developments with cubic EoS

The major capabilities and limitations of cubic EoS using the vdW1f mixing rules (from the practical point of view) which we have just discussed are summarized in Table 3.11. Due to these serious limitations, much research over the past 20 years has focused on the developments of advanced, theoretically derived EoS which can correct for these deficiencies. These models will be discussed in the third part of the book (Chapters 7–14).

Still, due to the great success of cubic EoS, especially for problems of interest to the petrochemical industry, the field of research on cubic EoS has continued to grow during the past 20 years, to some extent in parallel

Table 3.11 *Capabilities and limitations of cubic EoS, using the vdW1f mixing rules*

Capabilities	Limitations
Simple models, easy to program, fast calculations	Computational problems at the near-critical area
Excellent results for low- and high-pressure VLE for mixtures with hydrocarbons, gas–hydrocarbons and other non-polar compounds	Not very accurate when 'condensed phases' dominate, e.g. LLE, SLE, VLLE, etc. Serious problems for polar and/or associating compounds, especially for multicomponent, multiphase equilibria Impossible to correlate LLE for highly immiscible systems, e.g. water or glycols with alkanes
They can be predictive in some cases: (1) small or zero k_{ij} for mixtures with hydrocarbons; (2) use of same k_{ij} at different T in some cases or for similar systems; (3) correlation of binary mixtures and prediction for multicomponent systems	In many cases, the k_{ij} are not known a priori for the binary systems
Many years' experience and familiarity, e.g. extensive databases and correlations of the interaction parameters (k_{ij})	Results can be highly sensitive to k_{ij}, e.g. for gas–hydrocarbons
High correlation flexibility when two interaction parameters are used, e.g. for solid–gas equilibria or polar VLE (H_2S–water)	The interaction parameters must always be fitted to experimental phase equilibria data and cannot be typically generalized
Allow for implementation of the critical point, thus allowing vapor pressure calculations up to the critical point	Critical properties are not available for heavy, thermally unstable and/or complex compounds Calculations can be sensitive to the critical properties used Liquid volumes are not predicted well, and a separate treatment is needed via the so-called volume translation[49] Problems close to the critical point or at very low temperatures (close to the triple point)
They can be used for many properties beyond phase equilibria, e.g. excess enthalpies, heat capacities, excess volumes, speed of sound, etc.	Often not very successful representation of excess and derived properties

Table 3.12 Some (mostly recent) developments of cubic EoS (with the vdW1f or related mixing rules)

Development	Reference
Estimation of EoS parameters based on vapor pressures and liquid densities	Yakoumis *et al.*,[15] Kontogeorgis *et al.*,[54] Oliveira *et al.*,[16] Ting *et al.*,[13] Voutsas *et al.*[14]
A group contribution scheme for estimating the interaction parameter, k_{ij}, for mixtures with hydrocarbons (and also CO_2)	Jaubert and Mutelet[55], Mutelet *et al.*,[56] Jaubert *et al.*[57]
Use of the activity coefficient derived from the EoS for evaluating mixing and combining rules	Kontogeorgis *et al.*,[33] Kontogeorgis and Coutsikos,[58] Sacomani and Brignole[59]
Use of EoS for hydrate calculations, combined with the vdW–Platteeuw model (mainly for the gas and liquid hydrocarbon phase, but with excess Gibbs energy models or mixing rules for the aqueous phase)	Munck *et al.*,[53] Hendriks *et al.*,[60] Parrish and Prausnitz,[61] Anderson and Prausnitz,[62] Ng and Robison,[63] Madsen *et al.*,[64] Lundgaard and Mollerup,[65,66] Sloan,[67]
Mixing and combining rules based on extension of the vdW1f theory and/or molecular theory/London	Smith,[68] Plocker *et al.*,[69] Radosz *et al.*,[70] Ungerer *et al.*[71]

with the developments on non-cubic EoS (especially SAFT-type models). Much of the research on cubic EoS has focused on improvements in the mixing rules via the development of the so-called EoS/G^E mixing rules. We will discuss these in detail in Chapter 6, after presenting the local composition activity coefficient models in Chapter 5. We will see that cubic EoS with these advanced mixing rules extended the range of applicability of cubic EoS like SRK and PR to many polar systems of interest to the chemical industry. Another field in cubic EoS research which flourished over the last few years is the extension to polymers (solutions and blends). This subject will also be discussed in Chapter 6, as many of these developments include EoS/G^E mixing rules as well.

Moreover, cubic EoS have been extensively applied over the past 10–15 years to systems/types of phase equilibria, which can be considered beyond their original scope. Mixtures with refrigerants, solid–gas equilibria, solubility parameters and gas hydrates are some of these applications.[50–53] However, in this section we will present some advances which we consider significant within the framework of cubic EoS and vdW1f (and related) mixing rules. The presentation is limited to developments that we are aware of and feel are of significance, and may highlight future applications and advances in cubic EoS. They are summarized in Table 3.12 and some of them are briefly discussed hereafter.

3.4.1 Use of liquid densities in the EoS parameter estimation

We have already seen (Table 3.2) that one way to estimate the parameters (a and b; alternatively T_c, P_c) of cubic EoS is to 'sacrifice' the critical point and fit both parameters to vapor pressures and liquid densities. A few parameter values for cubic EoS estimated in this way were presented in Tables 3.4 and 3.5, while extensive tables of such estimated parameters are available in the literature for SRK[15,54] and PR.[13,14]

As Figures 3.12 and 3.13 illustrate, and unlike when T_c and P_c are fitted to the critical point, when the cubic EoS parameters are obtained from vapor pressure and liquid density data, smooth (linear for the co-volume) trends against size parameters (carbon number or vdW volume) are obtained.

Such linear plots, especially for the co-volume, should be attributed to the way the parameters are estimated, i.e. inclusion of the liquid density data, as they are observed for both SRK and PR. The trend is not linear if experimental T_c and P_c are used ('traditional' way of estimating the parameters). Moreover, as both Ting *et al.*[13] and Voutsas *et al.*[14] show, PR using these fitted pure compound parameters performs very well in correlating VLE for asymmetric alkane mixtures, much better than when the critical properties are used,

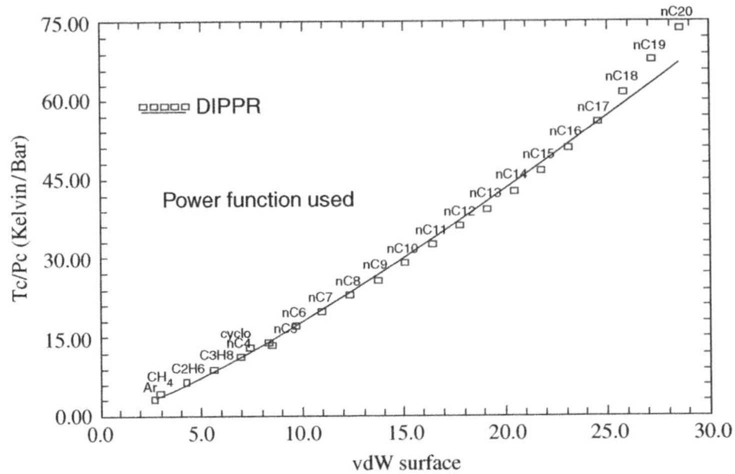

Figure 3.12 *The ratio T_c/P_c of alkanes against the vdW surface area, Q_w. The co-volume parameter of cubic EoS is not a linear function of the size (molecular weight, vdW volumes or surface areas) when the T_c and P_c are only estimated based on the critical point*

and actually as satisfactorily as molecular theoretically derived models such as SAFT (discussed in the third part of the book, Chapters 7–14).

The performance of the SRK EoS for highly asymmetric mixtures of hydrocarbons is presented in Figures 3.14 and 3.15 for the methane–C_{10} and methane–C_{16} binary systems. In this case, the approach discussed above (fitting the EoS parameters to vapor pressures and liquid densities) is used for SRK only for the heavy hydrocarbon (this is denoted as CPA in these figures). For methane the classical T_c, P_c and acentric factor approach is used, as presented in Table 3.1. For both systems, the use of fitted parameters to vapor pressures and liquid densities (referred to as SRK-fitted) for the heavy alkane significantly influences the

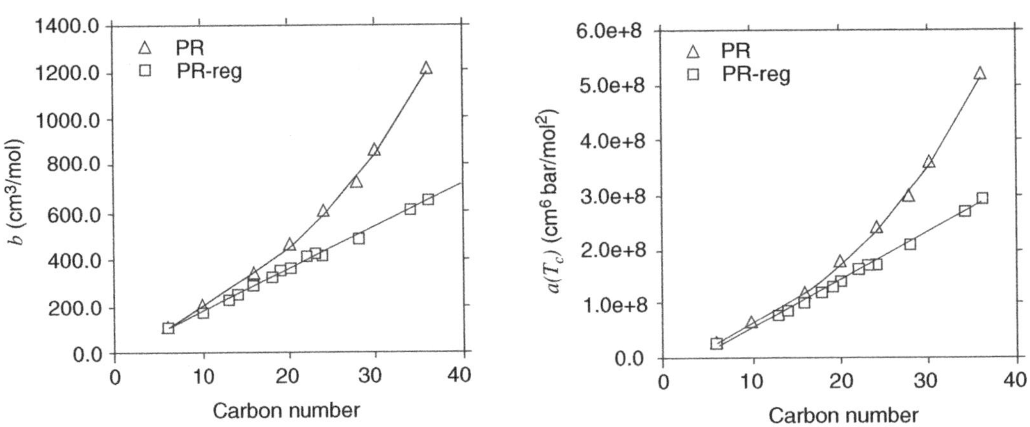

Figure 3.13 *Left: The co-volume parameter of PR, either obtained from the critical properties (PR) or fitted to vapor pressures and liquid densities (PR-reg). Right: The energy parameter of PR, either obtained from the critical properties (PR) or fitted to vapor pressures and liquid densities (PR-reg). Reprinted with permission from Fluid Phase Equilibria, Phase Equilibrium modeling of mixtures of long-chain and short-chain alkanes using Peng-Robinson and SAFT by D.P. Ting, P.C. Joyce et al., 206, 1–2, 267 Copyright (2003) Elsevier*

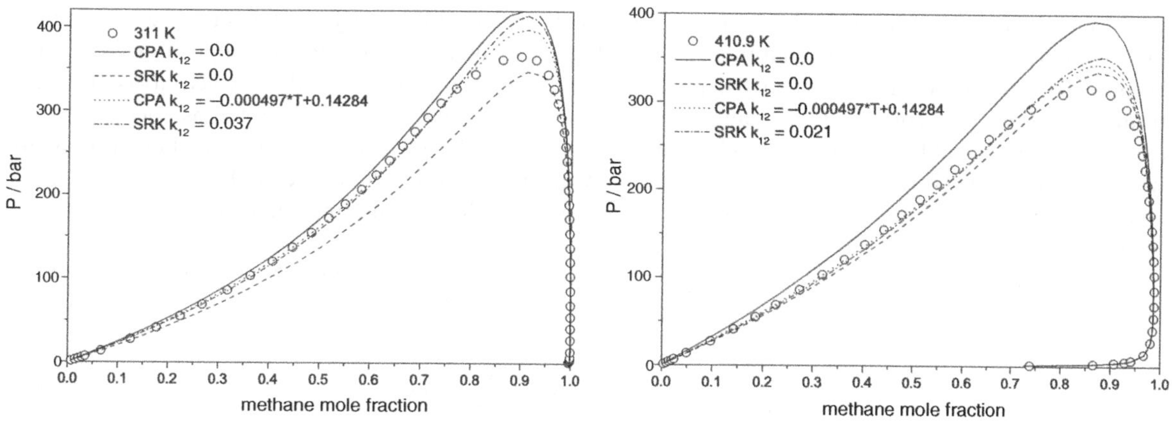

Figure 3.14 *VLE for methane–alkanes with SRK and CPA. Left: VLE of methane–n-C_{10} at 310.9 K. Right: VLE of methane–n-C_{10} at 410.9 K. Experimental data from Reamer et al., Ind. Eng. Chem.,* ***1942****, 34, 1526*

calculations, while the use of a linear temperature-dependent binary interaction parameter provides satisfactory results over an extended temperature range.

Liquid density data are sometimes more readily available than critical properties, especially for heavy compounds. Other efforts for estimating parameters of cubic EoS avoiding use of critical data have been reported in the literature.[72,73]

3.4.2 Activity coefficients for evaluating mixing and combining rules

Experience with activity coefficient models (see Chapters 4 and 5) is extensive and the earliest developments in this field are as old as the first cubic EoS (vdW), since the beginning of the twentieth century. To some extent, as will be discussed in Chapters 4 and 5, it can be argued that developments with activity coefficient models progressed somewhat faster than those of EoS and highly successful models for polar mixtures were available already during the 1970s (Chapter 5), while the advanced successful EoS (see third part of the book) had to

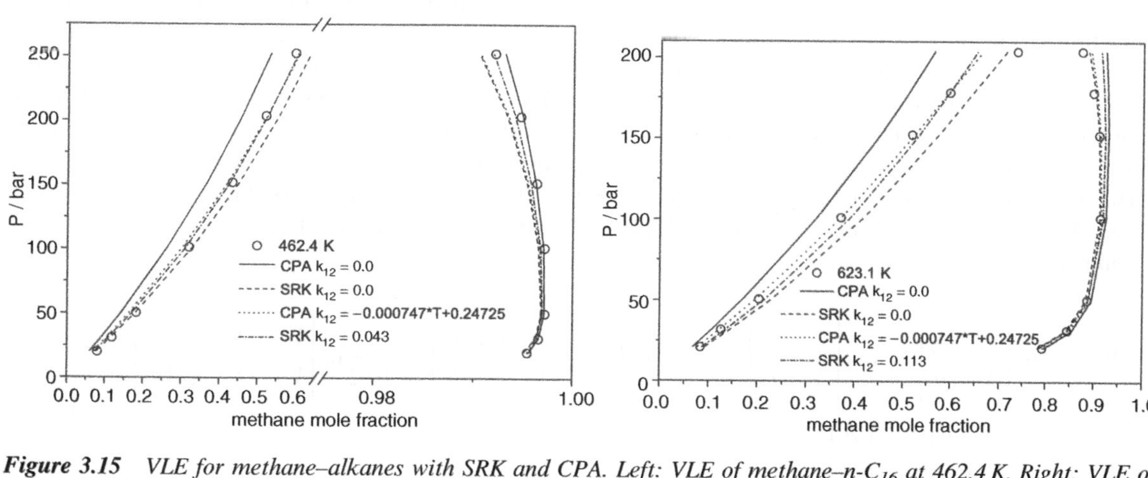

Figure 3.15 *VLE for methane–alkanes with SRK and CPA. Left: VLE of methane–n-C_{16} at 462.4 K. Right: VLE of methane–n-C_{16} at 623.1 K. Experimental data from Lin et al., J. Chem. Eng. Data,* ***1980****, 25, 252*

wait for one or two more decades. Due to this extensive experience with activity coefficient models, their clear link, i.e. derivation from well-defined theories (e.g. lattice, regular solutions, local compositions, etc.) and because activity coefficients are sensitive derivative properties, it is of interest to calculate activity coefficients from cubic EoS. It is equally interesting to develop the expressions for the excess Gibbs energy and for the activity coefficient from cubic EoS and to compare them to well-known activity coefficient models. In this way, as, among others, Kontogeorgis *et al.*[33] have shown, an analysis can be carried out of the various terms and an understanding of the assumptions behind EoS and especially their mixing and combining rules can be obtained.

Calculation of the excess Gibbs energy, g^E, and the activity coefficients, γ, from cubic EoS is straightforward and relies on well-known expressions from thermodynamics (see Chapter 1):

$$\frac{g^E}{RT} = \ln \varphi - \sum_i x_i \ln \varphi_i = \sum_i x_i \ln \gamma_i \tag{3.12}$$

$$\ln\gamma_i = \ln \frac{\hat{\varphi}_i}{\varphi_i} \tag{3.13}$$

Notice that the excess Gibbs energy expression only requires knowledge of the fugacity (of pure compound or mixture, φ_i, φ), and is thus independent of mixing and combining rules. On the other hand, the activity coefficient expression depends on the fugacity of the compound in the mixture ($\hat{\varphi}_i$) and is thus dependent on the mixing and combining rules used. For example, for the PR EoS, the excess Gibbs energy expression is:

$$\begin{aligned} \frac{g^{E,PR}}{RT} &= \left(\sum_i x_i \ln\left(\frac{V_i - b_i}{V - b}\right)\right) + \left(\frac{PV^E}{RT}\right) \\ &+ \left\{\frac{1}{RT2\sqrt{2}}\left[\sum_i x_i \frac{a_i}{b_i} \ln\left(\frac{V_i + (1+\sqrt{2})\, b_i}{V_i + (1-\sqrt{2})\, b_i}\right) - \frac{a}{b}\ln\left(\frac{V + (1+\sqrt{2})\, b}{V + (1-\sqrt{2})\, b}\right)\right]\right\} \end{aligned} \tag{3.14}$$

We observe that g^E can be divided into three parts, from left to right in Equation (3.14): a term which can be attributed solely to size ('combinatorial and free-volume') effects (containing only the volumes and co-volumes); one due to excess volume; and the last contribution which is largely due to energetic effects (and includes all EoS parameters, a,b). This division is largely based on the similarity of the combinatorial–free-volume part to that of well-known polymer models (Flory–Huggins and the like, see Chapter 4), but it is rather artificial. Alternative definitions of the combinatorial–free-volume part stemming from an EoS will be provided later in this section (see point 4, Equation (3.17)).

The expression for the activity coefficient depends on the mixing and combining rules. For PR using the vdW1f mixing rules (Equations (3.2) and (3.3)), the expression for the activity coefficient of compound i is:

$$\begin{aligned} \ln \gamma_i &= \ln\left[\frac{V_i - b_i}{V - b}\right] + 1 - \left[\frac{V_i - \bar{b}_i}{V - b}\right] + \frac{\alpha_i}{2\sqrt{2}}\ln\left[\frac{V_i + (1+\sqrt{2})b_i}{V_i + (1-\sqrt{2})b_i}\right] - \frac{\bar{\alpha}_i}{2\sqrt{2}}\ln\left[\frac{V + (1+\sqrt{2})b}{V + (1-\sqrt{2})b}\right] \\ &+ \frac{\alpha(bV_i - V\bar{b}_i)}{V(V+b) + b(V-b)} \end{aligned} \tag{3.15}$$

where:

$$\begin{aligned}
\alpha &= \frac{a}{bRT} \\
\bar{b}_i &= -b + 2\sum_j x_j b_{ij} \\
\bar{a}_i &= -a + 2\sum_j x_j a_{ij} \\
\bar{\alpha}_i &= \frac{a}{bRT}\left[\frac{\bar{a}_i}{a} + 1 - \frac{\bar{b}_i}{b}\right]
\end{aligned} \tag{3.16}$$

It can be argued, though with some degree of approximation, that the first three terms of Equation (3.15), which do not contain any energy parameters but only volumes and co-volumes (*V*,*b*), correspond to the combinatorial–free-volume part of the EoS and the rest is due to energetic interactions ('residual' term), though an alternative distinction is possible, see the discussion in the text. Tables 3.9 and 3.10 and Figures 3.8 and 3.9 show some activity coefficient values obtained with PR using two different mixing rules. Problems 6 and 7 on the companion website at www.wiley.com/go/Kontogeorgis illustrate various aspects of these derivations and the final result for the vdW and SRK EoS, while this methodology will be discussed and used again throughout Chapters 4–6, in the presentation of the activity coefficient models (random-mixing-based and local composition ones) and the EoS/G^E mixing rules.

We will now report some of the most important findings of the investigations of mixing and combining rules of cubic EoS (from the references listed in Table 3.12):

1. An approximate separation of the activity coefficient from EoS into a combinatorial–free-volume (*comb-FV*) and an energetic (residual, *res*) term permits to some extent a separate evaluation of repulsive and attractive terms of EoS and combining rules, as the combining rule of the cross-energy term only appears in the residual term. Activity coefficient data of asymmetric athermal alkane systems permit such an evaluation, as the residual contributions for these systems should be small.
2. Various combining rules proposed for the cross-energy and cross co-volume parameters (see Section 3.4.3 and Appendix 3.B) have been investigated. By far the best activity coefficient and VLE results are obtained with the 'classical' combining rules (Equation (3.3)). Moreover, using the arithmetic mean (AM) rule for b_{12}, the combinatorial–FV term of cubic EoS resembles that of well-known polymer models (Flory–Huggins, entropic FV), as will also become apparent in Chapters 4 and 5, where these activity coefficient models will be presented.
3. As Figures 3.16 and 3.17 and Table 3.13 illustrate, use of an l_{ij} interaction parameter improves the performance of asymmetric athermal systems much more than use of k_{ij}. For example, excellent VLE results are obtained with PR for the whole ethane–alkane series (up to n-C_{44}) using a single small, positive l_{ij} value for all mixtures. Both VLE, activity coefficients and the separate contributions, combinatorial–FV and residual, are improved in the right direction (i.e. the combinatorial–FV part dominates and the residual part becomes close to unity). Naturally, k_{ij} becomes progressively more important as the energetic effects become more important, e.g. with CO_2 or nitrogen mixtures (Table 3.13).
4. As shown by Sacomani and Brignole[59] and Kontogeorgis and Coutsikos,[58] the mixing rule of Equation (3.11) best isolates the non-residual, 'true' combinatorial–FV contributions from the EoS,

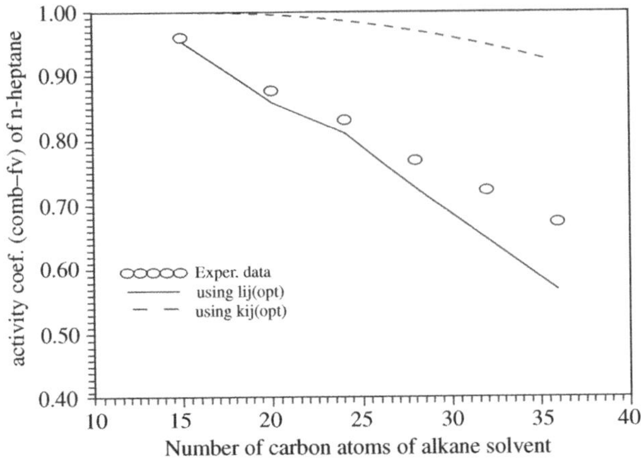

Figure 3.16 *The combinatorial–FV part of the infinite dilution activity coefficient of n-heptane in n-heptane/n-alkane mixtures at 373.15 K, as estimated from PR with the vdW1f mixing rules (Equation (3.2)). The classical combining rules are used, Equation (3.3), i.e. the arithmetic mean (AM) rule for the cross co-volume parameter and the geometric mean (GM) for the cross-energy parameter. The interaction parameters are estimated so that the total activity coefficient value is reproduced. Reprinted with permission from Chemical Engineering Science, A novel method for investigating the repulsive and attractive parts of cubic equations of state and the combining rules used with the vdW-1f theory by G. M. Kontogeorgis, P. Coutsikos et al., 53, 3, 541 Copyright (1998) Elsevier*

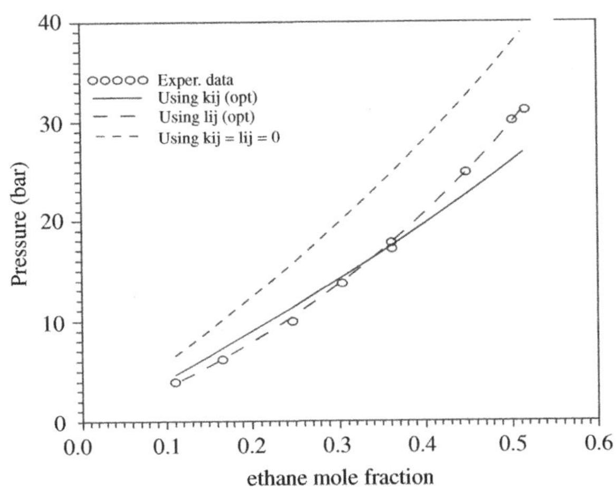

Figure 3.17 *Px plot for ethane/n-C_{44} ($T = 373$ K) with the PR EoS using the vdW1f mixing rules and the classical combining rules, Equations (3.2)–(3.3). Results are shown without interaction parameters and when one optimum interaction parameter is used either in the cross-energy (k_{ij}) or in the cross co-volume parameter (l_{ij}). Reprinted with permission from Chemical Engineering Science, A novel method for investigating the repulsive and attractive parts of cubic equations of state and the combining rules used with the vdW-1f theory by G. M. Kontogeorgis, P. Coutsikos et al., 53, 3, 541 Copyright (1998) Elsevier*

Table 3.13 VLE with the PR EoS using the vdW1f mixing rules (Equations (3.2) and (3.3)) and either one k_{ij} or one l_{ij} interaction parameter. The average absolute deviations in pressure (DP%) are given and the interaction parameter values are given in parentheses. C_x indicates an alkane with x carbon atoms (C_1 is methane, C_2 is ethane and C_3 is propane). Reprinted with permission from Chemical Engineering Science, A novel method for investigating the repulsive and attractive parts of cubic equations of state and the combining rules used with the vdW-1f theory by G. M. Kontogeorgis, P. Coutsikos et al., 53, 3, 541 Copyright (1998) Elsevier

System		DP% ($k_{ij}=l_{ij}=0$)	DP% (k_{ij} only)	DP% (l_{ij} only)
C_2–C_{20}	$T=373.8$ K	12.5	5.3 (−0.03)	2.5 (0.015)
	$T=572.9$ K	8.4	1.8 (−0.05)	1.1 (0.02)
C_2–C_{28}	$T=373.3$ K	17.7	5.8 (−0.05)	2.1 (0.016)
	$T=573.2$ K	13.2	2.0 (−0.08)	0.8 (0.021)
C_2–C_{36}	$T=373.2$ K	33	7.6 (−0.08)	1.0 (0.019)
C_2–C_{44}	$T=373$ K	45	10.7 (−0.11)	1.0 (0.019)
	$T=423.2$ K	41	5.4 (−0.14)	2.6 (0.0206)
C_3–C_{60}	$T=370$–374.6 K	18.5	9.6 (−0.037)	4.7 (0.038)
C_1–C_{28}	$T=373.3$ K	14	2.8 (−0.056)	1.4 (0.0075)
C_1–C_{36}	$T=373.2$ K	14.3	2.8 (−0.070)	1.6 (0.007)
CO_2–C_{24}	$T=373.15$ K	19	3.5 (0.0678)	4.7 (−0.015)
N_2–C_{16}	$T=323.15$ K	35	7.0 (0.179)	14.4 (−0.025)

possibly better than the distinction discussed previously in the text following Equations (3.15) and (3.16). It can be argued that at infinite pressure, the combinatorial–FV term should be zero and only the residual (energetic term) should remain (a concept that will be used again in Chapter 6). Using the mixing rule of Equation (3.11), the 'residual term' of SRK and PR disappears and only the combinatorial–FV part remains (which includes more terms than just those containing V and b, see Problem 6 on the companion website at www.wiley.com/go/Kontogeorgis for PR and Equation (3.17)). In the case of SRK, when the arithmetic mean rule is used for the cross co-volume and Equation (3.11) is used for the energy parameter mixing rule, then the infinite dilution activity coefficient of compound *1* in a binary mixture is given by the equation:

$$\ln \gamma_1^{\infty} = \ln\left(\frac{V_1 - b_1}{V_2 - b_2}\right) + 1 - \left(\frac{V_1 - b_1}{V_2 - b_2}\right) + \frac{a_1}{b_1 RT} \ln \frac{\left(1 + \frac{b_1}{V_1}\right)}{\left(1 + \frac{b_2}{V_2}\right)} + \frac{a_2}{b_2 RT}\left(\frac{V_1 b_2 - V_2 b_1}{V_2^2 + b_2 V_2}\right) \tag{3.17}$$

The very promising results shown with this mixing rule (Equation (3.11)) in Figures 3.8 and 3.9 highlight that, despite all odds, size-asymmetric systems can be well represented with classical cubic EoS and whatever problems with their representation should be attributed to mixing rules and in particular to the combining rule for the cross-energy parameter, a_{12}.

3.4.3 Mixing and combining rules – beyond the vdW1f and classical combining rules

The vdW1f mixing rules (Equations (3.2)) and the classical combining rules (Equations (3.3)) are indeed the most widely used choice in cubic EoS. They are by far not the *only* choice. Moreover, we wish to emphasize:

1. The link to intermolecular potential functions (Chapter 2) and molecular simulation studies, which are based on the molecular energy and size (diameter) parameters, ε, σ.

2. The connection to statistical mechanics theories like SAFT, which also use the molecular parameters ε, σ.
3. Other mixing rules employed in EoS and corresponding states theories and their similarity to the classical ones (Equations (3.2)–(3.3)).

For these reasons, we present the extension of vdW1f theory[68] and the types of mixing rules based on this extension, but first of all we will discuss the link between the cubic EoS (a,b) and the molecular parameter ε, σ (an issue briefly presented in Chapter 2).

The vdW1f mixing rules, Equations (3.2), can be written in terms of the molecular parameters and in terms of the critical properties as:

$$\begin{aligned} \varepsilon\sigma^3 &= \sum_i \sum_j x_i x_j \varepsilon_{ij} \sigma_{ij}^3 \\ \sigma^3 &= \sum_i \sum_j x_i x_j \sigma_{ij}^3 \end{aligned} \tag{3.18}$$

and:

$$\begin{aligned} T_c V_c &= \sum_i \sum_j x_i x_j T_{cij} V_{cij} \\ V_c &= \sum_i \sum_j x_i x_j V_{cij} \end{aligned} \tag{3.19}$$

Upon comparing these equations with Equations (3.2), the following relationships (proportionalities) between the 'microscopic' ε, σ and 'macroscopic' properties – parameter (T_c, V_c, a, b) – are obtained:

$$\begin{aligned} &\sigma^3 \propto b \propto V_c \\ &\varepsilon\sigma^3 \propto T_c V_c \propto a \Rightarrow \varepsilon \propto \frac{a}{b} \propto T_c \end{aligned} \tag{3.20}$$

T_c, V_c, a, b are respectively the critical temperature, the critical volume and the energy and co-volume parameters of cubic EoS. In principle, the choice of combining rules does not depend on the form of the mixing rules, and thus any combining rule for $\varepsilon_{ij}, \sigma_{ij}$ can be used in Equation (3.18), exactly as various combining rules for a_{ij}, b_{ij} can be used in the vdW1f mixing rules (Equations (3.2)). Using the relationships of Equation (3.20), we can express both the 'classical for cubic EoS' combining rules and the 'classical for simulation/theoretical EoS' combining rules in terms of the various parameters (macroscopic and microscopic).

For example, using the well-known combining rules for cubic EoS, Equation (3.3), and ignoring interaction parameters, we obtain the following combining rules in terms of ε, σ:

$$\begin{aligned} a_{ij} &= \sqrt{a_i a_j} \Rightarrow \varepsilon_{ij} = \sqrt{\varepsilon_i \varepsilon_j} \left(\frac{\sqrt{\sigma_i \sigma_j}}{\sigma_{ij}} \right)^3 \\ b_{ij} &= \frac{b_i + b_j}{2} \Rightarrow \sigma_{ij} = \left(\frac{\sigma_i^3 + \sigma_j^3}{2} \right)^{1/3} \end{aligned} \tag{3.21}$$

Similarly, using the well-known Lorentz–Berthelot rules, typically used in molecular simulation studies and in statistical-mechanics-based EoS, e.g. SAFT, we can derive the equivalent combining rules in terms of the cubic EoS parameters:

$$\varepsilon_{ij} = \sqrt{\varepsilon_i \varepsilon_j} \Rightarrow a_{ij} = \sqrt{a_i a_j}\frac{b_{ij}}{\sqrt{b_i b_j}}$$

$$\sigma_{ij} = \left(\frac{\sigma_i + \sigma_j}{2}\right) \Rightarrow b_{ij} = \frac{1}{8}\left(b_i^{1/3} + b_j^{1/3}\right)^3 \tag{3.22}$$

Note that different possibilities are available and differences exist, even among these well-known combining rules. Appendix 3.B discusses the extension of vdW1f mixing rules and a systematic classification of mixing and combining rules, together with some of the recent developments.

3.5 Concluding remarks

Cubic equations of state (EoS), especially SRK and PR, using the quadratic (van der Waals one fluid (vdW1f)) mixing rules and the classical combining rules have found widespread use, especially in the oil industry. They represent very well both low- and high-pressure phase equilibria (VLE) for mixtures with hydrocarbons and also hydrocarbon–gas mixtures, although in the latter case an interaction parameter (k_{ij}) is required. Various correlations for k_{ij} have been reported for different families of gas–hydrocarbon mixtures. Simple approaches for predicting the k_{ij} have been developed based on the Hudson–McCoubrey and other theories.

Three-parameter EoS (i.e. based on T_c, P_c and ω) are representations of the three-parameter corresponding states principle and their parameters are often estimated from the critical point constraints and vapor pressure data. The quadratic concentration dependency of second virial coefficients is maintained but densities are not very well represented, unless a volume translation is used. Physically more correct EoS parameters can be obtained if these are fitted simultaneously to vapor pressures and liquid densities, but then the critical point is overestimated. Cubic EoS with the classical mixing and combining rules can also moderately represent polar mixtures using the k_{ij} interaction parameters, but they often fail for highly polar and hydrogen bonding mixtures, especially for VLLE, LLE and SGE. Highly immiscible mixtures, e.g. water–alkanes or glycol–alkanes, cannot be satisfactorily represented.

EoS can be used for calculating various other properties including activity coefficients. While the excess Gibbs energy expression as derived from EoS does not depend on mixing and combining rules, the activity coefficients do! Deriving the activity coefficient expressions from cubic EoS can be used as a way to study the mixing and combining rules employed and to investigate strengths and weaknesses of the repulsive and attractive parts of the EoS. Such an analysis has shown that by using a simple mixing rule which essentially isolates the combinatorial free volume, i.e. the size and shape effects of EoS, models like SRK and PR can predict phase equilibria without any interaction parameters for size-asymmetric mixtures. Alternative (to vdW1f) mixing and combining rules are presented in Appendix 3.B, while the most widely used EoS/G^E mixing rules will be presented in Chapter 6, after the activity coefficient models are introduced in Chapters 4 and 5.

Appendix 3.A Free-volume theories

Free volume is often defined as the difference between the volume of the liquid and the minimum volume it would occupy if the molecules were firmly packed to each other (in other words, if they were close-packed spheres). Alternatively, the free volume is defined as the molar volume minus the hard-core (or excluded or inaccessible) volume of the molecules. In the context of cubic EoS of the vdW type, such as SRK and PR, the free volume is simply defined as $V_f = V - b$, where b is the co-volume, which can be estimated from the critical point, or combined vapor pressure/liquid density data or other approaches (Tables 3.2–3.4). The values of the thermodynamic properties will depend on the free volumes calculated from the EoS.

In a general formulation, however, the concepts of free volume and hard-core volumes are rather complex and different possibilities exist, as shown in Tables 3.14 and 3.15. There are so many variations that Bondi[74] stated that 'each author defines free-volume as what he wants it to mean'. As discussed previously and as observed from the various possibilities for the hard-core volume (in Table 3.15), it is widely accepted that the hard-core volume values (V^*) should be higher than the vdW volume (which is a measure of the true size of the molecule). The values (ratio V^*/V_w) will naturally depend on the expected expansion of the compound in the liquid state, but for most 'ordinary' liquids this ratio should be around 1.2–1.3, while for certain polymers it can be higher.[24] We can expect that similar ratios for the co-volume parameter of EoS (i.e. b/V_w between 1.2 and 1.5) would result in a good representation of the liquid state.

Table 3.14 *Various expressions for the free volume (V_f) and hard-core volumes (V^*)*

Theory	Free-volume expression, V_f	Comments
Van der Waals ($V_f = V - b$)	$V - V^{excl} = V - 4V^{mol} = V - 4\left(\frac{\pi\sigma^3}{6}\right)$ $= V - \frac{2}{3}\pi\sigma^3$	For many fluids, the excluded (or hard-core) volume is about 2–6 times higher than the van der Waals volume
Hard core	$V - V^* = V - V^{mol} = V - \left(\frac{\pi\sigma^3}{6}\right) \approx V - V_w$	The molecular volume is approximately equal to the van der Waals volume
Wanderer	$\left(V^{1/3} - (V^*)^{1/3}\right)^3 = \frac{4\pi\gamma}{3}\left(V^{1/3} - (V^*)^{1/3}\right)^3$ The parameter γ depends on packing, i.e. the average number of nearest neighbors surrounding a given molecule, which for solids is identical with the lattice coordination number, Z. $\gamma = 1.414$ for face-centered cubic ($Z = 12$) $\gamma = 1.299$ for body-centered cubic ($Z = 8$) $\gamma = 1.000$ for simple cubic ($Z = 6$) $\gamma = 0.650$ for diamond structure ($Z = 4$)	For $Z = 12$, face- centered cubic, the free volume is equal to $6(V^{1/3} - 1.1V_w^{1/3})^3$
Flory–FV	$\left(V^{1/3} - (V^*)^{1/3}\right)^{3c}$	Typical values for $c = 1.1$

Superscripts: *excl* = excluded volume; *mol* = molecular volume.

Table 3.15 *Relationships between the hard-core (V^*) volume and the van der Waals volume used in various thermodynamic models. 'Comments' indicate the main applications of the models and, when available, how the V^*/V_W coefficient is estimated*

Model	V^*/V_W	Comments
UNIFAC–FV[78]	1.28	Activity coefficient model for polymer solutions. The value *1.28* is obtained by fitting to phase equilibrium data
GC–Flory[79]	1.448	EoS for the liquid phase – applied to polymers
Flory EoS[23]	1.4–1.5	EoS for polymers. The value *1.4–1.5* is calculated from liquid density data
Sako EoS[80]	1.377	The first cubic EoS proposed for polymer solutions
Guggenheim ($V^*=0.286V_c$)[81]	1.53	Using the relationship $V_c=5.36V_w$[74,83]
Entropic–FV[75,82]	1.0	Activity coefficient model for polymers
Entropic–FV[84]	1.2	Activity coefficient model for polymers
Van der Waals[77]	1.3–1.45	Cubic EoS for polymers. The value *1.3–1.45* is obtained by fitting to liquid density data
PR[85]	1.3–1.4	Cubic EoS for polymers. The value *1.3–1.4* is obtained by fitting to volumetric data
PR[86]	1.3	Volume estimation method, value at 0 K
GC–VOL[83]	1.22	Volume estimation method, value for polyethylene at 0 K
PR[87]	1.65	Solid–gas equilibrium data

Moreover, a number of important conclusions should be mentioned:

1. The b/V_w ratios calculated from cubic EoS (Table 3.4) are smaller when the co-volume is estimated from vapor pressures and liquid densities compared to values based on critical data, thus approaching values which should better represent liquid state properties (based on the above discussion).
2. Free-volume concepts are very important in polymer–solvent systems, with polymers typically having lower free-volume percentages than solvents. Water with its very low free volume is a notable exception,[75] having a free-volume percentage close to that of polymers.
3. Wong and Prausnitz[76] state that an effective co-volume parameter, when used in vdW-type EoS, results in a repulsive term which is closer to the Carnahan–Starling one (of hard spheres), compared to the original vdW repulsive term. This b^{eff} is about $b/2$, when b is estimated in the 'vdW way', see first row of Table 3.14. This corresponds to a value of about $2V_w$, or even lower for small 'hard-sphere' molecules, e.g. Wong and Prausnitz's arguments when applied to argon result in a b^{eff}/V_w ratio of about 1.324.[77]

Appendix 3.B Alternative to the classical vdW1f mixing and combining rules

3.B.1 Beyond the vdW1f theory

Smith[68] extended the vdW1f theory and proposed the following general form of mixing rules from which the vdW1f ones (Equations (3.2)) can be obtained as a special case:

$$\varepsilon^m \sigma^n = \sum_i \sum_j x_i x_j \varepsilon_{ij}^m \sigma_{ij}^n$$
$$\varepsilon^p \sigma^q = \sum_i \sum_j x_i x_j \varepsilon_{ij}^p \sigma_{ij}^q \tag{3.23}$$

Equivalently, in terms of the cubic EoS parameters:

$$a^m b^{\frac{n}{3}-m} = \sum_i \sum_j x_i x_j a_{ij}^m b_{ij}^{\frac{n}{3}-m}$$
$$a^p b^{\frac{q}{3}-p} = \sum_i \sum_j x_i x_j a_{ij}^p b_{ij}^{\frac{q}{3}-p} \tag{3.24}$$

Many mixing and combining rules have been proposed based on these generalized expressions and some of the most popular examples are presented in Table 3.16. A few comments about these mixing rules are in order:

1. Of special importance are the mixing rules that, while using the vdW1f rule for the co-volume parameter b, employ a co-volume-dependent mixing rule for the energy parameter of the form a/b^n, with different values of the exponent n. Rules such as those of Radosz *et al.*[70] or Plöcker *et al.*[69] have found applications in asymmetric gas-containing mixtures and those by Rao and co-workers[87] in solid–gas equilibria.
2. The Kay rule[88] together with the Berthelot rule for ε_{ij} results in the mixing rule of Equation (3.11).

3.B.2 Beyond the classical combining rules

The magnitude and trends of the interaction parameters in cubic EoS may depend a lot on the choice not only of the mixing but also of the combining rules, especially those for the cross-energy parameter. It is, thus, of interest to develop combining rules based on the theory of intermolecular forces, as outlined in Chapter 2. We will discuss a few of the most relevant approaches. One useful starting point[90] is the geometric mean rule for the cross-intermolecular potential, which also contains the ionization potentials of the molecules:

$$\Gamma_{ij} = \sqrt{\Gamma_i \Gamma_j}\left(\frac{2\sqrt{I_i I_j}}{I_i + I_j}\right) \tag{3.25}$$

Using the general potential energy function of Mie (Equation (2.14), Chapter 2), we can see that the attractive potential of each compound and of the cross-interaction is proportional to ε/σ^n where n is the

Table 3.16 *Various mixing and combining rules corresponding to different (m,n,p,q) values in Equations (3.23) and (3.24)*

(*m*,*n*,*p*,*q*)	Mixing rules in form of ε, σ	Mixing rules in form of *a*,*b*	Examples of combining rules
1,3,0,3 vdW1f, classical	$\varepsilon\sigma^3 = \sum_i \sum_j x_i x_j \varepsilon_{ij} \sigma_{ij}^3$ $\sigma^3 = \sum_i \sum_j x_i x_j \sigma_{ij}^3$ Equation (3.18)	$a = \sum_{i=1}^{n} \sum_{j=1}^{n} x_i x_j a_{ij}$ $b = \sum_{i=1}^{n} \sum_{j=1}^{n} x_i x_j b_{ij}$ Equations (3.2)	GM/AM for a_{ij} and b_{ij}, Equations (3.3) and (3.21) Lorentz–Berthelot for $\sigma_{ij}, \varepsilon_{ij}$, Equation (3.22)
2,0,0,1 – Kay rules	$\varepsilon^2 = \sum_i \sum_j x_i x_j \varepsilon_{ij}^2$ $\sigma = \sum_i \sum_j x_i x_j \sigma_{ij}$	$\left(\frac{a}{b}\right)^2 = \sum_i \sum_j x_i x_j \left(\frac{a_{ij}}{b_{ij}}\right)^2$ $b^{1/3} = \sum_i \sum_j x_i x_j b_{ij}^{1/3}$	Lorentz–Berthelot, which give: $\varepsilon = \sum_i x_i \varepsilon_i$ $\sigma = \sum_i x_i \sigma_i$
1,*n*,1,*q* – Mie potential	$\varepsilon\sigma^n = \sum_i \sum_j x_i x_j \varepsilon_{ij} \sigma_{ij}^n$ $\varepsilon\sigma^q = \sum_i \sum_j x_i x_j \varepsilon_{ij} \sigma_{ij}^q$	$a\, b^{\frac{n}{3}-1} = \sum_i \sum_j x_i x_j a_{ij} b_{ij}^{\frac{n}{3}-1}$ $a\, b^{\frac{q}{3}-1} = \sum_i \sum_j x_i x_j a_{ij} b_{ij}^{\frac{q}{3}-1}$	$\frac{a_{ij}}{b_{ij}^{1-\frac{n}{3}}} = \sqrt{\frac{a_i a_j}{b_i^{1-\frac{n}{3}} b_j^{1-\frac{n}{3}}}}$ $b_{ij} = \frac{b_i + b_j}{2}$
1,0.75,0,3[69]	$\varepsilon\sigma^{0.75} = \sum_i \sum_j x_i x_j \varepsilon_{ij} \sigma_{ij}^{0.75}$ $\sigma^3 = \sum_i \sum_j x_i x_j \sigma_{ij}^3$	$ab^{-0.75} = \sum_i \sum_j x_i x_j a_{ij} b_{ij}^{-0.75}$ $b = \sum_i \sum_j x_i x_j b_{ij}$	$\frac{a_{ij}}{b_{ij}^{0.75}} = \sqrt{\frac{a_i a_j}{b_i^{0.75} b_j^{0.75}}}$ $b_{ij} = \frac{b_i + b_j}{2}$
1,2.25,0,3[70]	$\varepsilon\sigma^{2.25} = \sum_i \sum_j x_i x_j \varepsilon_{ij} \sigma_{ij}^{2.25}$ $\sigma^3 = \sum_i \sum_j x_i x_j \sigma_{ij}^3$	$ab^{-0.25} = \sum_i \sum_j x_i x_j a_{ij} b_{ij}^{-0.25}$ $b = \sum_i \sum_j x_i x_j b_{ij}$	$\frac{a_{ij}}{b_{ij}^{-0.25}} = \sqrt{\frac{a_i a_j}{b_i^{-0.25} b_j^{-0.25}}}$ $b_{ij} = \frac{b_i + b_j}{2}$
1,4.5,0,4.5[89]	$\varepsilon\sigma^{4.5} = \sum_i \sum_j x_i x_j \varepsilon_{ij} \sigma_{ij}^{4.5}$ $\sigma^{4.5} = \sum_i \sum_j x_i x_j \sigma_{ij}^{4.5}$	$ab^{0.5} = \sum_i \sum_j x_i x_j a_{ij} b_{ij}^{0.5}$ $b^{1.5} = \sum_i \sum_j x_i x_j b_{ij}^{1.5}$	$\frac{a_{ij}}{b_{ij}^{-0.5}} = \sqrt{\frac{a_i a_j}{b_i^{-0.5} b_j^{-0.5}}}$ $b_{ij} = \frac{b_i + b_j}{2}$
Rao[87]	$\varepsilon\sigma^{8.1} = \sum_i \sum_j x_i x_j \varepsilon_{ij} \sigma_{ij}^{8.1}$ $\sigma^3 = \sum_i \sum_j x_i x_j \sigma_{ij}^3$	$a\, b^{1.7} = \sum_i \sum_j x_i x_j a_{ij} b_{ij}^{1.7}$ $b = \sum_i \sum_j x_i x_j b_{ij}$	

GM = geometric mean rule, AM = arithmetic mean rule.

attractive tail of the potential. Substituting this relationship (Mie potential function) into Equation (3.25) and also utilizing the relationships of Equation (3.20), we arrive at the following expression for the combining rule of the cross-energy (microscopic and macroscopic) parameters:

$$\varepsilon_{ij} = \sqrt{\varepsilon_i \varepsilon_j}\left(\frac{\sqrt{\sigma_i \sigma_j}}{\sigma_{ij}}\right)^n \left(\frac{2\sqrt{I_i I_j}}{I_i + I_j}\right) \tag{3.26}$$

$$a_{ij} = \sqrt{a_i a_j}\left(\frac{\sqrt{b_i b_j}}{b_{ij}}\right)^{\frac{n}{3}-1} \left(\frac{2\sqrt{I_i I_j}}{I_i + I_j}\right) \tag{3.27}$$

An often utilized assumption is that the term with the ionization potentials can be ignored as the values of I_i of many compounds are close to each other. With this assumption, the classical geometric mean (GM) rule for a_{ij} is recovered for the 'empirical' value of $n = 3$ (which 'eliminates' the size dependence of the cross-energy term).

However, Coutinho *et al.*[91,92] showed that for many systems (*1,2* are used here instead of *i,j*):

$$I \propto \frac{1}{\sigma^3} \Rightarrow \frac{2\sqrt{I_1 I_2}}{I_1 + I_2} \cong \left(\frac{\sqrt{b_1 b_2}}{b_{12}}\right)^{-1} \tag{3.28}$$

Inserting Equation (3.28) into (3.27), we obtain:

$$a_{12} = \sqrt{a_1 a_2}\left(\frac{\sqrt{b_1 b_2}}{b_{12}}\right)^{\frac{n}{3}-2} \tag{3.29}$$

The GM rule for a_{12} is now recovered for the Lennard-Jones value ($n = 6$) of the general Mie potential. Moreover, using this equation, the interaction parameter correction to the GM rule, $a_{12} = \sqrt{a_1 a_2}(1 - k_{12})$, can be estimated as:

$$k_{12} = 1 - \left(\frac{\sqrt{b_1 b_2}}{b_{12}}\right)^{\frac{n}{3}-2} \tag{3.30}$$

The exponent n of the Mie potential function is not known a priori and should be considered a fitting parameter. It decreases with chain length for various families of compounds, e.g. N_2 and CO_2–alkanes.[92]

Table 3.17 Comparison of the I and C terms in Equations (3.26) and (3.31). The C values are from Singh et al.[93] while the I terms have been computed using the ionization potential values from Pesuit[95]

Mixture	$\left(\frac{C_{ij}}{\sqrt{C_i C_j}}\right)$	$\left(\frac{2\sqrt{I_i I_j}}{I_i + I_j}\right)$
CO_2–C_2H_6	0.9952	0.996 5
CO_2–C_3H_8	0.9956	0.994 1
CO_2–water	0.9999	0.999
CO_2–methanol	0.9986	0.992 6
CO_2–acetone	0.9969	0.988
CO_2–benzene	0.9891	0.980 4
N_2–C_2H_6	0.9951	0.989 5
Methanol–benzene	0.9951	0.996 9
Methanol–C_3H_8	0.9992	0.999 92
Methanol–H_2S	0.9895	0.999 8

Thus, although it has been observed that k_{ij} exhibits different trends with the chain length of n-alkanes for the two gases (decreasing for CO_2 and increasing for N_2), the exponent n decreases with increasing chain length in both cases. This may indicate that the more general combining rule, Equation (3.29), incorporates the size effects and provides a physically meaningful insight into EoS behavior. Thus, Equation (3.30) is a useful way for estimating interaction parameters for gas–alkane systems.

Recently, Singh *et al.*[93] and Leonhard *et al.*[94] have developed for application in the SAFT EoS (see Chapters 8 and 16) the following combining rule (based on a more rigorous solution of the London theory than that of Hudson and McCoubrey)[90]:

$$\varepsilon_{ij} = \sqrt{\varepsilon_i \varepsilon_j}\left(\frac{\sqrt{\sigma_i \sigma_j}}{\sigma_{ij}}\right)^6 \left(\frac{C_{ij}}{\sqrt{C_i C_j}}\right) \qquad (3.31)$$

The C_i values are the dispersion coefficients (of the potential function) and the term containing them can be rigorously calculated from quantum chemistry calculations. The resemblance of Equation (3.31) to (3.26) is striking, for $n = 6$ (the Lennard-Jones value). Their difference lies essentially in the use of ionization potentials or the C parameters. As Table 3.17 shows, the differences are small for many mixtures.

3.B.3 On the combining rule for the cross co-volume parameters

Compared to the combining rule for the cross-energy parameter, relatively less attention has been paid to the combining rule for the cross co-volume. This can be attributed to the great familiarity and success of the arithmetic mean combining rule for b_{12} (in cubic EoS), Equation (3.3), and the Lorentz rule for σ_{12} (in molecular simulation and theoretical EoS).

There are a couple of notable exceptions. One of them is the combining rule of Kong:[96]

$$\varepsilon_{ij}\sigma_{ij}^{12} = \frac{\varepsilon_i\sigma_i^{12}}{2^{13}}\left[1 + \left(\frac{\varepsilon_j\sigma_j^{12}}{\varepsilon_i\sigma_i^{12}}\right)^{\frac{1}{13}}\right]^{13} \tag{3.32}$$

These rules are discussed extensively by Ungerer *et al.*[71] Kong used Equation (3.26) with $n = 6$ (and without the ionization potential term) for the cross-energy parameter, ε_{ij}. Ungerer *et al.*[71] have showed that the use of the Kong rule results, in many cases, in improved representation of simulation results compared to the well-known Lorentz–Berthelot rules ($\varepsilon_{ij} = (\varepsilon_i\varepsilon_j)^{1/2}$, $\sigma_{ij} = (\sigma_i + \sigma_j)/2$). This has been illustrated, for example, for excess volumes and enthalpies of C_2–CO_2 and H_2S–CO_2.

Inspired by simulation studies, various alternatives to the arithmetic mean combining rule (Equation (3.3)) have been proposed for the co-volume parameter:

$$\text{Lorentz}^{97}\ \text{rule}: \quad b_{12} = \frac{1}{8}\left(b_1^{1/3} + b_2^{1/3}\right)^3 \tag{3.33}$$

$$\text{Lee and Sandler}^{98}\ \text{rule}: \quad b_{12} = \left(\frac{b_1^{2/3} + b_2^{2/3}}{2}\right)^{\frac{3}{2}} \tag{3.34}$$

$$\text{Good and Hope}^{99}\ \text{rule}: \quad b_{12} = \sqrt{b_1 b_2} \tag{3.35}$$

As previously indicated, the Lorentz rule is well known, typically used in molecular simulation studies and derived from the arithmetic mean combining rule for the molecular diameter:

$$\sigma_{12} = \frac{\sigma_1 + \sigma_2}{2} \tag{3.36}$$

These rules (Equations (3.32)–(3.35)) have been of limited success when used in cubic EoS, illustrating that we do not always get successful results by employing theoretical equations in semi-empirical models such as the well-known cubic EoS.

References

1. C.-C. Chen, P.M. Mathias, *AIChE J.*, **2002**, *48(2)*, 194.
2. J.C. De Hemptinne, E. Behar, *Oil Gas Sci. Technol. – Rev. IFP*, **2006**, *61(6)*, 303.
3. J.D. Van der Waals, On the continuity of the gas and liquid state. Doctoral Dissertation, Leiden, 1873.
4. O. Redlich, J.N.S. Kwong, *Chem. Rev.*, **1949**, *44*, 233.
5. G. Soave, *Chem. Eng. Sci.*, **1972**, 1197.
6. D.-Y. Peng, D.B. Robinson, *Ind. Eng. Chem. Fundam.*, **1976**, *15(1)*, 59.

7. M.S. Graboski, Th.E. Daubert, *Ind. Eng. Chem. Process Des. Dev.*, **1979**, *18*(2), 300.
8. G. Soave, *Chem. Eng. Sci.*, **1979**, *34*, 225.
9. K. Magoulas, D. Tassios, *Fluid Phase Equilib.*, **1990**, *56*, 119.
10. P.M. Mathias, T.W. Copeman, *Fluid Phase Equilib.*, **1983**, *13*, 91.
11. J.H. Vera, J.M. Prausnitz, *Chem. Eng. J.*, **1972**, *3*, 1.
12. K.J. Laidler, J.H. Meiser, *Physical Chemistry* (3rd edition). Houghton Mifflin, 1999.
13. P.D. Ting, P.C. Joyce, P.K. Jog, W.G. Chapman, M.C. Thies, *Fluid Phase Equilib.*, **2003**, *206*, 267.
14. E.C. Voutsas, G.D. Pappa, K. Magoulas, D.P. Tassios, *Fluid Phase Equilib.*, **2006**, *240*, 127.
15. I. Yakoumis, G.M. Kontogeorgis, E. Voutsas, D. Tassios, *Fluid Phase Equilib.*, **1997**, *130*, 31.
16. M.B. Oliveira, I.M. Marrucho, J.A.P. Coutinho, A.J. Queimada, *Fluid Phase Equilib.*, **2008**, *267*, 83.
17. W.D. Callister Jr, *Materials Science and Engineering: An Introduction* (4th edition). John Wiley & Sons, Inc., 1997.
18. B.E. Poling, J.M. Prausnitz, J.P. O'Connell, *The Properties of Gases and Liquids* (5th edition). McGraw-Hill, 2001.
19. G.M. Kontogeorgis, I.V. Yakoumis, Ph. Coutsikos, D.P. Tassios, *Fluid Phase Equilib.*, **1997**, *140*, 145.
20. I.V. Yakoumis, E. Nikitin, G.M. Kontogeorgis, *Fluid Phase Equilib.*, **1998**, *153*, 23.
21. A. Zbogar, V. da Silva Lopes, G.M. Kontogeorgis, *Ind. Eng. Chem. Res.*, **2006**, *45*, 476.
22. D.P. Tassios, *Applied Chemical Engineering Thermodynamics*. Springer-Verlag, 1993.
23. J.M. Prausnitz, R.N. Lichtenthaler, E.G. de Azevedo, *Molecular Thermodynamics of Fluid-Phase Equilibria* (3rd edition). Prentice Hall International, 1999.
24. J. N. Israelachvili, *Intermolecular and Surface Forces – With Applications to Colloidal and Biological Surfaces*. Academic Press, 1985.
25. K.S. Pedersen, Aa. Fredendslund, P. Thomassen, *Properties of Oils and Natural Gases*. Gulf Publishing, 1989.
26. A. Kordas, K. Tsoutsouras, S. Stamataki, D.P. Tassios, *Fluid Phase Equilib.*, **1994**, *93*, 141.
27. A. Kordas, K. Magoulas, S. Stamataki, D.P. Tassios, *Fluid Phase Equilib.*, **1995**, *112*, 33.
28. G. Avlonitis, G. Mourikas, S. Stamatakis, D. Tassios, *Fluid Phase Equilib.*, **1994**, *101*, 53.
29. S. Stamataki, K. Magoulas, *Oil Gas Sci. Technol. – Rev. IFP*, **2000**, *55*(*5*), 511.
30. J.J. Carroll, A.E. Mather, *Fluid Phase Equilib.*, **1995**, *105*, 221.
31. S. Stamataki, K. Magoulas, *Oil Gas Sci. Technol. – Rev. IFP*, **1998**, *53*(*3*), 367.
32. S. Stamataki, K. Magoulas, C.J. Boukouvalas, D.P. Tassios, *Oil Gas Sci. Technol. – Rev. IFP*, **1998**, *53*(*1*), 59.
33. G.M. Kontogeorgis, Ph. Coutsikos, V.I. Harismiadis, Aa. Fredenslund, D.P. Tassios, *Chem. Eng. Sci.*, **1998**, *53*(*3*), 541.
34. P. Ghosh, *Chem. Eng. Technol.*, **1999**, *22*(*5*), 379.
35. J.M. Prausnitz, F.W. Tavares, *AIChE J.*, **2004**, *50*(*4*), 739.
36. J.O. Valderrama, *Ind. Eng. Chem. Res.*, **2003**, *42*, 1603.
37. R. Agarwal, Y.-K. Li, O. Santollani, M.A. Satyro, A. Vieler, *Chem. Eng. Prog.*, **2001**, 64.
38. C. Tsonopoulos, J.L. Heidman, *Fluid Phase Equilib.*, **1986**, *29*, 391.
39. G.M. Kontogeorgis, D.P. Tassios, *Chem. Eng. J.*, **1997**, *66*, 35.
40. K. Knudsen, E.H. Stenby, Aa. Fredenslund, *Fluid Phase Equilib.*, **1993**, *82*, 361.
41. H. Orbey, S.I. Sandler, *Modeling Vapor-Liquid Equilibria: Cubic Equations of State and Their Mixing Rules*. Cambridge Series in Chemical Engineering, Cambridge University Press, 1998.
42. J.R. Elliott, A. Vahid, A.D. Sans, *Fluid Phase Equilib.*, **2007**, *256*, 4.
43. P.M. Mathias, H.C. Klotz, *Chem. Eng. Prog.*, **1994**, 67.
44. M. Yarrison, W.G. Chapman, *Fluid Phase Equilib.*, **2004**, *226*, 195.
45. P. Coutsikos, K. Magoulas, G.M. Kontogeorgis, *J. Supercrit. Fluids*, **2003**, *25*, 197.
46. B.L. Larsen, P. Rasmussen, Aa. Fredenslund, *Ind. Eng. Chem. Res.*, **1987**, *26*, 2274.
47. S. Gupta, J.D. Olson, *Ind. Eng. Chem. Res.*, **2003**, *42*(*25*), 6359.
48. S.I. Sandler, *Chemical and Engineering Thermodynamics* (3rd edition). John Wiley & Sons, Ltd, 1999.

49. A.E. Peneloux, E. Rauzy, R. Freeze, *Fluid Phase Equilib.*, **1982**, *8*, 7.
50. M. Teodorscu, L. Lugo, J. Fernandez, *Int. J. Thermophys.*, **2003**, *24(4)*, 1043.
51. M.D. Gordillo, M.A. Blanco, A. Molero, E.M. de la Ossa, *J. Supercrit. Fluids*, **1999**, *15*, 183.
52. S. Verdier, S.I. Andersen, *Fluid Phase Equilib.*, **2005**, *231*, 125.
53. J. Munck, S. Skjold-Jørgensen, P. Rasmussen, *Chem. Eng. Sci.*, **1988**, *43(10)*, 2661.
54. G.M. Kontogeorgis, M.L. Michelsen, G.K. Folas, S.D. Derawi, N. von Solms, E.H. Stenby, *Ind. Eng. Chem. Res.*, **2006**, *45(14)*, 2855.
55. J.-N. Jaubert, F. Mutelet, *Fluid Phase Equilib.*, **2004**, *224*, 285.
56. F. Mutelet, S. Vitu, R. Privat, J.-N. Jaubert, *Fluid Phase Equilib.*, **2005**, *238*, 157.
57. J.-N. Jaubert, S. Vitu, F. Mutelet, J.-P. Corriou, *Fluid Phase Equilib.*, **2005**, *237*, 193.
58. G.M. Kontogeorgis, Ph. Coutsikos, *Ind. Eng. Chem. Res.*, **2005**, *44(9)*, 3374.
59. P.A. Sacomani, E.A. Brignole, *Ind. Eng. Chem. Res.*, **2003**, *42*, 4143.
60. E.M. Hendriks, B. Edmonds, R.A.S. Moorwood, R. Szczepanski, *Fluid Phase Equilib.*, **1996**, *117*, 193.
61. W.R. Parrish, J.M. Prausnitz, *Ind. Eng. Chem. Process Des. Dev.*, **1972**, *11*, 26.
62. F.E. Anderson, J.M. Prausnitz, *AIChE J.*, **1986**, *32(8)*, 1321.
63. H.-J. Ng, D.B. Robinson, *AIChE J.*, **1977**, *23(4)*, 477.
64. J. Madsen, K.S. Pedersen, M.L. Michelsen, *Ind. Eng. Chem. Res.*, **2000**, *39*, 1111.
65. L. Lundgaard, J.M. Mollerup, *Fluid Phase Equilib.*, **1991**, *70*, 199.
66. L. Lundgaard, J.M. Mollerup, *Fluid Phase Equilib.*, **1992**, *76*, 141.
67. E.D. Sloan, *Clathrate Hydrates of Natural Gases* (2nd edition). Marcel Dekker, 1998.
68. W.R. Smith, *Can. J. Chem. Eng.*, **1972**, *50*, 271.
69. U. Plöcker, H. Knapp, J.M. Prausnitz, *Ind. Eng. Chem. Process Des. Dev.*, **1978**, *17*, 324.
70. M. Radosz, H.-M. Lin, K.-C. Chao, *Ind. Eng. Chem. Process Des. Dev.*, **1982**, *21*, 653.
71. Ph. Ungerer, B. Tavitian, A. Boutin, *Applications of Molecular Simulation in the Oil and Gas Industry: Monte Carlo Methods.* TECHNIP, IFP Publications, 2005.
72. G.M. Kontogeorgis, I. Smirlis, I.V. Yakoumis, V.I. Harismiadis, D.P. Tassios, *Ind. Eng. Chem. Res.*, **1997**, *36*, 4008.
73. G.S. Soave, A. Bertucco, M. Sponchiado, *AIChE J.*, **1995**, *41(8)*, 1964.
74. A. Bondi, *Physical Properties of Molecular Crystals, Liquid and Glasses.* John Wiley & Sons, Ltd, 1968.
75. H.S. Elbro, Aa. Fredenslund, P. Rasmussen, *Macromolecules*, **1990**, *23*, 4707.
76. J.O. Wong, J.M. Prausnitz, *Chem. Eng. Commun.*, **1985**, *37*, 41.
77. G.M. Kontogeorgis, V.I. Harismiadis, Aa. Fredenslund, D.P. Tassios, *Fluid Phase Equilib.*, **1994**, *96*, 65.
78. T. Oishi, J.M. Prausnitz, *Ind. Eng. Chem. Process Des. Dev.*, **1978**, *17(3)*, 333.
79. G. Bogdanic, Aa. Fredenslund, *Ind. Eng. Chem. Res.*, **1994**, *33*, 1331.
80. T. Sako, A.H. Wu, J.M. Prausnitz, *J. Appl. Polym. Sci.*, **1989**, *38*, 1839.
81. G.M. Kontogeorgis, I.A. Kouskoumvekaki, M.L. Michelsen, *Ind. Eng. Chem. Res.*, **2002**, *41*, 4686.
82. G.M. Kontogeorgis, Aa. Fredenslund, D.P. Tassios, *Ind. Eng. Chem. Res.*, **1993**, *32*, 362.
83. H.S. Elbro, Aa. Fredenslund, P. Rasmussen, *AIChE J.*, **1991**, *37*, 1107.
84. I.A. Kouskoumvekaki, M.L. Michelsen, G.M. Kontogeorgis, *Fluid Phase Equilib.*, **2002**, *202*, 325.
85. H. Orbey, S.I. Sandler, *AIChE J.*, **1994**, *40(7)*, 1203.
86. D.W. Van Krevelen, *Properties of Polymers: Their Correlation with Chemical Structure, Their Numerical Estimation and Prediction from Additive Group Contributions.* Elsevier, 1990.
87. V.S.G. Rao, M. Mukhopadhyay, *J. Supercrit. Fluids*, **1990**, *3*, 66.
88. W.B. Kay, D.B. Brice, *Ind. Eng. Chem.*, **1953**, *45*, 615.
89. T.J. Lee, L.L. Lee, K.E. Starling, Three-parameter corresponding states conformal solution mixing rules for mixtures of heavy and light hydrocarbons. In: K.C. Chao, R.L. Robinson Jr, Eds, *Equations of State in Engineering and Research*, Advances in Chemistry Series 182. American Chemical Society, 1979.
90. G.H. Hudson, J.C. McCoubrey, *Trans. Faraday Soc.*, **1960**, *56*, 761.
91. J.A.P. Coutinho, G.M. Kontogeorgis, E.H. Stenby, *Fluid Phase Equilib.*, **1994**, *102*, 31.
92. J.A.P. Coutinho, P.M. Vlamos, G.M. Kontogeorgis, *Ind. Eng. Chem. Res.*, **2000**, *39*, 3076.
93. M. Singh, K. Leonhard, K. Lucas, *Fluid Phase Equilib.*, **2007**, *258(1)*, 16.

94. K. Leonhard, V.N. Nguyen, K. Lucas, *Fluid Phase Equilib.*, **2007**, *258(1)*, 41.
95. D.R. Pesuit, *Ind. Eng. Chem. Fundam.*, **1978**, *17*(*4*), 235.
96. C.L. Kong, *J. Chem. Phys.*, **1973**, *59*, 2464.
97. H.A. Lorentz, *Annu. Phys.*, **1871**, *12*, 127.
98. K.H. Lee, S.I. Sandler, *Fluid Phase Equilb.*, **1987**, *34*, 113.
99. R.J. Good, Ch.J. Hope, *J. Chem. Phys.*, **1970**, *53*(2), 540.
100. L.C. Constantinou, R. Gani, *AIChE J.* **1994**, *40*, 1697.

4

Activity Coefficient Models, Part 1: Random-Mixing Models

4.1 Introduction to the random-mixing models

We have seen (Chapter 3) that the well-known cubic equations of state (EoS) using the van der Waals one-fluid (vdW1f) mixing rules are typically used with success mostly for mixtures containing non-polar/slightly polar compounds, e.g. gas–hydrocarbons or mixtures of hydrocarbons. For many years (to some extent this is still the case today) this limited use of cubic EoS led to the widespread use of the so-called *gamma–phi* ($\gamma - \varphi$) approach, where the vapor phase is described via an EoS (ideal gas, virial, SRK or PR) and the liquid phase is described via an activity coefficient model (excess Gibbs energy, g^E) specifically suitable for liquid solutions.

Two basic categories of models exist: the so-called 'random-mixing models' (e.g. the Margules and van Laar equations discussed in this chapter) and the advanced, theoretically based local composition (LC) models, like Wilson, UNIQUAC and UNIFAC (Chapter 5). The random-mixing models dominated up to about 1965, but are today, in many respects, replaced by the LC models, since the development of the LC concept by Grant Wilson in the mid 1960s.[1]

Thus, an evident question is, 'why bother to present *also* these classical random-mixing models, i.e. are there any practical reasons for doing so besides historical necessity?' We believe that the answer to this question must be affirmative for the following reasons:

- Simple models (Margules, van Laar) can satisfactorily correlate in many cases complex polar system vapor–liquid equilibria (VLE).
- Margules and van Laar parameters are available for many systems in the DECHEMA database and the models are available in commercial simulators.
- They are useful for fast pocket 'calculator-based' estimations (in a few cases these may be needed even in today's computer-oriented world).
- Their parameters can be estimated easily and from few data, e.g. a single point from infinite dilution activity coefficients or azeotropic data.
- These simple models are useful for illustrating some of the fundamental principles in model developments as well as some of the important interrelations between various models (theories).

Thermodynamic Models for Industrial Applications: From Classical and Advanced Mixing Rules to Association Theories
Georgios M. Kontogeorgis and Georgios K. Folas

- Regular solution theory is a widely used predictive tool, with applications to polymers (including paints and coatings), solids and pharmaceuticals.

We will start this chapter with a discussion of the activity coefficients, how they can be obtained experimentally and their trends, and then continue with a discussion first of the Margules equation and then of the 'van der Waals' family models, in which the van Laar and Flory–Huggins models and the regular solution theory are included.

4.2 Experimental activity coefficients

Data from experimental activity coefficients are useful in model development as well as in understanding the non-ideality of mixtures. There are no 'activimeters' and the term 'experimental' activity coefficient refers to values calculated from experimental VLE (*PTxy*) or SLE data.

4.2.1 VLE

$$y_i \hat{\varphi}_i^V P = x_i \gamma_i P_i^{sat} \varphi_i^{sat} \exp\left[\frac{V_i(P - P_i^{sat})}{RT}\right] \Rightarrow$$
$$\gamma_i = \frac{y_i P}{x_i P_i^{sat}} \left[\frac{\hat{\varphi}_i^V}{\varphi_i^{sat} PE}\right] \tag{4.1}$$
$$\text{where } PE = \exp\left[\frac{V_i(P - P_i^{sat})}{RT}\right]$$

In Equation (4.1), a separate model should be used for the vapor phase, e.g. the virial equation of state or a cubic EoS. Fortunately, the activity coefficients are not very sensitive to the choice of model used for the vapor phase. In most cases (unless acids are present), even if we assume that the vapor phase is completely ideal, this introduces little error in the 'experimental' activity coefficient values. At low pressures (below 10 bar), the activity coefficient is by far the most important quantity, contributing to the non-ideality of the solutions. Typically, the quantity in brackets in the last part of Equation 4.1 (including the fugacity coefficients) is between 0.9 and 1.1 up to pressures of a few bars for most substances (except when acids are present). This is true even in cases where the fugacity in the vapor, $\hat{\varphi}_i^V$, and the saturated fugacity, φ_i^{sat}, are different from each other; it is fortunate that their ratio is often close to unity.

4.2.2 SLE (assuming pure solid phase)

$$\ln(\gamma_i x_i) = -\frac{\Delta H_{fus,i}}{RT}\left(1 - \frac{T}{T_{m,i}}\right) + \frac{\Delta C_{p,i}}{R}\left(\frac{T_{m,i} - T}{T}\right) - \frac{\Delta C_{p,i}}{R}\ln\frac{T_{m,i}}{T} \tag{4.2}$$

or in a simpler form:

$$\ln(\gamma_i x_i) = -\frac{\Delta S_{fus,i}}{R}\left(\frac{T_{m,i}}{T} - 1\right) \tag{4.3}$$

where $\Delta C_{p,i}$ is the difference between the heat capacity of the liquid and the solid.

Equation (4.3) is derived from Equation (4.2) if the terms containing the heat capacity differences can be neglected. The first term in Equation (4.2) is much more important than the other two and the heat of fusion is equal to the entropy of fusion divided by the melting point. The triple-point temperature (which should be rigorously used in Equations (4.2) and (4.3)) is almost equal to the normal melting temperature, T_m. As Equation (4.2) illustrates, the solubility of solids in liquids is determined not only by the intermolecular forces between solute and solvent, but also by the melting point and the enthalpy of fusion of the solute. SLE data of aqueous organic pollutants can be used for back-calculating their activity coefficient. The values are very high due to their low solubility in water. Such calculations are important in environmental applications where we are interested in the distribution of chemicals between air, water, soil, etc. For example, from data on the solubility of benzo[a]pyrene in water at 25 °C ($=3.37 \times 10^{-10}$) we can estimate an activity coefficient for benzo[a] pyrene equal to 3.76×10^8.

The ideal solubility at every temperature T is calculated from Equation (4.2) or (4.3) by setting the activity coefficient equal to unity, and it therefore depends solely on the solute's melting temperature and its enthalpy or entropy of fusion:

$$x_i = \frac{x_i^{ideal}}{\gamma_i} \quad \text{with} \quad \gamma_i = 1 \tag{4.4}$$

The ideal solubility does not depend on the solvent's properties and is typically higher than the experimental solubility in liquids: for example, for white phosphorus in n-C_7 at 25 °C, the ideal solubility is 0.942 while the experimental value is 0.0124; and for naphthalene in n-C_6 at 20 °C, the ideal solubility is 0.269 and the experimental value 0.09.

4.2.3 Trends of the activity coefficients

Typically, activity coefficients for most systems have values above unity (positive deviations from Raoult's law); they are below unity (negative deviations) for solvating systems, e.g. chloroform–acetone, for asymmetric 'athermal systems' such as solutions of alkanes and for polymer solutions. Activity coefficients are strong functions of concentration and can also depend on temperature, but they are only very weak functions of pressure. Plots of activity coefficients against concentration present minima/maxima in very rare cases of 'strange' combinations of physical and chemical forces, e.g. for methanol–acetone and methanol–chloroform (and other chloroform–alcohol systems). A useful way to assess the non-ideality of mixtures is via the so-called infinite dilution activity coefficients, γ_i^∞, the limiting value of the activity coefficient when the concentration is close to zero ($\gamma_i^\infty = \lim_{x_i \to 0} \gamma_i$). Figure 4.1 shows one example. Infinite dilution activity coefficients are widely important in chemical, biochemical and environmental engineering.[2]

Typically, as the dissimilarity (chemical or size) between the two components increases, so do the activity coefficients. Activity coefficients are measures of the non-ideality of solutions. With reference to Figure 4.1, consider for example the binary systems of aliphatic ketones with n-heptane. The polar character of ketones which is due to the presence of the carbonyl (CO) group decreases with increasing values of the carbon atom number (N_c) in the ketone. Thus, the ketones become progressively more hydrocarbon-like, which explains the observed decrease in the activity coefficient values of ketones in n-heptane as N_c increases. The same phenomenon, but with the opposite effect, occurs in the case of alcohols in water, where the main factor is the hydrogen bonding among the hydroxyl (OH) groups. As N_c increases in the alcohols, hydrocarbon behavior is approached and the hydrogen bonding effect diminishes. However, alcohols become progressively more dissimilar to water, which again explains the trend of the

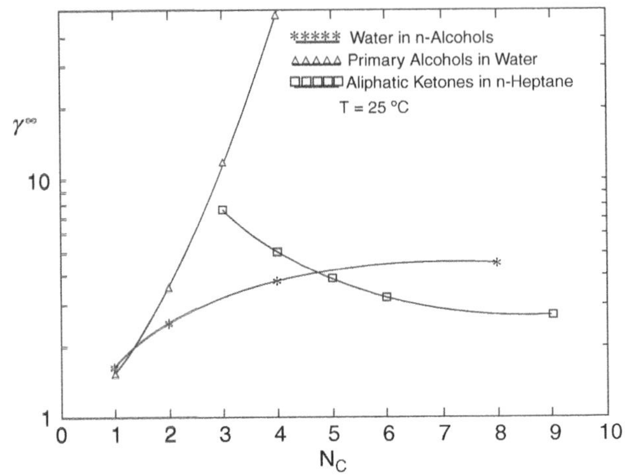

Figure 4.1 *Infinite dilution activity coefficients as function of the chain length, N_c, for three families of compounds illustrating the effect of chemical dissimilarity between the components of mixtures on the activity coefficients. Reprinted with permission from Applied Chemical Engineering Thermodynamics by Dimitrios P. Tassios, Copyright (1993) Springer Science + Business Media*

infinite dilution activity coefficients. This dissimilarity leads to partial miscibility with higher alcohols (butanol and higher).

Activity coefficients are linked to the excess Gibbs energy:

$$g^E = h^E - TS^E \tag{4.5}$$

$$RT \ln \gamma_i = \left(\frac{\partial n g^E}{\partial n_i} \right)_{T,P,n_{j \neq i}} \tag{4.6}$$

Development of an excess Gibbs energy model and thus of an activity coefficient model requires an understanding of the different effects involved in the mixing of the molecules, the molar excess enthalpy (energetic effect due to differences in intermolecular forces) and the entropic effect represented by the excess entropy, resulting from a lack of complete randomness in the distribution of molecules in the mixtures. It is well known that it is easier to develop a model for the 'sum' of the two (i.e. the excess Gibbs energy) than separately for the two other effects (excess enthalpy and excess entropy), which for many systems are rather complex and, moreover, may be complex functions of temperature (especially the excess enthalpy). Thus, the various developments for activity coefficient models almost always start from an expression for g^E. In most of the models we discuss in this and the next chapter, it is correctly assumed that the dominating factor is the first term of Equation (4.5), i.e. h^E (at least for mixtures containing molecules which do not differ much in size). This leads to the concept of 'regular solutions'.

4.3 The Margules equations

The Margules equations were the first important random-mixing models and the two known forms (one- and two-parameter versions) are presented in Table 4.1.

Table 4.1 The Margules equation for binary systems

Model	g^E equation	Expressions for the activity coefficient (binary)	Comments
One-parameter Margules	$g^E = Ax_1x_2$	$\ln\gamma_i = \frac{A}{RT}x_j^2$	$A = z\left[\Gamma_{12} - \frac{\Gamma_{11}+\Gamma_{22}}{2}\right]$ *A* value from the chemical theory
Two-parameter Margules	$\frac{g^E}{RT} = (A_{ji}x_i + A_{ij}x_j)x_ix_j$	$\ln\gamma_i = x_j^2[A_{ij} + 2(A_{ji} - A_{ij})x_i]$	

The parameters of the Margules and other activity coefficient models are typically obtained by regressing activity coefficient or VLE (*PTxy*) data at constant *T* or *P*. Alternatively, in the case of lack of plentiful data, they can be estimated from:

- a single activity coefficient point at a specific concentration (for both components);
- infinite dilution activity coefficients (many databases are available, e.g. Reid *et al.*[4]);
- azeotropic data ($\gamma = P/P^{sat}$).

Max Margules (1856–1920) was an Austrian meteorologist (see Wisniak,[5] for an interesting biography of this exciting scientist). The one-parameter (two-suffix) version is a symmetric activity coefficient model, which provides good results only for very simple symmetric systems like Ar/Xe, Ar/oxygen, benzene/CC_6 and benzene/2,2,4-trimethyl pentane. The activity coefficients at infinite dilution of both compounds are the same in this case. The coefficient *A* may be positive or negative and is in general a function (often decreasing) of temperature. For simple systems and over a small temperature range *A* is nearly constant. As Table 4.1 shows, there is some theoretical significance behind the single parameter version of the model based on random lattice theory. It can be shown[6] that the parameter *A* is related to the intermolecular potentials of the compounds. An ideal solution is obtained in the case where the cross-potential is given by the arithmetic mean (AM) average of the potentials ($\Gamma_{12} = (\Gamma_1 + \Gamma_2)/2$). However, we have seen (Chapters 2 and 3) that for simple non-polar molecules where dispersion forces dominate, the cross-potential is given by the geometric mean (GM) average of the potentials of the pure compounds ($\Gamma_{12} = (\Gamma_1\Gamma_2)^{1/2}$). As the GM is always less than the AM average and since all potentials are negative in sign, it follows that for simple molecules, the one-parameter Margules equation predicts positive deviations from ideal solution behavior, in agreement with experiment (for many binary mixtures).

The two-parameter (three-suffix) Margules model is, to some extent, developed in an empirical way. It has been observed[7] that for certain moderately non-ideal systems, e.g. MEK–toluene, the function g^E/x_1x_2RT plotted against the concentration yields an almost linear plot. Thus, having established a useful relationship for the excess Gibbs energy, the equations of the activity coefficients can be simply obtained by differentiation (Equation (4.6)). The final equations are shown in Table 4.1 for binary mixtures and it can be easily shown that the two parameters of the model are equal to the logarithms of the infinite dilution activity coefficients of the two components ($\ln\gamma_i^\infty = A_{ij}$, $\ln\gamma_j^\infty = A_{ji}$). The two-parameter Margules equation can be extended to multicomponent systems but has been mostly applied up to ternary mixtures and a few quaternary ones. Extension to ternary and quaternary mixtures is possible using only binary parameters but under certain simplifying assumptions.[6] Better results are obtained for multicomponent mixtures using a single ternary parameter.

As Figure 4.2 illustrates, the two-parameter Margules equation gives very good results in many cases, often even for highly non-ideal systems, and in the case of both positive and negative deviations from Raoult's law. This model can even describe (as one of the very few models) the maxima/minima in plots of activity coefficient against concentration observed for some systems. Of course, the two parameters have to be obtained from experimental data, not necessarily over the entire concentration range, but some data, e.g. the

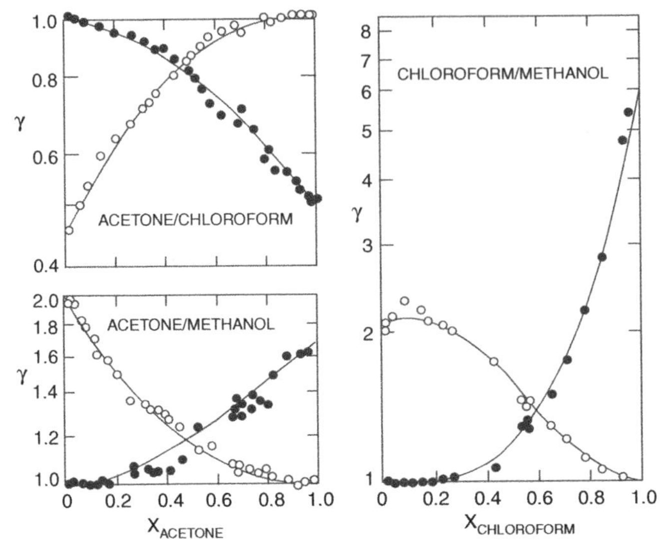

Figure 4.2 *Activity coefficients from three binary systems at 50 °C, using the two-parameter Margules equation. Reprinted with permission from AIChE, Estimation of Ternary Vapor-liquid equilibrium by W.H. Strevens, A. Sesonske et al., 1, 3, 401–409 Copyright (1955) John Wiley & Sons Ltd*

infinite dilution activity coefficients, must be available. There are no methods for predicting the parameters of the Margules equation if no data are available. One model with an improved theoretical basis, built on the van der Waals equation, is the van Laar activity coefficient, especially its improved version known as the regular solution theory, discussed next.

4.4 From the van der Waals and van Laar equation to the regular solution theory

4.4.1 From the van der Waals EoS to the van Laar model

All three models (van der Waals EoS, van Laar model and regular solution theory) belong to the van der Waals family which can be seen if the van der Waals EoS is written in terms of excess Gibbs energy and activity coefficients. As discussed in Chapter 3, and assuming the validity of the vdW1f mixing rules (for the activity coefficient expression), the van der Waals equation of state (Chapter 3, Table 3.1) can be written as (ignoring the excess volume term):

$$\frac{g^{E,vdW}}{RT} = \left[\sum_i x_i \ln\left(\frac{V_i - b_i}{V - b}\right)\right] + \left[\frac{1}{RT}\left(\sum_i x_i \frac{a_i}{V_i} - \frac{a}{V}\right)\right]$$
$$= \sum_i x_i \ln\left(\frac{\Phi_i^{fv}}{x_i}\right) + \frac{V}{RT}\Phi_i\Phi_j(\delta_i - \delta_j)^2 \qquad (4.7)$$

The latter part of Equation (4.7) holds for binary mixtures, assuming the validity of vdW1f mixing rules, classical combining rules (Equations (3.2) and (3.3) in Chapter 3) and all interaction parameters equal to zero.

The expression for the activity coefficient from the vdW EoS is equivalently (using the vdW1f mixing rules):

$$\begin{aligned}
\ln\gamma_i &= \ln\gamma_i^{comb\text{-}fv} + \ln\gamma_i^{res} = \left(\ln\frac{\Phi_i^{fv}}{x_i} + 1 - \frac{\Phi_i^{fv}}{x_i}\right) + \left(\frac{V_i}{RT}(\delta_i - \delta_j)^2\Phi_j^2\right) \\
\Phi_i^{fv} &= \frac{x_i(V_i - b_i)}{\sum_j x_j(V_j - b_j)} \\
\Phi_i &= \frac{x_i V_i}{\sum_j x_j V_j} \\
\delta_i &= \frac{\sqrt{a_i}}{V_i}
\end{aligned} \tag{4.8}$$

Although it is possible to derive the van Laar equation empirically, similar to Margules, by noticing that g^E/x_1x_2RT is inversely proportional to the composition, a better derivation which illustrates its physical meaning and the approach used by van Laar himself is based on the vdW EoS. Van Laar was a student of van der Waals and was naturally inspired by his EoS, despite his not always harmonic relationship with his teacher.[8] Such a derivation of the van Laar equation permits us also to see the clear interrelation between the two models. We will illustrate several such interrelations later; they are useful for an understanding of the various models, also in realizing that there are in reality 'fewer' really independent models, and, moreover, for comparing the various models to each other. Starting from the vdW EoS, e.g. in the form of Equations (4.7) or (4.8), van Laar used the same vdW1f mixing rules (without interaction parameters, Equations (3.2) and (3.3) in Chapter 3) but furthermore assumed that the volume can be approximated for liquids (far from the critical point) by their co-volume (i.e. $V_i = b_i$) and that the excess entropy and excess volume are zero. In this way, he neglected the first term of Equations (4.7) or (4.8), i.e. the so-called 'combinatorial–free-volume' term, stemming largely from the repulsive term of the EoS. Thus, the van Laar equation can be written in the following two forms (for a binary mixture):

$$\begin{aligned}
\ln\gamma_i &= \left(\frac{b_i}{RT}(\delta_i - \delta_j)^2\Phi_j^2\right) \\
\Phi_i &= \frac{x_i b_i}{\sum_j x_j b_j} \\
\delta_i &= \frac{\sqrt{a_i}}{b_i}
\end{aligned} \tag{4.9}$$

and:

$$\begin{aligned}
\ln\gamma_1 &= \frac{A}{\left(1 + \frac{A\,x_1}{B\,x_2}\right)^2} \\
A &= \frac{b_1}{RT}(\delta_1 - \delta_2)^2 \\
B &= \frac{b_2}{RT}(\delta_1 - \delta_2)^2
\end{aligned} \tag{4.10}$$

Activity coefficients of compound 2 can be obtained by simply setting B in place of A and A in place of B, as well as x_2 in place of x_1.

Van Laar himself presented his equation in the form shown in Equation (4.9), but Equation (4.10) is the form typically used in practice and presented in the thermodynamics literature and textbooks.[3,7] The two expressions are essentially identical. In some respects, the 'empirical' second form (Equation (4.10)) is somewhat more convenient in practice as, unlike the first form, it can give either positive or negative deviations from Raoult's law, depending on the values of the parameters.

As in the case of the two-parameter Margules equation, the two parameters of the van Laar equation are equal to the logarithms of the two activity coefficients (of component 1 and 2) at infinite dilution ($\ln\gamma_i^\infty = A$, $\ln\gamma_j^\infty = B$). The equations for the activity coefficient presented here (Equations (4.9) and (4.10)) are for binary systems. The van Laar equation can be extended to multicomponent systems (using solely binary parameters), although most applications involve binary or ternary systems.[6]

Despite its theoretical origin, the two parameters of the van Laar equation must still be treated as empirical parameters to be fitted to experimental data; they cannot be estimated from the parameters of the vdW EoS. Unfortunately, according to the first 'theoretically derived' form of the van Laar equation (Equation (4.9)), non-idealities should be attributed to the difference in the square root of the critical pressures ($\delta = \sqrt{a}/b \propto \sqrt{P_c}$) but this is, in the general case, not correct (see Problem 4 on the companion website at wiley.co.uk). Still, when van Laar's parameters are treated as adjustable parameters, as Figure 4.3 shows for a characteristic system, the model provides very satisfactory results even for complex (polar or size-asymmetric) binary systems typically having the same accuracy as the two-parameter Margules equation.

4.4.2 From the van Laar model to the Regular Solution Theory (RST)

Scatchard and Hildebrand (in Hildebrand and Scott)[9] soon realized the value of the van Laar equation and how much more useful it could be if it were free from the assumptions of the vdW equation. Their work led to the

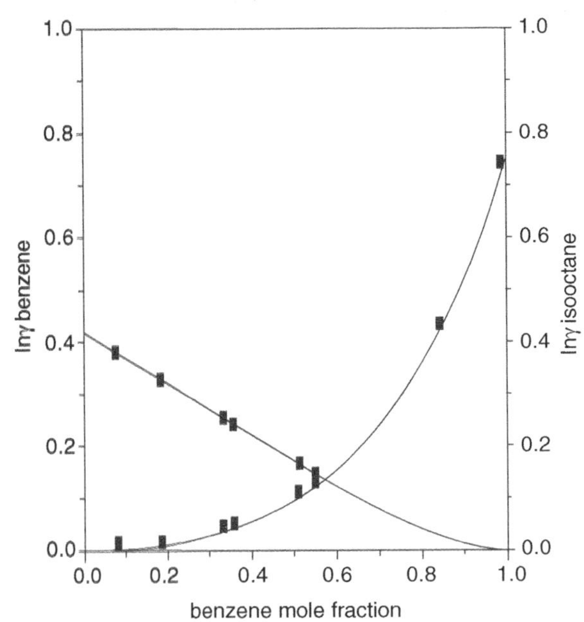

Figure 4.3 Application of the van Laar equation to a non-polar mixture (components differ slightly in molecular size): benzene–isooctane at 45 °C

famous (and still much used today) regular solution theory (RST). They termed 'regular' those solutions which obey $g^E = h^E$, $V^E = S^E = 0$, i.e. the same assumption as van Laar used. This is an assumption that, with the exception of polymer solutions, seems to be reasonably correct for many systems. Indeed, excess enthalpy is by far more important than the excess entropy for solutions having molecules similar in size.

The final equations of the RST derived are shown below for both binary and multicomponent systems, with and without interaction coefficients (where l_{ij} is the correction to the GM rule for the cross-solubility parameter $\delta_{ij} = (\delta_i\delta_j)^{1/2}(1 - l_{ij})$):

Binary systems ($l_{ij} = 0$):

$$\ln\gamma_i = \left(\frac{V_i}{RT}(\delta_i - \delta_j)^2\Phi_j^2\right)$$

$$\Phi_i = \frac{x_i V_i}{\sum_j x_j V_j}$$

Binary systems ($l_{ij} \neq 0$):

$$\ln\gamma_i = \left(\frac{V_i}{RT}\Phi_j^2\left[(\delta_i - \delta_j)^2 + 2l_{ij}\delta_i\delta_j\right]\right) \tag{4.11}$$

Multicomponent systems ($l_{ij} = 0$):

$$\ln\gamma_i = \left(\frac{V_i}{RT}(\delta_i - \delta^{av})^2\right)$$

$$\delta^{av} = \sum_{i=1}^{m}\Phi_i\delta_i$$

The top equation for activity coefficients is valid for binary systems, while the last one is the general form for multicomponent mixtures; in both equations it is assumed that the interaction parameter $l_{ij} = 0$. The middle equation is the general one for binary systems, in case a non-zero interaction parameter l_{ij} is used. A slightly more complex equation is needed for multicomponent systems when l_{ij} is different than zero:

Multicomponent systems ($l_{ij} \neq 0$):

$$\ln\gamma_k = \frac{V_k}{RT}\sum_i\sum_j\Phi_i\Phi_j\left(D_{ik} - \frac{1}{2}D_{ij}\right) \tag{4.12}$$

$$D_{ij} = (\delta_i - \delta_j)^2 + 2l_{ij}\delta_i\delta_j$$

where for a pure component i, $l_{ii} = D_{ii} = 0$.

The key concept of the theory is the solubility parameter, δ, which is defined as:

$$\delta = \sqrt{c} = \sqrt{\frac{\Delta H^{vap} - RT}{V}} \tag{4.13}$$

Table 4.2 'Liquid' volumes and solubility parameters for gaseous solutes at 25 °C to be used in Equation (4.16).[6] Given in parentheses are the values used by Thorlaksen et al.[17] in the Hildebrand–free-volume model, Equation (4.23), for gas solubilities in polymers

Gas	Volume (cm^3/mol)	δ $(J/cm^3)^{1/2}$
N_2	32.4 (70.8)	5.30 (9.08)
CO	32.1	6.40
O_2	33.0 (43.5)	8.18 (8.18)
Ar	57.1 (107.9)	10.9 (14.13)
CH_4	52.0 (54.6)	11.6 (11.6)
CO_2	55.0 (82.4)	12.3 (14.56)
C_2H_6	70.0	13.5

The quantity under the square root, c, is called cohesive energy density and can be readily measured for most compounds (except for polymers). Solubility parameters are reported, usually at 25 °C, in $(J/cm^3)^{1/2}$ or $(cal/cm^3)^{1/2}$. Table 4.3 shows the solubility parameter values for a few compounds.

Upon comparing Equations (4.9) and (4.11), we notice the striking resemblance of the RST with the van Laar equation. Unlike van Laar, though, an important advantage of the RST is its easy extension to multicomponent systems. Like the van Laar equation (in its 'theoretical formulation'), the RST with $l_{ij} = 0$ predicts only positive deviations from Raoult's law.

4.5 Applications of the Regular Solution Theory

4.5.1 General

The Regular Solution Theory (RST) has been used, in its original formulation (Equation (4.11)) or with modifications, in a wide spectrum of applications:

- low-pressure VLE and LLE;
- SLE, e.g. selecting solvents for pharmaceuticals;
- polymer solutions (in combination with the Flory–Huggins or other models for the combinatorial term);
- gas solubilities in liquids and polymers;
- study of controlled release of drugs in polymers.

Table 4.3 Solubility parameters for selected solvents and polymers (in $(J/cm^3)^{1/2}$)

Solvent	δ	Polymer	δ
Methyl Ethyl Ketone	18.5	Teflon	12.7
Hexane	14.9	Poly(dimethyl siloxane)	14.9
Styrene	19.0	Polyethylene	16.2
Cyclohexanone	19.0	Polystyrene	18.6
Acetone	19.9	Poly(methyl methacrylate)	19.4
Carbon tetrachloride	17.6	Poly(vinyl chloride)	19.8
Water	47.9	Poly(ethylene terephthalate)	21.9
Toluene	18.2	Poly(acrylonitrile)	25.3

These are some of the many applications of the RST, which has been literally applied to almost every type of phase equilibria and applications including metal solutions,[9] wax precipitation from oil[10,11] and asphaltenes.[12] It is one of the most widely used theories of thermodynamics and is attractive because of its simplicity and ease in application even in fields far different from those it was originally developed for. The RST is especially useful for qualitative calculations and interpreting data, as the examples in this section will illustrate.

4.5.2 Low-pressure VLE

It is fair to say that, for mixtures of non-polar liquids, whereas Raoult's law gives a zeroth approximation, the RST usually gives a first-order approximation to VLE. The results are often reasonably accurate for non-polar systems (especially those showing appreciable non-ideality) and in these cases the RST provides a useful guide; less satisfactory results are expected for polar/associating substances. The only major failure of the theory for non-polar fluids appears to be when it is applied to certain solutions containing fluorocarbons; the reasons are not fully understood.[6]

For low-pressure VLE, activity coefficients will be estimated using Equation (4.11). When no interaction parameters are used ($l_{ij} = 0$), the model is purely predictive, i.e. activity coefficients can be estimated from pure compound data alone. Very good and often excellent results are obtained for 'relatively simple' non-polar systems such as benzene–heptane, CO–methane and neo-pentane–carbon tetrachloride, even with zero l_{ij}. Better results can be achieved, though, if l_{ij} is used as an adjustable parameter. The following general comments can be made regarding the interaction parameter of the RST:

1. The results can be quite sensitive to the l_{ij} values employed, especially when the solubility parameters are close to each other. For this reason, the RST performs best and is recommended for non-polar solutions exhibiting moderate deviations from ideality rather than for nearly ideal solutions (with solubility parameters being very close to each other).
2. The l_{ij} parameter is essentially a correction to the GM rule for the cohesive energy density (or to the solubility parameter, $c_{ij} = (c_i c_j)^{1/2}(1 - l_{ij})$ or $\delta_{ij} = (\delta_i \delta_j)^{1/2}(1 - l_{ij})$), which is strictly valid for non-polar molecules according to the London theory. Thus, the RST's interaction parameter is essentially similar to our familiar k_{ij} from (cubic) EoS (also a correction to the GM combining rule of the EoS cross-energy parameter, $a_{ij} = (a_i a_j)^{1/2}(1 - k_{ij})$).
3. In most cases, the l_{ij} parameter cannot be correlated with the physical properties of the compounds in mixtures; some rough approximations have been proposed for aromatic–saturated hydrocarbon mixtures.[6]
4. The l_{ij} parameter is often a weak function of temperature.

Even if we cannot always make quantitative calculations with the RST, solubility parameters can help us to arrive at qualitative statements about deviations from ideality for certain mixtures. We recall that the logarithm of the activity coefficients varies directly with the difference in solubility parameters. Based on that, we can see why a mixture of carbon disulfide (CS_2), with solubility parameter equal to 20.5, exhibits with n-hexane ($\delta = 14.9$) large positive deviations from Raoult's law, whereas a mixture of carbon tetrachloride (CCl_4, $\delta = 17.6$) and cyclohexane ($\delta = 16.8$) is nearly ideal. (All solubility parameter values are in $(\text{J/cm}^3)^{1/2}$.) The difference in solubility parameters of the mixture components provides a measure of the solution non-ideality.

The application of the RST to VLE is straightforward in the sense that no additional assumptions are needed. Solubility parameters were primarily developed for applications of mixtures containing low-molecular-weight liquids. On the other hand, numerous other applications of the RST which we discuss further require assumptions and approximations because solubility parameters for solids, gases or polymers are not available in the same straightforward way as for liquids. Similarly it could be stated that the foundations of the RST do not, in principle, justify use of the model for strongly polar and especially hydrogen bonding liquids, but the model has been applied also for such compounds, especially in combination with the Hansen solubility parameters (see Equation (4.22) later).

4.5.3 Solid-liquid equilibria (SLE)

The starting point is Equation (4.3), and thus the melting point and heat of fusion of the solid must be known. The basic equations for the activity coefficient of the solid (2) in a liquid are:

$$\ln\gamma_2 = \frac{V_2^L}{RT}(\delta_1 - \delta_2)^2\Phi_1^2$$

$$\delta_2 = \sqrt{\frac{\Delta U^{vap}}{V_2^L}} = \sqrt{\frac{\Delta H^{sub} - \Delta H^{fus} - \Delta C_p(T - T_m) - RT}{V_2^L}} \tag{4.14}$$

$$V_2^L = V_2^S + \Delta V^{fus}$$

where ΔV^{fus} is the volume change of fusion at the melting temperature, T_m.

The difficulty in using the RST in this case is due to the fact that we must be able to estimate both the solubility parameter and the liquid molar volume for substances whose pure compound state is solid at the mixture temperature. Thus, we need to estimate the molar volumes and solubility parameters of subcooled liquids (compound 2) and how this is done is shown in Equation (4.14). The energy of vaporization of the subcooled liquid (ΔU^{vap}) depends on the enthalpies of fusion (*fus*) of the solid at T_m and the enthalpy of sublimation of the solid at temperature *T*. Rigorously we should be using the triple-point temperature, but the melting-point temperature (T_m) is a good approximation. Equation (4.14) is approximate but valid in the (realistic) cases where the temperature *T* is not far removed from the triple- or melting-point temperature.

The RST has found various applications in SLE[6]:

1. Using the more general form (Equation (4.11) or (4.12)) which employs a correction l_{ij} parameter, Preston and Prausnitz[38] have used the RST to correlate the solubilities of non-polar solids in non-polar liquids at low temperatures. They used generalized corresponding states charts for estimating the solubility parameters and liquid volumes of the subcooled liquids involved in these calculations. The l_{ij} parameter has been fitted from a single experimental data point.
2. Myers and Prausnitz[39] used the RST to study the solubility of solid CO_2 in liquefied light hydrocarbons. They accounted for both dispersion and quadrupolar effects of the cohesive energy density of the CO_2. Their analysis has shown that the quadrupolar effects, though lower than the dispersion ones, cannot be ignored.
3. The RST can be used for mixed solids as well.

One exciting application of the RST is the solvent screening for pharmaceuticals – an application published by researchers from Mitsubishi Chemicals.[13] As Figure 4.4 illustrates, it is shown that when the RST is used for calculating the very small solid solubilities of pharmaceuticals in various liquid solvents, the solubility has a parabolic plot with respect to the solvent solubility parameter with a maximum at the point where the solute and solvent solubility parameters are equal (see Problem 10 on the companion website at wiley.com/go/kontogeorgis). This technique has been used by Kolar *et al.*[13] and Abildskov and O'Connell[14] for selecting solvents for pharmaceuticals. For very complex pharmaceuticals, the solute activity coefficient can be much lower than one and thus the solubility can be much higher than the ideal solubility, as can be seen from the plot (b) of Figure 4.4. The RST cannot be used in such cases as it does not predict negative deviations from Raoult's

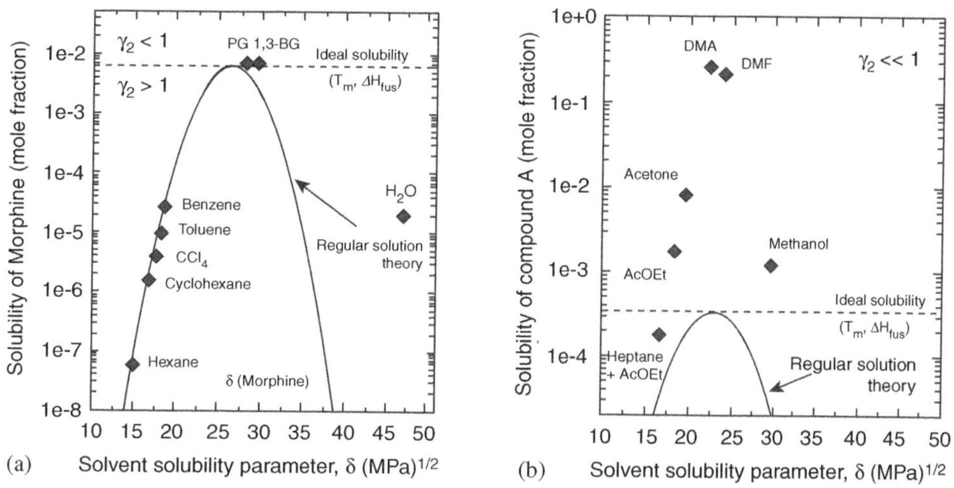

Figure 4.4 *Solubility prediction by regular solution theory for (a) morphine at 308 K and (b) a pharmaceutical intermediate containing N, O and S (298 K). Solvent abbreviations: CCl_4 = carbon tetrachloride, PG = propylene glycol, 1,3-BG = 1,3-butanediol, AcOEt = ethyl acetate, DMA = dimethylacetamide, DMF = dimethylformamide. The RST accurately predicts the solubility behavior of many solvents in morphine, but not in cases such as for the second complex pharmaceutical, shown here (b). This is due to the enhanced interactions which lead to negative deviations from Raoult's law, and result in solubility values which are much higher than the ideal solubility, a behavior which cannot be represented by the RST. Reprinted with permission from Fluid Phase Equilibria, Solvent selection for pharmaceuticals by P. Kolar, J.-W. Shen A. Tsuboi and T. Ishikawa; 194–197, 2, 771 Copyright (2002) Elsevier*

law. Other activity coefficient methods, e.g. UNIFAC with suitable parameters (discussed in Chapter 5), and theoretically based techniques could be used instead, e.g. COSMO-RS (see Chapter 16).

Alternatively, the extension of the RST based on the Flory–Huggins model and the Hansen solubility parameters,[15] as employed for polymers (Equations (4.20) and (4.22) with $\alpha = 1$, below), has been used for screening solvents for pharmaceuticals.[16]

4.5.4 Gas-liquid equilibrium (GLE)

GLE are very important in many applications, and as in the case of ideal solid solubility in liquids, the ideal solubility of a gas (2) in a liquid (1) is given by:

$$f_2^{gas} = f_2^{liquid} \Rightarrow y_2 P = x_2 P_2^{sat} \tag{4.15}$$

The ideal gas solubility depends only on the gas and is, at fixed T and P, the same for all solvents. This is unfortunately not correct in many cases, e.g. the solubility of CO_2 in various solvents (acetone, CS_2, CCl_4, C_7H_{16}, C_7F_{16}) varies from 22–209 × 10^{-4} (at 25 °C), while the ideal solubility is 160. The RST can be used also in this context, to provide estimates of the gas solubility in liquids:

$$\begin{aligned} f_2^{gas} &= y_2 \varphi_2 P = x_2 \gamma_2 f_2^L \\ \ln \gamma_2 &= \frac{V_2^L}{RT}(\delta_1 - \delta_2)^2 \Phi_1^2 \end{aligned} \tag{4.16}$$

'Liquid' volumes and solubility parameters of gases at 25 °C have been estimated from experimental mixture (gas solubility) data and some values are given in Table 4.2, while the hypothetical liquid fugacity can be obtained from corresponding states plots or correlations, provided that the critical properties of the gaseous solute (T_c,P_c) are known:[6]

$$\ln\frac{f_2^l}{P_c} = 3.548\,11 - \frac{4.745\,47}{T_r} + 1.601\,51T_r - 0.874\,66T_r^2 + 0.109\,71T_r^3 \tag{4.17}$$

where T_r ($= T/T_c$) is the reduced temperature of the gas.

4.5.5 Polymers

The RST and the concept of solubility parameters (including the Hansen solubility parameters) have been extensively applied in polymer technology in many contexts:

- identifying solvents for polymers (useful for example in the paint and coatings industry);
- activity coefficients in polymer–solvent solutions of importance (e.g. for solvent emission studies from paints);
- gas solubilities in polymers (e.g. choosing suitable polymeric adhesives to be used in thermopane windows);
- transport of pharmaceuticals via polymeric membranes.

The last three applications require coupling of the RST (which is an 'energetic' model) with an additional term which accounts for the size ('free-volume') differences between polymers and solvents. One suitable choice is the well-known Flory–Huggins (FH) 'combinatorial' model,[18,19] for binary mixtures given as:

$$\ln\gamma_1^{comb} = \ln\frac{\Phi_1}{x_1} + 1 - \frac{\Phi_1}{x_1} = \ln\frac{\Phi_1}{x_1} + \left(1 - \frac{1}{r}\right)\Phi_2 \tag{4.18}$$

The volume fraction Φ_i is defined in Equation (4.11), and $r = V_2/V_1$. For very high-molecular-weight polymers, the parameter $1/r$ is close to zero.

Another suitable approach is to use one of the free-volume theories, e.g. the entropic-FV model,[20,21] which are similar to FH but use free-volume (Φ_i^{fv}) instead of volume fractions (Φ_i):

$$\begin{aligned} \ln\gamma_i^{comb\text{-}fv} &= \ln\left(\frac{\Phi_i^{fv}}{x_i}\right) + 1 - \left(\frac{\Phi_i^{fv}}{x_i}\right) \\ \Phi_i^{fv} &= \frac{x_i V_{f_i}}{\sum\limits_j x_j V_{f_j}} \\ V_{f_i} &= V_i - V_{w,i} \end{aligned} \tag{4.19}$$

The free volume (V_{f_i}) is defined as the difference between the molar volume V_i and the vdW volume $V_{w,i}$, the latter calculated from the Bondi group contributions, e.g. as available in the UNIFAC tables.[22] Equation (4.19) is generally valid for multicomponent mixtures. We will outline next how the RST can be used for various polymer-related applications.

Solvent choice

The difference between the solubility parameters of a chemical and a polymer provides an indication of whether the polymer will be dissolved or not, and this is a manifestation of the 'like dissolves like' rule of chemistry. Table 4.3 lists solubility parameter values for a few polymers and low-molecular-weight compounds. It can be appreciated why polystyrene (PS) is soluble in the chemically similar toluene, as can be verified by their solubility parameters, while most polymers are *not* soluble in water, which has a very high solubility parameter value. This is, evidently, unfortunate, since water is a cheap and environmentally friendly solvent.

Activity coefficients using the FH model

The well-known FH model is based on the lattice theory and captures many of the essential characteristics of polymer solutions, especially with respects to size differences (entropic effects). The expression for the activity coefficient contains a combinatorial term (Equation (4.18)) and an energetic van-Laar-type term, with one interaction parameter (χ parameter), which can be estimated roughly from the solubility parameters (RST). For a binary solvent(1)–polymer(2) system, the expression is (assuming that the χ parameter depends only on temperature but not on concentration; see Problem 22 on the companion website at www.wiley.com/go/kontogeorgis):

$$\begin{aligned} \ln\gamma_1 &= \ln\left(\frac{\Phi_1}{x_1}\right) + 1 - \left(\frac{\Phi_1}{x_1}\right) + \chi_{12}\Phi_2^2 \\ \chi_{12} &= \frac{V_1}{RT}(\delta_1 - \delta_2)^2 \end{aligned} \qquad (4.20)$$

where Φ_i is the volume fraction (Equation (4.11)).

Appendix 4.A presents an extension of the FH model to multicomponent systems. Due to limitations of the FH combinatorial formula, especially the fact that it does not explicitly account for free-volume effects, a constant factor (approximately 0.35) is sometimes added to the RST-based equation (Equation (4.20)) for estimating the FH parameter. That is:

$$\chi_{12} = 0.35 + \frac{V_1}{RT}(\delta_1 - \delta_2)^2 \qquad (4.21)$$

The FH model can, in combination with the RST, be used for estimating activity coefficients, including both VLE and LLE calculations for polymer solutions and blends. Many values of the FH parameter exist in the literature.[36] The FH parameter is typically estimated from experimental phase equilibrium (pressure) data but it can be related to the solubility parameters, as shown in Equation (4.20). A useful rule of thumb based on the FH parameter is that values of χ (usually much) below 0.5 indicate miscibility, while those above 0.5 indicate non-miscibility (see Problem 16 on the companion website at www.wiley.com/go/kontogeorgis). Other rules of thumb for predicting polymer–solvent miscibility are presented in Appendix 4.B, while Appendix 4.C shows the various concentration scales of relevance to polymer solutions.

The FH model is:

- The oldest and one of the most widely used models in polymer thermodynamics.
- A semi-empirical tool since the FH parameter is typically a function of temperature and (unfortunately) often also of concentration.
- Basically of correlative value (qualitative estimation of LLE phase diagrams).

Due to the above limitations, many improvements of the FH model have been proposed, most of which include incorporation of group contributions, free-volume effects and LC concepts. They will be discussed in Chapter 5. One way to develop a predictive version of the FH model, while retaining both the volume-based combinatorial term and the van Laar/RST-type energetic term, is by using the Hansen instead of the Hildebrand parameters, as suggested by Lindvig *et al.*:[23]

$$\chi_{12} = \alpha \frac{V_1}{RT}\left[(\delta_{d1} - \delta_{d2})^2 + 0.25(\delta_{p1} - \delta_{p2})^2 + 0.25(\delta_{hb1} - \delta_{hb2})^2\right] \tag{4.22}$$

where d, p, h indicate dispersion, polar and hydrogen bonding contributions.

The α parameter in Equation (4.22) was optimized from various polymer–solvent systems (using also different choices for the combinatorial term of FH). A universal value of α was found, equal to 0.6, when the classical form of the FH equation is used (i.e. based on volume fractions, Equation (4.20)). Excellent results are obtained (about 20% average deviation for many systems). The deviations from the experimental data when α is equal to 0 or 1 are about 40%. The accuracy of this approach is similar to (actually a bit better than) the entropic–FV and UNIFAC–FV models, discussed in Chapter 5 (which for the same systems show a deviation of about 35–40%).

Gas solubilities in polymers

An interesting application of the RST combined with free-volume concepts is in the estimation of gas solubilities in polymers which are important in many contexts, e.g. they are relevant to the design of thermopane windows (Figure 4.5). This product design problem is discussed by Cussler and Moggridge.[24]

Sealant design requires a predictive method for estimating the permeability of various gases in various polymers. In order to be able to use the model for design purposes, it is important that reliable estimates can be made with a minimum of experimental data. Thus, a predictive model for gas solubilities in polymers is desirable.

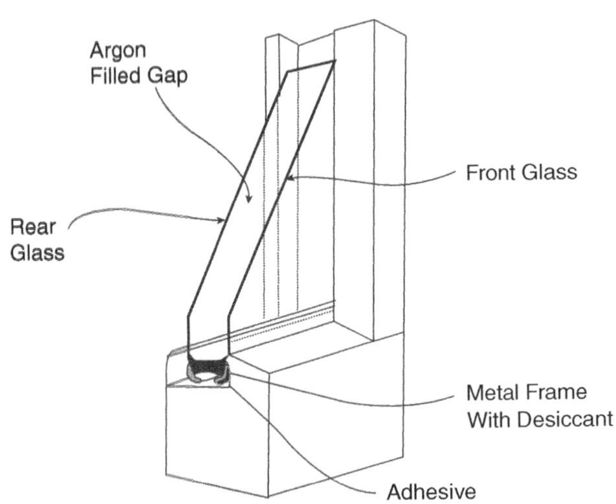

Figure 4.5 *A thermopane window. The space between the panes is filled with argon or other gases, which may leak out, causing problems. Reproduced with permission from Chemical Product Design by E. L. Cussler and G. D. Moggridge. Copyright (2001) Cambridge University Press*

The RST application to gas solubilities in elastomers related to thermopane windows has been presented by Thorlaksen *et al.*,[17] who combined the entropic–FV term (Equation (4.19)) with Hildebrand's RST for estimating the FH parameter (Equation (4.20)). This results in the following equation:

$$\frac{1}{x_2} = \frac{f_2^l}{f_2^g} \exp\left(\frac{V_2^L(\delta_1 - \delta_2)^2 \Phi_1^2}{RT} + \ln \frac{\Phi_2^{fv}}{x_2} + 1 - \frac{\Phi_2^{fv}}{x_2} \right) \tag{4.23}$$

Free-volume fractions are calculated as in Equation (4.19), the hypothetical liquid fugacity from Equation (4.17), while the 'liquid' volumes of some gases and their solubility parameters have been regressed from experimental data and they are shown to be largely independent of the polymer employed (some values are given in Table 4.2). The 'liquid' volumes can be correlated to the critical volume of the gases (V_c) via:

$$V_2^L = 1.7553 V_c - 85.093 \tag{4.24}$$

Very satisfactory results are obtained as shown in Figure 4.6 for the solubility of various gases in polyisoprene. The average deviations between experimental and calculated gas solubilities for four different gases (nitrogen, oxygen, argon and carbon dioxide) in five different elastomers is 17% using the 'liquid' volume estimated from Equation (4.24), or 11% using the average hypothetical volume of each gas (Table 4.2). The results are better than other 'standard' approaches or correlations tailored to specific polymers.

Controlled drug release

An interesting approach for estimating the controlled drug release (Figure 4.7) via polymers using the RST combined with the FH model has been developed by Michaels *et al.*[25] and discussed by Prausnitz[26].

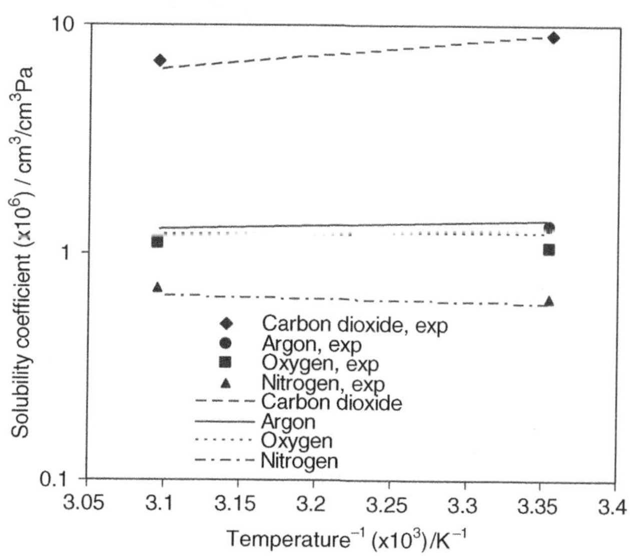

__Figure 4.6__ The solubility of several gases in polyisoprene (PIP) as a function of temperature predicted by the Hildebrand entropic–FV model, Equation (4.23). Reprinted with permission from Fluid Phase Equilibria, Prediction of gas solubilities in elastomeric polymers for the design of thermopane windows by Peter Thorlaksen, Jens Abildskov and Georgios M. Kontogeorgis, 211, 1, 17 Copyright (2003) Elsevier

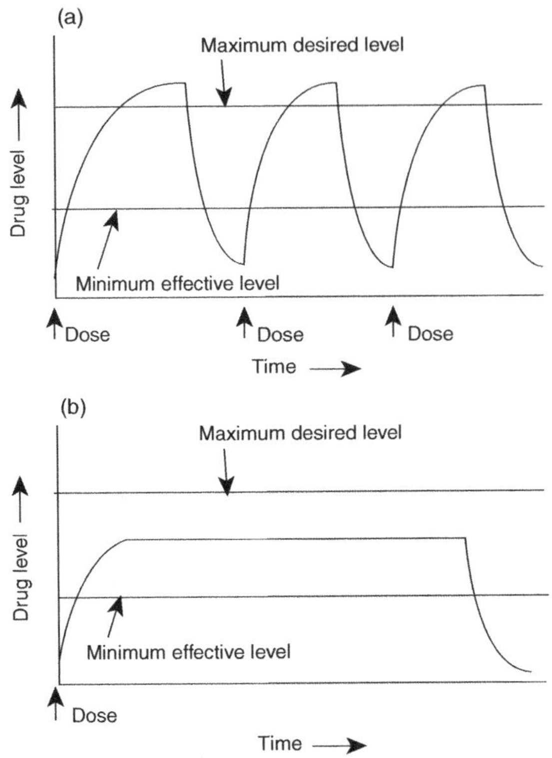

Figure 4.7 *Controlled drug release against traditional drug administration. The traditional drug administration curve may cause rhythmic changes (pulse release effects) in absorption, distribution, excretion and bioavailability and problems related to drug effect: minimum and maximum value. In contrast, the controlled drug delivery curve can achieve long-term administration, constant drug concentration in blood and stable rate between minimum and maximum value. Reprinted with kind permission from Medical Plastics and Biomaterials, Polymers in Controlled Drug Delivery by Lisa Brannon-Peppas Copyright (1997) Canon Communications LLC http://www.devicelink .com/mpb/archive/97/11/003.html*

By combining the SLE equation (Equation (4.2)) with the FH/RST equations (Equation (4.20)), a simple model can be developed for estimating the controlled release of pharmaceuticals from polymeric matrices (see Problem 9 on the companion website at www.wiley.com/go/kontogeorgis):

$$\ln\left[J_{\max}\cdot l\cdot \exp(1+\chi)\right] = -\frac{\Delta S_f}{R}\left(\frac{T_m}{T}-1\right)+\ln(\rho D) \tag{4.25}$$

where $J_{\max}$ is the maximum flux and l is the thickness of the polymer, ρ is the density of the pharmaceutical and D its diffusion coefficient through the polymer. The FH parameter is calculated from the volume and solubility parameters, as in Equation (4.20). Equation (4.25) suggests that a logarithmic plot of $J_{\max}\cdot l\cdot \exp(1+\chi)$ against $(T_m/T)-1$ should be linear for different drugs through the same polymer. Indeed, successful results are obtained in some cases, as can be seen for several steroids in the example of Figure 4.8.

However, Equation (4.25) should be used cautiously, as there are many assumptions involved; e.g. the validity of the FH equation with Hildebrand ('total') solubility parameters, uncertainty related to the estimation/values of the properties of pharmaceuticals (solubility parameters and densities). These limitations become apparent in

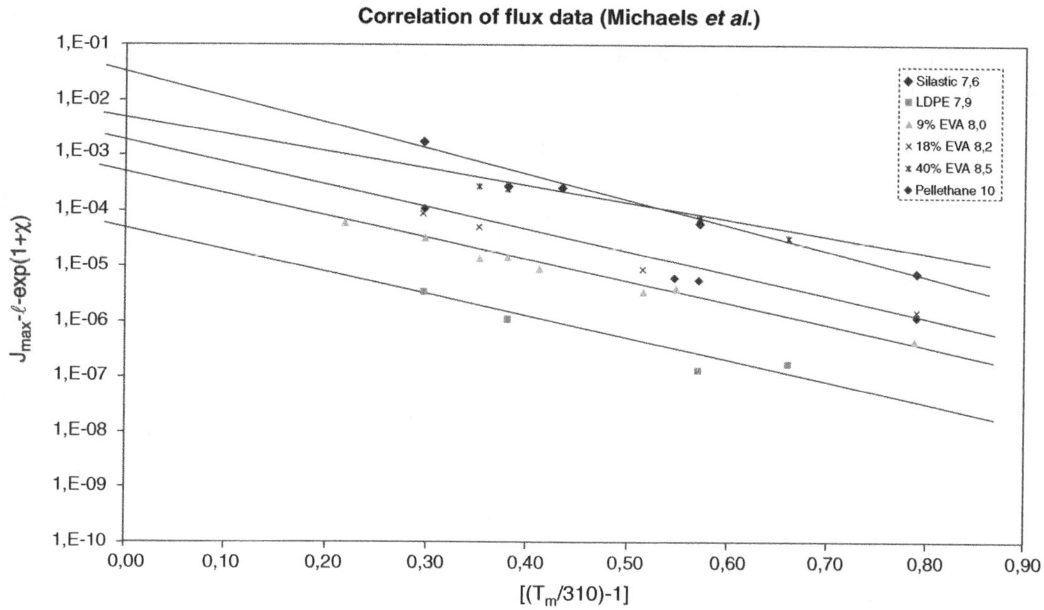

Figure 4.8 *Correlation of flux (permeability) data of 11 steroids diffusing through six polymers against a function of the steroid melting temperature. The data are plotted according to Equation (4.25). Calculations redone with the method of Michaels et al.*[25,27] *The calculated* $\Delta S_f/R$ *value is between 7 and 10.3, close to the values reported by Michaels et al.*[25] *and the experimental value (about 8.43 for all polymers and steroids)*

another case study. Only fair agreement is obtained between the experimental permeability data and values estimated from Equation (4.25) in the case of the permeability of nine steroids in two other polymers (PEU 2000E, a poly(etherurethane) copolymer and Elvax 40, a poly(ethylene vinylacetate) copolymer).[27] The experimental data have been reported by Lee *et al.*[28] The solubility parameters have been estimated by a group contribution method and the density of all steroids is assumed constant. The linearity is only approximate and much lower $\Delta S_f/R$ values are obtained (2.3 and 3.3) compared to the calculations by Michaels *et al.*[25]

4.6 SLE with emphasis on wax formation

The RST (alone or in combination with FH-type terms) has been used for predicting wax precipitation, which may take place at temperatures higher than the freezing point of water. By the term 'wax' we define mixtures of heavy paraffinic and/or microcrystalline (branched/cycloparaffinic) hydrocarbons. Wax precipitation can be a serious problem when oil is transported in pipelines as it can cause plugging of pipes or equipment.[11] Of special interest are the conditions under which wax precipitation takes place and the amount of wax formed.

The starting point is the SLE equation:

$$\frac{x_i^S}{x_i^L} = \frac{\gamma_i^L}{\gamma_i^S} \exp\left[\frac{\Delta H_i^f}{RT}\left(1 - \frac{T}{T_{m,i}}\right) + \frac{\Delta C_{p,i}}{R}\left(1 + \ln\frac{T_{m,i}}{T} - \frac{T_{m,i}}{T}\right)\right] \tag{4.26}$$

where x_i is the mole fraction of component i, S or L refers to solid or liquid phase, respectively, γ_i is the activity coefficient of component i, T_m is the melting temperature and ΔH^{fus} is the enthalpy of fusion.

Two RST-type models applied to wax precipitation are those of Won[10,29] and Hansen *et al.*[30] In Won's model, both the activity coefficients of compound *i* in the solid and liquid phases are expressed via the RST (without use of interaction parameters):

$$\ln\gamma_i = \frac{V_i}{RT}(\delta^{av} - \delta_i)^2$$
$$\delta^{av} = \sum_i \Phi_i \delta_i \tag{4.27}$$

where the volume fractions in the liquid and solid are given as:

$$\Phi_i^L = \frac{x_i^L V_i^L}{\sum_j x_j^L V_j^L} \quad \text{and} \quad \Phi_i^S = \frac{x_i^S V_i^S}{\sum_j x_j^S V_j^S} \tag{4.28}$$

The liquid and solid molar volumes are given as functions of the molecular weight (MW) by:

$$V_i^L = V_i^S = \frac{MW_i}{d_{i,25}^L}$$
$$d_{i,25}^L = 0.8155 + 0.6273 \times 10^{-4} MW_i - 13.06/MW_i \tag{4.29}$$

Won's values for the liquid phase solubility parameters (δ_i^L) for CO_2 and all alkanes from C_1 to C_{40} are available in his 1986 publication[10] and those above C_{40} can be calculated from the following equation as a function of the *MW*:[37]

$$\delta_i^L = 7.971 + 0.361\,32 \ln MW_i \tag{4.30}$$

The solubility parameter of the solid can be expressed as a function of the liquid solubility parameter and the enthalpy and temperature of melting:

$$(\delta_i^s)^2 = (\delta_i^L)^2 + \frac{\Delta H_i^f}{V_i}$$
$$\Delta H_i^f = 0.1426 MW_i T_i^f \tag{4.31}$$

Expressions for the temperature of melting as a function of the molecular weight have been presented by Won[10] and Hansen *et al.*[30]

The cloud points or wax appearance temperatures (the temperature at which the first small amount of wax is formed) are predicted to be rather high with the original version of Won's model. Pedersen *et al.*[11] offered a solution by implementing the ΔC_p term in Equation (4.26) as well as new expressions for the solubility parameter of the solid. An alternative solution is offered by the model of Hansen *et al.*[30] They assumed the ideal solid phase ($\gamma_i^s = 1$), but they accounted for nucleation phenomena. The liquid phase activity coefficients are calculated by a multicomponent form of the FH model, which is essentially identical to the equations shown in Appendix 4.A, Equations (4.38)–(4.42), but with two important differences compared to the classical FH:

1. The volume fractions are not based on volumes but on carbon numbers:

$$\Phi_i = \frac{x_i \zeta_i}{\sum_j x_j \zeta_j} \tag{4.32}$$

where ζ_i is the carbon number of component i. Thus (see also Equation (4.38)):

$$\chi_{21} = \chi_{12} \frac{\zeta_2}{\zeta_1} \tag{4.33}$$

2. The FH interaction parameters are functions of $|\zeta_i - \zeta_j|$ and three different equations are proposed for PN–PN, PN–A and A–A interactions (P, N, A are paraffinic, naphthenic and aromatic hydrocarbons). Instead of the carbon number, Pedersen *et al.*[11] used the *MW* of hydrocarbons in these correlations and in Equation (4.32).

4.7 Asphaltene precipitation

As discussed in the literature,[31–34] the simple RST, alone or in combination with the FH combinatorial term, can be used to describe asphaltene precipitation. Asphaltenes are heavy organics of polar, polynuclear character, containing several heteroatoms (S, O, N) which can precipitate during the production, transportation, refining and processing of crude oil. Asphaltenes are the *most polar* part of petroleum and they are often stabilized in oil in the presence of the equally polar but smaller resin molecules. Asphaltenes may precipitate depending on changes in *T,P*, but also because of the addition of light gases used for enhancing oil recovery.

At first glance, it is rather surprising that RST/FH approaches can be used for modeling asphaltene precipitation, considering the complexity of asphaltene molecules, possible association phenomena with resins and even the uncertainty as to how asphaltene–oil equilibria should be described (liquid–liquid, solid–liquid, colloidal instability, etc.). However, even though the number of unknown parameters and uncertain factors is rather high, e.g. the structure and molecular weight of asphaltenes, the RST may in some cases provide a good 'descriptive' tool and this is why it is still used, as indicated in the literature.[31–34]

The three most used approaches are illustrated below, in Equations (4.34)–(4.37), where in the last two the RST is combined with the FH equation, thus accounting for the polymeric character of asphaltene molecules. The mole or volume fraction of asphaltene is calculated assuming flocculation to be modeled as LLE in a two-constituent asphaltene–oil (a–s) system, assuming, moreover, that the activity of asphaltene is equal to one (i.e. the asphaltene-rich phase is assumed to be pure asphaltenes). It is also assumed that the amount of asphaltenes in the oil phase is very small.

In Equations (4.34)–(4.37), subscript a = asphaltenes and s = solvent (oil).

1. Hildebrand equation (RST), Equation (4.11), $l_{ij} = 0$:

$$\begin{aligned} \ln x_a &= -\frac{V_a}{RT}\Phi_s^2(\delta_s - \delta_a)^2 \\ V_a &= \frac{M_a}{\rho_a} \end{aligned} \tag{4.34}$$

2. FH/RST, Equation (4.20), (see also Problem 22 on the companion website at www.wiley.com/go/kontogeorgis for the 'polymer activity') assuming that the volume fraction of the oil is much higher than that of asphaltenes, i.e. negligible quantities of asphaltenes are dissolved in oil:

$$\ln \Phi_a^{\max} = \left[\left(\frac{V_a}{V_s} - 1 - \frac{V_a}{RT}(\delta_a - \delta_s)^2\right)\right] \tag{4.35}$$

3. Scott–Magat equation (extended FH), essentially a FH/RST model, Equation (4.20), but with the FH parameter estimated from the RST including also the entropic parameter of the FH model:

$$\chi_{12} = \frac{1}{Z} + \frac{V_1}{RT}\left[(\delta_1 - \delta_2)^2 + 2l_{12}\delta_1\delta_2\right] \tag{4.36}$$

$$\ln \Phi_a = \frac{V_a}{V_s} - 1 - \left(\frac{V_a}{V_s}\right)\left(\frac{1}{Z}\right) - \frac{V_a}{RT}\left[(\delta_a - \delta_s)^2 + 2l_{12}\delta_a\delta_s\right] \tag{4.37}$$

In all cases we assume insoluble asphaltenes, i.e. the activity of asphaltene is equal to one.

Satisfactory correlation of existing data is achieved in several cases, especially at ambient conditions, fitting the asphaltene parameters (molecular weight or volume, solubility parameter) as well as the l_{12} parameter to experimental precipitation data. The FH-type models for asphaltenes are not predictive and the parameters of these models must always be obtained from experimental data. In most studies the solubility parameter of asphaltenes is around 20 $MPa^{1/2}$, similar to the value of certain polar polymers (PET, PVAC, PVC). Numerous variations of Equations (4.34)–(4.37) have been proposed. For example, the FH parameter can be assumed to be composition dependent or a function of the molecular weight of oil and asphaltene. All these variations add a few more parameters to be adjusted to experimental data.

4.8 Concluding remarks about the random-mixing-based models

- They are basically empirical, except for the regular solution theory.
- They are simple, easy-to-use models for fast calculations, even using 'pocket' calculators.
- The one-parameter Margules equation is useful only for symmetric systems.
- The two-parameter Margules and the van Laar equations are generally of the same accuracy for systems of different asymmetries and energies (polarity).
- Clear interrelations exist between four models: the van der Waals equation of state, the van Laar and Flory–Huggins/free-volume activity coefficient models and the regular solution theory. An overview of the models is provided in Table 4.4, while their connections and derivations from the van der Waals equation are shown graphically in Figure 4.9.
- Most random-based models (like the local composition ones discussed in Chapter 5) typically have two adjustable parameters per binary mixture.
- They perform better for VLE than for LLE.
- Problems often occur for complex systems, e.g. alcohol–hydrocarbons (where false phase split may be predicted).
- Parameters are somewhat temperature dependent (thus the models are less useful, e.g. for constant pressure distillations).

Table 4.4 *Overview of the 'van der Waals family models', i.e. the van Laar, RST, as well as the FH, entropic–FV models for polymers combined with the RST. The g^E expression for SRK is also given for comparison purposes*

Theory	$\frac{g^E}{RT}$ (general/multicomponent), ignoring the excess volume terms	$\frac{g^E}{RT}$ (binary/vdW1f mixing rules/classical combining rules)	$\ln \gamma_i$ (binary mixtures)
van der Waals	$\left[\sum_i x_i \ln\left(\frac{V_i - b_i}{V - b}\right)\right] + \left[\frac{1}{RT}\left(\sum_i x_i \frac{a_i}{V_i} - \frac{a}{V}\right)\right]$	$\sum_i x_i \ln\frac{\Phi_i^{fv}}{x_i} + \frac{V}{RT}\Phi_i\Phi_j(\delta_i - \delta_j)^2$ $\Phi_i^{fv} = \frac{x_i V_{f_i}}{\sum_j x_j V_{f_j}}$ $V_{f_i} = V_i - b_i$ $\quad \Phi_i = \frac{x_i V_i}{\sum_j x_j V_j}$ $\delta_i = \frac{\sqrt{a_i}}{V_i}$	$\ln\left(\frac{\Phi_i^{fv}}{x_i}\right) + 1 - \left(\frac{\Phi_i^{fv}}{x_i}\right) + \frac{V_i}{RT}(\delta_i - \delta_j)^2\Phi_j^2$ $\Phi_i = \frac{x_i V_i}{\sum_j x_j V_j}$
van Laar	$\frac{1}{RT}\left(\sum_i x_i \frac{a_i}{b_i} - \frac{a}{b}\right)$	$\frac{b}{RT}\Phi_i\Phi_j(\delta_i - \delta_j)^2$ $\Phi_i = \frac{x_i b_i}{\sum_j x_j b_j}$ $\delta_i = \frac{\sqrt{a_i}}{b_i}$	$\frac{b_i}{RT}(\delta_i - \delta_j)^2\Phi_j^2$ $\Phi_i = \frac{x_i b_i}{\sum_j x_j b_j}$
RST	$\sum_i x_i \frac{V_i}{RT}(\delta_i - \delta^{av})^2$ $\Phi_i = \frac{x_i V_i}{\sum_j x_j V_j}$ $\delta^{av} = \sum_i \Phi_i \delta_i$ $\delta = \sqrt{c} = \sqrt{\frac{\Delta H^{vap} - RT}{V}}$	$\frac{V}{RT}\Phi_i\Phi_j(\delta_i - \delta_j)^2$	$\frac{V_i}{RT}(\delta_i - \delta_j)^2\Phi_j^2$
FH + RST	$\sum_i x_i \ln\left(\frac{\Phi_i}{x_i}\right) + V\sum_i\sum_j \Phi_i\Phi_j a_{ij}$ $\chi_{ij} = 2a_{ij}V_i$	$\sum_i x_i \ln\frac{\Phi_i}{x_i} + \frac{V}{V_i}\Phi_i\Phi_j\chi_{ij}$	$\ln\left(\frac{\Phi_i}{x_i}\right) + 1 - \left(\frac{\Phi_i}{x_i}\right) + \chi_{ij}\Phi_j^2$ $\chi_{ij} = \frac{V_i}{RT}(\delta_i - \delta_j)^2$ or via HSP, Equation (4.22)

(continued)

Table 4.4 (*Continued*)

Theory	$\frac{g^E}{RT}$ (general/multicomponent), ignoring the excess volume terms	$\frac{g^E}{RT}$ (binary/vdW1f mixing rules/classical combining rules)	$\ln \gamma_i$ (binary mixtures)
Entropic–FV + RST (Hildebrand–FV)	$\left(\sum_i x_i \ln \frac{\Phi_i^{fv}}{x_i}\right) + \sum_i x_i \frac{V_i}{RT}(\delta_i - \delta^{av})^2$ $\delta^{av} = \sum_i \Phi_i \delta_i$ $\Phi_i^{fv} = \frac{x_i V_{f_i}}{\sum_j x_j V_{f_j}}$ $V_{f_i} = V_i - V_{w,i}$	$\sum_i x_i \ln \frac{\Phi_i^{fv}}{x_i} + \frac{V}{RT}\Phi_i \Phi_j (\delta_i - \delta_j)^2$	$\ln\left(\frac{\Phi_i^{fv}}{x_i}\right) + 1 - \left(\frac{\Phi_i^{fv}}{x_i}\right) + \frac{V_i}{RT}(\delta_i - \delta_j)^2 \Phi_j^2$ $\Phi_i = \frac{x_i V_i}{\sum_j x_j V_j}$
SRK EoS $\delta_1 = 1, \delta_2 = 0$ or PR EoS $\delta_1 = 1 + \sqrt{2}$ $\delta_2 = 1 - \sqrt{2}$	$\left[\sum_i x_i \ln\left(\frac{V_i - b_i}{V - b}\right)\right] +$ $\left\{\frac{1}{(\delta_1 - \delta_2)RT}\left[-\sum_i x_i \frac{a_i}{b_i} \ln\left(\frac{V_i + \delta_2 b_i}{V_i + \delta_1 b_i}\right)\right.\right.$ $\left.\left. + \frac{a}{b}\ln\left(\frac{V + \delta_2 b}{V + \delta_1 b}\right)\right]\right\}$	Same as in previous column	$\ln\gamma_1 = \ln\left[\frac{V_1 - b_1}{V - b}\right] + 1 - \left[\frac{V_1 - b_1}{V - b}\right]$ $+ \frac{a_1}{b_1 RT}\ln\left[\frac{V_1 + b_1}{V_1}\right]$ $+ \frac{a}{RT(V + b)}\left(\frac{V_1}{V} - \frac{b_1}{b}\right)$ $- \frac{a}{bRT}\left[\frac{2\sum_j x_j a_{ji}}{a} - \frac{b_1}{b}\right]\ln\left[\frac{V + b}{V}\right]$ (vdW1f mixing rules, linear mixing rule for the co-volume parameter). See also Chapter 3, Equations (3.15) and (3.16) (for PR)

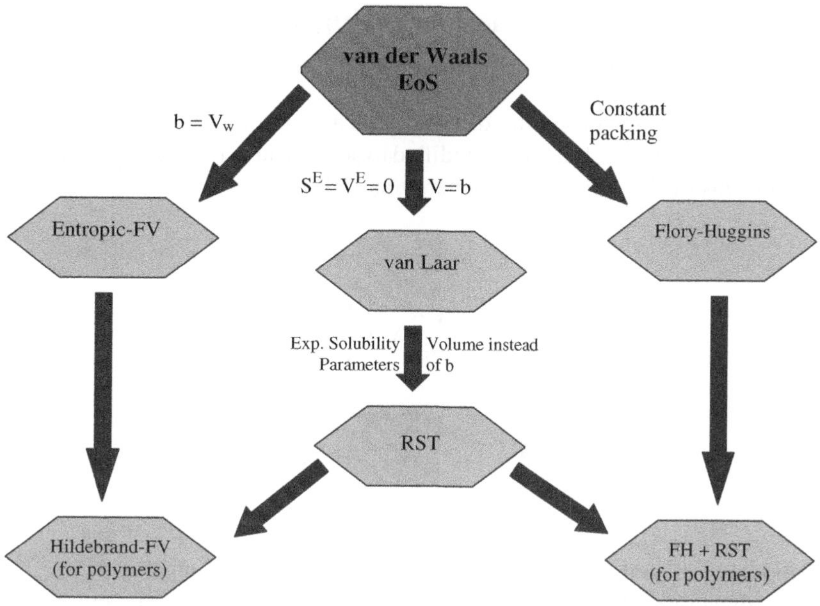

Figure 4.9 *Derivation of the van Laar, RST, FH and entropic–FV models from the vdW EoS. The RST energetic term can be combined with either the FH or the entropic–FV combinatorial terms, resulting in models suitable for polymer solutions*

- The van Laar and Margules models are applicable up to ternary mixtures but difficult to extend to multicomponent systems, so not very good ternary predictions are obtained.
- The regular solution theory is easily extended to multicomponent mixtures.
- The regular solution theory has a wide spectrum of applications which includes besides VLE for non-polar/slightly polar mixtures (original scope of the theory), polar/hydrogen bonding compounds, gases, solids, pharmaceuticals and polymers.

Table 4.4 provides an overview of the 'van der Waals family models'. The interrelations between the various models are clear with the understanding that:

1. Experimental solubility parameters are used in the RST and its combinations with FH or entropic–FV combinatorial terms, while the definitions of 'solubility parameters' in connection with vdW and van Laar are given in Table 4.4 ($\delta = \sqrt{a}/V$ or $\delta = \sqrt{a}/b$, respectively).
2. The van Laar and RST models have no 'combinatorial' terms, i.e. they only account for differences in the energetic interaction between the molecules.
3. Volume fractions, Φ_i, are, in all cases except for the van Laar equation, based on volumes, while for the van Laar equation they are based on the co-volume parameter, b.
4. Free-volume fractions, Φ_i^{fv}, are always based on the difference between the volume and the hard-core volume of the molecules ($V_f = V - V^*$). The hard-core volume, V^*, is the co-volume parameter in the case of vdW and SRK EoS ($V^* = b$), or it is considered to be equal to the vdW volume in the entropic–FV model ($V^* = V_w$).
5. Volumes are calculated by solving the EoS (vdW, SRK), while the experimental volume values are used as input in the activity coefficient models (RST, FH + RST, entropic–FV + RST).

Appendix 4.A An expression for the Flory–Huggins model for multicomponent mixtures

The Flory-Huggins (FH) model was originally developed as a model for the entropy of mixing for mixtures containing molecules of different size, but it was modified to account also for energetic interactions. The model can be formulated in terms of the excess Gibbs energy as follows:[35]

$$
\begin{aligned}
g^E &= g^{E,comb} + g^{E,res} \\
\frac{g^{E,comb}}{RT} &= \sum_i x_i \ln\left(\frac{\Phi_i}{x_i}\right) \\
\frac{g^{E,res}}{RT} &= V \sum_i \sum_j \Phi_i \Phi_j a_{ij} \\
\chi_{ij} &= 2a_{ij} V_i
\end{aligned}
\tag{4.38}
$$

where all $a_{ii} = 0$ and $a_{ij} = a_{ji}$ and the volume fraction Φ_i is defined in Equation (4.8).

Using basic thermodynamics, the following expression for the activity coefficient can be obtained:

$$\ln \gamma_i = \ln \gamma_i^{comb} + \ln \gamma_i^{res} \tag{4.39}$$

where the combinatorial term is given by:

$$\ln \gamma_i^{comb} = \ln \frac{\Phi_i}{x_i} + 1 - \frac{\Phi_i}{x_i} \tag{4.40}$$

and the residual term (NC is the number of components) is:

$$\ln \gamma_i^{res} = 2V_i \sum_{j=1}^{NC} \Phi_i a_{ij} - V_i \sum_{j=1}^{NC} \sum_{k=1}^{NC} \Phi_j \Phi_k a_{jk} \tag{4.41}$$

The above formulation of the FH model is slightly different from the conventionally used formulation using the FH interaction parameter (χ_{12}), although there is an interrelationship based on the simple equation shown above (Equation (4.38)). For a binary mixture, the multicomponent equation reduces to the traditional FH residual term:

$$\ln \gamma_i^{res} = \chi_{12} \Phi_2^2 \tag{4.42}$$

Appendix 4.B Rules of thumb for predicting polymer–solvent miscibility

Table 4.5 presents some rules of thumb for predicting polymer–solvent miscibility.

Table 4.5 *Rules of thumb for predicting solvent (1) and polymer (2) miscibility. $\delta_d, \delta_p, \delta_h$ are, respectively, the dispersion, polar and hydrogen bonding parts of the Hansen solubility parameter, which are related to the total solubility parameter as $\delta = (\delta_d^2 + \delta_p^2 + \delta_h^2)^{1/2}$*

Rule	Good solvent (1) in a polymer (2) if ...	Poor solvent if ...
Hildebrand solubility parameters (non-polar/slightly polar mixtures)	$\lvert\delta_1 - \delta_2\rvert \leq 1.8\ (\text{cal/cm}^3)^{1/2}$	$\lvert\delta_1 - \delta_2\rvert > 1.8\ (cal/cm^3)^{1/2}$
Hansen solubility parameters	$Ha = \sqrt{4(\delta_{d1} - \delta_{d2})^2 + (\delta_{p1} - \delta_{p2})^2 + (\delta_{h1} - \delta_{h2})^2} \leq R$ R = radius of solubility	$Ha > R$
FH model	$\chi_{12} \leq 0.5$ $\chi^{crit} = \frac{1}{2}\left[1 + \frac{1}{\sqrt{r}}\right]^2 = 0.5$ (for high MW polymers)	$\chi_{12} > 0.5$
Infinite dilution activity coefficients	$\Omega_1^\infty \leq 6$ $\Omega_i^\infty = \lim_{w_i \to 0}\left(\frac{x_i \gamma_i}{w_i}\right)$	$\Omega_1^\infty > 10$

Appendix 4.C Concentration scales (useful in polymer thermodynamics)

Mole fraction of compound i:

$$x_i = \frac{n_i}{\sum_j n_j} = \frac{n_i}{n} \tag{4.43}$$

where x_i is the mole fraction of compound i, n_i is the number of moles of compound i and n is the total number of moles in the solution.

Weight fraction of compound i:

$$w_i = \frac{m_i}{\sum_j m_j} = \frac{m_i}{m} \tag{4.44}$$

where w_i is the weight fraction of compound i, m_i is the mass (e.g. in g) of compound i and m is the total mass of the compound.

Volume fraction of compound i:

$$\Phi_i = \frac{V_i}{\sum_j V_j} = \frac{V_i}{V} \tag{4.45}$$

where Φ_i is the volume fraction of compound i, V_i is the volume (e.g. in cm^3) of compound i and V is the total volume of the compound.

4.C.1 Some useful interrelations between concentration scales

Weight and mole fractions:

$$w_i = \frac{x_i M_i}{\sum_j x_j M_j} \tag{4.46}$$

where M_i is the molar mass (molecular weight) of compound i.

Volume and mole fractions and volume and weight fractions:

$$\Phi_i = \frac{x_i V_i}{\sum_j x_j V_j} = \frac{w_i/\rho_i}{\sum_j w_j/\rho_j} \tag{4.47}$$

where ρ_i is the density of compound i.

References

1. G. Wilson, *J. Am. Chem. Soc.*, **1964**, *86*, 127.
2. S.I. Sandler, *Fluid Phase Equilib.*, **1996**, *116*, 343.
3. D.P. Tassios, *Applied Chemical Engineering Thermodynamics*. Springer-Verlag, 1993.
4. R.C. Reid, J.M. Prausnitz, B.E. Poling, *The Properties of Gases and Liquids* (4th edition). McGraw-Hill International, 1988.
5. J. Wisniak, *J. Phase Equilib.*, **2003**, *24* (2), 103.
6. J.M. Prausnitz, R.N. Lichtenthaler, E.G. de Azevedo, *Molecular Thermodynamics of Fluid-Phase Equilibria* (3rd edition). Prentice Hall International, 1999.
7. J.M. Smith, H.C. van Ness, M.M. Abbott, *Introduction to Chemical Engineering Thermodynamics* (7th edition). McGraw-Hill International, 2005.
8. J. Wisniak, *Chem. Educ.*, **2000**, *5*, 1.
9. J.H. Hildebrand, R.L. Scott, *The Solubility of Non-electrolytes*. Dover, 1964.
10. K.W. Won, *Fluid Phase Equilib.*, **1986**, *30*, 265.
11. K.S. Pedersen, P. Skovborg, H.P. Rønningsen, *Energy Fuels*, **1991**, *5*, 924.
12. S.I. Andersen, J.G. Speight, *J. Pet. Sci. Eng.*, **1999**, *22*, 53.
13. P. Kolar, J.-W. Shen, A. Tsuboi, T. Ishikawa, *Fluid Phase Equilib.*, **2002**, *194–197*, 771.
14. J. Abildskov, J.P. O'Connell, *Ind. Eng. Chem. Res.*, **2003**, *42*, 5622.
15. C.M. Hansen, *Hansen Solubility Parameters: A User's Handbook*. CRC Press, 2000.
16. T.C. Franck, J.R. Downey, S.K. Gupta, *Chem. Eng. Prog.*, **1999**, 41.
17. P. Thorlaksen, J. Abildskov, G.M. Kontogeorgis, *Fluid Phase Equilib.*, **2003**, *211*, 17.
18. P.J. Flory, *J. Chem. Phys.*, **1941**, *9*, 660.
19. M.L. Huggins, *J. Chem. Phys.*, **1941**, *15*, 225.
20. H.S. Elbro, Aa. Fredenslund, P. Rasmussen, *Macromolecules*, **1990**, *23*, 4707.
21. G.M. Kontogeorgis, Aa. Fredenslund, D.P. Tassios, *Ind. Eng. Chem. Res.*, **1993**, *32*, 362.
22. B.E. Poling, J.M. Prausnitz, J.P. O'Connell, *The Properties of Gases and Liquids* (5th edition). McGraw-Hill, 2001.
23. Th. Lindvig, M.L. Michelsen, G.M. Kontogeorgis, *Fluid Phase Equilib.*, **2002**, *203*, 247.
24. E.L. Cussler, G.D. Moggridge, *Chemical Product Design*. Cambridge University Press, 2001.
25. A.S. Michaels, P.S.L. Wong, R. Prather, R.M. Gale, *AIChE J.*, **1975**, *21*(6), 1073.
26. J.M. Prausnitz, *Fluid Phase Equilib.*, **1999**, *158–160*, 95.

27. A.P.D. Sattar, Controlled drug release and a test of Michaels et al. approach. Report written for a special course at the Department of Chemical Engineering, DTU, Denmark, 2005.
28. E.K.L. Lee, H.K. Lonsdale, R.W. Baker, E. Drioli, P.A. Bresnahan, *J. Membrane Sci.*, **1985**, *24*, 125.
29. K.W. Won, *Fluid Phase Equilib.*, **1989**, *53*, 377.
30. J.H. Hansen, Aa. Fredenslund, K.S. Pedersen, H.P. Rønningsen, *AIChE J.*, **1988**, *34*, 1937.
31. A. Hirschberg, L.N.J. de Jong, B.A. Schipper, J.G. Meijer, *Soc. Pet. Eng. J.*, **1984**, June, 283.
32. G.R. Pazuki, M. Nikookar, M.R. Omidkhah, *Fluid Phase Equilib.*, **2007**, *254*, 42.
33. G.R. Pazuki, M. Nikookar, *Fuel*, **2006**, *85*, 1083.
34. S.J. Park, G.A. Mansoori, *Ind. J. Energy Sources*, **1988**, *10*, 109.
35. Th. Lindvig, I.G. Economou, R.P. Danner, M.L. Michelsen, G.M. Kontogeorgis, *Fluid Phase Equilib.*, **2004**, *220*, 11.
36. A.F.M. Barton, *Handbook of Polymer–Liquid Interaction Parameters and Solubility Parameters*. CRC Press, 1990.
37. K.S. Pedersen, A. Fredenslund, P. Thomassen, *Properties of Oils and Natural Gases*. Gulf Publishing Company, 1989.
38. G.T. Preston, J.M. Prausnitz, *Ind. Eng. Chem. Process Des. Dev.*, **1970**, *9*, 264.
39. A.L. Myers, J.M. Prausnitz, *Ind. Eng. Chem. Fundam.*, **1965**, *4*, 209.

5

Activity Coefficient Models, Part 2: Local Composition Models, from Wilson and NRTL to UNIQUAC and UNIFAC

5.1 General

Traditional cubic EoS using the classical vdW1f mixing rules and activity coefficient models like the Margules and van Laar equations use 'average' or 'overall' compositions. They are models based on 'random mixing'. However, due to intermolecular forces, the mixing of molecules is never entirely random and a way to account for the non-randomness can lead to improved models and better descriptions of phase behavior. Since their advent with the Wilson equation in 1964,[1] local composition (LC) activity coefficient models have drastically changed the range of applicability of liquid phase models.

There exist several models which employ the LC concept, which is illustrated in Figure 5.1. All of these models are based on the physical picture that the mixing of molecules is 'non-random' and this is accomplished by using the so-called 'local compositions' which are, in the general case, different from the average concentrations due to the short-range nature of intermolecular forces. Due to their basis in this different principle, LC models differ drastically from random-mixing-based models (Chapter 4). LC models allow for a certain degree of non-randomness and they can thus be expected to represent more realistically the phase behavior of complex mixtures. Of course, we need to know the distribution of the local fractions, which is given by a Boltzmann factor expression, and different functions are employed in the various LC models, see Table 5.1.

As we will see, in most cases two interaction parameters per binary mixture in LC models are sufficient for obtaining good VLE results. The interaction parameters in LC models have, as seen from their derivation (see Appendix 5.A), an in-built temperature dependency and some theoretical significance. Moreover, LC models can be readily extended to multicomponent systems, easier than, for example, the van Laar and Margules equations. Two of the LC models, namely Wilson and UNIQUAC, have been further developed into predictive group contribution (GC) versions (ASOG and UNIFAC), suitable for preliminary design in the absence of

Thermodynamic Models for Industrial Applications: From Classical and Advanced Mixing Rules to Association Theories
Georgios M. Kontogeorgis and Georgios K. Folas

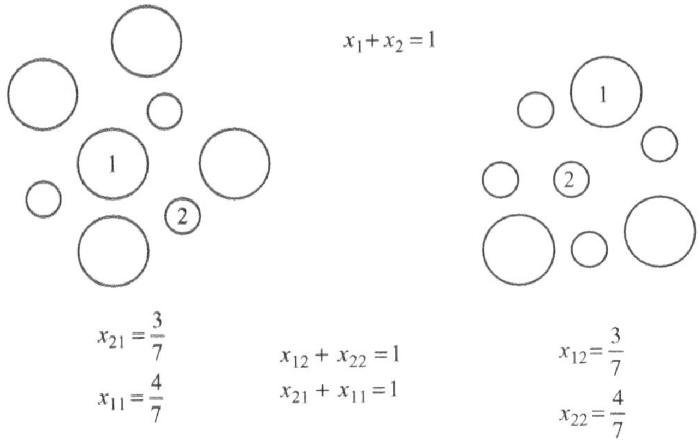

Figure 5.1 *The concept of local compositions, where x_{ij} indicates the mole fraction of compound i (number of molecules i over total number of molecules) around a central molecule j. Note that the local compositions differ from the overall ones ($x_i = 0.5$)*

experimental data. All the LC models suffer from a number of deficiencies, which we will discuss in depth. Some deficiencies are common for all or most LC models while others are specific; some of them are of theoretical origin while others have practical implications. Despite these limitations, the LC models are powerful tools, especially for low-pressure multicomponent VLE calculations (and to a lesser degree LLE) using solely binary parameters estimated from binary mixture data.

5.2 Overview of the local composition models

The expressions

Although there are many successful LC models, it could be reasonably argued that the most well known and widely used ones are the Wilson,[1] NRTL[2] and UNIQUAC[3] equations. The expressions for the excess Gibbs energy, g^E, for binary mixtures together with the expression for the 'local fractions' used in their derivations are shown in Table 5.1, while Table 5.2 presents the interaction parameters used in the models, which are all 'energy differences' (i.e. cross-energy interaction minus the energy between like molecules). Moreover, we can see that the Wilson and UNIQUAC models typically have two binary interaction parameters (which can be temperature dependent), while the NRTL equation has three parameters, where the additional parameter in NRTL, a_{ij}, is the so-called non-randomness factor. Details about the parameters of the models are given in the remaining of this section.

5.2.1 NRTL

The variables

The interaction energy parameter g_{ij}, referring to the i–j interaction, is expressed as:

$$\tau_{ji} = (g_{ji} - g_{ii})/RT \tag{5.1}$$

Table 5.1 *g^E/RT for binary mixtures and local mole fractions for LC models, given as Boltzmann factors of the energy differences. The UNIQUAC model is based on local area fractions, ϑ_{ij}, as the concentration variable. Parameters for all models are available in various collections, especially the DECHEMA series. Although different symbols are used for the interaction energies in the various models ($\lambda_{ij}, g_{ij}, U_{ij}$ in Wilson, NRTL and UNIQUAC, respectively), they are essentially equivalent and per definition: $\lambda_{ij} = \lambda_{ji}$ (and similarly for the cross-interaction energies for the other models)*

Model	g^E/RT	x_{21}/x_{11} or $\vartheta_{21}/\vartheta_{11}$
Wilson	$-x_1\ln(x_1+x_2\Lambda_{12})-x_2\ln(x_2+x_1\Lambda_{21})$	$\frac{x_{21}}{x_{11}}=\frac{x_2\exp(-\lambda_{21}/RT)}{x_1\exp(-\lambda_{11}/RT)}$ $=\frac{x_2}{x_1}\exp\left(-\frac{\Delta\lambda_{21}}{RT}\right)$ $\Delta\lambda_{ij}=\lambda_{ij}-\lambda_{jj}$
NRTL	$x_1x_2\left[\frac{\tau_{21}G_{21}}{x_1+x_2G_{21}}+\frac{\tau_{12}G_{12}}{x_2+x_1G_{12}}\right]$	$\frac{x_{21}}{x_{11}}=\frac{x_2\exp(-a_{12}g_{21}/RT)}{x_1\exp(-a_{12}g_{11}/RT)}$ $=\frac{x_2}{x_1}\exp\left(-\frac{a_{12}\Delta g_{21}}{RT}\right)$ $\Delta g_{ij}=g_{ij}-g_{jj}$
UNIQUAC	$x_1\ln\frac{\Phi_1}{x_1}+x_2\ln\frac{\Phi_2}{x_2}-\frac{Z}{2}\left(x_1q_1\ln\frac{\Phi_1}{\vartheta_1}+x_2q_2\ln\frac{\Phi_2}{\vartheta_2}\right)$ $-x_1q_1\ln(\vartheta_1+\vartheta_2\tau_{21})-x_2q_2\ln(\vartheta_1\tau_{12}+\vartheta_2)$	$\frac{\vartheta_{21}}{\vartheta_{11}}=\frac{\vartheta_2}{\vartheta_1}\exp\left(-\frac{Z}{2}\frac{U_{21}-U_{11}}{RT}\right)$ $=\frac{\vartheta_2}{\vartheta_1}\exp\left(-\frac{\Delta U_{21}}{RT}\right)$ $\Delta U_{ij}=U_{ij}-U_{jj}$
	$Z=10$ (the coordination number)	$\vartheta_{21}+\vartheta_{11}=1$ $\vartheta_{12}+\vartheta_{22}=1$

$$G_{ij}=\exp(-a_{ij}\tau_{ij}) \text{ and } a_{ij}=a_{ji} \tag{5.2}$$

Basics about the non-randomness parameter, a_{12}

NRTL has three parameters, i.e. two interaction parameters and the so-called non-randomness parameter, a_{12}, which has values between 0.20 and 0.47 in most cases. In the absence of data it can be set (somewhat arbitrarily) equal to 0.3. Values below 0.426 predict phase immiscibility. A value equal to zero means that the mixture is completely random. The non-randomness parameter has a clear physical meaning, as can be seen from a comparison of NRTL to the quasi-chemical theory of Guggenheim. The non-randomness parameter is equal to 2/*Z*, where *Z* is the coordination factor (*Z* values for liquids vary between 8 and 12), but it is most often treated as an adjustable parameter.

Table 5.2 Interaction parameters used in the LC models. All 'cross' parameters are equal to each other, e.g. for Wilson $\lambda_{ij} = \lambda_{ji}$. The interaction parameters of LC models have some physical significance (see Section 5.4) and an in-built temperature dependency via the Boltzmann factors

Model	Interaction parameters
Wilson	$\Lambda_{ij} = \frac{V_j}{V_i}\exp\left(-\frac{\Delta\lambda_{ji}}{RT}\right)$ $\Delta\lambda_{ij} = \lambda_{ij} - \lambda_{jj}$ $\Lambda_{ii} = \Lambda_{jj} = 1$
NRTL	$G_{ij} = \exp(-a_{ij}\tau_{ij})$ $\tau_{ji} = (g_{ji} - g_{ii})/RT$ $G_{ii} = G_{jj} = 1$ $\tau_{ii} = \tau_{jj} = 0$
UNIQUAC	$\tau_{ij} = \exp\left(-\frac{\Delta U_{ij}}{RT}\right)$ $\Delta U_{ij} = U_{ij} - U_{jj}$ or $\Delta U_{ij} = \frac{Z}{2}(U_{ij} - U_{jj})$ $\tau_{ii} = \tau_{jj} = 1$

More about the non-randomness parameter, a_{12}

NRTL is often used in its 'full' form, i.e. using a_{12} as the third adjustable parameter of the model. Care should be exercised, though, to make sure that a_{12} is indeed used to provide better correlation of the data rather than to fit the errors in them. In general, values greater than 0.5 are questionable; negative values are also questionable (not in agreement with the physical meaning of a_{12}), although some authors[4] have found that a single value of $a_{12} = -1.0$ provides good results for a large variety of mixtures. Renon and Prausnitz[2] have proposed some general rules depending on the family of compounds; however, these rules cannot be easily applied and thus often a_{12} is treated as an extra adjustable parameter, which is determined from binary data. Renon and Prausnitz's recommendations for a_{12} are summarized as follows:

- 0.2 for hydrocarbons–polar non-associated compounds;
- 0.3 for non-polar compounds, polar mixtures with slight negative deviations from Raoult's law or moderate positive deviations, water–polar components;
- 0.4 for hydrocarbons–perfluorocarbons;
- 0.47 for alcohols–non-polars, water–butyl alcohols, pyridine, CCl_4–acetonitrile and nitromethane.

5.2.2 UNIQUAC

The variables

The surface area and volume fractions used in UNIQUAC (binary mixtures) are:

$$\vartheta_i = \frac{x_i q_i}{x_i q_i + x_j q_j} = \frac{x_i q_i}{q} \tag{5.3}$$

$$\Phi_i = \frac{x_i r_i}{x_i r_i + x_j r_j} = \frac{x_i r_i}{r} \tag{5.4}$$

where r_i and q_i are the molecular volume and surface area (called 'van der Waals volume and area'), which in UNIQUAC are estimated using the group contribution values of Bondi,[5] R and Q:

$$r_i = \sum_k v_k^{(i)} R_k$$
$$q_i = \sum_k v_k^{(i)} Q_k \tag{5.5}$$

where $v_k^{(i)}$ is the number of functional groups of type k in molecule i and R_k and Q_k are the volume and surface area parameters of each functional group k. In UNIQUAC (and in UNIFAC, discussed later) there are 'dimensionless' values of R_k and Q_k which are based on the van der Waals volume and surface values of Bondi[5] but normalized using the volume and external area of the CH_2 unit in polyethylene:

$$R = \frac{V_w}{15.17}$$
$$Q = \frac{A_w}{2.5 \times 10^9} \tag{5.6}$$

where the units of V_w and A_w are cm^3/mol and cm^2/mol, respectively.

The basics

UNIQUAC has two contributions to the excess Gibbs energy and the activity coefficient: a combinatorial term accounting for differences in size and shape between the components; and a residual (energetic) term accounting for energy differences between the molecules. The r and q parameters ('van der Waals volume and areas') are measures of the molecular volume and area. The only fitted parameters to experimental phase equilibrium data are, in most cases, the energy interactions. However, a comment should be made about the r and especially the q parameters of alcohols and water. The q parameters appearing in the combinatorial and residual terms are typically the same, but this is not always the case for alcohols and water, where better results are obtained if different q parameters are employed in the combinatorial and residual terms (this model is sometimes called modified UNIQUAC). Without these special q parameters for alcohols, UNIQUAC may give false phase splits for alcohol–hydrocarbons.[6]

5.2.3 On UNIQUAC's energy parameters

UNIQUAC can be derived from the two-fluid theory by adopting (Boltzmann-type) LCs similar to those used in the Wilson and NRTL equations. In this case, however, it is assumed that the interactions take place using the areas of molecules. ϑ is the so-called area fraction, calculated using the Bondi surface areas. In the 'proper' derivation of UNIQUAC from the two-fluid theory (see Appendix 5.A), a term $Z/2$ appears in the exponential part containing the interaction energies (Z is the coordination factor, e.g. $Z = 10$ for the liquid phase). This $Z/2$ term provides a 'far too strong' correction to non-randomness for the UNIQUAC equation and is thus, somewhat arbitrarily, ignored (in the expression with the interaction energies, but kept in the combinatorial term). When, however, as is typically the case, the interaction parameters are fitted to experimental data (VLE or LLE), then the Z factor can be considered to be incorporated into the values for the energy parameters, as discussed later.

Table 5.3 *The g^E and activity coefficient expressions of LC models for multicomponent mixtures*

Model	g^E/RT	$\ln\gamma_i$
Wilson	$-\sum_i x_i \ln \sum_j \Lambda_{ij} x_j$	$1-\ln\sum_j x_j\Lambda_{ij}-\sum_k \frac{x_k\Lambda_{ki}}{\sum_j x_j\Lambda_{kj}}$
NRTL	$\sum_i x_i \frac{s_i}{r_i}$ $r_i = \sum_j x_j G_{ji}$ $s_i = \sum_j x_j \tau_{ji} G_{ji}$	$\frac{s_i}{r_i}+\sum_j \frac{x_j G_{ij}}{r_j}\left(\tau_{ij}-\frac{s_j}{r_j}\right)$
UNIQUAC	$\frac{g^E(\text{combinatorial})}{RT}+\frac{g^E(\text{residual})}{RT}$ $=\sum_i x_i \ln\frac{\Phi_i}{x_i}-\frac{Z}{2}\sum_i q_i x_i \ln\frac{\Phi_i}{\vartheta_i}-\sum_i x_i q_i \ln S_i$ $S_i=\sum_j \vartheta_j \tau_{ji}$ $\Phi_i=\frac{r_i x_i}{\sum_j r_j x_j}\quad \vartheta_i=\frac{q_i x_i}{\sum_j q_j x_j}$ $Z=10$	$\ln\gamma_i^{comb}+\ln\gamma_i^{res}$ $\ln\gamma_i^{comb}=\ln\left(\frac{\Phi_i}{x_i}\right)+1-\left(\frac{\Phi_i}{x_i}\right)$ $-\frac{Z}{2}q_i\left(\ln\frac{\Phi_i}{\vartheta_i}+1-\frac{\Phi_i}{\vartheta_i}\right)$ $\ln\gamma_i^{res}=q_i\left(1-\ln S_i-\sum_j\frac{\tau_{ij}\vartheta_j}{S_j}\right)$

All LC models can be readily extended to multicomponent systems and the expressions for g^E and activity coefficients are provided in Table 5.3. It should be mentioned that the fact that the energetic parameters of the models are 'energy differences' (Table 5.2), which is evident from their derivation, is 'silently' neglected in this extension of the models to multicomponent mixtures. When the LC models are used with the expressions of Table 5.3, it is assumed that all binary interaction parameters are determined independently, which is apparently not the case from a theoretical point of view (when their derivation is considered). For example, for a ternary system using the Wilson equation only five out of the six binary parameters can be chosen freely (as $\Delta\lambda_{12}=\lambda_{12}-\lambda_{22}$, $\Delta\lambda_{13}=\lambda_{13}-\lambda_{33}$, $\Delta\lambda_{23}=\lambda_{23}-\lambda_{33}$, $\Delta\lambda_{21}=\lambda_{21}-\lambda_{11}$, $\Delta\lambda_{31}=\lambda_{31}-\lambda_{11}$, $\Delta\lambda_{32}=\lambda_{32}-\lambda_{22}$), while for a quaternary system, only nine out of twelve binary parameters should be chosen independently, if the relationships of Table 5.2 are used.

5.2.4 On the Wilson equation parameters

The Λ_{ij} parameters are all equal to unity in the limiting case of an ideal solution. If both are greater than unity, we have negative deviations from ideality, while if both are less than unity, we have positive deviations from ideality. In some cases with only medium deviations from ideality, one of the parameters can be higher than unity and the other can be lower than unity. Volumes are dependent on temperature, but the effect is much smaller compared to the exponential term.

5.3 The theoretical limitations

We have already discussed in the previous section the important limitation, common to all LC models, related to 'the parameter interrelation', which refers to their extension to multicomponent systems. This is, however,

Table 5.4 *Theoretical limitations of the LC models*

Model	Entropic term	Problems with Z	Problems with *R*, *Q*	Interrelation of parameters
Wilson	Yes (hidden)	Yes	No	Yes
NRTL	No	Yes	No	Yes
UNIQUAC	Yes	Yes	Yes (many!)	Yes

not the only theoretical limitation of the LC models. Table 5.4 summarizes several more problems, due to the presence of the coordination number and the size (R and Q) parameters, which are discussed hereafter (for details see also a short discussion on the derivation of LC models in Appendix 5.A):

1. *Division into an entropic and an energetic term*: NRTL has no entropic term (it is essentially an 'enthalpic' or H^E model), while the other two models do have such an entropic contribution. For the Wilson equation this is not apparent when the model is written in the forms shown in Tables 5.1 and 5.3, but the entropic term is the same as in the Flory–Huggins model (i.e. as can be seen by setting all interaction parameters equal to each other; see also Section 5.6). Only UNIQUAC 'clearly' has two distinct contributions to the activity coefficient, one due to size and shape effects (combinatorial) and one due to energetic interactions (residual).
2. *The coordination number*: All models have problems arising from the presence of the coordination number, Z, though of different type. For UNIQUAC ($Z = 10$), it is mostly related to the renormalization (scale) problem: if we change Z (and q), e.g. via a new normalization way, we need to re-estimate the interaction parameters. Moreover, the original derivation of UNIQUAC includes a term $Z/2$ in the exponential factor with the interaction parameters, which is a rather extreme correction to non-randomness,[7] as verified by molecular simulation and quantum mechanics calculations. For NRTL, the Z problem is mostly related to the fact that the non-randomness parameter a_{12} is proportional to $2/Z$ according to theory, thus values of a_{12} are expected to be between 0.33 and 0.17 (Z between 6 and 12), which are indeed close to the values typically used (often 0.2–0.3). However, the a_{12} parameter is fitted to experimental data rather than being calculated from this expression. For Wilson, if the model is derived from the two-fluid theory (see Appendix 5.A), we need to employ $Z = 2$, which is evidently a very low value for the liquid state.
3. *UNIQUAC's many problems*: UNIQUAC has more problems due to the presence of the normalized surface area, Q, parameter. First of all, R and Q are normalized to a specific value, where Q is a multiplication factor to the residual term, thus they suffer from the same (normalization) problem as described above for UNIQUAC and Z. Moreover, the best results for mixtures with water or alcohols require fitted (to VLE data) R and Q values, especially in the residual term. In the case of alcohols, often different Q values are used in the residual and combinatorial terms. The same is true for one of the UNIFAC variants, the Dortmund modified UNIFAC (see later), for which the interaction parameters as well as the R and Q values are fitted simultaneously to experimental mixture data. Naturally, the need to use fitted R and Q values is a limitation of the approach.

Unfortunately, the above discussion does not cover all the 'theoretical' limitations of LC models. Vidal[8] explains that the Boltzmann-type LCs, x_{ij}, shown in Table 5.1 and used in LC models, do not in general respect the material balance imposed on the local mole fractions:

$$\begin{aligned} x_1x_{11} + x_2x_{12} &= x_1 \\ x_1x_{21} + x_2x_{22} &= x_2 \end{aligned} \tag{5.7}$$

This is not the end 'of the problems', either! Additional assumptions inherent in the derivation of the LC models are discussed in Appendix 5.A. It should, however, be emphasized that despite this extensive list of theoretical limitations, LC models have found widespread use in engineering calculations, especially multicomponent VLE, and in addition their interaction parameters do possess a theoretical significance. These are discussed in Sections 5.4 and 5.5, immediately after a short historical note on the development of the LC models.

5.3.1 Necessity for three models

We have presented the three 'famous', widely used LC models and some of their theoretical limitations, while their practical range of applicability will be discussed in the next section. However, at this stage, a short historical note is of interest in order to appreciate the background and practical necessity behind the three models, all of which are in use today.

The LC era essentially began with the pioneering work of Grant Wilson in 1964.[1] He derived the Wilson equation based on the Flory–Huggins equation using 'local composition fractions' instead of the traditional segment or volume fractions used in the Flory–Huggins model. That was essentially a 'one-fluid' derivation of a 'two-fluid' model, but the most serious limitation of the Wilson equation was its inability to represent LLE, no matter what the values of the parameters. When the Wilson equation is substituted into the classical equations of thermodynamic stability, it is found that the Wilson equation cannot account for a system of limited miscibility. Wilson's LC model is suitable only for binary and multicomponent VLE.

John M. Prausnitz has been the inspiring force behind almost all subsequent major developments in the LC field: NRTL in 1968[2] (with H. Renon), UNIQUAC in 1975[3] (with Abrams) and UNIFAC[9] shortly after UNIQUAC (with R. Jones and especially the late Aage Fredenslund). Unlike Wilson, all three subsequent models are suitable for both VLE and LLE. NRTL offered a solution to the LLE problem of Wilson, while maintaining good results for VLE, but in some cases also for heats of mixing. However, NRTL is indeed essentially an H^E rather than G^E model, as understood by its lack of a combinatorial term (S^E and V^E are zero, as in the regular solution theory). It has, moreover, three adjustable parameters and even though the non-randomness factor can sometimes be set to a constant value prior to parameter estimation, experimental data may not be sufficient for adjusting all three parameters with good accuracy.

Thus, despite the great success with the NRTL equation, it was soon realized that there was a need for another model, one that in some ways combines the positive features of Wilson and NRTL. The answer appeared with the development of the UNIQUAC model. Indeed UNIQUAC has 'elements' from both Wilson (it has a combinatorial term; two adjustable parameters) and from NRTL (it can describe both LLE and VLE).

5.4 Range of applicability of the LC models

The basic characteristics of the three most important LC models are summarized in Table 5.5:

1. All models have parameters with an in-built temperature dependency and can be applied to multicomponent VLE, but Wilson cannot be applied to LLE (irrespective of the values of the energy parameters).
2. NRTL and UNIQUAC have some success in simultaneously representing VLE and LLE, the latter especially for binary mixtures, less so for multicomponent mixtures.
3. NRTL can be applied, in most cases, with some success also to excess enthalpies (at the cost of an extra parameter). UNIQUAC with temperature-dependent parameters (more than two interaction parameters per binary) often yields good excess enthalpies.
4. The Wilson model cannot predict maxima/minima in activity coefficient–concentration plots of alcohol–chloroform, while UNIQUAC can describe such behavior (though not always quantitatively correct).

Table 5.5 Applicability range of the three most important LC models

Model	Number of parameters	Multicomponent VLE	Multicomponent LLE	H^E	In-built T dependency
Wilson	2	Yes	No	Yes*	Yes
NRTL	3	Yes	Yes	Yes	Yes
UNIQUAC	2	Yes	Yes	Yes*	Yes

*When temperature-dependent parameters are used.

A few illustrative examples of the behavior of LC models for various types of phase equilibria are shown in Figures 5.2–5.7, first for binary mixtures.

The following points summarize our conclusions:

1. For VLE of binary systems, Wilson and the other LC models perform at least as well as van Laar and Margules, and often better. For non-polar/slightly polar systems, there is little to gain by using the LC models over the random-mixing-based models discussed in Chapter 4.
2. For highly polar (hydrogen bonding) systems such as alcohol–alkanes, van Laar may result in false phase splits, while LC models perform very well as shown for example in Figure 5.2. In the example of Figure 5.2, the interaction parameters were obtained from the azeotropic point. Lacking extensive

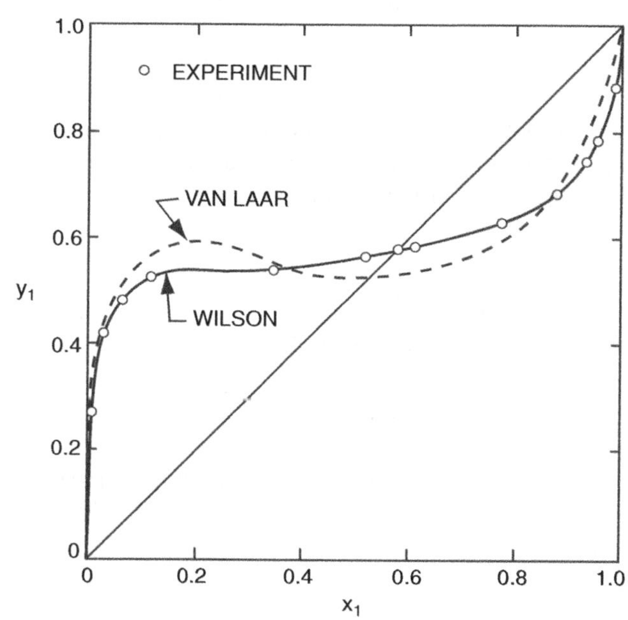

Figure 5.2 Comparison of Wilson and van Laar for ethanol(1)–isooctane(2) VLE at 50 °C. Wilson performs much better than van Laar, which erroneously predicts a phase split. Van Laar and other semi-empirical two-parameter models often have problems for solutions of alcohols with hydrocarbons (and may erroneously predict phase split when the values of activity coefficients exceed 7)[10]. In the example shown here, both the van Laar and the Wilson parameters have been estimated from the azeotropic point, which is a reliable approach as the azeotrope is formed near the middle of the concentration range. Reprinted with permission from Ind. Eng. Chem., Multicomponent Equilibria: The Wilson Equation by R. V. Orye et al., 57, 5, 18–26 Copyright (1965) American Chemical Society

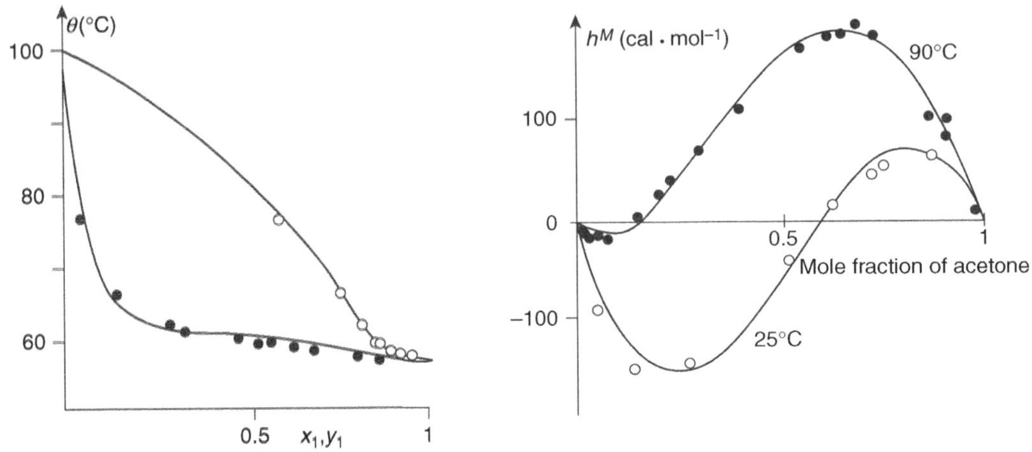

Figure 5.3 *Use of the NRTL model for the calculation of VLE at atmospheric pressure (left) and heat of mixing (right) for acetone(1)–water(2). Reprinted with permission from Editions Technip, Thermodynamics. Applications in chemical engineering and the petroleum industry by J. Vidal, Copyright (1997) Editions Technip*

experimental data, LC model parameters can be estimated from a single point, e.g. from infinite dilution activity coefficient data. Such estimated parameters often yield good results over extensive concentrations.

3. NRTL can be successfully used for correlating both VLE and excess enthalpies for several mixtures, as can be seen in Figure 5.3 for water–acetone. Excess enthalpies (heats of mixing) are usually difficult to describe satisfactorily with models having parameters based on phase equilibria (e.g. VLE) data. Heats of mixing are often complex functions of temperature and thus represent a stringent test for many thermodynamic models, especially their temperature dependency.

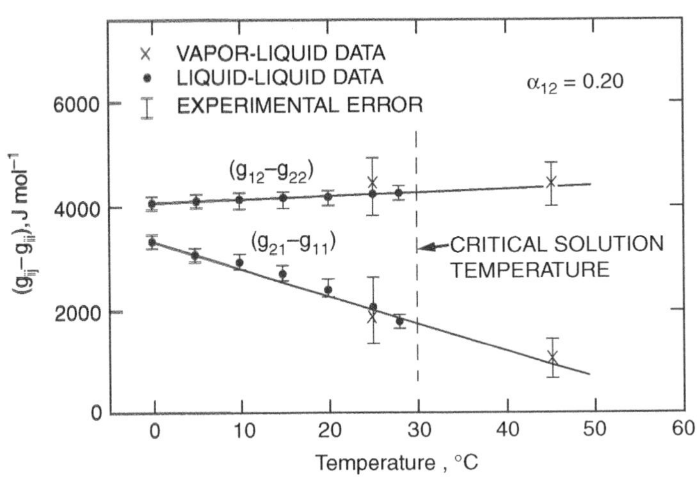

Figure 5.4 *NRTL parameters for nitroethane(1)–isooctane(2) calculated from VLE and LLE data, illustrating that the same parameters can be used for both equilibrium types. Reproduced with permission from AIChE J., Local compositions in thermodynamic excess functions for liquid mixtures by H. Renon and J. M. Prausnitz, 14, 1, 135–144 Copyright (1968) John Wiley and Sons, Inc.*

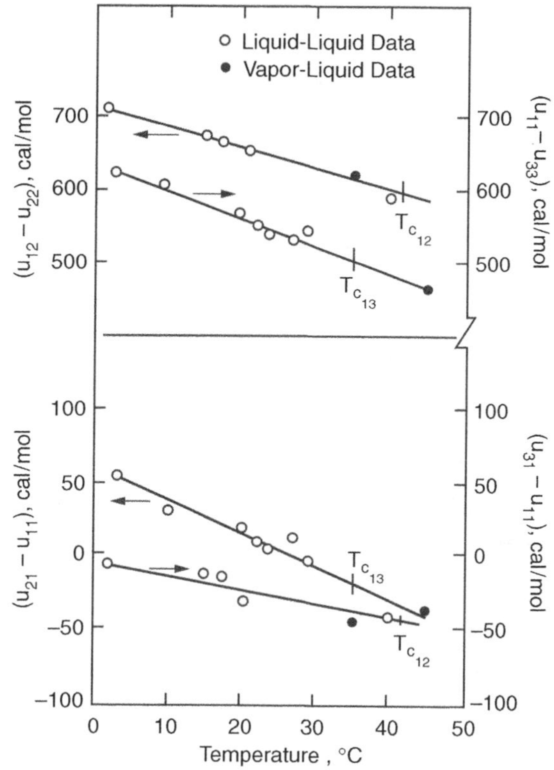

Figure 5.5 *UNIQUAC parameters for nitroethane(1)–octane(2) and nitroethane(1)–hexane(3) calculated from VLE and LLE data. T_c denotes the upper consolute temperature. The same UNIQUAC parameters can be used for VLE and LLE. Reproduced with permission from AIChE J., Statistical thermodynamics of liquid mixtures: a new expression for the excess Gibbs energy of partly or completely miscible systems by D. S. Abrams and J. M. Prausnitz, 21, 1, 116–128 Copyright (1975) John Wiley and Sons, Inc.*

4. NRTL and UNIQUAC (Figures 5.4 and 5.5) can, for several binary mixtures (not extremely non-ideal ones), successfully describe VLE and LLE with the same interaction parameters.
5. UNIQUAC is a very successful model, widely applicable to a variety of non-electrolyte liquid mixtures containing especially VLE of non-polar or polar and associating fluids such as hydrocarbons, nitriles, ketones, alcohols, aldehydes, acids, water, etc. Two examples are shown in Figure 5.6. It provides, in general, excellent correlation of binary VLE, even for highly complex systems, including those having associating substances. In some cases, however, two parameters are not enough to represent high-quality data with high accuracy, but for most practical applications the representation is satisfactory. When organic acids are present, however, it is important to correct for the deviations from ideality also in the vapor phase.
6. LC models have been applied with success to SLE of various types of mixtures including pharmaceuticals.[11,12] SLE data can be useful for obtaining the parameters of the LC models for compounds, e.g. heavy complex ones for which experimental VLE data are scarce. It is not always possible, however, to describe satisfactorily phase equilibria (VLE and SLE) with LC models over extensive temperature ranges.[7]

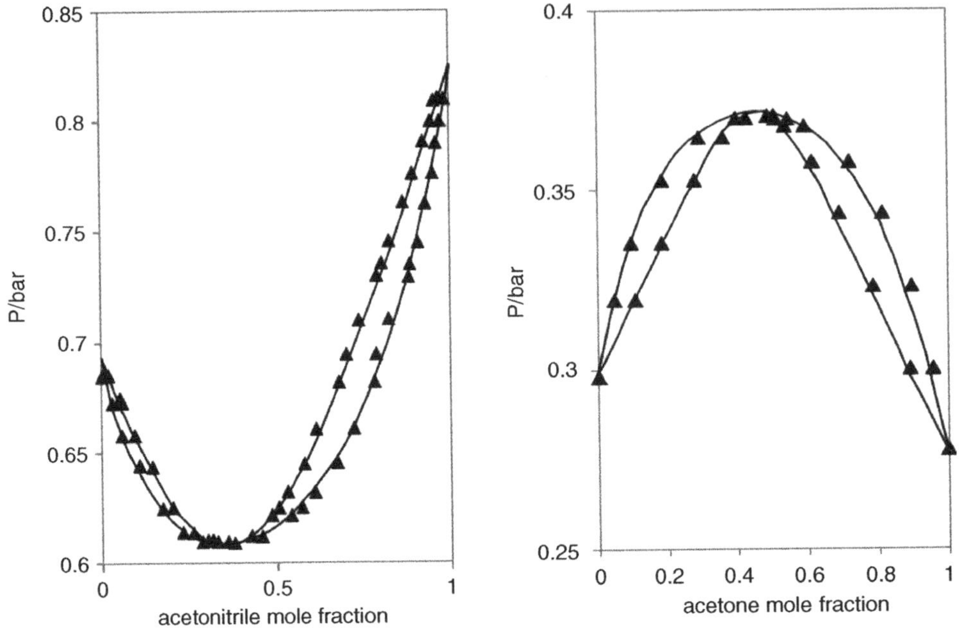

Figure 5.6 *VLE correlation with UNIQUAC for one mixture with positive (right: acetonitrile(1)–benzene(2) at 45 °C) and another one with negative deviations from Raoult's law (left: acetone(1)–chloroform(2) at 50 °C). Excellent correlation is achieved in both cases*

7. UNIQUAC has been successfully used for hydrate formation calculations combined with the van der Waals–Platteeuw framework, in order to calculate activity coefficients of condensable components via a modified form where the parameter ΔU_{ij} varies linearly with temperature,[13] or in its original form for mixtures in the presence of inhibitors.[14] Furthermore, a modification of UNIQUAC (extended UNIQUAC developed by Sander *et al.*,[15] see Appendix 5.C) was used by Munck *et al.*[14] for hydrate formation calculations is the presence of electrolyte inhibitors. Typical results for a seven-component mixture in the presence of methanol as hydrate inhibitor are presented in Figure 5.7.

One of the most important applications of LC models, which contributed to their extensive use, is their capability to provide (often very) satisfactory prediction of the VLE of multicomponent systems, as shown in Figure 5.8 for a few examples.

Wilson and LC models, in general, perform clearly better than van Laar and Margules equations, based on binary parameters. The results shown here are predictions, i.e. they are based on binary parameters. For 'simpler' ternary systems, e.g. argon–oxygen–nitrogen, the performance of Wilson and van Laar is similar but in the cases of polar and hydrogen bonding compounds the Wilson equation performs better.

The results with LC models are less convincing for LLE, especially for multicomponent LLE (important, for example, for liquid–liquid extraction design), where the calculations are highly sensitive to the values chosen for the interaction parameters. The performance (accuracy of prediction) of UNIQUAC in multicomponent LLE depends greatly on the type of system considered, the accuracy of the binary data, but also on the method used to obtain the binary parameters. The method of data reduction is far more important for LLE than for VLE. Especially for the 'difficult' Type I systems (those with only one binary mixture which is partially miscible, e.g. water–ethanol–benzene) some ternary data, preferably in the form of tie lines, away from the plait point, are useful for obtaining good 'predictions' of ternary LLE using UNIQUAC (and NRTL).

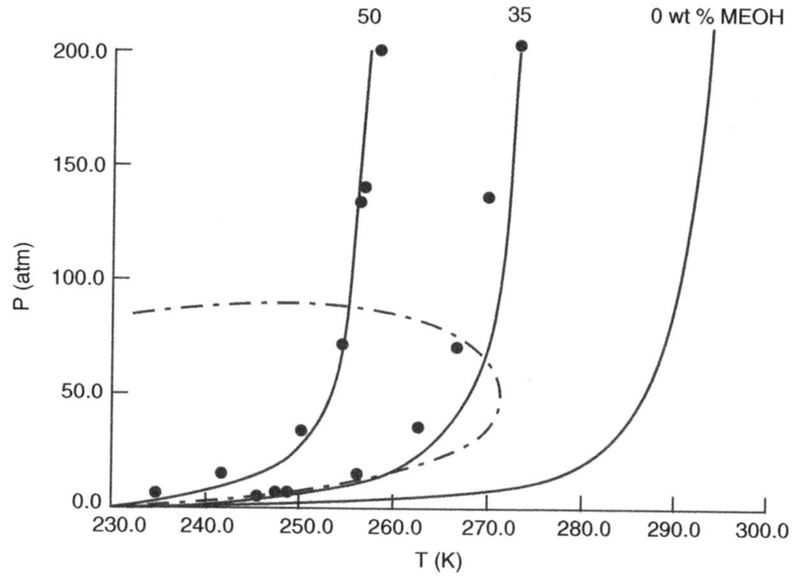

Figure 5.7 *Calculated coexistence curves for a seven-component mixture and aqueous solutions of methanol (MEOH). Reprinted with permission from Chemical Engineering Science, Computations of the formation of gas hydrates by J. Munck, S. Skjold–Jørgensen and P. Rasmussen, 43, 10, 2661 Copyright (1988) Elsevier*

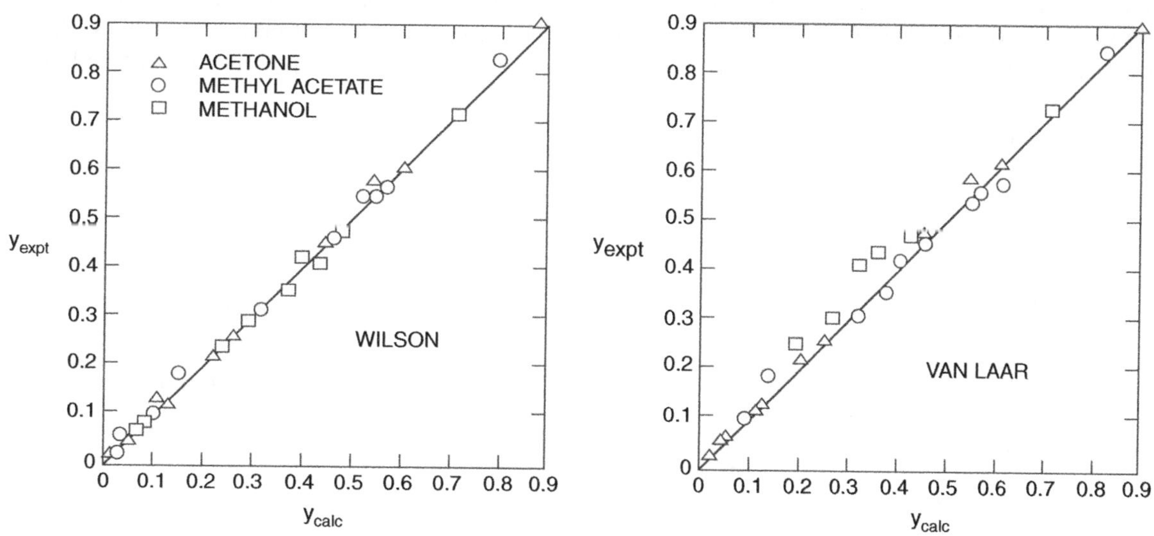

Figure 5.8 *Experimental and calculated with the Wilson and van Laar models vapor compositions for the ternary systems acetone/methanol/chloroform at 50°C (lower figure) and acetone/methyl acetate/methanol at 50°C (upper figure). Calculations use only binary data. Reprinted with permission from Ind. Eng. Chem., Multicomponent Equilibria: The Wilson Equation by R.V. Orye et al., 57, 5, 18–26 Copyright (1965) American Chemical Society*

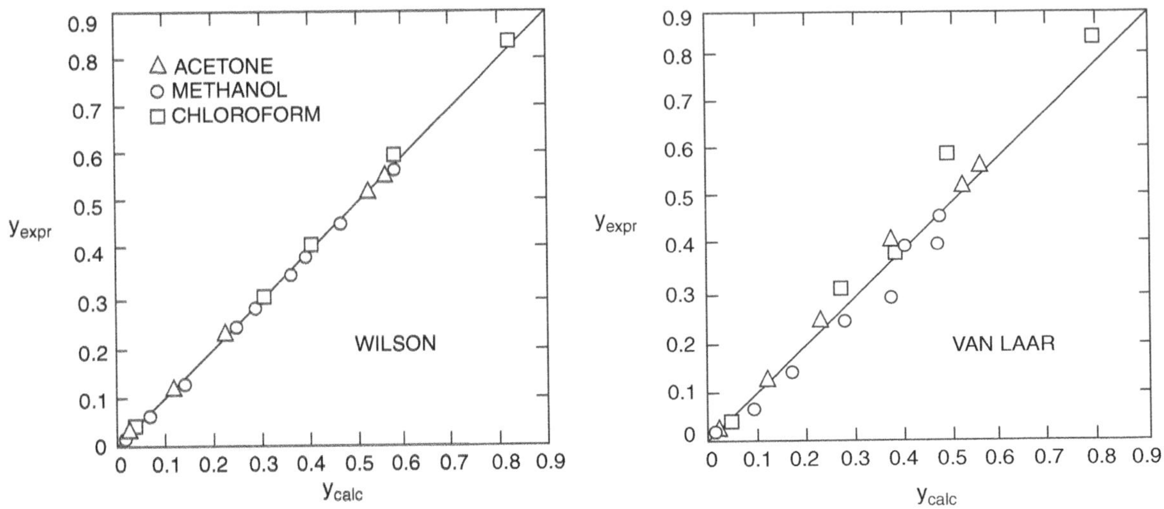

Figure 5.8 (Continued)

Regression of experimental data using UNIQUAC (and NRTL) typically results in multiple sets of parameters, which can represent the binary data equally well. A certain set of parameters may be best for VLE data, but other sets could be regressed from VLE data which can significantly improve ternary LLE prediction, while only slightly decreasing the accuracy of binary VLE representation.[6]

In these cases, it is, generally, best to fit binary parameters simultaneously to binary VLE and ternary LLE data, rather than fitting to either binary or ternary data only. Fitting to different data (both VLE and LLE) simultaneously may result, as Prausnitz *et al.*[6] discuss, in 'providing effectively constraints on the binary parameters, preventing them from attaining values of little physical significance'. Certain distillations (three-phase ones) depend very much on a single set of parameters, which can adequately represent both VLE and LLE.

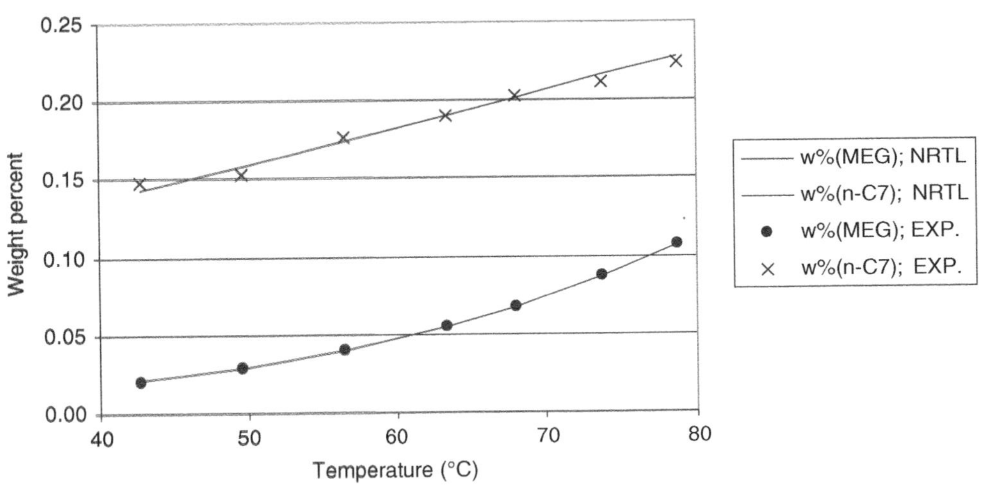

Figure 5.9 LLE for MEG–heptane with NRTL using linearly T-dependent parameters. Reprinted with permission from J. Chem Eng Data, Liquid–Liquid Equilibria for Glycols + Hydrocarbons: Data and Correlation by Samer O. Derawi, Georgios M. Kontogeorgis, et al., 47, 2, 169–173 Copyright (2002) American Chemical Society

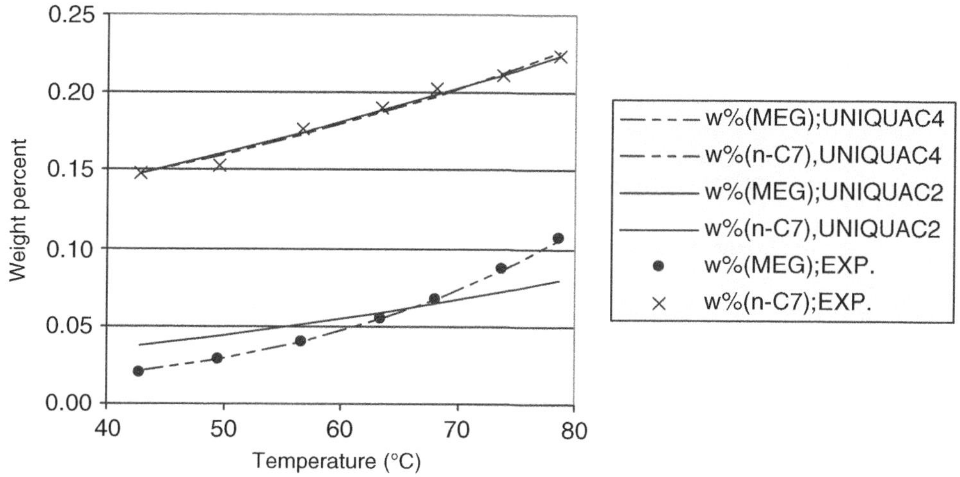

Figure 5.10 *LLE for MEG–heptane with UNIQUAC using T-independent (UNIQUAC2; as in Table 5.2) and linearly T-dependent parameters (UNIQUAC4, i.e. ΔU_{ij} of Table 5.2 is a linear function of temperature). Reprinted with permission from J. Chem Eng Data, Liquid–Liquid Equilibria for Glycols + Hydrocarbons: Data and Correlation by Samer O. Derawi, Georgios M. Kontogeorgis, et al., 47, 2, 169–173 Copyright (2002) American Chemical Society*

The problems of the LC models for multicomponent LLE can be seen in the study of binary LLE, especially for highly immiscible mixtures. Despite the successful representation of binary LLE in a few cases discussed previously (Figures 5.4 and 5.5), LLE of highly immiscible systems is much more difficult to represent using LC models, at least when temperature-independent parameters are used. Two examples are shown in Figures 5.9 and 5.10 for NRTL and UNIQUAC (for MEG–heptane) where accurate correlation of LLE over an extensive temperature range requires temperature-dependent interaction parameters.[55] In the case of NRTL this is a total of *five* and for UNIQUAC a total of *four* interaction parameters for this mixture.

5.5 On the theoretical significance of the interaction parameters

Despite the difficulties described in Section 5.3, the interaction parameters of the LC models do have a theoretical significance or physical meaning and this will be illustrated in many ways briefly discussed here.

5.5.1 Parameter values for families of compounds

Figure 5.11 illustrates that Wilson's parameters (always negative as the potential energy in the liquid phase is less than that of the ideal gas whose potential energy is zero) have some theoretical significance and they represent the interactions between molecules. Values obtained from experimental VLE data for different compound families are shown to follow well-expected trends based on the nature of the compounds involved.

5.5.2 One-parameter LC models

If we take the physical meaning of the interaction parameters seriously (see Appendix 5.A), it should be possible to develop ways to estimate the interaction energies between like molecules independently of the interactions between unlike molecules. In this way we would fit only one instead of two interaction parameters per binary, similar to the k_{ij} of cubic equations of state, or even predict all parameters from theory. Indeed such one-parameter LC models have been proposed, at first for Wilson and UNIQUAC and later also for NRTL.

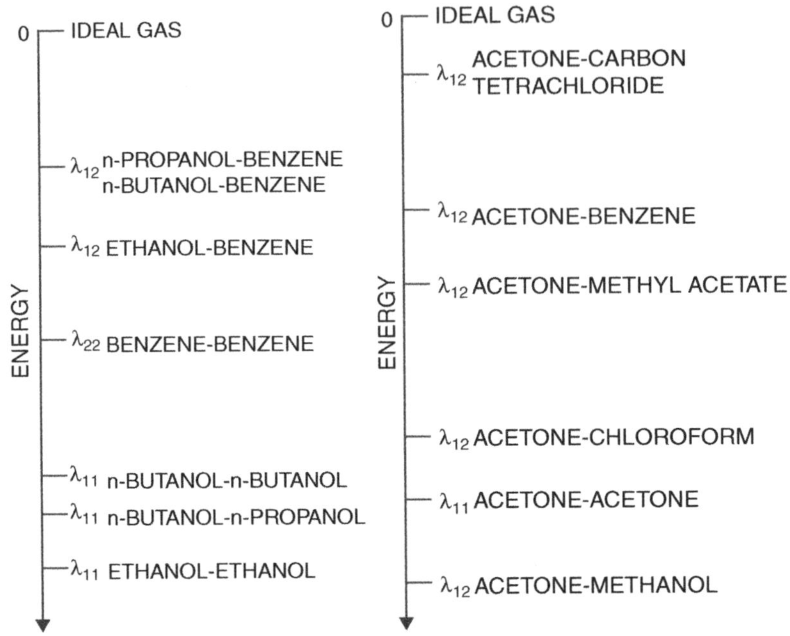

Figure 5.11 *Wilson interaction energies of alcohol(1)–benzene(2) and acetone–solvent ((1)–(2)) systems. For the acetone(1)–solvent(2) systems, the interaction increases as we proceed from the physical solvent CCl_4 (880), through methyl acetate (576) and benzene (535) (where weak interaction of acetone with π electrons is believed to take place) to the components which form hydrogen bonds with acetone, like chloroform (44) and methanol (−128). Given in parentheses are the $\lambda_{12}-\lambda_{11}$ values in cal/mol. Reprinted with permission from Ind. Eng. Chem., Multicomponent Equilibria: The Wilson Equation by R. V. Orye et al., 57, 5, 18–26 Copyright (1965) American Chemical Society*

First, for Wilson, Tassios[17] and Wong and Eckert[18] have proposed ways to estimate the pure compound interaction energies in the Wilson equation as a function of the enthalpy of vaporization, ΔH^{vap}. These methods, when combined with the Wilson equation, result in a Wilson-type model having only one adjustable parameter per binary:

Tassios approach, 1971:

$$\lambda_{ii} = -\Delta U_i^{VAP} = -(\Delta H_i^{VAP} - z_i RT) \tag{5.8}$$

z is the compressibility factor of saturated vapor at temperature T (z can be considered close to unity).

Wong and Eckert, 1971:

$$\lambda_{ii} = -\frac{2}{Z}\Delta U_i^{VAP} \tag{5.9}$$

Z is the coordination number (typically around 10).

Wong and Eckert's method is possibly theoretically more correct than Tassios's (see Appendix 5.A, Equation (5.36) where $-Zq_iU_{ii}/2 \approx \Delta U_i^{VAP}$). Both perform equally well in practice when combined with the Wilson equation. Note also (again see Appendix 5.A) that Wilson's derivation from the two-fluid theory implies that $Z=2$, which is consistent with Tassios's approach. When λ_{ii} is estimated using either Equation (5.8) or (5.9), there is only one adjustable parameter per binary system, λ_{ij}. The agreement is satisfactory for many binary and multicomponent systems (one example is shown in Figure 5.12), but

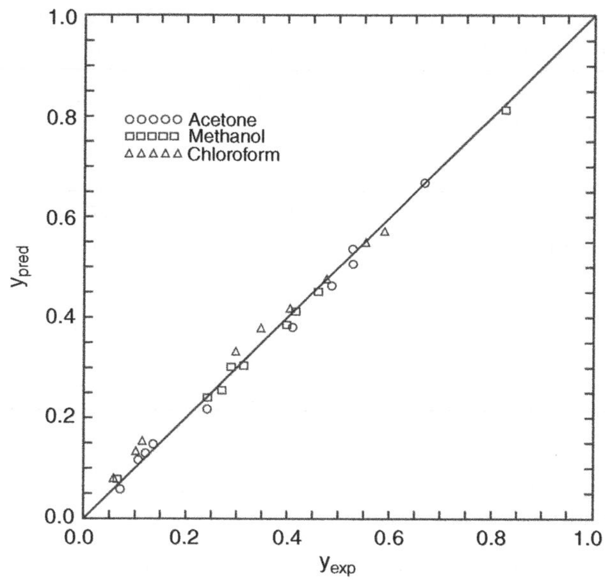

Figure 5.12 Comparison of predicted and experimental VLE for acetone–methanol–chloroform at 50 °C using the single-parameter Wilson equation. Reprinted with permission from Applied Chemical Engineering Thermodynamics by Dimitrios P. Tassios, Copyright (1993) Springer Science + Business Media

sometimes unreliable results are obtained for strongly non-ideal systems (e.g. those with infinite dilution activity coefficients above 10). Due to the latter reason, these one-parameter Wilson equations have gained limited acceptance but are useful when the quality and quantity of the data do not permit estimation of two parameters. They are also useful when a single data point is available, e.g. a value for the infinite dilution activity coefficient. They have also regained interest the last years, with the advent of quantum chemistry methods for estimating the interaction parameters (see Section 5.5.3).

One-parameter versions of the NRTL equation have been presented by Vetere in a series of publications,[20–22] as discussed in Appendix 5.B.

It is possible to develop one-parameter expressions also for UNIQUAC, as shown in the equations below, based on the energies of vaporization:

$$U_{ii} = -\frac{\Delta U_i^{VAP}}{q_i} \approx -\frac{\Delta H_i^{VAP} - RT}{q_i} \tag{5.10}$$

$$U_{ij} = \sqrt{U_{ii}U_{jj}}(1 - c_{ij}) \tag{5.11}$$

Notice that if the 'correct' U values are used, then the right-hand part of Equation (5.10) should be multiplied by 2/Z. From Appendix 5.A, Equation (5.36), we can see that $-Zq_iU_{ii}/2 \approx \Delta U_i^{VAP}$. The geometric mean rule, Equation (5.11), can be used for estimating the cross-energetic UNIQUAC parameter, but typically a correction c_{ij} parameter is required. For non-polar molecules, c_{ij} is positive and small, while it is positive and large (up to 0.5) for mixtures of polar and non-polar molecules. Finally, it can be negative in cases of solvating molecules, e.g. chloroform and acetone. Like the one-parameter Wilson model, in a typical case, NRTL and UNIQUAC are used with two (or three) parameters fitted to experimental data. One parameter is often not adequate for highly non-ideal (complex) mixtures.

Table 5.6 *Energies of interactions (in kJ/mol) estimated from quantum mechanics and from equations (5.8)–(5.10). The quantum mechanics values are from Jonsdottir.*[24] *The enthalpies and energies of vaporization are estimated at 298.15 K*

Compound	Energy parameter, U (quantum mechanics)	$\Delta H^{vap} - RT = \Delta U^{vap}$	$\frac{2}{Z}\Delta U^{vap}$	$\frac{\Delta U^{vap}}{q}$	$\frac{2}{Z}\frac{\Delta U^{vap}}{q}$
			$Z=10$		$Z=4$
Butane	3.37	18.5	3.7	6.65	3.32
Pentane	3.74	24.2	4.84	7.30	3.65
Hexane	3.84	29.5	5.9	7.65	3.82
Cyclohexane	3.54	30.6	6.12	9.44	4.72
Acetone	6.55	28.8	5.76	12.30	6.15
Butylamine	3.73	32.4	6.48	10.24	5.12
Ethylamine	3.62	24.3	4.86	11.67	5.83

5.5.3 Comparison of LC model parameters to quantum chemistry and other theoretically determined values

Recent data from quantum mechanics calculations for the UNIQUAC energy interactions for several pure compounds[23–25] can be used for testing the validity of various equations (Equations (5.8)–(5.10)) for estimating these parameters independently, i.e. without using phase equilibrium data. Table 5.6 presents the results with these methods, based on the energy of vaporization alone, the Wong–Eckert method or when the q and Z parameters are also included. The first two methods have been developed for the Wilson equation and only the last two methods are 'theoretically' correct for UNIQUAC, but for comparison reasons the results are shown with all four methods. We can see that using $Z=4$ results in relatively good agreement, although such a Z value is rather low for the liquid phase. This conclusion verifies, though, what was stated previously: that the more realistic value of $Z=10$ for the liquid phase would present a large and actually erroneous correction for non-randomness in the UNIQUAC model.

Using quantum mechanics calculations to obtain the interaction parameters, Sandler and co-workers[26,27] have shown that UNIQUAC (without the $Z/2$ term in the exponential) provides much better results than the Wilson model. One example is shown in Figure 5.13. These conclusions about the Z parameter in UNIQUAC verify the analysis of Table 5.6 regarding the overcorrection of the non-randomness with UNIQUAC when the Z parameter is retained. The Z term in the exponential has been dropped by other researchers as well (e.g. Kemeny and Rasmussen[28]). Fischer[29] also states that based on perturbation theory, inclusion of Z in the Boltzmann factors is not correct. Moreover, the results from Sandler's work (e.g. Figure 5.13) possibly indicate that the energy parameters in the UNIQUAC model are of greater physical meaning than those of the Wilson equation.

In addition, we can mention that simulation data provide confidence in the in-built temperature dependency of the interaction parameters of LC models, i.e. the Boltzmann factors,[30] at least over narrow temperature ranges. The above discussion is not conclusive but a useful indication about the physical significance of the LC interaction parameters, and possibly more of UNIQUAC, despite the significant theoretical limitations discussed in Section 5.3.

5.6 LC models: some unifying concepts

The remainder of the chapter will present the transformation of UNIQUAC into a fully predictive model via the UNIFAC equation as well as the extension of LC models to polymers. Prior to that, however, a unifying discussion including some conclusions on the advantages and shortcomings of LC models is provided.

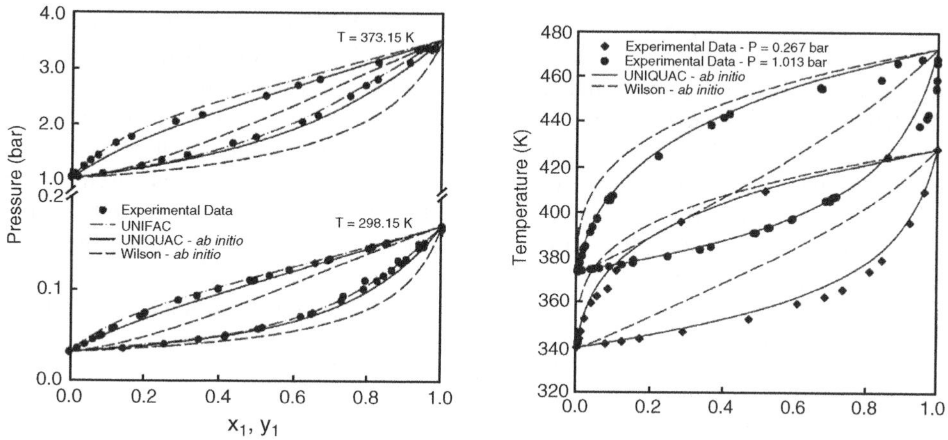

Figure 5.13 VLE for methanol(1)–water(2) (left) and m-methylformamide(1)–water(2) (right) using various LC models where the parameters have been estimated using molecular orbital ab initio methods. UNIQUAC performs much better than Wilson when these theoretically determined parameters are used. Reprinted with permission from Ind. Eng. Chem. Res., A Novel Approach to Phase Equilibria Predictions Using Ab Initio Methods by Amadeu K. Sum and Stanley I. Sandler, 38, 7, 2849–2855 Copyright (1999) American Chemical Society

5.6.1 Wilson and UNIQUAC

Equations (5.13) and (5.14) give a comparative presentation of the Wilson and UNIQUAC equations, which illustrates the similarities and differences between the two models. Equation (5.14) for UNIQUAC is identical to that shown in Table 5.1, but to distinguish it from the volume fraction, $\Phi_i = x_i V_i / V$, used in Wilson (Equation (5.13)), we use the symbol Φ_i^s for UNIQUAC's segment fraction, defined in Equation (5.4). Equation (5.13) for Wilson is of course mathematically identical to that shown in Table 5.1 but is written here in a way so that the combinatorial and residual terms are clearly shown:

Wilson written in a somewhat unusual form:

$$\frac{g^E}{RT} = \frac{g^{E,comb}}{RT} + \frac{g^{E,res}}{RT} \tag{5.12}$$

$$\frac{g^E}{RT} = x_1 \ln \frac{\Phi_1}{x_1} + x_2 \ln \frac{\Phi_2}{x_2} - x_1 \ln\left[\Phi_1 + \Phi_2 \exp\left(-\frac{\Delta\lambda_{21}}{RT}\right)\right] - x_2 \ln\left[\Phi_2 + \Phi_1 \exp\left(-\frac{\Delta\lambda_{12}}{RT}\right)\right] \tag{5.13}$$

Compare to UNIQUAC:

$$\frac{g^E}{RT} = x_1 \ln \frac{\Phi_1^s}{x_1} + x_2 \ln \frac{\Phi_2^s}{x_2} + \frac{Z}{2}\left(x_1 q_1 \ln \frac{\vartheta_1}{\Phi_1^s} + x_2 q_2 \ln \frac{\vartheta_2}{\Phi_2^s}\right) - x_1 q_1 \ln\left[\vartheta_1 + \vartheta_2 \exp\left(-\frac{\Delta U_{21}}{RT}\right)\right] - x_2 q_2 \ln\left[\vartheta_2 + \vartheta_1 \exp\left(-\frac{\Delta U_{12}}{RT}\right)\right] \tag{5.14}$$

We can, thus, see that both models have a combinatorial and a residual term: in the case of Wilson it is a Flory–Huggins term based on volume fractions, while in the case of UNIQUAC it is a modified Flory–Huggins term based on segment and surface area fractions. The second part of the combinatorial term of UNIQUAC (Staverman–Guggenheim term) does not usually contribute much to the activity coefficient compared to the Flory–Huggins term. The residual terms of the two models look similar, but again the Wilson one is based on volume fractions, while the UNIQUAC one is based on surface area fractions. Both models have two interaction parameters that, in a typical case, have to be estimated from experimental phase equilibrium data. When the cross-interaction parameters are equal to those between like compounds or, in other words, when the exponential factors are equal to one in the two models, then the residual terms disappear. Even in this case, there are deviations from ideal solution behavior due to size and shape differences between the molecules, which are approximately accounted for via the combinatorial terms of the LC models. Recall that NRTL has no combinatorial term.

5.6.2 The interaction parameters of the LC models

- They are not very sensitive to the quality of binary VLE data.
- They are (somewhat) sensitive to temperature; they can be used for narrow extrapolations (30–40 °C) and can be considered rather independent of temperature over narrow temperature ranges.
- Parameters from binary data are often sufficient for predicting ternary (in general multicomponent) VLE.
- They are not sensitive to the method used (minimization function) in the case of VLE (the opposite is true for LLE!).
- Usually two interaction parameters per binary mixture are sufficient for successful VLE results and in many cases experimental data are not of sufficient quantity and quality to justify use of more than two interaction parameters.
- If we wish to estimate temperature-dependent parameters, extensive data must be used e.g., either VLE data over a wide temperature range or VLE and heat of mixing data.

5.6.3 Successes and limitations of the LC models

- They often correlate binary and multicomponent VLE much better than the random-mixing-based models (Margules and van Laar) for mixtures of non-polar and polar/complex compounds.
- Successful VLE representation is obtained at various temperatures, due to their in-built temperature dependency, but to obtain good results for heats of mixing typically requires temperature-dependent interaction parameters.
- Wilson, NRTL and UNIQUAC perform similarly for VLE but only NRTL and UNIQUAC can be applied to LLE.
- LC models are easily extended to multicomponent systems and yield satisfactory multicomponent VLE in many cases, based solely on binary data. This is of great importance in separation design, e.g. of distillation columns.
- They exhibit (often serious) problems in representing LLE well, especially for simultaneous descriptions of VLE and LLE with the same interaction parameters and multiphase equilibria (VLLE), and when highly polar and hydrogen bonding compounds (water, acids, etc.) are present. LLE prediction for ternary systems is improved when a few ternary LLE data are used in the parameter estimation.

- The parameters of the LC models are strongly intercorrelated and often several sets of parameter pairs may represent VLE data equally well. Such intercorrelation may be somewhat eliminated if extensive data and possibly also some ternary data are included in the parameter estimation.[8]
- Prediction of multicomponent VLE requires quite a lot of fitting – even an 'ordinary' 10-component mixture contains 45 binaries! Despite the extensive databases available, it may be the case that for one or more of those binaries experimental data are lacking. If we do not/cannot measure them and if they are *not* crucial for the process, we could use a predictive group contribution method, such as UNIFAC, discussed in the next section.

5.7 The group contribution principle and UNIFAC

Group contribution (GC) methods have been applied with success for pure compound properties, e.g. critical constants, acentric factors, densities and solubility parameters.[31,32] In these methods the properties of the compounds are considered to be approximated by the sum of the contributions of the functional groups into which the compounds are divided. Evidently, such GC methods would be of enormous value for mixtures, since the number of possible mixtures of interest is enormous, but some 50–60 functional groups may be sufficient to describe thousands of mixtures. GC methods for activity coefficients would thus be a valuable tool for designing processes in the absence of data or only when few data are available. In such methods the (thousands of different) molecular interactions will be estimated by appropriate averages of (much fewer) group–group interactions.

There exist several GC versions of LC models. The first one was ASOG (Analytical Solution of Groups; Derr and Deal[33]), which was based on the Wilson equation, but a FH–combinatorial term is added, based on the number of carbon atoms, i.e. functionally similar to that which the Wilson model already has, but with a different concentration scale. The model has had some success and extensive parameters tables exist,[34,35] but it has not received the widespread acceptance of UNIFAC, which is based on UNIQUAC, so we will thus limit our discussion here to UNIFAC. Comparative evaluations of ASOG and UNIFAC have been published.[36,37]

The UNIFAC (Universal quasi-chemical Functional group Activity Coefficient) method was proposed by Aage Fredenslund, Russel Jones and John Prausnitz in 1975.[9] UNIFAC is a widely used GC activity coefficient model based on UNIQUAC. When using UNIFAC we need first to identify groups of a compound using available suitable tables, e.g. in Poling *et al.*[31] Numerous modifications (see later) of the original version of the model have been proposed, but the basic equations of the original UNIFAC are for the activity coefficient of compound *i*:

$$\ln \gamma_i = \ln \gamma_i^{comb} + \ln \gamma_i^{res} \tag{5.15}$$

where the combinatorial (*comb*) and residual (*res*) contributions are given by the equations:

$$\begin{aligned} \ln \gamma_i^{comb} &= \ln\left(\frac{\Phi_i}{x_i}\right) + 1 - \left(\frac{\Phi_i}{x_i}\right) - \frac{Z}{2} q_i \left(\ln \frac{\Phi_i}{\vartheta_i} + 1 - \frac{\Phi_i}{\vartheta_i}\right) \\ \ln \gamma_i^{res} &= \sum_k \nu_k^{(i)} \left(\ln \Gamma_k - \ln \Gamma_k^{(i)}\right) \end{aligned} \tag{5.16}$$

ϑ_i, Φ_i are defined in Equations (5.3) and (5.4). Note that the combinatorial term is identical to that of UNIQUAC ($Z = 10$ also in Equation (5.16)), while the residual term is evaluated from group contributions. The summation in the residual term is over all groups in the mixture.

The variables of the residual term are explained as follows: $\nu_k^{(i)}$ is the number of k groups present in compound i and $\ln \Gamma_k^{(i)}$ is the residual contribution to the activity coefficient of group k in the pure fluid i. This term is needed so that the residual term of a pure compound i (which is still a mixture of groups) goes to zero. The residual contribution to the logarithm of the activity of group k is given (for both the mixture and the pure compound) by the following equation (which is functionally similar to that of UNIQUAC, see Table 5.3, but involves group quantities):

$$\ln \Gamma_k = Q_k \left[1 - \ln\left(\sum_m \Theta_m \Psi_{mk} \right) - \sum_m \frac{\Theta_m \Psi_{km}}{\sum_n \Theta_n \Psi_{nm}} \right] \tag{5.17}$$

where the surface area fraction of group m in the mixture is defined similarly to ϑ_i for molecules as:

$$\Theta_m = \frac{X_m Q_m}{\sum_n X_n Q_n} \tag{5.18}$$

X_m is the mole fraction of the group m in the mixture:

$$X_m = \frac{\sum_j \nu_m^{(j)} x_j}{\sum_j \sum_n \left(\nu_n^{(j)} x_j \right)} \tag{5.19}$$

The group interaction parameters a_{mn} between groups m and n enter into the function Ψ_{mn}:

$$\Psi_{mn} = \exp\left(-\frac{a_{mn}}{T} \right) \tag{5.20}$$

$$a_{mn} = U_{mn} - U_{nn}$$

In the original UNIFAC, the interaction parameters a_{mn} are assumed to be temperature independent and the only temperature dependency is that of the Boltzmann factors, as shown in Equation (5.20).

The UNIFAC and UNIQUAC models have many features in common. They have the same combinatorial term based on the FH/Staverman–Guggenheim theory, with the same structural parameters, r and q, which for both models are estimated from the van der Waals volumes and surface areas from Bondi.[5] The same normalization method is used, Equations (5.5) and (5.6). Both UNIFAC and UNIQUAC have residual terms (which look alike) but in the case of UNIFAC the residual term is estimated from GCs. While energetic interactions are present for both models in the residual term, for UNIQUAC these are molecular interactions, while for UNIFAC they are group–group interactions. In both models, the energetic interactions are obtained from fitting experimental phase equilibrium data, usually VLE, although other types of data are occasionally also used (see Table 5.7). UNIFAC is usually highly successful in estimating VLE, as shown for one binary mixture in Figure 5.14. However, UNIFAC is essentially a family of GC models and various variations have

Table 5.7 Some of the most important UNIFAC variants

UNIFAC variant	Reference	Temperature dependency of the parameters	Data used in the parameter estimation	Comments
Original–VLE	Fredenslund *et al.*[9] Hansen *et al.*[38]	Independent of temperature	VLE	Recommended over narrow temperature ranges
Original–LLE	Magnussen *et al.*[39]	Independent of temperature	LLE	Limited use today (Very) narrow temperature range, around room temperature
'Linear' UNIFAC	Hansen *et al.*[40]	Linearly dependent on temperature: $a_{mn} = a_{mn,1} + a_{mn,2}(T - T_0)$	VLE	Parameters not published in the 'open' literature but are widely available via a report from IVC-SEP, Institut for Kemiteknik, Technical University of Denmark[40]
Modified UNIFAC (Lyngby)	Larsen *et al.*[41]	Logarithmic dependency: $a_{mn} = a_{mn,0} + a_{mn,1}(T - T_0) + a_{mn,2}\left(T \ln \frac{T_0}{T} + T - T_0\right)$	VLE, H^E	Modified combinatorial term
Modified UNIFAC (Dortmund)	Weidlich and Gmehling[42] Gmehling and co-workers[43–45]	$a_{mn} = a_{mn,0} + a_{mn,1}T + a_{mn,2}T^2$	VLE, H^E Infinite dilution activity coefficients	Modified combinatorial term Fitted r and q parameters (not obtained from Bondi's method) For most recent update, see UNIFAC consortium: http://www.uni-oldenburg.de/tchemie/consortium

(*continued*)

Table 5.7 *(Continued)*

UNIFAC variant	Reference	Temperature dependency of the parameters	Data used in the parameter estimation	Comments
Water–UNIFAC	Chen *et al.*[46]	Independent of temperature	VLE Infinite dilution activity coefficients, water data	Used for octanol–water partition coefficient calculations, see Chapter 17
Water–UNIFAC	Hooper *et al.*[47]	$a_{mn} = a_{mn,0} + a_{mn,1}T + a_{mn,2}T^2$		Suitable for water–hydrocarbon systems
Second order or KT–UNIFAC	Kang *et al.*[48]			Changes in equations and new first- and second-order group parameter tables

been proposed, as seen in Table 5.7, which illustrates some of the most important UNIFAC variants which are available.

The most important differences compared to the original UNIFAC are as follows:

1. Use of different combinatorial terms (modified UNIFAC versions):[41,42]
   Lyngby version:[41]

$$\ln \gamma_i = \ln \frac{\Phi_i}{x_i} + 1 - \frac{\Phi_i}{x_i}$$
$$\Phi_i = \frac{x_i r_i^{2/3}}{\sum_j x_j r_j^{2/3}} \tag{5.21}$$

2. *Dortmund version:*[42]

$$\ln\gamma_i = \ln\frac{\varphi'_i}{x_i} + 1 - \frac{\varphi'_i}{x_i} - \frac{Z}{2} q_i \left(\ln\frac{\Phi_i}{\vartheta_i} + 1 - \frac{\Phi_i}{\vartheta_i} \right) \tag{5.22}$$

where:

$$\varphi'_i = \frac{x_i r_i^{3/4}}{\sum_j x_j r_j^{3/4}} \text{ and } \Phi_i = \frac{x_i r_i}{\sum_j x_j r_j} \tag{5.23}$$

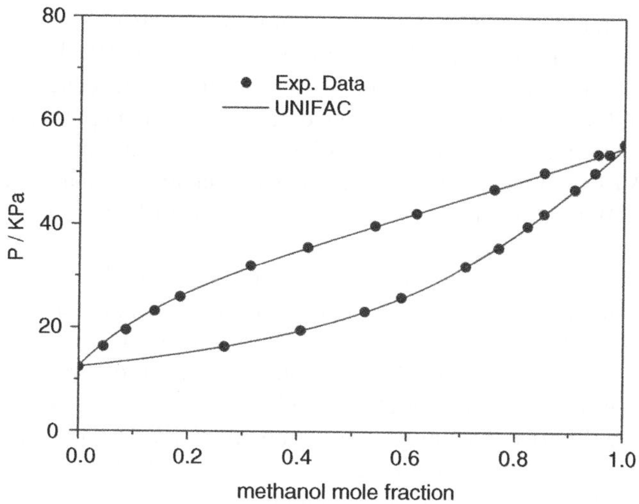

Figure 5.14 *VLE for methanol–water at 50 °C with UNIFAC. Excellent agreement is obtained in this case and for many more systems, but for several mixtures the agreement is only fair. Experimental data are from McGlashan and Williamson, J. Chem. Eng. Data,* ***1976****, 21, 196*

3. These modified combinatorial terms perform better for asymmetric alkane systems but unfortunately they cannot be applied to nearly athermal polymer solutions (see Problems 7 and 9 on the companion website at www.wiley.com/go/Kontogeorgis).
4. Use of temperature-dependent parameters and thus more adjustable parameters (Table 5.7) but also greater flexibility in data fitting.
5. Use of other/more than VLE data for fitting the parameters, e.g. LLE, infinite dilution activity coefficients, heats of mixing, etc.
6. The Dortmund version of modified UNIFAC uses R and Q parameters which are fitted to the experimental mixture data together with the energetic parameters. Unfortunately, this leads to rather large Q values that are often higher than the R ones and thus the Staverman–Guggenheim part of the combinatorial term often reaches such high values that the whole combinatorial activity coefficients become (much) greater than unity. Such high values of γ^{comb} are not realistic (see the discussion by Fredenslund and Rasmussen[49] and more recently by Abildskov *et al.*[50]).

5.7.1 Why there are so many UNIFAC variants

The original UNIFAC (1975, 1977 variants) is a successful model (at least qualitatively, often also quantitatively) for VLE calculations but within a narrow temperature range and extrapolations above 425 K should be avoided. Excess enthalpies and LLE are poorly predicted as well as representation of dilute systems and phase equilibria for mixtures containing certain complex compounds like water as well as multifunctional chemicals. The latter represent an inherent limitation of 'first-order' GC methods, which do not account for 'proximity' effects (effect of several polar groups close to each other). The most recent versions of the original UNIFAC[38] still suffer from the same limitations but the group parameters have been re-estimated based on a broader database of experimental data. The modified UNIFAC versions (Lyngby and Dortmund) 'correct' both the combinatorial and residual terms compared to the original UNIFAC. The former is accomplished by using an exponent-type term (Equations (5.21)–(5.23)) and the latter by adding a temperature dependency in the interaction parameters. Due to this temperature dependency, these modified UNIFAC versions are quite successful for both VLE and to some extent excess enthalpies and infinite dilution activity coefficients, especially the Dortmund version, which is based on more experimental data. Modified UNIFAC models can extrapolate VLE at higher temperatures better than the original UNIFAC. Moreover, these temperature-dependent UNIFAC versions employ more than just VLE data in the parameter estimation, typically H^E and γ^∞ as well.

The reason 'linear UNIFAC' was developed seems at first to be a step backwards, i.e. the combinatorial term of Kikic *et al.*[51] in the modified UNIFAC is dropped and a linear temperature dependency is adopted for the interaction parameters. The reason for changing back to the 'older' combinatorial is that (as shown by Kontogeorgis *et al.*;[52] see also Problems 7 and 9 on the companion website at www.wiley.com/go/Kontogeorgis) the exponent-type combinatorials, though satisfactory for asymmetric alkane mixtures, extrapolate poorly to athermal polymer solutions. The 'original UNIFAC' combinatorial is an FH-type term which extrapolates better to polymer solutions and this is why it was adopted in the '1992' linear UNIFAC.[40] Its parameter table has never been published in the open literature but is available in a technical report[40] which is freely provided by Institut for Kemiteknik, DTU. As the purpose of 'linear UNIFAC' was simply to fit VLE data up to high temperatures, the logarithmic part of the modified UNIFAC has also been dropped, and no excess enthalpies were used in the parameter estimation. The special UNIFAC versions for LLE/water systems/K_{ow} aim at specific applications where VLE-based UNIFAC are not well suited (they are discussed further in Chapter 17, devoted to environmental applications). Naturally, the presence of various UNIFAC variants can be considered to be at the same time both a strength ('flexibility') and weakness ('confusion, no universal parameters') of the method. The Dortmund modified UNIFAC is today the UNIFAC version that has the most extensive parameter table, although some of its parameters are not yet published in the open literature.

5.7.2 UNIFAC applications

UNIFAC has been widely used since its appearance and this is clear from the numerous citations that the articles referring to these models have received (at 14/10/2009 – in ISI Web of Knowledge): 1301 for Fredenslund *et al.*,[9] 560 for Larsen *et al.*,[41] 302 for Weidlich and Gmehling,[42] 407 for Magnussen *et al.*[39] (cf. 2158 for Soave's[53] article on SRK). Thus, citing all useful applications of UNIFAC would maybe require a whole book, and definitely much more than a section in a chapter. In order to illustrate that UNIFAC has indeed been applied to much more than VLE and LLE (which were the original targets of the model), Table 5.8 summarizes some of the applications of UNIFAC.

Due to its extensive parameter table and flexibility, UNIFAC has been widely used in the chemical engineering community. Table 5.9 illustrates some of the most important strengths and weaknesses of the approach. Its major application remains the prediction of binary and multicomponent VLE, when experimental data are not available, and thus it is primarily useful for preliminary design calculations and not for the very difficult separations. UNIFAC has been successfully used for the design of distillation columns[79] including azeotropic and extractive distillation, where there is often a lack of some data for the multicomponent mixtures involved. UNIFAC is not recommended for use in the last stages of the design, where very accurate calculations are needed.

However, UNIFAC can be used together with UNIQUAC for generating 'pseudo-experimental' data for one or more of the binary mixtures of a multicomponent system (not one of the crucial ones) for which real data

Table 5.8 *Some applications of UNIFAC (beyond 'ordinary VLE and LLE')*

UNIFAC variant	Applications	Comments/reference
Various UNIFAC, water UNIFAC	Octanol–water partition coefficients	Derawi *et al.*[54]
		Chen *et al.*[46]
		Li *et al.*[56]
Modified	SLE	Prausnitz *et al.*[7]
		Gmehling *et al.*[57]
		Mahmoud *et al.*[58]
	Pharmaceuticals	Prausnitz *et al.*[7]
		Abildskov and O'Connell[59,60]
		Kolar *et al.*[61]
		Franck *et al.*[62]
Original	Vapor pressures	Jensen *et al.*[63]
Modified (Lyngby)	Surface tension	Suarez *et al.*[64]
	Solubility of antibiotics in mixed solvents	Gupta and Heidemann[65]
	Flash points of flammable liquid mixtures	Tegtmeier and Misselhorn[66]
	Flavor sorption in packaging polymers	Li and Paik[67]
	Solvent selection for extractions	Brignole *et al.*[68]
		Ruzicka *et al.*[69]
	Viscosities of liquid mixtures	Wu[70]
		Cao *et al.*[71]
Various	Reid vapor pressure of gasoline	Hatzioannidis *et al.*[72]
	Partition coefficients of biochemicals	Kuramochi *et al.*[78]
Original, UNIFAC–FV	Infinite dilution activity coefficients	Paksoy *et al.*[73]
Modified (Lyngby, Dortmund)	Henry's law constants	Ornektekin *et al.*[74]
		Voutsas and Tassios[75,76]
		Zhang *et al.*[77]

are not available. Then, UNIQUAC is used for correlating all binary mixtures and finally predicting the multicomponent mixture VLE.

5.8 Local-composition–free-volume models for polymers

5.8.1 Introduction

Low-pressure activity coefficient models for polymers are very popular tools as many applications in the polymer industry are at low pressures, e.g. those related to paints and coatings. Of course, there are also industrial processes, e.g. the production of polyolefins, which occur at high pressures and for which equations of state are preferred (cubic EoS applied to polymers are discussed in Chapter 6, while advanced EoS for polymers in the SAFT family are discussed in Part C of the book, especially Chapter 14). We saw in Chapter 4 that a popular activity coefficient model for polymers is a combination of the FH combinatorial term together with a van-Laar-type residual term, containing an interaction parameter. If this parameter is independent of

Table 5.9 *Advantages and disadvantages of UNIFAC*

Advantages	Disadvantages
Reasonably accurate VLE predictions at moderate pressures and complex (non-associating) fluids	Poor accuracy for aqueous systems. e.g. H_2O–hydrocarbons
Predictive tool when no experimental data are available	Narrow temperature range (273–373 K) – caution with extrapolations required
It can be used to generate 'pseudo-experimental data'	'Pseudo-experimental' data should not be used for 'key binary mixtures' in a design problem (latest stage)
Excellent tool for multicomponent VLE but for preliminary design calculations (not for very difficult separations)	Often poor results for multicomponent, multifunctional systems with strong interactions, e.g. H_2O–alcohol–alkanes
Simple, fast calculations	Inability to distinguish between isomers Inability to handle proximity effects Difficulties in extrapolating to dilute systems
Continuous development, new parameters for missing groups appear, old parameters are revised based on new data	Need for different sets of parameters for estimating different phase equilibria types (VLE, LLE, GLE, SLE) Certain group interaction parameters are missing due to lack of experimental data Only qualitative agreement for excess enthalpies Extension to polymers requires corrections to account for free-volume effects

composition (but may depend on temperature) and for a binary polymer–solvent system, the solvent activity coefficient is given by the equation:

$$\ln\gamma_1 = \ln\left(\frac{\Phi_1}{x_1}\right) + 1 - \left(\frac{\Phi_1}{x_1}\right) + \chi_{12}\Phi_2^2 \tag{5.24}$$

where Φ_i is the volume fraction of compound i and the FH interaction parameter, χ_{12}, is typically fitted to polymer–solvent VLE data or the critical solution temperature (LLE data). Alternatively, it can be estimated from the regular solution theory and using the Hildebrand or the Hansen solubility parameters (see also Chapter 4), according to one of the following equations:

$$\chi_{12} = 0.35 + \frac{V_1}{RT}(\delta_1 - \delta_2)^2 \tag{5.25}$$

$$\chi_{12} = \alpha\frac{V_1}{RT}\left[(\delta_{d1} - \delta_{d2})^2 + 0.25(\delta_{p1} - \delta_{p2})^2 + 0.25(\delta_{hb1} - \delta_{hb2})^2\right] \tag{5.26}$$

Despite the success of the FH approach and the availability of solubility parameters for many compounds (solvents and polymers),[80,81] there are serious limitations on the FH/RST combinations, as also outlined in Chapter 4:

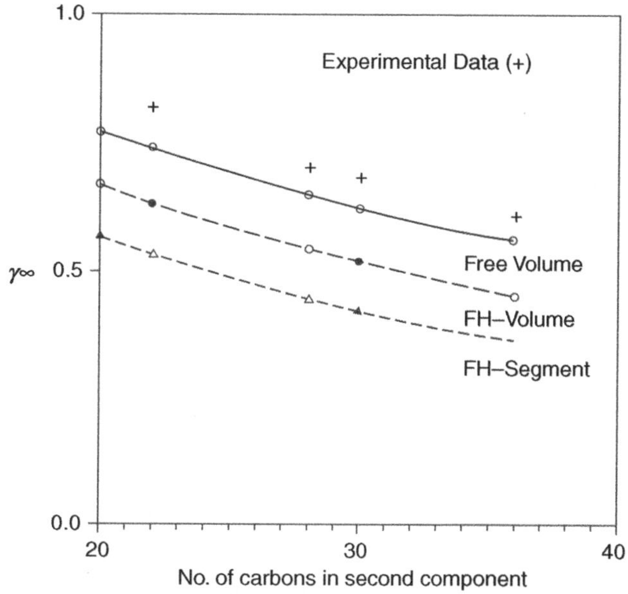

Figure 5.15 *Activity coefficients at infinite dilution at 100 °C for pentane + alkanes with the entropic–FV combinatorial term, with $\chi_{12} = 0$, Equation (5.27) and the FH combinatorial, Equation (5.24) (using either the volume or segment fractions). Reprinted with permission from Macromolecules, A new simple equation for the prediction of solvent activities in polymer solutions by H. S. Elbro, A. Fredenslund, P. Rasmussen, 23, 21, 4707–4714 Copyright (1990) American Chemical Society*

1. As Figure 5.15 illustrates, the FH combinatorial (based on either volume or segment fractions) systematically underestimates the activities in athermal alkane solutions. This is because it does not account for the free-volume (FV) effects. It is well known (Figure 5.16) that polymers and solvents have different FVs, with polymers typically exhibiting lower ones. FV differences have a profound effect on phase behavior, both VLE and LLE (Figure 5.17).
2. Although FV differences are a dominant factor of non-idealities between polymers and solvents, clearly the description of energetic effects could also be improved (compared to regular solutions) by using one of the non-random-mixing models which we have seen in this chapter.
3. The UNIQUAC model does contain a term for accounting for the differences in size and shape of various molecules, which is essentially an FH-segment-type term with an extra contribution, a Staverman–Guggenheim shape term. Thus, in principle, UNIQUAC can be used for polymer solutions, as shown for example in Figure 5.18. Actually, numerous polymer–solvent VLE systems have been correlated well with UNIQUAC using two system-specific interaction parameters.[82] However, as Figure 5.15 illustrates, the FH–segment combinatorial does not adequately account for the FV differences between polymers and solvents (the Staverman–Guggenheim term has only a small effect). This implies that an error due to this incorrect representation of size effects in UNIQUAC will be incorporated or absorbed by the interaction parameters of the residual term.

5.8.2 FV non-random-mixing models

FV and non-random-mixing effects have been combined in a simple and successful way in the entropic–FV model (using either UNIQUAC or UNIFAC in the residual term)[83,85] and in the UNIFAC-FV model:[86]

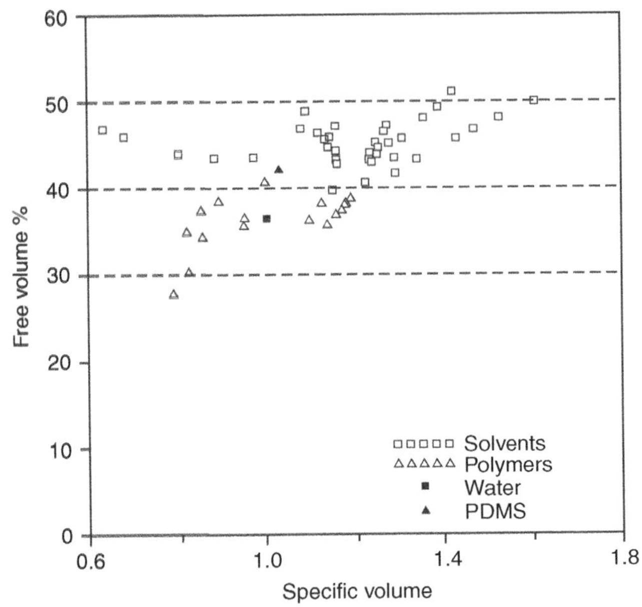

Figure 5.16 *Calculated free-volume percentages for polymers and solvents using the FV definition used in the entropic–FV model (Equations (5.27) and (5.28)). Polymers have typically lower FV percentages than solvents. Poly (dimethyl siloxane) (PDMS) and water are two notable exceptions, with this polymer being as 'flexible' as most solvents, while water has a very low FV percentage. Reprinted with permission from Macromolecules, A new simple equation for the prediction of solvent activities in polymer solutions by H. S. Elbro, A. Fredenslund, P. Rasmussen, 23, 21, 4707–4714 Copyright (1990) American Chemical Society*

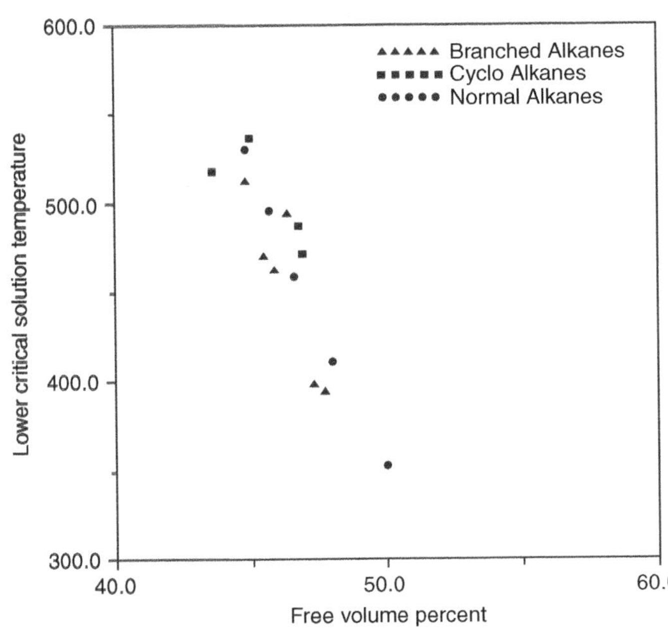

Figure 5.17 *Lower critical solution temperature (LCST) for polyethylene mixed with a series of hydrocarbons as a function of the FV percentage (at 298.15 K). From Elbro*[84]

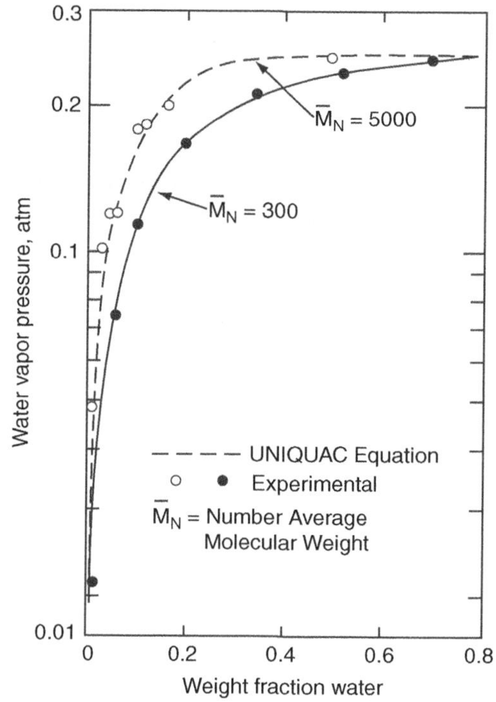

Figure 5.18 Vapor pressure of water in solutions of polyethylene glycol at 65 °C with the UNIQUAC equation. Reproduced with permission from AIChE J., Statistical thermodynamics of liquid mixtures: a new expression for the excess Gibbs energy of partly or completely miscible systems by D. S. Abrams and J. M. Prausnitz, 21, 1, 116–128 Copyright (1975) John Wiley and Sons, Inc.

Entropic–FV:

$$\ln\gamma_i = \ln\gamma_i^{comb\text{-}fv} + \ln\gamma_i^{res} = \ln\left(\frac{\Phi_i^{fv}}{x_i}\right) + 1 - \left(\frac{\Phi_i^{fv}}{x_i}\right) + \ln\gamma^{res} \tag{5.27}$$

$$\Phi_i^{fv} = \frac{x_i V_{f_i}}{\sum_j x_j V_{f_j}} \qquad V_{f_i} = V_i - V_{w,i} \tag{5.28}$$

where the residual term, γ^{res}, can be that of UNIQUAC (correlative model; Tables 5.1 and 5.3) or UNIFAC (predictive model; Equations (5.16) and (5.17)). When γ^{res} is calculated from UNIQUAC, the energetic parameters of polymer–solvents can be estimated from VLE data of the solvent plus a small molecule homologue of the polymer, e.g. ethylbenzene–CCl_4 for polystyrene–CCl_4. When UNIFAC is used for γ^{res}, the residual term based on the so-called 'new or linear UNIFAC' model is employed, which uses a linear-dependent parameter table:[40]

$$a_{mn} = a_{mn,1} + a_{mn,2}(T - T_o) \tag{5.29}$$

T_o is a reference temperature.

Notice that combinatorial and FV effects are combined into a single term in the entropic–FV model.

UNIFAC-FV:

$$\ln \gamma_i = \ln \gamma_i^{comb} + \ln \gamma_i^{res} + \ln \gamma_i^{FV} \tag{5.30}$$

The combinatorial and residual terms are, for this model, identical to those of UNIFAC (Equations (5.16) and (5.17)) but the FV term is given by the equation:

$$\ln \gamma_i^{fv} = 3c_i \ln\left[\frac{(\bar{v}_i^{1/3}-1)}{(\bar{v}_m^{1/3}-1)}\right] - c_i\left[\left(\frac{\bar{v}_i}{\bar{v}_m}-1\right)\left(1-\frac{1}{\bar{v}_i^{1/3}}\right)^{-1}\right] \tag{5.31}$$

where the reduced volumes are defined as:

$$\begin{aligned} \bar{v}_i &= \frac{v_i}{bV_{w,i}} \\ \bar{v}_m &= \frac{\sum w_i v_i}{b\sum w_i V_{w,i}} \end{aligned} \tag{5.32}$$

In Equation (5.31), the volumes v_i and the van der Waals volumes are all expressed in cm^3/mol and w_i is the weight fraction. In the UNIFAC–FV model as suggested by Oishi and Prausnitz[86] the parameters c_i ($3c_i$ is the number of external degrees of freedom) and b are set to constant values for all polymers and solvents ($c_i = 1.1$ and $b = 1.28$). Even though in the case of both entropic–FV and UNIFAC–FV, the combinatorial–FV contributions are not identical to that of UNIFAC, in the residual term the (original) UNIFAC parameters (developed from the Staverman–Guggenheim combinatorial) have been used without re-estimation.

Both entropic–FV and UNIFAC–FV have been applied with success to polymer–solvent VLE, including the infinite dilution activity coefficient and ternary VLE.[87–89] Entropic–FV has also been applied with qualitatively good results to polymer–solvent LLE (but not polymer blends where energetic interactions dominate),[90–92] SLE for polymer–solvents,[93] paints and dendrimers.[94] If more accurate representation of LLE is required, then the UNIQUAC residual instead of UNIFAC should be used and the interaction parameters should be fitted to low-molecular-weight mixtures (of the solvent and the low-molecular-weight homologue of the polymer). Table 5.10 and Figures 5.19–5.27 illustrate some of the results. The entropic–FV/UNIQUAC (results shown in Figures 5.23–5.27) is of particular interest,[83,95,96] as it offers additional flexibility over the predictive entropic–FV model (which is based on UNIFAC's residual term) and can be used with success in polymer–solvent LLE for both polymer solutions and blends, including high-pressure LLE (as the volumes are functions of both temperature and pressure).

5.9 Conclusions: is UNIQUAC the best local composition model available today?

This is indeed a very interesting question! UNIQUAC (and its GC version, UNIFAC) appear to represent the 'end of the road', with respect to LC models. They came after the Wilson equation and NRTL and with the aim

Table 5.10 Prediction of infinite dilution activity coefficients for polyisoprene (PIP) systems with two predictive GC models. Experimental values and calculations are at 328.2 K. The values in parentheses are the average absolute percentage deviations between experimental and predicted activity coefficients

PIP systems	Exp. value	Entropic–FV	UNIFAC–FV
+ acetonitrile	68.6	47.7 (31%)	52.3 (24%)
+ acetic acid	37.9	33.5 (12%)	17.7 (53%)
+ cyclohexanone	7.32	5.4 (27%)	4.6 (38%)
+ acetone	17.3	15.9 (8%)	13.4 (23%)
+ MEK	11.4	12.1 (6%)	10.1 (12%)
+ benzene	4.37	4.5 (2.5%)	4.4 (0%)
+ 1,2 dichloroethane	4.25	5.5 (29%)	6.5 (54%)
+ CCl_4	1.77	2.1 (20%)	1.8 (0%)
+ 1,4 dioxane	6.08	6.3 (4%)	5.9 (2%)
+ tetrahydrofurane	4.38	4.9 (14%)	3.9 (10%)
+ ethylacetate	7.47	7.3 (2%)	6.6 (11%)
+ *n*-hexane	6.36	5.1 (20%)	4.6 (27%)
+ chloroform	2.13	3.00 (41%)	2.6 (20%)

of resolving several problems in those previous models. Indeed UNIQUAC did solve several of the 'practical' limitations of previous LC models while creating a few new ones. Let us summarize the most important 'practical' and 'theoretical' aspects of UNIQUAC compared to the other two LC models:

1. It can be used for both multicomponent VLE and LLE (unlike Wilson) and does have separate combinatorial and energetic (residual) terms (unlike NRTL). The results are at least as satisfactory as for the other models (if not better). UNIQUAC can account for non-ideality due to size and shape and energetic effects.

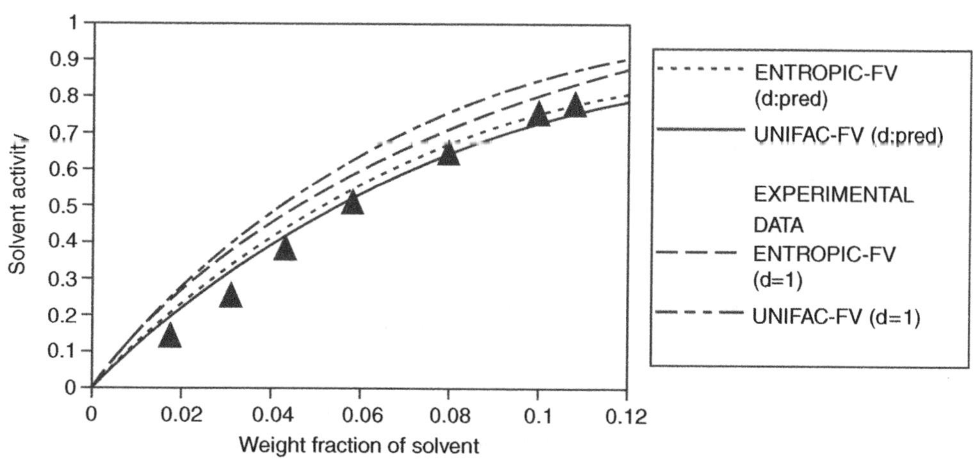

Figure 5.19 Experimental and predicted activities of acetone in the dendrimer A4 with the entropic–FV and UNIFAC–FV models. Reprinted with permission from Ind. Eng. Chem. Res., Free-Volume Activity Coefficient Models for Dendrimer Solutions by Irene A. Kouskoumvekaki, Georgios M. Kontogeorgis et al., 41, 19, 4848–4853 Copyright (2002) American Chemical Society

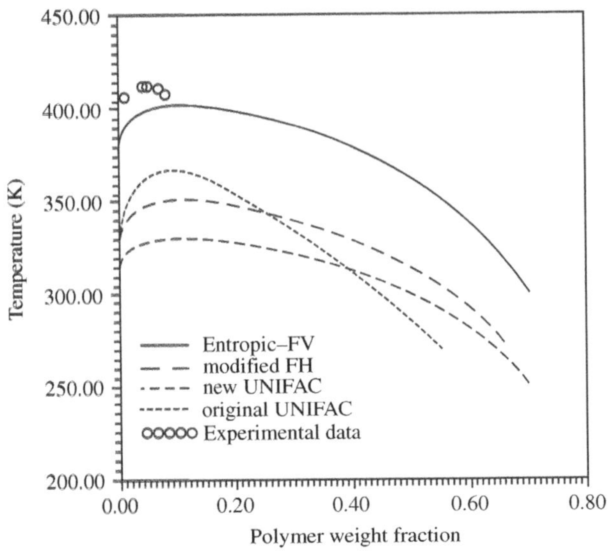

Figure 5.20 Experimental and predicted LLE (UCST-type phase diagrams) for HDPE (13 600)/n-butyl acetate using entropic–FV, modified FH and UNIFAC (either with T-independent parameters, indicated as 'original', or linearly T-dependent parameters, indicated as 'new'). Modified FH is the Flory–Huggins combinatorial based on volume fraction, Equation (5.24), combined with the 'linear' UNIFAC residual term. Reprinted with permission from Ind. Eng. Chem. Res., Prediction of Liquid-Liquid Equilibrium for Binary Polymer Solutions with Simple Activity Coefficient Models by Georgios M. Kontogeorgis, Dimitrios P. Tassios et al., 34, 5, 1823–1834 Copyright (1998) American Chemical Society

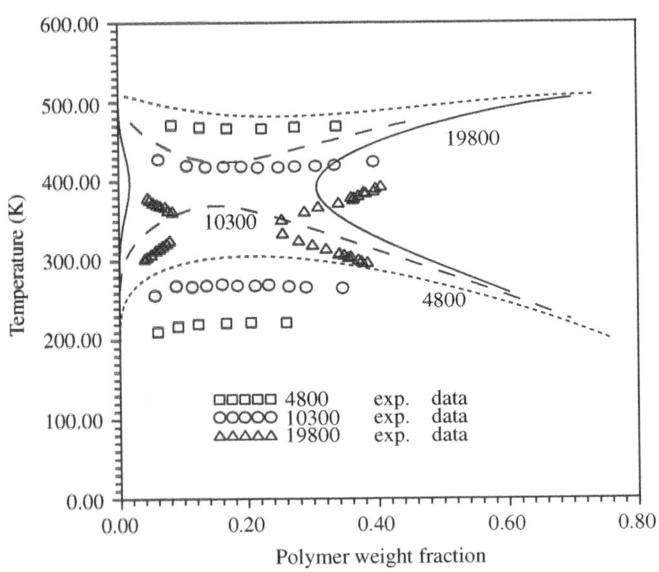

Figure 5.21 Experimental and predicted LLE with the entropic–FV model for the system polystyrene–acetone at three polymer molecular weights (4800, 10 300, 19 800). The points are the experimental data and the lines are the predictions with the entropic–FV model. Reprinted with permission from Ind. Eng. Chem. Res., Prediction of Liquid-Liquid Equilibrium for Binary Polymer Solutions with Simple Activity Coefficient Models by Georgios M. Kontogeorgis, Dimitrios P. Tassios et al., 34, 5, 1823–1834 Copyright (1998) American Chemical Society

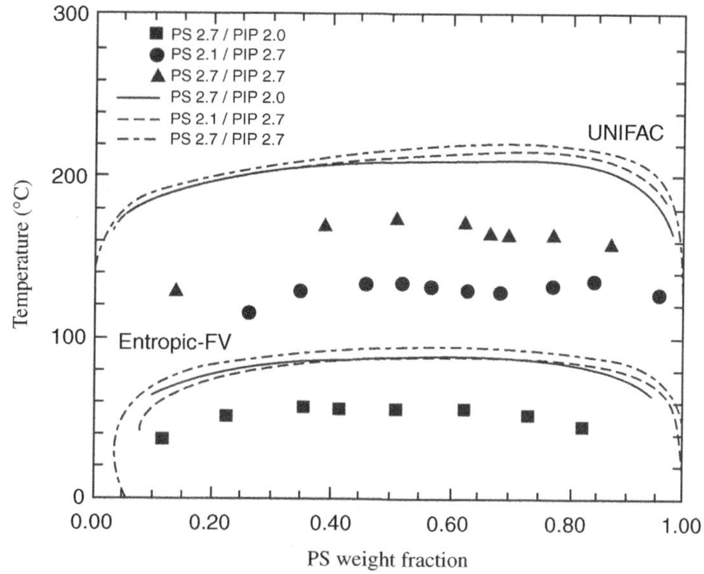

Figure 5.22 *Experimental (symbols) and calculated (lines) cloud point curves for three polyisoprene (PIP) and polystyrene (PS) blends with entropic–FV and UNIFAC. The molecular weights of the polymers are indicated in the figure in kg/mol. Reproduced with permission from AIChE J, Miscibility of polymer blends with Engineering models by V.I. Harismiadis, A.R.D. van Bergen, A. Saraiva, G.M. Kontogeorgis, Aa. Fredenslund and D.P. Tassios, 42, 11, 3170–3180 Copyright (1996) John Wiley and Sons, Inc*

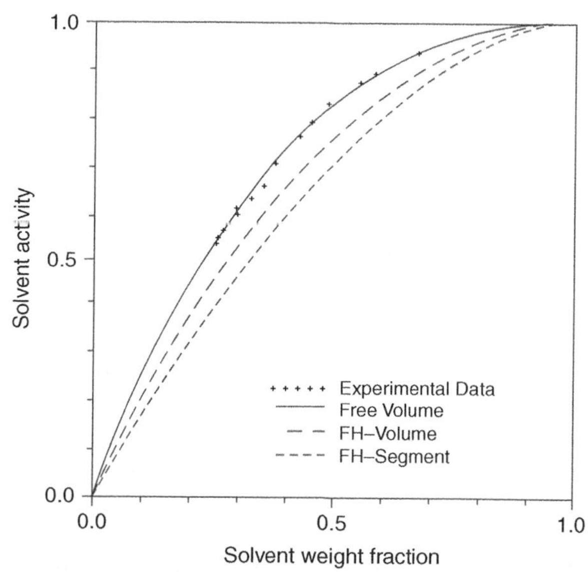

Figure 5.23 *Comparison of three versions of UNIQUAC combined either with the FV or FH combinatorial terms based on either volume or segment fractions for the system PS–CCl_4. Reprinted with permission from Macromolecules, A new simple equation for the prediction of solvent activities in polymer solutions by H. S. Elbro, A. Fredenslund, P. Rasmussen, 23, 21, 4707–4714 Copyright (1990) American Chemical Society*

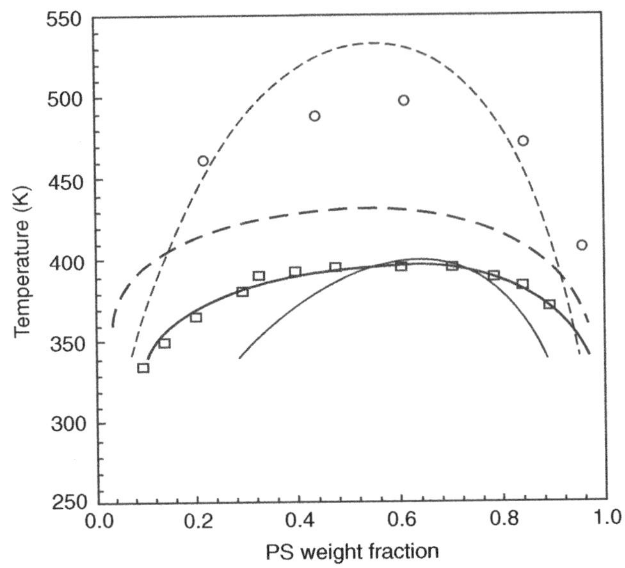

Figure 5.24 *Correlation and molecular weight effect prediction results with the entropic–FV/UNIQUAC model for the PS–PBD blend. Comparison between T-independent and T-dependent interaction parameters (IPs). The points are the experimental data (circles PS(3300)–PBD(2350) and squares PS(1900)–PBD(2350)) and the lines are the calculations. The thin solid line is correlation with T-independent IPs and the thin dashed line is prediction using T-independent IPs. Solid thick line is correlation using T-dependent IPs and dashed thick line is prediction using T-independent IPs. Reprinted with permission from Ind. Eng. Chem. Res., Miscibility in Binary Polymer Blends: Correlation and Prediction by Epaminondas C. Voutsas, Georgia D. Pappa et al., 43, 5, 1312–1321 Copyright (2004) American Chemical Society*

2. Multicomponent LLE is a challenging task, but this is also the case for NRTL. Temperature-dependent interaction parameters may improve LLE description (even though they are not justified based on the derivation of LC models).
3. UNIQUAC's derivation from two-fluid or other theories is quite problematic: the coordination number Z is set to 10 in the combinatorial term, but must be equal to 2 (or some low value, much lower than 10) in the residual term, otherwise the non-randomness of the mixtures is greatly overestimated. The coordination numbers of all compounds are assumed to be equal, which cannot be rigorously true for very asymmetric mixtures. Moreover, the energetic parameters for multicomponent mixtures cannot be chosen independently from the binaries. However, all three LC models suffer from 'coordination number' and 'parameter interrelation' problems.
4. A great advantage of UNIQUAC over NRTL and Wilson is that interactions take place via the surface of the molecules. It is, unfortunately, exactly this new characteristic which creates an inconvenience, which only UNIQUAC has among the three LC models: the size (R and Q) parameters are normalized to the polyethylene segment value and thus available energetic values are 'tightened to' these normalized R and Q values. Nonetheless, in the 'classical' UNIQUAC approach these parameters are estimated from the van der Waals volume and surface area and thus contain a physical significance. This physical significance becomes somewhat deteriorated when these parameters, in certain applications and for the sake of achieving better results, are fitted to mixture data together with the energy parameters.

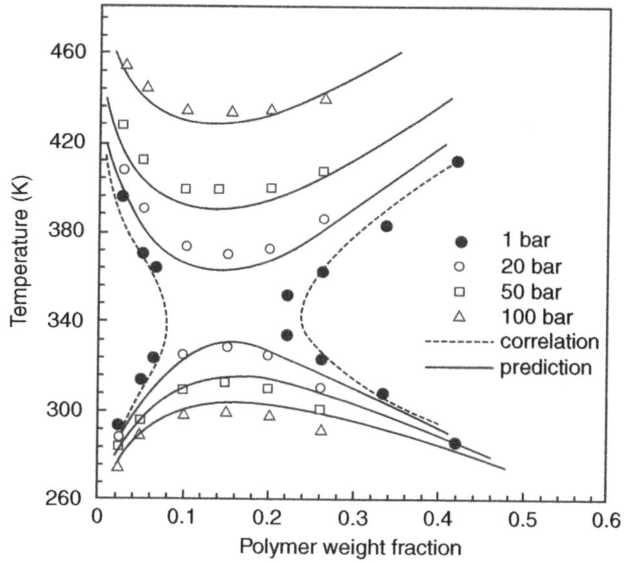

Figure 5.25 *Correlation and prediction results of the pressure effect on LLE for the system PS(20 400)–acetone with the entropic–FV/UNIQUAC model. Reprinted with permission from Ind. Eng. Chem. Res., Liquid–Liquid Phase Equilibrium in Polymer–Solvent Systems: Correlation and Prediction of the Polymer Molecular Weight and the Pressure Effect by Georgia D. Pappa, Epaminondas C. Voutsas, and Dimitrios P. Tassios, 40, 21, 4654–4663 Copyright (2001) American Chemical Society*

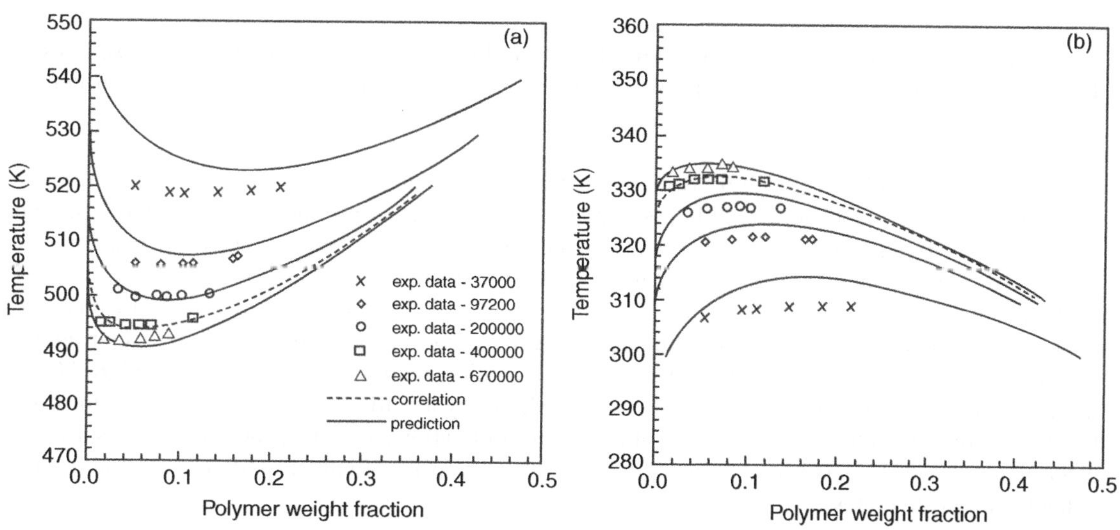

Figure 5.26 *Correlation (dashed lines) and prediction (full lines) results of the molecular weight (MW) effect on LLE UCST (a) and LCST (b) for the system PS–methylcyclohexane using the entropic-FV/UNIQUAC model. The numbers in the caption denote the polymer MW. Reprinted with permission from Ind. Eng. Chem. Res., Liquid–Liquid Phase Equilibrium in Polymer–Solvent Systems: Correlation and Prediction of the Polymer Molecular Weight and the Pressure Effect by Georgia D. Pappa, Epaminondas C. Voutsas, and Dimitrios P. Tassios, 40, 21, 4654–4663 Copyright (2001) American Chemical Society*

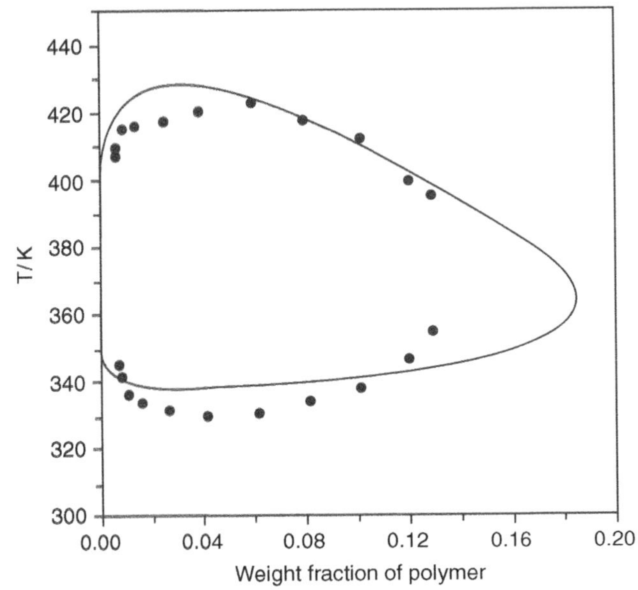

Figure 5.27 *Correlation of LLE for the PVAL–water system with the entropic–FV/UNIQUAC model: •, experimental data ($M_n = 140\,000$ g/mol); ——, correlation. Reprinted with permission from Fluid Phase Equilibria, A segmental interaction model for liquid-liquid equilibrium calculations for polymer solutions by Grozdana Bogdanic and Jean Vidal, 173, 2, 241 Copyright (2000) Elsevier*

5. There have been numerous molecular simulation studies of the LC concept with mixed results. There is some support for the use of the Boltzmann factors, but in general the agreement between LC models and simulation data is moderate, though somewhat better for UNIQUAC over the other equations.
6. Despite the above problems, UNIQUAC's parameters have been estimated from 'theoretical principles' (quantum chemistry/mechanics) with good results, better than with the other LC models. Other considerations (one-parameter versions etc.) provide some confidence about the theoretical significance of UNIQUAC parameters, but the same can be said for Wilson and NRTL.
7. By replacing its combinatorial term with a free-volume one, UNIQUAC has been extended with success to polymers and (see Appendix 5.C) and it has also been extensively used for electrolytes (upon addition of a Debye–Hückel term) and solids. The latter application requires no changes in UNIQUAC and, moreover, the resulting model is fully predictive for SLE of hydrocarbon mixtures (wax formation), as all the interaction parameters are estimated from heats of sublimation and other properties of the solids. The electrolyte UNIQUAC has been combined with the van der Waals–Platteeuw framework and has been extended to gas hydrates.

So, what should the answer be to the question we stated at the beginning of this section? We can conclude that, considering the balance between the practical advantages of UNIQUAC and its theoretical strengths/weaknesses, it does indeed represent the best 'stand-alone' activity coefficient model for low-pressure applications, i.e. in the context of the γ–φ approach. Especially over Wilson, its advantages are overwhelming, considering that calculations today with such models will typically be done via computer programs and thus the 'simpler' function of the Wilson equation does not offer a significant advantage. As we will see in the next chapter about EoS/G^E mixing rules, NRTL's 'true value' today may be more appreciated when it is used in connection with (i.e. as a mixing rule for) a cubic EoS. In this way, as we will see in the coming chapter, the fact that NRTL is essentially an H^E rather than the G^E model can, in some contexts, turn out to be an advantage.

Appendix 5.A A note on the derivation of the local-composition models

According to the physical picture shown in Figure 5.1, one might have expected that all local composition (LC) models would have been developed from some variation of a 'two-fluid' theory (for a binary mixture; *n*-fluid for a multicomponent one). That was originally done only for NRTL (see Problem 3 on the companion website at www.wiley.com/go/Kontogeorgis) but not for Wilson or UNIQUAC. The original derivations[1,3] are based on 'one-fluid' lattice theories (Flory-Huggins; quasi-chemical of Guggenheim)[98] combined with expressions for the local fractions. A detailed guide to solution of the problem illustrating the original derivation of the Wilson equation is shown in Problem 1 on the companion website at www.wiley.com/go/Kontogeorgis. Due to the 'conceptual' difficulties in these derivations, it has been shown that all three known LC models can be 'more rigorously' derived from the two-fluid theory (see Maurer and Prausnitz[99] and Prausnitz *et al.*[7] for UNIQUAC; Vidal[8] and Elliott and Lira[100] for Wilson).

The starting point for developing an expression for the excess Gibbs energy or excess Helmholtz energy, i.e. a G^E or A^E expression, is:

$$\begin{aligned} A^E &= G^E - PV^E \\ G^E &= H^E - TS^E = U^E + PV^E - TS^E \end{aligned} \quad (5.33)$$

All models assume that $V^E = 0$ (reasonable assumption for the liquid phase), and thus the problem is reduced to developing expressions for the excess energy and entropy, U^E and S^E, which can yield finally the equations for A^E or G^E. In the case of NRTL, we have made clear from the outset the assumption that, similar to that of the regular solution theory, $S^E = 0$, and thus 'all excess energies' are equal to each other:

$$\text{NRTL: } A^E = G^E = H^E = U^E \quad (5.34)$$

Thus the expression for U^E from the two-fluid theory (NRTL and Wilson):

$$U^E = \frac{Z}{2}[x_1 x_{21}(\varepsilon_{21} - \varepsilon_{11}) + x_2 x_{12}(\varepsilon_{12} - \varepsilon_{22})] \quad (5.35)$$

combined with expressions for the LCs x_{ij} yield an equation for the excess Gibbs energy. See Problem 3 on the companion website at www.wiley.com/go/Kontogeorgis (for NRTL). Note that, in principle, the coordination number, Z, can be incorporated into the parameter values, at least when the latter are estimated from experimental data. Equation (5.35) is valid for both Wilson and NRTL but for UNIQUAC, which assumes interaction via surface areas, the U^E expression is:

$$U^E = \frac{Z}{2}[x_1 q_1 \vartheta_{21}(U_{21} - U_{11}) + x_2 q_2 \vartheta_{12}(U_{12} - U_{22})] \tag{5.36}$$

An important assumption in the derivations of both equations (5.35) and (5.36) is that the coordination number is the same for all compounds and, moreover, that it is not modified during the mixing process. These are indeed rather serious assumptions! To proceed further we need: (1) expressions for the local mole fractions, x_{12} or ϑ_{12}; and (2) a way to derive A^E from U^E.

The latter relationship is provided by thermodynamics, where we can see that the temperature dependency of U^E is very important:

$$\frac{A^E}{RT} = \int_{1/T_0}^{1/T} \frac{-U^E}{RT}\frac{dT}{T} + C = \int_{1/T_0}^{1/T} \frac{-U^E}{R} d\left(\frac{1}{T}\right) + C \tag{5.37}$$

The constant of integration C is the value of A^E/RT at very high temperatures ($1/T_0 \to 0$). As illustrated in Tables 5.1 and 5.2, the expressions for the local fractions are similar for all LC models (variations of Boltzmann expressions), but differ in details, e.g. for Wilson (in the framework of two-fluid theory):

$$\frac{x_{21}}{x_{11}} = \frac{x_2}{x_1}\Omega_{21}$$

$$\Omega_{21} = \frac{V_2}{V_1}\exp\left[-\frac{Z}{2}\left(\frac{\varepsilon_{21} - \varepsilon_{11}}{RT}\right)\right] \tag{5.38}$$

$$\frac{x_{21}}{x_{11}} = \frac{\Phi_2}{\Phi_1}\exp\left[-\frac{Z}{2}\left(\frac{\varepsilon_{21} - \varepsilon_{11}}{RT}\right)\right]$$

Note that this definition of LC fractions in the Wilson equation is different from that shown in Table 5.1. Using Equations (5.35) and (5.38), we arrive at the well-known expressions of Table 5.1, if we assume that $Z = 2$ (which is yet another drastic assumption!).

For UNIQUAC:

$$\frac{\vartheta_{21}}{\vartheta_{11}} = \frac{x_2}{x_1}\Omega_{21}$$

$$\Omega_{21} = \frac{q_2}{q_1} \exp\left[-\frac{Z}{2}\left(\frac{U_{21}-U_{11}}{RT}\right)\right] \tag{5.39}$$

$$\frac{\vartheta_{21}}{\vartheta_{11}} = \frac{\vartheta_2}{\vartheta_1} \exp\left[-\frac{Z}{2}\left(\frac{U_{21}-U_{11}}{RT}\right)\right]$$

Having expressions now for both U^E and the local fractions, we can (1) substitute Equation (5.38) into (5.35) for Wilson and Equation (5.39) into (5.36) for UNIQUAC and then (2) integrate Equation (5.37) in order to get an expression for A^E. The final results are shown in Tables 5.1 and 5.3 (binary case and multicomponent mixtures). Two final important points of caution are:

1. The 'combinatorial term' (size and shape effects) is the constant C, the non-ideality at very high temperatures where intermolecular forces are not present. For Wilson, we set $C = 0$, but for UNIQUAC, Abrams and Prausnitz[3] used the Staverman–Guggenheim term, which is very similar to the FH term. There are no restrictions from the two- or n-fluid theory on which equation should be used for the combinatorial term, i.e. the constant C. In principle, any other term could have been used.
2. The final expressions for G^E shown in Tables 5.1 or 5.3 are derived from the integration of Equation (5.37) assuming that the energy differences, e.g. $\varepsilon_{ji} - \varepsilon_{ii}$ are independent of temperature and that the only temperature dependency is that of the Boltzmann factors. This implies that using temperature-dependent interaction coefficients in the LC models is not consistent with their derivation. It also implies that constant volume values should be used in the Wilson equation.

Appendix 5.B Modifications of the NRTL equation

The NRTL equation has been proved to be a successful model for correlating phase equilibrium data. For example, as Vetere[20,101] illustrated, it can successfully correlate VLE for numerous systems and the interaction parameters can be related to solubility parameter differences or other properties. However, as mentioned previously, NRTL is in reality an expression for H^E as it does not have an entropic term. As is apparent from the model's equation (here for binary systems):

$$\frac{g^E}{RT} = x_1 x_2 \left[\frac{\tau_{21} G_{21}}{x_1 + x_2 G_{21}} + \frac{\tau_{12} G_{12}}{x_2 + x_1 G_{12}}\right] \tag{5.40}$$

$$\frac{x_{21}}{x_{11}} = \frac{x_2}{x_1} \exp\left(-\frac{a_{12}(g_{21} - g_{11})}{RT}\right) \tag{5.41}$$

$$G_{ij} = \exp(-a_{ij}\tau_{ij})$$

$$\tau_{ij} = \frac{g_{ij} - g_{jj}}{RT} \tag{5.42}$$

when the non-randomness parameter is zero and irrespective of the values of the interaction parameters, we obtain the symmetrical one-parameter Margules expression:

$$\frac{g^E}{RT} = x_1 x_2 (\tau_{21} + \tau_{12}) \tag{5.43}$$

which seems to be a rather serious simplification (if we consider that g_{ij} are in general different from those of the pure compounds). At dilute conditions and for systems with large differences in size, the Wilson equation actually performs better than NRTL.[21] In order to accommodate these problems, to develop a predictive one-parameter NRTL form, but also in order to have an NRTL equation which can be conveniently used in combination with cubic EoS (as will be explained in the next chapter), various researchers have proposed modifications of the G_{ij} parameters in Equation (5.42) which still can be used in Equation (5.40). We present two of these modifications below:

Modified NRTL-1:[102]

$$G_{ij} = b_i \exp(-a_{ij}\tau_{ij}) \tag{5.44}$$

Modified NRTL-2:[20,101]

$$G_{ij} = \frac{V_i}{V_j} \exp(-a_{ij}\tau_{ij}) \tag{5.45}$$

When either equation (5.44) or (5.45) is implemented in NRTL, Equation (5.40), and if the non-randomness parameter a_{ij} is equal to zero, we obtain the following expressions which resemble the regular solution theory and could thus call them 'quasi-regular solutions':

Modified NRTL-1:

$$\begin{aligned} \frac{g^E}{RT} &= b\Phi_1\Phi_2\left(\frac{\tau_{12}}{b_2} + \frac{\tau_{21}}{b_1}\right) \\ \Phi_i &= \frac{x_i b_i}{b} \end{aligned} \tag{5.46}$$

Modified NRTL-2:

$$\begin{aligned} \frac{g^E}{RT} &= V\Phi_1\Phi_2\left(\frac{\tau_{12}}{V_2} + \frac{\tau_{21}}{V_1}\right) \\ \Phi_i &= \frac{x_i V_i}{V} \end{aligned} \tag{5.47}$$

The benefits of Equation (5.44) will be illustrated in Chapter 6. Equation (5.45) has been used together with NRTL by Vetere[20,101] in a series of publications. He showed that NRTL in this form yields much better results than the original NRTL, Equation (5.40) and can, moreover, be used in a one-parameter form, where g_{ii} is estimated using Equation (5.8) (the Tassios approach) and g_{ij} is obtained by the geometric mean rule $(=(g_{ii}g_{jj})^{1/2})$ or via suitable rules of thumb. Vetere also showed that Equation (5.9) (the Wong and Eckert approach) did not improve the results, indicating again that a low Z value ($=2$) provides better results, in accordance with previous findings for other models (Wilson and UNIQUAC).

Appendix 5.C Extension of UNIQUAC to electrolytes and solids

5.C.1 UNIQUAC for solids and waxes

Coutinho and co-workers[103–107] have proposed a predictive UNIQUAC model for solid–liquid calculations suitable especially for predicting wax formation in petroleum fluids. UNIQUAC was used within a framework based on the fact that macrocrystallinic waxes can be considered to be composed only of aliphatic alkanes, while for degassed oils one may consider the oil to be an average ideal fluid in which the non-ideal distribution of wax components is dispersed. Hence UNIQUAC is used for the non-ideality. The starting point for calculations is the 'full form' of the SLE equation, including not only the activity and heat of fusion term, but also the transition temperatures ($T_{tr,i}$) and heat capacity difference contributions ($\Delta C_{p,i}$):

$$\ln\left(\frac{f^{L,0}}{f^{S,0}}\right) = \ln\left(\frac{x_i^S\gamma_i^S}{x_i^L\gamma_i^L}\right) = \frac{\Delta H_{fus,i}}{RT}\left(1-\frac{T}{T_{t,i}}\right)+\frac{\Delta H_{tr,i}}{RT_{tr,i}}\left(\frac{T_{tr,i}}{T}-1\right)-\frac{\Delta C_{p,i}}{R}\left(\frac{T_{t,i}-T}{T}\right)+\frac{\Delta C_{p,i}}{R}\ln\frac{T_{t,i}}{T} \quad (5.48)$$

where t indicates the triple point properties (very close to melting point).

Coutinho assumes that the liquid phase activity coefficient can be calculated from the Flory–FV model (Equation (5.27) with $V_f = (V^{1/3}-V_w^{1/3})^{3c}$, $c=1.1$) or the liquid phase can be assumed ideal (activity coefficient of 1) for the alkane mixtures of relevance to wax formation. The UNIQUAC equation is used for the estimation of the solid activity coefficient. Originally, the Wilson model was used but UNIQUAC is preferred because – unlike Wilson – it can also predict multisolid phase equilibria and thus account for the non-miscibility of solid paraffins for which there is now experimental evidence via spectroscopy. The UNIQUAC model for solid activity coefficients is fully predictive. The equations are the same as before (Tables 5.1–5.3), where, in agreement with previous discussions, the Z term in the exponentials is dropped. As it is assumed that solid precipitation occurs as n-alkanes having an orthorhombic structure, the pure compound and cross-energetic parameters of UNIQUAC are estimated as (using $Z=6$ for this structure):

$$\begin{aligned} U_{ii} &= -\frac{2}{Z}(\Delta H_i^{subl}-RT) \\ U_{ij} &= U_{ii}(\text{shorter alkane}) \end{aligned} \quad (5.49)$$

Equation (5.49) is essentially similar to the Tassios approach, where the heat of sublimation is used instead of the heat of vaporization. The heat of sublimation of the orthorhombic solid phase is essentially a measure of the average molecular interactions in a crystal. The heat of sublimation can be obtained as the sum of the heats of vaporization, fusion and solid–solid transition of alkanes, all of which are given by correlations (see Coutinho[103] and Coutinho *et al.*[108]). In the same references the correlations for the melting and transition temperatures are provided. The UNIQUAC structural parameters are different from those of the original UNIQUAC and are based on a chain containing 10 CH_2 groups:

$$
\begin{aligned}
r &= 0.1C_n + 0.0672 \\
q &= 0.1C_n + 0.1141
\end{aligned}
\tag{5.50}
$$

Very good prediction results are obtained for the fraction of crystallized paraffins from different hydrocarbon mixtures including diesel fuels. One example is shown in Figure 5.28.

5.C.2 UNIQUAC for electrolytes

Extension of UNIQUAC to mixtures containing strong electrolytes requires significant changes to the original model. The most important one is the addition of a Debye–Hückel[109] term which explicitly accounts for the electrostatic interactions (for dilute solutions). In the original versions of the UNIQUAC extension to electrolytes, which is called extended UNIQUAC,[15] composition-dependent interaction parameters had been used, but these have now (fortunately!) been dropped and the version of the model discussed here has been developed by Thomsen and co-workers:[110,111]

$$
g^E = g^{E,comb} + g^{E,res} + g^{E,Debye\text{–}Hückel} \tag{5.51}
$$

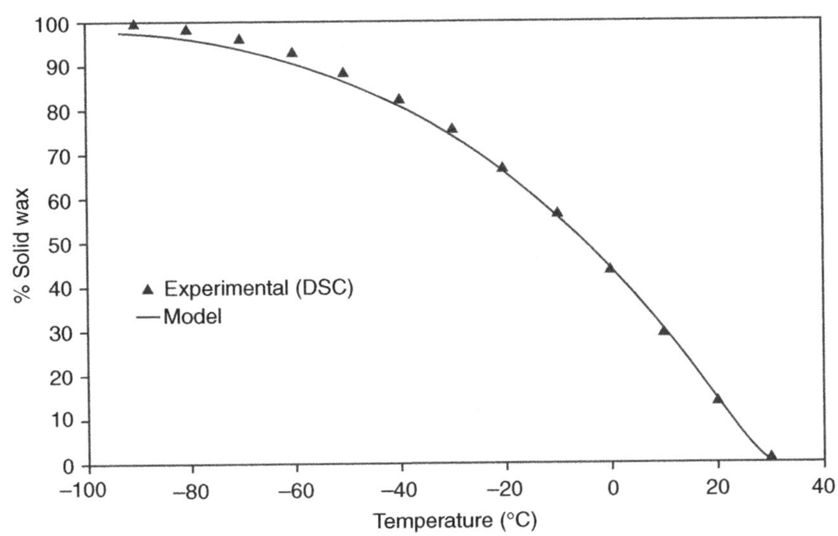

Figure 5.28 *Comparison between experimental measurements and model predictions for Calange's Brut X. Reprinted with permission from Energy & Fuels, Low-Pressure Modeling of Wax Formation in Crude Oils by J. A. P. Coutinho and J.-L. Daridon, 15, 6, 1454 Copyright (2001) American Chemical Society*

The combinatorial and residual terms are identical to that of UNIQUAC (Tables 5.1–5.3) and the interaction parameters are linear functions of temperature:

$$U_{ij} = U_{ij}^{0} + U_{ij}^{(1)}(T-298.15) \tag{5.52}$$

The interaction parameters and the volume and surface area parameters (R and Q) (of the combinatorial and residual term) are considered to be adjustable parameters and are fitted to mixture data. These are the sole adjustable parameters of the model, i.e. the Debye–Hückel term contains no additional adjustable parameters. Thomsen has used an extended Debye–Hückel equation:

$$g^{E,Debye-Hückel} = -x_w M_w \frac{4A}{b^3}\left[\ln\left(1+b\sqrt{I}\right) - b\sqrt{I} + \frac{b^2 I}{2}\right] \tag{5.53}$$

where x_w is the mole fraction of water, M_w is the molar mass of water (kg/mol), I is the ionic strength, which is equal to $0.5\sum_i m_i Z_i^2$ in mol/kg (water) where m is molality and Z is the ionic charge, and $b = 1.50\ (\text{kg/mol})^{1/2}$:

$$A = \sqrt{2\pi N_A d_0}\left(\frac{e}{\sqrt{4\pi\varepsilon_0 DkT}}\right)^3 \tag{5.54}$$

e.g. $A = 1.1717\ (\text{kg/mol})^{1/2}$ at 25 °C. The parameter A (in $(\text{kg/mol})^{1/2}$) can be approximated (in the range 273–383 K) by the equation:

$$A = 1.131 + 1.335 \times 10^{-3}(T-273.15) + 1.164 \times 10^{-5}(T-273.15)^2 \tag{5.55}$$

When calculating activity coefficients for aqueous (and other) electrolyte solutions, it is important to remember that the combinatorial and residual terms of UNIQUAC are based on the Lewis–Randall convention and the same is true for the Debye–Hückel term for water, but not for the ions for which the unsymmetrical convention is used. Thus, the activity coefficients of water and of ions are given by slightly different expressions (the combinatorial and residual terms are the same as for UNIQUAC, Table 5.3):

$$\gamma_w = \gamma_w^{comb}\gamma_w^{res}\gamma_w^{DH} \tag{5.56}$$

$$\gamma_i^{*} = \frac{\gamma_i^{comb}}{\gamma_i^{comb,\infty}}\frac{\gamma_i^{res}}{\gamma_i^{res,\infty}}\gamma_i^{*,DH} \tag{5.57}$$

and thus for the molal activity coefficient:

$$\gamma_i^{m} = x_w \gamma_i^{*} \tag{5.58}$$

where:

$$\ln\gamma_w^{DH} = M_w \frac{2A}{b^3}\left[1 + b\sqrt{I} - \frac{1}{1+b\sqrt{I}} - 2\ln\left(1+b\sqrt{I}\right)\right] \tag{5.59}$$

$$\ln\gamma_i^{*,DH} = -Z_i^2 \frac{A\sqrt{I}}{1+b\sqrt{I}} \tag{5.60}$$

Thomsen has presented parameters for water and the following ions: H^+, Na^+, K^+, $NH_4{}^+$, Cl^-, $SO_4{}^{2-}$, $NO_3{}^-$, OH^-, $HSO_4{}^-$, $CO_3{}^{2-}$, $HCO_3{}^-$, $S_2O_8{}^{2-}$. Extended UNIQUAC has been applied with success to gas solubilities and vapor–liquid–solid equilibria in aqueous electrolyte systems,[110,112,113] mixed solvent–salt mixtures,[114,115] heavy metal ions[116] and recently also pressure-dependent salt solubility.[117] The results are generally satisfactory using only pure compound and binary parameters.

Other LC models have been combined with versions of the Debye–Hückel equation and thus applied to electrolyte systems. Special attention has been given to electrolyte NRTL[118–120] as this model is also available in Aspen's process simulator. Despite the similarities with extended UNIQUAC, an important difference is that electrolyte NRTL has salt-specific parameters, while those of extended UNIQUAC are ion specific. More discussions about these models as well as presentation of other models suitable for electrolyte solutions are provided in Chapter 15.

References

1. G. Wilson, *J. Am. Chem. Soc.*, **1964**, *86*, 127.
2. H. Renon, J. M. Prausnitz, *AIChE J.*, **1968**, *14(1)*, 135.
3. D.S. Abrams, J.M. Prausnitz, *AIChE J.*, **1975**, *21(1)*, 116.
4. J.M. Marina, D.P. Tassios, *Ind. Eng. Chem. Process Des. Dev.*, **1973**, *12(1)*, 67.
5. A. Bondi, *Physical Properties of Molecular Crystals, Liquid and Glasses.* John Wiley and Sons Inc., 1968.
6. J.M. Prausnitz, T. Anderson, E. Grens, C.A. Eckert, J.P. O'Connell, *Computer Calculations for Multicomponent Vapour-Liquid and Liquid-Liquid Equilibria.* Prentice Hall International, 1980.
7. J.M. Prausnitz, R.N. Lichtenthaler, E.G. de Azevedo, *Molecular Thermodynamics of Fluid-Phase Equilibria.* (3rd edition), Prentice Hall International, 1999.
8. J. Vidal, *Thermodynamics: Applications in chemical engineering and the petroleum industry.* TECHNIP, IFP Publications, 1997.
9. Aa. Fredenslund, R. Jones, J.M. Prausnitz, *AIChE J.*, **1975**, *21*, 1086.
10. J.D. Seader, E.J. Henley, *Separation Process Principles.* (2nd edition), John Wiley & Sons Inc., 2006.
11. U. Domanska, *Ind. Eng. Chem. Res.*, **1987**, *26*, 1153.
12. F. Varanda, M.J.P. de Melo, A.I. Caco, R. Dohrn, F.A. Makrydaki, E.C. Voutsas, D.P. Tassios, I.M. Marrucho, *Ind. Eng. Chem. Res.*, **2006**, *45*, 6368.
13. F.E. Anderson, J.M. Prausnitz, *AIChE J.*, **1986**, *32(8)*, 1321.
14. J. Munck, S. Skjold-Jørgensen, P. Rasmussen, *Chem. Eng. Sci.*, **1988**, *43*, 2661.
15. B. Sander, P. Rasmussen, Aa. Fredenslund, *Chem. Eng. Sci.*, **1986**, *41*, 1197.
16. R.V. Orye, J.M. Prausnitz, *Ind. Eng. Chem.*, **1965**, *57(5)*, 18.
17. D.P. Tassios, *AIChE J.*, **1971**, *17(6)*, 1367.
18. K. Wong, C.A. Eckert, *Ind. Eng. Chem. Fundam.*, **1971**, *10(1)*, 23.
19. D.P. Tassios, *Applied Chemical Engineering Thermodynamics.* Springer-Verlag, 1993.
20. A. Vetere, *Fluid Phase Equilib.*, **1994**, *99*, 63.
21. A. Vetere, *Fluid Phase Equilib.*, **2000**, *173*, 57.
22. A. Vetere, *Fluid Phase Equilib.*, **2004**, *218*, 33.
23. S.O. Jonsdottir, K. Rasmussen, Aa. Fredenslund, *Fluid Phase Equilib.*, **1994**, *100*, 121.
24. S.O. Jonsdottir, Theoretical determination of UNIQUAC interaction parameters. PhD Thesis, Department of Chemical Engineering, Technical University of Denmark, 1995.
25. S.O. Jonsdottir, R.A. Klein, K. Rasmussen, *Fluid Phase Equilib.*, **1996**, *115*, 59.
26. A.K. Sum, S.I. Sandler, *Ind. Eng. Chem. Res.*, **1999**, *38*, 2849.
27. S.I. Sandler, *Chemical and Engineering Thermodynamics.* (3rd edition), John Wiley & Sons Inc., 1999.
28. S. Kemeny, P. Rasmussen, *Fluid Phase Equilib.*, **1981**, *7*, 197.

29. J. Fischer, *Fluid Phase Equilib.*, **1983**, *10*, 1.
30. S.I. Sandler, K.-H. Lee, *Fluid Phase Equilib.*, **1986**, *30*, 135.
31. B.E. Poling, J.M. Prausnitz, J.P. O'Connell, *The Properties of Gases and Liquids.* (5th edition), McGraw-Hill, 2001.
32. Van Krevelen, *Properties of Polymers: Their correlation with chemical structure; their numerical estimation and prediction from additive group contributions.* Elsevier, 1990.
33. E.L. Derr, C.H. Deal, *Instrum. Chem. Eng. Symp. Ser.*, **1969**, *32(3)*, 40.
34. K. Tochigi, D. Tiegs, J. Gmehling, K. Kojima, *J. Chem. Eng. Jpn*, **1990**, *23*, 453.
35. K. Tochigi, *Fluid Phase Equilib.*, **1998**, *144(1–2)*, 59.
36. R.P. Danner, P.A. Gupte, *Fluid Phase Equilib.*, **1986**, *29*, 415.
37. J. Gmehling, D. Tiegs, U. Knipp, *Fluid Phase Equilib.*, **1990**, *54*, 147.
38. H.K. Hansen, P. Rasmussen, Aa. Fredenslund, M. Schiller, J. Gmehling, *Ind. Eng. Chem. Res.*, **1991**, *30*, 2352.
39. T. Magnussen, P. Rasmussen, Aa. Fredenslund, *Ind. Eng. Chem. Process Des. Dev.*, **1981**, *26*, 2274.
40. H.K. Hansen, B. Coto, B. Kuhlmann,UNIFAC with linearly temperature-dependent group-interaction parameters. IVC-SEP Internal Report SEP 9212, Institut for Kemiteknik, Technical University of Denmark, 1992.
41. B.L. Larsen, P. Rasmussen, Aa. Fredenslund, *Ind. Eng. Chem. Res.*, **1987**, *26*, 2274.
42. U. Weidlich, J. Gmehling, *Ind. Eng. Chem. Res.*, **1987**, *26*, 1372.
43. J. Gmehling, K. Fischer, J. Li, M. Schiller, *Pure Appl. Chem.*, **1993**, *65(5)*, 919.
44. J. Gmehling, R. Wittig, J. Lohmann, R. Joh, *Ind. Eng. Chem. Res.*, **2002**, *41*, 1678.
45. J. Gmehling, *Fluid Phase Equilib.*, **2003**, *210*, 161.
46. F. Chen, J. Holten-Andersen, H. Tyle, *Chemosphere*, **1993**, *26(7)*, 1325.
47. H.H. Hooper, S. Michel, J.M. Prausnitz, *Ind. Eng. Chem. Res.*, **1988**, *27*, 2182.
48. J.W. Kang, J. Abildskov, R. Gani, J. Cobas, *Ind. Eng. Chem. Res.*, **2002**, *41*, 3260.
49. Aa. Fredenslund, P. Rasmussen,From UNIFAC to SUPERFAC – And Back? SEP 8419 (Technical Report), Institut for Kemiteknik, Technical University of Denmark, 1984.
50. J. Abildskov, G.M. Kontogeorgis, R. Gani, Models for liquid phase activity coefficients – UNIFAC. In: G.M. Kontogeorgis, R. Gani, *Computer-Aided Property Estimation for Process and Product Design.* Elsevier, 2004.
51. I. Kikic, P. Alessi, P. Rasmussen, Aa. Fredenslund, *Can. J. Chem. Eng.*, **1980**, *58*, 253.
52. G.M. Kontogeorgis, Ph. Coutsikos, D.P. Tassios, Aa. Fredenslund, *Fluid Phase Equilib.*, **1994**, *92*, 35.
53. G. Soave, *Chem. Eng. Sci.*, **1972**, 1197.
54. S.O. Derawi, G.M. Kontogeorgis, E.H. Stenby, *Ind. Eng. Chem. Res.*, **2001**, *40*, 434.
55. S.O. Derawi, G.M. Kontogeorgis, E.H. Stenby, T. Haugum, A.O. Fredheim, *J. Chem. Eng. Data*, **2002**, *47(2)*, 169.
56. A. Li, W.J. Doucette, A.W. Andren, *Chemosphere*, **1994**, *29(4)*, 657.
57. J. Gmehling, T. Anderson, J.M. Prausnitz, *Ind. Eng. Chem. Fundam.*, **1978**, *17(4)*, 269.
58. R. Mahmoud, R. Solimando, M. Rogalski, *Fluid Phase Equilib.*, **1998**, *148*, 139.
59. J. Abildskov, J.P. O'Connell, *Ind. Eng. Chem. Res.*, **2003**, *42*, 5622.
60. J. Abildskov, J.P. O'Connell, *Fluid Phase Equilib.*, **2005**, *228–229*, 395.
61. P. Kolar, J.-W. Shen, A. Tsuboi, T. Ishikawa, *Fluid Phase Equilib.*, **2002**, *194–197*, 771.
62. T.C. Franck, J.R. Downey, S.K. Gupta, *Chem. Eng. Prog.*, **1999**, 41.
63. T. Jensen, Aa. Fredenslund, P. Rasmussen, *Ind. Eng. Chem. Fundam.*, **1981**, *20*, 239.
64. J.T. Suarez, C. Torres-Marchal, P. Rasmussen, *Chem. Eng. Sci.*, **1989**, *44(3)*, 782.
65. R.B. Gupta, R.A. Heidemann, *AIChE J.*, **1990**, *36*, 333.
66. U. Tegtmeier, K. Misselhorn, *Chem. Ing. Tech.*, **1981**, *53*, 542.
67. S. Li, J.S. Paik, *Trans. ASAE*, **1996**, *39(3)*, 1013.
68. E.A. Brignole, S. Bottini, R. Gani, *Fluid Phase Equilib.*, **1986**, *29*, 125.
69. V. Ruzicka, Aa. Fredenslund, P. Rasmussen, *Ind. Eng. Chem. Process Des. Dev.*, **1983**, *17*, 333.
70. D.T. Wu, *Fluid Phase Equilib.*, **1986**, *30*, 149.

71. W. Cao, K. Knudsen, Aa. Fredendslund, P. Rasmussen, *Ind. Eng. Chem. Res.*, **1993**, *32*, 2088.
72. I. Hatzioannidis, E.C. Voutsas, E. Lois, D.P. Tassios, *J. Chem. Eng. Data*, **1998**, *43*, 386.
73. H.O. Paksoy, S. Ornektekin, B. Balcl, Y. Demirel, *Thermochim. Acta*, **1996**, *287*, 235.
74. S. Ornektekin, H.O. Paksoy, Y. Demirel, *Thermochim. Acta*, **1996**, *287*, 251.
75. E.C. Voutsas, D.P. Tassios, *Ind. Eng. Chem. Res.*, **1996**, *35*, 1438.
76. E.C. Voutsas, D.P. Tassios, *Ind. Eng. Chem. Res.*, **1997**, *36*, 4965.
77. S. Zhang, T. Hiaki, K. Kojima, *Fluid Phase Equilib.*, **2002**, *198*, 15.
78. H. Kuramochi, H. Noritomi, D. Hoshino, S. Kao, K. Nagahama, *Fluid Phase Equilib.*, **1998**, *144*, 87.
79. A. Fredenslund, J. Gmehling, M.L. Michelsen, P. Rasmussen, J.M. Prausnitz, *Ind. Eng. Chem. Process Des. Dev.*, **1977**, *16*(*4*), 450.
80. A.F.M. Barton, *CRC Handbook of Polymer-Liquid Interaction Parameters and Solubility Parameters*. CRC Press, 1990.
81. C.M. Hansen, *Hansen Solubility Parameters: A User's Handbook*. CRC Press, 2000.
82. H. Wen, H.S. Elbro, P. Alessi, *Data Collection on Polymer Containing Solutions and Blends*. Chemistry Data Series, DECHEMA, Frankfurt, 1991.
83. H.S. Elbro, Aa. Fredenslund, P. Rasmussen, *Macromolecules*, **1990**, *23*, 4707.
84. H.S. Elbro, Phase equilibria of polymer solutions – with special emphasis on free volumes. PhD Thesis, Department of Chemical Engineering, Technical University of Denmark, 1992.
85. G.M. Kontogeorgis, Aa. Fredenslund, D.P. Tassios, *Ind. Eng. Chem. Res.*, **1993**, *32*, 362.
86. T. Oishi, J.M. Prausnitz, *Ind. Eng. Chem. Process Des. Dev.*, **1978**, *17*(*3*), 333.
87. Th. Lindvig, M.L. Michelsen, G.M. Kontogeorgis, *Fluid Phase Equilib.*, **2002**, *203*, 247.
88. Th. Lindvig, I.G. Economou, R.P. Danner, M.L. Michelsen, G.M. Kontogeorgis, *Fluid Phase Equilib.*, **2004**, *220*, 11.
89. G.D. Pappa, E.C. Voutsas, D.P. Tassios, *Ind. Eng. Chem. Res.*, **1999**, *38*, 4975.
90. G.D. Pappa, G.M. Kontogeorgis, D.P. Tassios, *Ind. Eng. Chem. Res.*, **1997**, *36*, 5461.
91. G.M. Kontogeorgis, A. Saraiva, Aa. Fredenslund, D.P. Tassios, *Ind. Eng. Chem. Res.*, **1995**, *34*, 1823.
92. V.I. Harismiadis, A.R.D. van Bergen, A. Saraiva, G.M. Kontogeorgis, Aa. Fredenslund, D.P. Tassios, *AIChE J.*, **1996**, *42*, 3170.
93. V.I. Harismiadis, D.P. Tassios, *Ind. Eng. Chem. Res.*, **1996**, *35*, 4667.
94. I. Kouskoumvekaki, R. Giesen, M.L. Michelsen, G.M. Kontogeorgis, *Ind. Eng. Chem. Res.*, **2002**, *41*(*19*), 4848.
95. G.D. Pappa, E.C. Voutsas, D.P. Tassios, *Ind. Eng. Chem. Res.*, **2001**, *40*(*21*), 4654.
96. E.C. Voutsas, G.D. Pappa, C.J. Boukouvalas, K. Magoulas, D.P. Tassios, *Ind. Eng. Chem. Res.*, **2004**, *43*, 1312.
97. G. Bogdanic, J. Vidal, *Fluid Phase Equilib.*, **2000**, *173*, 241.
98. E.A. Guggenheim, *Mixtures*. Clarendon Press, 1952.
99. G. Maurer, J.M. Prausnitz, *Fluid Phase Equilib.*, **1978**, *2*, 91.
100. J.R. Elliott, C.T. Lira, *Introductory Chemical Engineering Thermodynamics*. Prentice Hall International, 1999.
101. A. Vetere, *Fluid Phase Equilib.*, **1993**, *91*, 265.
102. M.J. Huron, J. Vidal, *Fluid Phase Equilib.*, **1979**, *3*, 255.
103. J.A.P. Coutinho, *Ind. Eng. Chem. Res.*, **1998**, *37*, 4870.
104. J.A.P. Coutinho, *Fluid Phase Equilib.*, **1999**, *158–160*, 447.
105. J.A.P. Coutinho, J.-L. Daridon, *Energy Fuels*, **2001**, *15*(*6*), 1454.
106. J.A.P. Coutinho, J. Pauly, J.-L. Daridon, *Braz. J. Chem. Eng.*, **2001**, *18*(*4*), 1.
107. J.A.P. Coutinho, F. Mirante, J. Pauly, *Fluid Phase Equilib.*, **2006**, *247*, 8.
108. J.A.P. Coutinho, J. Pauly, J.-L. Daridon, Modelling phase equilibria in systems with organic solid solutions. In: G.M. Kontogeorgis, R. Gani, *Computer-Aided Property Estimation for Process and Product Design*. Elsevier, 2004.
109. P. Debye, E. Hückel, *Phys. Z.*, **1923**, *24*, 185.
110. K. Thomsen, P. Rasmussen, *Chem. Eng. Sci.*, **1999**, *54*, 1787.
111. K. Thomsen, *Pure Appl. Chem.*, **2005**, *77*, 531.
112. S. Pereda, K. Thomsen, P. Rasmussen, *Chem. Eng. Sci.*, **2000**, *55*, 2663.

113. S.G. Christensen, K. Thomsen, *Ind. Eng. Chem. Res.*, **2003**, *42*, 4260.
114. M. Iliuta, K. Thomsen, P. Rasmussen, *Chem. Eng. Sci.*, **2000**, *55*, 2673.
115. K. Thomsen, M. Iliuta, P. Rasmussen, *Chem. Eng. Sci.*, **2004**, *59*, 3631.
116. M. Iliuta, K. Thomsen, P. Rasmussen, *AIChE J.*, **2002**, *48*, 2664.
117. A.V. Garcia, K. Thomsen, E.H. Stenby, *Geothermics*, **2005**, *34*, 61.
118. C.C. Chen, H.I. Britt, J.F. Boston, L.B. Evans, *AIChE J.*, **1982**, *28*, 588.
119. C.C. Chen, L.B. Evans, *AIChE J.*, **1986**, *32*, 444.
120. C.C. Chen, Y. Song, *AIChE J.*, **2004**, *50*, 1928.

6

The EoS/G^E Mixing Rules for Cubic Equations of State

6.1 General

The range of applicability of cubic equations of state (EoS) (using the standard van der Waals one-fluid mixing rules) and g^E models (especially the local composition ones) is illustrated in Table 6.1 and schematically also in Figure 6.1. In a somewhat simplified way, Figure 6.1 illustrates the advantages which may be obtained by combining the strengths of both approaches (i.e. the cubic EoS and the activity coefficient models) and thus having a single model suitable for phase equilibria (at least VLE) of polar and non-polar mixtures and at both low and high pressures. This combination of EoS and g^E models has been possible via the so-called EoS/G^E models, which are essentially mixing rules for, typically, the energy parameter of cubic EoS. These mixing rules permit the incorporation of an expression for the excess Gibbs energy g^E (i.e. an activity coefficient model) inside the EoS, permitting thus a cubic EoS to be applied to polar compounds at high pressures as well.

The starting point for deriving *most* (but not all) EoS/G^E models is the equality of the excess Gibbs energies from an EoS and from a (successful) explicit activity coefficient model at a suitable reference pressure, P:

$$\left(\frac{g^E}{RT}\right)_P^{EoS} = \left(\frac{g^E}{RT}\right)_P^{\text{model},*} \tag{6.1}$$

where a suitable reference pressure P can be, for example, the infinite pressure (Huron–Vidal, Wong–Sandler models) or the zero pressure (e.g. MHV2, the PSRK models), while the superscript * refers to the specific activity coefficient model, e.g. NRTL. The right-hand part of Equation (6.1) is the g^E expression of an explicit activity coefficient model, e.g. Wilson, NRTL or UNIQUAC (see Chapters 4 and 5), whereas the g^E of the EoS is obtained from classical thermodynamics when the fugacity expression is known:

$$\frac{g^E}{RT} = \ln\varphi - \sum_i x_i \ln\varphi_i = \sum_i x_i \ln\gamma_i \tag{6.2a}$$

Thermodynamic Models for Industrial Applications: From Classical and Advanced Mixing Rules to Association Theories
Georgios M. Kontogeorgis and Georgios K. Folas

Table 6.1 *Range of applicability of cubic EoS (using the van der Waals one fluid mixing rules) and activity coefficient models*

Application	Cubic EoS	g^E (activity coefficient) models
Low pressures	×	×
High pressures	×	
'Simple' mixtures (hydrocarbons, gases)	×	×
Polar mixtures		×
Many more properties other than phase equilibria	×	
Predictive calculations	Seldom	× (using the GC models, e.g. UNIFAC)

$$RT \ln\gamma_i = \ln\frac{\hat{\varphi}_i}{\varphi_i} = \left(\frac{\partial n g^E}{\partial n_i}\right)_{T,P,n_{j\neq i}} \tag{6.2b}$$

where φ, φ_i are the fugacity coefficients of the mixture and of the pure compound, respectively, and their derivation from an EoS requires no mixing rules; $\hat{\varphi}_i$ is the fugacity coefficient of component i in the mixture. Thus, Equation (6.1) permits, under some assumptions, a solution to be obtained for the energy parameter of the mixture, a_{mix}. This results in a mixing rule for the EoS parameter which includes the activity coefficient model, e.g. Wilson, NRTL or UNIQUAC. The g^E expressions for three of the most well-known cubic EoS are shown in Table 6.2.

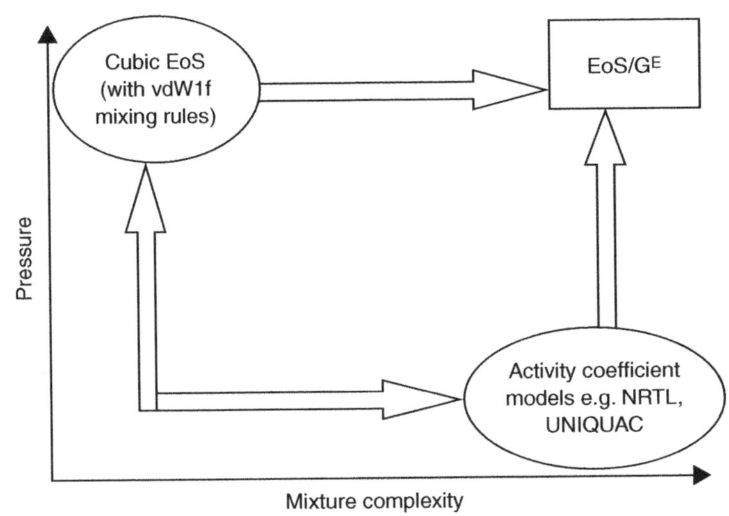

Figure 6.1 *Cubic EoS' major strength is their application at high pressures while that of explicit activity coefficient models (G^E) is their application to polar and non-polar mixtures. EoS/G^E models (mixing rules) combine the advantages of the two approaches, thus being able to describe satisfactorily both non-polar and polar system phase equilibria over the entire pressure range*

Table 6.2 The g^E expressions for vdW, SRK and PR, as well as the energy parameter mixing rule according to the Huron–Vidal approach (infinite reference pressure). The linear mixing rule is used for the co-volume parameter ($b = \sum_i x_i b_i$). $g^{E,\infty}$ is the excess Gibbs energy at infinite pressure

Theory	$\frac{g^E}{RT}$	a_{mix} (mixing rule for the energy parameter at infinite reference pressure)
van der Waals (vdW)	$\left[\sum_i x_i \ln\left(\frac{V_i-b_i}{V-b}\right)\right] + \frac{P}{RT}\left(V-\sum_i x_i V_i\right) + \left[\frac{1}{RT}\left(\sum_i x_i \frac{a_i}{V_i} - \frac{a}{V}\right)\right]$	$\frac{a}{b} = \sum_i x_i \frac{a_i}{b_i} - g^{E,\infty}$
Soave–Redlich–Kwong (SRK)	$\left[\sum_i x_i \ln\left(\frac{V_i-b_i}{V-b}\right)\right] + \frac{P}{RT}\left(V-\sum_i x_i V_i\right) + \left\{\frac{1}{RT}\left[\sum_i x_i \frac{a_i}{b_i}\ln\left(\frac{V_i+b_i}{V_i}\right) - \frac{a}{b}\ln\left(\frac{V+b}{V}\right)\right]\right\}$	$\frac{a}{b} = \sum_i x_i \frac{a_i}{b_i} - \frac{g^{E,\infty}}{\ln 2}$
Peng–Robinson (PR)	$\left[\sum_i x_i \ln\left(\frac{V_i-b_i}{V-b}\right)\right] + \frac{P}{RT}\left(V-\sum_i x_i V_i\right) + \left\{\frac{1}{RT2\sqrt{2}}\left[\sum_i x_i \frac{a_i}{b_i}\ln\left(\frac{V_i+(1+\sqrt{2})b_i}{V_i+(1-\sqrt{2})b_i}\right) - \frac{a}{b}\ln\left(\frac{V+(1+\sqrt{2})b}{V+(1-\sqrt{2})b}\right)\right]\right\}$	$\frac{a}{b} = \sum_i x_i \frac{a_i}{b_i} - \frac{g^{E,\infty}}{\lambda}$, $\lambda = \frac{1}{2\sqrt{2}}\ln\left(\frac{2+\sqrt{2}}{2-\sqrt{2}}\right)$

6.2 The infinite pressure limit (the Huron–Vidal mixing rule)

The first systematic successful effort in developing an EoS/G^E model is that of Huron and Vidal[1,2]. The basic assumption of the method is the use of the infinite pressure as the reference pressure. The linear mixing rule is used for the co-volume parameter ($b = \sum_i x_i b_i$). Huron and Vidal[2] used the SRK EoS but other EoS could be used as well. Table 6.2 provides the expressions for the mixing rule of the energy parameter for three cubic EoS. Appendix 6.A illustrates the derivation of the mixing rule for SRK. As shown, the 'combinatorial' and excess volume terms of the cubic EoS disappear at infinite pressure (thus $V \to b$), which simplifies the mathematics and the resulting mixing rule is explicit. The following observations can be made:

1. Although any activity coefficient model could, in principle, be used, it is clear from the derivation that a 'solely energetic' model should be preferred, i.e. a model which does not have an explicit combinatorial/free-volume term. For example, the NRTL equation (as Huron and Vidal did in their 1979 article)[2] or the residual term of UNIQUAC or UNIFAC.[3–9] Excellent correlation results are obtained.
2. The excess Gibbs energy at high pressures is, in general, different from the value at low pressures where the parameters of local composition models are typically estimated. Thus, the interaction parameters of the g^E expression must be re-estimated using an EoS with the Huron–Vidal mixing rule. As shown in Figures 6.2 and 6.3, excellent correlation is achieved for binary mixtures, and often satisfactory VLE (and VLLE) predictions are also obtained for multicomponent mixtures containing polar components and at high pressures.[10–12] For polar mixtures, the Huron–Vidal mixing rule offers much flexibility and improved correlation over the vdW1f mixing rules for mixtures containing polar and hydrogen bonding compounds.

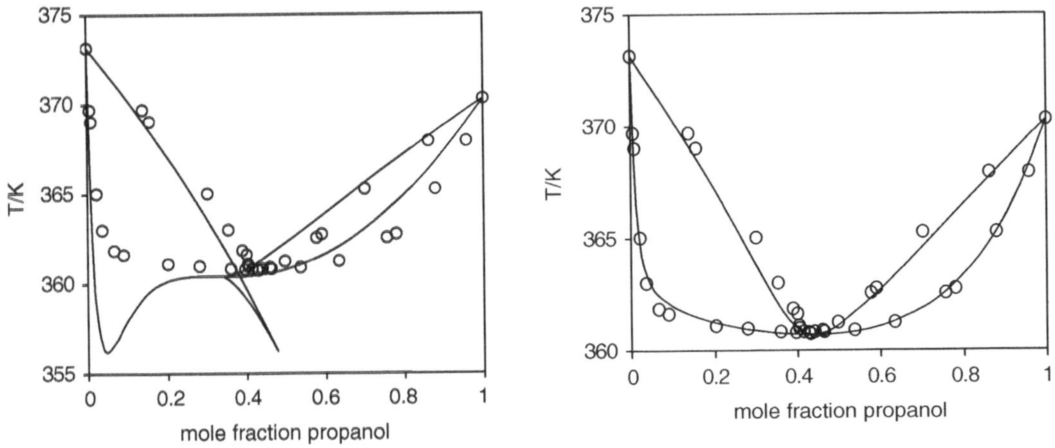

Figure 6.2 *VLE (P = 1 bar) for propanol–water with SRK using the vdW1f mixing rules with $k_{12} = -0.09$ (left) and the Huron–Vidal mixing rule (right). Experimental data are from Udovenko et al., Zh. Fiz. Khim.,* ***1972****, 46, 218*

Figures 6.2 and 6.3 illustrate that SRK can indeed satisfactorily describe polar mixtures when the Huron–Vidal mixing rule is used.

3. Other positive features of the Huron–Vidal mixing rule include the better correlation (compared to the vdW1f mixing rules) of water–hydrocarbon LLE[13] and the possibility to introduce salinity in a simple practical way suitable for calculations involving gas hydrate inhibitors.[14]
4. A useful feature of Huron–Vidal mixing rules is that, by using a special version of NRTL (see Chapter 5, Appendix 5.B, Equations (5.40) and (5.44)), SRK can result in the classical vdW1f mixing rules, which are useful for mixtures with hydrocarbons.

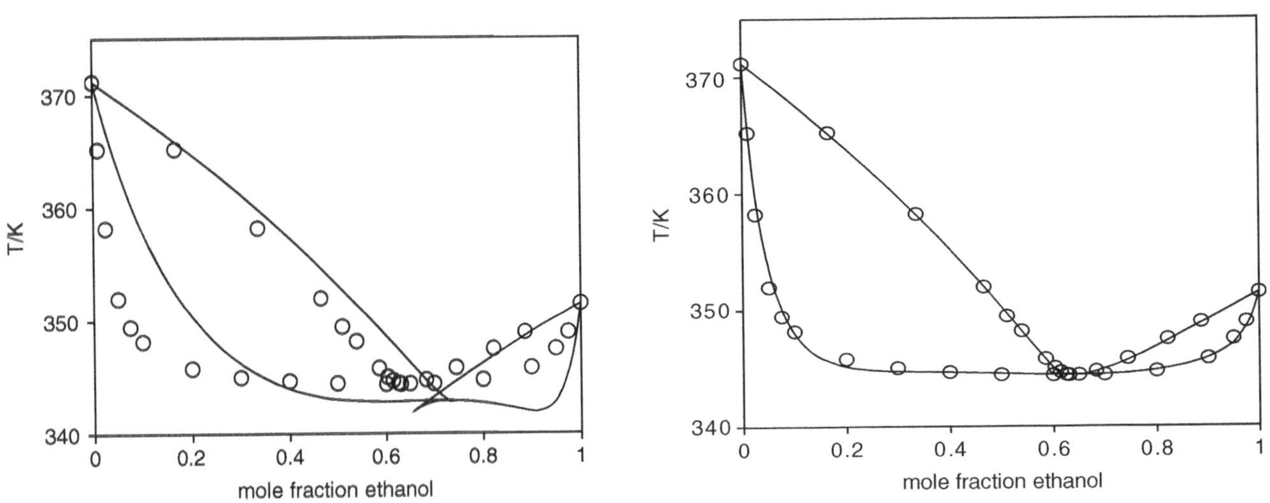

Figure 6.3 *VLE (P = 1 bar) for ethanol–heptane with SRK using the vdW1f mixing rules with $k_{12} = 0.085$ (left) and the Huron–Vidal mixing rule (right). Experimental data are from van Ness et al., J. Chem. Eng. Data,* ***1967****, 12, 346*

5. A limitation of the Huron–Vidal mixing rule is that it does not permit use of the large collections of interaction parameters of models such as Wilson, UNIQUAC or UNIFAC, which are based on low-pressure VLE data, e.g. the parameters available in the DECHEMA data collections or in the UNIFAC tables. An alternative solution is needed, and this is based on the Michelsen zero reference pressure mixing rules (within the concept originally introduced by Mollerup[15]), discussed in the next section.

6.3 The zero reference pressure limit (the Michelsen approach)

The zero reference pressure approach, originally introduced by Mollerup,[15] and then worked out further and applied to practical systems by Michelsen and co-workers[17,18], is a popular choice, as it permits direct use of interaction parameter tables, e.g. DECHEMA or UNIFAC. It is, moreover, consistent with using activity coefficient models with both combinatorial and residual terms like UNIFAC and UNIQUAC. There are some limitations (both theoretical and practical) which will be explained later. First, we will present the mixing rule in its various forms. It can be shown that by setting $P = 0$ in any vdW-type cubic EoS (e.g. vdW, SRK, PR), the 'exact' mixing rule obtained at zero pressure, when Equation (6.1) is employed, is:

$$\left(\frac{g^E}{RT}\right)_0^{\text{model},*} + \sum_i x_i \ln\left(\frac{b}{b_i}\right) = q^e(\alpha) - \sum x_i q_i^e(\alpha_i) \tag{6.3}$$

where $\alpha = \dfrac{a}{bRT}$.

The $q^e(\alpha)$ function depends on the EoS used and is defined only for values $\alpha > \alpha_{\text{lim}}$. The $q^e(\alpha)$ expressions and α_{lim} values for two cubic EoS are given in Table 6.3.

In practice this condition ($\alpha > \alpha_{\text{lim}}$) means that the $q^e(\alpha)$ function is defined for reduced temperature T_r values up to about 0.9, i.e. the $q^e(\alpha)$ function is not defined for most gas-containing mixtures like CO_2 or N_2 with alkanes.

Equation (6.3) is an implicit mixing rule for the energy parameter, which means that an iterative procedure is needed for calculating the energy parameter. In order to obtain an explicit (or simpler) mixing rule and to address the limitation introduced by the presence of α_{lim}, a number of approximate zero reference pressure mixing rules have been proposed. Most of them are based on a linear or a quadratic approximation of the $q(\alpha)$ function, and several of these mixing rules have been applied to both SRK and PR. The most well-known approximate zero reference pressure mixing rules are presented in Table 6.4.

Michelsen[16,118] illustrated originally the validity of the 'exact' approach, Equation (6.3), by combining the SRK EoS with the Wilson activity coefficient equation. Dahl and Michelsen[17] proposed the MHV1 and MHV2 mixing rules (modified Huron–Vidal first-order and second-order approximations) and applied them to SRK combined with the modified UNIFAC of Larsen *et al.*[20]. Gmehling and co-workers[21–24] proposed the

Table 6.3 *The $q^e(\alpha)$ expressions (Equation (6.3)) and the α_{lim} values in the Michelsen zero reference pressure mixing rule for two cubic EoS. u_0 is the value of V/b at zero pressure and is essentially the volume solution of a cubic EoS at zero pressure (the cubic EoS becomes quadratic at zero pressure)*

EoS	$q^e(\alpha)$, with $\alpha = \dfrac{a}{bRT}$	α_{lim}
SRK	$q(\alpha) = -1 - \ln(u_0 - 1) - \alpha \ln\left(\dfrac{u_0 + 1}{u_0}\right)$	$3 + 2\sqrt{2}$
PR	$q(\alpha) = -1 - \ln(u_0 - 1) - \dfrac{\alpha}{2\sqrt{2}} \ln\left(\dfrac{u_0 + (1+\sqrt{2})}{u_0 + (1-\sqrt{2})}\right)$	$4 + 2\sqrt{2}$

Table 6.4 *Well-known approximate zero reference pressure mixing rules. The linear mixing rule for the b parameter is used in all cases.* ($b=\sum_i x_i b_i$)

$q(\alpha)$	α_{mix} with $\alpha = \frac{a}{bRT}$	Name of the model
$q(\alpha) \approx q_0 + q_1\alpha$	$\alpha = \frac{1}{q_1}\left[\frac{g^{E,*}}{RT} + \sum_i x_i \ln\left(\frac{b}{b_i}\right)\right] + \sum_i x_i\alpha_i$ $q_1 = -0.593,\ -0.64663$ (SRK) – MHV1 and PSRK, respectively $q_1 = -0.53$ (PR) $q_1 = -0.85$ (vdW)	MHV1 or PSRK
$q(\alpha) \approx q_0 + q_1\alpha + q_2\alpha^2$	$q_1\left(\alpha - \sum_i x_i\alpha_i\right) + q_2\left(\alpha^2 - \sum_i x_i\alpha_i^2\right) = \left(\frac{G^E}{RT}\right)^* + \sum_i x_i \ln\left(\frac{b}{b_i}\right)$ $q_1 = -0.4783,\ q_2 = -0.0047$ for SRK (for α values between 10 and 13) $q_1 = -0.4347,\ q_2 = -0.003654$ for PR[19]	MHV2
$q = 3.365 - \frac{\alpha \ln 2 + \sqrt{(\alpha \ln 2)^2 + 69}}{2}$		Soave[7]

MHV1 = modified Huron–Vidal first order, MHV2 = modified Huron–Vidal second order, PSRK = predictive Soave–Redlich–Kwong.

PSRK model (predictive SRK) which is, essentially, the MHV1 combining rule with a slightly different q_1 value (see Table 6.4). In PSRK, SRK is combined with the original or the modified Dortmund UNIFAC versions. These mixing rules are generic, thus other cubic EoS and activity coefficient models can, in principle, be used. In all cases the linear mixing rule is used for the co-volume parameter ($b=\sum_i x_i b_i$), but this could be relaxed.

A number of observations can be made:

1. MHV1 and PSRK are the simplest mixing rules and they are explicit.
2. MHV2 provides a better match than MHV1/PSRK of the g^E model that the EoS is combined with and thus a better reproduction of the low-pressure VLE data, but it does not yield an explicit mixing rule.
3. Mathematically, the MHV1 and the Huron–Vidal mixing rules are rather similar, with the major difference being the presence in MHV1 and PSRK models of a Flory–Huggins term based on the co-volumes ($=\sum_i x_i \ln(b/b_i)$).
4. As shown first by Mollerup[15] and also by Orbey and Sandler[25–27], the MHV1 and PSRK models can be somewhat 'mechanistically' derived using the assumption of constant volume packing fraction b/v for pure compounds and mixtures (see also, later, Table 6.7 and Problem 3 on the companion website at www.wiley.com/go/Kontogeorgis).
5. For MHV2 and PSRK models, there are extensive parameters also available for gas-containing mixtures,[18,24] with PSRK having the most extensive parameter table. In the case of MHV2, the energy parameters for gas-containing mixtures are linearly dependent on temperature, although modified UNIFAC is used as the base model.

There are only a few measured systems (available VLE data) with polar compounds at high pressures where gases are *not* present. Such mixtures can be used for testing the EoS/G^E mixing rules, as no parameter

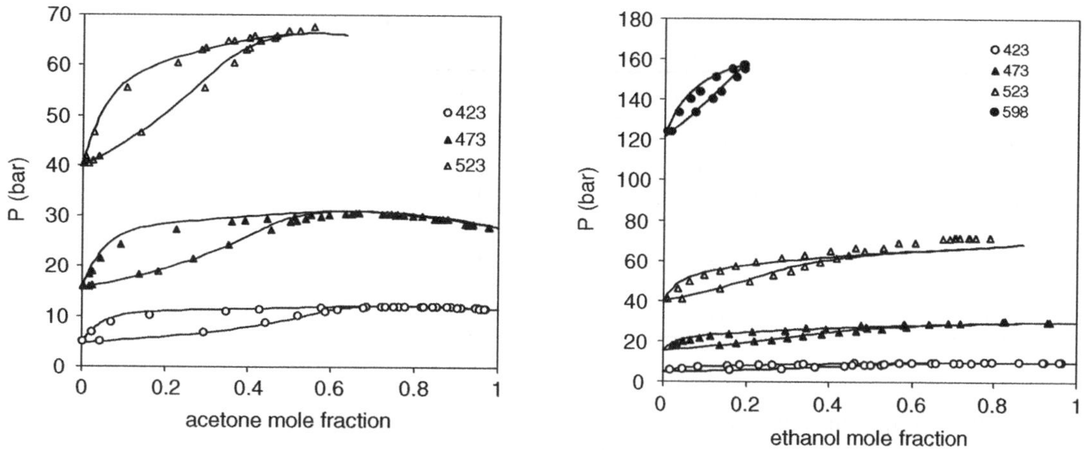

Figure 6.4 *Left: VLE (Pxy) diagram for acetone–water with SRK using the MHV2 mixing rule. Experimental data from Griswold and Wong, AIChE Symp. Ser.,* ***1952****, 48, 18. Right: VLE (Pxy) diagram for ethanol–water with SRK using the MHV2 mixing rule. Experimental data from Barr-David and Dodge, J. Chem. Eng. Data,* ***1959****, 4, 107*

estimation is involved (beyond those already available in the activity coefficient models, which are obtained from low-pressure data). Two of the most extensively studied mixtures are shown in Figure 6.4. Excellent results are obtained with MHV2 for acetone–water, ethanol–water and other similar systems over extended temperature and especially pressure ranges. This illustrates the great success of the zero reference pressure EoS/G^E models. They satisfactorily extend the use of UNIFAC to high pressures or, alternatively, they extend the applicability of cubic EoS to polar systems. The results shown in Figure 6.4 with MHV2 are obtained using the modified UNIFAC[20] parameter table which is based exclusively on low-pressure data.

6.4 Successes and limitations of zero reference pressure models

The great success and wide applicability of zero reference pressure models can be largely attributed to two reasons:

1. Successful representation of polar high- (and low-)pressure VLE, often in a predictive manner when UNIFAC is used as the incorporated activity coefficient model, e.g. as illustrated in Figure 6.4 and in Table 6.5. In this way, UNIFAC parameters which are estimated from low-pressure data can be used in a consistent way for high-pressure VLE calculations.
2. The approximate zero reference pressure models can be extended to gas-containing systems and both the MHV2 and PSRK models (mixing rules) have extensive parameter tables for many gases.[18,24] MHV2 has also been applied to multiphase equilibria (VLLE, LLLE) of difficult mixtures such as CO_2–ethanol–water and water–butanol–2-butane, with some success considering the complexity of such mixtures.[36]

Zero reference pressure EoS/G^E models have a number of important limitations. The first two, reported hereafter, are common for other EoS/G^E models, i.e. also for the Huron–Vidal mixing rule:

1. Following their derivation from Equation (6.1), the performance of cubic EoS using EoS/G^E mixing rules should not be expected to be different or better at low pressures than the g^E model they are combined with. Thus, their 'strongest' part (the local composition activity coefficient model) defines their limit of success at low pressures.

Table 6.5 *Average percentage deviations between experimental and predicted bubble point pressures with various EoS/G^E models. Modified after Boukouvalas et al.*[29]

System	*P* range (bar)	MHV2	PSRK	LCVM
Methanol–benzene	3–58	2.7	3.7	4.3
Ethanol–water	6–186	2.3	1.6	1.1
Acetone–CHI_3	0–1	4.7	1.2	1.1
Ethane–*n*-C_{12}	3–56	14	12	3.2
Ethane–*n*-C_{20}	2–168	34	28	3.6
Ethane–*n*-C_{44}	4–31	83	163	7.1
Acetone–water	1–63	2.4	3.0	2.5

2. The EoS/G^E mixing rules discussed so far do not satisfy the theoretically justified mixing rule for the second virial coefficient (see Chapter 2, Equation (2.18)).

In addition, the approximate (but not the 'exact', Equation (6.3)) zero reference pressure models, e.g. MHV1, PSRK and MHV2, suffer from two (not entirely connected) limitations:

1. They do not fully reproduce at low pressures the activity coefficient expression (G^E model) they are combined with.[28]
2. They have serious problems in representing VLE for size-asymmetric systems, as shown in Table 6.5 and Figure 6.5, e.g. mixtures of CO_2 or ethane with heavy hydrocarbons.

Efforts to address these limitations have resulted in some improved EoS/G^E mixing rules which are discussed in Section 6.6. We will first, however, present a somewhat different EoS/G^E mixing rule which has found

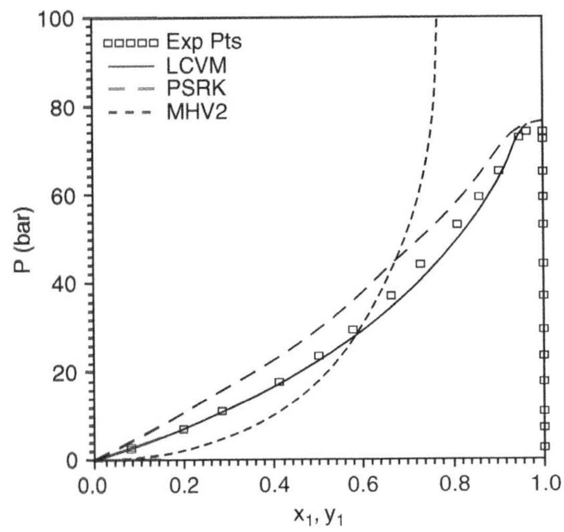

Figure 6.5 *Prediction of VLE for the system ethane–eicosane at 320 K with three EoS/G^E models. Reprinted with permission from Fluid Phase Equilibria, Prediction of vapor-liquid equilibrium with the LCVM model: a linear combination of the Vidal and Michelsen mixing rules coupled with the original UNIFAC by C. Boukouvalas, N. Spiliotis et al., 92, 75 Copyright (1995) Elsevier*

widespread use and which resolves one of the problems of this family of mixing rules – namely, that related to the concentration dependency of the second virial coefficient ($B = \sum_i \sum_j x_i x_j B_{ij}$).

6.5 The Wong–Sandler (WS) mixing rule

Wong and Sandler[30,31] have developed an EoS/G^E mixing rule which has found widespread use. Their model differs from the ones presented so far in the sense that the theoretically correct quadratic composition dependence of the second virial coefficient is satisfied. The mixing rule is derived by equating the excess Helmholtz energy A^E of an activity coefficient model to that of an EoS at infinite pressure. Low-pressure activity coefficient model parameters are found to be useful in this mixing rule because of the insensitivity of A^E to pressure. Unlike the other EoS/G^E models it contains one more binary interaction parameter in the cross second virial coefficient. Like the other mixing rules, it can be applied to any cubic EoS, but it has been proposed and used extensively in conjunction with the PR EoS (especially its modified form by Stryjek and Vera)[32].

For PR, the expression for the mixture energy parameter a is similar to that of MHV1:

$$a = b\left(\sum_i x_i \frac{a_i}{b_i} + \frac{g^{E,*}}{0.623}\right) \tag{6.4}$$

For any cubic EoS, the more general form is:

$$a = b\left(\sum_i x_i \frac{a_i}{b_i} - \frac{A^E_\infty}{\sigma}\right)$$

However, for the co-volume parameter, the following mixing rule is used:

$$b = \frac{\sum_i \sum_j x_i x_j \left(b - \frac{a}{RT}\right)_{ij}}{1 + \frac{A^E_\infty}{RT} - \sum_i x_i \frac{a_i}{b_i RT}} \tag{6.5}$$

The combining rule for the cross second virial coefficient is given by:

$$\left(b - \frac{a}{RT}\right)_{ij} = \frac{\left(b_i - \frac{a_i}{RT}\right) + \left(b_j - \frac{a_j}{RT}\right)}{2}(1 - k_{ij}) \tag{6.6}$$

Various approaches have been proposed for estimating the 'presumably' extra interaction parameter (k_{ij}) introduced in the WS mixing rule:

1. In the original publications, k_{ij} is obtained from fitting experimental VLE data (expressed as G^E–x) at one T and $x = 0.5$. Then, the model can be used to extrapolate to other conditions.
2. Orbey *et al.*[33] proposed a predictive WS model, using UNIFAC as the activity coefficient model, and calculated the cross virial binary interaction parameter k_{ij} using the following procedure. First, for the binary mixture of interest, the two infinite dilution activity coefficients are predicted at 298 K from UNIFAC. Next, these infinite dilution activity coefficients are used to obtain the interaction parameters of the UNIQUAC model. Finally, k_{ij} is calculated by matching the excess Gibbs energy of the mixture calculated from UNIQUAC and from the EoS at the mid-concentration point, $x_i = 0.5$, and $T = 298$ K. The

values of the UNIQUAC parameters and k_{ij} are then used at all other temperatures. Voutsas *et al.*[34] proposed the direct calculation of k_{ij} by matching the excess Gibbs energy of the mixture calculated from UNIFAC and from the EoS at $x_i = 0.5$ and $T = 298$ K.

The results with the WS mixing rules for polar (relatively symmetric) mixtures, e.g. water–acetone and water–propanol, are excellent, at least as good as with the other mixing rules presented previously. Often better results are obtained at high temperatures compared to the MHV2 model, although it is not entirely clear whether this should be attributed to the stronger theoretical foundation of the mixing rule (due to satisfying the correct limit for the concentration dependency of the second virial coefficient). Nevertheless, there are limitations to the WS mixing rule:[35]

1. It has be shown that the WS mixing rule with a composition-independent k_{ij} is not able to match the EoS-obtained g^E expression to that of the activity coefficient model for asymmetric systems.
2. A serious disadvantage of the WS predictive scheme is that its application is limited only to systems containing components that are condensable at the temperature of interest and, consequently, it cannot be used for phase equilibrium predictions for systems containing gases.

In general, the WS mixing rule has not been used extensively for mixtures with gases and unlike other mixing rules (MHV2, PSRK, LCVM) extensive parameter tables for gas-containing mixtures are not available. More applications of the WS and the other 'classical' mixing rules (MHV2, PSRK) and some examples are presented in Section 6.7, following a discussion in the next section of how one of the most serious problems of these models has been addressed: the extension of EoS/G^E mixing rules to size-asymmetric systems (with compounds differing significantly in size). We will present in the next section a general explanation of a framework which justifies most (if not all) of the developments related to the extension of the EoS/G^E mixing rules to size-asymmetric mixtures.

6.6 EoS/G^E approaches suitable for asymmetric mixtures

An inspection of the MHV1-type mixing rules (Table 6.4) shows that in reality, when they are combined with the UNIQUAC/UNIFAC models, as typically done, the resulting models contain two different 'combinatorial' terms. These are the combinatorial term which comes from the activity coefficient model that is used (denoted as * below) and the Flory–Huggins (FH) combinatorial term containing the b parameters originating from the EoS:

$$\alpha = \frac{1}{q_1}\left(\frac{g^{E,*,comb}}{RT} - \frac{g^{E,FH}}{RT} + \frac{g^{E,*,res}}{RT}\right) + \sum_i x_i \alpha_i \tag{6.7}$$

where the FH combinatorial term stemming from the EoS is:

$$\frac{g^{E,FH}}{RT} = \sum_i x_i \ln\left(\frac{b_i}{b}\right) \tag{6.8}$$

The combinatorial term from (the original) UNIFAC is (see Chapter 5):

$$\frac{g^{E,*,comb}}{RT} = \sum_i x_i \ln\frac{\Phi_i}{x_i} - \frac{Z}{2} q \sum_i \vartheta_i \ln\frac{\Phi_i}{\vartheta_i} \tag{6.9}$$

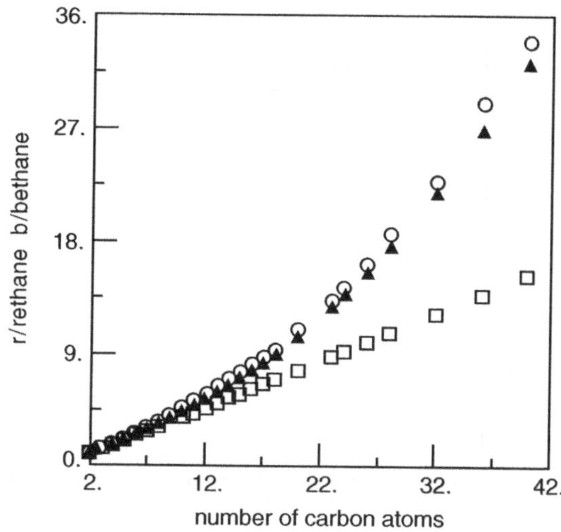

Figure 6.6 The ratios r/r_{ethane} and b/b_{ethane} for n-alkanes as a function of the number of carbon atoms of the n-alkane. The circles represent the calculations based on the co-volume parameters (b) and the squares those based on the vdW volume (r). The filled triangles represent the r/r_i calculations using the method of Li et al.[38] where the vdW volumes (and surface areas) for the alkane groups are essentially fitted to experimental data. Reprinted with permission from Ind. Eng. Chem. Res., Development of a Universal Group Contribution Equation of State. 2. Prediction of Vapor–Liquid Equilibria for Asymmetric Systems by Jens Ahlers and Jürgen Gmehling, 41, 14, 3489–3498 Copyright (2002) American Chemical Society

The last (Staverman–Guggenheim) contribution in Equation (6.9) (the one containing the q parameter) is often small, thus, ignoring this term, the combinatorial term of UNIFAC (or UNIQUAC) is also a FH-type term, using the vdW volume (r) values instead of co-volumes, b, or volumes, V:

$$\frac{g^{E,*,comb}}{RT} = \sum_i x_i \ln \frac{r_i}{r} \tag{6.10}$$

The two combinatorial terms of Equation (6.7) must be of similar magnitude, and indeed they would cancel out if $b_i = r_i$, but as Figure 6.6 illustrates, this is generally not the case. Moreover, the difference increases with asymmetry, when the co-volumes b_i are estimated in the 'usual' way, based on critical temperatures and pressures. The above 'inconsistency' (two combinatorials not being the same) does not create problems in practice for symmetric systems, because the two combinatorials are indeed numerically similar in this case. The 'inconsistency' is not a problem either when the activity coefficient model is fully reproduced at low pressures and the calculations are based on previously estimated parameters (used in the residual term of UNIFAC or UNIQUAC). Problems are, as might be anticipated, observed for size-asymmetric systems, especially as new group parameters have to be estimated for gas-containing systems. For example, problems are observed when the group interaction parameters between CO_2 and CH_2 must be estimated from mixtures of CO_2 with alkanes or varying molecular weight. In these cases, it could be argued that an 'increased' combinatorial difference might create problems in trying to optimize a single group gas/alkane (and similar) group interaction parameter. This problem with the 'increased combinatorial difference effect' appears in the classical MHV1, MHV2 and PSRK models, as discussed by Kontogeorgis and Vlamos[37], see also Figure 6.7.

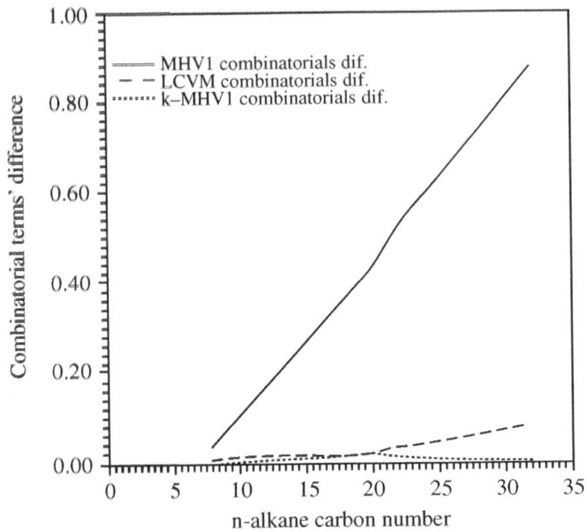

Figure 6.7 *The difference of combinatorial terms for the series of n-hexane/n-alkanes (up to n-C_{32}) with three EoS/G^E mixing rules using the modified UNIFAC of Larsen etal.*[20] *as the activity coefficient model. The k-MHV1 model is presented in Table 6.6. In all cases the PR is used as the EoS. Reprinted with permission from Chemical Engineering Science, An interpretation of the behavior of EoS/GE models for asymmetric systems by Georgios M. Kontogeorgis and Panayiotis M. Vlamos, 55, 13, 2351–2358 Copyright (2000) Elsevier*

In agreement with this explanation, numerous approaches have been developed which essentially eliminate this 'combinatorial terms' difference', by adding a constant to the FH-type term of Equation (6.8) or by fully eliminating the two combinatorial contributions altogether or by other approximations. Some of the approaches are presented in Table 6.6. It is important to emphasize that, even though we could classify all of the mixing rules of Table 6.6 as EoS/G^E ones (in the sense that they combine EoS with an activity coefficient model), not all of them can be derived from the fundamental Equation (6.1), thus not all of them have a specific reference pressure. Despite that, they are typically used as zero reference pressure models, i.e. using existing UNIFAC or other activity coefficient models with existing parameter tables. The approaches of Table 6.6 are not the only ones which provide solutions to the problems of EoS/G^E mixing rules for size-asymmetric mixtures. Other approaches have been used, also based on the elimination of the 'combinatorial terms' difference':

1. *Li et al.*[38] have presented a modified PSRK which performs better for size-asymmetric gas–alkane systems, using r and q values for alkanes fitted to VLE data. What is essentially accomplished in this way is that $b_i/b = r_i/r$, i.e. the elimination of the combinatorial terms' difference of Equation (6.7), as illustrated in Figure 6.6.
2. Knudsen and co-workers[40] and later also Jaubert *et al.*[41–43] have used MHV1 with a quadratic mixing rule for the co-volume parameter and a constant l_{ij} value (equal to 0.3) for a variety of asymmetric mixtures containing CO_2 and fatty acid esters (compounds present in fish oil) etc.

As Figures 6.7 and 6.8 illustrate for some of these models, the combinatorial terms' difference is minimized and excellent results are obtained for size-asymmetric mixtures.

Another recent approach which has been proved successful for asymmetric mixtures is the so-called universal mixing rule (UMR-PR), which is combined with a translated form of the Peng–Robinson EoS.[44,45]

Table 6.6 *EoS/G^E mixing rules suitable for size-asymmetric mixtures. Except for the last three models (new developments of PSRK and the UMR-PR mixing rule), all models use the linear mixing rule for the co-volume parameter ($b = \sum_i x_i b_i$)*

Model	Mixing rule for the energy parameter $\alpha = \frac{a}{bRT}$	EoS and activity coefficient typically used
LCVM[29]	$\alpha = C_1\left(\frac{g^{E,*}}{RT} - \frac{C_2}{C_1}\frac{g^{E,FH}}{RT}\right) + \sum_i x_i\alpha_i$	t-modified PR + original UNIFAC
	$C_1 = \frac{\lambda}{A_V} + \frac{1-\lambda}{A_M}$	$\lambda = 0.36 \Rightarrow C_2/C_1 = 0.68$
	$C_2 = \frac{1-\lambda}{A_M}$	For modified UNIFAC:[20]
	$A_V = -0.623, A_M = -0.52$	$\lambda = 0.7 \Rightarrow C_2/C_1 = 0.30$
κ-MHV1[46]	$\alpha = \frac{1}{q_1}\left(\frac{g^{E,*}}{RT} - \kappa\frac{g^{E,FH}}{RT}\right) + \sum_i x_i\alpha_i$	t-modified PR + original UNIFAC $\kappa = 0.65$ Modified UNIFAC $\kappa = 0.3$
	$q_1 = -0.553$	
GCVM-1[47]	$\alpha = C_1\left(\frac{g^{E,*}}{RT} - \frac{C_2}{C_1}\frac{g^{E,FH}}{RT}\right) + \sum_i x_i\alpha_i$	PR + UNIFAC VLE $C_2/C_1 = 0.715$
	$C_1 = \frac{1}{A_M}$	
	$C_2 = \frac{1-\mu}{A_M}$	
	$A_M = -0.53, \quad \mu = 0.285$	
GCVM-2[47]	$\alpha = C_1\left(\frac{g^{E,*}}{RT} - \frac{C_2}{C_1}\frac{g^{E,FH}}{RT}\right) + \sum_i x_i\alpha_i$	$C_2/C_1 = 0.4$ using the modified UNIFAC of Larsen *et al.*[20]
CHV[48]	$\alpha = \frac{1}{C^*}\left(\frac{g^{E,*}}{RT} - (1-\delta)\frac{g^{E,FH}}{RT}\right) + \sum_i x_i\alpha_i$	$1-\delta = 0.64$
	$C^* = -0.6931$	
CHV[49]	$\alpha = \frac{1}{C^*}\left(\frac{g^{E,*}}{RT} - (1-\delta)\frac{g^{E,FH}}{RT}\right) + \sum_i x_i\alpha_i$	$1-\delta = 0.715$
PSRK-new[50]	$\alpha = \frac{1}{A_1}\left(\frac{g^{E,res}}{RT}\right) + \sum_i x_i\alpha_i$	$b = \sum_i \sum_j x_i x_j b_{ij}$
	$A_1 = -0.646\,63$	$b_{ij} = \left(\frac{b_i^{3/4} + b_j^{3/4}}{2}\right)^{4/3}$
VTPR[a] [39,51–53,23]	$\alpha = \frac{1}{A_1}\left(\frac{g^{E,res}}{RT}\right) + \sum_i x_i\alpha_i$	$b = \sum_i \sum_j x_i x_j b_{ij}$
	$A_1 = -0.530\,87$	$b_{ij} = \left(\frac{b_i^{3/4} + b_j^{3/4}}{2}\right)^{4/3}$
UMR-PR[44,45]	$\alpha = \frac{1}{q_1}\left(\frac{g^{E,*}}{RT} + \sum_i x_i \ln\frac{b}{b_i}\right) + \sum_i x_i\alpha_i$	$b = \sum_i \sum_j x_i x_j b_{ij}$
	$q_1 = -0.53$	$b_{ij} = \left(\frac{b_i^{1/2} + b_j^{1/2}}{2}\right)^2$

[a] The only essential difference between PSRK-new and VTPR is the cubic EoS used, which is SRK in PSRK-new and a translated Peng–Robinson in VTPR. Otherwise, the mixing rules used (for both the energy and the co-volume parameters) are identical.

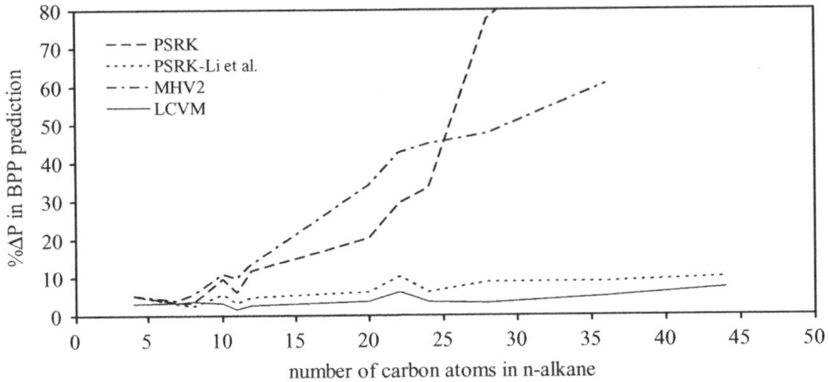

Figure 6.8 *Ethane–alkane VLE with various EoS/G^E models. As the size-asymmetry (size difference between the mixture components) increases, the results with MHV2 and PSRK become progressively worse and this is a serious problem for all types of gas and heavy alkane systems, e.g. ethane–alkanes, CO_2–alkanes, etc. The LCVM model performs well for size-asymmetric systems. Reprinted with permission from Computer-Aided Property Estimation for Process and Product Design by Gani and Kontogeorgis, Equations of state with emphasis on excess Gibbs energy mixing rules by E.C. Voutsas, Ph. Coutsikos and G.M. Kontogeorgis, Elsevier Ltd, Oxford, UK Copyright (2004) Elsevier*

This is essentially the MHV1 mixing rule (see Table 6.4) with the original UNIFAC used as the activity coefficient model (using either T-independent or linear T-dependent interaction parameters). The novel aspect of this approach is the mixing rule used for the co-volume parameter:

$$b = \sum_i \sum_j x_i x_j b_{ij}$$
$$b_{ij} = \left(\frac{b_i^{1/2} + b_j^{1/2}}{2} \right)^2 \tag{6.11}$$

The resulting model has, thus, an approximate zero reference pressure and the existing UNIFAC parameter table can be used. The authors have found that UMR-PR yields as satisfactory results as the PSRK, MHV2, etc., for polar mixtures and very satisfactory results for size-asymmetric systems. Some successful first applications have been presented for solid–gas equilibria, heats of mixing (with T-dependent parameters) and polymer–solvent VLE. The model can be used for LLE as well, but in this case the UNIFAC–LLE should be used instead or the UNIQUAC/UNIFAC model for correlating the parameters. Still, the model is flexible and it was shown that when UNIQUAC parameters were correlated to water–alkane LLE, then VLE could be well predicted for two water–alkane systems.

Some important observations can be made:

1. The most widely used of the models in Table 6.6 is LCVM, while the recent versions of PSRK-new and VTPR are also gradually gaining acceptance. LCVM, being the oldest among the successful models for size-asymmetric mixtures, has the most extensive parameter tables for mixtures with gases. LCVM was not originally developed according to the explanation presented previously in this section (elimination of the combinatorial terms' difference), but it was proposed as a rather empirical model based on a mixing rule which is a linear combination of the Vidal and Michelsen mixing rules:[29]

$$\alpha = \lambda \alpha_v + (1-\lambda)\alpha_M \tag{6.12}$$

$$\alpha_v = \frac{1}{A_v} \left(\frac{g^E}{RT} \right)^* + \sum_i x_i \alpha_i \tag{6.13}$$

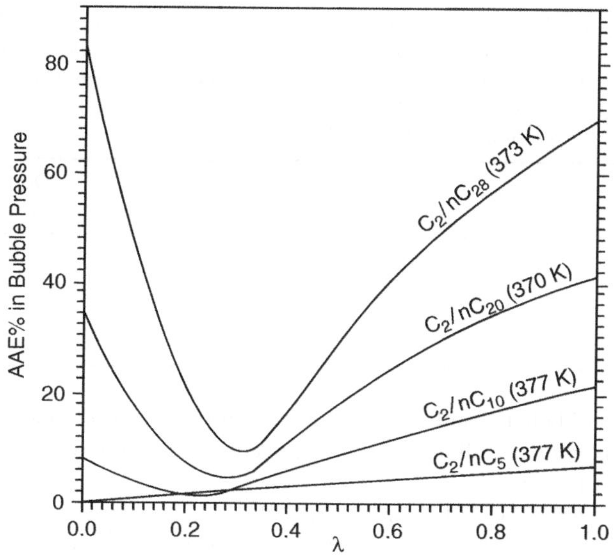

Figure 6.9 *Average absolute percentage error in bubble point pressure prediction for ethane–n-alkane systems as a function of the λ parameter of the LCVM mixing rule. The translated PR and original UNIFAC are used. An optimum λ value close to 0.36 is obtained, which is then used with LCVM for all mixtures. Reprinted with permission from Fluid Phase Equilibria, Prediction of vapor-liquid equilibrium with the LCVM model: a linear combination of the Vidal and Michelsen mixing rules coupled with the original UNIFAC by C. Boukouvalas, N. Spiliotis et al., 92, 75 Copyright (1995) Elsevier*

$$\alpha_M = \frac{1}{A_M}\left[\left(\frac{g^E}{RT}\right)^* + \sum_i x_i \ln\left(\frac{b}{b_i}\right)\right] + \sum_i x_i \alpha_i \tag{6.14}$$

PR EoS: $A_v = -0.623$, $A_M = 0.52$:

$$\alpha = \left(\frac{\lambda}{A_v} + \frac{1-\lambda}{A_M}\right)\left(\frac{g^E}{RT}\right)^* + \frac{1-\lambda}{A_M}\sum_i x_i \ln\left(\frac{b}{b_i}\right) + \sum_i x_i \alpha_i \tag{6.15}$$

$\lambda = 0.36$ (original UNIFAC).

The λ parameter was fitted from ethane–alkane data (Figure 6.9) and was used for all mixtures, but, as illustrated above, the model *does* have a theoretical explanation. It is clear from the previously presented explanation that the λ parameter is tightly connected to the activity coefficient and cubic EoS used (especially the former), and different values should be used for the original and modified UNIFAC models, which contain different forms for the combinatorial activity coefficient term. The LCVM as used today is based on the original UNIFAC equation and thus the value $\lambda = 0.36$ should be used in all applications (when existing UNIFAC and fitted parameters for gas-containing mixtures are used).

2. LCVM and the other EoS/G^E models of Table 6.6 suffer from several of the same limitations of the approximate zero reference pressure models:
   (a) they do not fully (only approximately) reproduce the g^E model they are combined with;
   (b) they do not satisfy the quadratic composition dependency for the mixture second virial coefficient;

Table 6.7 *The values for the volume packing fraction (equal for all pure compounds and the mixture) which can be used for deriving the PSRK, MHV1, VTPR and LCVM mixing rules. Noll and Fischer (Fluid Phase Equilib., **1998**, 53, 449) report average u values (based on 458 compounds at 2.3 bar) equal to 1.189 for SRK and 1.324 for PR. The LCVM value is estimated from the equation* $u = \lambda u^{Vidal} + (1-\lambda)u^{MHV1}$ *(with* $u^{Vidal} = 1$, $u^{Michelsen} = 1.257$*)*

Model	EoS	Derived based on the assumption $u_i = u = V/b$
PSRK	SRK	1.099
MHV1	SRK	1.234
VTPR	PR	1.225
LCVM	PR	1.164

(c) we should not expect better results than the g^E model they are combined with, although improved representation of asymmetric alkane systems has been observed in some cases, e.g. with LCVM and CHV (see Figure 6.13 below).

3. Many of the models of Table 6.6 can be derived (in a phenomenological way) using the constant volume packing fraction assumption as illustrated in Table 6.7. This is a rather reasonable assumption since, as shown by Fischer and Gmehling,[54] the 'experimental' V/b values lie between 0.9 (for small polar molecules) and 1.2 (for non-polar molecules) with most molecules having an average value around 1.1.
4. The PSRK-new and VTPR are essentially Huron–Vidal-type models, so all parameters of the residual term of UNIFAC have to be re-estimated.
5. The good results for size-asymmetric mixtures for LCVM and related models (see Table 6.6) can be attributed to the elimination from the mixing rule of the energy parameter of the cubic EoS of the combinatorial terms' difference (in Equation (6.7)), thus approximately resulting in the following mixing rule:

$$\alpha = \frac{1}{q_1}\left(\frac{g^{E,res}}{RT}\right) + \sum_i x_i\alpha_i \qquad (6.16)$$

which, when a UNIFAC-type model is used for alkane mixtures (where the residual term of the activity coefficient is unity), can be simplified as:

$$\alpha = \sum_i x_i\alpha_i \Rightarrow \frac{a}{b} = \sum_i x_i\frac{a_i}{b_i} \qquad (6.17)$$

We saw previously (Chapter 3) that such a simple mixing rule in combination with a cubic EoS can yield very good results for alkane-type size-asymmetric mixtures. This does not imply that the vdW repulsive term alone performs satisfactorily and is adequate for size-asymmetric systems, but that this classical repulsive term together with a residual term based on the above (a/b) mixing rule (Equation (6.17)) is a useful choice for asymmetric mixtures.

6.7 Applications of the LCVM, MHV2, PSRK and WS mixing rules

Three of the most widely used models which also have extensive parameter tables for gas-containing systems are MHV2, PSRK (especially the 'older version') and LCVM. A comparative evaluation of these models as well as a presentation of several of their application areas are given in this section. Applications of the WS mixing rule are also included, as well as some comparative evaluations of the models. Table 6.8 illustrates the wide variety of applications and Figures 6.10–6.16 present some typical results, illustrating both strengths and

Table 6.8 Some applications of EoS/G^E models. During the past 30 years, these mixing rules have found widespread use. This list presents a small number of some of the most important applications. Although not exhaustive, the selected references present comparisons of several mixing rules and/or calculations for several mixtures and various properties

Application	Model(s)	Reference
Multicomponent VLE containing acid gases, water, other gases and methanol	MHV2, HV and various mixing rules with k_{ij}/l_{ij} parameters	Knudsen *et al.*[60]
VLE for non-polar/weakly polar systems	MHV2, WS, pressure- and density-dependent mixing rules	Voros and Tassios[61]
High-pressure VLE	MHV2, WS	Huang and Sandler[19]
High-pressure VLE for polar systems	LCVM, PSRK, MHV2	Boukouvalas *et al.*[29]
VLE for gas-containing mixtures (CO_2, ethane, methane + alkanes)		
Hydrogen + hydrocarbons	WS	Huang *et al.*[62]
VLE for polar mixtures	LCVM, MHV2, WS, PSRK (and γ–Φ using UNIFAC)	Voutsas *et al.*[34]
High-pressure VLE for binary and ternary mixtures with SCF (CO_2, propane) and polar compounds (alcohols, water, heavy acids and esters)	LCVM, MHV2, MHV1, HV, WS and others	Knudsen *et al.*[63]
VLE of CO_2/fish-oil-related compounds (fatty esters and acids)	LCVM, MHV1, MHV2 also using quadratic mixing rule with l_{ij} in b_{ij} (MHV1QB and GCVM models)	Coniglio *et al.*[40,47]
Polar high-pressure VLE	PSRK, LCVM, WS, MHV2	Fischer and Gmehling,[54] Wong *et al.*[64]
VLE of asymmetric mixtures	MHV1, MHV2, PSRK, LCVM, HV type including WS	Orbey and Sandler[48] Orbey *et al.*[65]
V(L)LE of water–alcohol/phenol–hydrocarbons	WS	Abdel-Gani and Heidemann[66]
LLE	WS	Escobedo-Alvarado and Sandler[67]
VLE for gas-containing mixtures (CO_2, ethane, propane + alkanes)	LCVM and PHCT	Boukouvalas *et al.*[68]
VLE for a 10-component system	PSRK, MHV2, G^E–Henry's law models	Patel *et al.*[69]
Alcohol–alkanes (VLE, LLE)	WS	Seo *et al.*[70]
Synthetic gas condensate and oil systems	LCVM	Spiliotis *et al.*[71]
CO_2/alcohols	LCVM	Yakoumis *et al.*[72]
CO_2/esters, acids, ethers	LCVM	Yakoumis *et al.*[72]
Excess enthalpies	WS	Escobedo-Alvarado and Sandler[73]
Asymmetric mixtures, VLE/gas solubilities	PSRK, PSRK–Li, VTPR	Ahlers and Gmehling[39]
LLE	MHV1, WS	Ohta *et al.*[74]
	MHV2	Matsuda *et al.*[75]
Solid–gas equilibria	LCVM	Spiliotis *et al.*[76]
	WS	Coutsikos *et al.*[55]

(continued)

Table 6.8 (*Continued*)

Application	Model(s)	Reference
	VTPR	Escobedo-Alvarado *et al.*[77]
	UMR–PR	Ahlers *et al.*[53]
		Voutsas *et al.*[78]
		Bertakis *et al.*[79]
High-pressure wax formation in petroleum fluids	LCVM	Sansot *et al.*[80]
		Coutinho *et al.*[81]
Wax formation in diesel fuel		Pauly *et al.*[82]
Solid–fluid gas heavy paraffins		Pauly *et al.*[83]
Solid–liquid–vapor of North Sea Waxy Crude		Daridon *et al.*[84]
Heat capacities	VTPR	Diedrichs *et al.*[85]
Strong electrolytes	VTPR	Collinet and Gmehling[86]

shortcomings of the various mixing rules. Finally, Figures 6.17 and 6.18 present selected results using the UMR–PR model, illustrating its capabilities in calculating phase equilibria and other thermodynamic properties. The major conclusions are:

1. For about 10 of the well-known high-pressure VLE systems which do not include gases (e.g. acetone–-water, methanol–benzene, methanol–water, ethanol–water, acetone–methanol and propanol-2–water) excellent predictive performance is achieved with all four well-known EoS/G^E models (MHV2, PSRK, WS, LCVM).

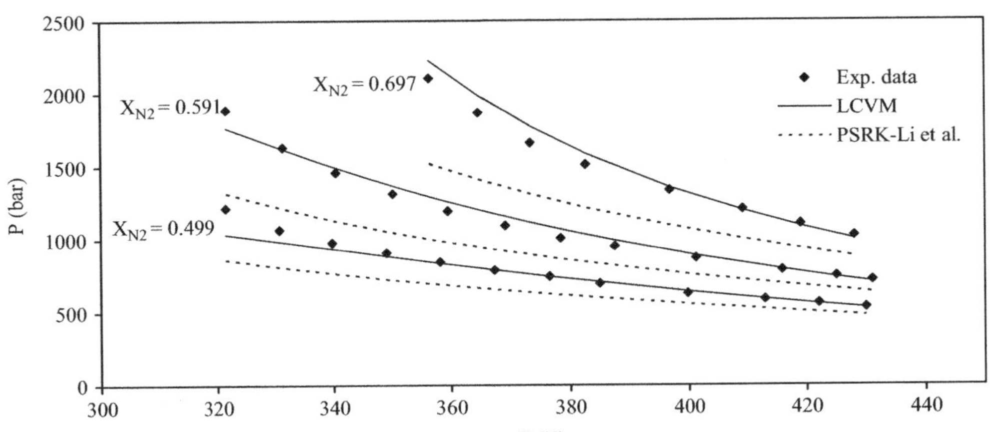

Figure 6.10 *High-pressure VLE for nitrogen–n-tetradecane. Only LCVM provides satisfactory VLE predictions for this asymmetric system for which there are data up to very high pressures. MHV2 yields very large errors and this is why the results with this model are not included. Reprinted with permission from Computer-Aided Property Estimation for Process and Product Design by Gani and Kontogeorgis, Equations of state with emphasis on excess Gibbs energy mixing rules by E.C. Voutsas, Ph. Coutsikos and G.M. Kontogeorgis, Elsevier Ltd, Oxford, UK Copyright (2004) Elsevier*

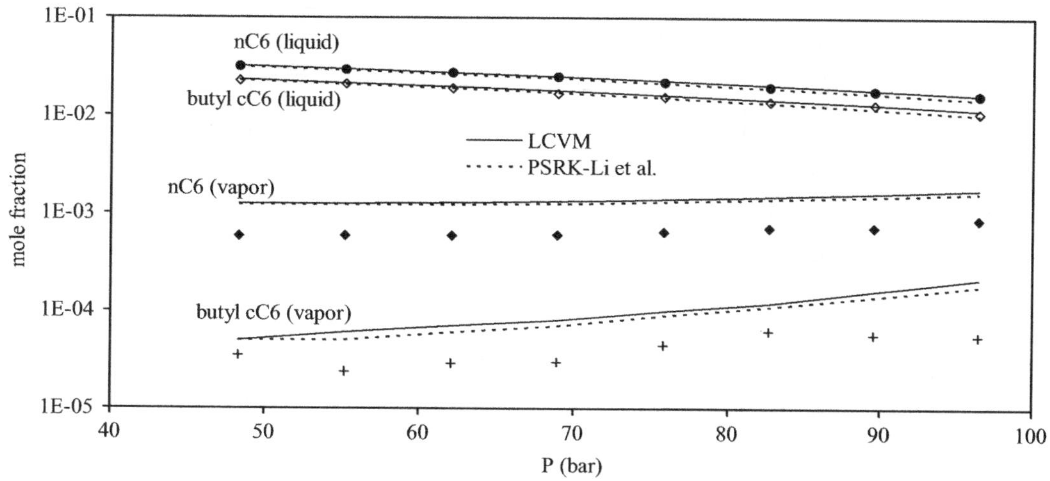

Figure 6.11 *VLE for a 19-component mixture with LCVM and PSRK/Li et al.*[38] *version. This 19-component mixture contains carbon dioxide, methane, paraffins (n-C_5 to n-C_{13}) and n-alkylcyclohexanes (cyclohexane–heptylcyclohexane). Very good results are obtained with LCVM and PSRK as modified by Li et al.*[38] *both for the heavy components shown here and for the gases. Somewhat higher deviations are observed in the vapor phase. Reprinted with permission from Computer-Aided Property Estimation for Process and Product Design by Gani and Kontogeorgis, Equations of state with emphasis on excess Gibbs energy mixing rules by E.C. Voutsas, Ph. Coutsikos and G.M. Kontogeorgis, Elsevier Ltd, Oxford, UK Copyright (Ref. 120) (2004) Elsevier*

2. For size-asymmetric systems, the PSRK-new/VTPR, UMR–PR and LCVM models are recommended. MHV2 and the original PSRK typically overestimate the experimental data. LCVM provides, in general, reliable VLE predictions also for multicomponent mixtures containing gases (H_2, CO_2, C_1, etc.) and non-polar/polar compounds. The WS mixing rule has not been extensively applied to asymmetric gas-containing systems.

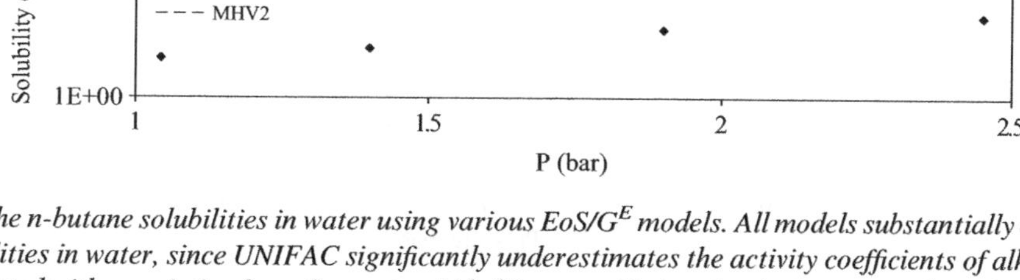

Figure 6.12 *The n-butane solubilities in water using various EoS/G^E models. All models substantially overestimate the n-butane solubilities in water, since UNIFAC significantly underestimates the activity coefficients of alkanes in aqueous mixtures. Reprinted with permission from Computer-Aided Property Estimation for Process and Product Design by Gani and Kontogeorgis, Equations of state with emphasis on excess Gibbs energy mixing rules by E.C. Voutsas, Ph. Coutsikos and G.M. Kontogeorgis, Elsevier Ltd, Oxford, UK Copyright (Ref. 120) (2004) Elsevier*

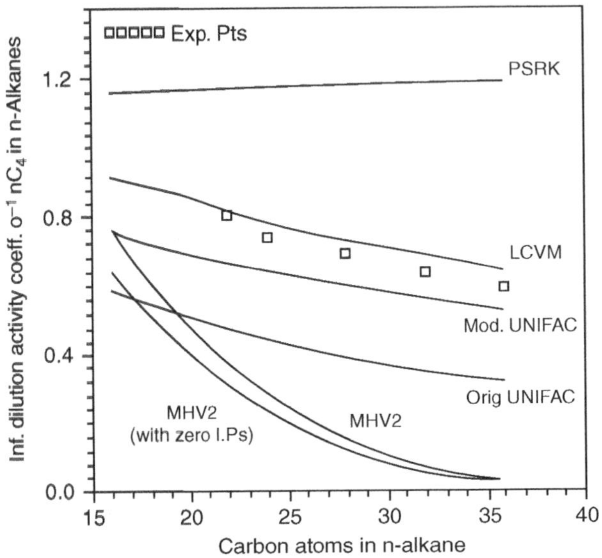

Figure 6.13 *Infinite dilution activity coefficient of n-butane in large n-alkanes (T = 373 K, P = 1 atm) as a function of the number of carbon atoms in the n-alkane. If the PSRK mixing rule were to reproduce the activity coefficient model it is combined with, the PSRK results should coincide with the original UNIFAC and the MHV2 results should coincide with the modified UNIFAC. None of these two happen. LCVM gives very good results (good agreement to the experimental data), but LCVM does not reproduce the activity coefficient model it is combined with either (which is the original UNIFAC). Reprinted with permission from Fluid Phase Equilibria, Prediction of vapor-liquid equilibrium with the LCVM model: a linear combination of the Vidal and Michelsen mixing rules coupled with the original UNIFAC by C. Boukouvalas, N. Spiliotis et al., 92, 75 Copyright (1995) Elsevier*

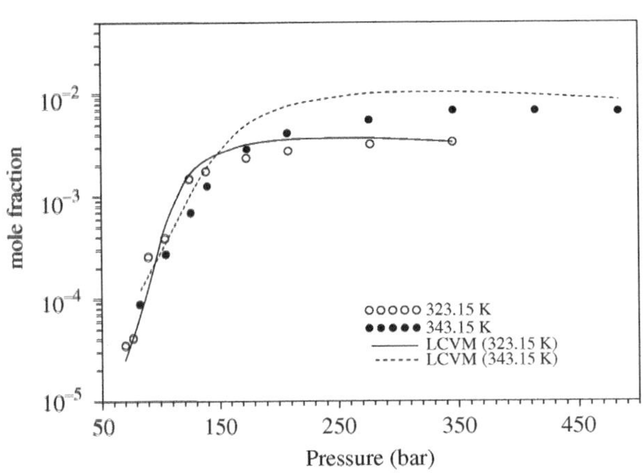

Figure 6.14 *Solid–gas equilibria with LCVM for CO_2–hexamethyl benzene. Reprinted with permission from Journal of Supercritical Fluids, Prediction of solid-gas equilibria with the Peng-Robinson equation of state by Coutsikos, Magoulas and Kontogeorgis, 25, 3, 197–212 Copyright (2003) Elsevier*

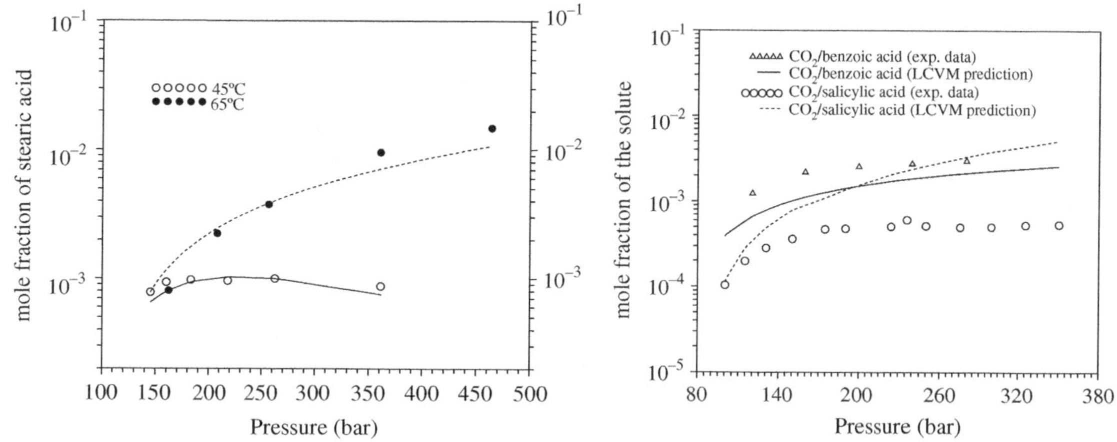

Figure 6.15 *Solid–gas equilibria with LCVM for CO_2–stearic acid (left) and CO_2–benzoic, salicylic acid (right). Reprinted with permission from Journal of Supercritical Fluids, Prediction of solid-gas equilibria with the Peng-Robinson equation of state by Coutsikos, Magoulas and Kontogeorgis, 25, 3, 197–212 Copyright (2003) Elsevier*

3. No EoS/G^E models can (or should, for that matter!) in principle perform (much) better than their 'best part', i.e. the local composition activity coefficient model they are combined with. Thus, EoS/G^E models do not perform well for LLE, especially for highly immiscible mixtures like water with n-alkanes (Figure 6.12).

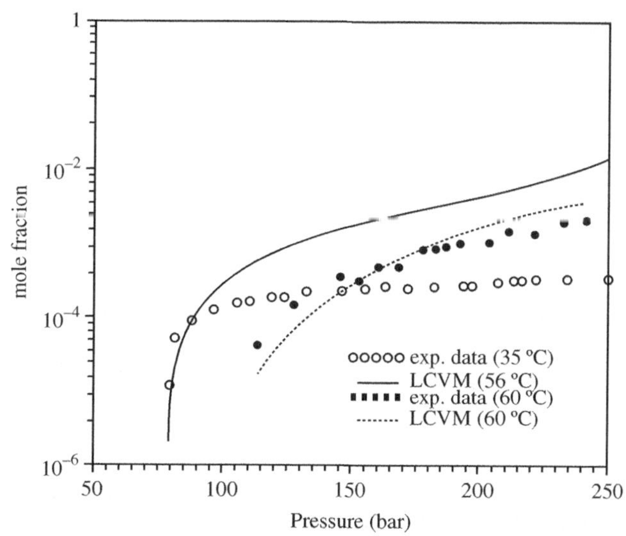

Figure 6.16 *Vapor–liquid equilibria and solid–gas equilibria for CO_2–phenol with LCVM. The error is 19% at 60 °C (liquid–gas) and 187% at 35 °C (solid–gas). Reprinted with permission from Journal of Supercritical Fluids, Prediction of solid-gas equilibria with the Peng-Robinson equation of state by Coutsikos, Magoulas and Kontogeorgis, 25, 3, 197–212 Copyright (2003) Elsevier*

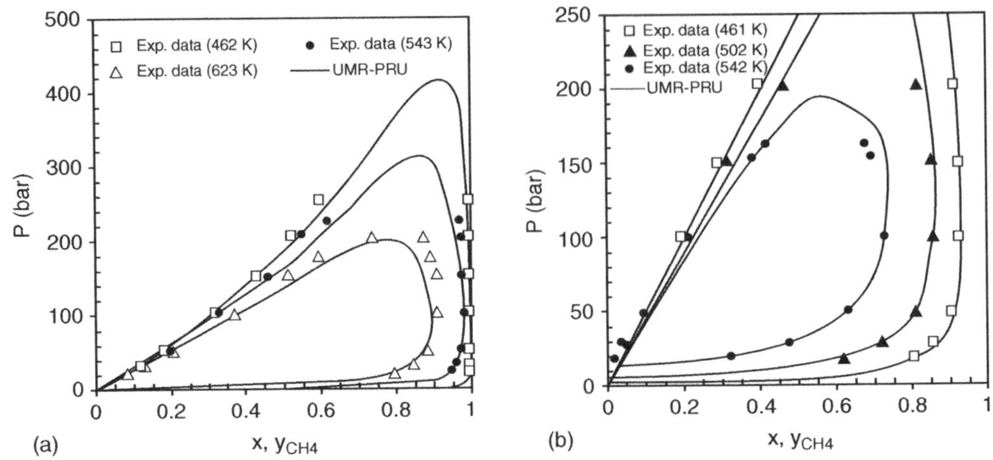

Figure 6.17 *VLE with UMR–PR model for (a) methane–n-hexadecane and (b) methane–m-xylene. Reprinted with permission from Fluid Phase Equilibria, Thermodynamic property calculations with the universal mixing rule for EoS/G^E models: Results with the Peng-Robinson EoS and a UNIFAC model by Epaminondas Voutsas, Vasiliki Louli et al., 241, 1-2, 216–228 Copyright (2006) Elsevier*

4. As Figure 6.13 illustrates, none of the approximate EoS/G^E models (MHV2, PSRK, LCVM) fully reproduces the activity coefficient model they are combined with, and this is greatly exemplified for asymmetric alkane mixtures (and in general as the size-asymmetry increases). LCVM performs best for such mixtures.
5. Figures 6.14–6.16 show a few typical solid–gas equilibrium results with LCVM. It can be seen (see also Coutsikos *et al.*[55]) that LCVM represents very accurately the solubilities of solid aromatic hydrocarbons in CO_2 and even for some polar compounds, e.g. organic acids, though better results are obtained for the aliphatic acids rather than for the aromatic ones. For even more complex compounds, e.g. cholesterol and pharmaceuticals, LCVM performs very poorly. In some cases, this poor performance can be attributed to the pure compound parameters (critical constants, solid vapor pressures), while in other cases it is due to the group energy interaction parameters. However, the poor performance of LCVM for solid–gas systems cannot always be attributed to the complexity of compounds and Figure 6.16 demonstrates the nature of the problem for a rather simple compound, phenol. Critical properties and the vapor pressure of phenol are accurately known. However, the percentage error with LCVM is only 19% at 60 °C (gas–liquid equilibria) and 187% a few degrees below (35 °C), i.e. the solid–gas equilibria area. The conclusion is that VLE-based parameters cannot always be used for solid–gas calculations. This is a general problem of the group contribution approach.

There have been several recent attempts to predict solid–gas equilibria with EoS/G^E mixing rules and two of the 'most recent' approaches (UMR–PR and VTPR) have been employed, though mostly for rather simple mixtures (CO_2 with aromatic hydrocarbons). Other recent approaches[56–59] involve use of the PR or the Patel–Teja cubic EoS together with the Huron–Vidal mixing rule and a modified UNIFAC activity coefficient model. In these models two adjustable parameters are used in the expression for the co-volume parameter which can be roughly estimated from the solid molar volumes.

6.8 Cubic EoS for polymers

6.8.1 High-pressure polymer thermodynamics

High-pressure phase equilibrium is important in many industrial processes involving polymers. For example, polymers are often dissolved in supercritical or liquid solvents and devolatilization and recovery of the solvents require information on the partial pressure of the solvent as a function of concentration. The aim in this case is to separate the polymer product from unreacted monomers, solvents and additives. Environmental regulations may set limits on the concentrations of these components in the product and thus these should be accurately known. Other applications of polymer thermodynamics directly involve the polymerization processes. For example, several processes such as the production of PET are carried out in

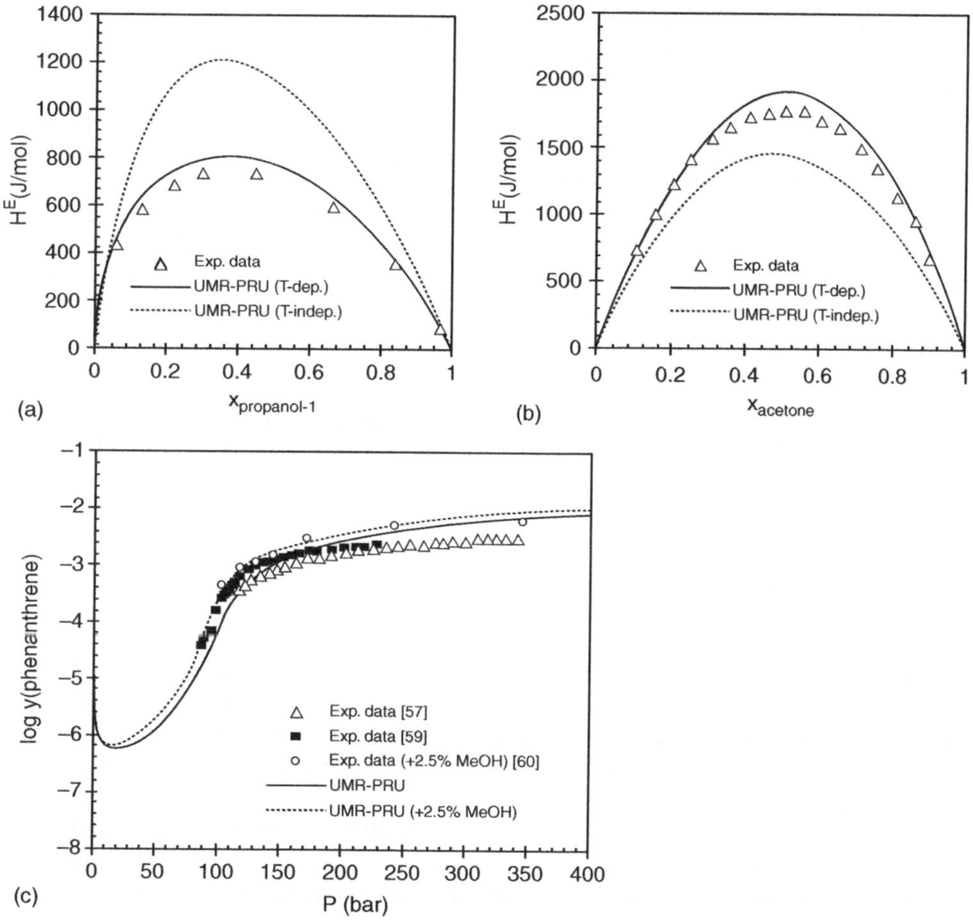

Figure 6.18 *Calculations with the UMR–PR for various properties: (a) and (b) heats of mixing for propanol–nonane (298 K) and acetone–heptane (323 K) respectively; (c) solubility of phenanthrene (323 K) in CO_2 with and without co-solvents. Reprinted with permission from Fluid Phase Equilibria, Thermodynamic property calculations with the universal mixing rule for EoS/G^E models: Results with the Peng–Robinson EoS and a UNIFAC model by Epaminondas Voutsas, Vasiliki Louli et al., 241, 1–2, 216–228 Copyright (2006) Elsevier*

two-phase (vapor–liquid) reactors. Phase equilibrium compositions of the reacting components will determine their phase concentrations and thus the outcome of the polymerization reaction. Another example is the case of LDPE made in autoclave reactors, where it is undesirable to perform the polymerization reaction in the two-liquid phase region but close to it, which makes accurate liquid–liquid equilibrium information at high pressures essential. As polymers are complex molecules, it is a challenging task to model phase equilibria involving polymers at both low and high pressures. Improved understanding of polymer properties (especially phase equilibrium) will assist in improving current operations, designing new installations, removing bottlenecks from existing plants and reducing time to market for new polymers.

For this reason, there is much interest in the polyolefin industry in developing and using thermodynamic models (EoS) for high-pressure applications involving polymers, e.g. mixtures containing polyolefins such as PE (LDPE, LLDPE and HDPE), PP and polyisobutene and chemicals like ethylene, propylene, butene, hexene, pentane, hexane at high T and P. Despite the widespread use of the SAFT-type approach for polymers (discussed in the third part of the book, Chapter 14), cubic EoS have often been shown to yield satisfactory results for many polymer solutions and blends, and some highlighted results and the way cubic EoS are applied to polymers will be illustrated here.

6.8.2 A simple first approach: application of the vdW EoS to polymers

The vdW EoS using the classical mixing and combining rules (Chapter 3, Equations (3.2) and (3.3)) written as an activity coefficient model is given by Equation (6.18) (where the excess volume term is ignored). Expressing the vdW EoS in this way and comparing this equation to 'classical' free-volume activity coefficient models (Chapters 4 and 5, Equations (4.11) and (5.27)) provides one explanation why cubic EoS could be used for polymers:

$$\ln \gamma_i = \ln \gamma_i^{comb\text{-}fv} + \ln \gamma_i^{res} = \left(\ln \frac{\Phi_i^{fv}}{x_i} + 1 - \frac{\Phi_i^{fv}}{x_i}\right) + \left(\frac{V_i}{RT}(\delta_i - \delta_j)^2 \Phi_j^2\right)$$

$$\Phi_i^{fv} = \frac{x_i(V_i - b_i)}{\sum_j x_j(V_j - b_j)} \qquad \delta_i = \frac{\sqrt{a_i}}{V_i} \tag{6.18}$$

1. Similarly to models such as UNIQUAC and UNIFAC, the vdW EoS contains a so-called 'combinatorial (free-volume)' term and a so-called 'residual–energetic' term. The first term can be considered to account for differences in sizes, shapes and free volumes between the molecules. The second term accounts for differences in the energetic effects between the molecules, thus the basic contributions to solution non-ideality are present in the model.
2. The first term is similar to the combinatorial terms of UNIQUAC/UNIFAC or of the Flory–Huggins/entropic–FV models. The second term is a modified form of the regular solution theory of Hildebrand and Scott. A vdW-based 'solubility parameter' is defined as shown in Equation (6.18).

In a series of articles published in the period 1994–1996, Kontogeorgis *et al.*[87] and Harismiadis *et al.*[88–90] have extended the vdW EoS to polymers. The pure polymer parameters were estimated from two volume–temperature data at $P = 0$ (a rather simplified approach). Excellent VLE correlation is obtained using the vdW1f mixing rules and the Berthelot combining rule for the cross-energy parameter

$(\varepsilon_{12} = (\varepsilon_1 \varepsilon_2)^{1/2} \quad \varepsilon = a/b)$:

$$a_{12} = \sqrt{a_1 a_2}\frac{b_{12}}{\sqrt{b_1 b_2}}(1 - l_{12})$$
$$b_{12} = \frac{b_1 + b_2}{2} \tag{6.19}$$

The most notable success of this first application of a cubic EoS (vdW) to polymers was the correlation of LLE for various polymer–solvent solutions and for polymer blends, as illustrated in Figures 6.19 and 6.20 for a few systems. The conclusions are:

1. Excellent LLE is obtained for numerous polymer solutions and blends, using in the case of UCST-type phase behavior a single temperature-independent interaction parameter, l_{ij} in Equation (6.19). Successful correlation of LCST, however, requires a temperature-dependent l_{ij} (for both polymer solutions and blends).
2. As can be seen, for example, for PBMA–octane, the vdW EoS can successfully capture the shape (flatness) of coexistence curves, while the calculated curves are not very sensitive to the interaction parameter values.
3. Excellent results are obtained for polymer blends as well, where other models from the literature including the entropic–FV, for example, fail (for quantitative calculations). The results shown in Figure 6.20 are

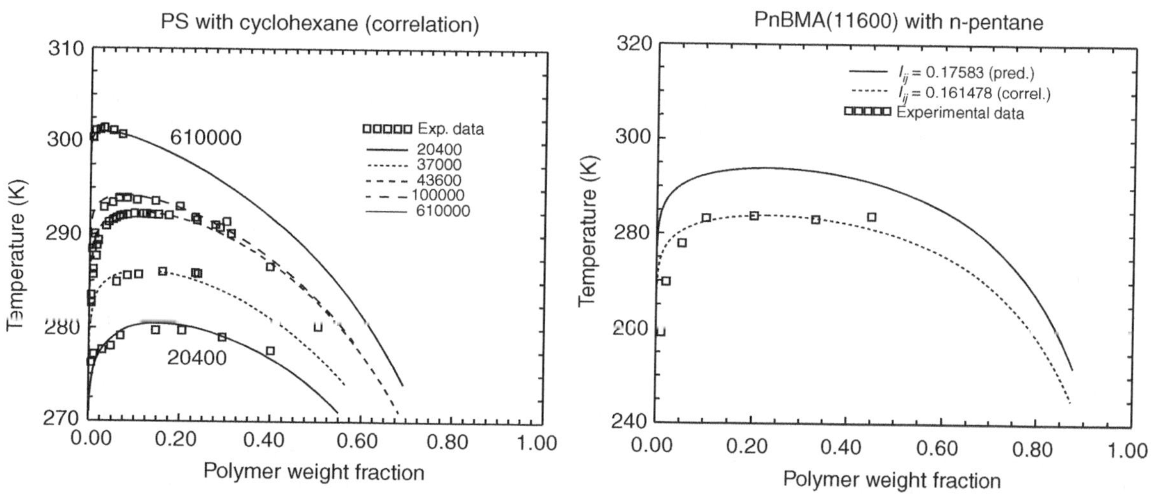

Figure 6.19 Left: LLE (UCST-type) correlation with the vdW EoS for polystyrene–cyclohexane at various molecular weights. The lines are correlations using a single (per molecular weight) interaction parameter l_{ij} (Equation (6.19)). The l_{ij} values vary from 0.122 (for polymer molecular weight 20 400) to 0.134 (molecular weight 610 000). Right: Predicted and correlated LLE (UCST-type) with the vdW EoS for PBMA(11 600)–n-pentane. A single interaction parameter is used. The predicted interaction parameter is based on a simple correlation of the interaction parameter with the solvent's molecular weight. Reprinted wih permission from Fluid Phase Equilibria, Application of the van der Waals equation of state to polymers III. Correlation and prediction of upper critical solution temperatures for polymer solutions by V. I. Harismiadis, G. M. Kontogeorgis et al., 100, 63–102 Copyright (Ref. 89) (1994) Elsevier

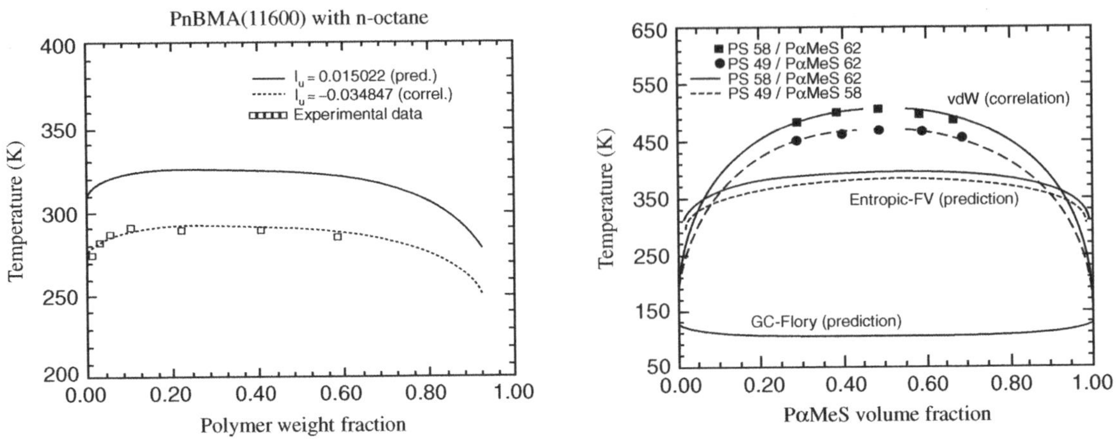

Figure 6.20 *Left: Predicted and correlated LLE (UCST-type) with the vdW EoS for PBMA(11 600)–n-octane. A single interaction parameter is used. The predicted interaction parameter is based on a simple correlation of the interaction parameter with the solvent's molecular weight. Reprinted wih permission from Fluid Phase Equilibria, Application of the van der Waals equation of state to polymers III. Correlation and prediction of upper critical solution temperatures for polymer solutions by V. I. Harismiadis, G. M. Kontogeorgis et al., 100, 63–102 Copyright (1994) Elsevier. Right: Predicted and correlated LLE (UCST-type) with the vdW EoS for a polymer blend. A single interaction parameter is used. Predictions with the entropic–FV and the GC–Flory models are also shown. Reproduced with permission from AIChE J, Miscibility of polymer blends with Engineering models by V.I. Harismiadis, A.R.D. van Bergen, A. Saraiva, G.M. Kontogeorgis, Aa. Fredenslund and D.P. Tassios, 42, 11, 3170–3180 Copyright (1996) John Wiley and Sons, Inc.*

correlations using a single interaction parameter (correction in the cross-energy term). Harismiadis *et al.*[90] presented a generalized equation for estimating the interaction parameters for several polystyrene-based blends. For correlating LCST, however, a temperature-dependent interaction parameter is required, as was the case for polymer solutions as well.

6.8.3 Cubic EoS for polymers

Following work by Kontogeorgis *et al.*,[87] numerous researchers developed cubic EoS for polymers. Such models are now used for polymer solutions and some are available in commercial simulators. See for example the review by Bokis *et al.*[91] and a review on cubic EoS for polymers by Orbey *et al.*[92] Various cubic EoS have been applied to polymers (vdW, SRK, PR) using different approaches for the estimation of pure polymer parameters and various mixing rules have been used as well, including EoS/G^E-type ones. For one of the first cubic EoS for polymers, that by Sako *et al.*[93] (SWP), several applications have appeared during recent years. Table 6.9 summarizes some of the most important approaches for cubic EoS for polymers. It can be seen that a variety of cubic EoS and mixing rules have been used, including various combinations of the EoS/G^E mixing rules which we have seen in this chapter together with the free-volume activity coefficient models discussed in Chapters 4 and 5. It is interesting to note that the combinatorial/chain term of the SWP EoS:

$$P = \frac{RT(V-b+bc)}{V(V-b)} - \frac{a}{V(V+b)} \tag{6.20}$$

Table 6.9 *Cubic EoS for polymers (1989 till now)*

Method/reference	Cubic EoS	Method for pure compound parameter estimation for polymers	Mixing rule	Major applications
Sako *et al.*[93] Tochigi *et al.*[95,96] Tork *et al.*[97] Browarzik *et al.*[98–100] Gharagheizi *et al.*[101]	SWP	*PVT* data VdW volume Segment-based approach	vdW1f	PE–ethylene high pressure *PVT*, VLE High-pressure polyolefin and copolymer phase equilibria PS–methylCC_6 LLE PS–CC_6–CO_2 Polydisperse mixtures
Tochigi *et al.*[95]	SWP		EoS/G^E (ASOG–FV)	Polymer–solvent VLE
Kontogeorgis *et al.*[87] Harismiadis *et al.*[88–90] Bithas *et al.*[102–103] Saraiva *et al.*[104]	vdW	Two volume–temperature data at low pressures	vdW1f	VLE, LLE and Henry's law constants for many polymer–solvent Polymer blends
Bertucco and Mio[105] Orbey *et al.*[106]	SRK	Default values for the 'critical' parameters	vdW1f	VLE Comparisons with SAFT, SL
Zhong and Masuoka[107]	SRK	Atmospheric density data	EoS/G^E (new UNIFAC)	Henry's law constants
Li *et al.*[108]	SRK	Default values for the 'critical' parameters	PSRK-new using: $b_{ij} = \left(\frac{b_i^{1/2} + b_j^{1/2}}{2}\right)^2$	VLE
Orbey *et al.*[109]	SRK		PSRK	
Louli and Tassios[110] Zhong and Masuoka[111,112] Voutsas *et al.*[119]	PR	Volumetric data (either extensive over wide *T*, *P* range or at low *P*)	vdW1f and related ones	VLE Gas solubilities Polymer blends
Orbey *et al.*[92] Orbey and Sandler[113] Kang *et al.*[114]	PR	Default values to account for very low fixed polymer vapor pressure + polymer *PVT* data	EoS/G^E (WS + FH)	VLE and comparisons with SAFT NLF–HB
Kalospiros and Tassios[28]	PR	Extensive volumetric data over wide *T*, *P* range	EoS/G^E (entropic–FV)	VLE
Tochigi *et al.*[115]	PR	Default values for the 'critical' parameters	EoS/G^E (UNIFAC–FV, ASOG–FV)	VLE
Tochigi[116]	PR	Polymer densities	EoS/G^E (ASOG–FV)	VLE

(continued)

Table 6.9 (*Continued*)

Method/reference	Cubic EoS	Method for pure compound parameter estimation for polymers	Mixing rule	Major applications
Wang *et al.*[117]	PR	Extensive volumetric data over wide T, P range	VTPR using: $b_{ij} = \left(\frac{b_i^{1/2} + b_j^{1/2}}{2}\right)^2$	VLE
Voutsas *et al.*[44,45]	PR	Extensive volumetric data over wide T, P range	UMR–PR	VLE

SWP = Sako–Wu–Prausnitz, SRK = Soave–Redlich–Kwong, PR = Peng–Robinson, SL = Sanchez–Lacombe, NLF–HB = lattice–fluid hydrogen bonding.

corresponds to the following activity coefficient expression which resembles the equation obtained from the vdW EoS (and indeed results in the vdW equation if all rotational parameters are equal to unity, $c_i = 1$):

$$\ln \gamma_i^{comb\text{-}fv} = \ln \frac{\Phi_i^{fv}}{x_i} + 1 - \frac{\Phi_i^{fv}}{x_i} + c_i \ln\left(\frac{\Phi_i^{fv}}{\Phi_i}\right) + c_m\left(\frac{\Phi_i}{x_i} - \frac{\Phi_i^{fv}}{x_i}\right) \tag{6.21}$$

The free volume, Φ_i^{fv}, and volume fractions, Φ_i, are defined in Equations (5.28) and (4.11).

The chain–FV model,[94]. Equation (6.21), can be derived from statistical mechanics (from the generalized vdW partition function) and includes rotational and vibrational effects. It is interesting to see that, as limiting cases, the chain–FV model results in the FH equation when the free-volume and volume fractions coincide and in Elbro's free-volume term used in the entropic–FV model when $c_i = 1$. Except for the applications with the vdW EoS, most of the applications with the other cubic EoS in the literature focused on low- and high-pressure VLE including gas solubilities. More specifically, it has been shown that:

1. The Sako *et al.*[93] EoS can satisfactorily correlate the polyethylene–ethylene high-pressure phase equilibria, describe the high-pressure binary and ternary polyolefin and co-polymer (poly(ethylene-co-propylene)) systems, and also can correlate high-pressure phase equilibria for certain polydisperse systems, e.g. ethylene–poly(ethylene-co-vinyl acetate) and CO_2–cyclohexane–polystyrene.
2. The SRK and PR EoS can correlate polymer–solvent VLE better than the Flory–Huggins model, when the comparison is made using a single binary interaction parameter. These cubic EoS can also correlate the high-pressure polyethylene–ethylene system and gas solubilities in various polymers.
3. Cubic EoS can predict low-pressure VLE and Henry's law constants using various EoS/G^E mixing rules combined with activity coefficient models suitable for polymer solutions. A better validation of cubic EoS and especially of the EoS/G^E mixing rules would be via high-pressure VLE, but few such data with non-gaseous solvents exist for polymers.

6.8.4 How to estimate EoS parameters for polymers

The development of EoS for polymers is still a very active area of research and it is difficult to recommend a specific approach (EoS). A serious problem with all EoS for polymers which, in our view, has not been adequately addressed as yet is the way that the EoS parameters should be estimated for polymers (see also the discussion in Chapter 14). Methods employed for low-molecular-weight compounds, e.g. based on the critical point and vapor pressures, cannot be used since critical properties and vapor pressure data are not available (have no meaning) for polymers. Numerous indirect methods have been employed using volumetric data and additional information (e.g. mixture data, the glass transition temperature, etc.), often including phase equilibria data for mixtures of polymers with low-molecular-weight compounds. Such methods may be a necessity since use of volumetric data alone does not seem to provide polymer EoS parameters useful for phase equilibrium calculations. Use of phase equilibria data, on the other hand, may render the parameters of pure polymers sensitive to the type of information employed. A thorough investigation of the methods to obtain meaningful polymer parameters for EoS will significantly improve and enhance their applicability to polymers. The subject of pure polymer parameter estimation for EoS is discussed for SAFT-type models in Chapter 14 (Section 14.2).

6.9 Conclusions: achievements and limitations of the EoS/G^E models

- EoS/G^E models such as MHV2, PSRK, WS and LCVM extend the applicability of cubic EoS to polar systems or, in other words, they extend the applicability of G^E models to high pressures.
- Excellent predictions of high-pressure VLE for polar systems, e.g. acetone–water, ethanol–water, etc., are obtained.
- Any combination of (cubic) EoS and G^E is possible (in many but not all the models).
- MHV2, PSRK and LCVM have extensive parameter tables for gases.
- The zero reference pressure approach is the best choice for use with 'existing' G^E models, i.e. when interaction parameters are estimated from experimental low-pressure VLE data.
- Vidal's infinite pressure mixing rule combined with especially the NRTL equation is a very useful model for correlation purposes.
- For asymmetric systems, only LCVM and the new PSRK provide successful results. Recently, the further development of PSRK (VTPR) and the UMR–PR models have been shown to be successful for such systems. The success is partially due to the 'cancellation' of the combinatorial terms from the G^E model and the EoS. A limitation of the recent approaches (PSRK-new, VTPR, UMR–PR) is that, despite their success for asymmetric systems, they still have a rather limited parameter table (for gas-containing systems). For this reason, classical models like LCVM are still more useful due to their larger parameter table. The same is the case for the original PSRK which, despite its limitations, is still being developed and new interaction parameters added.[24] The original PSRK is a well-known model among engineers, but the new approaches (VTPR, UMR–PR) are expected to gain more acceptance in the coming years (at least for high-pressure VLE and related calculations) as more interaction parameters become available.

The EoS/G^E models do have limitations. The most important ones are:

- Availability of UNIFAC interaction parameters, especially for gas-containing systems. The UNIFAC parameter table still has many gaps!
- None of the 'approximate' EoS/G^E models fully reproduces the G^E model they are combined with (especially for size-asymmetric systems).

Table 6.10 *Recommended models and their limitations*

Type of mixtures	Low pressures – recommended models	High pressures – recommended models	Limitations
Hydrocarbons	Ideal solution Simple act. coeff. models, Cubic EoS	Cubic EoS	Infinite dilution conditions High-molecular-weight compounds
Hydrocarbons/gases (CO_2, N_2, C_1, C_2, H_2S, ...)	(Cubic EoS)	Cubic EoS with k_{ij}	Need of k_{ij} databases/ correlations for solid–gas systems
Polar compounds			
General	Act. coeff. models (Margules, van Laar, LC)	EoS/G^E models, e.g. MHV2, LCVM, PSRK and WS	Systems with complex chemicals, water–HB compounds, solids, liquids, polymers, electrolytes, etc. LLE, VLLE Multicomponent, multiphase equilibria
Gas/polar compounds	—	MHV2, PSRK, LCVM	Systems and gases for which parameters are not available
Size/asymmetric systems	Act. coeff. models, modified UNIFAC	LCVM	Systems and gases for which parameters are not available
Complex and associating systems	UNIQUAC	EoS/G^E using UNIQUAC	Water systems LLE
Predictions	UNIFAC	EoS/G^E using UNIFAC	Often good only for preliminary design Many UNIFAC parameter tables for VLE, LLE, water systems, etc.

LC = local composition (Wilson, NRTL, UNIQUAC, UNIFAC models), HB = hydrogen bonding.

- The local composition g^E model is an underlying limitation of the approach. Associating compounds, solids, LLE and especially multicomponent and multiphase equilibria are still difficult to represent satisfactorily with the EoS/G^E models, as these mixtures are not well represented at low pressures by activity coefficient models.

Finally, considering the similarity of vdW-type EoS and similar models (e.g. SWP) to well-known free-volume and regular solution models, it is not surprising that numerous cubic EoS have been developed and applied to polymer–solvent and polymer–polymer phase equilibria. The user can now choose among various possibilities with different mixing rules, including the EoS/G^E ones, with free-volume models used as the g^E model incorporated, e.g. UNIFAC–FV or entropic–FV. Excellent correlative and often predictive results are obtained, but one of the most difficult aspects, common to all EoS, is to have available a reliable method for obtaining the EoS parameters for pure polymers. Parameters based on density data alone do not always yield good results when used in EoS, but they may be the only widely available pure compound data for polymers.

6.10 Recommended models – so far

Based on the models (and concepts) we have seen so far (Chapters 3–6), we can outline some recommendations on the applicability of the various models, depending on the type of compounds and pressure range. These recommended models are summarized in Table 6.10, including their limitations. The EoS/G^E models were, prior to the advent of association models, the best choice, suitable for wide applications over extended T, P ranges as well as for different types of phase equilibria. They are still much used today either in their 'predictive' form, e.g. MHV2, LCVM and PSRK/VTPR, or in their 'correlative' form, e.g. the Huron–Vidal mixing rule. For this reason, selective comparisons between the association models which will be discussed in the third part of the book (Chapters 7–14) and some EoS/G^E models will be presented in subsequent chapters.

Appendix 6.A Derivation of the Huron–Vidal mixing rule for the SRK EoS

The starting point is the expression for the excess Gibbs energy for SRK

$$\frac{g^E}{RT} = \sum_i x_i \ln\left[\frac{P(V_i-b_i)}{RT}\right] - \ln\left[\frac{P(V-b)}{RT}\right] + \frac{P}{RT}(V-\sum_i x_i V_i) - \left\{\frac{1}{RT}\left[-\sum_i x_i \frac{a_i}{b_i}\ln\left(\frac{V_i+b_i}{V_i}\right) + \frac{a}{b}\ln\left(\frac{V+b}{V}\right)\right]\right\}$$

$$= \left[\sum_i x_i \ln\left(\frac{V_i-b_i}{V-b}\right)\right] + \frac{P}{RT}(V-\sum_i x_i V_i) - \frac{1}{RT}\left[-\sum_i x_i \frac{a_i}{b_i}\ln\left(\frac{V_i+b_i}{V_i}\right) + \frac{a}{b}\ln\left(\frac{V+b}{V}\right)\right] \quad (6.22)$$

At the limit of infinite pressure, $P \to \infty$, the various terms of Equation (6.22) can be calculated as follows:

$$\lim_{P\to\infty}\left\{\ln\left[\frac{P(V-b)}{RT}\right]\right\} = \lim_{P\to\infty}\left\{\ln\left[1-\frac{a(V-b)}{V(V+b)RT}\right]\right\} = \ln 1 = 0 \quad (6.23)$$

Since:

$$\lim_{P\to\infty} V = b \text{ and } \lim_{P\to\infty} V_i = b_i$$

then:

$$\lim_{P\to\infty}\left[\frac{PV}{RT} - \frac{P}{RT}\sum_i x_i V_i\right] = \lim_{P\to\infty}\left\{\frac{P(V-b)}{RT} + \frac{Pb}{RT} - \sum_i x_i\left[\frac{P(V_i-b_i)}{RT} + \frac{Pb_i}{RT}\right]\right\}$$

$$= \frac{P}{RT}(b - \sum_i x_i b_i) = 0 \quad (6.24)$$

if we assume that the co-volume parameter can be given by the linear mixing rule:

$$b = \sum_i x_i b_i \tag{6.25}$$

Thus at $P \rightarrow \infty$, Equation (6.22) for the excess Gibbs energy becomes:

$$\frac{g^{E,\infty}}{RT} = -\frac{1}{RT}\frac{a}{b}\ln 2 - \sum_i \frac{x_i a_i}{RT b_i}\ln 2 \Rightarrow a = b\left(\sum_i \frac{x_i a_i}{b_i} - \frac{g^{E,\infty}}{\ln 2}\right) \tag{6.26}$$

This equation is actually a mixing rule for the energy parameter for the SRK EoS. It can be easily shown that for all vdW-type cubic EoS, the mixing rule has the form:

$$\alpha = \frac{1}{A_\nu}\left(\frac{g^E}{RT}\right)^{\infty} + \sum_i x_i \alpha_i \tag{6.27}$$

where the superscript ∞ implies infinite pressure and $\alpha = a/bRT$. For SRK, $A_\nu = -0.693$. Generally, A_ν depends on the attractive term of the cubic EoS (see Table 6.2). Originally, the Huron–Vidal model was employed with SRK and NRTL.

References

1. J. Vidal, *Chem. Eng. Sci.*, **1978**, *33*, 787.
2. M.J. Huron, J. Vidal, *Fluid Phase Equilib.*, **1979**, *3*, 255.
3. V. Feroiu, D. Geana, *Fluid Phase Equilib.*, **1996**, *120*, 1.
4. Ch. Lermite, J. Vidal, *Fluid Phase Equilib.*, **1988**, *42*, 1.
5. Ch. Lermite, J. Vidal, *Fluid Phase Equilib.*, **1992**, *72*, 111.
6. G. Soave, *Chem. Eng. Sci.*, **1984**, *39*(2), 357.
7. G. Soave, *Fluid Phase Equilib.*, **1992**, *72*, 325.
8. G.S. Soave, A. Bertucco, L. Vecchiato, *Ind. Eng. Chem. Res.*, **1994**, *33*, 975.
9. G.S. Soave, S. Sama, M.I. Oliveras, *Fluid Phase Equilib.*, **1999**, *156*, 35.
10. K.S. Pedersen, M.L. Michelsen, A.O. Fredheim, *Fluid Phase Equilib.*, **1996**, *126*, 13.
11. K.S. Pedersen, J. Milter, C.P. Rasmussen, *Fluid Phase Equilib.*, **2001**, *189*, 85.
12. E. Neau, C. Nicolas, J.N. Jaubert, F. Mutelet, *Pol. J. Chem.*, **2006**, *80*, 27.
13. I.G. Economou, C. Tsonopoulos, *Chem. Eng. Sci.*, **1997**, *52*(*4*), 511.
14. B. Edmonds, R.A.S. Moorwood, R. Szczepanski,SPE 35569, **1996**. Available at http://www.infochemuk.com/.
15. J. Mollerup, *Fluid Phase Equilib.*, **1986**, *25*, 323.
16. M.L. Michelsen, *Fluid Phase Equilib.*, **1990**, *60*, 47.
17. S. Dahl, M.L. Michelsen, *AIChE J.*, **1990**, *36*(*12*), 1829.
18. S. Dahl, Aa. Fredenslund, P. Rasmussen, *Ind. Eng. Chem. Res.*, **1991**, *30*, 1936.
19. H. Huang, S.I. Sandler, *Ind. Eng. Chem. Res.*, **1993**, *32*, 1498.
20. B.L. Larsen, P. Rasmussen, Aa. Fredenslund, *Ind. Eng. Chem. Res.*, **1987**, *26*, 2274.
21. T. Holderbaum, J. Gmehling, *Fluid Phase Equilib.*, **1991**, *70*, 251.
22. S. Horstmann, K. Fischer, J. Gmehling, *Fluid Phase Equilib.*, **2000**, *167*, 173.

23. J. Gmehling, *Fluid Phase Equilib.*, **2003**, *210*, 161.
24. S. Hortstmann, A. Jabloniec, J. Krafczyk, K. Fischer, J. Gmehling, *Fluid Phase Equilib.*, **2005**, *227*, 157.
25. H. Orbey, S.I. Sandler, *Fluid Phase Equilib.*, **1995**, *111*, 53.
26. H. Orbey, S.I. Sandler, *AIChE J.*, **1995**, *41(3)*, 683.
27. H. Orbey, S.I. Sandler, *AIChE J.*, **1996**, *42(80)*, 2327.
28. N. Kalospiros, D.P. Tassios, *Ind. Eng. Chem. Res.*, **1995**, *34*, 2117.
29. C. Boukouvalas, N. Spiliotis, P. Coutsikos, N. Tzouvaras, D.P. Tassios, *Fluid Phase Equilib.*, **1994**, *92*, 75.
30. D.S.H. Wong, S.I. Sandler, *Chem. Eng. Sci.*, **1992**, *33*, 787.
31. D.S.H. Wong, S.I. Sandler, *AIChE J.*, **1992**, *38(5)*, 671.
32. R. Stryjek, J.H. Vera, *Can. J. Chem. Eng.*, **1986**, *64*, 323.
33. H. Orbey, S.I. Sandler, D.S.H. Wong, *Fluid Phase Equilib.*, **1993**, *85*, 41.
34. E.C. Voutsas, N.S. Spiliotis, N.S. Kalospiros, D.P. Tassios, *Ind. Eng. Chem. Res.*, **1995**, *34*, 681.
35. Ph. Coutsikos, N.S. Kalospiros, D.P. Tassios, *Fluid Phase Equilib.*, **1995**, *108*, 59.
36. S. Dahl, Aa. Fredendlund, P. Rasmussen,The MHV2 model: prediction of phase equilibria at sub- and supercritical conditions. SEP 9105, Institut for Kemiteknik, Technical University of Denmark, 1991. Presented at the 2nd International Symposium on Supercritical Fluids, Boston, USA, 20–22 May 1991.
37. G.M. Kontogeorgis, P.M. Vlamos, *Chem. Eng. Sci.*, **2000**, *55*, 2351.
38. J. Li, K. Fischer, J. Gmehling, *Fluid Phase Equilib.*, **1998**, *143*, 71.
39. J. Ahlers, J. Gmehling, *Ind. Eng. Chem. Res.*, **2002**, *41*, 3489.
40. L. Coniglio, K. Knudsen, R. Gani, *Ind. Eng. Chem. Res.*, **1995**, *34(7)*, 2473.
41. J.-N. Jaubert, L. Coniglio, F. Denet, *Ind. Eng. Chem. Res.*, **1999**, *38*, 3162.
42. J.-N. Jaubert, L. Coniglio, C. Crampon, *Ind. Eng. Chem. Res.*, **2000**, *39*, 2623.
43. J.-N. Jaubert, P. Borg, L. Coniglio, D. Barth, *J. Supercrit. Fluids*, **2001**, *20*, 145.
44. E. Voutsas, K. Magoulas, D. Tassios, *Ind. Eng. Chem. Res.*, **2004**, *43*, 6238.
45. E. Voutsas, V. Louli, Ch. Boukouvalas, K. Magoulas, D. Tassios, *Fluid Phase Equilib.*, **2006**, *241*, 216.
46. N. Spiliotis, Ch. Boukouvalas, D.P. Tassios, The models κ- MHV1 and LCVM: prediction of phase equilibria in multicomponent systems. First Greek Chemical Engineering Conference, 29–31 May 1997, (pp. 325–330).
47. L. Coniglio, K. Knudsen, R. Gani, *Fluid Phase Equilib.*, **1996**, *116*, 510.
48. H. Orbey, S.I. Sandler, *Fluid Phase Equilib.*, **1997**, *132*, 1.
49. C. Zhong, H. Masuoka, *J. Chem. Eng. Jpn*, **1996**, *29*, 315.
50. J. Chen, K. Fischer, J. Gmehling, *Fluid Phase Equilib.*, **2002**, *200*, 411.
51. J. Ahlers, J. Gmehling, *Fluid Phase Equilib.*, **2001**, *191*, 177.
52. J. Ahlers, J. Gmehling, *Ind. Eng. Chem. Res.*, **2002**, *41*, 5890.
53. J. Ahlers, T. Yamaguchi, J. Gmehling, *Ind. Eng. Chem. Res.*, **2004**, *43*, 6569.
54. K. Fischer, J. Gmehling, *Fluid Phase Equilib.*, **1996**, *121*, 185.
55. Ph. Coutsikos, K. Magoulas, G.M. Kontogeorgis, *J. Supercrit. Fluids*, **2003**, *25(3)*, 197.
56. P.-C. Chen, M. Tang, Y.-P. Chen, *Ind. Eng. Chem. Res.*, **1995**, *34*, 332.
57. C.-C. Huang, M. Tang, W.-H. Tao, Y.-P. Chen, *Fluid Phase Equilib.*, **2001**, *179*, 67.
58. Y.-J. Sheng, P.-C. Chen, Y.-P. Chen, D.S.H. Wong, *Ind. Eng. Chem. Res.*, **1992**, *31*, 967.
59. K. Tochigi, T. Iizumi, H. Sekikawa, K. Kurihara, K. Kojima, *Ind. Eng. Chem. Res.*, **1998**, *37*, 3731.
60. K. Knudsen, E.H. Stenby, Aa. Fredenslund, *Fluid Phase Equilib.*, **1993**, *82*, 361.
61. N.G. Voros, D.P. Tassios, *Fluid Phase Equilib.*, **1993**, *91*, 1.
62. H. Huang, S.I. Sandler, H. Orbey, *Fluid Phase Equilib.*, **1994**, *96*, 143.
63. K. Knudsen, L. Coniglio, R. Gani, *Innovations in Supercritical Fluids. ACS Symposium Series*, **1995**, Vol. 608, pp. 140–153.
64. D.S.H. Wong, H. Orbey, S.I. Sandler, *Ind. Eng. Chem. Res.*, **1992**, *31*, 2033.
65. H. Orbey, C. Balci, G.A. Guruz, *Fluid Phase Equilib.*, **2002**, *41*, 963.
66. R.M. Abdel-Ghani, R.A. Heidemann, *Fluid Phase Equilib.*, **1996**, *116*, 495.
67. G.N. Escobedo-Alvarado, S.I. Sandler, *AIChE J.*, **1998**, *44(5)*, 1178.
68. C.J. Boukouvalas, K.G. Magoulas, D.P. Tassios, I. Kikic, *J. Supercrit. Fluids*, **2001**, *19*, 123.
69. N.C. Patel, V. Abovsky, S. Watanasiri, *Fluid Phase Equilib.*, **2001**, *185*, 397.

70. J. Seo, J. Lee, H. Kim, *Fluid Phase Equilib.*, **2001**, *182*, 199.
71. N. Spiliotis, C. Boukouvalas, N. Tzouvaras, D.P. Tassios, *Fluid Phase Equilib.*, **1994**, *101*, 187.
72. I. Yakoumis, K. Vlachos, G.M. Kontogeorgis, Ph. Coutsikos, N.S. Kalospiros, Fr. Kolisis, D. Tassios, *J. Supercrit. Fluids*, **1996**, *9*(2), 88.
73. G.N. Escobedo-Alvarado, S.I. Sandler, *Ind. Eng. Chem. Res.*, **2001**, *40*, 1261.
74. T. Ohta, H. Todoriki, T. Yamada, *Fluid Phase Equilib.*, **2004**, *225*, 23.
75. H. Matsuda, K. Kurihara, K. Ochi, K. Kojima, *Fluid Phase Equilib.*, **2002**, *203*(*1–2*), 269.
76. N. Spiliotis, K. Magoulas, D. Tassios, *Fluid Phase Equilib.*, **1994**, *102*, 121.
77. G.N. Escobedo-Alvarado, S.I. Sandler, A.M. Scurto, *J. Supercrit. Fluids*, **2001**, *21*, 123.
78. E.C. Voutsas, C. Boukouvalas, N. Kalospiros, D. Tassios, *Fluid Phase Equilib.*, **1996**, *116*, 480.
79. E. Bertakis, I. Lemomis, S. Katsoufis, E. Voutsas, R. Dohrn, K. Magoulas, D. Tassios, *J. Supercrit. Fluids*, **2007**, *41*, 238.
80. J.-M. Sansot, J. Pauly, J.-L. Daridon, J.A.P. Coutinho, *AIChE J.*, **2005**, *51*(*7*), 2089.
81. J.A.P. Coutinho, F. Mirante, J. Pauly, *Fluid Phase Equilib.*, **2006**, *247*, 8.
82. J. Pauly, J.-L. Daridon, J.-M. Sansot, J.A.P. Coutinho, *Fuel*, **2003**, *82*, 595.
83. J. Pauly, J.-L. Daridon, J.A.P. Coutinho, N. Lindeloff, S.I. Andersen, *Fluid Phase Equilib.*, **2000**, *167*(2), 145.
84. J.-L. Daridon, J. Paul, J.A.P. Coutinho, F. Montel, *Energy Fuels*, **2001**, *15*, 730.
85. A. Diedrichs, J. Rarey, J. Gmehling, *Fluid Phase Equilib.*, **2006**, *248*, 56.
86. E. Collinet, J. Gmehling, *Fluid Phase Equilib.*, **2006**, *246*, 111.
87. G.M. Kontogeorgis, V.I. Harismiadis, Aa. Fredenslund, D.P. Tassios, *Fluid Phase Equilib.*, **1994**, *96*, 65.
88. V.I. Harismiadis, G.M. Kontogeorgis, Aa. Fredenslund, D.P. Tassios, *Fluid Phase Equilib.*, **1994**, *96*, 93.
89. V.I. Harismiadis, G.M. Kontogeorgis, A. Saraiva, Aa. Fredenslund, D.P. Tassios, *Fluid Phase Equilib.*, **1994**, *100*, 63.
90. V.I. Harismiadis, A. Saraiva, G.M. Kontogeorgis, A.R.D. van Bergen, Aa. Fredenslund, D.P. Tassios, *AIChE J.*, **1996**, *42*(*11*), 3170.
91. C. Bokis, H. Orbey, C.-C. Chen, *Chem. Eng. Prog.*, **1999**, 39.
92. H. Orbey, C.-C. Chen, C.P. Bokis, *Fluid Phase Equilib.*, **1998**, *145*, 169.
93. T. Sako, A.H. Wu, J.M. Prausnitz, *J. Appl. Polym. Sci.*, **1989**, *38*, 1839.
94. G.M. Kontogeorgis, G. Nikolopoulos, Aa. Fredenslund, D.P. Tassios, *Fluid Phase Equilib.*, **1997**, *127*, 103.
95. K. Tochigi, K. Kojima, T. Sako, *Fluid Phase Equilib.*, **1996**, *117*, 55.
96. K. Tochigi, S. Kurita, T. Matsumoto, *Fluid Phase Equilib.*, **1999**, *158–160*, 313.
97. T. Tork, G. Sadowski, W. Arlt, A. de Haan, G. Krooshof, *Fluid Phase Equilib.*, **1999**, *163*, 61.
98. D. Browarzik, M. Kowalewski, *Fluid Phase Equilib.*, **1999**, *163*, 43.
99. C. Browarzik, M. Kowalewski, Calculation of the cloud-point and the spinodal curve for the system methylcyclohexane/polystyrene at high pressure. Presented at the 9th International Conference on Properties and Phase Equilibria for Product and Process Design, Kurashiki, Japan, 20–25 May 2001.
100. C. Browarzik, D. Browarzik, H. Kehlen, *J. Supercrit. Fluids*, **2001**, *20*(*1*), 73.
101. F. Gharagheizi, M. Mehrppoya, A. Vatani, *Braz. J. Chem. Eng.*, **2006**, *23*(*3*), 383.
102. S. Bithas, G.M. Kontogeorgis, N.S. Kalospiros, Aa. Fredenslund, D.P. Tassios, *Fluid Phase Equilib.*, **1995**, *113*, 79.
103. S. Bithas, N.S. Kalospiros, G.M. Kontogeorgis, D.P. Tassios, *Polym. Eng. Sci.*, **1996**, *36*(2), 254.
104. A. Saraiva, G.M. Kontogeorgis, V.I. Harismiadis, Aa. Fredenslund, D.P. Tassios, *Fluid Phase Equilib.*, **1996**, *115*, 73.
105. A. Bertucco, C. Mio, *Fluid Phase Equilib.*, **1996**, *117*, 18.
106. H. Orbey, C.P. Bokis, C.-C. Chen, *Ind. Eng. Chem. Res.*, **1998**, *37*, 4481.
107. C. Zhong, H. Masuoka, *Fluid Phase Equilib.*, **1996**, *126*, 1.
108. M. Li, L. Wang, J. Gmehling, *Chin. J. Chem. Eng.*, **2004**, *12*(*3*), 454.
109. H. Orbey, C.P. Bokis, C.-C. Chen, *Ind. Eng. Chem. Res.*, **1998**, *37*, 1567.
110. V. Louli, D.P. Tassios, *Fluid Phase Equilib.*, **2000**, *168*, 165.
111. C. Zhong, H. Masuoka, *Fluid Phase Equilib.*, **1996**, *123*, 59.
112. C. Zhong, H. Masuoka, *Fluid Phase Equilib.*, **1998**, *144*, 49.

113. H. Orbey, S.I. Sandler, *AIChE J.*, **1994**, *40(7)*, 1203.
114. J.W. Kang, J.H. Lee, K.-P. Yoo, C.S. Lee, *Fluid Phase Equilib.*, **2002**, *194–197*, 77.
115. K. Tochigi, H. Futakuchi, K. Kojima, *Fluid Phase Equilib.*, **1998**, *152*, 209.
116. K. Tochigi, *Fluid Phase Equilib.*, **1998**, *144*, 59.
117. L.-S. Wang, J. Ahlers, J. Gmehling, *Ind. Eng. Chem. Res.*, **2003**, *42*, 6205.
118. M.L. Michelsen, *Fluid Phase Equilib.*, **1990**, *60*, 213.
119. E. Voutsas, G. Pappa, Ch. Boukouvalas, K. Magoulas, D. Tassios, *Ind. Eng. Chem. Res.*, **2004**, *43*, 1312.
120. E.C. Voutsas, Ph. Coutsikos, G.M. Kontogeorgis, Equations of state with emphasis on excess Gibbs energy mixing rules. G.M. Kontogeorgis, R. Gani, *Computer-Aided Property Estimation for Process and Product Design*. Elsevier, 2004.

Part C

Advanced Models and Their Applications

7

Association Theories and Models: The Role of Spectroscopy

7.1 Introduction

This chapter introduces the third part of the book, which is devoted to the association theories. Emphasis will be placed on the chemical and perturbation theories. First, a short unified presentation of the different association theories (chemical, lattice and perturbation), highlighting similarities and differences, will be given. Then, the oldest approach, the chemical theory and first-order perturbation theory, will be outlined and the chapter will close with a discussion of spectroscopic data. Such data, especially for the monomer fraction of hydrogen bonding compounds (and their mixtures), are invaluable tools in the development and the evaluation of association theories. One of the most important association theories, belonging to the perturbation family, is SAFT (Statistical Associating Fluid Theory), which will be presented in Chapter 8. Chapters 9–12 will present another perturbation theory, the CPA (Cubic-Plus-Association) equation of state (EoS), a variation of SAFT which has found widespread use, especially in the petroleum and chemical industries. Applications of SAFT to associating compounds and polymers will be presented in Chapters 13 and 14, and applications of both CPA and SAFT to electrolytes will be discussed in Chapter 15.

7.2 Three different association theories

Over the last few years, a variety of models capable of explicitly accounting for the effects of hydrogen bonding in solutions have been proposed. They are called *association models*. The association models are particularly suitable for describing phase equilibria of mixtures containing highly polar and/or strongly associating compounds such as water, alcohols, acids, amines, phenols and others which have the capability of forming hydrogen bonds. The hydrogen bonding formation between molecules of the same kind is called self-association (e.g. in pure water or pure methanol), while the creation of hydrogen bonding complexes between two different molecules is called cross-association (e.g. water–methanol). Reviews of association models have been provided by Economou and Donohue[1] and Donohue and Economou.[2] We will not review

Thermodynamic Models for Industrial Applications: From Classical and Advanced Mixing Rules to Association Theories
Georgios M. Kontogeorgis and Georgios K. Folas

all the theories here, but most of them are classified in three main families: (1) chemical theories,[3] APACT;[4–6] (2) lattice–fluid theories;[7] and (3) perturbation theories (SAFT,[8,9] ESD[10] and CPA)[11].

When these theories are expressed in the form of a thermodynamic model, e.g. an EoS, two contributions are often presented: a so-called physical term accounting for the deviations from ideality due to physical forces; and an association term accounting for the effect of hydrogen bonding and other quasi-chemical interactions. The two contributions are not always entirely separable (see Section 7.3.1 for the chemical theory) but in general terms the compressibility factor, Z, of association EoS is often expressed as:

$$Z = Z^{phys} + Z^{assoc} \qquad (7.1)$$

where *phys*, *assoc* indicate the physical and association contributions to the compressibility factor, respectively.

The approach used for describing the hydrogen bonding is different in the three categories of association models:[2]

- Chemical theories are based on the formation of new species and the extent of association is determined by the number of oligomers formed, as a function of density, temperature and composition. The new species have essentially the same molecular properties as their constituent monomers.
- Lattice (often called quasi-chemical) theories account for the number of bonds formed between segments of different molecules that occupy adjacent sites in the lattice. The number of bonds determines the extent of association.
- In the perturbation theory, the total energy of hydrogen bonding is calculated from statistical mechanics and the important parameter for hydrogen bonding is, in this case, the number of bonding sites per molecule.

All these theories have been extensively developed since the late 1980s (although the chemical theory is older) and the engineer now has a number of alternatives to select in terms of 'closed-form' thermodynamic models (EoS), where the hydrogen bonding effect is included as a separate term. An important conclusion, independently reached by several investigators,[1,12,13] is that all three theories, despite their different physical origin and formulation, result in many cases, e.g. for acid or alcohol mixtures, in essentially identical mathematical expressions for the hydrogen bonding contribution.

7.3 The chemical and perturbation theories

7.3.1 Introductory thoughts: the separability of terms in chemical-based EoS

The chemical theory used in the development of thermodynamic models is an old concept discussed in several textbooks.[14,15] The earliest developments go back to the concepts of Dolezalek.[16] These earlier chemical models are often expressed in the form of activity coefficient models and different equations are derived depending on the 'chemical reaction' assumed, e.g. the formation of dimers, oligomers, or 1–1 complexes such as in the case of chloroform and acetone. The equilibrium constant or constants are adjustable parameters, fitted to experimental mixture data. Our discussion here will focus on the chemical theories expressed in the form of 'closed' EoS which can be used over extensive temperature and pressure ranges and for different types of reaction schemes.

Heidemann and Prausnitz[3] were the first to develop an association EoS in 'closed form' using the chemical theory. Two major types of chemical EoS have since been developed: those having a non-cubic physical term (APACT)[17,18] and those for which cubic EoS are used in the physical part.[5,6,19] Not all chemical-type EoS result in separable 'physical' and 'chemical' terms. Whether this will be the case or not depends on the combining rules used to relate the EoS parameters for the oligomers to those of monomers. These combining rules also affect the expression of the 'chemical term', $Z^{chem} = n_T/n_0$, where n_T is the true number of moles that exist after association and n_0 is the apparent (or superficial) number of moles, i.e. without association. The basic approaches will be outlined here, while more information and a detailed presentation of several of the well-known chemical theories (EoS) is given in Appendix 7.A. Anderko[5,6] assumed complete separability of the physical and chemical terms and expressed his EoS as:

$$Z = Z^{phys} + Z^{assoc} = Z^{phys} + Z^{chem} - 1 = Z^{attr} + Z^{rep} + \frac{n_T}{n_0} - 1 \tag{7.2}$$

The approach of Anderko

In Anderko's model, the attractive and repulsive terms are given by a cubic EoS.[20] Naturally the association term $(n_T/n_0) - 1$ in Equation (7.2) disappears if there are no associating molecules present, i.e. the true and apparent numbers of moles are identical ($n_T = n_0$). The chemical term, n_T/n_0, depends on the reaction scheme. In the case of pure compounds forming oligomers, which is a reasonable assumption for alcohols and phenols, it can be shown that:

$$Z^{chem} = \frac{n_T}{n_0} = \frac{2}{1 + \sqrt{1 + 4\rho KRT}} \tag{7.3}$$

where ρ is the density, T is the temperature (in K), R is the ideal gas constant and K is the equilibrium constant.

Equation (7.3) is essentially equivalent to X^A, the so-called site fraction used in perturbation theories (e.g. SAFT and CPA), i.e. the fraction of molecules non-bonded at site A, see also section 7.4 (Equation (7.22)). The key property in Equation (7.3) is the equilibrium constant K, which is related to the enthalpy (ΔH) and entropy of hydrogen bonding (ΔS):

$$\ln K = -\frac{\Delta H}{RT} + \frac{\Delta S}{R} \tag{7.4}$$

The number of monomers, n_1/n_0, is in this case equal to $(n_T/n_0)^2$.

The approach of Anderko, though justified via some unconventional combining rules for the oligomers (see Appendix 7.A), is not rigorous. In the general case, the choice of the combining rules for the oligomers affects the functional form of the EoS obtained from the chemical theory. Moreover, the physical term can be either a non-cubic or a cubic EoS.

The approach of Heidemann and Prausnitz

In the approaches of Heidemann and Prausnitz[3] and in the APACT EoS,[4] the most well-known and accepted combining rules for the relation between the EoS parameters of the oligomers (a_i, b_i) and those of the monomers (a_1, b_1) are used (here shown for the energy and the co-volume parameter of a cubic EoS):

$$\begin{aligned} a_i &= i^2 a_1 \\ b_i &= i b_1 \end{aligned} \tag{7.5}$$

where $i = n_0/n_T$ is the oligomer number index which gives the extent of association (it is the inverse of the chemical term defined previously, Equations (7.2) and (7.3)). In this case it can be shown (see Problem 2.ii on the companion website at www.wiley.com/go/Kontogeorgis) that:

$$Z = Z^{attr} + \frac{n_T}{n_0} Z^{rep} \tag{7.6}$$

where:

$$Z^{chem} = \frac{n_T}{n_0} = \frac{2}{1 + \sqrt{1 + 4\rho KRTe^g}} \tag{7.7}$$

with the g parameter depending on the repulsive term of the EoS:

$$g = \int_0^\eta \left(\frac{Z^{rep} - 1}{\eta} \right) d\eta \qquad \eta = \frac{b}{V} \tag{7.8}$$

Note 1: In the general case, the g factor depends on both the repulsive and attractive terms of the EoS via a rather complex function, but if the 'usual' combining rules, Equation (7.5), are used, then g is only a function of the repulsive term via Equation (7.8).

Note 2: In the case of either complete separability of physical and chemical terms or 'ideal chemical' behavior (the physical compressibility factor is unity), $g = 0$ and the e^g factor can be ignored.

Note 3: Many authors have discussed the importance of the e^g term in chemical theories but there is no overall consensus on its importance. Anderko[21] reports that chemical theories without this e^g term often perform best for mixture phase equilibria. Economou and Donohue[22] point out that, if this term is 'artificially' ignored, there are serious consistency problems.

7.3.2 Beyond oligomers and beyond pure compounds

Equations (7.3) and (7.7) are essentially the Mecke–Kempter equation solved when an infinite number of oligomers are assumed. They can be expected to be valid for a variety of associating compounds, e.g. alcohols, amines, phenols and pyridines.[5] They do not represent dimers (model used for organic acids) or three-dimensional networks, e.g. as present in water.

The chemical theory can be solved analytically in the case of dimers (organic acid model):

$$Z^{chem} = \frac{n_T}{n_0} = \frac{2(1 - KRT\rho)}{1 - 4KRT\rho + \sqrt{1 + 8KRT\rho}} \tag{7.9}$$

The chemical theory cannot be rigorously solved in the case of three-dimensional networks, a model used often for water. Anderko[19] proposed the following empirical equation for water:

$$Z^{chem} = \frac{n_T}{n_0} = \frac{1}{1 + KRT\rho + \alpha(KRT\rho)^2} \tag{7.10}$$

Equation (7.10) accounts, in an empirical way, for the three-dimensional structure of water. The parameter α is equal to 8.2 in Anderko's model. A better way to extend the chemical theory, e.g. APACT, to water and three-dimensional networks is to make use of the similarity and analogies with the perturbation theories, as discussed in Section 7.4 (see also Economou and Donohue[23]).

7.3.3 Extension to mixtures

The chemical theory can be extended to mixtures but analytical solutions cannot always be obtained. In the case of a multicomponent associating mixture containing any number of compounds following the infinite equilibria scheme (Equations (7.3) and (7.7)) and inert compounds, the general solution can be approximated by the equation:[24,5,6]

$$Z^{chem} = \sum_{i=1}^{n} \frac{2X_{A_i}}{1+\sqrt{1+4RT\rho\left(\sum_{j=1}^{n} K_{ij}X_{A_j}\right)}} + \sum_{k=1}^{\nu} X_{B_k} \tag{7.11}$$

where K_{ij} is either the self- or cross-association constant and X_{A_i} denotes the 'true' mole fraction of a species of stoichiometry A_i.

It can be seen from Equation (7.11) that extension to cross-associating systems, e.g. alcohol–alcohol, in the chemical theory requires a combining rule for the cross-equilibrium constant K_{ij}. One popular choice is the geometric mean rule:

$$K_{ij} = \sqrt{K_i K_j} \tag{7.12}$$

Equation (7.12) can be obtained from the definition of the equilibrium constant, Equation (7.4), and well-known theoretically accepted combining rules for the cross-enthalpy and entropy of hydrogen bonding:

$$\Delta H_{ij} = \frac{\Delta H_i + \Delta H_j}{2} \tag{7.13a}$$

$$\Delta S_{ij} = \frac{\Delta S_i + \Delta S_j}{2} \tag{7.13b}$$

The similarities between chemical and perturbation theory are discussed in Section 7.4.3 as well as in Chapters 8 and 9 where the SAFT and CPA EoS are presented.

Chemical EoS such as APACT, the Anderko EoS and others have been applied to alcohol–hydrocarbon and water–hydrocarbon VLE and LLE[25] as well as to mixtures containing organic acids.[26,27] Few results have been presented for multicomponent systems. Economou *et al.*[18] applied APACT to cross-associating systems exhibiting 'Lewis acid–Lewis base' interactions such as acetone–chloroform, alcohols–chloroform and alcohols–ketones.

7.3.4 Perturbation theories

The perturbation theory is based on the solution of integral equations using a potential function that mimics that of hydrogen bonds. Especially pertinent is Wertheim's contribution[28–31] which has been the basis for the association term of several EoS, such as the SAFT family. Wertheim developed a model for systems with a repulsive core and multiple attractive sites capable of forming chains and closed rings. Wertheim expanded the Helmholtz energy in a series of integrals of molecular distribution functions and showed that many integrals in the series must be zero and hence a cluster expansion was obtained. In this way an expression for the Helmholtz free energy was derived.

The key relation in Wertheim's theory is between the residual Helmholtz energy due to association and the fraction of molecules not bonded at a specific site, X^A. X^A is related to the association strength, $\Delta^{A_iB_j}$, which is given by the following equations for SAFT and CPA, respectively:

$$\Delta^{A_iB_j} = d_{ij}^3 g_{ij}(d_{ij})^{seg} \kappa^{A_iB_j} \left[\exp(\varepsilon^{A_iB_j}/kT) - 1\right] \quad \text{[SAFT]} \tag{7.14}$$

$$\Delta^{A_iB_j} = g(\rho)\left[\exp\left(\frac{\varepsilon^{A_iB_j}}{RT}\right) - 1\right] b_{ij}\beta^{A_iB_j} \text{ and } b_{ij} = \frac{b_i + b_j}{2} \quad \text{[CPA]} \tag{7.15}$$

The radial distribution function, g, is discussed in Chapters 8 and 9.

The association term is, in terms of the compressibility factor, expressed by the following equation:[13]

$$Z^{\text{assoc}} = -\frac{1}{2}\left(1 + \rho\frac{\partial \ln g}{\partial \rho}\right)\sum_i x_i \sum_{A_i}(1 - X_{A_i}) \tag{7.16}$$

Note 1: The first-order perturbation theory allows for chainlike and treelike associated clusters, but no closed rings.

Note 2: The activity of each site is independent of bonding at other sites on the same molecule. Hence, the effects of site hindrance are ignored.

Note 3: The repulsive cores of the molecules prevent more than two molecules from bonding at a single site, while no site on a specific molecule can bond simultaneously to two sites on another molecule. Finally, double bonding between molecules is not allowed.

Note 4: No intramolecular association is accounted for.

Note 5: The strength of association is represented by a square-well potential. The association strength is characterized by two parameters, the association energy ($\varepsilon^{A_iB_j}$, the well depth) and the association volume ($\kappa^{A_iB_j}$ or $\beta^{A_iB_j}$, the well width). These parameters can be related to spectroscopic properties as discussed in the next section.

7.4 Spectroscopy and association theories

7.4.1 A key property

A key property in all three types of association theories is the fraction of molecules present in monomeric form, hereafter simply called 'monomer mole fraction'. As shown in Section 7.3.4 and discussed in Chapter 8, the key property in perturbation theories such as SAFT is X^A, i.e. the fraction of molecules *not* bonded at site A. If we have two hydrogen bonding sites A and B and assuming that the activity of a site is independent of bonding at the other sites of the same molecule, the fraction of molecules present as monomers, i.e. the molecular monomer fraction, is:[8]

$$X_1 = X^A X^B = (X^A)^2 \tag{7.17}$$

assuming that $X^A = X^B$.

Table 7.1 *Bonding types in alcohols and water illustrating the basic association schemes, 2B, 3B and 4C, which are widely used in perturbation theories X_1 is the monomer mole fraction. The 2B scheme implies one proton donor and one proton acceptor per molecule, while the 4C scheme involves two proton donors and two proton acceptors per molecule. The 'asymmetric' 3B scheme has either two proton donors and one proton acceptor or one proton donor and two proton acceptors per molecule*

Species	Formula	Type	Site fractions
Alcohol	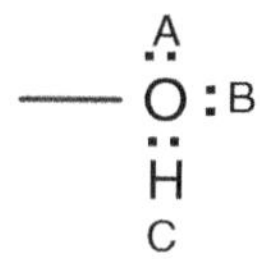	3B	$X^A = X^B; X^C = 2X^A - 1$ $X_1 = X^A X^B X^C$
	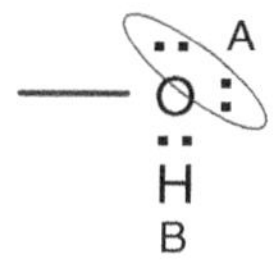	2B	$X^A = X^B$ $X_1 = X^A X^B$
Water	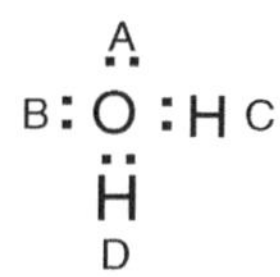	4C	$X^A = X^B = X^C = X^D$ $X_1 = X^A X^B X^C X^D$
	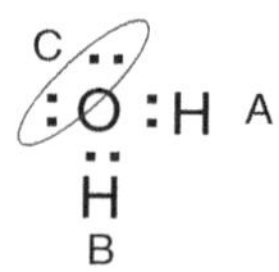	3B	$X^A = X^B; X^C = 2X^A - 1$ $X_1 = X^A X^B X^C$
	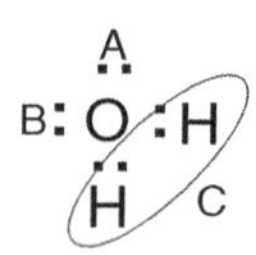	3B[a]	$X^A = X^B; X^C = 2X^A - 1$ $X_1 = X^A X^B X^C$
	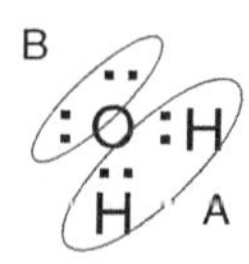	2B	$X^A = X^B$ $X_1 = X^A X^B$

[a] Assuming one proton donor and two proton acceptors for water results in the same mathematical expression as if two proton donors and one proton acceptor are assumed; however, this is physically not likely.

Equation (7.17) is valid, for example, for alcohols assuming two equivalent sites, which is the so-called 2B scheme in the SAFT terminology from Huang and Radosz;[9] see Table 7.1 and Chapter 8 (Tables 8.11 and 8.12). The equivalency of sites is a statistical consideration that all site types on a molecule are equally likely to participate in hydrogen bonding. A short explanation of Equation (7.17) is given in Figure 7.1 for an associating two-site molecule.

In general, the monomer mole fraction (fraction of molecules of compound 1 that are *not* bonded at any site) is given as:

$$X_1 = \prod_K X^K \tag{7.18}$$

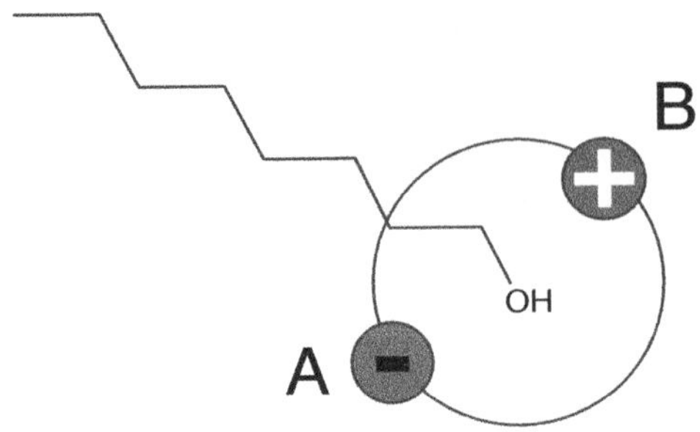

Figure 7.1 *An associating two-site molecule: the fraction of bonded molecules = fraction of molecules bonded at A + fraction of molecules bonded at B − fraction of molecules bonded at A and B = $(1 - X^A) + (1 - X^B) - (1 - X^A)(1 - X^B) = 1 - X^A X^B$. Thus, the fraction of non-bonded molecules = $1 - (1 - X^A X^B) = X^A X^B$*

For water, for which a realistic situation could be to assume four equivalent sites (A = B = C = D; Table 7.1), the monomer fraction is:

$$X_1 = X^A X^B X^C X^D = (X^A)^4 \tag{7.19}$$

The fraction of molecules present as m-mers is:[8]

$$X(m\text{-mer}) = m(X^A)^2(1 - X^A)^{m-1} \tag{7.20}$$

and the average chain length is given by:

$$m_{ave} = \frac{1}{X^A} \tag{7.21}$$

7.4.2 Similarity between association theories

It has been previously shown[1,11,32] that the lattice, chemical and perturbation theories result, in many cases, in mathematically similar expressions for the monomer fraction. This is a significant result, considering the different theoretical background, assumptions and derivations of these theories. These similarities have been illustrated for different cases, e.g. one- or two-site hydrogen bonding molecules, and for both pure compounds and when they are in mixtures with inert compounds. We will present these similarities here in the case of pure compounds following the two-site scheme (2B in Table 7.1), an assumption widely used for alcohols and amines, for example.

Perturbation theories

In SAFT-type theories, the site fraction in the 2B scheme is given by (see also Chapter 8):

$$X^A = \frac{-1 + \sqrt{1 + 4\rho\Delta}}{2\rho\Delta} \tag{7.22}$$

where ρ is the molar density and Δ is the association strength, defined for SAFT and CPA in Equations (7.14) and (7.15) (for more information, including a discussion about the radial distribution functions, g, see Chapters 8 and 9 where the SAFT and CPA EoS are presented). Thus, using equations (7.17) or (7.18), it can be shown that the monomer fraction of a pure alcohol molecule, following the 2B scheme, is given as:

$$X^{monomer} = \frac{2}{1+2\rho\Delta+\sqrt{1+4\rho\Delta}} \tag{7.23}$$

Chemical theories

In the chemical theories discussed earlier in the chapter, the monomer fraction is simply the ratio n_1/n_0 (see after Equation (7.4)). For example, in the case of the APACT chemical-based EoS and assuming we have linear oligomers, which represent the physically correct picture for alcohols corresponding to the 2B site molecules (in the SAFT terminology), we get for the monomer fraction (using Equation (7.7)):

$$X^{monomer} = \frac{2}{1+2\rho KRTe^{g}+\sqrt{1+4\rho KRTe^{g}}} \tag{7.24}$$

K is the equilibrium constant and Equation (7.24) is mathematically identical to Equation (7.23) under the assumption that:

$$\Delta = KRTe^{g} \tag{7.25}$$

This equivalency shows the link between chemical and perturbation theories and their 'key parameters', the association strength and the equilibrium constant, and is further discussed in Section 7.4.3. Other chemical theories, e.g. as shown by Prausnitz *et al.*,[15] result in similar expressions for the monomer fraction of oligomer-creating associating molecules.

Lattice–fluid theories, e.g. LFHB, NRHB

In the modeling of alcohols with lattice–fluid theories, usually two hydrogen bonding sites are assumed (i.e. one proton donor and one proton acceptor). Then, the final result for the fraction of unbonded molecules (monomer fraction) under certain assumptions (discussed in Appendix 7.B) is:

$$X_1 = \frac{2}{1+2\rho K+\sqrt{1+4\rho K}} \tag{7.26}$$

where ρ is the density and:

$$K = \frac{v^*}{r}\exp\left(-\frac{\Delta G^{hb}}{RT}\right) \tag{7.27}$$

This equation is mathematically similar to the expressions from the chemical and perturbation theories, i.e. Equations (7.23) and (7.24). It is interesting to note that several investigators, e.g. Brinkley and Gupta,[33] have used the non-bonded *site* fraction expression from the lattice–fluid hydrogen bonding (LFHB) theory to compare against experimental *monomer* fraction data. The results reported in their manuscripts do not compare favorably with the experimental data. When, however, Equation (7.26) is used, the performance of LFHB is satisfactory, as von Solms *et al.*[34,35] have shown both for pure alcohols and alcohol–alkane mixtures. Methanol is an exception, see Figure 7.10.

7.4.3 Use of the similarities between the various association theories

Equation (7.25) is obtained upon comparing the monomer fraction equations calculated from chemical and perturbation theories for two-site molecules (oligomer creation). In the general case (see Chapter 16, Equation (16.2)), a proportionality constant, C, should be used ($\Delta = CKRTe^{g}$).

Assuming that the density-dependent terms of the two expressions (e^{g} of the chemical theory and g of the perturbation theory) are also of similar magnitude or can be ignored, then it can be approximately written that (where component and site indicators have been dropped):

$$\Delta \propto KRT \Rightarrow \left[\exp\left(\frac{\varepsilon}{RT}\right) - 1\right]\beta b \frac{1}{RT} \propto \exp\left(\frac{-\Delta H}{RT}\right)\exp\left(\frac{\Delta S}{R}\right) \tag{7.28}$$

Further assuming that:

$$\exp\left(\frac{\varepsilon}{RT}\right) - 1 \approx \exp\left(\frac{\varepsilon}{RT}\right)$$

it is easily shown that:

$$\begin{aligned} &\exp\left(\frac{\varepsilon}{RT}\right) \propto \exp\left(\frac{-\Delta H}{RT}\right) \Rightarrow \varepsilon \propto -\Delta H \\ &\frac{\beta b}{RT} \propto \exp\left(\frac{\Delta S}{R}\right) \Rightarrow \beta \propto \exp\left(\frac{\Delta S}{R}\right) \end{aligned} \tag{7.29}$$

Equations (7.28) and (7.29) are written for the case of the CPA model, but similar equations can be written for SAFT as well. In the case of SAFT, the term βb should be replaced by $\kappa\sigma^3$ or κd^3.

Equation (7.28) shows that the association energy is proportional to the enthalpy of hydrogen bonding, while from Equation (7.29) we observe that the association volume parameter is related to the entropy of hydrogen bonding. Equivalent equations can be written between the lattice and the other theories.

These similarities between the various association theories have been extensively used in the literature, as follows:

- By Gupta and Johnston[36] in applying lattice–fluid theories using SAFT concepts.
- By Economou and Donohue[23] who used Equation (7.25) and the SAFT equations for water in order to extend the APACT chemical theory to aqueous mixtures.
- By Wolbach and Sandler[37,38] and Sandler *et al.*[39] who via molecular orbital theory (quantum chemistry) calculated theoretical values for the enthalpies and entropies of association of several alcohols, acids and

water and then used them in SAFT (see also Chapter 16). Yarrison and Chapman[40] compared the performance of SAFT for methanol using such theoretically based parameters and the parameters obtained from the 'classical' approach, i.e. optimizing vapor pressures and liquid densities.

Moreover, Wolbach and Sandler[41,42] used molecular-orbital-based enthalpies and entropies of association for testing the combining rules of the association parameters, in particular those shown in Equations (7.12) and (7.13). More information is provided in Chapter 16, especially Section 16.3.

7.4.4 Spectroscopic data and validation of theories

Spectroscopy (FTIR, NMR) can be used for obtaining experimental data for the monomer fractions of associating compounds. These data can be used for testing association theories, since as shown previously, Equation (7.18), the monomer fraction can be explicitly calculated from these theories. Monomer fraction data for a few compounds have been reported in the literature, e.g. for pure water, alcohols and for some mixtures of alcohols or glycolethers with alkanes, and Table 7.2 presents some references containing experimental monomer fraction data. Table 7.3 presents some investigations where these data were used for testing chemical, lattice or perturbation association theories.

Table 7.2 *Experimental spectroscopic data for monomer fractions*

Compound/mixture	Conditions (T_r is the reduced temperature)	Reference
Water, methanol, ethanol	$0.5 < T_r < 0.95$	Luck[43]
Propanol	$0.54 < T_r < 0.61$	Lien[44]
Octanol	$0.43 < T_r < 0.57$	Fletcher and Heller[45] Palombo *et al.*[46]
Propanol–heptane	15–55 °C	Lien[44]
Methanol–hexane	Ambient temperature	Martinez[47]
Methanol–CCl_4	20 °C	Prausnitz *et al.*[15]
2-Methoxyethanol–hexane	35, 45 °C	Brinkley and Gupta[33]
2-Butoxyethanol–hexane	Intra- and intermolecular hydrogen bonding	
2-Methoxyethanol–hexane 2-Ethoxyethanol–hexane	30–50 °C	Missopolinou *et al.*[64]
Pentanol–hexane Hexanol–hexane	25 and 35 °C	Gupta and Brinkley[48]
Hexanol–toluene, *m*-xylene	25–45 °C	Brinkley and Gupta[49]
Cyclohexanol–toluene	11, 61 °C	Brinkley and Gupta[49]
Methanol–hexane Ethanol–hexane Pentanol–hexane	Room temperature	Asprion *et al.*[50]
Propanol–hexane Hexanol–hexane	25–40 °C	Asprion *et al.*[50]
Methanol–hexane Ethanol–hexane, heptane Propanol–hexane, heptane Pentanol–hexane Hexanol–hexane	23.3 °C	Von Solms *et al.*[35]

Table 7.3 *Investigation of association theories against spectroscopic data for monomer fractions*

Reference	Systems studied	Models tested
Peschel and Wenzel[51]	Methanol	Chemical theory
Gupta *et al.*[52]	Water, methanol, ethanol	LFHB
Koh *et al.*[53]	Methanol	SAFT
Brinkley and Gupta[33]	Alkoxyalcohols–hexane	LFHB
Gupta and Brinkley[48]	Pentanol–hexane Hexanol–hexane	LFHB
Muthukumaran *et al.*[54]	Ethanol	LFHB
Clark *et al.*[55]	Water	SAFT–VR
Von Solms *et al.*[34]	Water, alcohols	CPA, sPC–SAFT
Aparicio-Martinez and Hall[56]	Water	CPA based on SRK CPA based on PR SAFT PC–SAFT
Von Solms *et al.*[35]	Alcohol–alkanes	CPA, sPC–SAFT
Grenner *et al.*[57]	Water	CPA, sPC–SAFT
Tsivintzelis *et al.*[32]	Alcohols (pure and mixtures)	sPC–SAFT, NRHB

[a]LFHB = lattice–fluid hydrogen bonding, NRHB = non-random hydrogen bonding.

Figures 7.2–7.9 present some typical results, while additional results will be presented in subsequent chapters for the perturbation theories. The following points summarize the major conclusions from these investigations.

Comments on the experimental data

1. Considering the importance of spectroscopic data for monomer fractions in testing association theories, it is somewhat surprising how few data are available in the literature. Some of them are rather old, e.g. the well-known data from Luck,[43] which are widely used for water, methanol and ethanol, are almost 30 years old (Figure 2.5 in Chapter 2). Note that Luck is one of the few authors who present site fractions; the majority of the investigations in the literature report monomer fractions. The fraction of monomeric molecules can be calculated from the spectroscopically determined fraction of free OH groups. Similarly the fraction of free OH groups can be calculated (open squares in Figure 2.5) from the experimentally determined monomeric fraction. The transformation between site and monomer fraction in Luck's work must follow the assumptions adopted by Luck, i.e. 4 site for water and 3 site for the alcohols.
2. Luck's well-known data are presented as % free OH groups, essentially site fractions (Figure 2.5). These must be transformed into monomer fractions if comparisons are to be made using Equation (7.18) for the monomer fractions from perturbation, chemical and lattice theories. Alternatively, the experimental site fraction data could be compared against the predictions of the models for the site fractions. This point is not always properly appreciated in the literature, where sometimes the concepts of monomer fractions and site fractions (from experiment and from models) are confused, thus comparing unequal quantities, e.g. see discussions by Gupta and Brinkley[48], Gupta *et al.*[52] and Kahl and Enders.[58] See a further discussion by von Solms *et al.*[34,35].

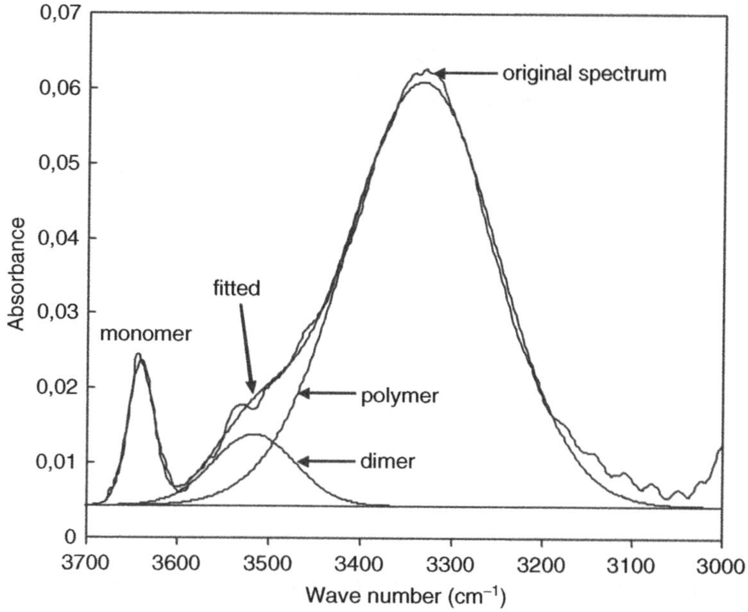

Figure 7.2 Deconvolution of an FTIR spectrum for the mixture ethanol–n-heptane with 1 wt% ethanol. The spectra are deconvoluted into monomer, dimer and higher mer. The peak on the left at around 3650 cm^{-1} represents the monomeric ethanol molecules (no hydrogen bonding), the peak at around 3500 cm^{-1} represents ethanol dimers and the large peaks with a maximum at around 3300 cm^{-1} represent ethanol molecules in longer chains (polymers). Reprinted with permission from Fluid Phase Equilibria, Measurement and modeling of hydrogen bonding in 1-alkanol + n-alkane binary mixtures by N. von Solms, L. Jensen et al., 261, 1–2, 272 Copyright (2007) Elsevier

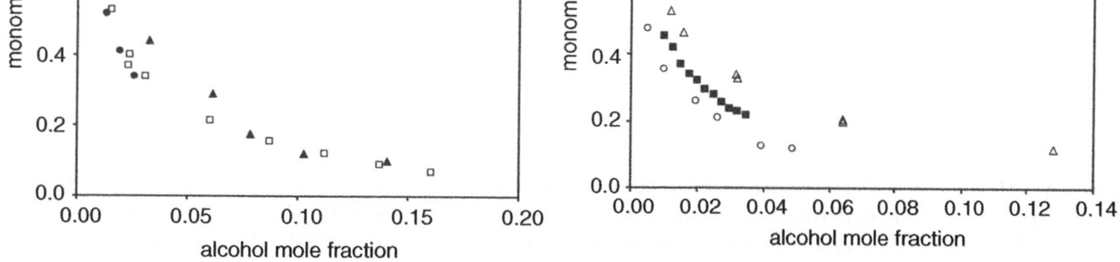

Figure 7.3 Comparisons of experimental monomer fractions for methanol + hexane (left) and pentanol + hexane (right). The data indicated as 'Jensen 2005' have been published by von Solms et al.[35]

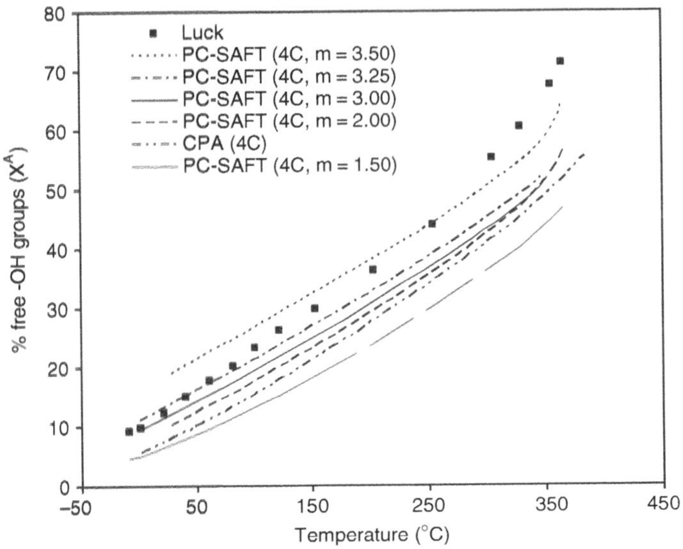

Figure 7.4 *Data and modeling of percentage free OH groups (X^A fraction) against temperature for water. Experimental data are from Luck.[43] The lines represent calculations with PC–SAFT and CPA and different parameter sets. CPA and PC–SAFT ($m = 2$–3.5) results and parameters are from von Solms et al.[34] PC–SAFT ($m = 1.5$) results are from Grenner et al.[59]*

3. The interpretation of spectroscopic data into monomer fractions is not entirely straightforward, see discussion in Figure 7.2. The spectroscopic measurements provide the monomer fraction, not the unbonded fraction. Wertheim's theory only differentiates between monomer and the 'other' fractions.

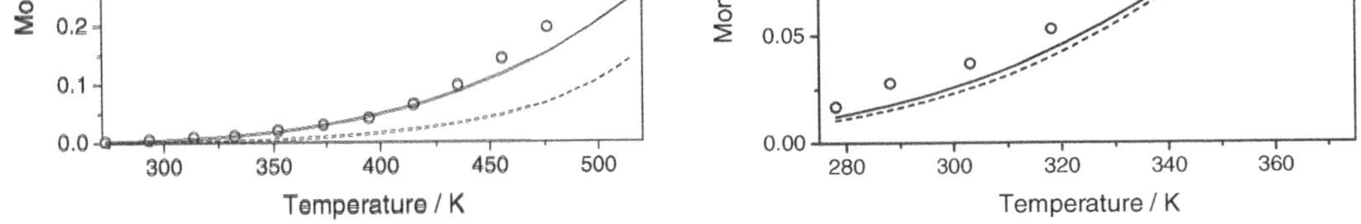

Figure 7.5 *Monomer fraction for pure alcohols with two association theories. Experimental data (points), NRHB (solid lines) and sPC–SAFT predictions (dashed lines) for: (a) methanol and (b) 1-octanol. Reprinted with permission from Ind. Eng. Chem. Res., Evaluation of the Nonrandom Hydrogen Bonding (NRHB) Theory and the Simplified Perturbed-Chain–Statistical Associating Fluid Theory (sPC-SAFT). 2. Liquid–Liquid Equilibria and Prediction of Monomer Fraction in Hydrogen Bonding Systems by I. Tsivintzelis, A. Grenner et al., 47, 15, 5651–5659 Copyright (2008) American Chemical Society and Errata, Ind. Eng. Chem. Res, 2009, 48: 7860*

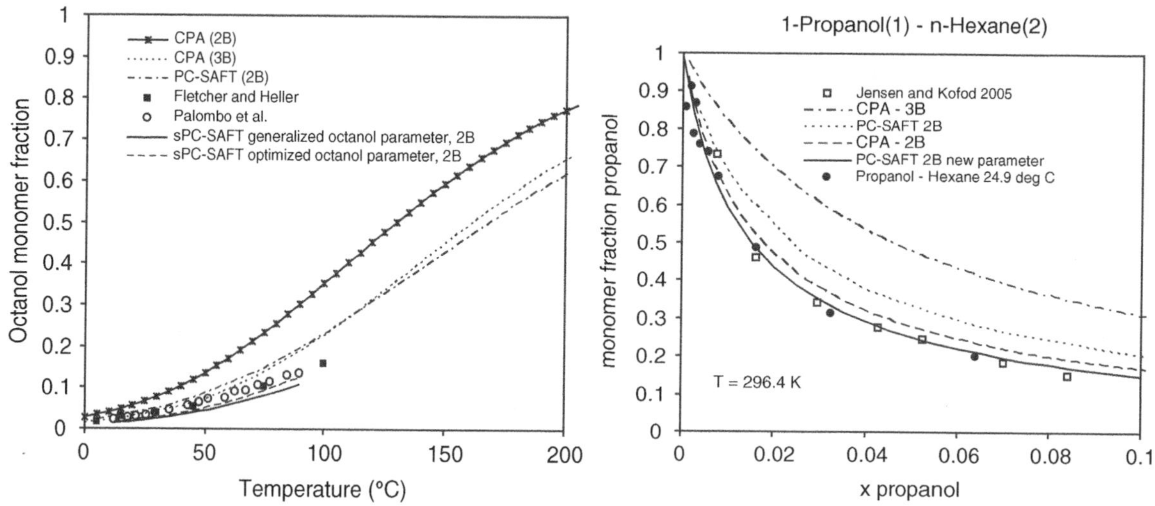

Figure 7.6 Left: Monomer fraction of 1-octanol calculated with CPA and PC–SAFT against experimental data from two sources. Right: Propanol monomer fraction for 1-propanol–hexane mixture at approximately 298 K. 'Jensen and Kofod 2005' data are published by von Solms et al.[35] The '2B new parameter set' for PC–SAFT is from Grenner et al.[60] and performs better than when the parameters from Gross and Sadowski[61] are used. For CPA, the 2B set performs better than the 3B set

4. There are more experimental monomer fraction data for mixtures than for pure compounds, possibly because the mixture measurements are often reported at room or low temperatures. There is fair but not excellent agreement between experimental monomer fraction data from different sources, as shown in Figure 7.3.

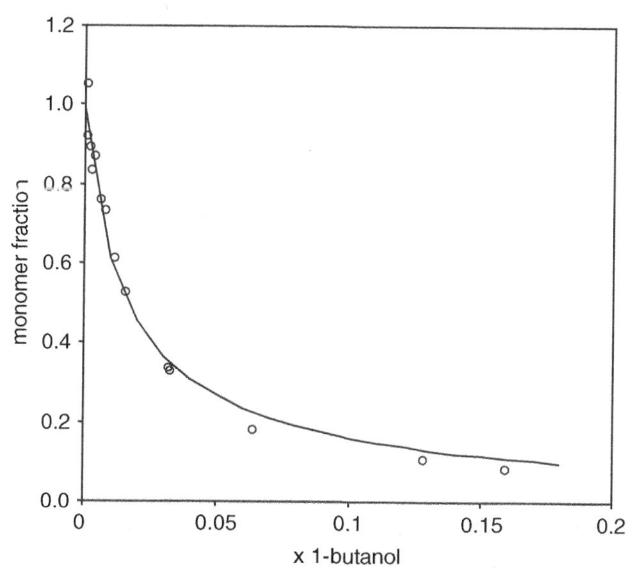

Figure 7.7 The 1-butanol monomer fraction in the 1-butanol–hexane mixture at 298 K with sPC–SAFT using the 2B parameters from Grenner et al.[60]

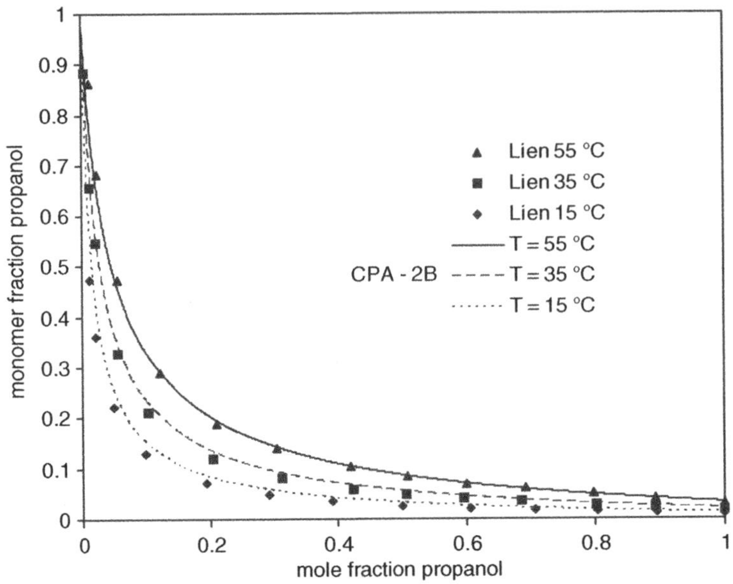

Figure 7.8 *Propanol monomer fraction in propanol–heptane mixture at three temperatures using the CPA EoS and the 2B scheme. Reprinted with permission from Fluid Phase Equilibria, Measurement and modeling of hydrogen bonding in 1-alkanol + n-alkane binary mixtures by N. von Solms, L. Jensen et al., 261, 1–2, 272 Copyright (2007) Elsevier*

Comparisons against monomer fraction data for pure compounds

1. Extensive comparisons between spectroscopic data and monomer fraction predictions from association models have been presented for various versions of lattice–fluid and perturbation theories. As shown in Figures 7.4–7.6 for a few typical compounds (water, methanol, octanol), the performance of the models against each other is similar. Similar results are obtained for other alcohols as well, as shown in the investigations by von Solms *et al.*[34] and Tsivintzelis *et al.*[32].
2. In terms of pure compounds' monomer fraction, the agreement is fair for methanol (underestimation of monomer fraction data; Figure 7.5, left) but satisfactory for the heavier alcohols (with all models studied). No significant differences are observed between the perturbation and lattice theories studied, sPC–SAFT and NRHB.
3. Different optimized parameter sets (representing equally well vapor pressure and liquid density data) may yield different results for the monomer fraction, as shown in Figure 7.4. Similar conclusions from their analysis with CPA and SAFT variants for water monomer fraction are drawn from the work of Clark *et al.*[55] and Grenner *et al.*[57].
4. Spectroscopic data are useful for investigating which association scheme is most suitable for specific hydrogen bonding compounds. The studies with CPA and PC–SAFT for water and alcohols resulted in the following conclusions: the 4C is the best scheme for water, while for alcohols the picture is less clear and distinction between the 2B and 3B schemes is not always possible. It seems that the best agreement is obtained for methanol using the 3B scheme, while the 2B scheme performs best for ethanol and the heavier alcohols – and for both CPA and PC–SAFT.

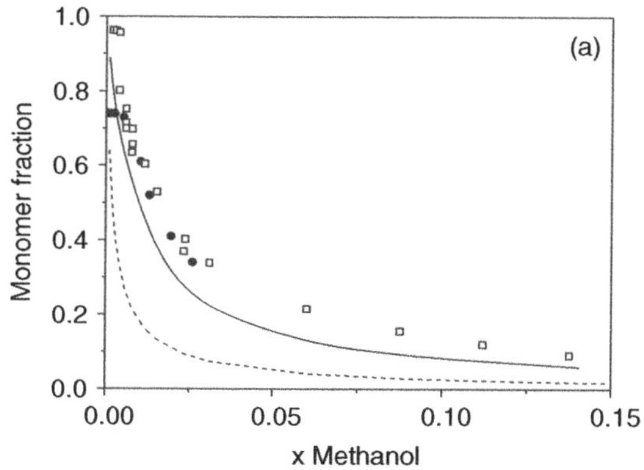

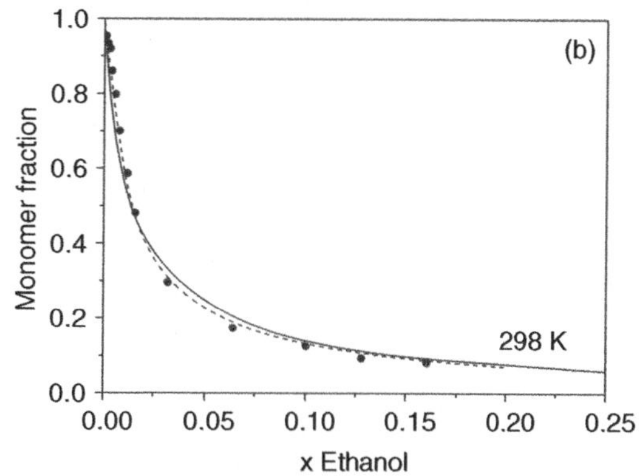

Figure 7.9 *Monomer fraction for hexane–alcohol mixtures. Experimental data (points), NRHB (solid lines) and sPC–SAFT predictions (dashed lines) for mixtures of hexane with: (a) methanol; (b) ethanol. Reprinted with permission from Ind. Eng. Chem. Res., Evaluation of the Nonrandom Hydrogen Bonding (NRHB) Theory and the Simplified Perturbed-Chain–Statistical Associating Fluid Theory (sPC-SAFT). 2. Liquid–Liquid Equilibria and Prediction of Monomer Fraction in Hydrogen Bonding Systems by I. Tsivintzelis, A. Grenner et al., 47, 15, 5651–5659 Copyright (2008) American Chemical Society*

Comparisons against monomer fraction data for mixtures

1. Lattice–fluid and perturbation theories predict qualitatively correctly the dependency of monomer fraction on temperature, concentration and alcohol chain length. The results shown in Figures 7.8–7.9 are with zero interaction parameters, i.e. they are pure predictions. In many cases, quantitative agreement between theory and experiment is achieved. In all cases, there is excellent qualitative agreement with experimental data and trends. For example, the monomer fraction of different alcohols with the same alkane appears to be roughly independent of the alcohol chain length in agreement with the experimental data from Asprion *et al.*[50]. One might have expected, however, that the monomer fraction would be somewhat higher for the heavier alcohols.

7.5 Concluding remarks

- Chemical, quasi-chemical (lattice) and perturbation theories account explicitly for hydrogen bonding in mixtures. Despite their different derivations and physical background, they yield similar expressions for the hydrogen bonding contributions in many cases, e.g. for dimers and oligomers, which are the association schemes often used for organic acids and alcohols, respectively.
- There are many chemical theory-based equation of state theories, many originating from the pioneering work of Heidemann and Prausnitz.[3]. The physical and chemical terms are, in these theories, not always separable.
- Experimental monomer fraction data can be extracted from spectroscopic measurements and such data are available for pure water and alcohols as well as for several alcohol–alkane mixtures. Most authors report them as 'monomer fraction', a few report them as 'site fraction'.
- There is some confusion in the literature about the difference between monomer fraction and site fraction as well as the interpretation of site fractions – correct interpretation enables these data to be reconciled. It is important to compare 'monomer fractions' from the spectroscopic data against the same property (and not site fractions) from the various association theories.

Appendix 7.A Chemical theories and underlying assumptions

Table 7.4 Chemical theories and underlying assumptions

Name of the EoS	EoS – compressibility factor term (Z)	EoS for the physical terms $\eta = \frac{b}{V}$	Combining rules for oligomers a_i, b_i	g function; with $\eta = \frac{b}{V}$
Heidemann and Prausnitz[3]	$Z^{attr} + \frac{n_T}{n_0} Z^{rep}$	CS expression for the repulsive term: $Z^{rep} = -\frac{\eta(4-2\eta)}{(1-\eta)^3}$ A vdW-type attractive term: $Z^{attr} = -\frac{a}{bRT}\eta(1-4\eta)$	$a_i = i^2 a_1$ $b_i = ib_1$	$\frac{4\eta-3\eta^2}{(1-\eta)^2}$
APACT[4]	$Z^{attr} + \frac{n_T}{n_0} Z^{rep}$	Repulsive term is the chain CS equation. It yields CS when $c=1$: $Z^{rep} = -\frac{\eta(4-2\eta)}{(1-\eta)^3}$ Attractive term based on Lennard-Jones potential + extra term for anisotropic effects: $Z^{attr} = Z^{LJ} + Z^{ani}$	$a_i = i^2 a_1$ $b_i = ib_1$ $c_i = ic_1$ (Later the c-rule was modified by Economou and Donohue)[1]	Zero
Anderko[5]	$Z^{attr} + Z^{rep} + \frac{n_T}{n_0} - 1$	A three-parameter cubic EoS:[20] $Z^{rep} = \frac{\eta}{1-\eta}$ $Z^{attr} = -\frac{a}{bRT}\frac{\eta}{1+\eta(d/b+3)+\eta^2 d/b}$	$a_i = i^2 a_1$ $b_i = ib_1$ $b'_i = i^2 b_1$ (repulsive term only!)	Zero

ESD[62,10]	$Z^{attr} + Z^{chain} + \frac{n_T}{n_0} Z^{rep}$	$Z^{rep} = \frac{1}{1-1.9\eta}$ $Z^{chain} = \frac{4c\eta - 1.9\eta}{1-1.9\eta}$ $Z^{attr} = -\frac{9.49\eta q Y}{1+1.7745\eta Y}$		$-\ln(1-1.9\eta)$
Twu *et al.*[26]	$Z^{attr} + \frac{n_T}{n_0} Z^{rep}$	SRK	$a_i = i^2 a_1$ $b_i = ib_1$	$-\ln(1-\eta)$
Hong and Hu[27]	$\frac{n_T}{n_0}(Z^{rep} + Z^{attr})$	SRK	$a_i = ia_1$ $b_i = ib_1$	$-\ln(1-\eta)$ $-\frac{a}{bRT}\ln(1+\eta)$
Junyang and Ying[63]	$\frac{n_T}{n_0}(Z^{rep} + Z^{attr})$	CS expression for the repulsive term A vdW-type attractive term	$a_i = ia_1$ $b_i = ib_1$	$\frac{4\eta - 3\eta^2}{(1-\eta)^2}$ $-\frac{a}{bRT}\eta$

CS = Carnahan–Starling.

- Chemical, lattice and perturbation theories can be used for calculating the monomer fraction in associating compounds and their mixtures and thus spectroscopic data can be used in the development and validation of these theories.
- As chemical, lattice and perturbation theories typically have at least five parameters for pure associating fluids which are often estimated solely based on vapor pressures and liquid densities, spectroscopic data can be used as an extra degree of freedom, i.e. as a property which can be used for identifying the best parameter set.
- Similar results for the water monomer fractions are obtained from the various theories.
- Spectroscopy can help guide the scheme selection. The investigations so far with some perturbation theories (CPA, PC–SAFT) point out that water is best represented as a four-site molecule (4C), methanol as a three-site (3B), while for ethanol and heavier alcohols there is little difference between two- and three-site schemes, with the former (2B) being possibly the best choice.
- Both CPA and PC–SAFT can predict monomer fractions with reasonably good accuracy – correct trends are observed with respect to chain length (for alcohols) and temperature.
- Perturbation theories such as SAFT and CPA ignore certain aspects of association such as steric effects and the hydrogen bonding cooperativity. The agreement of these theories with the spectroscopic-based monomer fraction data for alcohols is satisfactory and therefore there does not seem to be sufficient evidence that such limitations are very important for alcohol mixtures.

Appendix 7.B Hydrogen bonding monomer fractions in lattice–fluid hydrogen bonding (LFHB) theories – the effect of different assumptions

The hydrogen bonding term of the non-random hydrogen bonding (NRHB) equation of state (EoS) is identical to the hydrogen bonding term of the lattice–fluid hydrogen bonding (LFHB) model. The NRHB EoS is described in Chapter 19 (Appendix 19.B). In this approach the number of hydrogen bonds that occur in the system is calculated. However, when the number of hydrogen bonds is known, the number of bonded molecules cannot be obtained. In other words, for a certain number of hydrogen bonds, the number of bonded (or unbonded) molecules may vary. In order to obtain the number of bonded (or unbonded) molecules, the kind of oligomers that occur in the system must be known.

Usually the alcohols are modeled assuming two sites (one proton donor and one proton acceptor) per molecule. For these pure fluids, the fraction of hydrogen bonds in the system is:

$$r\nu_H = \frac{N_H}{N} = 1 + \frac{1 - \sqrt{1+4K}}{2K} \tag{7.30}$$

where:

$$K = \frac{\tilde{\rho}}{r}\exp\left(\frac{-\Delta G^{hb}}{RT}\right) \tag{7.31}$$

(Note that K in Equation (7.31) is equal to ρK, with K as in (7.27), since $\tilde{\rho} = \rho/\rho^* = \rho v^*$.) N_H is the number of hydrogen bonds and N is the total number of molecules (bonds) in the system. Gupta *et al.*[52] applied the LFHB theory to alcohols. They assumed that the number of hydrogen bonds is equal to the number of bonded molecules. In other words, there is no unbonded site in every bonded alcohol molecule. Consequently, the fraction of the monomers is obtained from the following equation:

$$X_1 = 1 - \frac{N_H}{N} \tag{7.32}$$

This is not the fraction of monomeric alcohol (as Gupta *et al.*[52] report it), but it corresponds exactly with the non-bonded site fraction X^C. This equivalency is also illustrated in the work of Economou and Donohue.[1]

From Equations (7.30) and (7.32) we obtain:

$$X_1 = \frac{\sqrt{1+4K}-1}{2K} = \frac{2}{1+\sqrt{1+4K}} \tag{7.33}$$

Note that in this approach (the approach of Gupta *et al.*[52]) the number of hydrogen bonds, N_H, is equal to the number of bonded molecules. Consequently, the extent of hydrogen bonding can be calculated by subtracting the number of free alcohol molecules from the total number of alcohol molecules in the system. It appears that Gupta and Brinkley[48] followed this approach when they presented experimental and theoretical results for the percentage hydrogen bonding in alcohol–alkane systems. The authors report that the percentage hydrogen bonding was obtained from the monomer fraction. Economou and Donohue[1] compared the hydrogen bonding term of lattice and perturbation theories. In their approach, they assumed linear chains of alcohol oligomers. Consequently, the number of hydrogen bonds in the system may be calculated from the following equation:

$$N_H = N - N_T \Leftrightarrow \frac{N_H}{N} = 1 - \frac{N_T}{N} \tag{7.34}$$

where N the total number of molecules in the system and N_T the 'true' number of molecules (HB oligomer is considered as a single molecule). By combining Equations (7.30) and (7.34), the following expression is obtained (presented by Economou and Donohue)[1]:

$$\frac{N_T}{N} = \frac{\sqrt{1+4K}-1}{2K} = \frac{2}{1+\sqrt{1+4K}} \tag{7.35}$$

Consequently the fraction of monomers is equal to the probability of finding a molecule with two unbonded sites:

$$X_1 = \left(1 - \frac{N_H}{N}\right)^2 \tag{7.36}$$

By combining Equations (7.34)–(7.36) the following equation is obtained:[1]

$$X_1 = \left(1 - \frac{N_H}{N}\right)^2 = \frac{2}{1+2K+\sqrt{1+4K}} \tag{7.37}$$

Note that in the latter approach of Economou and Donohue,[1] the number of hydrogen bonds, N_H, is not equal to the number of bonded molecules as it was assumed in the former approach of Gupta *et al.*[52] Consequently, the extent of hydrogen bonding cannot be calculated by subtracting the number of free alcohols from the

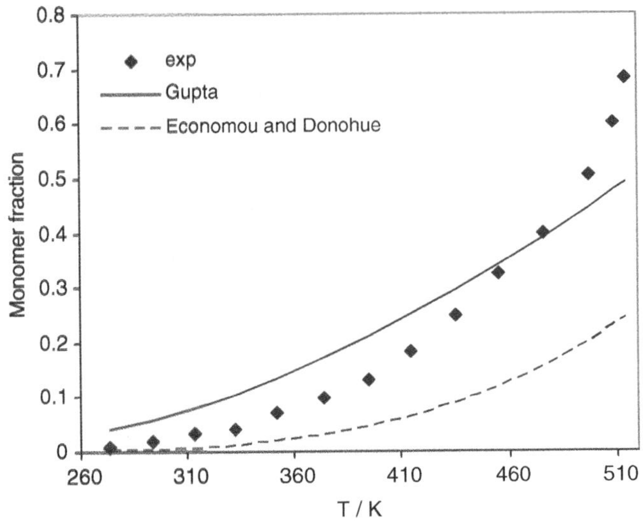

Figure 7.10 *Monomer fraction for methanol. Experimental data and NRHB predictions: approach of Gupta et al.*[52] *(solid line) and approach of Economou and Donohue*[1] *(dashed line)*

total number of alcohols in the system as was reported by Gupta and Brinkley.[48] Moreover, according to the first approach[52] the quantity $1 - (N_H/N)$ is equal to the monomer fraction, while according to the approach of Economou and Donohue[1] it is equal to the number of unbonded proton donors or acceptors per molecule. Figure 7.10 shows the predictions of the NRHB model according to the approach of Gupta *et al.*[52] (solid line) and the approach of Economou and Donohue[1] (dashed line).

References

1. I.G. Economou, M.D. Donohue, *AIChE J.*, **1991**, *37(12)*, 1875.
2. M.D. Donohue, I.E. Economou, *Fluid Phase Equilib.*, **1996**, *116(1–2)*, 518.
3. R.A. Heidemann, J.M. Prausnitz, *Proc. Natl Acad. Sci. USA*, **1976**, *73(6)*, 1773.
4. G.D. Ikonomou, M.D. Donohue, *Fluid Phase Equilib.*, **1988**, *39*, 129.
5. A. Anderko, *Fluid Phase Equilib.*, **1989**, *45*, 39.
6. A. Anderko, *Fluid Phase Equilib.*, **1989**, *50*, 21.
7. C. Panayiotou, I.C. Sanchez, *J. Phys. Chem.*, **1991**, *95*, 10090.
8. W.G. Chapman, K.E. Gubbins, G. Jackson, M. Radosz, *Ind. Eng. Chem. Res.*, **1990**, *29*, 1709.
9. S.H. Huang, M. Radosz, *Ind. Eng. Chem. Res.*, **1990**, *29*, 2284.
10. J.S. Suresh, J.R. Elliott, *Ind. Eng. Chem. Res.*, **1992**, *31*, 2783.
11. G.M. Kontogeorgis, E. Voutsas, I. Yakoumis, D.P. Tassios, *Ind. Eng. Chem. Res.*, **1996**, *35*, 4310.
12. E.M. Hendriks, J.M. Walsh, A.R.D. van Bergen, *J. Stat. Phys.*, **1997**, *87(5/6)*, 1287.
13. M.L. Michelsen, E.M. Hendriks, *Fluid Phase Equilib.*, **2001**, *180(1–2)*, 165.
14. J.H. Hildebrand, R.L. Scott, *The Solubility of Non-electrolytes*. Dover, 1964.
15. J.M. Prausnitz, R.N. Lichtenthaler, E.G. de Azevedo, *Molecular Thermodynamics of Fluid-Phase Equilibria* (3rd edition). Prentice Hall International, 1999.
16. F. Dolezalek, *Phys. Chem.*, **1908**, *64*, 727.
17. P. Vimalchand, G.D. Ikonomou, M.D. Donohue, *Fluid Phase Equilib.*, **1988**, *43*, 121.
18. I.G. Economou, G.D. Ikonomou, P. Vimalchand, M.D. Donohue, *AIChE J.*, **1990**, *36(12)*, 1851.

19. A. Anderko, *Fluid Phase Equilib.*, **1992**, *75*, 89.
20. J.M. Yu, B.C.-Y. Lu, *Fluid Phase Equilib.*, **1987**, *34*, 1.
21. A. Anderko, *Fluid Phase Equilib.*, **1991**, *65*, 89.
22. I.G. Economou, M.D. Donohue, *Ind. Eng. Chem. Res.*, **1992**, *31*, 1203.
23. I.G. Economou, M.D. Donohue, *Ind. Eng. Chem. Res.*, **1992**, *31*, 2388.
24. G.D. Ikonomou, M.D. Donohue, *Fluid Phase Equilib.*, **1986**, *39*, 129.
25. I.G. Economou, C. Tsonopoulos, *Chem. Eng. Sci.*, **1997**, *52(4)*, 511.
26. C.H. Twu, J.E. Coon, J.R. Cunningham, *Fluid Phase Equilib.*, **1993**, *82*, 379.
27. J. Hong, Y. Hu, *Fluid Phase Equilib.*, **1989**, *51*, 37.
28. M.S. Wertheim, *J. Stat. Phys.*, **1984**, *35*, 19.
29. M.S. Wertheim, *J. Stat. Phys.*, **1984**, *35*, 35.
30. M.S. Wertheim, *J. Stat. Phys.*, **1986**, *42*, 459.
31. M.S. Wertheim, *J. Stat. Phys.*, **1986**, *42*, 477.
32. I. Tsivintzelis, A. Grenner, G.M. Kontogeorgis, I.G. Economou, C. Panayiotou, *Ind. Eng. Chem. Res.*, **2008**, *47(15)*, 5651.
33. R.L. Brinkley, R.B. Gupta, *Ind. Eng. Chem. Res.*, **1998**, *37*, 4823.
34. N. von Solms, M.L. Michelsen, C.P. Passos, S.O. Derawi, G.M. Kontogeorgis, *Ind. Eng. Chem. Res.*, **2006**, *45*, 5368.
35. N. von Solms, L. Jensen, J.L. Kofod, M.L. Michelsen, G.M. Kontogeorgis, *Fluid Phase Equilib.*, **2007**, *261*, 272.
36. R.B. Gupta, K.P. Johnston, *Fluid Phase Equilib.*, **1994**, *99*, 135.
37. J.P. Wolbach, S.I. Sandler, *Ind. Eng. Chem. Res.*, **1997**, *36*, 4041.
38. J.P. Wolbach, S.I. Sandler, *Ind. Eng. Chem. Res.*, **1998**, *37*, 2917.
39. S.I. Sandler, J.P. Wolbach, M. Castier, G. Escobedo-Alvarado, *Fluid Phase Equilib.*, **1997**, *136*, 15.
40. M. Yarrison, W.G. Chapman, *Fluid Phase Equilib.*, **2004**, *226*, 195.
41. J.P. Wolbach, S.I. Sandler, *AIChE J.*, **1997**, *43(6)*, 1589.
42. J.P. Wolbach, S.I. Sandler, *AIChE J.*, **1997**, *43(6)*, 1597.
43. W.A.P. Luck, *Angew. Chem. Int. Ed. Engl.*, **1980**, *19*, 28.
44. T.R. Lien,A study of the thermodynamic excess functions of alcohol solutions by IR spectroscopy: applications to chemical solution theory. PhD Thesis, University of Toronto, 1972.
45. A.N. Fletcher, C.A. Heller, *J. Phys. Chem.*, **1967**, *71*, 3742.
46. F. Palombo, P. Sassi, M. Paolantoni, A. Morresi, R.S. Cataliotti, *J. Phys. Chem. B*, **2006**, *110*, 18017.
47. S. Martinez, *Spectrochim. Acta Part A: Mol. Spectrosc.*, **1986**, *42*, 531.
48. R.B. Gupta, R.L. Brinkley, *AIChE J.*, **1998**, *44(1)*, 207.
49. R.L. Brinkley, R.B. Gupta, *AIChE J.*, **2001**, *47(4)*, 948.
50. N. Asprion, H. Hasse, G. Maurer, *Fluid Phase Equilib.*, **2001**, *186*, 1. Supplementary material at http://www.itt.uni-stuttgart.de.
51. W. Peschel, H. Wenzel, *Ber. Bunsenges Phys. Chem.*, **1984**, *88*, 807.
52. R.B. Gupta, C.G. Panayiotou, I.C. Sanchez, K.P. Johnston, *AIChE J.*, **1992**, *38(8)*, 1243.
53. C.A. Koh, H. Tanaka, J.M. Walsh, K.E. Gubbins, J.A. Zollweg, *Fluid Phase Equilib.*, **1993**, *83*, 51.
54. P. Muthukumaran, R.L. Brinkley, R.B. Gupta, *AIChE J.*, **2002**, *48(2)*, 386.
55. G.N.I. Clark, A.J. Haslam, A. Galindo, G. Jackson, *Mol. Phys.*, **2006**, *104*, 3562.
56. S. Aparicio-Martinez, K.R. Hall, *Fluid Phase Equilib.*, **2007**, *254*, 112.
57. A. Grenner, G.M. Kontogeorgis, M.L. Michelsen, G.K. Folas, *Mol. Phys.*, **2007**, *105(13–14)*, 1797.
58. H. Kahl, S. Enders, *Fluid Phase Equilib.*, **2000**, *172*, 27.
59. A. Grenner, J. Schmelzer, N. von Solms, G.M. Kontogeorgis, *Ind. Eng. Chem. Res.*, **2006**, *45*, 8170.
60. A. Grenner, G.M. Kontogeorgis, N. von Solms, M.L. Michelsen, *Fluid Phase Equilib.*, **2007**, *258(1)*, 83.
61. J. Gross, G. Sadowski, *Ind. Eng. Chem. Res.*, **2002**, *41*, 5510.
62. J.R. Elliott, S.J. Suresh, M.D. Donohue, *Ind. Eng. Chem. Res.*, **1990**, *29*, 1476.
63. Z. Junyang, H. Ying, *Fluid Phase Equilib.*, **1990**, *57*, 89.
64. D. Missopolinou, K. Ioannou, I. Prinos, C. Panayiotou, *Z. Phys. Chem.*, **2002**, *216*, 1.

8

The Statistical Associating Fluid Theory (SAFT)

8.1 The SAFT EoS: a brief look at the history and major developments

The SAFT EoS is a theoretically derived model, based on perturbation theory, in particular as outlined in the four articles by Wertheim[1–4]. However, the model appeared in the form known today a few years later, first in two articles in *Molecular Physics*[5,6] and a bit later in two articles in the August and November issues of the journal *Industrial and Engineering Chemistry Research*, and for this reason both of these articles are often referred to as 'original SAFT'[7,8]. However, the term 'CK–SAFT' can be alternatively used for the Huang and Radosz version, for reasons which will become clear later. There are many more SAFT variants, some of which will be briefly discussed in this chapter.

The SAFT EoS and the reason for the several variations can be understood by referring to Figure 8.1. The final single molecule is shown in Figure 8.1(a). A fluid is first assumed to consist of equal-sized hard spheres (b), then a dispersive potential is added to account for attraction between the spheres (c), e.g. the square-well or Lennard-Jones potential. Next, each sphere is given two (or more) 'sticky' spots, which enables the formation of chains (d). Finally, specific interaction sites are introduced at certain positions in the chain which enable the chains to associate through some attractive interaction (hydrogen bonding) (e). The hydrogen bonding energy is often taken to be a square-well potential. Each of these steps contributes to the Helmholtz energy. The residual Helmholtz energy is given by:

$$a^{res} = a^{seg} + a^{chain} + a^{assoc} \tag{8.1}$$

where a^{seg} is the Helmholtz energy of the segment, including both hard-sphere reference and dispersion terms, a^{chain} is the contribution from chain formation and a^{assoc} is the contribution from association.

Wertheim's contributions are responsible for the expressions of the chain and the association term, which are essentially unchanged in the various versions of SAFT. However, due to the separation of the Helmholtz energy into additive components, various variations have been proposed, most of which use different expressions for the attractive contribution. Under the headline 'first major SAFT developments or variants', we could include the original SAFT[7] and the CK–SAFT,[8] both often called 'original SAFT' as mentioned above, the simplified SAFT,[9] the LJ–SAFT,[10,11] the soft–SAFT,[12] the SAFT–VR[13] and the PC–SAFT, in both

Thermodynamic Models for Industrial Applications: From Classical and Advanced Mixing Rules to Association Theories
Georgios M. Kontogeorgis and Georgios K. Folas

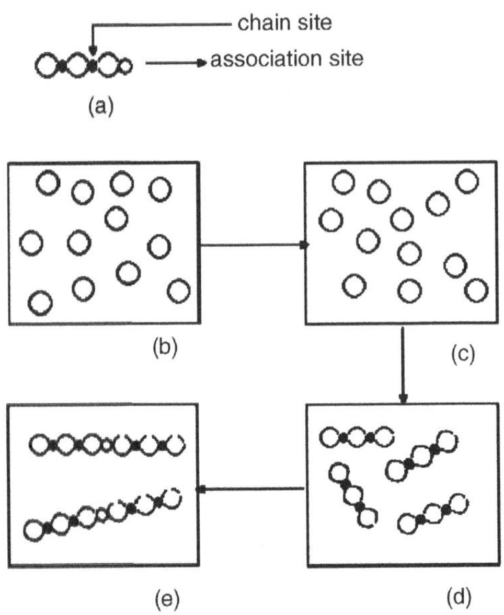

Figure 8.1 *Procedure to form a molecule in the SAFT model: (a) the proposed molecule containing chain and association sites; (b) initially the fluid is a hard-sphere fluid; (c) attractive forces are added; (d) chain sites are added and chain molecules appear; (e) association sites are added and molecules form association complexes through association sites (after Fu and Sandler)*[9]

its original[14] and simplified versions[15]. These are summarized in Table 8.1; their differences lie primarily in the different terms used for the dispersion term, as will be illustrated later (Section 8.2). Almost all of these SAFT variants include five pure compound parameters with well-defined physical meaning (the segment energy and size, the number of segments and the association volume and energy).

Table 8.1 *References of some of the most well-known SAFT variants*

SAFT variant	Reference	Comments
Original SAFT	Chapman *et al.*[7,16]	Mostly comparisons against simulation data Parameters for six hydrocarbons and two associating fluids are given
CK–SAFT	Huang and Radosz[8,17]	Parameters for 100 different fluids
Simplified SAFT	Fu and Sandler[9]	Parameters for ten non-associating and eight associating compounds
LJ–SAFT	Kraska and Gubbins[10,11]	Alkanes, alkanols, water (pure components)/mixtures of alkanes, alkanols, water
SAFT–VR	Gil-Vilegas *et al.*[13] McCabe *et al.*[18]	alkanes, perfluoroalkanes (pure components)/comparisons against simulation data
Soft SAFT	Blas and Vega[12]	
PC–SAFT	Gross and Sadowski[14]	The Gross and Sadowski[14] article contains parameters for 100 compounds and Tihic *et al.*[19] another 400 parameters. See Appendix A (on the companion website at www.wiley.com/go/Kontogeorgis) for a full parameter table
Simplified PC–SAFT	Von Solms *et al.*[15] Tihic *et al.*[19] (full table)	See Table 8.10 for a comparison between PC–SAFT and simplified PC–SAFT

Table 8.2 CPA – a special case of SAFT variants

Model	Reference	Main applications
CPA (based on SRK + SAFT)	Kontogeorgis *et al.*[20,21]	Water–alcohols(glycols)–hydrocarbons (V)LLE
	Kontogeorgis et al.[22,23]	
e-CPA	Ruffine *et al.*[24]	Acids, amines, etc.
	Lin *et al.*[25]	Acid gases (H_2S)
		Electrolytes (water–salts VLE, SLE)
ESD	Elliott *et al.*[26]	Water–alcohols–gases–hydrocarbons
	Suresh and Elliott[27,28]	
	Puhala and Elliott[29]	
PR–CPA (+ extra electrolyte term)	Wu and Prausnitz[30]	Water–hydrocarbons–salts
PR–CPA	Pfohl *et al.*[31]	CO_2–phenols/cresol
PR–CPA	Perakis *et al.*[32,33]	CO_2–water–ethanol and acetic acid VLE
	Voutsas *et al.*[34]	Water–alkanes
		CO_2–polars
PR–CPA	Huang *et al.*[35]	CO_2/alcohol–aspirin, naproxen SCFE

There have been numerous further developments as well as extensions and modifications of the SAFT approach which could be classified into four main directions:

1. A simplification by substituting the chain and dispersion terms with that of a cubic EoS (e.g. SRK and PR). This has led (Table 8.2) to a variety of 'cubic-plus-association' (CPA) EoS, and the version based on SRK which has so far been systematically applied to many types of mixtures will be presented in detail in Chapters 9–12.
2. Discussion of the computational aspects of SAFT (Table 8.3), with the most important one possibly being the analysis and simplification of the association term, in order to permit easy use and fast implementation[41].
3. Improving the SAFT model by adding polar and quadrupolar contributions (Table 8.4) or extending SAFT to electrolytes (Table 8.5), which requires the addition of one or two terms in the Helmholtz energy expression to account for electrostatic effects (long-range forces). Electrolyte versions of PC–SAFT, SAFT–VR and CPA have been presented and will be discussed in Chapter 15.
4. Development of group contribution (GC) versions of SAFT (Table 8.6), in the sense of developing GC schemes for estimating the pure compound parameters, thus permitting the estimation of the parameters

Table 8.3 Computational aspects of SAFT

Subject	Reference
Multiplicity of roots	Koak *et al.*[36]
Rescaling of PC–SAFT and sPHCT	Cismondi *et al.*[37]
Analytical solutions of association term	Kraska[38]
Comparison of computing times for various models and mixtures	von Solms *et al.*[15]
Simplified but rigorous expression of the Wertheim association term	Hendriks *et al.*[39]
	Yakoumis *et al.*[40]
	Michelsen and Hendriks[41]
	Michelsen[42,43]

Table 8.4 *Polar and quadrupolar SAFT variants*

SAFT variant	Additional term	Reference
CK–SAFT	Dipolar	Jog *et al.*[44]
PC–SAFT	Dipolar	Tumakaka and Sadowski[45,46]
		Sauer and Chapman[47]
PC–SAFT	Dipolar + induced polar	Karakatsani *et al.*[48–50]
PC–SAFT	Dipolar	Gross and Vrabec[51]
PC–SAFT	Quadrupolar	Gross[52]
PC–SAFT	Polarizable dipoles	Kleiner and Gross[53]
PC–SAFT	Dipolar + polarizable	Karakatsani and Economou[54]
		Kleiner and Sadowski[55]

Table 8.5 *Application of SAFT (and CPA) to electrolytes*

SAFT variant	Reference	Application
SAFT–VR	Behzadi *et al.*[56]	Water–alkanes
	Patel *et al.*[57]	
PC–SAFT	Cameretti *et al.*[58]	Water–salts
	Fuchs *et al.*[59]	Amino acids
CPA (Peng–Robinson)	Wu and Prausnitz[30]	Water–hydrocarbons–salts
CPA (SRK)	Lin *et al.*[25]	Water–salts VLE, SLE

Table 8.6 *Group contribution versions of SAFT*

SAFT variant	Application	Reference
Original SAFT	Alkanes, aromatics, olefins, alcohols	Tamouza *et al.*[60]
SAFT–VR		
Original SAFT	Alcohols	Tamouza *et al.*[61]
SAFT–VR	Alkanes	
Original SAFT	Esters	Thi *et al.*[62]
PC–SAFT + polar term		
SAFT–VR		
SAFT	General	Emami *et al.*[63]
PC–SAFT		
PC–SAFT	H_2, CO_2 + alkanes (including a GC method for the k_{ij} parameter)	Le Thi *et al.*[64]
SAFT–VR		
Original SAFT	Polycylic aromatic hydrocarbons	Huynh *et al.*[65]
PC–SAFT + quadrupolar		
SAFT–VR		
PC–SAFT	General (non-associating compounds)	Tihic *et al.*[66]
SAFT–VR	Hydrocarbons, alcohols	Lymberiadis *et al.*[103,119]

Table 8.7 *SAFT reviews*

Topic	Reference
Electrolytes, interfaces and polymers	Paricaud *et al.*[67]
General, including aqueous mixtures, electrolytes, liquid crystals, polymers, oil mixtures and high pressure	Müller and Gubbins[68]
General	Economou,[69] Wei and Sadus,[70] Prausnitz and Tavares,[71] Tan *et al.*[104]
PC–SAFT applications	Arlt *et al.*[72]
PC–SAFT applications (strengths and limitations)	von Solms *et al.*[73]
Oil applications (PC–SAFT, SAFT–VR and CPA)	De Hemptinne *et al.*[74]

when data for the pure compounds which are typically used for parameter estimation (vapor pressures and liquid densities, see Section 8.5) are not available.

Finally, Table 8.7 presents some of the most recent and useful reviews of the SAFT family of models; some are quite general, while others target specific families of mixtures.

8.2 The SAFT equations

In SAFT, each pure component is characterized by a chain length m (which is the number of segments in the molecule), a size parameter σ and a segment energy parameter, ε. If the molecule is self-associating, there are two further parameters which characterize the volume: ($\kappa^{A_iB_i}$) and energy ($\varepsilon^{A_iB_i}$) of association.

8.2.1 The chain and association terms

As almost all SAFT variants use the same chain and association terms, these will be presented first, in the form of the contributions to the Helmholtz energy.

The chain term is given by the equation:

$$\frac{a^{chain}}{RT} = \sum_i x_i(1 - m_i)\ln(g_{ii}(d_{ii})^{hs}) \tag{8.2}$$

while the association term is expressed as:

$$\frac{a^{assoc}}{RT} = \sum_i x_i \left[\sum_{A_i} \left(\ln X^{A_i} - \frac{X^{A_i}}{2} \right) + \frac{1}{2} M_i \right] \tag{8.3}$$

where X^{A_i} is the fraction of molecules i not bonded at site A and M_i is the number of association sites on molecule i, defined as:

$$X^{A_i} = \left[1 + \sum_j \sum_{B_j} \rho_j X^{B_j} \Delta^{A_iB_j} \right]^{-1} \tag{8.4}$$

where ρ_j is the molar density of j and $\Delta^{A_iB_j}$ is the association strength between two sites A and B belonging to two different molecules i and j, which is given by:

$$\Delta^{A_iB_j} = d_{ij}^3 g_{ij}(d_{ij})^{seg} \kappa^{A_iB_j} \left[\exp(\varepsilon^{A_iB_j}/kT) - 1\right] \tag{8.5}$$

where k is the Boltzmann factor and T is the temperature.

There are two important issues when using the association term in SAFT, Equations (8.3)–(8.5):

- Should the temperature-independent σ or the temperature-dependent diameter d be used in Equation (8.5)?
- Which expression should be used for the radial distribution function g_{ij}?

The various SAFT variants use different assumptions or answers to these two questions.

Temperature-dependent or temperature-independent diameter in the association strength Δ?

In the original SAFT,[7] the temperature-dependent diameter d_{ij} ($d_{ij} = (d_{ii} + d_{jj})/2$) is used and is related to the temperature-independent diameter σ_{ij} ($\sigma_{ij} = (\sigma_{ii} + \sigma_{jj})/2$), which for pure components is given as:

$$\frac{d}{\sigma} = \frac{1 + 0.2977kT/\varepsilon}{1 + 0.331\,63kT/\varepsilon + f(m)(kT/\varepsilon)^2} \tag{8.6}$$

where:

$$f(m) = 0.001\,047\,7 + 0.025\,337\frac{m-1}{m} \tag{8.7}$$

and m is the number of segments.

The temperature-dependent size diameter is also used in the other 'original SAFT', that of Huang and Radosz[8] (CK–SAFT), but a simpler temperature dependency is used, following Chen and Kreglewski,[75] hence the name CK–SAFT used for this SAFT variant:

$$d = \sigma\left[1 - 0.12\exp\left(\frac{-3u^0}{kT}\right)\right] \tag{8.8}$$

where $u^0 = \varepsilon$, and u (see Equation (8.19)) is the temperature-dependent energy parameter given by $u = u^0(1 + e/kT)$ and e/k is a constant set to 10, with a few exceptions. The u^0 parameter is, in CK–SAFT, the segment energy parameter. In most other SAFT variants, the symbol ε is used for the segment energy parameter. Other versions of SAFT (PC–SAFT in its original and simplified forms and SAFT–VR) employ the temperature-independent diameter in Equation (8.5), i.e. σ_{ij} is used instead of d_{ij} (see Table 8.10). The temperature dependency of the size parameter does have a physical significance: it accounts for the fact that real molecules are not hard spheres, but rather there is some degree of interpenetration between molecules, particularly at high temperatures, thus the 'effective' hard-sphere diameter of a segment is smaller at higher temperatures. This effect seems, however, to be small in practical applications.

On the radial distribution function

The other 'special' issue is the expression for the radial distribution function in Equation (8.5). Most SAFT variants (original and PC–SAFT, etc.) use the expression derived for a mixture of hard spheres from the Carnahan–Starling EoS:

$$g_{ij}^{seg}(d_{ij}) \approx g_{ij}^{hs}(d_{ij}^{+}) = \frac{1}{1-\zeta_3} + \left(\frac{d_i d_j}{d_i + d_j}\right)\frac{3\zeta_2}{(1-\zeta_3)^2} + \left(\frac{d_i d_j}{d_i + d_j}\right)^2 \frac{2\zeta_2^2}{(1-\zeta_3)^3} \tag{8.9}$$

where:

$$\zeta_k = \frac{\pi N_{AV}}{6}\rho \sum_i X_i m_i d_{ii}^k \tag{8.10}$$

However, in the simplified PC–SAFT,[15] inspired by developments with the CPA EoS[20] and the fact that the values of the segment diameters for various compounds are quite similar (see Appendix A on the companion website at www.wiley.com/go/Kontogeorgis), it has been set as $\eta \equiv \zeta_3$, thus Equation (8.9) is simplified to (see also Table 8.10 and Equation (8.41)):

$$g^{hs}(d^{+}) = \frac{2-\eta}{2(1-\eta)^3} \tag{8.11}$$

8.2.2 The dispersion terms

As previously mentioned, despite some small differences in the chain and association terms above, it is the different expressions for the dispersion (and repulsion) terms especially which characterize the various SAFT variants. We present here some of the popular SAFT expressions for a^{seg} of Equation (8.1). Not all SAFT variants are presented here and the expressions for LJ–SAFT, soft-SAFT and other SAFT equations, for example, are available in the original articles describing these models (see Table 8.1).

Original SAFT[7]

The segment term in Equation (8.1) is:

$$a^{seg} = a_0^{seg} \sum_i x_i m_i \tag{8.12}$$

where the zero subscript indicates a non-associated segment. The segment energy consists of a hard-sphere reference and a dispersion contribution:

$$a_0^{seg} = a_0^{hs} + a_0^{disp} \tag{8.13}$$

The Carnahan–Starling[76] equation is used for both pure components and mixtures to give:

$$\frac{a_0^{hs}}{RT} = \frac{4\eta - 3\eta^2}{(1-\eta)^2} \tag{8.14}$$

where for mixtures $\eta \equiv \zeta_3$ as defined by Equation (8.10). The dispersion term is given by:

$$a_0^{disp} = \frac{\varepsilon R}{k}\left(a_{01}^{disp} + \frac{a_{02}^{disp}}{T_R}\right) \tag{8.15}$$

where:

$$a_{01}^{disp} = \rho_R\left[-0.859\,59 - 4.5424\rho_R - 2.1268\rho_R^2 + 10.285\rho_R^3\right] \tag{8.16}$$

$$a_{02}^{disp} = \rho_R\left[-1.9075 + 9.9724\rho_R - 22.216\rho_R^2 + 15.904\rho_R^3\right] \tag{8.17}$$

The reduced quantities are given by $T_R = kT/\varepsilon$ and $\rho_R = \left[6/(2^{0.5}\pi)\right]\eta$.

CK–SAFT [8,1]

In this version, the full Carnahan–Starling equation for the hard-sphere mixtures reference system is used:

$$\frac{a_0^{hs}}{RT} = \frac{1}{\zeta_0}\left[\frac{3\zeta_1\zeta_2}{1-\zeta_3} + \frac{\zeta_2^3}{\zeta_3(1-\zeta_3)^2} + \left(\frac{\zeta_2^3}{\zeta_3^2} - \zeta_0\right)\ln(1-\zeta_3)\right] \tag{8.18}$$

The dispersion term is based on simulation data for square-well fluids:

$$\frac{a_0^{disp}}{RT} = \sum_i \sum_j D_{ij}\left[\frac{u}{kT}\right]^i\left[\frac{\eta}{\tau}\right]^j \tag{8.19}$$

where the D_{ij} are (in total 24, see Table 8.8) universal constants and u is the temperature-dependent energy parameter given by $u = u^0(1 + e/kT)$ and e/k is a constant set to 10, with a few exceptions. There are also some notational differences between the Huang and Radosz[8] CK–SAFT and other SAFT variants: (1) the symbol u is used instead of ε for the segment energy parameter; and (2) instead of a size parameter σ used in most SAFT variants, a volume parameter v^{00} is used :

$$v^{00} = \frac{\pi N_{AV}}{6\tau}\sigma^3 \tag{8.20}$$

Here $\tau = 0.740\,48$ is the highest possible packing fraction for a system of pure hard spheres.

Table 8.8 Chen and Kreglewski[75] constants (D_{ij}) used in Equation (8.19) for the CK–SAFT EoS

D_{ij}	$i=1$	$i=2$	$i=3$	$i=4$
$j=1$	– 8.8043	2.9396	– 2.8225	0.34
$j=2$	4.164 627	– 6.086 538 3	4.760015	– 3.187 501 4
$j=3$	– 48.203 555	40.13 796	11.257 18	12.2318
$j=4$	140.4362	– 76.230 797	– 66.382 743	– 12.110 681
$j=5$	– 195.233 39	– 133.700 55	69.248 79	0.0
$j=6$	113.515	860.2535	0.0	0.0
$j=7$	0.0	– 1535.3224	0.0	0.0
$j=8$	0.0	1221.426	0.0	0.0
$j=9$	0.0	– 409.105 39	0.0	0.0

Simplified SAFT

The dispersion term is given by the expression:

$$\frac{a^{disp}}{RT} = mZ_M \ln\left(\frac{v_s}{v_s + \langle v^* Y \rangle}\right) \tag{8.21}$$

where the 'average' chain length $m = \sum_i x_i m_i$, $Z_M = 36$ is the maximum coordination number, $v_s = 1/\rho m$ is the total molar volume of a segment and:

$$\langle v^* Y \rangle = \frac{N_{AV} \sum_i \sum_j x_i x_j m_i m_j (d_{ij}^3/\sqrt{2}) \left[\exp(u_{ij}/kT) - 1\right]}{\sum_i \sum_j x_i x_j m_i m_j} \tag{8.22}$$

Otherwise, simplified SAFT is identical to CK–SAFT.

SAFT–VR[13]

SAFT–VR is also identical to CK–SAFT, except for the dispersion term, which is based on a square-well potential. Thus, in addition to a segment being characterized by a size and an energy parameter, the square-well width (λ) is also included as a pure component parameter. Changing the parameter λ changes the range of attraction of the segment (hence the name VR for 'variable range'). It is the introduction of this extra term that gives SAFT–VR greater flexibility, since an extra pure component parameter is available. Although it is generally desirable to describe pure component liquid densities and vapor pressures with the minimum number of parameters, the extra variable range parameter may be necessary for the description of certain anomalous behaviors in systems containing water. The Helmholtz energy for the dispersion energy is given by:

$$a^{disp} = \frac{a_1}{kT} + \frac{a_2}{(kT)^2} \tag{8.23}$$

where:

$$a_1 = -\rho_s \sum_i \sum_j x_{s,i} x_{s,j} \alpha_{ij}^{VDW} g^{HS}\left[\sigma_x; \zeta_x^{eff}\right] \tag{8.24}$$

The subscript s refers to segment rather than molecule properties. α_{ij}^{VDW} in Equation (8.24) is given as:

$$\alpha_{ij}^{VDW} = 2\pi\varepsilon_{ij}\sigma_{ij}^3(\lambda_{ij}^3 - 1)/3 \tag{8.25}$$

where g^{HS} is the radial distribution function for hard spheres as before except that the arguments are different:

$$\zeta_x^{eff} = c_1\zeta_x + c_2\zeta_x^2 + c_3\zeta_x^3 \tag{8.26}$$

$$\sigma_x^3 = \sum_i \sum_j x_{s,i}x_{s,j}\sigma_{ij}^3 \tag{8.27}$$

$$\zeta_x = \frac{\pi}{6}\rho_s\sigma_x^3 \tag{8.28}$$

The constants c_i in Equation (8.26) are given by:

$$\begin{pmatrix} c_1 \\ c_2 \\ c_3 \end{pmatrix} = \begin{pmatrix} 2.258\,55 & -1.503\,49 & 0.249\,434 \\ -0.669\,27 & 1.400\,49 & -0.827\,739 \\ 10.1576 & -15.0427 & 5.308\,27 \end{pmatrix} \begin{pmatrix} 1 \\ \lambda_{ij} \\ \lambda_{ij}^2 \end{pmatrix} \tag{8.29}$$

The second-order term in Equation (8.23) is given by:

$$a_2 = \sum_{i=1}^{n}\sum_{j=1}^{n} x_{s,i}x_{s,j}\frac{1}{2}K_{HS}\varepsilon_{ij}\rho_s\frac{\partial a_1}{\partial \rho_s} \tag{8.30}$$

where:

$$K_{HS} = \frac{\zeta_0(1-\zeta_3)^4}{\zeta_0(1-\zeta_3)^2 + 6\zeta_1\zeta_2(1-\zeta_3) + 9\zeta_2^3} \tag{8.31}$$

PC–SAFT [14,7]

Also, in this case, most of the terms are similar to CK–SAFT but with a different dispersion term. There is, however, a fundamental difference between PC–SAFT and the previous versions of SAFT. The dispersion term attempts to account for dispersion attraction between whole chains. Referring to Figure 8.1 should make this clear. Instead of adding the dispersion to hard spheres and then forming chains, we first form hard-sphere chains and then add a chain dispersion term, so the route in Figure 8.1 would be (b)–(d)–(c)–(e). To do this we require interchain rather than intersegment radial distribution functions. These are given by O'Lenick *et al.*[78] The Helmholtz energy for the dispersion term is given as the sum of a first-order and second-order term:

$$\frac{a^{disp}}{kTN} = \frac{A_1}{kTN} + \frac{A_2}{kTN} \tag{8.32}$$

where:

$$\frac{A_1}{kTN} = -2\pi\rho m^2\left(\frac{\varepsilon}{kT}\right)\sigma^3\int_1^\infty \tilde{u}(x)g^{hc}(m;x\sigma/d)x^2dx \tag{8.33}$$

$$\frac{A_2}{kTN} = -\pi\rho m\left(1+Z^{hc}+\rho\frac{\partial Z^{hc}}{\partial\rho}\right)^{-1} m^2\left(\frac{\varepsilon}{kT}\right)^2\sigma^3\frac{\partial}{\partial\rho}\left[\rho\int_1^\infty \tilde{u}(x)^2g^{hc}(m;x\sigma/d)x^2dx\right] \tag{8.34}$$

where $x = r/\sigma$ and $\tilde{u}(x) = u(x)/\varepsilon$ is the reduced intermolecular potential.

Two important points are now: (1) the radial distribution function chains (rather than segment functions as before); and (2) expressions for the two integrals. These are given from the following expressions :

$$\left(1+Z^{hc}+\rho\frac{\partial Z^{hc}}{\partial\rho}\right) = \left(1+m\frac{8\eta-2\eta^2}{(1-\eta)^4}+(1-m)\frac{20\eta-27\eta^2+12\eta^3-2\eta^4}{((1-\eta)(2-\eta))^2}\right) \tag{8.35}$$

$$I_1 = \int_1^\infty \tilde{u}(x)g^{hc}(m;x\sigma/d)x^2dx = \sum_{i=0}^{6} a_i\eta^i \tag{8.36}$$

$$I_2 = \frac{\partial}{\partial\rho}\left[\rho\int_1^\infty \tilde{u}(x)^2g^{hc}(m;x\sigma/d)x^2dx\right] = \sum_{i=0}^{6} b_i\eta^i \tag{8.37}$$

with the power series in reduced density being given by the equations:

$$a_i = a_{0i}+\frac{m-1}{m}a_{1i}+\frac{m-1}{m}\frac{m-2}{m}a_{2i} \tag{8.38}$$

$$b_i = b_{0i}+\frac{m-1}{m}b_{1i}+\frac{m-1}{m}\frac{m-2}{m}b_{2i} \tag{8.39}$$

Equations (8.36)–(8.39) were developed using the Lennard-Jones potential and the radial distribution function of O'Lenick *et al.*[78] for the series of *n*-alkanes and the integrals were fitted as a power series; 42 constants were adjusted to fit experimental pure component data of *n*-alkanes. These constants are presented in Table 8.9.

Simplified PC–SAFT[15]

This is not a new EoS, rather a simplified version in terms of mixing rules of the original PC–SAFT EoS. Thus, the pure compound parameters of the original and simplified PC–SAFT are the same. The equation for the

Table 8.9 Universal model constants for Equations (8.38) and (8.39) in the PC–SAFT EoS. After Gross and Sadowski[14]

i	a_{0i}	a_{1i}	a_{2i}	b_{0i}	b_{1i}	b_{2i}
0	0.910 563 144 5	−0.308 401 691 8	−0.090 614 835 1	0.724 094 694 1	−0.575 549 807 5	0.097 688 311 6
1	0.636 128 144 9	0.186 053 115 9	0.452 784 280 6	2.238 279 186 1	0.699 509 552 1	−0.255 757 498 2
2	2.686 134 789 1	−2.503 004 725 9	0.596 270 072 8	−4.002 584 948 5	3.892 567 339 0	−9.155 856 153 0
3	−26.547 362 491 0	21.419 793 629 0	−1.724 182 913 1	−21.003 576 815 0	−17.215 471 648 0	20.642 075 974 0
4	97.759 208 784 0	−65.255 885 330 0	−4.130 211 253 1	26.855 641 363 0	192.672 264 470 0	−38.804 430 052 0
5	−159.591 540 870 0	83.318 680 481 0	13.776 631 870 0	206.551 338 410 0	−161.826 461 650 0	93.626 774 077 0
6	91.297 774 084 0	−33.746 922 930 0	−8.672 847 036 8	−355.602 356 120 0	−165.207 693 460 0	−29.666 905 585 0

Table 8.10 Modifications of the simplified PC–SAFT EoS[15] compared to the PC–SAFT EoS[14]

PC–SAFT[14]	Simplified PC–SAFT[15]
$g_{ij}^{hs} = \frac{1}{1-\zeta_3} + \left(\frac{d_i d_j}{d_i + d_j}\right)\frac{3\zeta_2}{(1-\zeta_3)^2} + \left(\frac{d_i d_j}{d_i + d_j}\right)^2 \frac{2\zeta_2^2}{(1-\zeta_3)^3}$	$g^{hs}(\eta) = \frac{1-\eta/2}{(1-\eta)^3}$
$\frac{a_0^{hs}}{RT} = \frac{1}{\zeta_0}\left[\frac{3\zeta_1\zeta_2}{1-\zeta_3} + \frac{\zeta_2^3}{\zeta_3(1-\zeta_3)^2} + \left(\frac{\zeta_2^3}{\zeta_3^2} - \zeta_0\right)\ln(1-\zeta_3)\right]$	$\tilde{a}^{hs} = \frac{4\eta - 3\eta^2}{(1-\eta)^2}$
$\Delta^{A_iB_j} = N_{AV} d_{ij}^3 g_{ij}^{hs} \kappa^{A_iB_j}[\exp(\varepsilon^{A_iB_j}/kT) - 1]$	$\Delta^{A_iB_j} = N_{AV}\frac{\pi}{6}\sigma_{ij}^3 g^{hs}(\eta)\kappa^{A_iB_j}[\exp(\varepsilon^{A_iB_j}/kT) - 1]$

residual Helmholtz energy is:

$$a^{res} = m a^{hs} + a^{chain} + a^{disp} + a^{assoc} \tag{8.40}$$

By setting $\eta \equiv \zeta_3$, i.e. using an average diameter:

$$d = \left(\frac{\sum_i x_i m_i d_i^3}{\sum_i x_i m_i}\right)^{1/3} \tag{8.41}$$

the hard-sphere and association equations are simplified, thus Equations (8.9) and (8.18) reduce to:

$$g^{hs}(d^+) = \frac{2-\eta}{2(1-\eta)^3} \tag{8.42}$$

and:

$$\tilde{a}^{hs} = \frac{4\eta - 3\eta^2}{(1-\eta)^2} \tag{8.43}$$

respectively. The remaining terms are the same, except, as mentioned previously, the simpler radial distribution function will affect both the chain and association terms, since the radial distribution function appears in both of them. Appendix 8.A presents the equations necessary for the fugacity calculation with the simplified PC–SAFT. Table 8.10 summarizes the main differences between the original and simplified PC–SAFT EoS.

8.3 Parameterization of SAFT

8.3.1 Pure compounds

As mentioned previously, most SAFT variants have five pure compound parameters:

- the chain length number or number of segments m;
- the segment diameter σ;
- a segment energy parameter, ε;

- the volume ($\kappa^{A_iB_i}$) of association;
- the energy ($\varepsilon^{A_iB_i}$) of association.

The last two parameters are needed only if the molecule is self-associating, e.g. water, alcohols and amines.

SAFT–VR has a sixth pure compound parameter, the square-well width (λ), while several of the polar and quadrupolar SAFT variants contain additional parameters, e.g. a value of the dipole moment adjusted to experimental data. All these parameters have to be estimated from available experimental pure compound data and, in the typical case, vapor pressures and liquid densities over extensive temperature range are used. Two SAFT variants (the CK–SAFT by Huang and Radosz[8] and PC–SAFT) have extensive parameter tables for numerous compounds including several self-associating ones, and Appendix A on the companion website at www.wiley.com/go/Kontogeorgis contains the list of the PC–SAFT parameters compiled from various publications. Parameters for over 500 non-associating and associating compounds are currently available for PC–SAFT.

For associating compounds, prior to using Equations (8.3) and (8.4), the number and type of association sites, i.e. the association scheme of the compound, should be established. Tables 8.11 and 8.12, first presented by Huang and Radosz,[8] represent a frequently used starting point.

Due to their physical meaning (see also Chapter 7) and similarity to the enthalpy and entropy of hydrogen bonding, the association parameters of SAFT ($\varepsilon^{A_iB_i}$ and $\kappa^{A_iB_i}$) could be estimated from molecular orbital calculations or based on experimental values of the enthalpy and entropy of hydrogen bonding,[79–81] thus leaving only the three remaining parameters (ε, σ, m) to be fitted to vapor pressures and liquid densities.

Table 8.11 *Unbonded site fractions* X^A for different bonding types (after Huang and Radosz)[8]

Type	Δ approximations	X^A approximations	X^A
1	$\Delta^{AA} \neq 0$		$\dfrac{-1+\sqrt{1+4\rho\Delta}}{2\rho\Delta}$
2A	$\Delta^{AA} = \Delta^{AB} = \Delta^{BB} \neq 0$	$X^A = X^B$	$\dfrac{-1+\sqrt{1+8\rho\Delta}}{4\rho\Delta}$
2B	$\Delta^{AA} = \Delta^{BB} = 0$ $\Delta^{AB} \neq 0$	$X^A = X^B$	$\dfrac{-1+\sqrt{1+4\rho\Delta}}{2\rho\Delta}$
3A	$\Delta^{AA} = \Delta^{AB} = \Delta^{BB} = \Delta^{AC} = \Delta^{BC} = \Delta^{CC} \neq 0$	$X^A = X^B = X^C$	$\dfrac{-1+\sqrt{1+12\rho\Delta}}{6\rho\Delta}$
3B	$\Delta^{AA} = \Delta^{AB} = \Delta^{BB} = \Delta^{CC} = 0$ $\Delta^{AC} = \Delta^{BC} \neq 0$	$X^A = X^B$ $X^C = 2X^A - 1$	$\dfrac{-(1-\rho\Delta)+\sqrt{(1+\rho\Delta)^2+4\rho\Delta}}{4\rho\Delta}$
4A	$\Delta^{AA} = \Delta^{AB} = \Delta^{BB} = \Delta^{AC} = \Delta^{BC} = \Delta^{CC}$ $= \Delta^{AD} = \Delta^{BD} = \Delta^{CD} = \Delta^{DD} \neq 0$	$X^A = X^B = X^C = X^D$	$\dfrac{-1+\sqrt{1+16\rho\Delta}}{8\rho\Delta}$
4B	$\Delta^{AA} = \Delta^{AB} = \Delta^{BB} = \Delta^{AC} = \Delta^{BC} = \Delta^{CC} = \Delta^{DD} = 0$ $\Delta^{AD} = \Delta^{BD} = \Delta^{CD} \neq 0$	$X^A = X^B = X^C$ $X^D = 3X^A - 2$	$\dfrac{-(1-2\rho\Delta)+\sqrt{(1+2\rho\Delta)^2+4\rho\Delta}}{6\rho\Delta}$
4C	$\Delta^{AA} = \Delta^{AB} = \Delta^{BB} = \Delta^{CC} = \Delta^{CD} = \Delta^{DD} = 0$ $\Delta^{AC} = \Delta^{AD} = \Delta^{BC} = \Delta^{BD} \neq 0$	$X^A = X^B = X^C = X^D$	$\dfrac{-1+\sqrt{1+8\rho\Delta}}{4\rho\Delta}$

Table 8.12 *Types of bonding in real associating fluids (after Huang and Radosz[8]). The recommendations can be used as guidance in association scheme selection, while more information on parameter selection can be found in Chapters 9, 12 and 14 (in connection with discussions related to CPA and SAFT EoS)*

Species		Formula	Rigorous	Assigned type
Acids		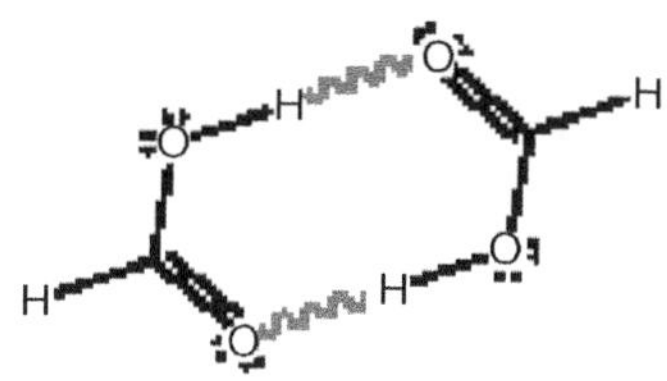	1A	**1A**, 2B
Alcohols		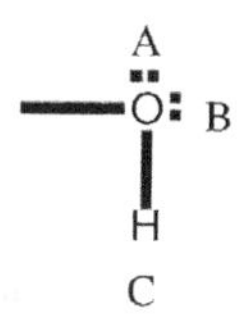	3B	**2B**/3B
Water		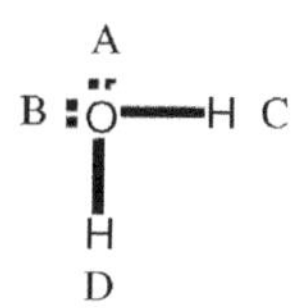	4C	3B/**4C**
Amines	Ternary	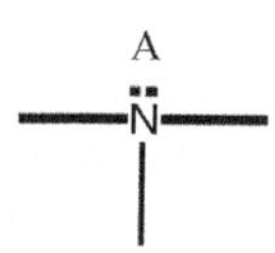	1A	Non-self-associating
	Secondary	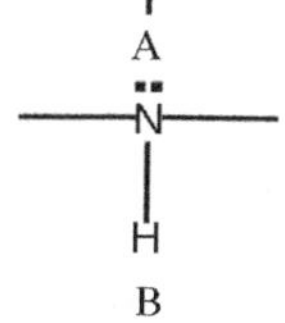	2B	2B
	Primary	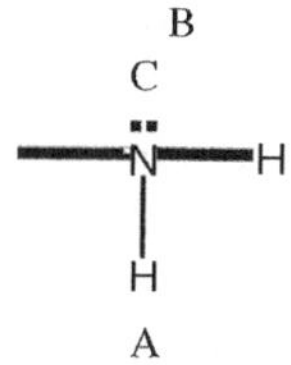	3B	3B
Ammonia		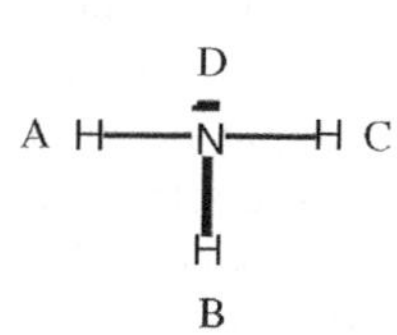	4B	3B

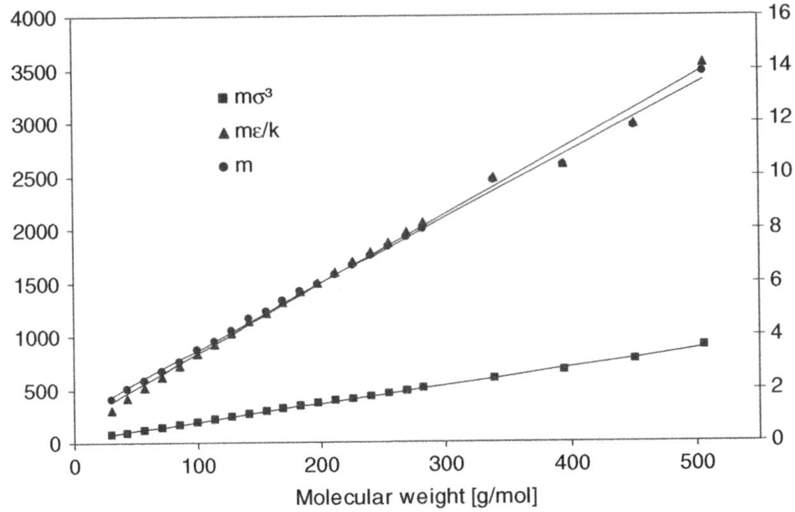

Figure 8.2 *The groups m, $m\sigma^3$ and $m\varepsilon/k$ against molecular weight for linear alkanes up to n-C_{36}. Points are PC–SAFT parameters, lines are linear fits to these points, excluding methane. The equations are: $m = 0.0249M_w + 0.9711$, $m\varepsilon/k = 6.5446M_w + 177.92$ and $m\sigma^3 = 1.6947M_w + 23.27$. Reprinted with permission from Fluid Phase Equilibria, Applications of the simplified perturbed-chain SAFT equation of state using an extended parameter table by A. Tihic, G. M. Kontogeorgis et al., 248, 1, 29 Copyright (2006) Elsevier*

The following points summarize the most important observations related to the estimation of pure compounds with various SAFT variants:

1. Five parameters for associating fluids and three for non-associating ones result in most cases in excellent vapor pressure and liquid density correlation. The critical point is, however, overestimated. Cismondi *et al.*[37] present a rescaling of PC–SAFT so that the critical point is matched, while several 'crossover SAFT' versions have been proposed for describing the critical area,[82,83] but at the cost of more adjustable parameters.
2. The parameters estimated based on vapor pressures and liquid densities follow well-defined trends with the molecular weight, as shown in Figures 8.2–8.5 for PC–SAFT. Similar plots are obtained for other SAFT variants as well. It can be seen that the parameters m, $m\sigma^3$ and $m\varepsilon/k$ are all linear with molecular weight, which can be used for extrapolating to higher molecular weight compounds (e.g. polymers) from knowledge of similar shorter chain compounds. It can be seen that not only for *n*-alkanes but also for other families of compounds there are well-defined trends of PC–SAFT (and other SAFT variant) parameters observed.
3. The segment diameter is rather constant for different molecules (see Appendix A on the companion website at www.wiley.com/go/Kontogeorgis), while the segment energy often increases with molecular weight but becomes constant for the heavier members of a homologous series.
4. When data for one of the key properties used in parameter estimation, i.e. vapor pressure and liquid density, are not available, the parameters of PC–SAFT must be estimated using other considerations. For example, for polyethylene the parameters can in principle be predicted by extrapolating from the properties of the *n*-alkanes, from Figure 8.2. For other polymers, this is more difficult and the estimation of pure polymer parameters to be used in SAFT will be discussed in Chapter 14.
5. Various association schemes have been used for pure self-associating compounds, but in most successful applications water is modeled as 4C, alcohols as 2B or 3B and acids as 1A. More details on the parameter estimation for associating compounds and related applications are given in Chapter 13.
6. Recently, some researchers[84,85] have investigated the possibility of using additional data to vapor pressure and liquid density data in the parameter estimation of pure compounds. In particular, enthalpies of vaporization

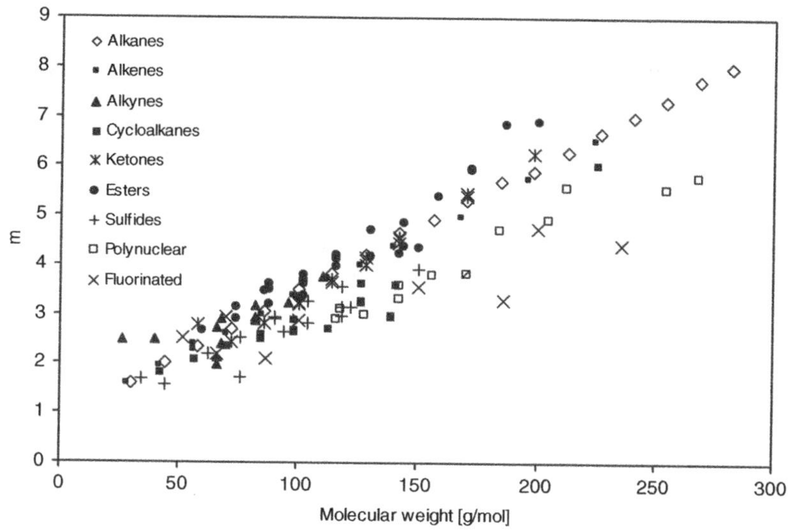

Figure 8.3 *The parameter m of PC–SAFT against molecular weight for different families of compounds. Reprinted with permission from Fluid Phase Equilibria, Applications of the simplified perturbed-chain SAFT equation of state using an extended parameter table by A. Tihic, G. M. Kontogeorgis et al., 248, 1, 29 Copyright (2006) Elsevier*

and monomer fractions were used. However, as discussed in Chapter 7, monomer fraction data are only available for water and a few alcohols. Enthalpies of vaporization are available for many compounds.

7. Recently, Grenner *et al.*[86,87] have proposed a generalized way for estimating parameters for two families of self-associating compounds (alcohols, glycols). They applied this generalized method to simplified PC–SAFT but the approach could equally well be used with other SAFT variants as well (with different coefficients). One unique set of associating parameters for all alcohols and one for all glycols provide

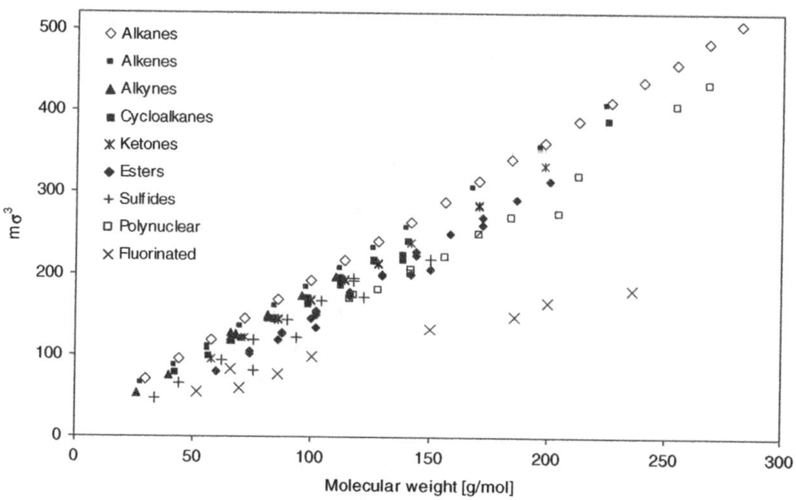

Figure 8.4 *The parameter $m\sigma^3$ of PC–SAFT against molecular weight for different families of compounds. Reprinted with permission from Fluid Phase Equilibria, Applications of the simplified perturbed-chain SAFT equation of state using an extended parameter table by A. Tihic, G. M. Kontogeorgis et al., 248, 1, 29 Copyright (2006) Elsevier*

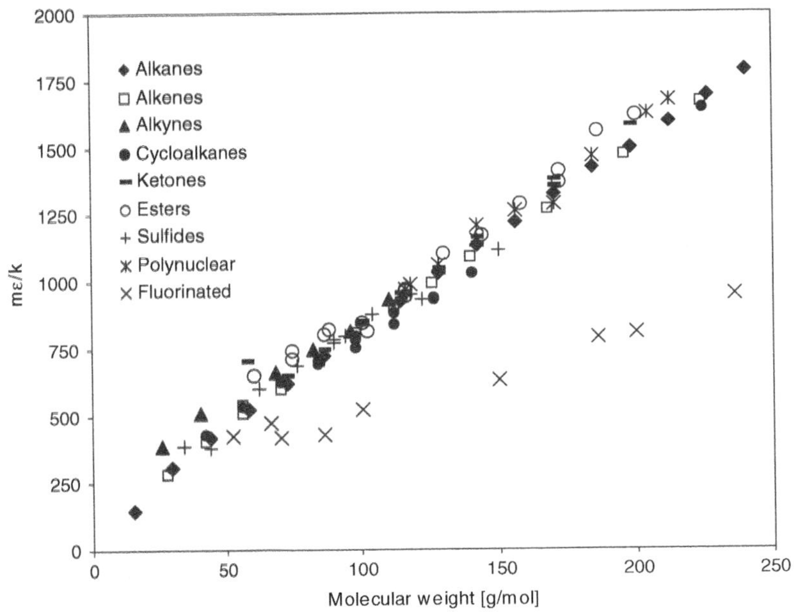

Figure 8.5 *The parameter $m\varepsilon/k$ of PC–SAFT against molecular weight for different families of compounds. Reprinted with permission from Fluid Phase Equilibria, Applications of the simplified perturbed-chain SAFT equation of state using an extended parameter table by A. Tihic, G. M. Kontogeorgis et al., 248, 1, 29 Copyright (2006) Elsevier*

excellent pure compound properties as well as mixture phase equilibria (see Chapter 13). The generalized equations for sPC–SAFT are:

Glycols:

$$\begin{aligned} m &= 0.0192M + 0.7924 \\ m\sigma^3 &= 1.3121M + 8.5441 \\ m\varepsilon/k &= 6.1866M + 224.49 \end{aligned} \tag{8.44a}$$

$$\begin{aligned} \varepsilon^{HB} &= 2080.03\ \text{K} \\ \kappa^{HB} &= 0.0235 \end{aligned}$$

Alcohols:

$$\begin{aligned} m &= 0.0287M + 0.0749 \\ m\sigma^3 &= 1.6906M + 5.5449 \\ m\varepsilon/k &= 7.3017M + 91.577 \end{aligned} \tag{8.44b}$$

$$\begin{aligned} \varepsilon^{HB} &= 2811.02\ \text{K} \\ \kappa^{HB} &= 0.0033 \end{aligned}$$

where M is the molar mass in g/mol, m denotes the number of segments, σ represents the segment diameter in Å and ε/k is the dispersion energy in K. ε^{HB}, κ^{HB} are respectively the association energy and volume.

8.3.2 Mixtures

One of the especially attractive features of SAFT-type approaches, which stems from their theoretical origin and can be seen from Equations (8.2) and (8.3), is that no mixing rules are needed in the chain and association terms. These terms are thus rigorously extended to mixtures. Mixing rules are needed, however, in the dispersion term of the equation for all SAFT variants. Moreover, combining rules are needed for the segment energy and volume (or diameter) parameters and the Lorentz–Berthelot rules are typically used. As with cubic EoS, a correction–interaction parameter k_{ij} is often used in the combining rule for the cross-energy parameter. Typically, the van der Waals one-fluid mixing rules are used, which involve only one temperature-independent interaction parameter k_{ij}, i.e. for the CK–SAFT:

$$\frac{u}{kT}=\frac{\sum_i\sum_j x_ix_jm_im_j\left[\frac{u_{ij}}{kT}\right](v^*)_{ij}}{\sum_i\sum_j x_ix_jm_im_j(v^*)_{ij}} \tag{8.45}$$

Recall that $u_{ij}=\varepsilon_{ij}$ in other SAFT variants.

The cross hard-core volume is given by the Berthelot rule:

$$(v^*)_{ij}=\left\{\frac{1}{2}\left[(v^*)_i^{1/3}+(v^*)_j^{1/3}\right]\right\}^3 \tag{8.46}$$

and the cross-energy parameter is given by the geometric mean rule for u_{ij} or ε_{ij} in most SAFT variants:

$$u_{ij}=\sqrt{u_iu_j}(1-k_{ij}) \tag{8.47a}$$

The Lorentz–Berthelot combining rules, as used for PC–SAFT and other SAFT variants which employ the ε_{ij} parameter and the cross-diameter σ_{ij} instead of the hard-core volume, are:

$$\varepsilon_{ij}=\sqrt{\varepsilon_i\varepsilon_j}(1-k_{ij}) \qquad \sigma_{ij}=\frac{\sigma_i+\sigma_j}{2} \tag{8.47b}$$

where again the interaction parameter k_{ij} is introduced to correct for the dispersion energies of unlike molecules. Other mixing rules have been proposed[17,88] for asymmetric systems or in order to represent better the critical area.

The interaction parameter k_{ij} should be typically optimized from experimental data, but for relatively simple systems, e.g. mixtures of hydrocarbons or gases with hydrocarbons, the interaction parameter can be estimated from the following equation, derived from the Hudson–McCoubrey theory (see Chapter 3):

$$k_{ij}=1-\left[2^7\left(\frac{(I_iI_j)^{1/2}}{I_i+I_j}\right)\left(\frac{\sigma_i^3\sigma_j^3}{(\sigma_i+\sigma_j)^6}\right)\right] \tag{8.48a}$$

I_i is the ionization potential of compound i in eV and the molecular-size diameters are expressed in Å. The Lennard-Jones value of the exponent in the attractive potential ($n=6$) has been used in Equation (8.48a).

The values for the segment diameters of the different molecules are quite similar in SAFT and thus, with some degree of approximation, the last part of Equation (8.48a) could be ignored. This has been done recently[108,109] and the following way to estimate the interaction parameter has been proposed:

$$k_{ij} = 1 - \left(\frac{2\sqrt{I_i I_j}}{I_i + I_j} \right) \tag{8.48b}$$

Equation (8.48b) has been used in connection with the polar/quadrupolar GC PC–SAFT (see below, Equation (8.52)), where the (pseudo-)ionization potentials I_i can also be obtained from GC values:

$$I_i = \sqrt[n]{\prod_i I_i^{n_i}} \tag{8.48c}$$

The estimated I_i values from Equation (8.48c) are close to, but systematically lower compared to, the experimental values and should thus be considered of semi-empirical nature.

The polar GC PC–SAFT using interaction parameters estimated from Equations (8.48b) and (8.48c) has been applied with success to VLE of several gases (CO_2, N_2, H_2S, methane, ethane) with hydrocarbons[108,109].

Equation (8.48a) has been derived from the Hudson–McCoubrey theory assuming the validity of the Lennard-Jones potential. If a different potential function is used, then the equation should be modified. For example, as Haslam *et al.*[110] showed for the square-well potential used in SAFT–VR, the following expression should be used:

$$k_{ij} = 1 - \left(\frac{\sqrt{\sigma_i \sigma_j}}{\sigma_{ij}} \right)^6 \left(\frac{2\sqrt{I_i I_j}}{I_i + I_j} \right) \left(\frac{\sqrt{\lambda_i^3 - 1}\sqrt{\lambda_j^3 - 1}}{\lambda_{ij}^3 - 1} \right) \tag{8.48d}$$

where the cross-parameter λ_{ij} is given by the following equation:

$$\lambda_{ij} = \frac{\lambda_i \sigma_i + \lambda_j \sigma_j}{\sigma_i + \sigma_j} \tag{8.48e}$$

All the above equations (8.48a,b,d) are expressions for estimating the interaction parameter in the cross-energy term. It is of course also possible to include an interaction parameter in the combining rule for the cross-diameter (in Equation (8.47b)), and this has been done for some SAFT variants, e.g. the soft-SAFT EoS, but in most cases the value is close to zero and the importance of this adjustable parameter compared to k_{ij} is smaller.

Although mixing rules are not needed in the association term, combining rules are needed for the association parameters for cross-associating mixtures. There are various choices and two successful ones are given in Equations (8.49) and (8.50):[77,80,86,87]

$$\varepsilon^{A_i B_j} = \frac{\varepsilon^{A_i B_i} + \varepsilon^{A_j B_j}}{2} \quad \text{and} \quad \kappa^{A_i B_j} = \sqrt{\kappa^{A_i B_i} \kappa^{A_j B_j}} \tag{8.49}$$

$$\varepsilon^{A_i B_j} = \frac{\varepsilon^{A_i B_i} + \varepsilon^{A_j B_j}}{2} \quad \text{and} \quad \kappa^{A_i B_j} = \sqrt{\kappa^{A_i B_i} \kappa^{A_j B_j}} \left(\frac{\sqrt{\sigma_i \sigma_j}}{\sigma_{ij}} \right)^3 \tag{8.50}$$

We call Equation (8.49) the CR-1 rule, while it can be shown (Problem 2 on the companion website at www.wiley.com/go/Kontogeorgis) that Equation (8.50) is essentially identical to the square-root rule for the cross-association strength:

$$\Delta^{A_iB_j} = \sqrt{\Delta^{A_iB_i}\Delta^{A_jB_j}} \tag{8.51}$$

Equation (8.51) hereafter will be called 'Elliott's rule', as it was presented by Suresh and Elliott[27] in connection with the ESD (Elliott–Suresh–Donohue) EoS. The difference between Equations (8.49) and (8.50) is expected to be small for SAFT variants (except for CPA discussed in Chapter 9), as the segment diameter values are similar even for rather different molecules. Equations (8.49)–(8.51) cannot be used for mixtures where solvation between two molecules is present, but one of them (or both!) are non-self-associating, e.g. water or glycols with aromatic hydrocarbons and chloroform–alcohols. If one molecule is self-associating but the other is not, two possibilities which have been proposed[55,89,90] are based on Equation (8.49) or (8.50) for the cross-association energy, with the value for the cross-association volume either being equal to that of the associating molecule[55] ($\kappa^{A_iB_j} = \kappa^{assoc}$) or fitted to mixture phase equilibrium data[89,90] ($\kappa^{A_iB_j} = fitted$).

8.4 Applications of SAFT to non-polar molecules

As can be expected, much of the strength of the SAFT family of EoS is revealed in applications involving associating or polymeric fluids, because in these cases the association and chain terms from Wertheim's theory will often dominate. Results for such mixtures will be presented in Chapters 13 and 14. In this chapter, we highlight some applications for non-polymeric, non-polar, size-asymmetric systems. Such mixtures can serve as a way to test some of the terms of SAFT EoS, i.e. the repulsive, chain and dispersion contributions, thus testing the reference terms (repulsive, chain) and part of the perturbation (dispersion). Various SAFT variants, especially PC–SAFT, have been applied to size-asymmetric mixtures, e.g. those containing alkanes with different sizes and gas–alkane mixtures. Table 8.13 summarizes some of the applications and Figures 8.6–8.10

Table 8.13 *Applications of SAFT to size-asymmetric systems*

Systems	SAFT variants used	Reference
CO_2–alkanes VLE	Original	Pasarello *et al.*[93] Benzaghou *et al.*[94]
Hydrocarbon mixtures VLE	PC–SAFT	Tihic *et al.*[19]
CO_2–solids, alkanes multicomponents (SGE at high pressures)	CK–SAFT	Seiler *et al.*[95]
C_1, H_2, CO, C_2 + alkanes	PC–SAFT	Ghosh *et al.*[96] Tihic *et al.*[19]
Alkanes with varying size length (short- and long-chain molecules)	PC–SAFT, CK–SAFT PC–SAFT	Ting *et al.*[91] Voutsas *et al.*[92] Tihic *et al.*[19]
Alkanes including critical properties and chain length dependency of the critical density	Soft-SAFT	Pamies and Vega[97,98]
Gas–perfluorocarbons	Soft-SAFT	Dias *et al.*[99]
H_2S, N_2 + alkanes	PC–SAFT	Tihic *et al.*[19]

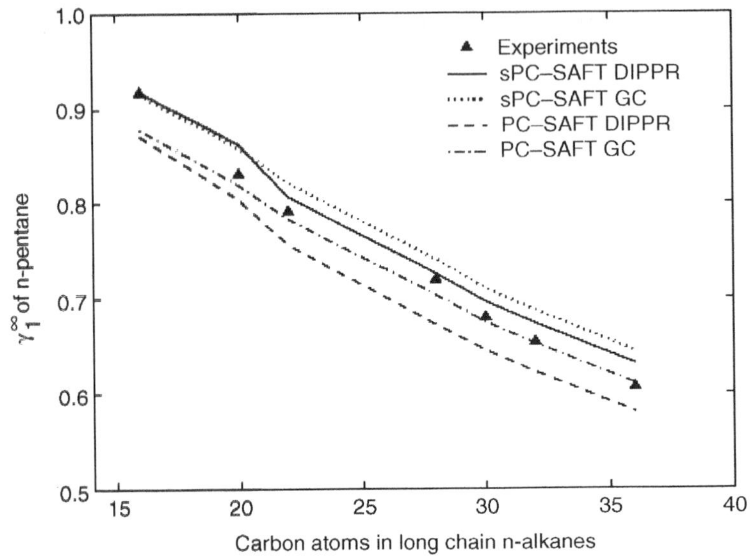

Figure 8.6 *Experimental and predicted PC–SAFT activity coefficients at infinite dilution for n-pentane in various n-alkanes as a function of the carbon atoms in long-chain n-alkanes. All calculations have been performed with k_{ij} equal to zero. Results are shown with both original and simplified PC–SAFT (sPC–SAFT) and with alkane parameters estimated from vapor pressures/liquid densities (indicated as DIPPR) or the GC method by Tihic et al.[66] (indicated as GC). Reprinted with permission from Fluid Phase Equilib. by Tihic et al., 281, 1, 60 Copyright (2009) Elsevier*

present a few typical results. Two of the investigations[91,92] present comparisons for alkane mixtures with the PR EoS.

The most important conclusions are:

1. PC–SAFT can describe phase equilibria for a large variety of non-associating mixtures due to the extensive parameter tables available[19].

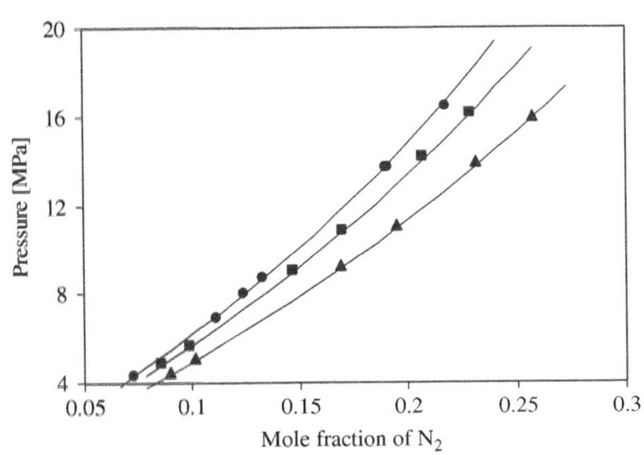

Figure 8.7 *Solubility of nitrogen in octacosane at 323.2 K (circles), 373.2 K (squares) and 423.3 K (triangles). Lines correspond to the simplified PC–SAFT calculations with $k_{12}=0.095$. Experimental data from J. Tong, W. Gao, R.L. Robinson, K.A.M. Gasem, J. Chem. Eng. Data, **1999**, 44, 784–787*

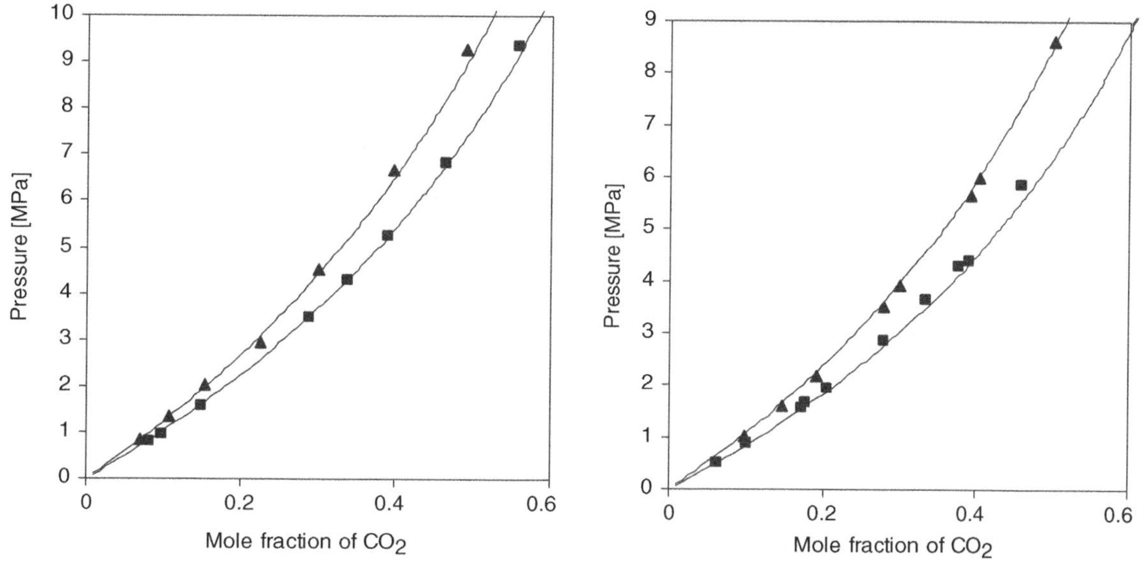

Figure 8.8 *VLE for CO_2–octacosane (left) and CO_2–hexatriacontane (right) mixtures at two temperatures, 373.2 K and 423.2 K, respectively. Comparison of simplified PC–SAFT correlation results ($k_{12} = 0.065$) to experimental data. Reprinted with permission from Fluid Phase Equilibria, Applications of the simplified perturbed-chain SAFT equation of state using an extended parameter table by A. Tihic, G. M. Kontogeorgis et al., 248, 1, 29 Copyright (2006) Elsevier*

2. The correlation of VLE using PC–SAFT (and other SAFT variants) is excellent for numerous mixtures of varying complexity and asymmetry, using a single system-dependent interaction parameter.
3. The performance of SAFT is satisfactory up to very asymmetric mixtures, including both VLE and infinite dilution activity coefficients. This has been illustrated via extensive studies for mixtures containing

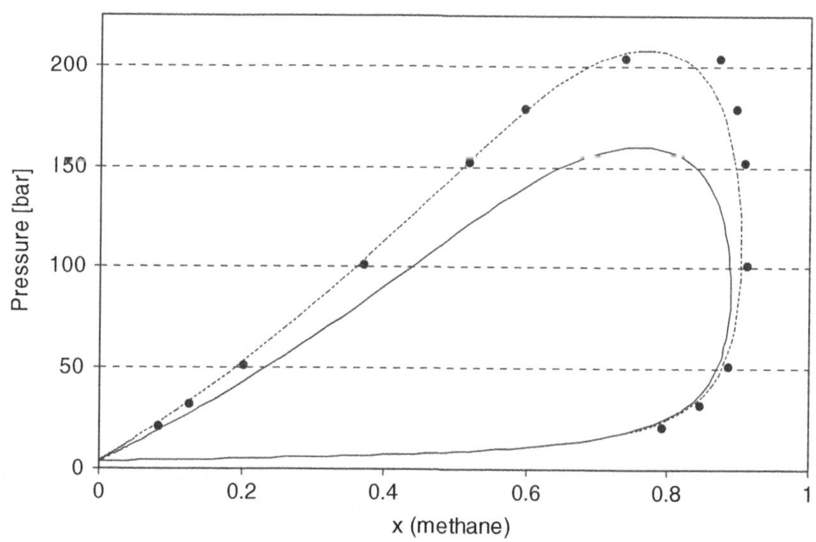

Figure 8.9 *VLE correlations for methane–hexadecane at 623.15 K, where symbols represent experimental data, and dotted lines correspond to the correlations from simplified PC–SAFT with $k_{12} = 0.013$, while solid lines are predictions ($k_{12} = 0$). Experimental data from H.M. Lin, H.M. Sebastian, K.C. Chao, J. Chem. Eng. Data, **1970**, 15, 82–91*

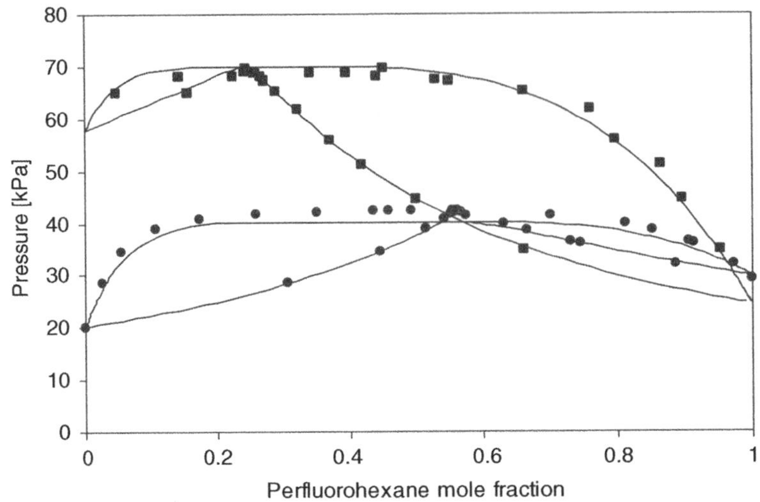

Figure 8.10 *Vapor pressure of perfluorohexane + pentane (squares) at 293.15 K and perfluorohexane + hexane (circles) at 298.65 K, where symbols represent experimental data, and lines correspond to the correlations from simplified PC–SAFT with* $k_{12} = 0.077$. *Experimental data from C. Duce, M.T. Tiné, L. Lepori, E. Matteoli, Fluid Phase Equilib.,* ***2002****, 199, 197–212. The occurrence and location of the azeotrope are well captured by the model using a single interaction parameter. Fluorinated compounds have recently found extensive applications in a variety of fields, ranging from use as substitutes of the toxic chlorinate solvents up to biomedical ones, e.g. oxygen carriers in artificial blood substitutes. Reprinted with permission from Fluid Phase Equilibria, Applications of the simplified perturbed-chain SAFT equation of state using an extended parameter table by A. Tihic, G. M. Kontogeorgis et al., 248, 1, 29 Copyright (2006) Elsevier*

various gases (nitrogen, methane, ethane, CO_2, CO, ethylene, H_2S) with heavy alkanes. Very low deviations are obtained, typically using low values of a single interaction parameter (k_{ij}).

4. A small, positive interaction parameter (k_{ij}) value is needed in most cases for obtaining an excellent correlation. There is some sensitivity to the k_{ij} value but, in most cases it is not extreme, at least for VLE calculations. Small k_{ij} values are required, e.g. for CO_2–alkanes, when a quadrupolar PC–SAFT version is used.[52] For PC–SAFT, the same k_{ij} value can be used for different gas–alkanes, depending only on the gas but not on the alkane used.
5. The performance of PC–SAFT is, for non-polar compounds, similar to cubic EoS, e.g. PR if the parameters of the cubic EoS are obtained in the same way as SAFT, i.e. based on vapor pressures and liquid densities. For non-polar systems and VLE, a similar performance between PC–SAFT and cubic EoS such as PR has been reported.[14] In some cases, e.g. CO_2–dodecane LLE, a similar performance between PC–SAFT and SRK (CPA) or PR is obtained when the parameters of the models are estimated from vapor pressures and liquid densities.[92] This indicates the importance of using vapor pressure and liquid density data in pure compound parameter estimation.
6. Compared to the large amount of results available for phase equilibrium calculations, relatively less attention has been paid with SAFT-type models to derivative properties such as heat capacity and speed of sound. Llovell and Vega[120] have applied soft-SAFT to the calculation of heat capacities, speed of sound and Joule–Thomson coefficients of three families of pure compounds; alkanes, alkenes and alcohols. With parameters based on vapor pressures and liquid densities, promising and semi-quantitatively correct results are obtained. The maxima of residual isochoric heat capacities

with density are qualitatively correctly represented for alkanes. The deviations are about 20–30% for the residual heat capacities but they are reduced when a crossover version of the soft-SAFT EoS is used.

7. An extensive investigation of derivative properties has been carried out by Lafitte *et al.*[121–123] with a specific variant of SAFT–VR, called SAFT–VR Mie. The authors considered isothermal compressibility, isobaric thermal expansivity, residual and isobaric heat capacities and speed of sound of *n*-alkanes, hydrofluoroethers and alcohols as well as alcohol–alkane mixtures. SAFT–VR Mie is a modification of the SAFT–VR equation based on the Mie *m*–*n* potential. Satisfactory representation of phase behavior, densities and derivative properties is achieved with a single set of molecular parameters, if all properties are included in the pure component parameter estimation. The authors emphasized especially the need to use these 'additional derivative properties' in the parameter estimation so that SAFT-type EoS can provide reliable values of derivative properties over an extensive temperature and pressure range.

8.5 GC SAFT approaches

Various research groups are currently developing GC versions of SAFT, but three approaches which have recently appeared in the literature are those by a French research group[60], DTU[66] and Imperial College for SAFT–VR[103]. They differ in both the GC schemes used and the range of applicability. The approaches are briefly described below.

8.5.1 French method[60,61]

This method is based on GC schemes for the three non-associating parameters and tables have been presented for original SAFT, SAFT–VR and PC–SAFT. The GC equations used for m, σ and ε/k are (for original SAFT and PC–SAFT):

$$m = \sum_i (n_i m_i)$$

$$\sigma = \frac{\sum_i n_i \sigma_i}{\sum_i n_i} \tag{8.52}$$

$$\varepsilon = \sum_i \sqrt[n]{\prod_i \varepsilon_i^{n_i}}$$

where m_i, σ_i, ε/k_i are the contributions of the individual groups and n_i the number of times they appear in the molecule.

For SAFT–VR, the GC scheme for the λ parameter is identical to that for the segment diameter σ.

Tamouza *et al.*[60,61] presented group parameters for hydrocarbon families and alcohols and the method was applied to alcohol–alkane VLE. Thi *et al.*[62] extended the approach to esters but an additional polar term is used. The most recent extensions[65] considered polycyclic aromatic hydrocarbons where an additional quadrupolar term is included and gas (H_2, CO_2)–alkanes were considered. Moreover, an attempt to develop a GC method for the interaction parameter k_{ij} has been presented[64]. Finally, the polar GC PC–SAFT has been recently applied to alcohol–alkanes and aromatic hydrocarbons[105] as well as mixtures containing, alcohols, aromatic or aliphatic hydrocarbons and esters.[106] The polar versions of the SAFT EoS are discussed in Chapter 13.

8.5.2 DTU method[66]

Unlike the French group, Tihic *et al.*[66] developed GC parameters for the three non-associating parameters of simplified PC–SAFT based on the equations:

$$
\begin{aligned}
m &= \sum_i (n_i m_i)_{FOG} + \sum_j (n_j m_j)_{SOG} \\
m\sigma^3 &= \sum_i (n_i m_i \sigma_i^3)_{FOG} + \sum_j (n_j m_j \sigma_j^3)_{SOG} \\
m\varepsilon/k &= \sum_i (n_i m_i \varepsilon_i/k)_{FOG} + \sum_j (n_j m_j \varepsilon_j/k)_{SOG}
\end{aligned}
\tag{8.53}
$$

which are based on the functionalities of SAFT EoS which are approximately linear with respect to molecular weight. Moreover, both first-order groups (FOG) and second-order groups (SOG) were considered. An extensive GC parameter table (for numerous FOG and SOG but only for the three non-associating parameters) has been developed. The parameters are presented in Appendix 8.B (Tables 8.14 and 8.15). In total, parameters for 45 FOG and 26 SOG have so far been estimated. Successful results are presented for polymers where the method has been primarily applied to: polymer densities, polymer–solvent VLE and LLE. Some results are shown in Chapter 14. Figure 8.11 shows one result for the system CO_2–methyl oleate, where the parameters of

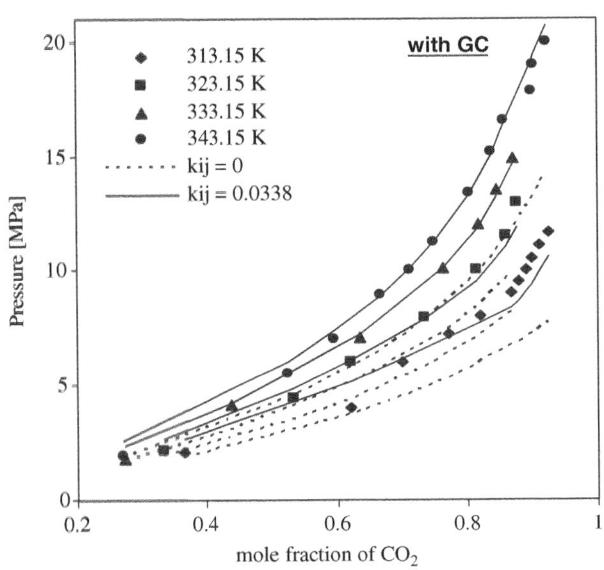

Figure 8.11 *VLE for CO_2–methyl oleate with sPC–SAFT using EoS parameters estimated from the GC method of Tihic et al.*[66]

the ester are obtained from the GC method. Equally satisfactory results with the GC method have been obtained for other esters as well as other heavy or complex compounds such as polyaromatics, sulfides, thiols, phytochemicals, etc.[107]

8.5.3 Other methods

The SAFT–γ method

Lymperiadis *et al.*[103,119] developed a GC EoS based on SAFT–VR, called the SAFT–γ EoS, assuming that the molecule is constructed by fused heteronuclear united-atom groups. This is a fundamental difference from the French and DTU methods which can both be considered to be 'homonuclear methods' (all of the segments making up the chain are identical). Each segment type in SAFT–γ EoS is characterized by a set of potential parameters (size, energy, range, hydrogen bonding energy and range) for like and unlike interactions that are used in a transferable fashion regardless of the molecules that contain the segment. Within this formalism, an extra parameter (shape parameter S_k) is introduced per group for non-associating components, which essentially characterizes the portion of the group that contributes to the overall molecular properties. For hydrogen bonding groups two additional parameters (the site–site energy and volume parameters) for sites on the various groups are required. The authors present optimized group parameters for CH_3, CH_2, CH_3CH, ACH, $ACCH_2$, CH_2=, CH=, C=O, COOH, NH_2 and OH with average deviations of 3.6% for vapor pressure and 0.9% for liquid densities for various families of organic compounds (*n*-alkanes and other hydrocarbons as well as alcohols). Alcohols are modeled as three-site (3B) molecules, which is found to represent them better than the simpler 2B (two-site) scheme. Besides the shape parameter (and the corresponding changes in the underlying theory in order to implement this parameter), another difference of SAFT–γ EoS over the previously described GC methods is the inclusion of cross-energy interaction parameters between various groups. Thus a group–energy interaction parameter table similar to UNIFAC tables has been developed, for the time being limited to the groups mentioned previously. The SAFT–γ EoS has been tested with both pure compounds not included in the parameter estimation (heavy alkanes and other hydrocarbons and heavy alcohols) as well as VLE of hydrocarbon and alcohol–alkane mixtures. The results are satisfactory and no adjustable parameters have been used in the case of mixtures.

GC SAFT and ESD models

Elliott and co-workers[63] proposed a generalized GC approach which was applied to SAFT and PC–SAFT EoS (as well as to the ESD EoS), based on the following methodology:

1. Pure component parameters of each EoS were obtained by matching their boiling point temperatures at 10 or 760 mmHg, their estimated solubility parameter, and estimated liquid density, while applying standard hydrogen bonding parameters.
2. GCs were regressed for the shape factor parameters of each EoS. Following this methodology, the authors obtained parameters for 86 first-order functional groups, covering 19 different families of compounds (hydrocarbons, alcohols, amines, thiols, nitriles, aldehydes, esters, ethers, silicones, etc.).

The strength of this approach is that no experimental data are required for optimizing the group pure component parameters, since the required values for the compressibility factors, liquid molar volume and heat of vaporization at the conditions of interest are obtained via other GC methods. This method allows for vapor pressure predictions based solely on the chemical structure and very satisfactory results are reported (deviations around 30% for vapor pressure). The method has not yet been applied to mixtures.

GC association terms: the GCA EoS

Several approaches have been presented in the literature which, although they do not use the SAFT EoS as a basis, include a GC version of the association term of SAFT. Fu *et al.*[111] have presented such a GC association term (essentially that of Wertheim's theory) coupled with UNIFAC leading to an 'association UNIFAC model'. A more well-known approach which is, moreover, in the form of an EoS is the GCA EoS developed by Brignole and co-workers since 1996.[112–118] GCA stands for 'group contribution plus association' and is essentially an extension of an older EoS (that of Skjold–Jorgensen) to associating fluids. The Skjold–Jorgensen EoS has a density-dependent local composition expression based on the NRTL equation, with the interactions expressed between functional groups, and typically contains four parameters for each group interaction. The new element in the GCA EoS is the inclusion of an association term, which is essentially a GC version of Wertheim's term used in SAFT. The radial distribution function is ignored (it is set equal to unity). Originally the GCA EoS was applied only to mixtures containing alcohols and water as the initial emphasis was on supercritical systems related to near or supercritical extraction and dehydration of oxygenated compounds from aqueous solutions.[112,113] Even though a very simple association scheme was used (a single OH group with two sites to characterize both water and alcohols), excellent results were reported for water–alcohol–alkanes and water–alcohol–gases VLE, including accurate prediction of the alcohol distribution coefficients. More recently the focus and application areas of the model have been expanded to include organic acids,[114,115] as well as polar compounds such as esters and ketones.[116–118] Esters and ketones are considered to be non-self-associating but capable of hydrogen bonding with, for example, alcohols or acids. The CR-1 rule is used between two self-associating compounds[115] but both cross-association parameters are fitted to data for the interaction between the ester or ketone group, for example, with OH or COOH. The association energy parameters of OH and COOH are equal to 2700 and 6300 K (in accordance with the high degree of association of carboxylic acids). The values obtained from the fitting procedure for the cross-association energies with esters and ketones do not always agree with the CR-1 rule (eq. 8.49). The value reported for the COOH–ester group (= 3249 K) is very close to that given by the CR-1 rule, while the values reported for OH–ester (= 2105 K) and OH–ketone (= 2485 K) are very close to the association energy of the OH group.

Overall the performance of the model is very satisfactory for the extensive database that has been tested (mostly VLE), including highly asymmetric mixtures, e.g. CO_2 with heavy acids, esters or glycerides and complex mixtures with multiple interactions such as methyl oleate–methanol–glycerol, which exhibits VLLE.[118] It has, moreover, been shown that the GCA EoS performs better than MHV2 and other models which do not explicitly account for association interactions.

8.6 Concluding remarks

The SAFT (Statistical Associating Fluid Theory) is a family of theoretically based equations of state, originating in the theory of Wertheim. SAFT accounts explicitly for dispersion, chain and association effects due to, for example, hydrogen bonding. Except for the dispersion term (which gives rise to the different SAFT variants), the other terms (chain, association) can be rigorously extended to mixtures without the need for mixing rules. Additional contributions to the model are needed for polar molecules and electrolytes and these are discussed in Chapters 13 and 15, respectively.

SAFT-type models contain a number of (typically five to six) parameters for pure associating compounds (three to four parameters for non-associating ones) which are typically estimated from vapor pressure and liquid density data, but other data can be occasionally used, e.g. monomer fractions or enthalpies of vaporization. The parameters show well-defined trends against the molecular weight within homologous series, trends which can be used e.g. in estimating parameters for polymers.

For associating compounds, a choice for the association scheme (number and type of association sites) is needed prior to the parameter estimation, e.g. two sites for alcohols and amines and four equal sites for water and glycols. For cross-associating mixtures, e.g. water–methanol or when induced association is present as in water–benzene, combining rules are required for the two cross-association parameters.

Numerous SAFT variations exist today, of which the CK–SAFT, PC–SAFT and SAFT–VR are possibly the most well known; the first two have extensive parameter tables, and for all three of them have been recently developed group contribution methods for estimating the pure compound parameters.

SAFT today is available in existing process simulators from Aspen Tech[100] and Simulation Sciences (the non-association part) and several other companies offer SAFT versions especially for polymers (Infochem, VLXE). SAFT is also available in many other in- house simulators and is used by many industries, especially in the polymer field, although other applications, e.g. asphaltenes, have gained some attention recently.[101]

We have shown in this chapter that SAFT can correlate satisfactorily phase equilibria for mixtures of different polarity and complexity, including size-asymmetric systems using a single interaction parameter.

As will be illustrated in subsequent chapters (Chapters 13 and 14), SAFT variants, especially PC–SAFT and SAFT–VR, have been applied to a variety of polar and associating mixtures (containing water, alcohols, glycols, organic acids, etc.) as well as to low- and high-pressure phase equilibria for polymers. It is therefore fair to conclude this section with a statement by Jog *et al.*[102] which seems to be still valid today:

> *SAFT offers advantages to industry today, but SAFT has not been fully optimized. In a sense, SAFT exists in a Redlich-Kwong form prior to transformation by Soave or Peng & Robinson. As recent theoretical extensions are applied and as researchers fine-tune the SAFT model, the predictive capabilities and range of applicability of the model will increase.*

This statement has indeed been verified in the years since it was published.

Appendix 8.A Calculation of fugacity coefficients with the simplified PC–SAFT EoS

This appendix presents the required equations for the calculation of the fugacity coefficient with the simplified PC–SAFT EoS. The fugacity coefficient ϕ_i of a component i in a mixture is given by:

$$RT \ln \phi_i = \left(\frac{\partial A^r}{\partial n_i}\right)_{T,V,n_j} - RT \ln Z \tag{8.54}$$

A^r is the residual Helmholtz energy for the mixture and Z is the compressibility factor, defined as:

$$Z = \frac{PV}{nRT} \tag{8.55}$$

The expression of the residual Helmholtz energy for the mixture (A^r) with the simplified PC–SAFT EoS is given by:

$$A^r(T,V,n) = A^r_{hard\text{-}sphere\ chain}(T,V,n) + A^r_{dispersion}(T,V,n) + A^r_{association}(T,V,n) \tag{8.56}$$

where the hard-sphere chain reference-fluid contribution $A^r_{hard\text{-}sphere\ chain}(T,V,n)$ is made up of the hard-sphere and the chain-formation contributions:

$$\frac{A^r_{hard\text{-}sphere\ chain}(T,V,n)}{RT} = \tilde{m}\frac{A^r_{hard\text{-}sphere}(T,V,n)}{RT} + \frac{A^r_{chain}(T,V,n)}{RT} \tag{8.57}$$

where $\tilde{m}$ is the average number of segments per chain given by:

$$\tilde{m} = \sum_i n_i m_i \tag{8.58}$$

For the simplified PC–SAFT EoS:

$$\frac{A^r_{hard\text{-}sphere}(T,V,n)}{RT} = \frac{4\eta - 3\eta^2}{(1-\eta)^2} \tag{8.59}$$

$$\frac{A^r_{chain}(T,V,n)}{RT} = \sum_i n_i(1-m_i)\ln g^{hs}(d) \tag{8.60}$$

$$g^{hs}(d) = \frac{2-\eta}{2(1-\eta)^3} \tag{8.61}$$

where the reduced packing fraction is given by:

$$\eta = \frac{\pi\rho d^3}{6}\tilde{m} \tag{8.62}$$

with:

$$d = \left(\frac{\sum_i n_i m_i d_i^3}{\sum_i n_i m_i}\right)^{1/3} \tag{8.63}$$

while d_i is the Chen and Kreglewski temperature-dependent segment diameter of component i given by:

$$d_i = \sigma_i \left[1 - 0.12 \exp\left(-3 \frac{\varepsilon_i}{kT} \right) \right] \tag{8.64}$$

The residual Helmholtz energy of the dispersion term is given by the following equation:

$$\frac{A^r_{dispersion}(T, V, n)}{RT} = -\left(2\pi\rho I_1 m^2 \varepsilon\sigma^3 + \pi\rho C_1 \tilde{m} I_2 m^2 \varepsilon^2 \sigma^3 \right) \tag{8.65}$$

where:

$$C_1 = \left(1 + \tilde{m} \frac{8\eta - 2\eta^2}{(1-\eta)^4} + (1 - \tilde{m}) \frac{20\eta - 27\eta^2 + 12\eta^3 - 2\eta^4}{(1-\eta)^2 (2-\eta)^2} \right)^{-1} \tag{8.66}$$

$$\begin{aligned} I_1 &= \sum_{i=0}^{6} a_i(\tilde{m}) \eta^i \\ I_2 &= \sum_{i=0}^{6} b_i(\tilde{m}) \eta^i \end{aligned} \tag{8.67}$$

$$\begin{aligned} a_i(\tilde{m}) &= a_{0i} + \frac{\tilde{m} - 1}{\tilde{m}} a_{1i} + \frac{\tilde{m} - 1}{\tilde{m}} \frac{\tilde{m} - 2}{\tilde{m}} a_{2i} \\ b_i(\tilde{m}) &= b_{0i} + \frac{\tilde{m} - 1}{\tilde{m}} b_{1i} + \frac{\tilde{m} - 1}{\tilde{m}} \frac{\tilde{m} - 2}{\tilde{m}} b_{2i} \end{aligned} \tag{8.68}$$

$$m^2 \varepsilon^y \sigma^3 = \sum_i \sum_j n_i n_j m_i m_i \left(\frac{\varepsilon_{ij}}{kT} \right)^y \sigma_{ij}^3 \tag{8.69}$$

with $y = 1, 2$.

The combining rules applied are:

$$\begin{aligned} \varepsilon_{ij} &= \sqrt{\varepsilon_i \varepsilon_j} (1 - k_{ij}) \\ \sigma_{ij} &= \frac{\sigma_i + \sigma_j}{2} \end{aligned} \tag{8.70}$$

The values of the universal constants are presented by Gross and Sadowski[14] (see also Table 8.9). The residual Helmholtz free energy for the association term is similar to the CPA EoS, and is presented in Chapter 9. Consequently the fugacity coefficient for the simplified PC–SAFT EoS is calculated as follows:

$$RT\ln\phi_i = \left(\frac{\partial A^r_{hard\text{-}sphere\ chain}}{\partial n_i}\right)_{T,V,n_j} + \left(\frac{\partial A^r_{dispersion}}{\partial n_i}\right)_{T,V,n_j} + \left(\frac{\partial A^r_{association}}{\partial n_i}\right)_{T,V,n_j} - RT\ln Z \qquad (8.71)$$

where the derivative with respect to n for the hard-sphere chain term is given by the following equations:

$$\begin{aligned} \frac{\partial}{\partial n_i}\left(\frac{A^r_{hard\text{-}sphere\ chain}}{RT}\right)_{T,V,n_j} &= m_i\left[\frac{4\eta-3\eta^2}{(1-\eta)^2} - \ln g^{hs}(d)\right] + \ln g^{hs}(d) + F \\ F &= \frac{m_i d_i^3 \rho\pi}{6}\left[\left(\tilde{m}\frac{4-2\eta}{(1-\eta)^3} - \tilde{m}\frac{5-2\eta}{(1-\eta)(2-\eta)}\right) + \frac{5-2\eta}{(1-\eta)(2-\eta)}\right] \end{aligned} \qquad (8.72)$$

For the dispersion term:

$$\frac{\partial}{\partial n_i}\left(\frac{A^r_{dispersion}}{RT}\right)_{T,V,n_j} = -\rho F_{dispersion} \qquad (8.73)$$

$$\begin{aligned} F_{dispersion} = {} & 2\pi\frac{\partial I_1}{\partial n_i}m^2\varepsilon\sigma^3 + 2\pi I_1\frac{\partial(m^2\varepsilon\sigma^3)}{\partial n_i} + \pi C_1 m_i I_2 m^2\varepsilon^2\sigma^3 + \pi I_2\tilde{m}\frac{\partial C_1}{\partial n_i}m^2\varepsilon^2\sigma^3 \\ & + \pi C_1\tilde{m}\frac{\partial I_2}{\partial n_i}m^2\varepsilon^2\sigma^3 + \pi C_1\tilde{m}I_2\frac{\partial(m^2\varepsilon^2\sigma^3)}{\partial n_i} \end{aligned} \qquad (8.74)$$

with:

$$\begin{aligned} \frac{\partial(m^2\varepsilon\sigma^3)}{\partial n_i} &= 2m_i\sum_j n_j m_j\left(\frac{\varepsilon_{ij}}{kT}\right)\sigma_{ij}^3 \\ \frac{\partial(m^2\varepsilon^2\sigma^3)}{\partial n_i} &= 2m_i\sum_j n_j m_j\left(\frac{\varepsilon_{ij}}{kT}\right)^2\sigma_{ij}^3 \end{aligned} \qquad (8.75)$$

$$\frac{\partial C_1}{\partial n_i} = C_2\frac{\partial\eta}{\partial n_i} - C_1^2\left[m_i\frac{8\eta-2\eta^2}{(1-\eta)^4} - m_i\frac{20\eta-27\eta^2+12\eta^3-2\eta^4}{(1-\eta)^2(2-\eta)^2}\right] \qquad (8.76)$$

$$C_2 = \frac{\partial C_1}{\partial\eta} = -C_1^2\left[\tilde{m}\frac{-4\eta^2+20\eta+8}{(1-\eta)^5} + (1-\tilde{m})\frac{2\eta^3+12\eta^2-48\eta+40}{(1-\eta)^3(2-\eta)^3}\right] \qquad (8.77)$$

$$\frac{\partial I_1}{\partial n_i} = \sum_{j=0}^{6}\left(\frac{\pi\rho m_i d_i^3}{6} j a_j(\tilde{m})\eta^{j-1} + \frac{\partial a_j}{\partial n_i}\eta^j\right)$$
$$\frac{\partial I_2}{\partial n_i} = \sum_{j=0}^{6}\left(\frac{\pi\rho m_i d_i^3}{6} j b_j(\tilde{m})\eta^{j-1} + \frac{\partial b_j}{\partial n_i}\eta^j\right) \tag{8.78}$$

$$\frac{\partial a_j}{\partial n_i} = \frac{m_i}{\tilde{m}^2}a_{1j} + \frac{m_i}{\tilde{m}^2}\left(\frac{3-4}{\tilde{m}}\right)a_{2j}$$
$$\frac{\partial b_j}{\partial n_i} = \frac{m_i}{\tilde{m}^2}b_{1j} + \frac{m_i}{\tilde{m}^2}\left(\frac{3-4}{\tilde{m}}\right)b_{2j} \tag{8.79}$$

For the association term, as will be further discussed for CPA (Chapter 9):

$$\frac{\partial}{\partial n_i}\left(\frac{A^r_{association}}{RT}\right)_{T,V,n_j} = \sum_{A_i}\ln X_{A_i} - \frac{1}{2}\sum_i n_i\sum_{A_i}(1-X_{A_i})\frac{\partial \ln g^{hs}}{\partial n_i} \tag{8.80}$$

8.A.1 Calculation of volume

The volume corresponding to a specific pressure, temperature and mixture composition can be calculated from the pressure equation:

$$P = \frac{nRT}{V} - \left(\frac{\partial A^r}{\partial V}\right)_{T,n} = \frac{nRT}{V} - \left(\frac{\partial A^r_{hard\text{-}sphere\ chain}}{\partial V}\right)_{T,n} - \left(\frac{\partial A^r_{dispersion}}{\partial V}\right)_{T,n} - \left(\frac{\partial A^r_{association}}{\partial V}\right)_{T,n} \tag{8.81}$$

For the hard-sphere chain term:

$$\left(\frac{\partial\left(\frac{A^r_{hard\text{-}sphere\ chain}}{RT}\right)}{\partial V}\right)_{T,n} = \frac{\partial \eta}{\partial V}\left[\left(\tilde{m}\frac{4-2\eta}{(1-\eta)^3} - \tilde{m}\frac{5-2\eta}{(1-\eta)(2-\eta)}\right) + \frac{5-2\eta}{(1-\eta)(2-\eta)}\right] \tag{8.82}$$

$$\frac{\partial \eta}{\partial V} = -\frac{1}{V}\eta \tag{8.83}$$

For the dispersion term:

$$\left(\frac{\partial\left(\frac{A^r_{dispersion}}{RT}\right)}{\partial V}\right)_{T,n} = \frac{1}{V}\left(F_v - \frac{A^r_{dispersion}}{RT}\right) \tag{8.84}$$

with:

$$F_v = -\frac{\partial \eta}{\partial V}\left(2\pi\frac{\partial I_1}{\partial \eta}m^2\varepsilon\sigma^3 + \pi C_1\tilde{m}\frac{\partial I_2}{\partial \eta}m^2\varepsilon^2\sigma^3 + \pi\tilde{m}\frac{\partial C_1}{\partial V}I_2 m^2\varepsilon^2\sigma^3\right) \tag{8.85}$$

$$\begin{aligned}\frac{\partial I_1}{\partial \eta} &= \sum_{i=0}^{6} i a_i(\tilde{m})\eta^{i-1}\\ \frac{\partial I_2}{\partial \eta} &= \sum_{i=0}^{6} i b_i(\tilde{m})\eta^{i-1}\end{aligned} \tag{8.86}$$

and $\partial C_1/\partial\eta = C_2$ as already given by Equation (8.77). Finally, for the association term as discussed in detail in Chapter 9:

$$\left(\frac{\partial\left(\frac{A^r_{association}}{RT}\right)}{\partial V}\right)_{T,n} = RT\left(\frac{\partial Q_{SP}}{\partial V}\right)_{T,n} \tag{8.87}$$

$$\left(\frac{\partial Q_{SP}}{\partial V}\right) = \frac{1}{2V}\left(1 - V\frac{\partial \ln g^{hs}}{\partial V}\right)\sum_i n_i \sum_{A_i}(1 - X_{A_i}) \tag{8.88}$$

Appendix 8.B The group parameters of the GC sPC–SAFT by Tihic *et al.*[66]

Table 8.14 *First-order group contributions for the m, $m\sigma^3$* and *$m\varepsilon/k$ parameters.* Reprinted with permission from Ind. Eng. Chem. Res., A Predictive Group-Contribution Simplified PC-SAFT Equation of State: Application to Polymer Systems by Amra Tihic, Georgios M. Kontogeorgis et al., 47, 15, 5092-5101 Copyright (2008) American Chemical Society

First-order groups (FOG)	Contributions			Sample group assignment (occurrences)
	m (–)	$m\sigma^3$ (Å^3)	$m\varepsilon/k$ (K)	
'–CH_3'	0.644 362	34.169 55	129.3866	Propane (2)
'–CH_2'	0.384 329	24.339 81	102.3238	Butane (2)
'–CH<'	0.043 834	13.953 91	68.2084	Isobutene (1)
'>C<'	– 0.492 080	2.3254 15	– 10.9830	Neopentane (1)
'CH_2=CH – '	1.031 502	52.086 40	240.7577	Propylene (1)
'–CH=CH–'	0.900 577	39.686 39	257.6914	*cis*-2-butane (1)
'CH_2=C<'	0.739 720	40.857 26	214.9811	Isobutene (1)
'–CH=C<'	0.513 621	29.695 89	208.6538	2-Methyl-2-butene (1)
'>C=C<'	0.377 151	17.070 79	212.3236	2,3-Dimethyl-2-butene (1)
'CH_2=C=CH–'	1.588 361	67.460 85	412.7788	1,2-Butadiene (1)
'CHEC–'	1.172 342	41.800 54	287.8396	Propyne (1)
'CEC'	0.715 504	33.171 67	276.7768	2-Butyne (1)
'ACH'	0.366 330	19.817 53	106.9481	Benzene (6)
'AC'	0.010 721	10.514 68	87.9300 8	Naphthalene (2)

Table 8.14 *(Continued)*

First-order groups (FOG)	Contributions			Sample group assignment (occurrences)
	m (–)	$m\sigma^3$ (Å^3)	$m\varepsilon/k$ (K)	
'ACCH$_3$'	0.836 861	43.574 97	235.0636	Toluene (1)
'ACCH$_2$ – '	0.389 760	34.498 23	171.3898	*m*-Ethyl toluene (1)
'ACCH<'	– 0.036 170	27.226 17	109.9270	*sec*-Butylbenzene (1)
'CH$_3$CO'	1.793 821	61.903 47	496.6067	Methyl ethyl ketone (1)
'CH$_2$CO'	1.552 067	52.185 21	444.9888	Cyclopentanone (1)
'CHO'	1.889 630	31.066 93	436.2165	1-Butanal (1)
'CH$_3$COO'	2.362 557	65.594 42	538.0440	Ethyl acetate (1)
'CH$_2$COO'	1.952 796	54.809 90	462.7927	Methyl propionate (1)
'HCOO'	1.745 841	43.392 09	426.2767	*n*-Propyl formate (1)
'COO'	1.439 110	32.513 28	351.1344	Ethyl acetate (1)
'CH$_3$O'	1.527 003	38.226 02	330.7321	Methyl ethyl ether (1)
'CH$_2$O'	1.226 298	28.979 37	277.2849	Ethyl vinyl ether (1)
'CHO'	1.544 171	16.263 32	321.8993	Diisopropyl ether (1)
'CH$_2$O (cyclic)'	1.101 871	32.855 69	307.6912	1,4-Dioxane (2)
'CH$_3$S'	1.391 979	61.468 31	428.3381	Methyl ethyl sulfide (1)
'CH$_2$S'	0.970 886	51.070 98	341.3012	Diethyl sulfide (1)
'CHS'	0.928 117	40.554 89	363.4252	Diisopropyl sulfide (1)
'I'	0.890 179	49.922 70	341.5094	Isopropyl iodide (1)
'Br'	0.888 904	35.942 99	276.4405	2-Bromopropane (1)
'CH$_2$Cl'	1.209 673	53.658 76	368.3897	*n*-Butyl chloride (1)
'CHCl'	0.810 132	43.140 92	272.4527	Isopropyl chloride (1)
'ACCl'	0.713 602	41.178 43	273.3549	*m*-Dichlorobenzene (2)
'ACF'	1.458 050	49.349 37	254.5195	Fluorobenzene (1)
'CF$_3$'	0.615 750	46.220 00	118.2550	Perfluorohexane (2)
'CH$_2$NO$_2$'	2.210 123	65.583 54	702.2415	1-Nitropropane (1)
'ACNO2'	2.340 985	48.053 76	687.6930	Nitrobenzene (1)
'CF$_2$'	0.933 850	35.658 25	153.7675	Perfluoromethyl cyclohexane (5)
'CH$_2$=C=C<'	1.328 356	57.764 90	375.6868	3-Methyl-1,2-butadiene (1)
'CH=C=CH–'	1.479 486	56.356 90	408.9468	2,3-Pentadiene (1)
'CHCO'	1.085 097	15.635 89	347.6752	Diisopropyl ketone (1)
'O (except as above)'	– 0.088 640	14.502 39	– 21.2312	Divinyl ether (1)

Table 8.15 *Second-order group contributions for the m, mσ³ and mε/k parameters.* Reprinted with permission from Ind. Eng. Chem. Res., A Predictive Group-Contribution Simplified PC-SAFT Equation of State: Application to Polymer Systems by Amra Tihic, Georgios M. Kontogeorgis et al., 47, 15, 5092–5101 Copyright (2008) American Chemical Society

Second-order groups (SOG)	Contributions			Sample group assignment (occurrences)	Total occurrences
	m (–)	$m\sigma^3$ (Å^3)	$m\varepsilon/k$ (K)		
'(CH$_3$)$_2$–CH–'	0.016 263	0.280 872	– 9.836 15	Isobutane (1)	38
'(CH$_3$)$_3$–C–'	0.041 437	1.472 296	– 6.895 16	Neopentane (1)	18
'–CH(CH$_3$)-CH(CH$_3$)–'	– 0.046 340	– 2.464 521	– 6.814 56	2,3-Dimethylbutane (1)	9

(continued)

Table 8.15 *(Continued)*

Second-order groups (SOG)	Contributions			Sample group assignment (occurrences)	Total occurrences
	m (–)	$m\sigma^3$ (Å^3)	$m\varepsilon/k$ (K)		
'–CH(CH_3)-C(CH_3)$_2$–'	– 0.101 480	– 1.913 722	– 4.680 34	2,2,3-Trimethylpentane (1)	4
'–C(CH_3)$_2$-C(CH_3)$_2$–'	– 0.183 290	– 7.464 243	23.727 92	2,2,3,3-Tetramethylpentane (1)	3
'ring of 3 carbons'	0.672 813	5.826 574	121.088 7	Cyclopropane (1)	3
'ring of 4 carbons'	0.506 911	2.643 507	101.789 8	Cyclobutane (1)	3
'ring of 5 carbons'	0.309 498	1.015 915	61.402 57	Cyclopentane (1)	8
'ring of 6 carbons'	0.064 237	– 0.636 912	29.246 73	Cyclohexane (1)	12
'ring of 7 carbons'	0.121 577	– 4.990 554	52.689 76	Cycloheptane (1)	9
'–C=C–C=C–'	– 0.112 200	– 0.955 843	– 19.753 90	1,3-Butadiene (1)	9
'CH_3–C='	0.013 538	0.342 635	1.528 91	Isobutene (2)	27
'–CH_2–C='	– 0.043 500	0.658 982	– 5.937 27	1-Butene (1)	54
'>C{H or C}–C='	0.087 414	– 0.806 591	– 0.989 85	3-Methyl-1-butene (1)	6
'string in cyclic'	– 0.145 510	– 0.220 100	– 5.719 82	Ethylcyclohexane (1)	6
'>CHCHO'	0.254 933	4.322 626	24.838 24	2-Methyl butyraldehyde (1)	5
'CH_3(CO)CH_2–'	– 0.038 690	– 1.234 490	– 6.906 54	2-Pentanone (1)	3
'C(cyclic)=O'	– 0.030 920	– 0.169 280	32.299 44	Cyclopentanone (1)	4
'CH_3(CO)OC{H or C}<'	0.088 960	– 1.218 040	– 18.939 80	Isobutyric acid (1)	4
'(CO)C{H_2}COO'	– 0.010 910	3.553 918	47.566 69	Ethyl acetoacetate (1)	3
'(CO)O(CO)'	– 0.077 070	3.242 018	– 13.212 50	Propanoic anhydride (1)	4
'ACHO'	– 0.242 220	– 0.310 420	– 8.358 46	Benzaldehyde (1)	4
'ACBr'	– 0.123 261	0.424 762	– 37.976 00	Bromotoluene (1)	5
'ACCOO'	– 0.080 001	1.303 279	– 23.830 30	Benzoic acid ethyl ester (1)	18
'AC(ACHm)$_2$AC(ACHn)$_2$'	– 0.000 133	0.665 582	5.950 01	Naphthalene (1)	4
'Ocyclic–Ccyclic=O'	0.363 529	3.423 997	– 0.922 22	Diketene (1)	2

References

1. M.S. Wertheim, *J. Stat. Phys.*, **1984**, *35*, 19.
2. M.S. Wertheim, *J. Stat. Phys.*, **1984**, *35*, 35.
3. M.S. Wertheim, *J. Stat. Phys.*, **1986**, *42*, 459.
4. M.S. Wertheim, *J. Stat. Phys.*, **1986**, *42*, 477.
5. W.G. Chapman, G. Jackson, K.E. Gubbins, *Mol. Phys.*, **1988**, *65*, 1057.
6. G. Jackson, W.G. Chapman, K.E. Gubbins, *Mol. Phys.*, **1988**, *65*, 1.
7. W.G. Chapman, K.E. Gubbins, G. Jackson, M. Radosz, *Ind. Eng. Chem. Res.*, **1990**, *29*, 1709.
8. S.H. Huang, M. Radosz, *Ind. Eng. Chem. Res.*, **1990**, *29*, 2284.
9. Y.-H. Fu, S.I. Sandler, *Ind. Eng. Chem. Res.*, **1995**, *34*, 1897.
10. T. Kraska, K.E. Gubbins, *Ind. Eng. Chem. Res.*, **1996**, *35*, 4727.
11. T. Kraska, K.E. Gubbins, *Ind. Eng. Chem. Res.*, **1996**, *35*, 4738.
12. F.J. Blas, L.F. Vega, *Ind. Eng. Chem. Res.*, **1998**, *37*, 360.
13. A. Gil-Vilegas, A. Galindo, P.J. Whitehead, S.J. Mills, G. Jackson, A.N. Burgess, *J. Chem. Phys.*, **1997**, *106*, 4168.
14. J. Gross, G. Sadowski, *Ind. Eng. Chem. Res.*, **2001**, *40*, 1244.
15. N. von Solms, M.L. Michelsen, G.M. Kontogeorgis, *Ind. Eng. Chem. Res.*, **2003**, *42*, 1098.
16. W.G. Chapman, K.E. Gubbins, G. Jackson, M. Radosz, *Fluid Phase Equilib.*, **1989**, *52*, 31.

17. S.H. Huang, M. Radosz, *Ind. Eng. Chem. Res.*, **1991**, *30*, 1994.
18. C. McCabe, A. Gil-Vilegas, G. Jackson, *Chem. Phys. Lett.*, **1999**, *303*, 27.
19. A. Tihic, G.M. Kontogeorgis, N. von Solms, M.L. Michelsen, *Fluid Phase Equilib.*, **2006**, *248*, 29.
20. G.M. Kontogeorgis, E. Voutsas, I. Yakoumis, D.P. Tassios, *Ind. Eng. Chem. Res.*, **1996**, *35*, 4310.
21. G.M. Kontogeorgis, I.V. Yakoumis, H. Meijer, E.M. Hendriks, T. Moorwood, *Fluid Phase Equilib.*, **1999**, *158–160*, 201.
22. G.M. Kontogeorgis, M.L. Michelsen, G.K. Folas, S. Derawi, N. von Solms, E.H. Stenby, *Ind. Eng. Chem. Res.*, **2006**, *45(14)*, 2855.
23. G.M. Kontogeorgis, M.L. Michelsen, G.K. Folas, S. Derawi, N. von Solms, E.H. Stenby, *Ind. Eng. Chem. Res.*, **2006**, *45(14)*, 4869.
24. L. Ruffine, P. Mougin, A. Barreau, *Ind. Eng. Chem. Res.*, **2006**, *45*, 7688.
25. Y. Lin, K. Thomsen, J.-C. de Hemptinne, *AIChE J.*, **2007**, *53(4)*, 989.
26. J.R. Elliott, S.J. Suresh, M.D. Donohue, *Ind. Eng. Chem. Res.*, **1990**, *29*, 1476.
27. J.S. Suresh, J.R. Elliott, *Ind. Eng. Chem. Res.*, **1992**, *31*, 2783.
28. J.S. Suresh, J.R. Elliott, *Ind. Eng. Chem. Res.*, **1991**, *30*, 524.
29. A.S. Puhala, J.R. Elliott, *Ind. Eng. Chem. Res.*, **1993**, *32*, 3174.
30. J. Wu, J.M. Prausnitz, *Ind. Eng. Chem. Res.*, **1998**, *37*, 1634.
31. O. Pfohl, A. Pagel, G. Brunner, *Fluid Phase Equilib.*, **1999**, *157*, 53.
32. C. Perakis, E. Voutsas, K. Magoulas, D. Tassios, *Fluid Phase Equilib.*, **2006**, *243(1–2)*, 142.
33. C.A. Perakis, E.C. Voutsas, K.G. Magoulas, D.P. Tassios, *Ind. Eng. Chem. Res.*, **2007**, *46(3)*, 932.
34. E. Voutsas, C. Perakis, G. Pappa, D. Tassios, *Fluid Phase Equilib.*, **2007**, *261*, 343.
35. Z. Huang, Y.C. Chiew, M. Feng, H. Miao, J.-H. Li, L. Xu, *J. Supercrit. Fluids*, **2007**, *43*, 259.
36. N. Koak, Th.W. de Loos, R.A. Heidemann, *Ind. Eng. Chem. Res.*, **1999**, *38*, 1718.
37. M. Cismondi, E.A. Brignole, J. Mollerup, *Fluid Phase Equilib.*, **2005**, *234*, 108.
38. T. Kraska, *Ind. Eng. Chem. Res.*, **1998**, *37(12)*, 4889.
39. E.M. Hendriks, J.M. Walsh, A.R.D. van Bergen, *J. Stat. Phys.*, **1997**, *87(5/6)*, 1287.
40. I.V. Yakoumis, G.M. Kontogeorgis, E.C. Voutsas, E.M. Hendriks, D.P. Tassios, *Ind. Eng. Chem. Res.*, **1998**, *37*, 4175.
41. M.L. Michelsen, E.M. Hendriks, *Fluid Phase Equilib.*, **2001**, *180(1–2)*, 165.
42. M.L. Michelsen, *Ind. Eng. Chem. Res.*, **2004**, *43(19)*, 6262.
43. M.L. Michelsen, *Ind. Eng. Chem. Res.*, **2006**, *45(25)*, 8449.
44. P.K. Jog, S.G. Sauer, J. Blaesing, W.G. Chapman, *Ind. Eng. Chem. Res.*, **2001**, *40*, 4641.
45. F. Tumakaka, G. Sadowski, *Fluid Phase Equilib.*, **2004**, *217*, 233.
46. F. Tumakaka, J. Gross, G. Sadowski, *Fluid Phase Equilib.*, **2005**, *228–229*, 89.
47. S.G. Sauer, W.G. Chapman, *Ind. Eng. Chem. Res.*, **2003**, *42*, 5687.
48. E.K. Karakatsani, T. Spyriouni, I.G. Economou, *AIChE J.*, **2005**, *51*, 2328.
49. E.K. Karakatsani, I.G. Economou, *J. Phys. Chem. B*, **2006**, *110*, 9252.
50. E.K. Karakatsani, G.M. Kontogeorgis, I.G. Economou, *Ind. Eng. Chem. Res.*, **2006**, *45 (17)*, 6063.
51. J. Gross, J. Vrabec, *AIChE J.*, **2006**, *52*, 1194.
52. J. Gross, *AIChE J.*, **2005**, *51*, 2556.
53. M. Kleiner, J. Gross, *AIChE J.*, **2006**, *52*, 1951.
54. E.K. Karakatsani, I.G. Economou, *Fluid Phase Equilib.*, **2007**, *261*, 265.
55. M. Kleiner, G. Sadowski, *J. Phys. Chem. C*, **2007**, *111*, 15544.
56. B. Behzadi, B.H. Patel, A. Galindo, C. Ghotbi, *Fluid Phase Equilib.*, **2005**, *236*, 241.
57. B.H. Patel, P. Paricaud, A. Galindo, G.C. Maitland, *Ind. Eng. Chem. Res.*, **2003**, *42*, 3809.
58. L.F. Cameretti, G. Sadowski, J.M. Mollerup, *Ind. Eng. Chem. Res.*, **2005**, *44*, 3355.
59. D. Fuchs, J. Fischer, F. Tumakaka, G. Sadowski, *Ind. Eng. Chem. Res.*, **2006**, *45*, 6578.
60. S. Tamouza, J.-Ph. Passarello, P. Tobaly, J.-Ch. de Hemptinne, *Fluid Phase Equilib.*, **2004**, *222–223*, 67.
61. S. Tamouza, J.-Ph. Passarello, P. Tobaly, J.-Ch. de Hemptinne, *Fluid Phase Equilib.*, **2005**, *228–229*, 409.
62. T.X.N. Thi, S. Tamouza, P. Tobaly, J.-Ph. Passarello, J.-Ch. de Hemptinne, *Fluid Phase Equilib.*, **2005**, *238*, 254.
63. F. Emami, A. Vahid, J.R. Jr.Elliott, F. Feyzi, *Ind. Eng. Chem. Res.*, **2008**, *47*, 8401.

64. C. Le Thi, S. Tamouza, J.P. Passarello, P. Tobaly, J.-C. de Hemptinne, *Ind. Eng. Chem. Res.*, **2006**, *45*, 6803.
65. D.N. Huynh, M. Benamira, J.-Ph. Passarello, P. Tobaly, J.-Ch. de Hemptinne, *Fluid Phase Equilib.*, **2007**, *254*, 60.
66. A. Tihic, G.M. Kontogeorgis, N. von Solms, M.L. Michelsen, L. Constantinou, *Ind. Eng. Chem. Res.*, **2008**, *47*, 5092.
67. P. Paricaud, A. Galindo, G. Jackson, *Fluid Phase Equilib.*, **2002**, *194–197*, 87.
68. E.A. Müller, K.E. Gubbins, *Ind. Eng. Chem. Res.*, **2001**, *40*, 2193.
69. I.G. Economou, *Ind. Eng. Chem. Res.*, **2002**, *41*, 953.
70. Y.S. Wei, R.J. Sadus, *AIChE J.*, **2000**, *46(1)*, 169.
71. J.M. Prausnitz, F.W. Tavares, *AIChE J.*, **2004**, *50(4)*, 739.
72. W. Arlt, O. Spuhl, A. Klamt, *Chem. Eng. Process.*, **2004**, *43*, 221.
73. N. Von Solms, I. Kouskoumvekaki, M.L. Michelsen, G.M. Kontogeorgis, *Fluid Phase Equilib.*, **2006**, *241*, 344.
74. J.C. De Hemptinne, P. Mougin, A. Barreau, L. Ruffine, S. Tamouza, R. Inchekel, *Oil Gas Sci. Technol. – Rev. IFP*, **2006**, *61(3)*, 363.
75. S.S. Chen, A. Kreglewski, *Ber. Bunsenges. Phys. Chem.*, **1977**, *81*, 1048.
76. N.F. Carnahan, K.E. Starling, *J. Chem. Phys.*, **1970**, *53*, 600.
77. J. Gross, G. Sadowski, *Ind. Eng. Chem. Res.*, **2002**, *41*, 5510.
78. R. O'Lenick, X.J. Li, Y.C. Chiew, *Mol. Phys.*, **1995**, *86*, 1123.
79. J.P. Wolbach, S.I. Sandler, *Ind. Eng. Chem. Res.*, **1997**, *36*, 4041.
80. J.P. Wolbach, S.I. Sandler, *Ind. Eng. Chem. Res.*, **1998**, *37*, 2917.
81. S.I. Sandler, J.P. Wolbach, M. Castier, G. Escobedo-Alvarado, *Fluid Phase Equilib.*, **1997**, *136*, 15.
82. F. Llovell, L.F. Vega, *J. Chem. Phys.*, **2004**, *121(21)*, 10715.
83. F. Llovell, L.F. Vega, *J. Phys. Chem. B*, **2006**, *110*, 1350.
84. G.N.I. Clark, A.J. Haslam, A. Galindo, G. Jackson, *Mol. Phys.*, **2006**, *104*, 3561.
85. A. Grenner, G.M. Kontogeorgis, M.L. Michelsen, G.K. Folas, *Mol. Phys.*, **2007**, *105(13–14)*, 1797.
86. A. Grenner, G.M. Kontogeorgis, N. Von Solms, M.L. Michelsen, *Fluid Phase Equilib.*, **2007**, *261(1–2)*, 248.
87. A. Grenner, G.M. Kontogeorgis, N. Von Solms, M.L. Michelsen, *Fluid Phase Equilib.*, **2007**, *258 (1)*, 83.
88. I.G. Economou, C. Tsonopoulos, *Chem. Eng. Sci.*, **1997**, *52 (4)*, 511.
89. A. Grenner, I. Tsivintzelis, G.M. Kontogeorgis, I.G. Economou, C. Panayiotou, *Ind. Eng. Chem. Res.*, **2008**, *47 (15)*, 5636.
90. I. Tsivintzelis, A. Grenner, G.M. Kontogeorgis, G. Ioannis, I.G. Economou, C. Panayiotou, *Ind. Eng. Chem. Res.*, **2008**, *47 (15)*, 5651.
91. P.D. Ting, P.C. Joyce, P.K. Jog, W.G. Chapman, M.C. Thies, *Fluid Phase Equilib.*, **2003**, *206 (1–2)*, 267.
92. E.C. Voutsas, G.D. Pappa, K. Magoulas, D.P. Tassios, *Fluid Phase Equilib.*, **2006**, *240*, 127.
93. J.P. Pasarello, S. Benzaghou, P. Tobaly, *Ind. Eng. Chem. Res.*, **2000**, *39*, 2578.
94. S. Benzaghou, J.P. Pasarello, P. Tobaly, *Fluid Phase Equilib.*, **2001**, *180*, 1.
95. M. Seiler, J. Gross, B. Bungert, G. Sadowski, W. Arlt, *Chem. Eng. Technol.*, **2001**, *24 (6)*, 607.
96. A. Ghosh, W.G. Chapman, R.N. French, *Fluid Phase Equilib.*, **2003**, *209*, 229.
97. J.C. Pamies, L.F. Vega, *Ind. Eng. Chem. Res.*, **2001**, *40*, 2532.
98. J.C. Pamies, L.F. Vega, *Mol. Phys.*, **2002**, *100 (15)*, 2519.
99. A.M.A. Dias, J.C. Pamies, J.A.P. Coutinho, I.M. Marrucho, L.F. Vega, *J. Phys. Chem. B*, **2004**, *108*, 1450.
100. C.C. Chen, P.M. Mathias, *AIChE J.*, **2002**, *48* (2), 194.
101. K.S. Pedersen, P.L. Christensen, *Phase Behavior of Petroleum Reservoir Fluids*. CRC Press/Taylor & Francis, **2007**.
102. P.K. Jog, A. Garcia-Cuellar, W.G. Chapman, *Fluid Phase Equilib.*, 1999, *158–160*, 321.
103. A. Lymperiadis, C.S. Adjiman, A. Galindo, G. Jackson, *J. Chem. Phys.*, **2007**, *127*, 234903.
104. S.P. Tan, H. Adidharma, M. Radosz, *Ind. Eng. Chem. Res.*, **2008**, *47*, 8063.
105. D.N. Huynh, J.-Ph. Passarello, P. Tobaly, J.-Ch. de Hemptinne, *Fluid Phase Equilib.*, **2008**, *264*, 62.
106. D.N. Huynh, A. Falaix, J.-Ph. Passarello, P. Tobaly, J.-Ch. de Hemptinne, *Fluid Phase Equilib.*, **2008**, *264*, 182.
107. A. Tihic, N. von Solms, M.L. Michelsen, G.M. Kontogeorgis, L. Constantinou, *Fluid Phase Equilib.*, **2009**, *281 (1)*, 60.

108. D.N. Huynh, J.-Ph. Passarello, P. Tobaly, J.-Ch. de Hemptinne, *Ind. Eng. Chem. Res.*, **2008**, *47*, 8847.
109. D.N. Huynh, T.K.S. Tran, S. Tamouza, J.-Ph. Passarello, P. Tobaly, J.-Ch. de Hemptinne, *Ind. Eng. Chem. Res.*, **2008**, *47*, 8859.
110. A.J. Haslam, A. Galindo, G. Jackson, *Fluid Phase Equilib.*, **2008**, *266*, 105.
111. Y.-F. Fu, H. Orbey, S.I. Sandler, *Ind. Eng. Chem. Res.*, **1998**, *35*, 4656.
112. H.P. Gros, S. Bottini, E.A. Brignole, *Fluid Phase Equilib.*, **1996**, *116*, 537.
113. H.P. Gros, S. Bottini, E.A. Brignole, *Fluid Phase Equilib.*, **1997**, *139*, 75.
114. O. Ferreira, T. Fornari, E.A. Brignole, S.B. Bottini, *Lat. Am. Appl. Res.*, **2003**, *33*, 307.
115. O. Ferreira, E.A. Brignole, E.A. Macedo, *J. Chem. Thermodyn.*, **2004**, *36*, 1105.
116. O. Ferreira, E.A. Macedo, E.A. Brignole, *J. Food Eng.*, **2005**, *70*, 579.
117. P. Hegel, A. Andreatta, S. Pereda, S. Bottini, E.A. Brignole, *Fluid Phase Equilib.*, **2008**, *266*, 31.
118. A.E. Andreatta, L.M. Casas, P. Hegel, S. Bottini, E.A. Brignole, *Ind. Eng. Chem. Res.*, **2008**, *47*, 5157.
119. C.A. Lymperiadis, S. Adjiman, G. Jackson, A. Galindo, *Fluid Phase Equilib.*, **2008**, *274*, 85.
120. F. Llovell, L.F. Vega, *J. Phys. Chem. B*, **2006**, *110*, 11427.
121. T. Lafitte, D. Bessieres, M.M. Pineiro, J.-L. Daridon, *J. Chem. Phys.*, **2006**, *124*, 024509.
122. T. Lafitte, F. Plantier, M.M. Pineiro, J.-L. Daridon, D. Bessieres, *Ind. Eng. Chem. Res.*, **2007**, *46*, 6998.
123. T. Lafitte, M.M. Pineiro, J.-L. Daridon, D. Bessieres, *J. Phys. Chem. B*, **2007**, *111*, 344.

9

The Cubic-Plus-Association Equation of State

9.1 Introduction

9.1.1 The importance of associating (hydrogen bonding) mixtures

Associating systems are those which contain compounds capable of hydrogen bonding, e.g. alcohols, water, amines, acids, etc. Phase equilibria of complex associating systems are important in many practical cases. For example, in several hydrocarbon and chemical processes as well as in order to meet environmental and low-emissions legislation, systems involving a highly polar compound, a non-polar one and a co-solvent, with both polar and non-polar functional groups, are typically present, e.g. mixtures involving water, alcohols or glycols and hydrocarbons. The role of the alcohol or glycol is that of a co-solvent: they improve significantly the mutual solubility of water and hydrocarbon in the hydrocarbon-rich and water-rich phases, respectively. Without such co-solvents, these solubilities are very low. Some examples of cases where phase equilibria of associating systems are important are as follows:

- In the oil industry:
  - formation of gas hydrates, calculation of the amount of hydrate inhibitors;
  - partitioning of alcohols and other chemicals between water and oil;
  - use of alcohols as additives in gasoline for *Octane Number* improver and for environmental reasons;
  - heterogeneous azeotropic separation in alcohol manufacturing;
  - glycol regeneration units (use of TEG for gas dehydration);
    removal of S-contaminants (thiols etc.) from oil and gas streams;
  - Reid vapor pressure of gasoline in the presence of additives.[1]
- For environmental calculations:
  - octanol–water and other partition coefficients;
  - modeling of waste streams.
- In biotechnology and biochemical engineering.[23]
- In the food product and process design.[4]

Thermodynamic Models for Industrial Applications: From Classical and Advanced Mixing Rules to Association Theories
Georgios M. Kontogeorgis and Georgios K. Folas

- In the chemical industry:[5]
  - novel separation methods for pharmaceuticals (e.g. supercritical fluid extraction);
  - recovery of alcohols from aqueous solutions using high-pressure propane (or other gases); propane is often selected as solvent since no azeotropes are expected in mixtures of ethanol with propane and due to its high-pressure extracting capability.
- In the polymer industry:
  - polymer blends;
  - phase equilibria in coatings of mixed water–organic solvents.

9.1.2 Why specifically develop the CPA EoS?

CPA combines the classical simple Soave–Redlich–Kwong (SRK) equation with an advanced association term, which is similar to that of SAFT (presented in Chapter 8). The purpose of the development of the cubic-plus-association (CPA) EoS was to arrive at a model which is based on the SRK (or other cubic) EoS, widely used in the petroleum industry (e.g. for mixtures with gases and hydrocarbons), and on the SAFT theory for describing associating mixtures in a theoretically correct way. The CPA model reduces to SRK, the EoS often used for oil and gas applications, in the absence of hydrogen bonding compounds (water, alcohols, acids, etc.), thus achieving a balance between accuracy and simplicity and gaining acceptance in the oil, gas and chemical industries. Besides simplicity and accuracy, the numerical implementation of the association term ensures that the computation time is not much higher than that of SRK and other simple models.

From a practical point of view, the target in the CPA project was to develop a thermodynamic model capable of describing complex (multicomponent, multiphase) equilibria of mixtures containing hydrocarbons and polar/associating chemicals like water, alcohols, glycols, esters and organic acids. VLE and LLE of water–alcohol–hydrocarbons and other chemicals (sulfolane, heavy alcohols, glycols, etc.) were given priority at first. Moreover, the following targets were considered:

- accurate multicomponent calculations using solely parameters estimated from binary data;
- use of a simple mathematical formulation, but with theoretical background in order to describe complex compounds;
- the model should reduce to the classical SRK equation since the representation of hydrocarbons' phase equilibria via SRK is considered satisfactory and successful characterization procedures are available for cubic EoS for petroleum fluids.

The model developed from those principles, the so-called CPA EoS, was first presented in the literature in 1996.[6] Applications up to 1999 included mostly mixtures with alcohols, water and alkanes, while results for other polar chemicals (glycols, acids, amines, alkanolamines, glycolethers, etc.) and applications with aromatic–olefinic hydrocarbons were presented after 1999.

The reason that a new EoS was considered necessary was because widely used existing models (e.g. cubic EoS) were found to be inadequate for V(L)LLE applications, especially for mixtures containing highly immiscible compounds, e.g. water–hydrocarbons LLE or water–alcohol(glycol)–hydrocarbons VLLE. Of course, EoS/G^E models like SRK with the Huron–Vidal mixing rule sometimes provide satisfactory results which are highly dependent on the accuracy of the underlying activity coefficient model at low pressures and models like NRTL and UNIQUAC. Such local composition models often fail to describe well mixtures with hydrogen bonding compounds, especially for multiphase, multicomponent equilibria, as discussed in Chapters 5 and 6. A more 'direct' approach was desired. Thus, CPA (SRK + association term of SAFT) was considered to be a compromise between simplicity and accuracy as the use of the 'full' SAFT EoS was not considered necessary, at least for applications not containing polymers.

9.2 The CPA EoS

9.2.1 General

The CPA EoS, proposed by Kontogeorgis *et al.*,[6,7] can be expressed for mixtures in terms of pressure P as:

$$P = \frac{RT}{V_m - b} - \frac{a(T)}{V_m(V_m + b)} - \frac{1}{2}\frac{RT}{V_m}\left(1 + \rho\frac{\partial \ln g}{\partial \rho}\right)\sum_i x_i \sum_{A_i}(1 - X_{A_i}) \tag{9.1}$$

where ρ is the molar density ($= 1/V_m$).

The key element of the association term is X_{A_i}, which represents the fraction of sites A on molecule i that do not form bonds with other active sites ('site monomer fraction'), and x_i is the mole fraction of component i. X_{A_i} is related to the association strength $\Delta^{A_iB_j}$ between two sites belonging to two different molecules, e.g. site A on molecule i and site B on molecule j, determined from:

$$X_{A_i} = \frac{1}{1 + \rho\sum_j x_j \sum_{B_j} X_{B_j}\Delta^{A_iB_j}} \tag{9.2}$$

where the association strength $\Delta^{A_iB_j}$ in CPA is expressed as:

$$\Delta^{A_iB_j} = g(\rho)\left[\exp\left(\frac{\varepsilon^{A_iB_j}}{RT}\right) - 1\right] b_{ij}\beta^{A_iB_j} \tag{9.3}$$

with the radial distribution function:

$$g(\rho) = \frac{1}{1 - 1.9\eta} \quad \text{and} \quad \eta = \frac{1}{4}b\rho \quad \text{while } b_{ij} = \frac{b_i + b_j}{2}$$

The radial distribution of CPA is discussed further in Appendix 9.A. Finally, the energy parameter of the EoS is given by a Soave-type temperature dependency, while b is temperature independent:

$$a(T) = a_0(1 + c_1(1 - \sqrt{T_r}))^2 \tag{9.4}$$

where $T_r = T/T_c$ and T_c is the critical temperature.

In the expression for the association strength $\Delta^{A_iB_j}$, the parameters $\varepsilon^{A_iB_j}$ and $\beta^{A_iB_j}$ are called the association energy and the association volume, respectively. These two parameters are only used for associating components, and together with the three additional parameters in the SRK term (a_0, b, c_1) are the five pure compound parameters of the model. They are typically obtained by fitting vapor pressure and liquid density data. For inert (not self-associating) components, e.g. hydrocarbons, only the three parameters of the SRK term are required, which can be either obtained from vapor pressures and liquid densities or calculated in the conventional manner (critical data, acentric factor). Appendix A on the companion website at www.wiley.com/go/Kontogeorgis contains an extensive list with all pure compound parameters of CPA available to date, while Appendix 9.B discusses the parameterization of CPA in more detail.

9.2.2 Mixing and combining rules

Physical term

When the CPA EoS is used for mixtures, the conventional mixing rules are employed in the physical term (SRK) for the energy and co-volume parameters. The geometric mean rule is used for the energy parameter a_{ij}. The interaction parameter k_{ij} is, for mixtures containing only self-associating compounds (e.g. alcohol, water, glycol or acid with n-alkanes or other 'inert' compounds), the only binary adjustable parameter of CPA:

$$a = \sum_i \sum_j x_i x_j a_{ij}, \quad \text{where} \quad a_{ij} = \sqrt{a_i a_j}(1 - k_{ij}) \tag{9.5}$$

$$b = \sum_i x_i b_i \tag{9.6}$$

Appendix B on the companion website at www.wiley.com/go/Kontogeorgis includes a comprehensive list of interaction parameters for the mixtures considered with CPA so far.

Association term: cross-associating mixtures

As can be seen from Equation (9.1), no mixing rules are needed in the association term, but combining rules are sometimes needed for the two association parameters. More specifically, for extending the CPA EoS to mixtures of two associating compounds, e.g. alcohols or glycols with water, combining rules for the association energy ($\varepsilon^{A_iB_j}$) and the association volume ($\beta^{A_iB_j}$) are required in order to calculate the value of the association strength in Equation (9.3). Over the years different combining rules have been suggested, but only the CR-1 rule (Equation (9.7)) and Elliott's combining rule (ECR, Equation (9.8)) have been found successful in applications so far. Furthermore, as shown by Derawi *et al.*[8] when the CR-1 combining rule is used, the arithmetic mean for the cross-association energy is proportional to the enthalpy of hydrogen bonding (ΔH_{12}) and the geometric mean for the cross-association volume is also related to the cross-entropy of the hydrogen bonding (ΔS_{12}) (see Chapter 7, eqs. 7.12 and 7.13). Equivalent theoretical justifications can be made for ECR because both combining rules are functionally similar as shown below (see Chapter 7, eqs. 7.12 and 7.13). Other combining rules previously tested lack theoretical explanation.

The expressions for the cross-association energy and cross-association volume parameters with CR-1 are:

$$\varepsilon^{A_iB_j} = \frac{\varepsilon^{A_iB_i} + \varepsilon^{A_jB_j}}{2} \quad \text{and} \quad \beta^{A_iB_j} = \sqrt{\beta^{A_iB_i}\beta^{A_jB_j}} \tag{9.7}$$

The expression of the cross-association strength ($\Delta^{A_iB_j}$) with ECR is:

$$\Delta^{A_iB_j} = \sqrt{\Delta^{A_iB_i}\Delta^{A_jB_j}} \tag{9.8}$$

Assuming that the radial distribution function in Equation (9.3) is $g(\rho) \cong 1$, as well as the term $\exp(\varepsilon^{AB}/RT) - 1 \cong \exp(\varepsilon^{AB}/RT)$, it can be shown that the equivalent expressions for the cross-association energy and cross-association volume parameters with ECR in Equation (9.3) are:

$$\varepsilon^{A_iB_j} = \frac{\varepsilon^{A_iB_i} + \varepsilon^{A_jB_j}}{2} \quad \text{and} \quad \beta^{A_iB_j} = \sqrt{\beta^{A_iB_i}\beta^{A_jB_j}}\,\frac{\sqrt{b_i b_j}}{b_{ij}} \tag{9.9}$$

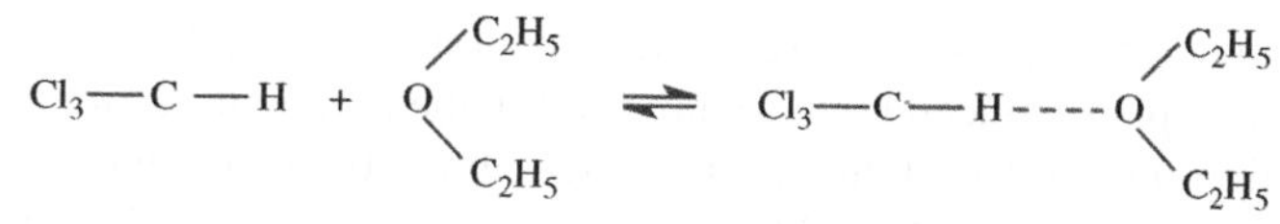

Figure 9.1 Solvation between chloroform and diethyl ether. Similar types of solvation exist between chloroform and other ethers as well

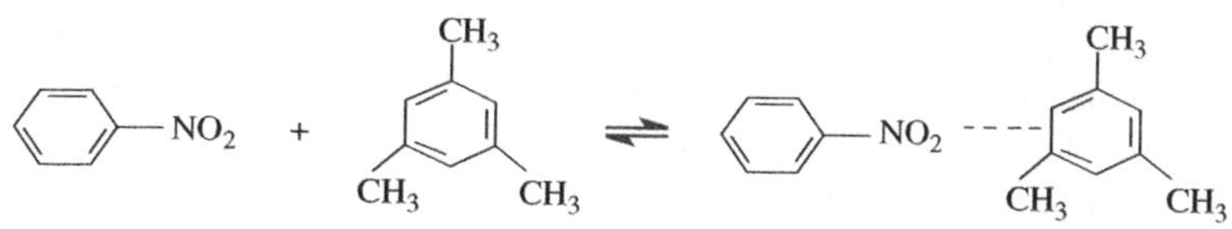

Figure 9.2 Solvation between nitrobenzene and mesitylene

Thus ECR and CR-1 are similar, the only difference being that the second term in equation (9.9) contains the co-volume parameters in the expression for the cross-association volume. Appendix 9.C provides the expressions for the calculation of the fugacity coefficient with the CPA EoS.

Association term: mixtures with solvation (induced association)

CPA can be applied to mixtures with one self-associating compound and one inert compound where there is the possibility for cross-association (solvation) between the two compounds, e.g. water with ethers or aromatic hydrocarbons. In this book such mixtures will be called 'solvating mixtures'. For this purpose, particularly useful is the so-called modified CR-1 rule proposed by Folas *et al.*,[9] which has been applied to mixtures with water or glycols and aromatic or olefinic hydrocarbons. In the modified CR-1 combining rule, the cross association volume $\beta^{A_iB_j}$ is optimized from the experimental data, while the cross-association energy parameter is equal to the value of the associating compound (e.g. water, alcohol or glycol) divided by two (from Equation (9.7) since $\varepsilon^{inert} = 0$):

$$\varepsilon^{A_iB_j} = \frac{\varepsilon_{associating}}{2} \quad \text{and} \quad \beta^{A_iB_j} \text{ (fitted)} \tag{9.10}$$

Then, the association strength will be estimated by Equation (9.3) and in this way the in-built temperature dependency of the cross-association strength $\Delta^{A_iB_j}$ is retained for solvating systems. Calculations have showed that this is important in order to obtain satisfactory results, e.g. for water or glycols with aromatic hydrocarbons over large temperature ranges (see Table 9.3 and Figure 9.17).

The modified CR-1 combining rule is useful for the solvation cases where one of the two compounds is not self-associating. However, there are several solvating mixtures, such as those illustrated in Figures 9.1 and 9.2, with hydrogen bonding between one electron donor and one electron acceptor (e.g. ketones or chloroform with ethers) or simply solvation of the charge-transfer type. The modeling of such cases cannot be formally described using Equations (9.7) and (9.10) or other similar approaches, as none of the involved compounds are self-associating. Different approaches are required, which are discussed in Chapters 11 and 12.

9.3 Parameter estimation: pure compounds

CPA has three parameters (like SRK) for 'inert', i.e. non-self-associating, compounds but five pure component parameters for associating compounds. They are typically estimated from vapor pressure and liquid density data over a specific temperature range, e.g. reduced temperature range between 0.5 and 0.9 or 0.95. Michelsen[50] showed that an upper reduced temperature limit in the parameter estimation of 0.95 is a suitable choice. However, extending the reduced temperature range above 0.95 to values close to one is of minor importance, since association models (CPA, SAFT, etc.) overpredict the critical temperature. It is of crucial importance that if DIPPR or similar correlations of experimental data are used, then they should be based on experimental data covering the same temperature range. Care should be taken since, especially for heavy compounds like glycols or alkanolamines, this is often not the case.

Before estimating pure component parameters for an associating compound, a suitable association scheme should be chosen, e.g. among those shown in Tables 8.11 and 8.12 (Chapter 8) (from Huang and Radosz)[10]. These association schemes have been very useful for the development of CPA for various associating compounds. As can be seen from these tables, a choice should first be based on the physical nature of the compound and the expected form of self-association, although experimental mixture data can assist in the final choice. The choice of the association scheme is a crucial step as the expression for the monomer fraction X_A (and thus the functional form of the EoS) will be different and depend on the choice of the association schemes. Spectroscopic and other external data can also assist in the final choice.

There is a link between the expected picture from a chemical theory viewpoint and the above-mentioned association schemes (of Tables 8.11 and 8.12); for example, the cyclic dimerization of organic acids like acetic acid is expected to be described via the 1A scheme (see Table 8.12). Alcohols and phenols form linear oligomers and this is equivalent to the 2B scheme, as presented in Figure 9.3. The 4C choice corresponds to a three-dimensional cluster structure like that expected for water and glycols.

Mixture data are useful for the final choice of association schemes and sites. We have found that it is particularly useful to use binary LLE data of the associating compound (under investigation) with inert compounds (e.g. *n*-alkanes) as in this way other effects (cross-association etc.) are not present and, moreover, LLE data offer a stringent test compared to VLE data. Such an approach has been used for water, glycols and alcohols (as water–alkane, glycol–alkane and alcohol–alkane LLE data are available) and has assisted in establishing that 4C is the best scheme for water and glycols, while most alcohols can best be represented with the 2B scheme. All these conclusions are in agreement with the aforementioned physical picture of these three types of hydrogen bonding compounds. In some cases, however, it is difficult to distinguish between association schemes if LLE data with inert compounds are not available, as for amines which are miscible with hydrocarbons. The 2B scheme has been found to be the best one for amines,[11] although 3B provides similar results in most cases. Water, organic acids, glycols, alcohols and amines have been modeled with CPA so far (see Appendix A on the companion website at www.wiley.com/go/Kontogeorgis, for a comprehensive parameter table). Recently, glycolethers and alkanolamines have been considered and selected results are presented in Chapter 11.

9.3.1 Testing of pure compound parameters

Although the parameter estimation of CPA and other SAFT-type approaches is most often based on vapor pressures and liquid densities, the parameters can be validated based on the following:

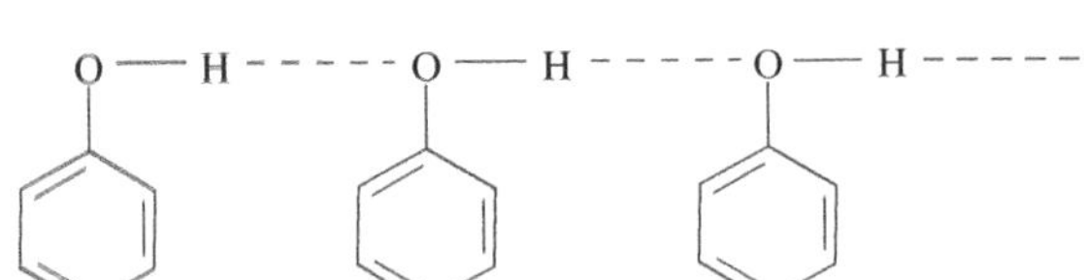

Figure 9.3 *Linear oligomers of alcohols corresponding to the 2B association scheme, in SAFT terminology*[10] *(Tables 8.11 and 8.12)*

- Trends of the parameters, e.g. against the van der Waals volume;.
- Prediction of other properties like second virial coefficient, enthalpy of vaporization, *dV/dP* and speed of sound.
- Monomer fraction calculations when such spectroscopic experimental data are available.
- Phase equilibria over extensive temperature ranges, especially when VLE, LLE (and sometimes also SLE) data are available for the same system.
- Prediction of multicomponent, multiphase equilibria based exclusively on binary parameters estimated from binary data.

The parameters of CPA follow well-defined trends which can be used in parameter estimations for new associating compounds, e.g. for selecting appropriate initial values needed in the parameter estimation of new (not previously studied) associating (and inert) compounds. One parameter that exhibits well-defined trends is the co-volume parameter, b, when plotted against the van der Waals volume (as shown in Figure 9.4). The co-volume is the only parameter of the model that is present in all three terms of CPA, i.e. the attractive, repulsive and association contributions. Among the five pure compound parameters the co-volume is the one that changes the least in the regression and an initial guess from Figure 9.4 is likely to be close to the final value. Even cubic EoS show similar trends for the co-volume parameter,[12,13] provided the co-volume parameter is estimated from vapor pressure and liquid density data and not set to the value dictated from the critical point constraints (see Chapter 3). The energy parameter also shows some trend against the size of the molecule, as can be seen in Figure 9.5, although, as might have been expected, this is somewhat less clear compared to the trend of the co-volume parameter. In the case of the parameter c_1, a trend can also be observed (Figure 9.5, right), but the different values are more disperse than in the previous two figures (Figures 9.4 and 9.5, left). Moreover, the glycols seem to follow a different trend. The outlier in Figure 9.5 (right) corresponds to Tetra-EG, a compound which has rather strange association parameters.[8] This problem for Tetra-EG has been resolved in a recent work,[54] as discussed further in Chapter 11.

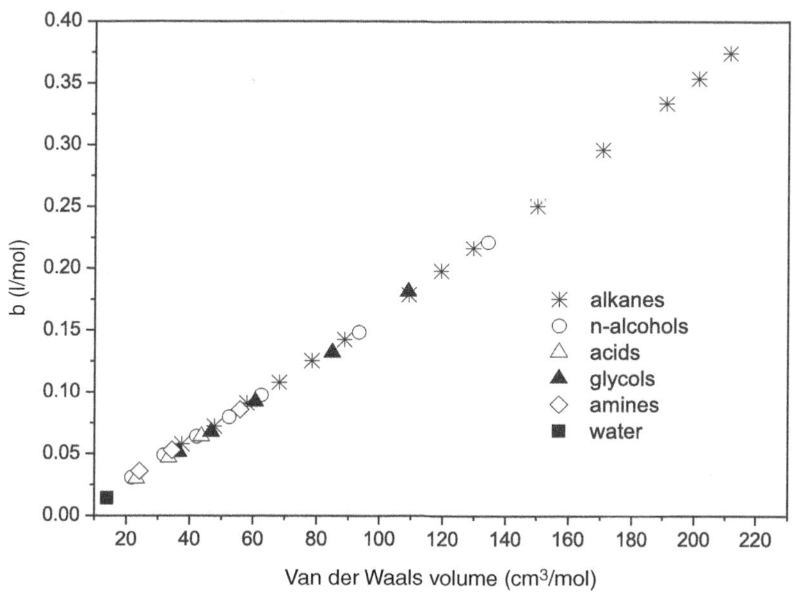

Figure 9.4 The co-volume parameter b of CPA for various associating and inert compounds plotted against the van der Waals volume, V_w (expressed in cm^3/mol). The co-volume parameters can be fitted to the linear correlation: $b = 0.00179V_w - 0.0132$ (b in l/mol, V_w in cm^3/mol). After Kontogeorgis et al.[16]

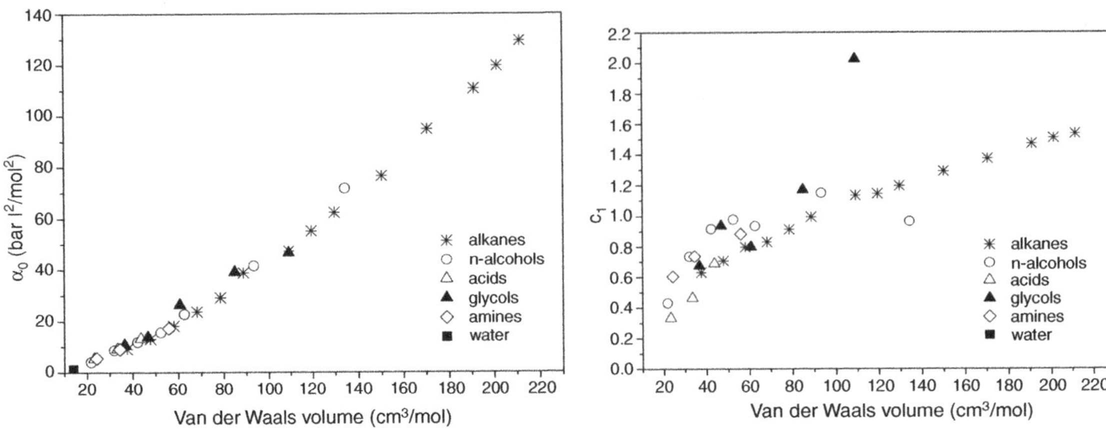

Figure 9.5 *Left: The energy parameter a_0 of CPA (Equation (9.4)) for various associating and inert compounds plotted against the van der Waals volume, V_w. Right: The c_1 parameter of CPA (temperature dependency of the energy term, Equation (9.4)) for various associating and inert compounds plotted against the van der Waals volume, V_w*

The two parameters of the association term are also related to externally measured quantities based on the link between association strength and the equilibrium constant, presented in Section 7.4.3 (Equations (7.28) and (7.29)). The clearest link is that of the association energy being (almost) equal to the enthalpy of the hydrogen bonding, while the association volume parameter depends on both the entropy of hydrogen bonding and the co-volume parameter. The hydrogen bonding enthalpy is available for some compounds from spectroscopic data and a comparison with CPA-obtained values reveals rather good agreement, as Table 9.1 indicates.

An even more direct test for the association term (and for the EoS as a whole) is the use of experimental data for the monomer fractions which are available for a few alcohol–alkane systems and for pure alcohols and water. One typical example is shown in Figure 9.6 for propanol–hexane. More results were presented in Chapter 7 by von Solms *et al.*[14,15] These results reveal good agreement between the monomer fractions calculated from CPA (for the 2B scheme, the monomer fraction is simply calculated

Table 9.1 *Comparison of spectroscopic values for the hydrogen bonding energies with the association energies from CPA ($\varepsilon^{A_iB_i}$), both expressed in K, for various hydrogen bonding compounds for which CPA parameters are available. Notice the high values of acids and the rather low values of the weakly associating amines. Alcohol association energy values are between the values of acids and amines*

Compound	Spectroscopy	CPA
Methanol (2B)	2630	2957
Ethanol up to hexanol (2B)	2526–3007	2590 (ethanol)
		2526 (butanol, pentanol)
		3219 (octanol)
Methylamine (2B)	1413	1379
Ethylamine (2B)	1141	1121
Diethylamine (2B)	552	445
Formic acid (1A)	7453	7478
Acetic acid (1A)	6949–8266	5463
Propanoic acid (1A)	7503	5223
Water (4C)	1813	2003

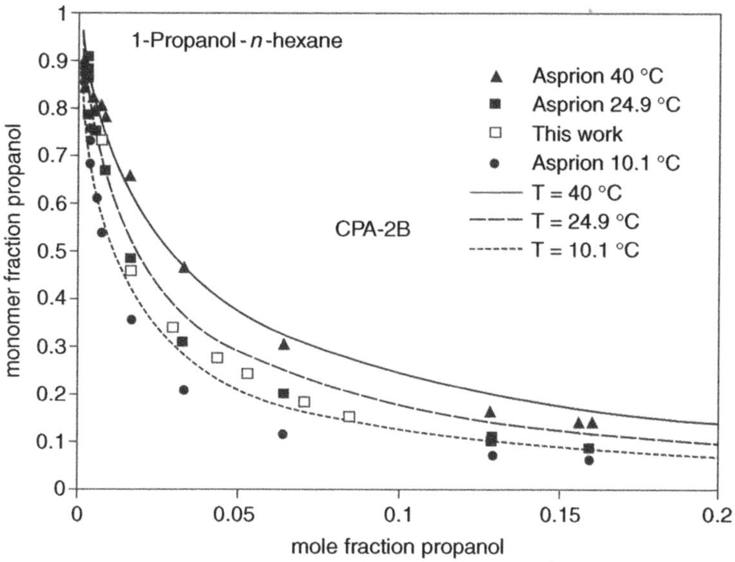

Figure 9.6 *Spectroscopic (experimental) and monomer fractions calculated from CPA for propanol–hexane. The Asprion data are from N. Asprion, H. Hasse, G. Maurer, Fluid Phase Equilib.,* ***2001****, 186, 1. Supplementary material at http://www.itt.uni-stuttgart.de. Reprinted with permission from Fluid Phase Equilibria, Measurement and modeling of hydrogen bonding in 1-alkanol + n-alkane binary mixtures by N. von Solms, L. Jensen et al., 261, 1–2, 272 Copyright (2007) Elsevier*

as X_A^2) and the experimental monomer fraction data obtained from spectroscopy. The 2B scheme performs best for most alcohols. For methanol, however, better agreement is obtained when the 3B scheme is used for this alcohol and the same conclusion is extracted from second virial coefficients. Even though monomer fraction data are not available for many compounds, when including those properties in the pure component parameter regression, physically more correct parameters may be obtained while maintaining good phase equilibrium results. There is ongoing research of this topic by several research groups.[19,20]

Testing the performance of the EoS against second virial coefficient data provides a stringent test of its capabilities, especially at low temperatures where the effect of intermolecular forces is greater. Cubic EoS, as known, fail to represent such virial coefficients at low temperatures. As can be seen from Figure 9.7, CPA represents very well the second virial coefficients of ethanol and the 2B scheme performs better than 3B, in both agreement with spectroscopic data and mixture calculations. As mentioned above, a similar analysis demonstrates that methanol is best represented with three sites (3B scheme).

Even though, based on monomer fraction and second virial coefficient calculations, the three-site (3B) scheme should be chosen for methanol, the 2B scheme is found to be the most appropriate choice overall. As will be discussed later, a major reason for this choice is the superiority of the 2B scheme for the prediction of methanol partition coefficients and in general multicomponent, multiphase equilibria calculations for water–methanol–hydrocarbon mixtures. Phase equilibrium calculations are of great importance in testing the applicability of engineering models like CPA, but theoretically based properties such as monomer fractions are still useful in model development.

Monomer fractions, energies of association and second virial coefficients offer alternative independent tests of the capability of CPA (and its parameters, which are typically estimated from vapor pressures and liquid

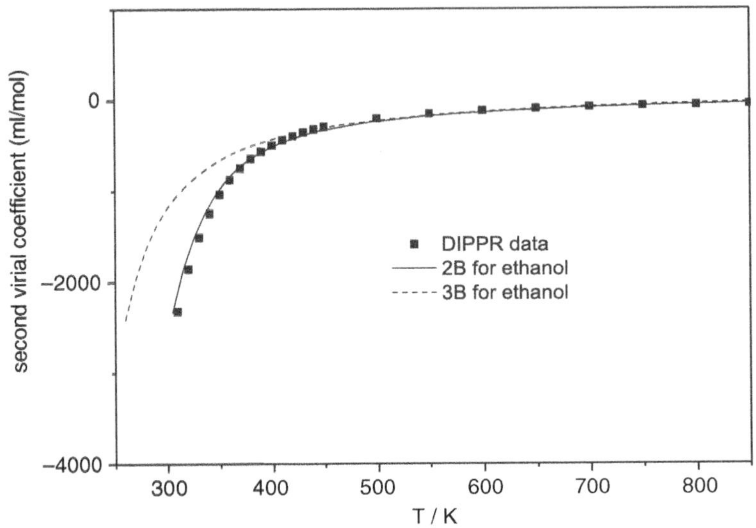

Figure 9.7 Ethanol second virial coefficients with CPA using the 2B and 3B association schemes. Reprinted with permission from Ind. Eng. Chem. Res., Ten Years with the CPA (Cubic-Plus-Association) Equation of State. Part 1. Pure Compounds and Self-Associating Systems by Georgios M. Kontogeorgis, Georgios K. Folas et al., 45, 14, 4855–4868 Copyright (2006) American Chemical Society

densities) in representing properties other than those used in the parameter estimation. For some molecules extensive data are available for other properties (density over extensive temperature and pressure ranges, enthalpy, entropy, speed of sound, etc.) and thus a more thorough test of the model can be carried out. Such an extensive investigation has been performed for pure water and pure methanol with good agreement as shown in Figures 9.8 and 9.9 for the density and speed of sound of methanol as a function of pressure. SRK (with or

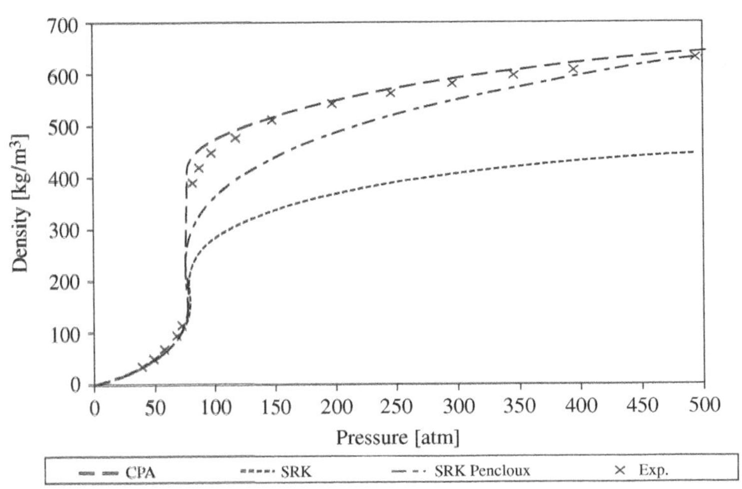

Figure 9.8 Density of methanol at 513 K with the CPA and SRK EoS. Reprinted with permission from Fluid Phase Equilibria, Comparison of the SRK and CPA equations of state for physical properties of water and methanol by C. Lundstrøm, M. L. Michelsen et al., 247, 1–2, 149 Copyright (2006) Elsevier

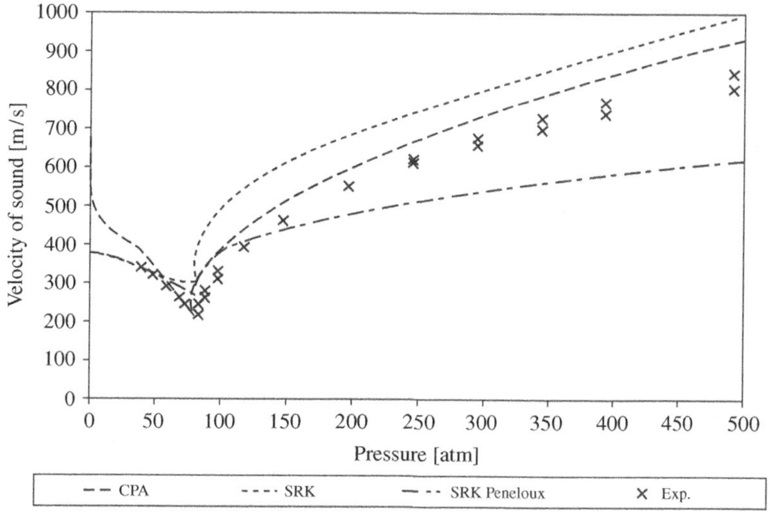

Figure 9.9 *Speed of sound in methanol at 513 K with the SRK and CPA EoS. Reprinted with permission from Fluid Phase Equilibria, Comparison of the SRK and CPA equations of state for physical properties of water and methanol by C. Lundstrøm, M. L. Michelsen et al., 247, 1–2, 149 Copyright (2006) Elsevier*

without the use of the Peneloux translation parameter) does not perform quite as satisfactorily. Additional results for other properties are reported by Lundstrøm *et al.*[17]

When extensive enthalpy of vaporization data are available, e.g. for water and methanol, these can be used for testing the parameters of association models, as illustrated in Figure 9.10, where calculations with three

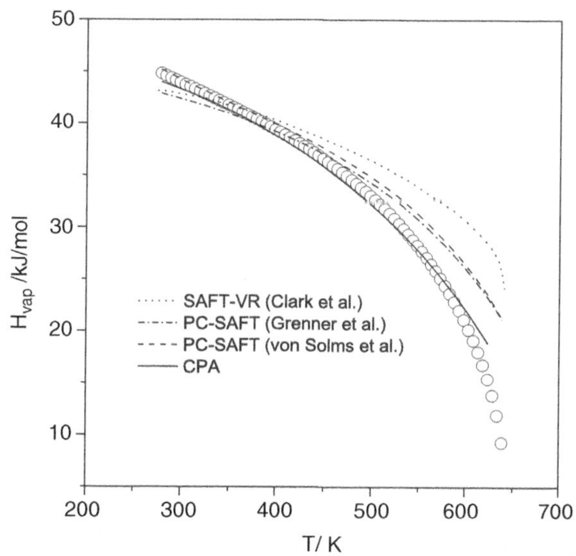

Figure 9.10 *Enthalpy of vaporization of water calculated with various EoS. Calculations with CPA, PC–SAFT (using various parameter sets, see von Solms et al.,*[14] *Grenner et al.*[18]*) and the SAFT–VR model (Clark et al.*[19]*). After Grenner et al.*[20]

SAFT-type models (CPA, PC–SAFT with various sets and SAFT–VR) are shown. CPA provides the best overall performance in this case.

9.4 The first applications

9.4.1 VLE, LLE and SLE for alcohol–hydrocarbons

One of the first applications of CPA was the alcohol–hydrocarbon VLE and LLE.[21,22] The results are very satisfactory with a single small value of the interaction parameter (k_{12}), but more importantly the model can predict details of phase behavior such as the azeotrope in the methanol–propane system at low methanol concentrations for which there is industrial evidence (but no measurements), as illustrated in Figure 9.11.

The lower alcohols are partially miscible with alkanes at low temperatures. For example, methanol is partially miscible with several alkanes (hexane, heptane, cyclohexane, etc.) at low temperatures and CPA can correlate LLE with a single small value of the interaction parameter, as shown in Figure 9.12 (left) for methanol–decane. Similar results are obtained for mixtures with methanol and other hydrocarbons, as well as for ethanol–alkane LLE (ethanol is partially miscible with heavy alkanes, e.g. tetradecane and hexadecane), as shown in several publications.[22,23]

Since for several alcohol–alkane mixtures both VLE and LLE (and sometimes also SLE) data are available, CPA can be validated over extensive temperature ranges. It can also be investigated whether different types of phase equilibria can be correlated using the same value of the interaction parameter. Several methanol–alkane (and other alcohol–alkane) systems have been studied[24] (with hexane, heptane, cyclohexane and heavier hydrocarbons) and a typical result is shown in Figure 9.12 (right). A number of observations can be made:

1. Excellent agreement between experimental and correlated VLE and LLE is achieved using a small value of a temperature-independent interaction parameter.
2. Prediction results ($k_{ij} = 0$) are acceptable (in most cases for VLE, less so for LLE).

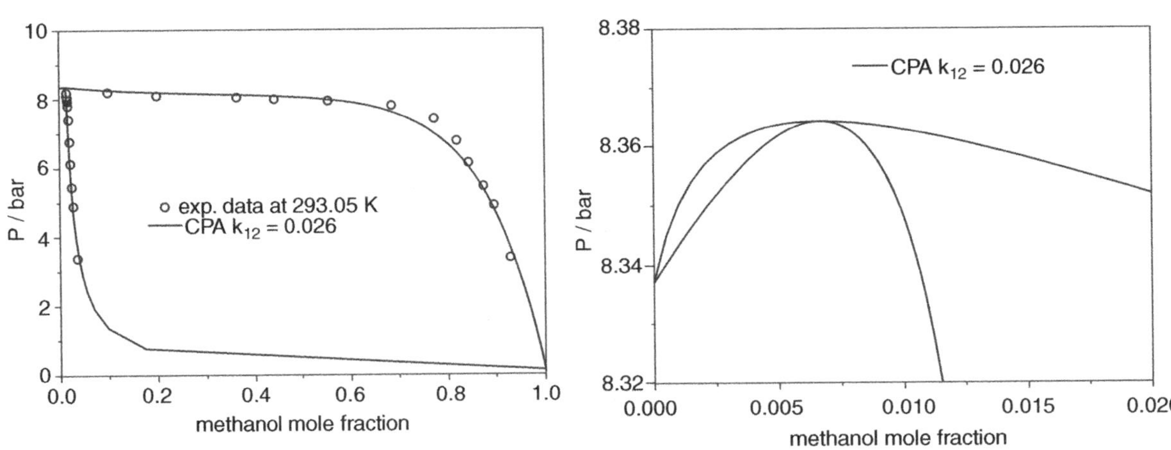

Figure 9.11 VLE of methanol–propane at 293.05 K with the CPA EoS. The right-hand figure is a magnification of the methanol dilute area

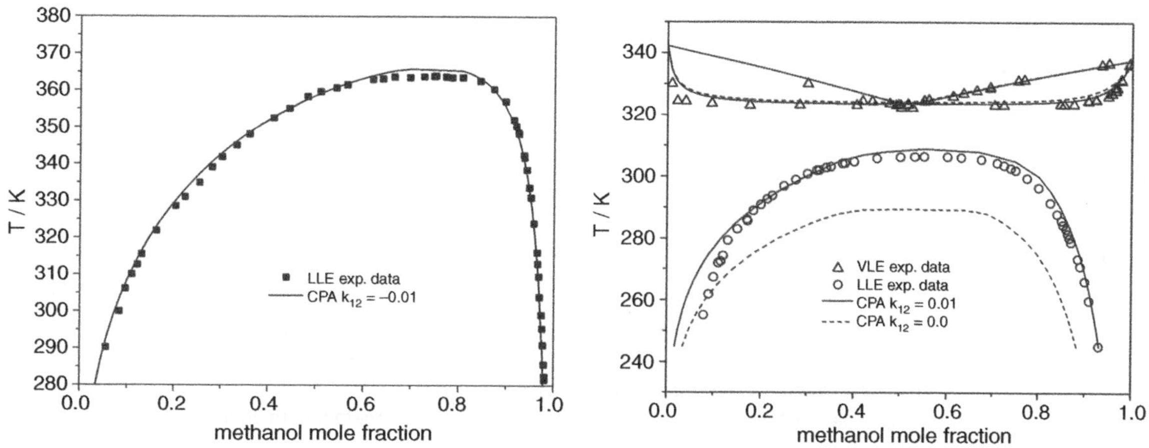

Figure 9.12 *Left: LLE correlation of methanol–n-decane with the CPA EoS using $k_{12} = -0.01$. Experimental data are from Higashiuchi et al., Fluid Phase Equilib., **1987**, 36, 35, Hiroyuki et al., Fluid Phase Equilib., **2004**, 224, 31 and Holscher et al., Fluid Phase Equilib., **1986**, 27, 153. Right: Methanol–hexane LLE and VLE using zero and optimum interaction parameters. Reprinted with permission from Fluid Phase Equilibria, Recent applications of the cubic-plus-association (CPA) equation of state to industrially important systems by G. Folas, G. M. Kontogeorgis et al., 228–229, PPEPPD 2004 Proceedings, 121–126 Copyright (2005) Elsevier*

3. LLE is, as could be anticipated, more sensitive to k_{ij} than VLE. Similar results are obtained also for other methanol–alkane systems studied. Systems like those of Figure 9.12 (right) have been studied with various SAFT versions (Gupta and Olson)[25] and the performance of, for example, PC-SAFT is similar to that of CPA (see also Chapter 13).

In the alcohol applications with CPA discussed above, all alcohols were assumed to have two active sites (i.e. 2B scheme; Tables 8.11 and 8.12), which is in agreement with the physical picture of alcohols known to form linear oligomers. Figures 9.13 and 9.14 show a few additional typical results for alcohol–alkane VLE and SLE where more features of CPA have been investigated:

- Excellent agreement is obtained with a single temperature independent interaction parameter (k_{12}) over extensive temperature ranges.
- The two association schemes (3B and 2B) typically yield similar results for alcohols, though the performance is a bit better when the simpler 2B (two-site) association scheme is used. The somewhat better performance of 2B over 3B for alcohol–alkane VLE results in a better performance of CPA/2B over CPA/3B for water–alcohol–alkane V(L)LE and thus no attempt has been made to change the 2B scheme for alcohols.
- Correlation of SLE for alcohol–alkanes is excellent at both low and high pressures,[26] using as in the case of VLE and LLE a small value of the interaction parameter.

9.4.2 Water–hydrocarbon phase equilibria

For alcohols, the 2B scheme has been found adequate. Water, however, is known to form three-dimensional clusters and a more appropriate scheme should be, within the CPA/SAFT framework, the 4C one, where we

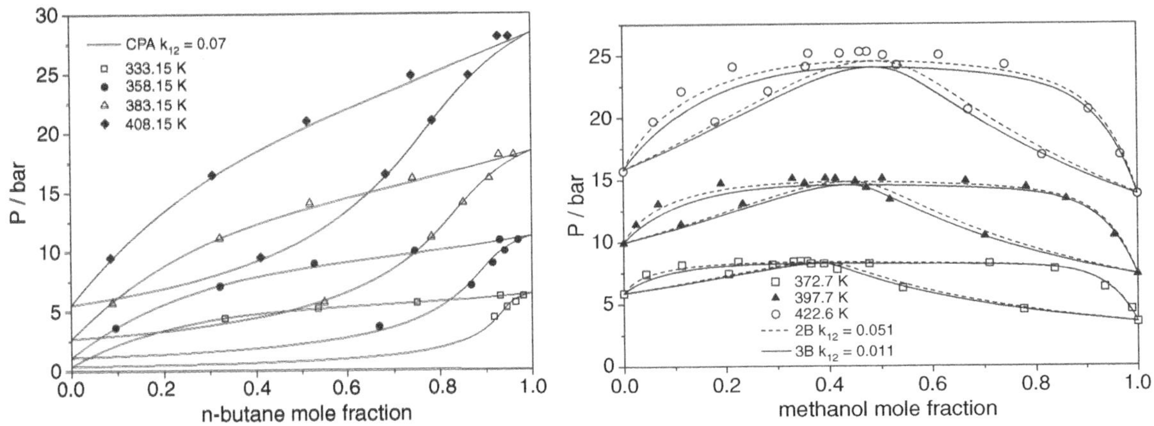

Figure 9.13 *VLE for alcohol-alkane with the CPA EoS. Left: VLE of t-butanol-butane with a single interaction parameter at four temperatures. Experimental data are from Melpolder, Fluid Phase Equilib.,* ***1986****, 26, 279. Right: VLE for methanol–pentane with a single k_{12} and two different association schemes for methanol (2B and 3B). Experimental data are from Wilsak et al., Fluid Phase Equilib.,* ***1987****, 33, 157. Reprinted with permission from Ind. Eng. Chem. Res., Ten Years with the CPA (Cubic-Plus-Association) Equation of State. Part 1. Pure Compounds and Self-Associating Systems by Georgios M. Kontogeorgis, Georgios K. Folas et al., 45, 14, 4855–4868 Copyright (2006) American Chemical Society*

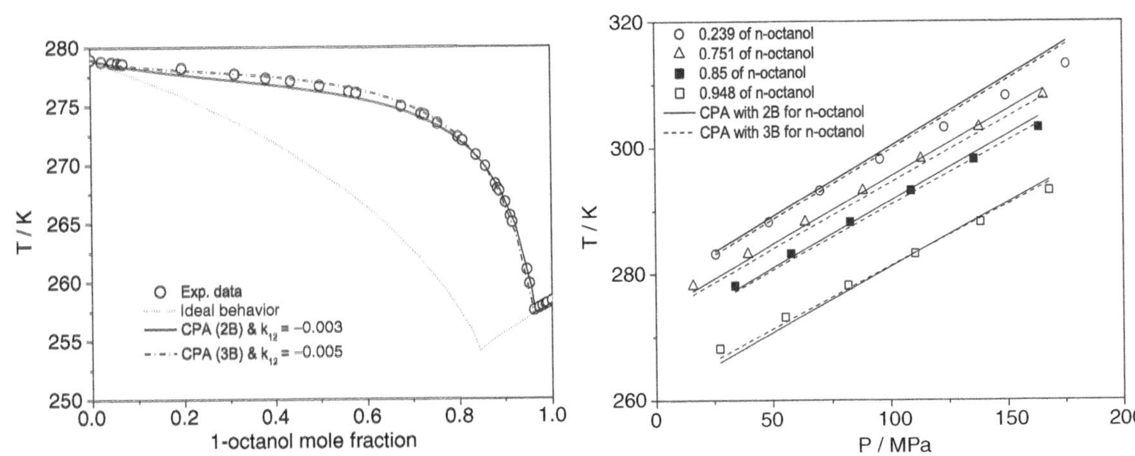

Figure 9.14 *SLE for n-octanol–tetradecane with the CPA EoS. Left: SLE correlation of n-octanol–tetradecane at 1 bar with the 3B and the 2B schemes. Experimental data are from Liu et al., J. Solution Chem.,* ***1991****, 20, 39. Right: SLE correlation of n-octanol–tetradecane system at high pressures with the 3B scheme for n-octanol and an interaction parameter $k_{12} = -0.005$ obtained from SLE data at P = 1 bar. The performance of the model with the 2B scheme for n-octanol and a binary interaction parameter $k_{12} = -0.003$ is also presented. Experimental data are from Yang et al., Fluid Phase Equilib.,* ***2002****, 194–197, 1119. Reprinted with permission from Ind. Eng. Chem. Res., Application of the Cubic-Plus-Association Equation of State to Mixtures with Polar Chemicals and High Pressures by Georgios K. Folas, Georgios M. Kontogeorgis et al., 45, 4, 1516–1526 Copyright (2006) American Chemical Society*

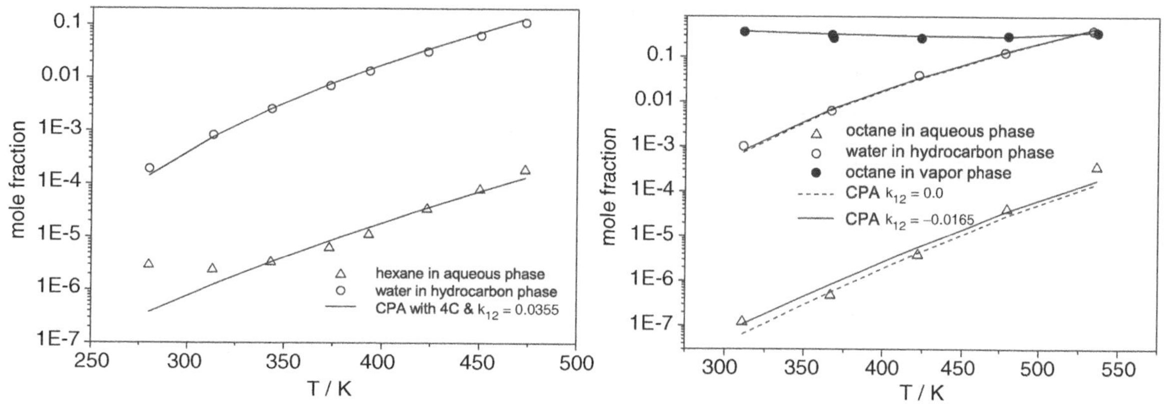

Figure 9.15 *LLE and VLLE for water–alkanes with CPA. Left: LLE for water–hexane. Right: VLLE for water–octane. The 4C scheme is used for water and k_{ij} is obtained from Equation (9.11). Reprinted with permission from Ind. Eng. Chem. Res., Ten Years with the CPA (Cubic-Plus-Association) Equation of State. Part 1. Pure Compounds and Self-Associating Systems by Georgios M. Kontogeorgis, Georgios K. Folas et al., 45, 14, 4855–4868 Copyright (2006) American Chemical Society*

assume that all atoms in water have the possibility of creating hydrogen bonds. There is, moreover, spectroscopic evidence for this. Figure 9.15 (left) shows that CPA can, with a single small value of a temperature-independent parameter, correlate very satisfactorily both the water solubility in the hydrocarbon phase and the very low hydrocarbon solubilities in the aqueous phase, for the water–hexane mixture. It is well known that classical cubic EoS fail to represent the phase behavior of mixtures of hydrocarbons and water. Water is often handled by assuming binary interaction coefficients of the order of 0.5 (see Chapter 3); this assumption will somewhat underestimate the solubility of water in hydrocarbon liquid phases and give a completely incorrect picture of the solubility of hydrocarbons in the aqueous phase.

Equally good results were obtained with CPA for other aliphatic hydrocarbons over extensive temperature ranges.[9] The only significant deficiency is the inability of the model to describe the minimum in the hydrocarbon solubility near room temperature which is often attributed to the so-called hydrophobic effect. Results like those of Figure 9.15 and other extensive comparisons in the literature[27–29] show that CPA correlates water–alkane LLE better than many more complex variants of the SAFT EoS. This improved better performance could be attributed to the better parameterization of CPA for water as CPA and SAFT have similar association terms.

A systematic study of many water–alkane LLE systems has been made. Excellent results are obtained in all cases, e.g. for multiphase equilibria such as for VLLE of water–octane (Figure 9.15, right) and for the water content of methane over an extensive temperature range[48] (Figure 9.16). A single k_{ij} is used but the results are satisfactory even with $k_{ij} = 0$, as can be seen in Figures 9.15 (right) and 9.16. It can be shown that the optimum k_{ij} parameters for water–alkanes (from water–propane up to water–decane) can be described by the following generalized equation:

$$k_{ij} = -0.026 \times (\text{carbon number}) + 0.1915 \tag{9.11}$$

CPA equally satisfactorily correlates the phase equilibria of aqueous systems with aromatic hydrocarbons or alkenes. The solubility of the aromatic hydrocarbons in water is two orders of magnitude higher than the aliphatic ones with the same carbon number. This is a physical indication of the importance of solvation for

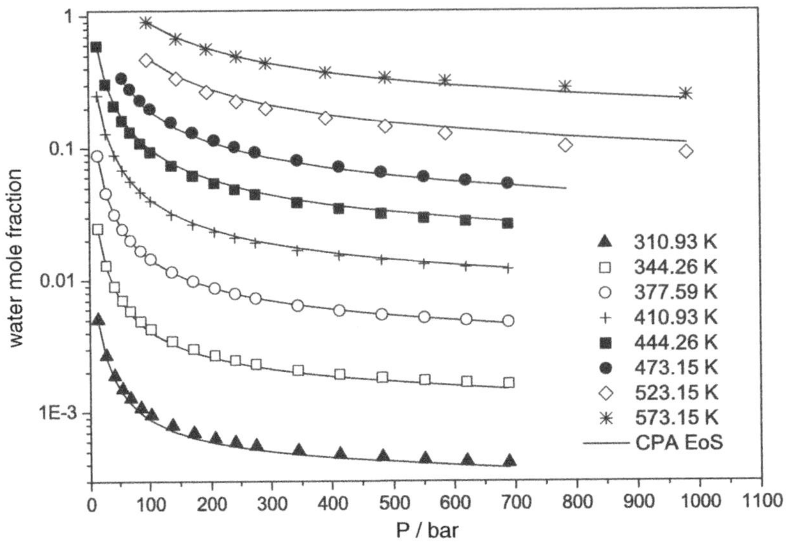

Figure 9.16 *Prediction of water content of methane with the CPA EoS. The interaction parameter is set to zero at all temperatures. Experimental data from Odds et al., Ind. Eng. Chem.,* ***1942****, 34, 1223 and Sultanov et al., Gazov. Prom.,* ***1971****, 4, 6. Reprinted with permission from Fluid Phase Equilibria, Data and prediction of water content of high pressure nitrogen, methane and natural gas by G. K. Folas, G. M. Kontogeorgis et al., 252, 1–2, 162–174 Copyright (2007) Elsevier*

such systems. Solvation is important for water–alkenes as well. The solubility of 1-alkenes in water is an order of magnitude higher than the solubility of the corresponding aliphatic hydrocarbon with the same carbon number. This is due to the double bond between the first and the second carbon in the carbon chain, which results in an increased electronegativity of 1-alkenes; hence it can act as electron donor, similar to the case of the aromatics. It is concluded that only when accounting for the solvation, using the modified CR-1 combining rule (Equation (9.10)), can both solubilities be adequately correlated with the CPA EoS.[30] Solvation implies that the hydrocarbons do not self-associate but are only able to solvate with water. Although this introduces an additional parameter in the cross-association term (see Equation (9.10), the $\beta^{A_iB_j}$), excellent correlation is obtained over the entire temperature range using this modified CR-1 rule and for both VLE and LLE, as Figure 9.17 demonstrates for water–benzene. A comparison to the correlative performance of SRK, using the Huron–Vidal mixing rule with the modified NRTL equation, is also presented. For aromatic hydrocarbon aqueous mixtures, the interaction parameter of the physical term of the model (k_{12}) can be obtained from the corresponding 'homomorph' alkane (e.g. the k_{12} of water–benzene is assumed to be that of water–hexane etc.) and thus only $\beta^{A_iB_j}$ is fitted to experimental data. Tables 9.2 and 9.3[9] provide an overview of the correlative performance of CPA for both aliphatic and olefinic or aromatic hydrocarbon aqueous mixtures. The results shown in Table 9.3 are based on the homomorph approach, i.e. k_{12} is taken from the corresponding alkane homomorph using Equation (9.11), and only the cross-association volume $\beta^{A_iB_j}$ is fitted to the LLE data.

9.4.3 Water–methanol and alcohol–alcohol phase equilibria

In many oil and gas applications, together with water and hydrocarbons, methanol or other gas hydrate inhibitors are present. Thus, a satisfactory description of water–methanol phase equilibria is important over

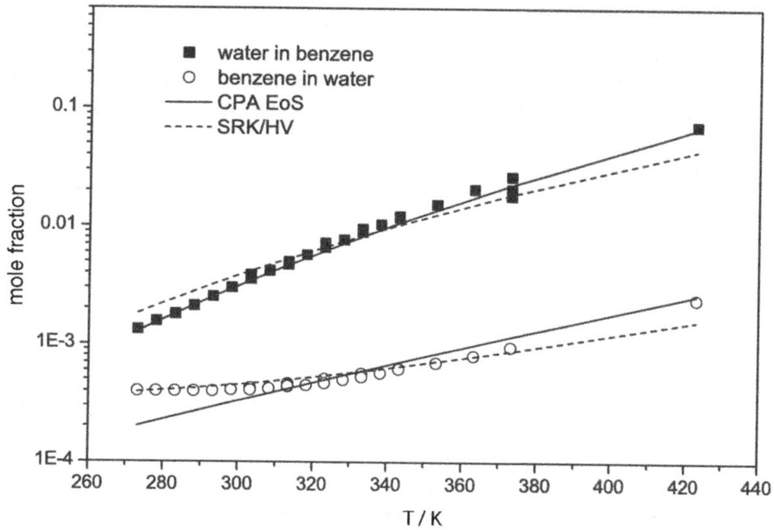

***Figure 9.17** LLE of water–benzene with the CPA ($k_{12}=0.0355$ from Equation (9.11) and $\beta^{A_iB_j}=0.079$) and SRK EoS using the Huron–Vidal (SRK/HV) mixing rule ($g_{12}-g_{22}/R=591$ K, $g_{21}-g_{11}/R=1998$ K, $\alpha_{12}=0.175$). Reprinted with permission from Fluid Phase Equilibria, Vapor–liquid, liquid–liquid and vapor–liquid–liquid equilibrium of binary and multicomponent systems with MEG: Modeling with the CPA EoS and an EoS/G^E model by G. K. Folas, G. M. Kontogeorgis et al., 249, 1–2, 67–74 Copyright (2006) Elsevier*

extensive temperature ranges. Both VLE and SLE data are available for water–methanol. Alcohol–alcohol mixtures (for which typically only VLE data are available) are also of significance in certain applications. From the modeling point of view, a way to account for the cross-association is needed. In the case of water with methanol (and other low-molecular-weight alcohols like ethanol and propanol) ECR (Equation (9.8)) has been found to be adequate. The solid complex formed at low temperatures between water and polar compounds like methanol and glycols can be described using the chemical reaction model presented in Appendix 9.D.

***Table 9.2** Percentage average absolute deviation (% AAD) between experimental water solubilities and calculated values from CPA in the hydrocarbon phase (X_{water}) or the vapor phase (y_{water}) and hydrocarbon (HC) solubilities in the aqueous phase (X_{HC}) using the generalized expression for the interaction parameter, Equation (9.11). Reprinted with permission from Ind. Eng. Chem. Res., Application of the Cubic-Plus-Association (CPA) Equation of State to Complex Mixtures Georgios K. Folas, Georgios M. Kontogeorgis, 45, 4, 1527–1538 Copyright (2006) American Chemical Society*

Hydrocarbon	T range (K)	k_{12}	% AAD in X_{water}	% AAD in X_{HC}	% AAD in y_{water}
Propane	278–366	0.1135	3.4	35.9	4.1
Butane	310–420	0.0875	11.7	26.5	9.5
n-Pentane	280–420	0.0615	13.4	28.4	—
n-Hexane	280–473	0.0355	11.9	31.1	—
n-Heptane	280–420	0.0095	11.5	63.3	—
n-Octane	310–550	−0.0165	9.7	44.1	1.9
n-Decane	290–566	−0.0685	8.2	264	—

Table 9.3 Percentage average absolute deviation (% AAD) between experimental water solubilities and calculated values from CPA in the hydrocarbon phase (X_{water}) or the vapor phase (y_{water}) and hydrocarbon (HC) solubilities in the aqueous phase (X_{HC}) using the generalized expression for the interaction parameter, Equation (9.11). The $\beta^{A_iB_j}$ parameter is fitted to experimental LLE data. Reprinted with permission from Ind. Eng. Chem. Res., Application of the Cubic-Plus-Association (CPA) Equation of State to Complex Mixtures Georgios K. Folas, Georgios M. Kontogeorgis, 45, 4, 1527–1538 Copyright (2006) American Chemical Society

Hydrocarbon	T range (K)	k_{12}	$\beta^{A_iB_j}$	% AAD in X_{water}	% AAD in X_{HC}	% AAD in y_{water}
Benzene	273–473	0.0355	0.079	5.3	19.5	—
Toluene	273–473	0.0095	0.06	5.1	23.5	—
Ethylbenzene	303–568	−0.0165	0.051	6.5	47.1	1.1
Propylbenzene	280–420	−0.0425	0.041	14.3	38.5	—
m-Xylene	373–473	−0.0165	0.039	3.7	8.3	—
1-Hexene	310–496	0.0355	0.021	7.6	29.3	1.2
1-Octene	310–540	−0.0165	0.021	4.7	23.4	1.1
1-Decene	310–550	−0.0685	0.021	12.7	288	—

The performance of CPA with ECR is excellent for both methanol–water and the azeotropic ethanol–water system over extended temperature and pressure ranges using a single temperature-independent interaction parameter ($k_{12} = -0.09$ for methanol–water and $k_{12} = -0.11$ for ethanol–water between 298 and 623 K), as shown for one of the two systems in Figure 9.18 (left). SLE of methanol–water can also be correlated very satisfactorily but a slightly different k_{ij} is required (−0.153), as shown in Figure 9.18 (right).

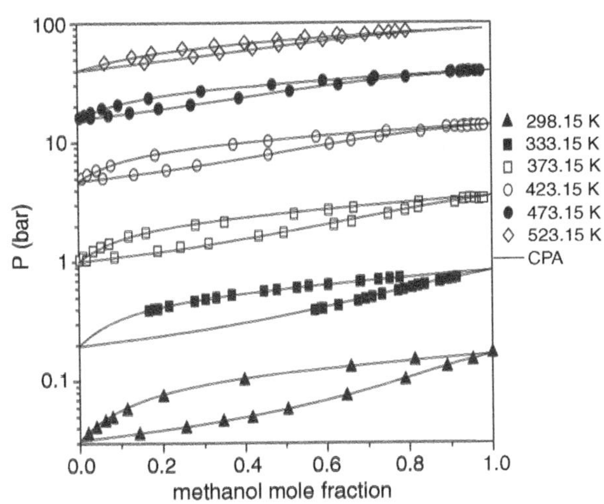

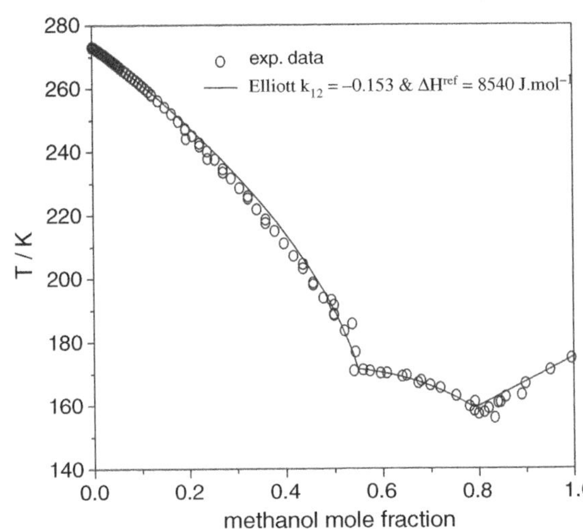

Figure 9.18 Left: VLE for methanol–water with CPA and ECR using $k_{12} = -0.09$. Right: SLE for methanol–water with CPA using ECR with a common interaction parameter $k_{12} = -0.153$. Reprinted with permission from Ind. Eng. Chem. Res., Application of the Cubic-Plus-Association (CPA) Equation of State to Cross-Associating Systems by Georgios M. Kontogeorgis, Georgios Folas et al., 44, 10, 3823–3833 Copyright (2005) American Chemical Society

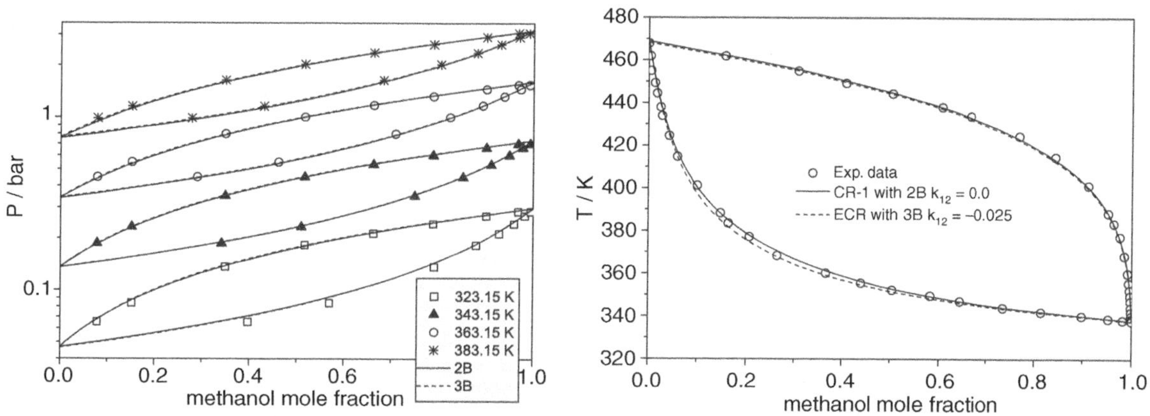

Figure 9.19 *VLE for alcohol–alcohol mixtures with CPA. Left: VLE calculations of ethanol–butanol using ECR and the 2B scheme with $k_{12} = 0.0$ (solid line) or ECR and the 3B scheme with $k_{12} = 0.0$ (dashed line). Experimental data are from Kharin et al., Khim. Khim. Tekhnol.,* ***1969****, 12, 424. Right: VLE calculations for methanol–octanol using CR1 and the 2B scheme with $k_{12} = 0.0$ (solid line) or ECR and the 3B scheme with $k_{12} = -0.025$ (dashed line). Experimental data are from Arce et al.[31] Reprinted with permission from Ind. Eng. Chem. Res., Ten Years with the CPA (Cubic-Plus-Association) Equation of State. Part 2. Cross-Associating and Multicomponent Systems by Georgios M. Kontogeorgis, Georgios Folas et al., 45, 14, 4869–4878 Copyright (2006) American Chemical Society*

Figure 9.19 illustrates that the ECR is sufficient for correlating alcohol–alcohol VLE using a small value of the interaction parameter[49]. As for alcohol–alkanes, small differences are observed in using the 2B or 3B schemes for alcohols. Moreover, the ECR and CR-1 rules perform equally well.

9.4.4 Water–methanol–hydrocarbons VLLE: prediction of methanol partition coefficient

Methanol is possibly the most important gas hydrate inhibitor. It suppresses gas hydrates but it is often injected at a higher rate than necessary, due to uncertainties as to what the correct rate should be. Two important reasons for optimizing the amount of methanol used for inhibition (and keeping it to a minimum) are the cost of providing this chemical, especially on offshore platforms, and the fact that methanol is a toxic substance. The injection rate is directly related to the phase equilibria of mixtures containing oil, water and methanol. Moreover, the loss of hydrate inhibitor in the gas or condensate phases should be accurately known.

The ultimate purpose of any thermodynamic model is the prediction of multicomponent, multiphase equilibria, based solely on interaction parameters obtained from binary mixtures and preferably over extended temperature and pressure ranges. Figure 9.20 demonstrates, for example, the prediction of gas phase methanol content for the methanol–water–methane ternary system over an extended temperature and pressure range. Accurate results for the loss of methanol in the gas phase are of major importance for the oil and gas industry (note that methane is typically at very high concentrations in natural gas mixtures), and as can be seen CPA provides very accurate predictions at various temperatures.

Some other typical applications related to VLLE for water–alcohol–hydrocarbon (aliphatic alkane) mixtures are presented in Figures 9.21 and 9.22, for one quaternary and one five-component mixture. In these applications, a single interaction parameter per binary system is used, obtained from the corresponding binary mixtures. Good results are achieved for the partition coefficients over an extended temperature and

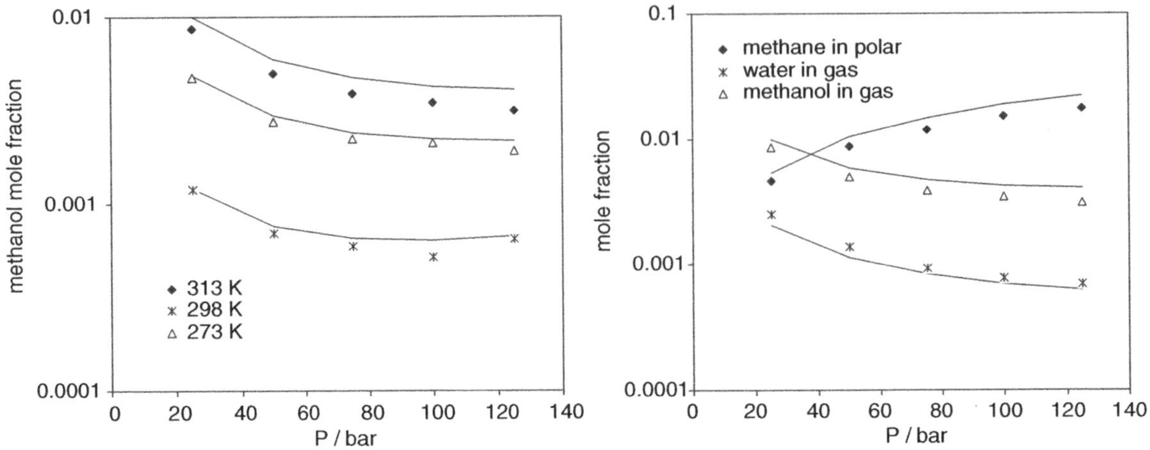

Figure 9.20 *Left: Prediction with CPA of the gas phase methanol content for the methanol–water–methane system at three different temperatures. Right: Prediction with CPA of methanol and water concentration in the gas phase and methane concentration in the liquid polar phase for the methanol–water–methane system at 313 K. In both figures the binary interaction parameters are: −0.09 for water–methanol using ECR, 0.0134 for methanol–methane and −0.045 for water–methane. Experimental data are from Sinyavskaya et al., Gazov. Prom-st,* ***1984****, 1, 39 and Sinyavskaya et al., Gazov. Prom-st,* ***1985****, 2, 26*

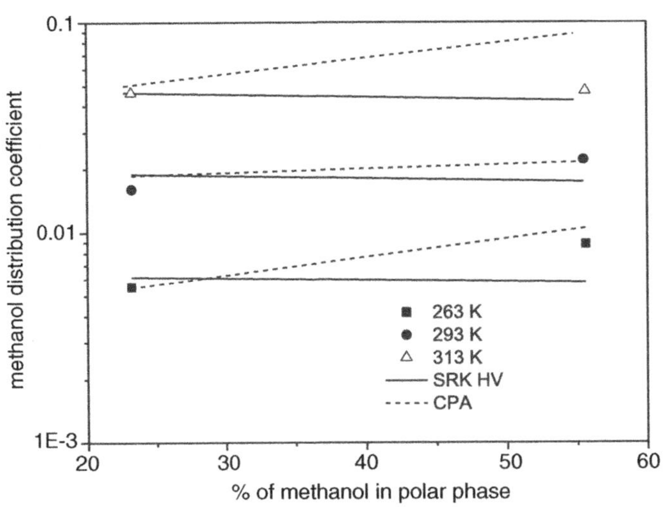

Figure 9.21 *Prediction of methanol partition coefficient (mole fraction of methanol in hydrocarbon phase/mole fraction of methanol in polar phase) at 69 bar and different temperatures with the CPA EoS and SRK Huron–Vidal (SRK HV) for the quaternary system methanol–water–methane–heptane. The interaction parameters are: ECR and* $k_{ij} = -0.09$ *for water–methanol, Equation (9.11) for water–heptane and water–propane, 0.0134 for methanol–methane, −0.045 for water–methane and 0.026 for methanol–propane. Experimental data are from Chen et al., GPA Research Report RR-117, 1988*

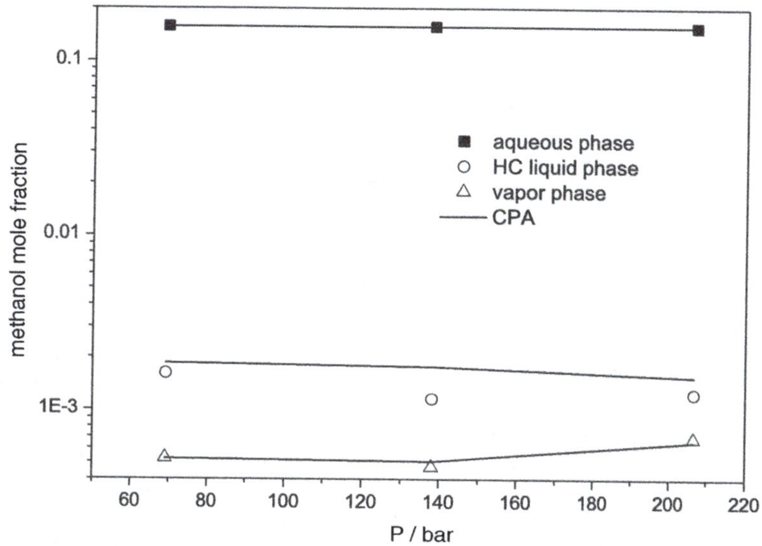

Figure 9.22 *VLLE prediction with CPA for methanol(1)–water(2)–methane(3)–propane(4)–n-heptane(5) at 284.14 K. The values of the interaction parameters are $k_{12}=-0.09$ (ECR), $k_{13}=0.0134$, $k_{14}=0.026$ and $k_{15}=0.005$. Water–alkane interaction parameters are estimated from Equation (9.11). All binary interaction parameters for hydrocarbon–hydrocarbon systems are set equal to zero. Experimental data are from Chen and Ng, GPA Research Report RR-149, 1995. Reprinted with permission from Ind. Eng. Chem. Res., Application of the Cubic-Plus-Association (CPA) Equation of State to Cross-Associating Systems by Georgios M. Kontogeorgis, Georgios Folas et al., 44, 10, 3823–3833 Copyright (2005) American Chemical Society*

pressure range while using the same binary parameters. Equally satisfactory results are obtained for the methanol partition coefficients in the water–methanol–hexane and water–methanol–propane ternary mixtures.[7] The good performance of CPA for systems like those of Figures 9.21 and 9.22, which contain some of the main types of compounds of a typical petroleum mixture (water, gases, oil and a hydrate inhibitor, e.g. methanol), is important for practical applications. The accurate prediction of the partitioning of methanol between the two liquid phases for such a mixture is very important for the oil industry. A difference of the order of 10 in the calculated partitioning coefficients may result in up to a 50% increase in the amount of methanol injected for hydrate inhibition. To realize the economical importance of such a prediction, it has been stated that each oil platform (in the North Sea) spends approximately £4 million ($6.5 million) on a yearly basis for the addition of methanol to the oil.

Models like SRK (using vdW1f mixing rules) typically overestimate by almost an order of magnitude the amount of alcohol in the organic phase. This would yield an overestimation of the amount of added methanol of almost 50%; the use of CPA provides more reasonable estimates of the required amount of methanol. On the other hand, classical cubic EoS with advanced mixing rules (e.g. Huron–Vidal types using a modified NRTL activity coefficient model, see Chapter 6) can provide satisfactory prediction of the methanol partition coefficient as well. Typically, however, five interaction parameters are required in NRTL for each binary mixture of an associating and an inert component, e.g. water–alkanes, in order to accurately represent the temperature dependency of phase behavior.

The study of multicomponent mixtures with aromatic hydrocarbons leads to the conclusion that accounting for the solvation between water and the aromatic hydrocarbons is crucial for obtaining satisfactory predictions. Solvation does exist also between alcohols and aromatic hydrocarbons (e.g. methanol is miscible with benzene

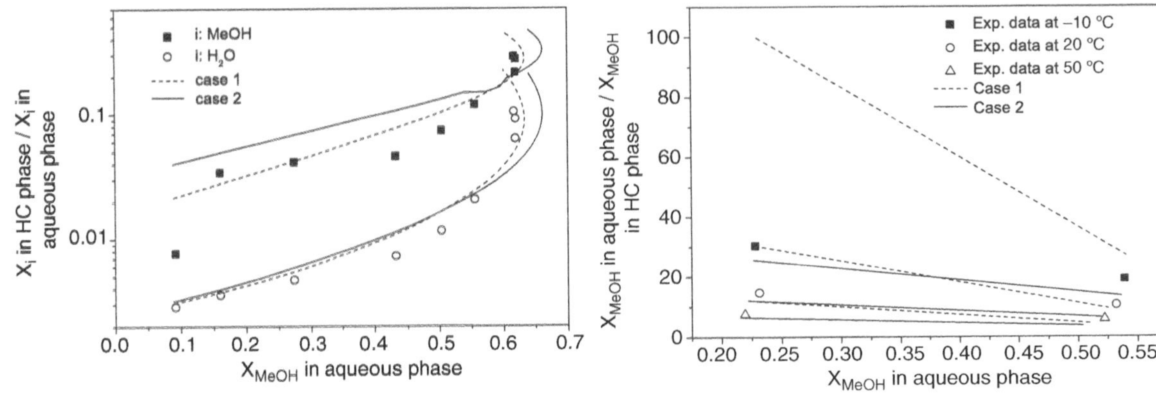

Figure 9.23 *Left: Prediction with CPA of methanol and water partition coefficient for the ternary system water(1)–methanol(2)–benzene(3) at 293.15 K, using two cases. Case 1:* $k_{23} = 0.006$*; Case 2:* $k_{23} = 0.02$ *and* $\beta^{A_iB_j} = 0.01$*. In both cases the model accounts for the solvation between water and benzene (parameters in Table 9.3). Experimental data are from Triday et al., J. Chem. Eng. Data,* ***1984****, 29, 231. Right: Prediction with CPA of methanol partition coefficients for the quaternary system water(1)–methanol(2)–toluene(3)–methane(4). In Case 1 the model does not account for the solvation between methanol and toluene, while in Case 2 solvation is accounted for. Only at low temperature does Case 2 perform better, otherwise the results are similar. Interaction parameters: water–hydrocarbons from Tables 9.2 and 9.3,* $k_{12} = -0.09$*,* $k_{23} = 0.0$ *(or* $k_{23} = 0.034$ *and* $\beta^{A_iB_j} = 0.029$*),* k_{23} *0.0134,* $k_{14} = -0.045$*. Experimental data are from Chen and Ng, GPA Research Report RR-149, 1995. Reprinted with permission from Ind. Eng. Chem. Res., Application of the Cubic-Plus-Association (CPA) Equation of State to Complex Mixtures Georgios K. Folas, Georgios M. Kontogeorgis, 45, 4, 1527–1538 Copyright (2006) American Chemical Society*

and toluene but only partially miscible with hexane and heptane). It has, however, been established that the importance of solvation in CPA is smaller for alcohol–aromatic hydrocarbons compared to mixtures with water.[9] For practical purposes and also for multicomponent calculations, it can be ignored since the use of a binary interaction parameter in the physical term (k_{ij}) is sufficient for satisfactory results. The lack of LLE data for such systems is possibly, from a practical point of view, the reason why additional interaction parameters are not needed, since VLE representation is often easier than LLE. Solvation between alcohols and aromatic hydrocarbons can be appreciated only when we look at 'sensitive' properties, e.g. infinite dilution activity coefficients[9] or calculations at low temperatures. Some typical prediction results are presented in Figure 9.23.

Finally, Figure 9.24 shows the predicted results for water–methanol–butane at three temperatures, where the effect of using different association schemes for methanol is also illustrated. When predicting multicomponent, multiphase equilibria based solely on binary interaction parameters, it might be the case that no experimental data are available for all the binary systems, or no experimental data might be available at the desired temperature and pressure. A study of the influence of the binary interaction parameters on the prediction of the alcohol partition coefficient reveals that for the system methanol–water–butane the amount of water in the hydrocarbon phase is very low, thus the correct representation of the *n*-butane–methanol system is crucial. The concentrations of both water and butane are rarely high in the polar phase, thus the accurate representation of the water–alkane system is not very important for this type of calculation (see case 3 in Figure 9.24). On the other hand, a successful correlation of the water–methanol and methanol–*n*-butane systems is required for a satisfactory prediction of the ternary system. Similar observations are found to be valid for other ternary systems both with methanol and ethanol.

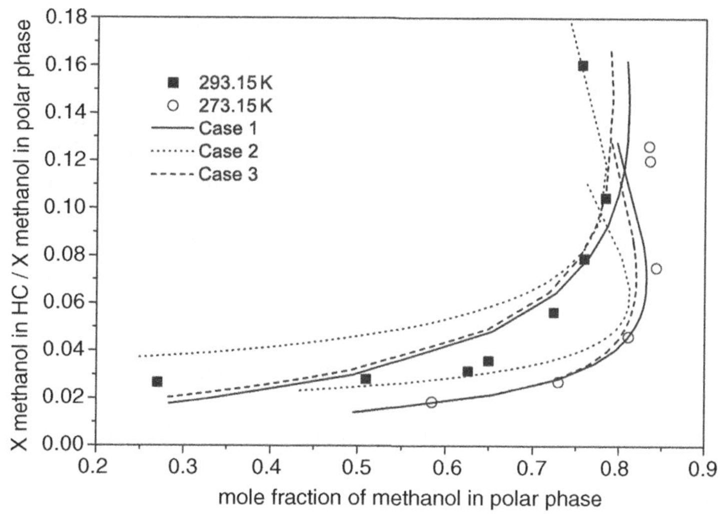

Figure 9.24 *Prediction with CPA of the partition coefficient of methanol between the hydrocarbon (HC) and the polar phase for the ternary system of methanol(1)–water(2)–n-butane(3) at 273.15 K and 293.15 K. The three different cases are: Case 1: $k_{12} = -0.09$ (ECR), $k_{13} = 0.035$ and $k_{23} = 0.0875$ and 2B scheme for methanol (solid line); Case 2: $k_{12} = 0.035$ (ECR), $k_{13} = -0.015$ and $k_{23} = 0.0875$ and 3B scheme for methanol (dotted line); Case 3: $k_{12} = -0.09$ (ECR), $k_{13} = 0.035$ and $k_{23} = 0.0$ and 2B scheme for methanol (dashed line). Experimental data are from Noda et al., J. Chem. Eng. Jpn,* ***1975****, 8, 492. Reprinted with permission from Ind. Eng. Chem. Res., Application of the Cubic-Plus-Association (CPA) Equation of State to Cross-Associating Systems by Georgios M. Kontogeorgis, Georgios Folas et al., 44, 10, 3823–3833 Copyright (2005) American Chemical Society*

Cinar and Orr[53] have presented three-phase liquid–liquid–liquid calculations with CPA for the system water–hexadecane–*n*-butanol–isopropanol. These calculations have been performed as part of their study on the effect of low interfacial tensions between two of the three phases on the three-phase relative permeabilities. Except for the area around the critical point and the critical tie line, CPA provides a reasonably good representation of the three-phase behavior.

9.5 Conclusions

- CPA is a combination of SRK with the association term of Wertheim, used in SAFT. The model has five pure compound parameters for associating compounds, typically estimated from vapor pressures and liquid densities. Parameters are available for a wide variety of compounds (alcohols, glycols, acids, amines, esters, ethers, etc.) and are available in Appendix A on the companion website at www.wiley.com/go/Kontogeorgis.
- Alcohols and amines are best modeled using the 2B scheme, while for water and glycols the 4C is used and acids are modeled using the 1A scheme. This chapter presents applications for methanol and few other alcohols, as well as water, while the next (three) chapters include a variety of applications, e.g. glycols, organic acids, amines, ketones and heavy alcohols.
- Monomer fractions, second virial coefficients and enthalpies of vaporization can be used for model validation. A challenging issue is to consider those properties for pure component parameter estimation together with vapor pressures and liquid densities.

- Hydrocarbons are always treated as 'inert' compounds, while polar compounds like ketones, ethers and esters may be treated as 'pseudo' self-associating compounds. This approach occasionally yields improved phase equilibria results, as will be illustrated in a subsequent chapter.
- Mixtures with one self-associating and one inert compound, e.g. alcohol or water with *n*-alkanes, require a single binary interaction parameter and often satisfactory results (especially VLE) are obtained even with $k_{ij}=0$ (prediction). Mixtures with two self-associating compounds, e.g. mixtures containing water or alcohols and glycols, also require a single k_{ij}. For these cross-associating mixtures, the interaction parameter is always required to obtain satisfactory correlation of phase equilibria. Mixtures where induced solvation is present, e.g. systems containing water or glycols with aromatic or olefinic hydrocarbons, require typically two interaction parameters, the k_{ij} and the cross-solvating parameter $\beta^{A_iB_j}$, using the modified CR-1 rule. Sometimes, the k_{ij} parameter can be obtained from the corresponding 'homomorph' system, e.g. from water–hexane in the case of water–benzene. This approach leaves a single adjustable interaction parameter ($\beta^{A_iB_j}$).
- Accounting for the solvation between water and aromatic or olefinic hydrocarbons is always crucial. For phase equilibrium calculations between alcohols and aromatic hydrocarbons the results suggest that the weak solvation might be ignored.
- Excellent binary VLE, LLE, SLE is obtained for mixtures of alcohols or water with alkanes as well as VLLE for water–methanol–alkanes. CPA provides reliable predictions of the methanol partition coefficients for multicomponent, multiphase oil and gas relevant mixtures, over an extended temperature and pressure range based solely on binary temperature-independent interaction parameters.
- CPA in combination with other frameworks has been used for predicting diffusion[51] and thermodiffusion (Soret) coefficients[52] for polar and associating mixtures, but further validation of these approaches is required.

Appendix 9.A Some comments on the radial distribution function (RDF)

9.A.1 RDF and repulsive terms

As Suresh and Elliott[32] point out, referring to Boublik,[33] there is a relationship between the radial distribution function (RDF) at contact and the repulsive compressibility factor of an equation of state (EoS) as follows:

$$g=\frac{Z^{rep}-1}{4y} \qquad y=\frac{b}{4V} \tag{9.12}$$

It can be easily shown that for the van der Waals EoS repulsive term:

$$g=Z^{rep}=V/(V-b)=1/(1-4y) \tag{9.13}$$

In the case of the Carnahan–Starling hard-sphere equation (CS)[55]:

$$Z^{rep}=\frac{1+y+y^2-y^3}{(1-y)^3} \tag{9.14}$$

which yields (by using Equation (9.12)):

$$g=\frac{2-y}{2(1-y)^3} \tag{9.15}$$

This is the well-known expression for SAFT presented in the various SAFT publications (and the original CPA version). For the Elliott EoS which (after some simplifications) can be written as:

$$Z^{rep} = \frac{1+2y}{1-2y} \tag{9.16}$$

the RDF at contact is given by:

$$g = \frac{1}{1-2y} \tag{9.17}$$

This is essentially the sCPA expression.

Equations (9.16) and (9.17) are essentially equivalent to the equation proposed by Scott[34], who also proposed a second CS-type equation:

$$Z^{rep} = \frac{1+1.5y}{1-2.5y} \tag{9.18}$$

There are many similar simplified CS expressions, e.g. that of Lin *et al.*[35]:

$$Z^{rep} = \frac{1+3.08y}{1-1.68y} \tag{9.19}$$

or Soave:[36]

$$Z^{rep} = \frac{1+2.4y}{1-4.0y} \tag{9.20}$$

9.A.2 The role of b/4V in the RDF expressions

In all CS-type EoS including all CS/empirical combinations, i.e. CS-type repulsive term and an empirical attractive[36–39] term, the reduced packing factor is defined as $y = b/4V$, where b is the co-volume parameter of the EoS and V is the molar volume. This definition of the packing fraction y is approximately equal to the ratio of molecular volume to molar volume. This is because b is defined in the van der Waals way as:

$$b = \frac{2}{3}\pi N_A \sigma^3 = 4V^{molecule} \tag{9.21}$$

The argumentation of such a large b value lies in the original free-volume definition of the free volume by van der Waals. Note that the b value in Equation (9.21) is the second virial coefficient of the hard-sphere potential. Alternative definitions of the packing fraction are essentially identical to that mentioned above, like $y = 0.7405V_o/V$, where V_o is the close-packed volume ($=1.35V_w$; V_w is the van der Waals volume for the FCC model). The original definition of the co-volume (Equation (9.21)) is 'forgotten' when such an EoS and the packing fraction are used in various EoS; b is simply the co-volume parameter of EoS to be obtained from various types of data (critical, vapor pressures, etc.) but not via the molecular diameter and Equation (9.21).

9.A.3 Justification of the sCPA RDF

As CPA uses the van der Waals repulsive term we would expect that the RDF derived from the vdW EoS should be used (see Equation (9.13)). Wong and Prausnitz[40] explain why vdW-type EoS can approach CS-type behavior if $b^{eff} = b/2$ (this is about $2V_w$). Note that the finally obtained CPA co-volume parameters estimated from vapor pressures and liquid densities turn out to be between 1.5 and $1.7V_w$ for most compounds studied, while similar values have been suggested in many other models, e.g. in the Sako–Wu–Prausnitz and GC–Flory EoS b is set to 1.4–$1.5V_w$, see also Table 3.15 (Chapter 3).

Thus, using the 'effective' co-volume (instead of the original co-volume b) and their relationship in the vdW RDF we get:

$$g = \frac{1}{1 - \frac{b}{2V}} = \frac{1}{1 - 2y} \tag{9.22}$$

$y = b/4V$ is defined the usual way, which is essentially the RDF of CPA. This analysis justifies in a way the use of the RDF (Equations (9.17) or (9.22)) in sCPA.

Appendix 9.B Parameterization of CPA

The energy parameter of the CPA EoS $a(T)$ is given by a Soave-type temperature dependency, while b (hereafter called b_{CPA}) is temperature independent:

$$a(T) = a_0[1 + c_1(1 - \sqrt{T_r})]^2 \tag{9.23}$$

where the reduced temperature is defined in the 'conventional' way as T/T_c.

The CPA model has five pure compound parameters: three for non-associating compounds (a_0, b_{CPA}, c_1) and two additional parameters for associating compounds ($\varepsilon^{A_iB_j}$, $\beta^{A_iB_j}$). All pure compound parameters are typically obtained by fitting experimental vapor pressure and saturated liquid density data. For non-associating compounds, e.g. n-alkanes, the three parameters (a_0, b_{CPA}, c_1) can be obtained either from vapor pressures and liquid densities or alternatively with the conventional methodology using critical data and acentric factors. The above procedure is somewhat inconvenient as it requires knowledge of the experimental critical temperature (T_c) which is used in the calculations. Thus, the exact value of the experimental critical temperature as used in the parameter estimation is required when using CPA together with Equation (9.23).

Alternatively, only three 'monomer' parameters can be estimated using the conventional SRK expressions:

$$a(T) = \Omega_A \frac{R^2 T_{cm}^2}{P_{cm}} \left[1 + m_m(1 - \sqrt{T/T_{cm}})\right]^2$$

$$b = \Omega_B \frac{R T_{cm}}{P_{cm}} \tag{9.24}$$

In Equation (9.24):

$$\Omega_A = 0.427\,48 \quad \text{and} \quad \Omega_B = 0.086\,64 \tag{9.25}$$

i.e. the usual SRK values, but the 'critical' parameters and m correspond to the 'monomer' and can be calculated from the energy and co-volume parameters of CPA. That is, they are based on optimizing vapor pressures and liquid density data. Comparing Equation (9.24) to the co-volume (b_{CPA}) CPA parameter, we get:

$$b_{CPA} = \Omega_B \frac{RT_{cm}}{P_{cm}} \Rightarrow P_{cm} = \Omega_B \frac{RT_{cm}}{b_{CPA}} \tag{9.26}$$

From Equations (9.23) and (9.24) the following expression can be obtained:

$$\begin{aligned} a(T) &= a_0[1 + c_1(1 - \sqrt{T_r})]^2 = \Omega_A \frac{R^2 T_{cm}^2}{P_{cm}} \left[1 + m_m(1 - \sqrt{T/T_{cm}})\right]^2 \Rightarrow \\ a &= K_1 - K_2\sqrt{T} \end{aligned} \tag{9.27}$$

where:

$$\begin{aligned} K_1 &= (1 + c_1)\sqrt{a_0} = (1 + m_m)\sqrt{\Omega_A \frac{R^2 T_{cm}^2}{P_{cm}}} \\ K_2 &= c_1 \frac{\sqrt{a_0}}{\sqrt{T_c}} = m_m \frac{\sqrt{\Omega_A \frac{R^2 T_{cm}^2}{P_{cm}}}}{\sqrt{T_{cm}}} \end{aligned} \tag{9.28}$$

Combining Equations (9.27) and (9.28) gives the equation derived for T_{cm}:

$$\sqrt{T_{cm}} = \frac{(1 + c_1)m_m}{c_1(1 + m_m)} \sqrt{T_c} \Rightarrow T_{cm} = \frac{\left(\frac{1 + c_1}{c_1}\right)^2}{\left(\frac{1 + m_m}{m_m}\right)^2} T_c \tag{9.29}$$

Finally, by combining Equation (9.26) and the K_2 expression of Equation (9.28):

$$m_m = c_1 \frac{\sqrt{a_0}\sqrt{T_{cm}}}{\sqrt{T_c}\sqrt{\Omega_A \frac{b_{CPA} R T_{cm}}{\Omega_B}}} \Rightarrow m_m = c_1 \sqrt{\frac{a_0 \Omega_B}{\Omega_A b_{CPA} R T_c}} \tag{9.30}$$

Using Equations (9.27)–(9.29) and the CPA parameters (a_0, b_{CPA}, c_1), the 'corresponding monomer' parameters can be calculated (a table is provided by Kontogeorgis *et al.*[16]). This implies that we only need to use the three conventional EoS parameters: the (monomer) critical temperature, pressure and m_m parameters.

Appendix 9.C Calculation of fugacity coefficients with CPA EoS

This appendix presents all the equations required for the calculation of the fugacity coefficient with the CPA EoS. The methodology presented by Michelsen and Mollerup[41] is followed, since the EoS is no longer cubic due to the association term. From classical thermodynamics the fugacity coefficient $\hat{\phi}_i$ of a component i in a mixture is given by:

$$RT \ln \hat{\phi}_i = \left(\frac{\partial A^r}{\partial n_i}\right)_{T,V,n_j} - RT \ln Z \tag{9.31}$$

A^r is the residual Helmholtz energy for the mixture and Z is the compressibility factor, defined as:

$$Z = \frac{PV}{nRT} \tag{9.32}$$

The CPA EoS combines the SRK EoS with the association term, derived from Wertheim's first-order perturbation theory, hence:

$$A^r(T, V, n) = A^r_{SRK}(T, V, n) + A^r_{association}(T, V, n) \tag{9.33}$$

According to Michelsen and Mollerup,[41] the fugacity coefficient for the SRK term can be calculated as follows:

$$\frac{A^r_{SRK}(T, V, n)}{RT} = -n \ln\left(1 - \frac{B}{V}\right) - \frac{D(T)}{RTB} \ln\left(1 + \frac{B}{V}\right) \tag{9.34}$$

where V is the total volume of the system, while:

$$nB = n^2 b_{mix} = \sum_i n_i \sum_j n_j b_{ij} \tag{9.35}$$

$$D(T) = n^2 a_{mix} = \sum_i n_i \sum_j n_j a_{ij}(T)$$
$$n = \sum_i n_i \tag{9.36}$$

The classical vdW1f mixing rules are used for the energy (α(T)) and co-volume parameter (b):

$$a_{ij}(T) = a_{ji}(T) = \sqrt{a_{ii}(T) a_{jj}(T)}(1 - k_{ij}) \tag{9.37}$$

$$b_{ij} = b_{ji} = \frac{1}{2}(b_{ii} + b_{jj}) \tag{9.38}$$

As a result of Equation (9.38), Equation (9.35) reduces to:

$$B = \sum_i n_i b_{ii} \tag{9.39}$$

Assuming that:

$$\frac{A^r_{SRK}(T,V,n)}{RT} = -n\ln\left(1-\frac{B}{V}\right) - \frac{D(T)}{RTB}\ln\left(1+\frac{B}{V}\right) = F^{SRK} \tag{9.40}$$

$$g(V,B) = \ln(1-B/V) \tag{9.41}$$

$$f(V,B) = \frac{1}{RB}\ln(1+B/V) \tag{9.42}$$

and inserting Equations (9.41) and (9.42) into (9.40), the resulting equation for F^{SRK} is:

$$F^{SRK} = -ng(V,B) - \frac{D(T)}{T}f(V,B) \tag{9.43}$$

Hence for the calculation of the fugacity coefficient $\hat{\phi}_i$ of a component i for the SRK term (Equation (9.31)) the derivative of the function F^{SRK} is required:

$$\left(\frac{\partial F^{SRK}}{\partial n_i}\right)_{T,V,n_j} = F_n + F_B B_i + F_D D_i \tag{9.44}$$

where:

$$\begin{aligned} F_n &= -g = -\ln(1-B/V) \\ F_B &= -ng_B - \frac{D(T)}{T}f_B \end{aligned} \tag{9.45}$$

with:

$$\begin{aligned} g_B &= -\frac{1}{V-B} \\ f_B &= -\frac{f+Vf_V}{B} \\ f_V &= -\frac{1}{RV(V+B)} \\ F_D &= -\frac{f}{T} \end{aligned} \tag{9.46}$$

B_i and D_i are the composition derivatives of the energy term (Equation (9.36)) and the co-volume term (Equation (9.39)), given by the following equation:

$$B_i = \frac{2\sum_j n_j b_{ij} - B}{n}$$

$$D_i = 2\sum_j n_j a_{ij} \tag{9.47}$$

9.C.1 The association term of the CPA EoS

The contribution of residual Helmholtz energy for the mixture to the association term $A^r_{association}(T, V, n)$ can be estimated, based on the approach proposed by Michelsen and Hendriks[42]. The authors introduced a Q function for the calculation of the derived properties of the association term, taking advantage of the fact that the association contribution to the Helmholtz energy is itself the result of a minimization. Consider the Q function given by:

$$Q(n, T, V, X) = \sum_i n_i \sum_{A_i} (\ln X_{A_i} - X_{A_i} + 1) - \frac{1}{2V}\sum_i \sum_j n_i n_j \sum_{A_i} \sum_{B_j} X_{A_i} X_{B_j} \Delta^{A_i B_j} \tag{9.48}$$

In Equation (9.48) X_{A_i} is the fraction of A sites on molecule i that do not form bonds with other active sites, n is the total composition of the mixture and V is the total volume. The association contribution of the CPA EoS equals the value of Q at a stationary point with respect to the site fractions X. The conditions that apply at a stationary point are:

$$\frac{\partial Q}{\partial X_{A_i}} = 0, \quad \text{for all sites} \tag{9.49}$$

By differentiating Equation (9.48):

$$n_i\left(\frac{1}{X_{A_i}} - 1\right) - \frac{1}{V} n_i \sum_j n_j \sum_{B_j} X_{B_j} \Delta^{A_i B_j} = 0 \tag{9.50}$$

which yields:

$$\frac{1}{X_{A_i}} = 1 + \frac{1}{V}\sum_j n_j \sum_{B_j} X_{B_j} \Delta^{A_i B_j} \tag{9.51}$$

The value of Q at a stationary point (sp) is:

$$Q_{sp} = \sum_i n_i \sum_{A_i} (\ln X_{A_i} - X_{A_i} + 1) - \frac{1}{2} \sum_i n_i \sum_{A_i} X_{A_i} \left(\frac{1}{V} \sum_j n_j \sum_{B_j} X_{B_j} \Delta^{A_i B_j} \right) \Rightarrow$$

$$Q_{sp} = \sum_i n_i \sum_{A_i} \left(\ln X_{A_i} - \frac{1}{2} X_{A_i} + \frac{1}{2} \right) = \frac{A^r_{association}(T, V, n)}{RT} \tag{9.52}$$

In order to proceed to the calculation of the fugacity coefficient from the association term, the chain rule should be used. According to the chain rule, the derivative of Q_{sp} with respect to n_i is given by the following equation:

$$\frac{\partial Q_{sp}}{\partial n_i} = \left. \frac{\partial Q}{\partial n_i} \right|_X + \sum_i \sum_{A_i} \frac{\partial Q}{\partial X_{A_i}} \frac{\partial X_{A_i}}{\partial n_i} \tag{9.53}$$

At the stationary point, however, the derivatives $\partial Q / \partial X_{A_i}$ are by definition zero, meaning that the fugacity coefficient for the association term can now be calculated using the explicit derivative of Q with respect to n_i:

$$\frac{\partial}{\partial n_i} \left(\frac{A^r_{association}}{RT} \right)_{T,V,n_j} = \frac{\partial}{\partial n_i} \left(\sum_i n_i \sum_{A_i} (\ln X_{A_i} - X_{A_i} + 1) - \frac{1}{2V} \sum_i \sum_j n_i n_j \sum_{A_i} \sum_{B_j} X_{A_i} X_{B_j} \Delta^{A_i B_j} \right)_{T,V,n_j}$$

$$= \sum_{A_i} (\ln X_{A_i} - X_{A_i} + 1) - \frac{1}{2V} \left. \frac{\partial}{\partial n_i} \right|_{\Delta} \left(\sum_i \sum_j n_i n_j \sum_{A_i} \sum_{B_j} X_{A_i} X_{B_j} \Delta^{A_i B_j} \right)$$

$$- \frac{1}{2V} \sum_i \sum_j n_i n_j \sum_{A_i} \sum_{B_j} X_{A_i} X_{B_j} \frac{\partial \Delta^{A_i B_j}}{\partial n_i}$$

When the yielding equation is combined with Equation (9.51) and after some algebra:

$$\frac{\partial}{\partial n_i} \left(\frac{A^r_{association}}{RT} \right)_{T,V,n_j} = \sum_{A_i} \ln X_{A_i} - \frac{1}{2V} \sum_i \sum_j n_i n_j \sum_{A_i} \sum_{B_j} X_{A_i} X_{B_j} \frac{\partial \Delta^{A_i B_j}}{\partial n_i} \tag{9.54}$$

The derivative $\partial \Delta^{A_i B_j} / \partial n_i$ is given by the following equation:

$$\frac{\partial \Delta^{A_i B_j}}{\partial n_i} = \Delta^{A_i B_j} \frac{\partial \ln g}{\partial n_i} \tag{9.55}$$

Therefore, combining Equation (9.54) with (9.55):

$$\frac{\partial}{\partial n_i}\left(\frac{A^r_{association}}{RT}\right)_{T,V,n_j} = \sum_{A_i} \ln X_{A_i} - \frac{1}{2V}\sum_i\sum_j n_i n_j \sum_{A_i}\sum_{B_j} X_{A_i} X_{B_j} \Delta^{A_i B_j} \frac{\partial \ln g}{\partial n_i} \tag{9.56}$$

Combining Equation (9.56) with (9.51) the resulting equation is:

$$\frac{\partial}{\partial n_i}\left(\frac{A^r_{association}}{RT}\right)_{T,V,n_j} = \sum_{A_i} \ln X_{A_i} - \frac{1}{2}\sum_i n_i \sum_{A_i}(1 - X_{A_i}) \frac{\partial \ln g}{\partial n_i} \tag{9.57}$$

Hence for the calculation of the contribution of the association term, the derivative of g with respect to n_i is required. Given that for the CPA EoS:

$$g(V,n) = \frac{1}{1 - 1.9\eta}, \quad \text{where} \quad \eta \frac{B}{4V} \tag{9.58}$$

and B is given by Equation (9.39), then:

$$\frac{\partial g}{\partial n_i} = \frac{\partial g}{\partial B} B_i$$

where B_i can be calculated from Equation (9.47), while:

$$\frac{\partial g}{\partial B} = 0.475V\left(\frac{1}{V - 0.475B}\right)^2 \tag{9.59}$$

9.C.2 Calculation of volume

For the calculation of the fugacity coefficient $\hat{\phi}_i$ of a component i in a mixture, the total volume is required, which can be calculated using the Newton–Raphson iteration method. The volume corresponding to a specific pressure, temperature and mixture composition can be calculated from the pressure equation:

$$P = \frac{nRT}{V} - \left(\frac{\partial A^r}{\partial V}\right)_{T,n} = \frac{nRT}{V} - \left(\frac{\partial A^r_{SRK}}{\partial V}\right)_{T,n} - \left(\frac{\partial A^r_{association}}{\partial V}\right)_{T,n} \tag{9.60}$$

From Equation (9.40) the expression for the SRK term is:

$$\left(\frac{\partial A^r_{SRK}}{\partial V}\right)_{T,n} = RT\left(\frac{\partial F^{SRK}}{\partial V}\right) \tag{9.61}$$

From Equation (9.52) the expression for the association term is:

$$\left(\frac{\partial A^r_{association}}{\partial V}\right)_{T,n} = RT\left(\frac{\partial Q_{SP}}{\partial V}\right)_{T,n} \tag{9.62}$$

Inserting Equations (9.61) and (9.62) into (9.60), the equation for the total pressure is:

$$P = \frac{nRT}{V} - RT\left(\frac{\partial F^{SRK}}{\partial V}\right)_{T,n} - RT\left(\frac{\partial Q_{SP}}{\partial V}\right)_{T,n} \tag{9.63}$$

Following the methodology presented by Michelsen and Mollerup,[41] the necessary equations for the estimation of the contribution of the physical term are:

$$\left(\frac{\partial F^{SRK}}{\partial V}\right)_{T,n} = F_V \tag{9.64}$$

$$F_V = -ng_v - \frac{D(T)}{T}f_v \tag{9.65}$$

$$g_V = \frac{B}{V(V-B)} \tag{9.66}$$

$$f_V = -\frac{1}{RV(V+B)} \tag{9.67}$$

Michelsen and Hendriks[42] showed, based on the Q function presented above and the chain rule:

$$\left(\frac{\partial Q_{sp}}{\partial V}\right) = \left(\frac{\partial Q}{\partial V}\right)_X + \sum_i \sum_{A_i} \frac{\partial Q}{\partial X_{A_i}} \frac{\partial X_{A_i}}{\partial V}$$

and since the derivatives $\partial Q/\partial X_{A_i}$ are by definition zero, the derivative $\partial Q_{sp}/\partial V$ is given by:

$$\left(\frac{\partial Q_{sp}}{\partial V}\right) = \frac{\partial}{\partial V}\left(\frac{A^r_{association}}{RT}\right)_{T,n} = \frac{1}{2V}\left(1 - V\frac{\partial \ln g}{\partial V}\right)\sum_i n_i \sum_{A_i}(1 - X_{A_i}) \tag{9.68}$$

Considering Equation (9.58), the required derivative $\partial \ln g/\partial V$ for the calculation of the contribution of the association term is given by the following expression:

$$\frac{\partial g}{\partial V} = -0.475B\left(\frac{1}{V - 0.475B}\right)^2 \tag{9.69}$$

For a Newton–Raphson variant for the calculation of the total volume V, the function that should be minimized is given by Equation (9.60) as follows:

$$H(V) = P - \left[\frac{nRT}{V} - \left(\frac{\partial A^r_{SRK}}{\partial V}\right)_{T,n} - \left(\frac{\partial A^r_{association}}{\partial V}\right)_{T,n}\right] \tag{9.70}$$

The derivative of the function H(V) with respect to volume is also required, meaning that second derivatives of A^r with respect to volume are required. The second derivative can, however, be calculated numerically from Equation (9.70); hence an analytical method for estimating the second derivatives is not required.

Appendix 9.D Modeling of the solid complex in water–alcohol or water–glycol mixtures

At atmospheric pressure and in a narrow composition range, an aqueous solution of MEG or methanol forms a solid complex when it freezes. Other water–alcohol and water–glycol mixtures are also shown to form such solid complexes. The complex is assumed to have a crystal lattice structure and the formation of the solid complex can thus be modeled as the product of the interaction between MEG or methanol and water. We consider in general the following interaction between two liquids A and B forming a solid A.B:

$$A \cdot B_{(s)} \leftrightarrow A_{(l)} + B_{(l)} \tag{9.71}$$

Assuming that the solid complex behaves as a pure solid, the activity of the complex will be unity. The chemical equilibrium constant (K) for the interaction can be calculated as:

$$K = \gamma_A \cdot x_A \cdot \gamma_B \cdot x_B \tag{9.72}$$

The temperature dependency of the chemical equilibrium constant, at constant pressure, can be derived from the Gibbs–Helmholtz equation:

$$\left[\frac{\partial}{\partial T}\left(\frac{\Delta G}{T}\right)\right] = -\frac{\Delta H}{T^2} \tag{9.73}$$

and given that:

$$\Delta G = -RT \ln K \tag{9.74}$$

the equation of the temperature dependence of K at constant pressure is:

$$\frac{d\ln K}{dT} = \frac{\Delta H}{RT^2} \tag{9.75}$$

Assuming that ΔH is a function of temperature, it can be described in the following manner:

$$\Delta H(T) = \Delta H^{ref} + \int_{T^{ref}}^{T} \Delta C_p dT \tag{9.76}$$

where ΔC_p is the difference in heat capacity and ΔH^{ref} is the enthalpy of fusion of the solid complex phase at a given reference temperature, T^{ref}. Furthermore, if it is assumed that ΔC_p is zero in the relatively small temperature area where the solid complex is formed, the following expression is obtained:

$$R \ln K = R \ln K^{ref} + \Delta H^{ref}\left(\frac{1}{T^{ref}} - \frac{1}{T}\right) \tag{9.77}$$

The mixture properties in the liquid phase are obtained from CPA using ECR (any suitable EoS that simultaneously correlates both freezing curves can be used) and the interaction parameters regressed from experimental data of the freezing curves of the pure MEG or methanol and water. ECR is preferred as it describes equally well the SLE and VLE of both systems. The interaction parameters are $k_{12} = -0.115$ for the MEG–water system (common from VLE and SLE data) and $k_{12} = -0.153$ for the methanol–water system (from SLE data), respectively (Figure 9.18). Modeling of the solid complex in this manner requires that the molar composition of the complex is known. The freezing diagram of the MEG–water system has been extensively studied in the literature with particular emphasis on the intermediate concentrations, where the solid complex occurs; the formation of a 1 : 1 molar composition complex is concluded.[43] The formation of a similar 1 : 1 solid complex has also been reported for the methanol–water system.[44,45] Finally, the enthalpy of fusion of the complex phase ΔH^{ref} is regressed from experimental data of the solid complex phase, whereas the value of the equilibrium constant K^{ref} at a chosen reference temperature T^{ref} is calculated from Equation (9.72). Table 9.4 presents the values of the equilibrium constant K^{ref} at a chosen reference temperature T^{ref} and the optimized value for the enthalpy of fusion of the complex phase ΔH^{ref} for each system studied.

Values in the literature for such complex solid phases are rare. A typical system forming such a solid complex phase is acetone–chloroform, which is also suggested to form a solid complex with 1 : 1 stoichiometric ratio.[46] Prausnitz *et al.*[46] reported the enthalpy of fusion for the solid complex of acetone–chloroform to be 11 370 J/mol, which is in fair agreement with the values suggested in Table 9.4 for MEG–water and methanol–water respectively.

Figures 9.18 and 10.6 present the complete freezing diagrams for the water–methanol and water–MEG systems, respectively. The calculated eutectics for the MEG–water system are at 223.95 K and 229.49 K and mole fractions of 0.30 and 0.54 of ethylene glycol, respectively. These results are in good agreement with the experimental values,[47] which are reported to be 224.12 K and 230.22 K and mole fractions 0.288 and 0.541 of ethylene glycol, respectively. Similar results are obtained for methanol. The calculated eutectics are at 171.25 K and 159.23 K and mole fractions of 0.55 and 0.79 of methanol, respectively, while an experimental eutectic is reported[45] to be at 157 K and a mole fraction of 0.807 of methanol.

Table 9.4 *Values of the parameters used in Equation (9.77)*

System	T^{ref} (K)	K^{ref}	ΔH^{ref} (J/mol)
MEG–water	223.95	0.168 048	144 50
Methanol–water	171.25	0.208 059	854 0

References

1. I. Hatzioannidis, E.C. Voutsas, E. Lois, D.P. Tassios, *J. Chem. Eng. Data*, **1998**, *43*, 386.
2. U. von Stockar, L.A.M. van der Wielen, *J. Biotechnol.*, **1997**, *59*, 25.
3. J.M. Prausnitz, *Fluid Phase Equilib.*, **1989**, *53*, 439.
4. S. Bruin, *Fluid Phase Equilib.*, **1999**, *158–160*, 657.
5. S. Zeck, *Fluid Phase Equilib.*, **1991**, *70*, 125.
6. G.M. Kontogeorgis, E. Voutsas, I. Yakoumis, D.P. Tassios, *Ind. Eng. Chem. Res.*, **1996**, *35*, 4310.
7. G.M. Kontogeorgis, I.V. Yakoumis, H. Meijer, E.M. Hendriks, T. Moorwood, *Fluid Phase Equilib.*, **1999**, *158–160*, 201.
8. S.O. Derawi, G.M. Kontogeorgis, M.L. Michelsen, E.H. Stenby, *Ind. Eng. Chem. Res.*, **2003**, *42*, 1470.
9. G.K. Folas, G.M. Kontogeorgis, M.L. Michelsen, E.H. Stenby, *Ind. Eng. Chem. Res.*, **2006**, *45(4)*, 1527.
10. S.H. Huang, M. Radosz, *Ind. Eng. Chem. Res.*, **1990**, *29*, 2284.
11. M. Kaarsholm, S.O. Derawi, M.L. Michelsen, G.M. Kontogeorgis, *Ind. Eng. Chem. Res.*, **2005**, *44(12)*, 4406.
12. E.C. Voutsas, G.D. Pappa, K. Magoulas, D.P. Tassios, *Fluid Phase Equilib.*, **2006**, *240*, 127.
13. P.D. Ting, P.C. Joyce, P.K. Jog, W.G. Chapman, M.C. Thies, *Fluid Phase Equilib.*, **2003**, *206(1–2)*, 267.
14. N. von Solms, M.L. Michelsen, C.P. Passos, S.O. Derawi, G.M. Kontogeorgis, *Ind. Eng. Chem. Res.*, **2006**, *45*, 5368.
15. N. von Solms, L. Jensen, J. Kofod, M.L. Michelsen, G.M. Kontogeorgis, *Fluid Phase Equilib.*, **2007**, *261(1–2)*, 272.
16. G.M. Kontogeorgis, M.L. Michelsen, G.K. Folas, S. Derawi, N. von Solms, E.H. Stenby, *Ind. Eng. Chem. Res.*, **2006**, *45*, 4855.
17. C. Lundstrøm, M.L. Michelsen, G.M. Kontogeorgis, K.S. Pedersen, H. Sørensen, *Fluid Phase Equilib.*, **2006**, *247 (1–2)*, 149.
18. A. Grenner, J. Schmelzer, N. von Solms, G.M. Kontogeorgis, *Ind. Eng. Chem. Res.*, **2006**, *45*, 8170.
19. G.N.I. Clark, A.J. Haslam, A. Galindo, G. Jackson, *Mol. Phys.*, **2006**, *104*, 3562.
20. A. Grenner, G.M. Kontogeorgis, M.L. Michelsen, G.K. Folas, *Mol. Phys.*, **2007**, *105(13–14)* 1797.
21. I. Yakoumis, G.M. Kontogeorgis, E. Voutsas, D. Tassios, *Fluid Phase Equilib.*, **1997**, *130*, 31.
22. E. Voutsas, I. Yakoumis, G.M. Kontogeorgis, D. Tassios, *Fluid Phase Equilib.*, **1997**, *132*, 61.
23. G.K. Folas, J. Gabrielsen, M.L. Michelsen, E.H. Stenby, G.M. Kontogeorgis, *Ind. Eng. Chem. Res.*, **2005**, *44*, 3823.
24. G.K. Folas, S.O. Derawi, M.L. Michelsen, E.H. Stenby, G.M. Kontogeorgis, *Fluid Phase Equilib.*, **2005**, *228–229*, 121.
25. S. Gupta, J.D. Olson, *Ind. Eng. Chem. Res.*, **2003**, *42(25)*, 6359.
26. G.K. Folas, G.M. Kontogeorgis, M.L. Michelsen, E.H. Stenby, *Ind. Eng. Chem. Res.*, **2006**, *45*, 1516.
27. I.G. Economou, C. Tsonopoulos, *Chem. Eng. Sci.*, **1997**, *52*, 511.
28. J. Wu, J.M. Prausnitz, *Ind. Eng. Chem. Res.*, **1998**, *37*, 1634.
29. E.C. Voutsas, G.C. Boulougouris, I.G. Economou, D.P. Tassios, *Ind. Eng. Chem. Res.*, **2000**, *39(3)* 797.
30. G.K. Folas, G.M. Kontogeorgis, M.L. Michelsen, E.H. Stenby, *Fluid Phase Equilib.*, **2006**, *249(1–2)*, 67.
31. A. Arce, A. Blanco, A. Soto, J. Tojo, *J. Chem. Eng. Data*, **1995**, *40*, 1011.
32. J.S. Suresh, J.R. Elliott, *Ind. Eng. Chem. Res.*, **1992**, *31*, 2783.
33. T. Boublik, *Mol. Phys.*, **1974**, *27*, 1415.
34. R.L. Scott, *Physical Chemistry: An advanced treatise*, Vol. 8A, Chapter 1. Academic Press. 1971.
35. H.M. Lin, H. Kim, T.M. Guo, K.C. Chao, *Fluid Phase Equilib.*, **1983**, *13*, 143.
36. G. Soave, *Fluid Phase Equilib.*, **1990**, *56*, 39.
37. J.S. Yanaki, V.F. Yesavage, *Ind. Eng. Chem. Res.*, **1990**, *29*, 1549.
38. N.F. Carnahan, K.E. Starling, *AIChE J.*, **1972**, *18(6)*, 1184.
39. I. Polishuk, J. Wisniak, H. Segura, *Ind. Eng. Chem. Res.*, **2002**, *41*, 4414.
40. J. Wong, J.M. Prausnitz, *Chem. Eng. Commun.*, **1984**, *37*, 41.
41. M.L. Michelsen, J.M. Mollerup, *Thermodynamic Models: Fundamentals & Computational Aspects*. Tie-Line Publications, 2004.
42. M.L. Michelsen, E.M. Hendriks, *Fluid Phase Equilib.*, **2001**, *180*, 165.
43. D.R. Cordray, L.R. Kaplan, P.M. Woyciesjes, T.F. Kozak, *Fluid Phase Equilib.*, **1996**, *117*, 146.
44. G.A. Miller, D.K. Carpenter, *J. Chem. Eng. Data*, **1964**, *9*, 371.

45. J.B. Ott, J.R. Goates, B.A. Waite, *J. Chem. Thermodyn.*, **1979**, *11*, 739.
46. J.M. Prausnitz, R.N. Lichtenthaler, E.G. de Azevedo, *Molecular Thermodynamics of Fluid-Phase Equilibria* (3rd edition). Prentice Hall International, 1999.
47. J.B. Ott, J.R. Goates, J.D. Lamb, *J. Chem. Thermodyn.*, **1972**, *4*, 123.
48. G.K. Folas, E.W. Froyna, J. Lovland, G.M. Kontogeorgis, E. Solbraa, *Fluid Phase Equilib.*, **2007**, *252*, 162.
49. G.M. Kontogeorgis, M.L. Michelsen, G. Folas, S. Derawi, N. von Solms, E.H. Stenby, *Ind. Eng. Chem. Res.*, **2006**, *45*(*14*), 4869.
50. M.L. Michelsen, Parameter estimation in association models. Proceedings of ESAT, France, 2008.
51. O.O. Medvedev, A.A. Shapiro, *Fluid Phase Equilib.*, **2008**, *208*, 291.
52. M.Z. Saghir, C.G. Jiang, S.O. Derawi, E.H. Stenby, M. Kawaji, *Eur. Phys. J. E*, **2008**, *15*, 241.
53. Y. Cinar, F.M. Orr Jr, *SPE 89419*, **2005**, *8*, 33.
54. M.P. Breil, G.M. Kontogeorgis, *Ind. Eng. Chem. Res.*, **2009**, *48*(*11*), 5472.
55. N.F. Carnahan, K.E. Starling, *J. Chem. Physics*, **1969**, *51*(2), 635.

10

Applications of CPA to the Oil and Gas Industry

10.1 General

There are various applications in the oil and gas industry where associating and/or oligomeric compounds are present. For these applications, it is expected that association models may be the preferred approach. Some examples are gas hydrate inhibitors, e.g. methanol and glycols, other flow assurance problems, e.g. related to asphaltenes, the variety of drilling, production and injection chemicals used in the oil industry, for which their environmental fate must be assessed and their use must be optimized, and finally the use of ethanol and other oxygenated polar compounds as additives in gasoline. Some of these examples are discussed below in more detail.

Methanol is possibly the most important gas hydrate inhibitor. It suppresses gas hydrates but is often injected at a higher rate than necessary, due to uncertainties as to what the correct rate should be. Two important reasons for optimizing the amount of methanol used for inhibition (and keeping it to a minimum) are the cost of providing this chemical, especially on offshore platforms, and the fact that methanol is a toxic substance. The injection rate is directly related to the phase equilibria of mixtures containing oil, water and methanol. Moreover, the loss of hydrate inhibitor in the gas or condensate phases should be accurately known.

Typically methanol is not being regenerated because it is relatively inexpensive. Common alternatives to methanol are the glycols, especially monoethylene glycol (MEG), which is added at rates up to 100% of the weight of water. Due to the high cost of MEG, regeneration is required, which, however, adds to the capital and operational cost as well as space requirements, especially for offshore installations. Hence, optimization of the MEG injection rate and minimizing the amount of MEG lost (mainly) in the hydrocarbon liquid and aqueous phases are directly related to the elimination of operational problems and decrease of cost.

Another important problem where associating compounds are involved in the petroleum industry is related to the production and processing of natural gas. As previously discussed, the freezing diagrams of MEG–water and methanol–water have two eutectic points and an intermediate phase which indicates the formation of a solid (complex) hydrate between the two compounds. When natural gas is being produced from offshore fields, e.g. in the North Sea, and then transported to the gas plant, it is brought to sales gas specifications by a series of separators. In order to prevent the formation of gas hydrates during transportation and further

Thermodynamic Models for Industrial Applications: From Classical and Advanced Mixing Rules to Association Theories
Georgios M. Kontogeorgis and Georgios K. Folas

processing, inhibitors such as ethylene glycol (MEG) are continuously added to the produced mixture (water and light hydrocarbons). Decreased efficiency has been observed in the separators at MEG concentrations, where according to the literature problem-free operation would be expected. This means that the concentration of ethylene glycol is high enough to inhibit formation of gas hydrates and low enough to prevent freezing of ethylene glycol. Experimental investigations indicate that ethylene glycol at increased pressures and in the presence of hydrocarbons may freeze at higher temperatures than expected, or a 'hydrate' (complex solid) phase of inhibitor–water might occur. The freezing of MEG causes the concentration of MEG in the liquid phase to drop, which lowers the inhibiting effect of MEG and causes gas hydrates to form. This narrows the operational limits of MEG concentration to about 75–80%.

Moreover, petroleum is a mixture of aliphatic, aromatic and naphthenic hydrocarbons, thus it is important that a model can describe the different thermodynamic behavior that is observed for aromatic/olefinic versus aliphatic hydrocarbons. In addition, the phase equilibria over extensive temperature and pressure ranges and in the presence of acid gases (CO_2, H_2S) are also important. Finally, a characterization procedure is needed for applying the thermodynamic models to reservoir fluids (oil and gas condensates).

CPA and PC–SAFT (as well as other SAFT variants) have been applied to many of the aforementioned applications, with CPA's strength lying so far mostly in describing mixtures with a variety of gas hydrate inhibitors, and SAFT's in asphaltenes. Some of the most characteristic applications of CPA in the oil industry are presented in this chapter, while for SAFT they are discussed in Chapter 13.

10.2 Glycol–water–hydrocarbon phase equilibria

10.2.1 Glycol–hydrocarbons

The extension of CPA to glycols was possible because of an extensive experimental program in the period 2002–2007 in collaboration with Statoil, where new experimental LLE data for glycol–alkanes and glycol–aromatics hydrocarbons were measured.[1,2] The loss of glycol in the gas phase is also important, and recently VLE measurements of MEG–water–methane at low temperature and high pressures[3] have been reported. Finally, measurements of TEG solubility in methane have also been reported.[50] Various glycols were considered (MEG, DEG, TEG, TeEG) and all were modeled using the 4C scheme (i.e. the same association scheme as for water), although the glycols which are heavier than MEG contain one or more oxygen atoms in their structure and there are indications of both inter- and intramolecular association. As shown, however, by Derawi *et al.*[4], even the simplified approach, i.e. CPA using four equal sites for all glycols, provides very satisfactory LLE correlation results for glycols–aliphatic hydrocarbon mixtures. Two characteristic results for MEG–hexane and TEG–heptane mixtures are shown in Figure 10.1. A single temperature-independent k_{12} is sufficient for satisfactory correlation over extensive temperature ranges. On the other hand, SRK with conventional mixing rules cannot simultaneously correlate both solubilities. With SRK, when the binary interaction parameter is fitted to the solubility of hydrocarbon in the glycol phase, the solubility of glycol in the hydrocarbon phase is erroneously calculated to be higher than the solubility of the hydrocarbon in the glycol phase. This opposes the physical behavior of the systems. The same holds for glycols–aromatic hydrocarbons, and Figure 10.2 (left) presents some results on the correlative performance of SRK for MEG–benzene. It has already been mentioned in Chapter 5 that local composition models like UNIQUAC and NRTL can also correlate LLE for such mixtures when temperature-dependent interaction parameters are used.

One important recent development is the extension of CPA to mixtures containing polar compounds (water, glycols) and aromatic or olefinic hydrocarbons. This permitted also the study of many multicomponent mixtures containing polar compounds (water, alcohols, glycols) and hydrocarbons, as BTEX compounds are

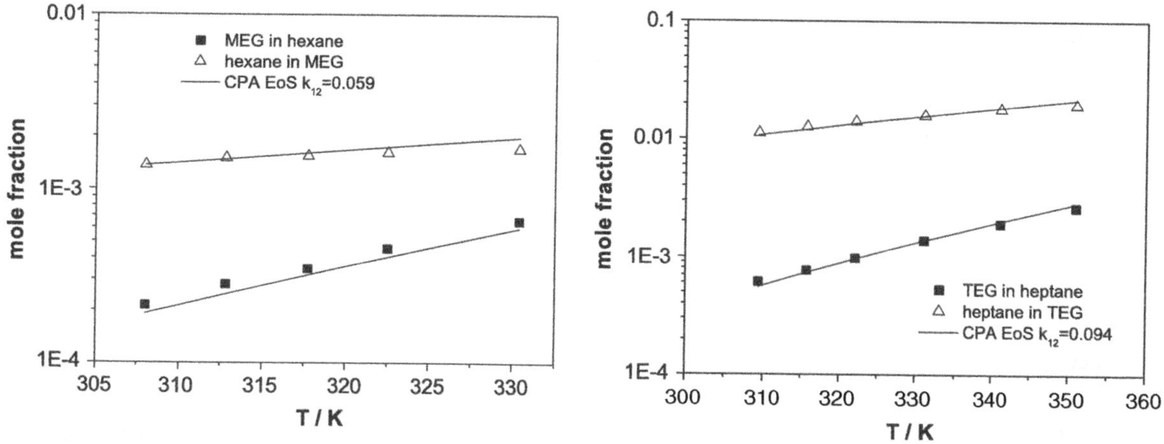

Figure 10.1 LLE for glycols–alkanes with CPA. Left: LLE of monoethylene glycol (MEG) and hexane with CPA using $k_{12}=0.059$. Right: LLE of triethylene glycol (TEG) and heptane with CPA using $k_{12}=0.094$. Experimental data are from Derawi et al.[1] Reprinted with permission from Fluid Phase Equilibria, Application of the CPA equation of state to glycol/hydrocarbons liquid-liquid equilibria by S. Derawi, M. L. Michelsen, G. M. Kontogeorgis et al., 209, 2, 163–184 Copyright (2003) Elsevier

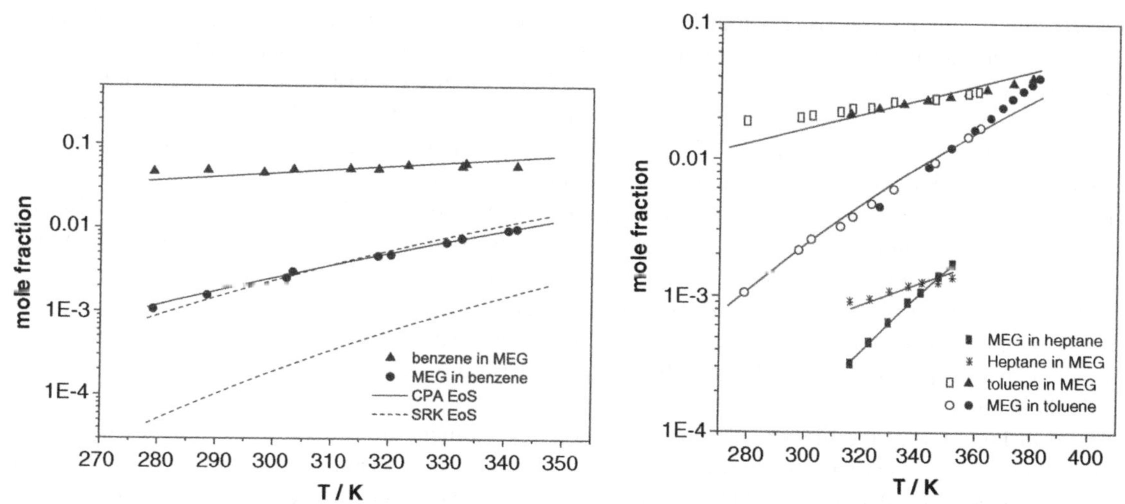

Figure 10.2 LLE for MEG–aromatic hydrocarbons with CPA. Left: LLE of MEG–benzene with the CPA ($k_{12}=0.049$ and $\beta^{A_iB_j}=0.04$) and the SRK equation of state ($k_{12}=0.21$). Right: LLE correlation of MEG–toluene ($k_{12}=0.051$ and $\beta^{A_iB_j}=0.042$) and MEG–heptane ($k_{12}=0.047$). Experimental data are from Folas et al.[2], Mandik and Lesek, Collect. Czech. Chem. Commun., **1982**, 47, 1686, while for MEG–heptane the experimental data are from Derawi et al.[1] Reprinted with permission from Ind. Eng. Chem. Res., Application of the Cubic-Plus-Association (CPA) Equation of State to Complex Mixtures Georgios K. Folas, Georgios M. Kontogeorgis, 45, 4, 1527–1538 Copyright (2006) American Chemical Society

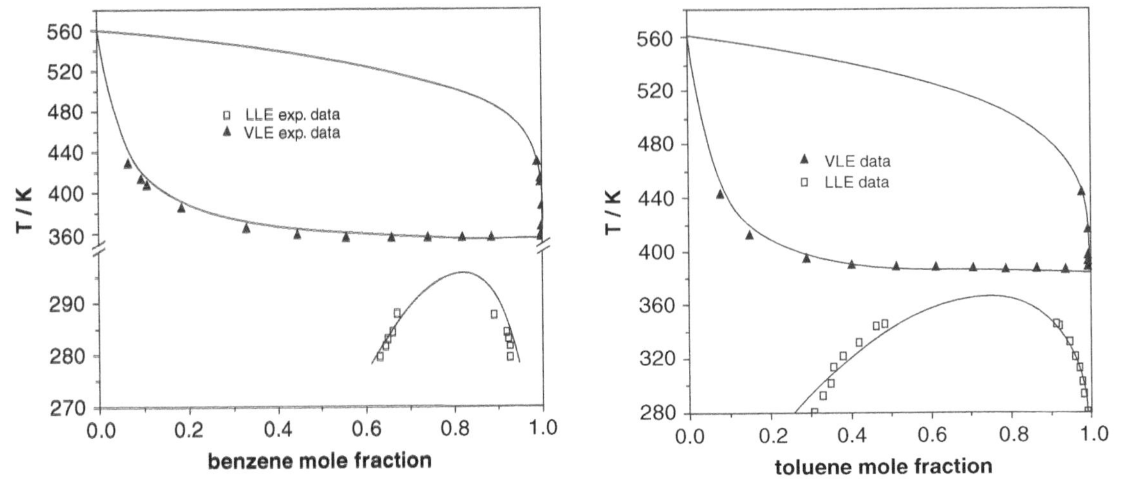

Figure 10.3 *LLE for TEG–aromatic hydrocarbons with CPA. Left: VLE and LLE for TEG–benzene with CPA ($k_{12}=0.032$ and $\beta^{A_iB_j}=0.083$). Right: VLE and LLE for TEG–toluene with the CPA ($k_{12}=0.038$ and $\beta^{A_iB_j}=0.048$). Reprinted with permission from J. Chem. Eng. Data, Liquid–Liquid Equilibria for Binary and Ternary Systems Containing Glycols, Aromatic Hydrocarbons, and Water: Experimental Measurements and Modeling with the CPA EoS by Georgios K. Folas, Georgios M. Kontogeorgis et al., 51, 3, 977–983 Copyright (2006) American Chemical Society*

often present in such mixtures (BTEX indicates benzene–toluene–ethylbenzene–xylene). A major obstacle in the development was the lack of experimental data, especially for glycols with aromatic hydrocarbons. For this reason, the second part of the experimental project with Statoil focused on LLE measurements of binary and ternary mixtures of glycols, aromatic hydrocarbons and water.[2] From the modeling point of view, CPA can model very satisfactorily phase equilibria of mixtures of glycols and BTEX if the solvation between polar compounds and such hydrocarbons is explicitly taken into account, via the modified CR-1 rule (see Chapter 9, Equation (9.10)). Typical results are presented in Figures 10.2 and 10.3, while more results are given by Folas *et al.*[2,5] Two interaction parameters are required for solvating systems (k_{ij} and $\beta^{A_iB_j}$), but the homomorph approach can be used for water–aromatic hydrocarbons, as shown in Tables 9.2 and 9.3. According to this approach, the k_{ij} parameter can be taken from the corresponding 'homomorph' system, e.g. water–hexane for water–benzene. Then, only $\beta^{A_iB_j}$ is adjusted to experimental data. Recent studies[62] indicate that the same approach can be applied to glycol–aromatic hydrocarbons, i.e. the k_{ij} parameter can be taken from glycol–alkane mixtures.

Notice in Figure 10.2 the higher solubilities in the MEG–toluene mixture compared to MEG–heptane. These higher solubilities are due to the solvation effects which must be accounted for by the model. Solvation does exist also between alcohols and BTEX compounds, e.g. methanol is miscible with benzene or toluene but only partially miscible with hexane or heptane. In the framework of CPA, it has been established that the alcohol–BTEX solvation is less important compared to the water–BTEX and glycol–BTEX solvation interactions. Thus, for practical purposes and for multicomponent calculations, the alcohol–BTEX solvation can be ignored and alcohol–BTEX phase equilibria can be described well using one k_{12} parameter. The lack of LLE data for alcohol–BTEX is possibly, from a practical point of view, the reason why additional interaction parameters are not needed. Explicit accounting for the solvation between alcohols and BTEX compounds is important for 'sensitive' properties, e.g. infinite dilution activity coefficients.

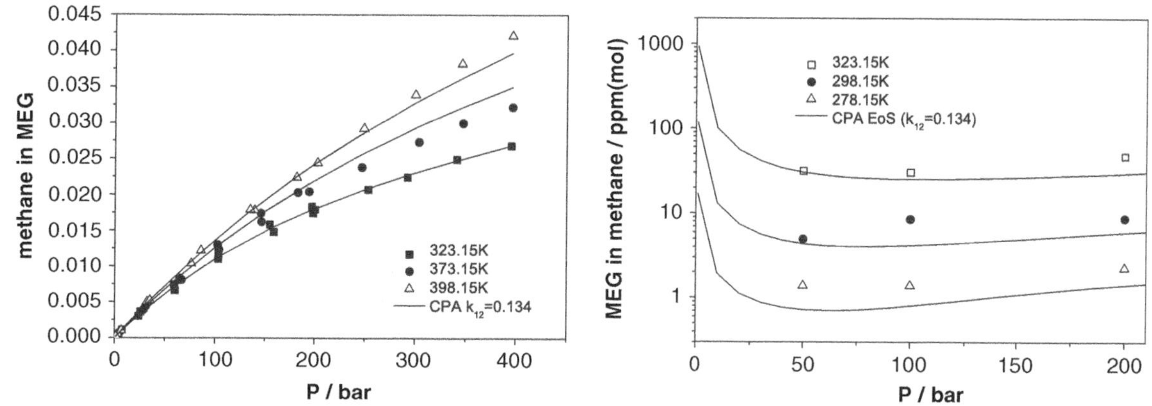

Figure 10.4 *CPA calculation for MEG–methane using a single interaction parameter $k_{12} = 0.134$. The interaction parameter is fitted to methane solubility in MEG (left plot). Reprinted with permission from Fluid Phase Equilibria, High-pressure vapor-liquid equilibria of systems containing ethylene glycol, water and methane: Experimental measurements and modeling by Georgios M. Kontogeorgis, Georgios Folas et al., 251, 1, 52–58 Copyright (2007) Elsevier*

VLE experimental measurements of MEG–methane and MEG–water–methane mixtures were recently published[3] with an emphasis on the MEG solubility in the gas phase. CPA was shown to provide very satisfactory correlation of the MEG solubility even when the binary interaction parameter is obtained from the methane solubility in MEG at different temperatures. As presented for example in Figure 10.4, the binary interaction parameter which is obtained based on methane solubility data in MEG within the temperature range of 323–390 K and a pressure range up to 400 bar provides very satisfactory calculations of the MEG solubility in the gas phase at lower temperatures. Since glycol measurements in the gas phase are both time consuming and expensive,[3] it is of interest to obtain reliable calculations based on the relatively inexpensive solubility measurements in the liquid phase.

10.2.2 Glycol–water and multicomponent mixtures

VLE and SLE data are available for MEG–water, whereas only reliable VLE data are available for other glycol–water mixtures. Some industrial data available for water–DEG and water–TEG SLE show significant scatter indicating the experimental difficulties associated with measurements involving the heavy (viscous) glycols. The VLE of water mixtures with MEG, TEG, DEG and other glycols are described well with CPA using the CR-1 rule and a rather large negative interaction parameter[6]. The negative values of the interaction parameters for these mixtures are higher in magnitude than the interaction parameters obtained for mixtures containing self-associating and inert compounds, e.g. alcohol–alkanes or water–hydrocarbons. These large negative interaction parameters may indicate that the extent of cross-association is somewhat underestimated with the combining rules used and the assigned association sites assumed for glycols, which many need revision. For example, the VLE correlation of DEG–water with the CR-1 rule requires a binary interaction parameter of $k_{12} = -0.115$, whereas the VLE of MEG–water requires a small binary interaction parameter $k_{12} = -0.012$. In all cases, however, the correlative performance of the model is excellent, as typically presented in Figure 10.5, for the VLE of DEG–water. ECR is a successful alternative to CR-1 for MEG–water VLE as can be seen in Figure 10.6, although correlation results with ECR are less accurate for the heavier glycols when compared to those with CR-1.

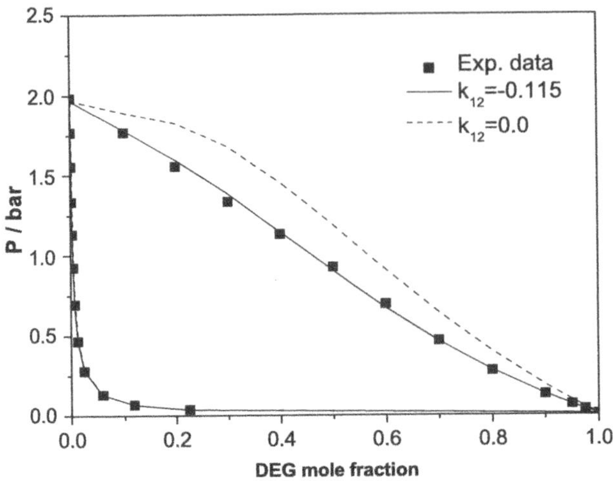

Figure 10.5 *CPA correlation and prediction of the DEG–water VLE at T = 393.15 K using the CR-1 rule with $k_{12} = -0.115$ (solid line) and $k_{12} = 0$ (dashed line). Reprinted with permission from Ind. Eng. Chem. Res., Extension of the Cubic-Plus-Association Equation of State to Glycol–Water Cross-Associating Systems by S. O. Derawi, G. M. Kontogeorgis, M. L. Michelsen, and E. H. Stenby, 42, 7, 1470–1477 Copyright (2003) American Chemical Society*

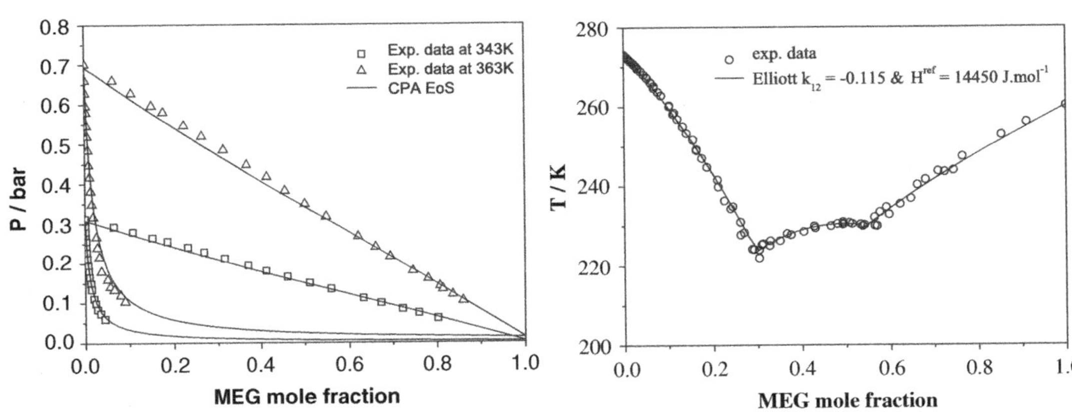

Figure 10.6 *SLE (right) and VLE (left) for water–MEG using ECR and the same binary interaction parameter $k_{12} = -0.115$.Experimental data are from Cordray et al., Fluid Phase Equilib.,* ***1996****, 117, 146, Ott et al., J. Chem. Thermodyn.,* ***1972****, 4, 123 and Chiavone-Filho et al., J. Chem. Eng. Data,* ***1993****, 38, 128. Reprinted with permission from Ind. Eng. Chem. Res., Application of the Cubic-Plus-Association (CPA) Equation of State to Cross-Associating Systems by Georgios M. Kontogeorgis, Georgios Folas et al., 44, 10, 3823–3833 Copyright (2005) American Chemical Society. Reprinted with permission from Fluid Phase Equilibria, Vapor–liquid, liquid–liquid and vapor–liquid–liquid equilibrium of binary and multicomponent systems with MEG: Modeling with the CPA EoS and an EoS/G^E model by G. K. Folas, G. M. Kontogeorgis et al., 249, 1–2, 67–74 Copyright (2006) Elsevier*

Especially for the MEG–water mixture, the performance of CR-1 and ECR can be tested over an extended temperature range, since both VLE and SLE data are available. We have seen in Chapter 9 that CPA with Elliott's rule (ECR) satisfactorily describes mixtures of water–methanol (and hydrocarbons), thus multicomponent aqueous mixtures with one of the most important gas hydrate inhibitors (methanol). The same holds for aqueous mixtures with the other important gas hydrate inhibitor, MEG. As can be seen in Figure 10.6, CPA can correlate VLE and SLE of MEG–water using ECR and the same k_{12} parameter (−0.115). CPA can correlate the freezing curves of water and glycol (or methanol) and the performance of the model depends on the combining rule as well. For example, the correlation of both freezing curves of MEG–water with ECR is superior to CR-1, when using the same value of the binary interaction parameter. To describe the solid complex at low temperatures a different approach is required. The chemical model[7] presented in Appendix 9.D (Chapter 9) is used. A 1:1 complex formation between MEG and water is assumed. The correlation is satisfactory using the enthalpy of complex formation as an additional adjustable parameter. The obtained enthalpies are, for MEG–water, equal to −14.5 kJ/mol and for methanol–water equal to −8.5 kJ/mol, which are in between the value reported for the solid complex between chloroform and acetone, which is −11.4 kJ/mol.[8]

CPA attracts the interest of the oil and gas industry, not just because of the successful simulations of glycol–hydrocarbon and glycol–water mixtures over extended temperature ranges, but also because of the capability of the model to predict multicomponent, multiphase equilibria. Predictions are successful even when the binary interaction parameters are extrapolated over extended ranges of temperatures and pressure. For example, CPA can accurately predict the MEG and water solubility in the gas phase at low temperatures and over an extended temperature range for the MEG–water–methane ternary system. Some results and comparison with the SRK EoS (using the Huron–Vidal combining rule) are presented in Figure 10.7. The VLLE predictive performance of CPA for several multicomponent mixtures containing MEG, water and hydrocarbons (both aromatic and aliphatics) was presented by Folas *et al.*[9] Predictions for multicomponent systems are based on binary interaction parameters, adjusted solely to binary data. A comparison with the SRK EoS using the Huron–Vidal mixing rule was also presented,[9] since the latter model is a popular (within the oil and gas industry) alternative approach for mixtures containing associating and polar compounds. Even

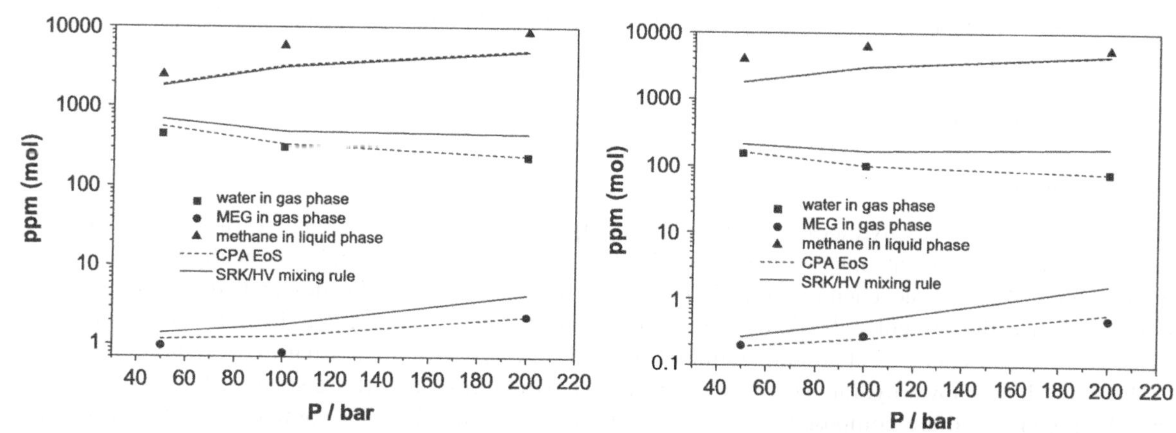

Figure 10.7 *VLE prediction for the MEG–water–methane system at 298.15 K (left) and 278.15 K (right) with the CPA EoS and the SRK EoS using the Huron–Vidal mixing rule (SRK/HV). Reprinted with permission from Fluid Phase Equilibria, High-pressure vapor-liquid equilibria of systems containing ethylene glycol, water and methane: Experimental measurements and modeling by Georgios M. Kontogeorgis, Georgios Folas et al., 251, 1, 52–58 Copyright (2007) Elsevier*

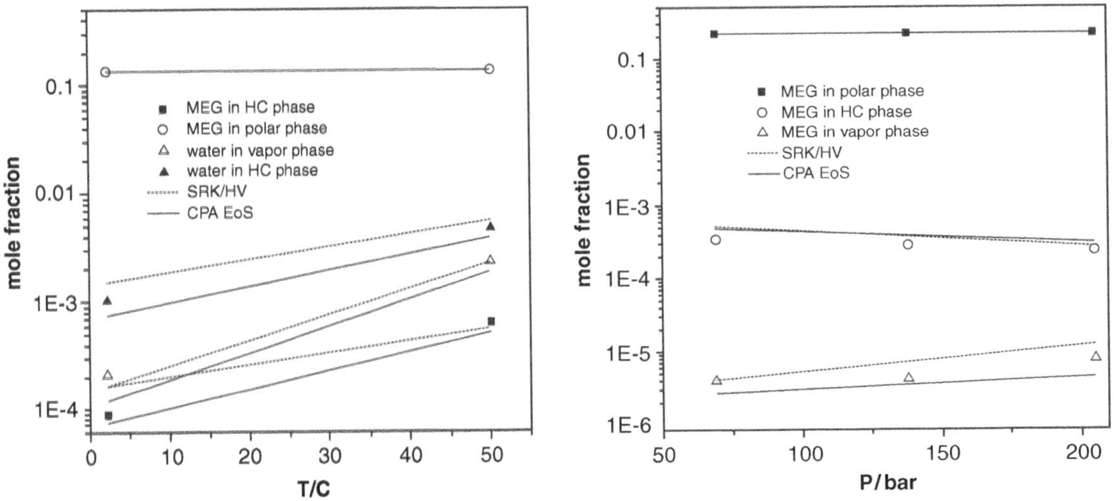

Figure 10.8 Prediction with CPA and SRK/Huron–Vidal (SRK/HV) of VLLE for multicomponent systems containing MEG, water and hydrocarbons. Left: Prediction of MEG solubility in the polar and liquid hydrocarbon phases and water solubility in vapor and liquid hydrocarbon phase with the CPA and SRK/HV EoS, for the quaternary system MEG–water–methane–toluene at 69 bar. Right: Prediction of MEG solubility in the liquid hydrocarbon, polar and gas phase with CPA and SRK/HV, for the five-component system MEG–water–methane–propane–toluene at 310.85 K. Reprinted with permission from Fluid Phase Equilibria, Vapor–liquid, liquid–liquid and vapor–liquid– liquid equilibrium of binary and multicomponent systems with MEG: Modeling with the CPA EoS and an EoS/G^E model by G. K. Folas, G. M. Kontogeorgis et al., 249, 1–2, 67–74 Copyright (2006) Elsevier

though the predictive performance of both models was satisfactory overall, CPA was found to be systematically superior regarding the prediction of the MEG and water solubility in the hydrocarbon phase. Typical results are presented in Figures 10.7 and 10.8. In the case of CPA, explicit accounting for the solvation between glycols or water and the aromatic hydrocarbons is crucial for obtaining satisfactory predictions.

10.3 Gas hydrates

10.3.1 General

Clathrate hydrates or gas hydrates are crystalline complexes, where water molecules are linked through hydrogen bonding and create interstitial cavities that can enclose 'guest' molecules, typically light gases and hydrocarbons. There is no chemical bonding between the host water molecules and the enclosed guest molecule; hence, the gas molecules do not occupy a position in the water lattice. Gas hydrates behave like wet snow; however, they may exist at temperatures below as well as above the normal freezing point of water.

There are three known hydrate structures: I, II and H. Structures I and II are small (12–17 Å) repeating crystal structures and are composed of two types of cages, one small cavity common for both sI and sII hydrates 5^{12} (i.e. pentagonal dodecahedron) and a larger one. For sI hydrates, this is a tetrakaidecahedron, $5^{12}6^2$, consisting of twelve pentagonal and two hexagonal faces), while for sII this is a hexakaidecahedron, $5^{12}6^4$, consisting of twelve pentagonal and four hexagonal faces. The structure H has three different types of cavities: three 5^{12} cavities, two $4^35^66^3$ cavities and one large $5^{12}6^8$ cavity. The $4^35^66^3$ cavity has three square faces, six pentagonal faces, and three hexagonal faces, whereas the $5^{12}6^8$ cavity has twelve pentagonal faces and eight hexagonal faces. The light gases typically form hydrate structure I, with the heavier gases forming structure II. Heavier components can also form hydrates, often with small 'help' gases, such as methane or nitrogen, in the smaller hydrate cavities. The majority of mixtures of interest to the oil and gas industry form structure II.

Although hydrates may be of potential benefit as a hydrocarbon resource and as a means of storing and transporting natural gas, traditionally they are known due to their nuisance behavior. They can form at pressures and temperatures found in natural gas and oil pipelines causing blockages typically during end tail reservoir production or unexpected shutdowns. Together with other solid depositions, such as waxes, and asphaltenes, hydrates are a serious potential problem for offshore technology, possibly the most important one. In order to inhibit the formation of hydrates, the major technologies used in the offshore industry are either continuous injection of chemicals or direct electrical heating, with the first being the preferred solution for long-distance multiphase flowlines, while the most common chemicals are methanol and MEG.

10.3.2 Thermodynamic framework

When computing hydrate equilibria, the values of the fugacities of all components present in the mixture at the different phases need to be calculated, typically vapor (V), aqueous liquid phase (L_W), liquid hydrocarbon phase (L_H), hydrate structure I (H_I), hydrate structure II (H_{II}) and ice (I):

$$f_i^H = f_i^a \tag{10.1}$$

e.g. for water:

$$f_w^H = f_w^{L_H} = f_w^V = f_w^{ice} \tag{10.2}$$

where α denotes the vapor (V), aqueous liquid phase (L_W), liquid hydrocarbon phase (L_H) or ice (I) phase and H is the hydrate phase. The fugacity f_i^a of the component i in the vapor or liquid phase is obtained typically from an EoS, e.g. SRK or PR, according to the following equation:

$$f_i^a = x_i \varphi_i^a P \tag{10.3}$$

where P is the total pressure of the system and x_i is the mole fraction of the component i in the vapor or liquid phase, respectively. For example:

$$f_w^V = y_w \varphi_w^V P \quad \text{or} \quad f_w^L = x_w \varphi_w^L P \tag{10.4}$$

Here x_w, y_w are, respectively, the mole fractions of water in the liquid and vapor phase. Applying classical thermodynamics, the fugacity of ice at the desired pressure can be calculated by correcting the saturation fugacity at the same temperature by the Poynting factor:

$$\ln f_w^{ICE} = \ln f_{w,P\_ref}^{ICE} + \frac{1}{RT} \int_{P\_ref}^{P} V_W^{ICE} dP \tag{10.5}$$

f_{w,P_ref}^{ICE} is the fugacity of water in the ice phase at the reference pressure P_ref (the atmospheric pressure) while V_W^{ICE} in the Poynting term correction (the term with the integral) is the molar volume of ice. Usually this

Poynting term is set to unity for relatively low pressures;[20] however, correlations may also be used (see e.g. Avlonitis[10]). The fugacity of water in the ice phase, f^{ICE}_{w,P_ref}, can be computed via a term involving the heat of fusion ΔH^{fus} and the difference of heat capacity Δc_p^{fus} between solid and liquid, similar to the way the fugacity of the solid phase is calculated in the case of SLE:

$$\ln(f^{ICE}_{w,P\_ref}) = \ln(f^{Lo}_{w}) + \frac{\Delta H^{fus}}{RT_m}\left(1 - \frac{T_m}{T}\right) + \frac{\Delta C_p^{fus}}{R}\left[\frac{T_m}{T} - 1 - \ln\left(\frac{T_m}{T}\right)\right] \tag{10.6}$$

The fugacity of water in the hydrate phase (*H*) is expressed, using standard thermodynamic relationships, as follows, utilizing the empty hydrate (*EH*) as reference state:

$$f_w^H = f_w^{EH} \exp\left(\frac{\mu_w^H - \mu_w^{EH}}{RT}\right) \tag{10.7}$$

where μ_w^H is the chemical potential of water in the hydrate phase, and μ_w^{EH}, f_w^{EH} are the properties (chemical potential and fugacity of water) in the hypothetical empty hydrate phase.

The difference of the chemical potentials (water in hydrate minus water in empty hydrate phase) is typically calculated using the solid solution theory of van der Waals and Platteeuw[11]. The model is essentially a combination of the Gibbs–Duhem equation with an expression for the multicomponent Langmuir adsorption equation (based on fugacities; for further discussion see Chapter 18, Section 18.8.3). The final expressions are:

$$\mu_W^H = \mu_W^{EH} + RT \sum_i v_i \ln\left(1 - \sum_{\text{guests } m} \Theta_{mi}\right) \tag{10.8}$$

where the summation is over all cavity types (1 and 2), R is the universal gas constant, ν_i is the number of type i cavities per water molecule ($\nu_1 = 1/23$ and $\nu_2 = 3/23$ for structure I hydrate and $\nu_1 = 2/17$ and $\nu_2 = 1/17$ for type II hydrates). The occupancy of cavity i by a component m, Θ_{mi}, is calculated from the Langmuir multicomponent expression:

$$\Theta_{mi} = \frac{C_{mi} f_m}{1 + \sum_{\text{guests } k} C_{ki} f_k} \tag{10.9}$$

where f_k is the fugacity of a component k in the equilibrium vapor phase (typically obtained from an EoS) and the summation is over all hydrate-forming components. C_{mi} are the Langmuir isotherm constants, which are assumed to be temperature dependent. The Langmuir multicomponent expression of Equation (10.9) is similar to the one typically presented in the colloid and surface literature, which uses partial pressures instead of fugacities.[12]

10.3.3 Calculation of hydrate equilibria

Equations (10.1)–(10.8) are general thermodynamic expressions. Equation (10.9), the Langmuir theory, is essentially the only new aspect of hydrate equilibria introduced so far. However, prior to performing hydrate equilibria calculations, we need expressions for two properties: (1) the Langmuir adsorption constants; and (2) the properties of the empty hydrate .

The Langmuir constants

A general expression for the Langmuir constants as a function of the intermolecular potential is:

$$C_{mi}(T) = \frac{4\pi}{kT} \int_0^{R_i-\alpha_i} \exp\left(-\frac{W(r)}{kT}\right) r^2 dr \tag{10.10}$$

For the computation of the Langmuir constants, a model potential experienced by the guest molecule in the cage, based on water–guest interactions, is rigorously introduced. The Kihara potential is commonly used:

$$W(r) = 2z\varepsilon\left[\frac{\sigma^{12}}{R^{11}r}\left(\delta^{10} + \frac{a}{R}\delta^{11}\right) - \frac{\sigma^6}{R^5 r}\left(\delta^4 + \frac{a}{R}\delta^5\right)\right] \tag{10.11}$$

where:

$$\delta^N = \frac{1}{N}\left[\left(1 - \frac{r}{R} - \frac{a}{R}\right)^{-N} - \left(1 + \frac{r}{R} - \frac{a}{R}\right)^{-N}\right] \tag{10.12}$$

k is Boltzmann's constant, R_i is the radius of cage i and α_i is the guest core radius. Three parameters are required for the guest molecules (a,σ,ε) which are simultaneously fitted to hydrate equilibrium data, including simple hydrates (i.e. gas–water) and mixtures. Lundgaard and Mollerup[13] suggest that the Kihara parameters can be obtained over the ice–gas–hydrate dissociation line of single gas-water systems (because the influence of the gas phase fugacities is negligible). Such an approach was shown to provide very satisfactory results also for hydrate–vapor–liquid calculations. Several authors[14–16] suggest that including all available data adds to the accuracy of the calculations. A detailed discussion of this issue is provided by Avlonitis[10]. Kihara parameters can be theoretically obtained also from second virial coefficient data, but this approach is not recommended as it results in rather poor results. Alternatively, an empirical expression was suggested by Parrish and Prausnitz[17], which has been successfully used by other authors as well:[14,16]

$$C_{mi}(T) = \frac{A_{mi}}{T}\exp\left(\frac{B_{mi}}{T}\right) \tag{10.13}$$

Equation (10.13) is similar to the one derived from statistical mechanics:[12]

$$C_{mi}(T) = A_{mi}\exp\left(\frac{B_{mi}}{T}\right) \tag{10.14}$$

The properties of empty hydrate: two approaches

There are two approaches for obtaining the properties of empty hydrate. The first one relates the empty hydrate to pure liquid water/ice at some reference conditions and the following expressions are used:

$$f_w^{EH} = f_w^{Ice/L_0} \exp\left(\frac{\Delta\mu_w^{EH-Ice/L_0}}{RT}\right) \tag{10.15}$$

$$\frac{\Delta\mu_w^{EH-Ice}}{RT} = \frac{\Delta\mu_w^0}{RT_0} - \int_{T_0}^{T} \frac{\Delta h_w^{EH-Ice}}{RT^2} dT + \frac{\Delta V_w^{EH-Ice} P}{RT} \tag{10.16}$$

$$\frac{\Delta\mu_w^{EH-L_0}}{RT} = \frac{\Delta\mu_w^0}{RT_0} - \int_{T_0}^{T} \frac{\Delta h_w^{EH-L_0}}{RT^2} dT + \frac{\Delta V_w^{EH-L_0} P}{RT} \tag{10.17}$$

T in the right-hand term of Equation (10.16) may in some cases be replaced by an average temperature $[(T + T_0)/2]$ as mentioned by Pedersen and Christensen[18]. $f_w^{L_0}$ is the fugacity of pure liquid water, $\Delta\mu_w^0$ the chemical potential difference between the empty hydrate and pure liquid water/ice at reference conditions usually taken to be the freezing point of water (273.15 K) and zero pressure, while $\Delta h_w^{EH-L_0}$ and $\Delta V_w^{EH-L_0}$ are the enthalpy and volume differences between the empty hydrate lattice and liquid water, respectively. Finally $\Delta h_w^{EH-L_0}$ is usually expressed by the following equation:

$$\Delta h_w^{EH-L_0} = \Delta h_w^0 + \int_{T_0}^{T} \left[\Delta Cp_w^0 + b(T - T_0)\right] dT \tag{10.18}$$

The method is dependent on the values of $\Delta\mu_w^0$, $\Delta h_w^{EH-L_0}$ and the expression for $\Delta V_w^{EH-L_0}$ used. Depending on these values and the calculated $f_w^{L_0}$ with the EoS used, Langmuir constants need to be optimized for reliable predictions of hydrate formation conditions. Unfortunately, there is no unique set of parameters used and reported values in the literature show a lot of scatter (see e.g. Parrish and Prausnitz[17], Holder *et al.*[19], Sloan[20], John *et al.*[21]). Besides, most of these parameters cannot be measured directly, e.g. $\Delta\mu_w^0$ is related to a hypothetical (empty hydrate) phase.

An alternative expression for the empty hydrate fugacity is a modification[22,23,20,16,24,25] which treats the empty hydrate phase as a solid phase:

$$f_W^{EH} = P_w^{EH} \varphi_w^{EH} \exp \int_{P_w^{EH}}^{P} \frac{V_w^{EH}}{RT} dP \tag{10.19}$$

Typically ϕ_w^{EH} is set to unity while the vapor pressure of the hypothetical empty hydrate structures I and II, P_w^{EH}, is calculated from empirical equations. For example, Folas *et al.*[16] used the expressions given by Sloan[20]:

$$\ln P_w^{EH} = 17.440 - 6003.9/T, \quad \text{for structure I} \tag{10.20}$$

$$\ln P_w^{EH} = 17.332 - 6017.6/T, \quad \text{for structure II} \tag{10.21}$$

Even though the two approaches for estimating the empty hydrate fugacity (Equations (10.15) and (10.19)) appear to be different, the results are similar. As discussed by Folas[26] using the CPA for estimating the

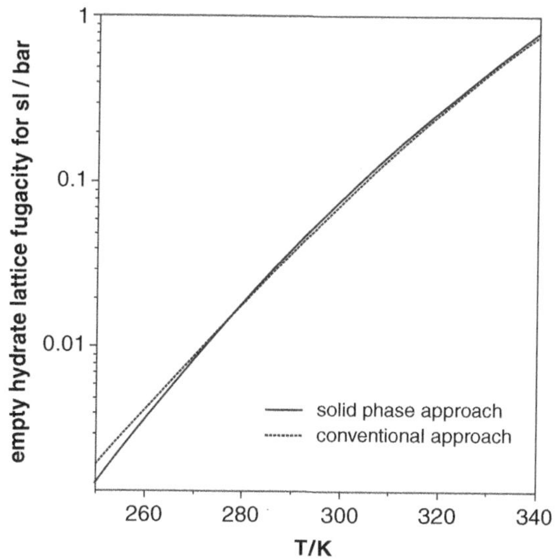

Figure 10.9 Comparisons of calculated fugacities of the hypothetical empty hydrate lattice for sI hydrate at 20 bar using Equation (10.15) (dashed line) or (10.19) (solid line) with the CPA EoS

fugacity f_k of a component k in the equilibrium vapor phase, assuming that Equation (10.19) is valid over the whole temperature and pressure range (and not limited over the *I–H–V* region) does not influence the overall performance of the hydrate model. This is because the calculated fugacity of water in the hypothetical empty hydrate phase (f_w^{EH}), when assuming the validity of Equation (10.19), is similar to the empty hydrate fugacity obtained from Equations (10.15)–(10.17) and CPA for calculating the $f_w^{L_0}$. This is elucidated in Figure 10.9.

Table 10.1 presents optimized values of A_{mi} and B_{mi} for calculating the Langmuir constants of Equation (10.13) using the CPA EoS. Typical temperature dissociation calculations for some simple water–gas mixtures are presented in Figure 10.10. For the dense phases, a zero binary interaction parameter is used for CPA.

Table 10.1 Optimized values of A_{mi} and B_{mi} for calculating the Langmuir constants in the CPA/van der Waals and Platteeuw model (Equations (10.9) and (10.13))

Component	Structure	Small cavity		Large cavity	
		$A_{mi} \times 10^3$ (K/bar)	B_{mi} (K)	$A_{mi} \times 10^3$ (K/bar)	B_{mi} (K)
Methane	I	0.621	2760	421.2	1963
	II	4.05	2637	295.2	900
Ethane	I	0.0	0.0	109.8	2855
	II	0.0	0.0	89.4	3363
Propane	II	0.0	0.0	79.9	3886
Iso-butane	II	0.0	0.0	81.6	4000
n-Butane	II	0.0	0.0	1053	2691
Nitrogen	I	11.64	2159	400.1	1037
	II	7.18	2091	300	1150

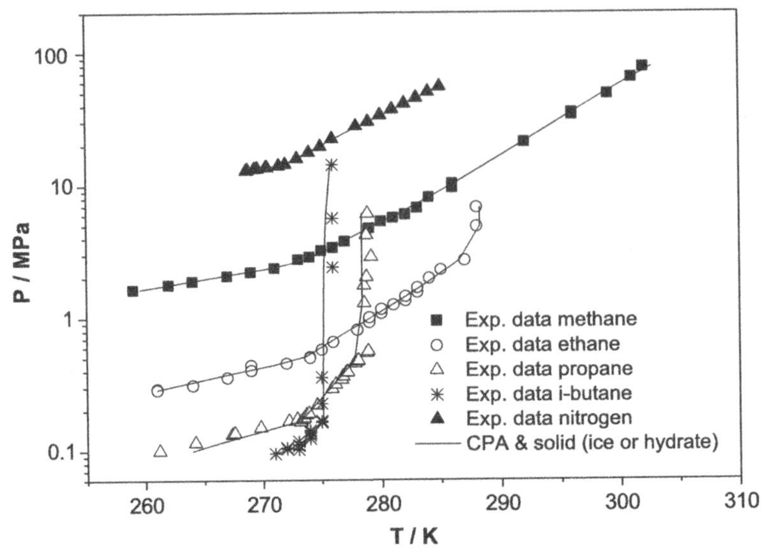

Figure 10.10 *Comparisons of experimental and calculated dissociation temperatures of methane, ethane, propane, iso-butane and nitrogen. Reprinted with permission from Fluid Phase Equilibria, Data and prediction of water content of high pressure nitrogen, methane and natural gas by G. K. Folas, G. M. Kontogeorgis et al., 252, 1–2, 162–174 Copyright (2007) Elsevier*

10.3.4 Discussion

The following points summarize the most important conclusions regarding the phase equilibria of fluid and solid/hydrate phases:

1. If the properties for all fluid phases are obtained from the same EoS and irrespectively of the combining rules used, the approach is thermodynamically consistent. In recent years, several authors have presented successful calculations within this framework.[15,16,27–29] Association theories, which are shown to be very promising tools for phase equilibria, especially provide an appealing method for modeling the fluid phases. Kontogeorgis *et al.*[30] presented hydrate dissociation calculations for simple gas mixtures in the presence of inhibitors. Figures 10.11–10.14 demonstrate the applicability of the CPA EoS for hydrate dissociation predictions for single gas and multicomponent mixtures with and without inhibitors. The important interaction parameter is that between water and the inhibitor. For MEG–water, this is $k_{12} = -0.115$, shown to provide satisfactory results for both VLE and SLE (ECR). For TEG–water, the CR-1 rule with $k_{12} = -0.211$ is used (from VLE, in absence of SLE data). For water–methanol, ECR with two values for the interaction parameter is tested ($k_{12} = -0.09$ from VLE and -0.153 from SLE), with the latter performing slightly better. Overall, the performance of the model is satisfactory in most of the cases over an extended pressure range. Equally good results have been presented by Haghighi *et al.*[61] for hydrate curves calculated with CPA using MEG as the gas hydrate inhibitor. Somewhat inferior results are obtained when very high amounts of inhibitor are used.
2. Mainly due to the inability of traditional EoS, e.g. SRK and PR (with the vdW1f mixing rules), to satisfactorily represent phase equilibria of aqueous mixtures, especially in the presence of inhibitors, combined EoS–activity coefficient approaches were previously employed for gas hydrate mixtures containing polar inhibitors (now being phased out and replaced by EoS/G^E or association models).

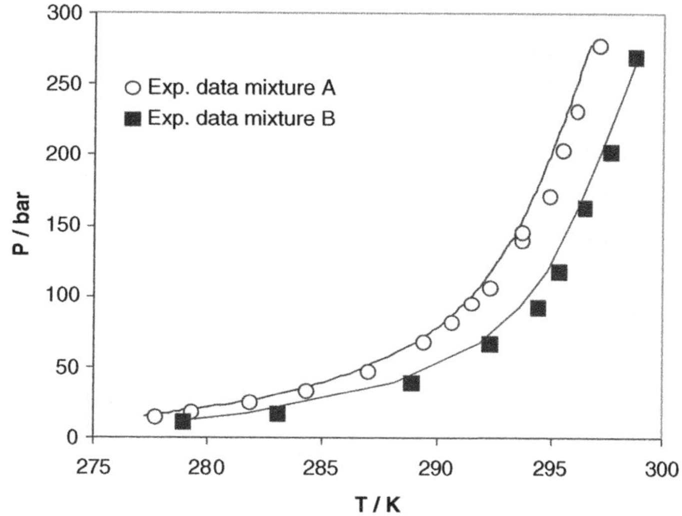

Figure 10.11 CPA predictions of hydrate dissociation temperatures for two natural gas mixtures (mixture A: 93.20 mol% CH_4, 4.25 mol% C_2H_6, 1.61 mol% C_3H_8, 0.51 mol% CO_2, 0.43 mol% N_2; mixture B: 86.41 mol% CH_4, 6.47 mol% C_2H_6, 3.57 mol% C_3H_8, 0.99 mol% i-C_4H_{10}, 1.14 mol% n-C_4H_{10}, 0.64 mol% N_2, 0.78 mol% C_{5+}) using the CPA EoS for the fluid phases. Experimental data are from Wilcox et al., *Ind. Eng. Chem.*, **1941**, 33, 662

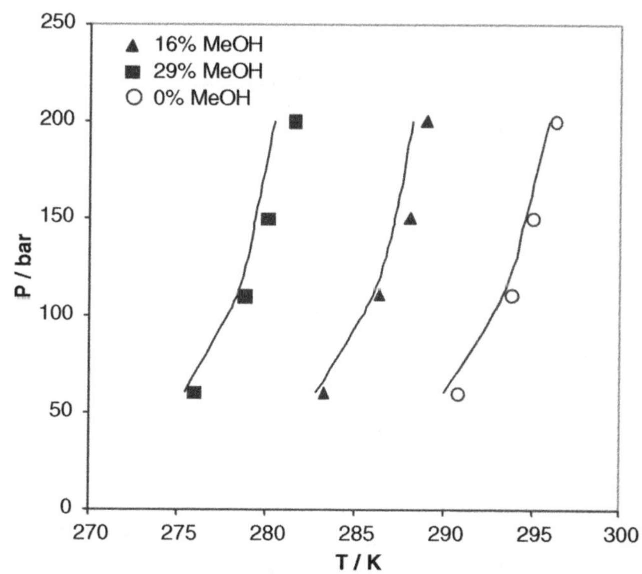

Figure 10.12 CPA predictions of hydrate dissociation temperatures for a natural gas mixture (93.20 mol% CH_4, 4.25 mol% C_2H_6, 1.61 mol% C_3H_8, 0.51 mol% CO_2, 0.43 mol% N_2) and different amount of methanol (MeOH) using the CPA EoS for the fluid phases. The interaction parameter between water and methanol is −0.153 taken from SLE data. All other interaction parameters are provided in Appendix B on the companion website at www.wiley.com/go/Kontogeorgis. Experimental data are from Ng et al., *Fluid Phase Equilib.*, **1987**, 36, 99

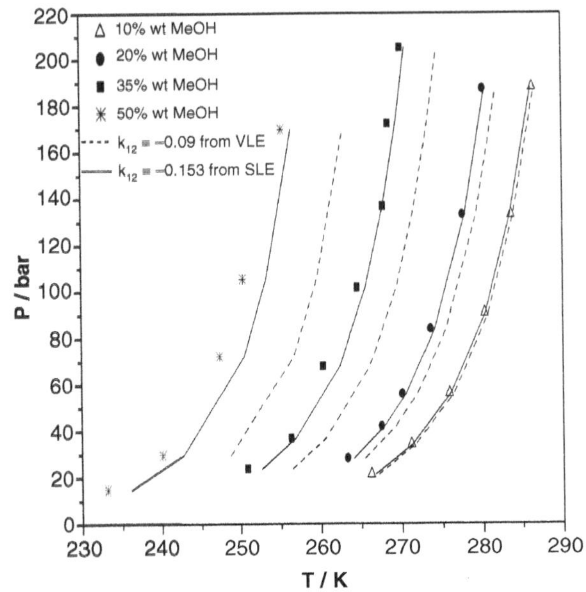

Figure 10.13 CPA prediction of methane hydrate formation in the presence of methanol (MeOH) as inhibitor. Results are shown with two values of k_{12} for water–methanol, one optimized from VLE (−0.115) and one optimized from SLE data (−0.153). ECR is used for water–methanol. The interaction parameter for methanol–methane is $k_{12}=0.0134$ (from VLE at 293 K). The water–methane interaction parameter is equal to zero. Reprinted with permission from Fluid Phase Equilib., by G. M. Kontogeorgis et al., 261, 1–2, 205. Copyright (2007) Elsevier

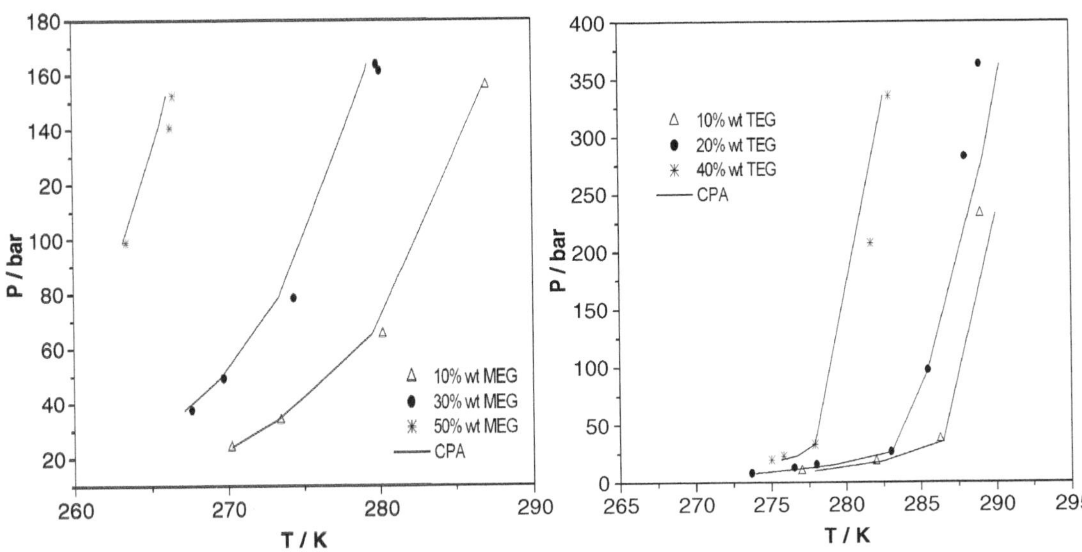

*Figure 10.14 Prediction of methane hydrate formation with CPA in the presence of MEG and TEG as inhibitor. Experimental data are from Robinson and Ng, J. Can. Pet. Technol., **1986**, 26 and Ross and Toczylkin, J. Chem. Eng. Data, **1992**, 37, 488. Reprinted with permission from Fluid Phase Equilibria, Modelling of associating mixtures for applications in the oil & gas and chemical industries by G. M. Kontogeorgis, G. Folas et al., 261, 1–2, 205–211 Copyright (2007) Elsevier*

According to this combined approach, cubic EoS are used for the vapor phase while an activity coefficient model is used for the liquid phase. In such a framework, it is well known[13] that SRK provides more accurate calculations of fugacities than PR and it should be preferred for representing the vapor phase. UNIQUAC, as also discussed in Chapter 5, has been successfully used by several researchers[31,14,32] for describing the liquid phase properties. The use of an activity coefficient model results in an accurate calculation of the water activity and of the way the water activity is influenced by the presence of inhibitors including salts. The role of water activity in hydrate calculations is obvious, since Equation (10.16) can also be written as (in equilibrium with an aqueous phase):

$$\frac{\Delta\mu_w^{EH-Ice/liq}}{RT} = \frac{\Delta\mu_w^0}{RT_0} - \int_{T_0}^{T} \frac{\Delta h_w^{EH-Ice/liq}}{RT^2} dT + \frac{\Delta V_w^{EH-Ice/liq} P}{RT} - \ln(x_w \gamma_w) \tag{10.22}$$

In the absence of inhibitors, Henry's law constants can be used to account for the influence of the gas solubility in the aqueous phase.[14]

3. Focusing on EoS, Hendriks *et al.*[15] provided a discussion on the importance of reproducing the vapor pressure of water accurately, over the whole temperature range (i.e. from the triple to the critical point), including the sublimation pressure of ice. Models based on associating theories are privileged in this case (not to an extreme degree of accuracy because of the temperature range), since the pure component parameters are usually fitted to vapor pressure data of pure liquid water. It is true, however, that Hendriks *et al.*'s approach provides even more accurate results for association models as well as for the CPA EoS when a Matthias–Copeman expression is used for the energy term (E. Solbraa, 2007, personal communication).
4. Hydrate phases (as well as other solid phases such as waxes) can be incorporated in the multiphase flash algorithm, as for example discussed by Michelsen and Mollerup[33], where a description of the method for tracing a phase boundary to any number of coexisting phases is presented. This provides a consistent way of incorporating hydrate calculations into industrial simulations.

Even though some of the models in the literature are shown to provide satisfactory results for hydrate dissociation calculations for natural gas mixtures, often in the presence of inhibitors or salts, there are several challenges. The shortcomings of the solid solution theory of van der Waals and Platteeuw, the role of a non-existing phase (the hypothetical empty hydrate phase) in the calculations, the use of Kihara parameters to model both hydrate structures and the fact that results are poor when those parameters are obtained from second virial coefficients are some of the limitations. Still, the most important element in hydrate equilibria calculations which can influence the calculations is the choice of the liquid phase model, as the calculations are greatly controlled by the change in the fugacity of water in the presence of inhibitors.

10.4 Gas phase water content calculations

The water content of natural gas creates problems during transportation and processing, and thus it is of great importance to natural gas processors to calculate accurately the equilibrium water content of the gas. Pipeline conditions are usually in the temperature range of 50 to −20 °C and a pressure range 50–250 bar, although it is occasionally of interest to predict water content at both higher and lower temperatures. GERG-water is the ISO standard model for calculations of natural gas water content, suggested by the European Gas Research Group GERG (Group Européen de Recherche Gazière; ISO 18453). GERG-water is the Peng–Robinson EoS, with the classical expression for the co-volume parameter and the van der Waals one-fluid mixing rules.

However, in order to ensure an accurate calculation of water vapor pressure above ice and liquid, the following energy term is used for water:

$$a = a_c a(T_r) \tag{10.23a}$$

$$\sqrt{a(T_r)} = 1 + A_1(1-\sqrt{T_r}) + A_2(1-\sqrt{T_r})^2 + A_3(1-\sqrt{T_r})^4 \tag{10.23b}$$

where a_c is given in Table 3.1.

The energy term of Equation (10.23b) is divided into two parts. In the temperature range of 223.15–273.16 K the energy term is fitted to vapor pressure data above ice (with parameters $A_1 = 0.106\,025$, $A_2 = 2.683\,845$ and $A_3 = -4.75638$), while in the temperature range of 273.16–313.15 K vapor pressure data over liquid water were used (with parameters $A_1 = 0.905\,436$, $A_2 = -0.213781$ and $A_3 = 0.260\,05$). Finally, for the water–methane, water–ethane and water–CO_2 systems, a temperature-dependent binary interaction parameter is used, which is of the form:

$$k_{ij}(T) = k_{ij,0} + k_{ij,1}\left(\frac{T}{273.15} - 1\right) \tag{10.24}$$

All required coefficients for the calculation of the binary interaction parameter from Equation (10.24) as well as the critical component properties that should be used are provided in ISO 18453. The model parameters are tailor-made specifically for water content calculations in natural gas limited to temperatures below 313 K, while the solid phases are treated as a 'pseudo' liquid phases.

Folas *et al.*[16] recently tested an alternative to the GERG-water approach, by using CPA to describe the dense phases (i.e. the gas phase or the liquid water phase) coupled with a solid phase model, which can be an ice or a hydrate model. The fugacity of ice can be expressed by Equations (10.5) and (10.6), while equation (10.7) is used for the hydrate fugacity. Gas phase water content calculations were performed for simple gases (nitrogen and methane) as well as for selected natural gas mixtures. Nitrogen was chosen because it only forms hydrates at very low temperatures in the actual pressure ranges, and methane because it is the dominant component in natural gas and also forms hydrates at high temperatures (e.g. at 25 °C). Overall the performance of this approach was found to be similar to the GERG-water EoS both for the single gases and for the mixtures, especially when ice or hydrate is the stable solid phase in equilibrium with gas (Figures 10.15 and 10.16). For high-pressure measurements, where water condenses (i.e. VLE), the CPA model performs systematically better compared to the GERG-water EoS. Finally, the modeling approach presented by Folas *et al.*[16] enables, also, differentiation between the various heavy phases (i.e. liquid, ice or hydrate phase) that could equilibrate with the gas phase, as for example demonstrated in Figure 10.15. The thermodynamically stable phase is the phase that corresponds to the higher temperature for the given water content.

10.5 Mixtures with acid gases (CO_2 and H_2S)

Mixtures with acid gases (CO_2 and H_2S) are very important in petroleum applications as well as in the chemical industry. EoS/G^E models have been partially successful for vapor–liquid equilibria, but less satisfactory results are obtained for VLLE and especially for multicomponent systems. While the results with the EoS/G^E approaches are predictive using UNIFAC-type models in the mixing rules, for mixtures with CO_2, H_2S, water and low-molecular-weight alcohols, essentially all parameters are obtained from the corresponding binaries and the 'group contribution character' of these models is lost.

There have been several modeling attempts in the literature with SAFT-type approaches, including variations of the CPA EoS. Pfohl *et al.*[34] used a PR–CPA for CO_2–ethanol/cresols, but instead of the van der Waals one fluid mixing rules with a single interaction parameter, they used a mixing rule for the energy

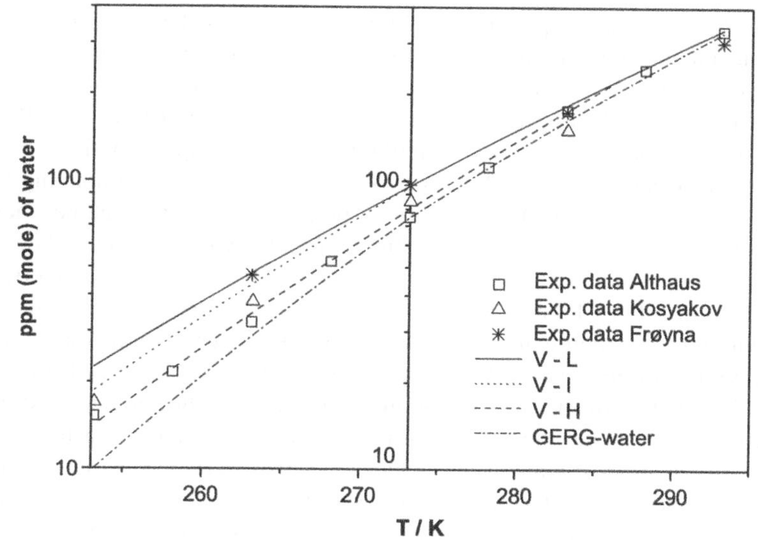

Figure 10.15 *Prediction with CPA and the GERG-water EoS of the water content in methane in equilibrium with stable (hydrate) and metastable (ice, liquid water) phases at 100 bar. The GERG-water EoS gives only the stable equilibrium. Reprinted with permission from Fluid Phase Equilibria, Data and prediction of water content of high pressure nitrogen, methane and natural gas by G. K. Folas, G. M. Kontogeorgis et al., 252, 1–2, 162–174 Copyright (2007) Elsevier*

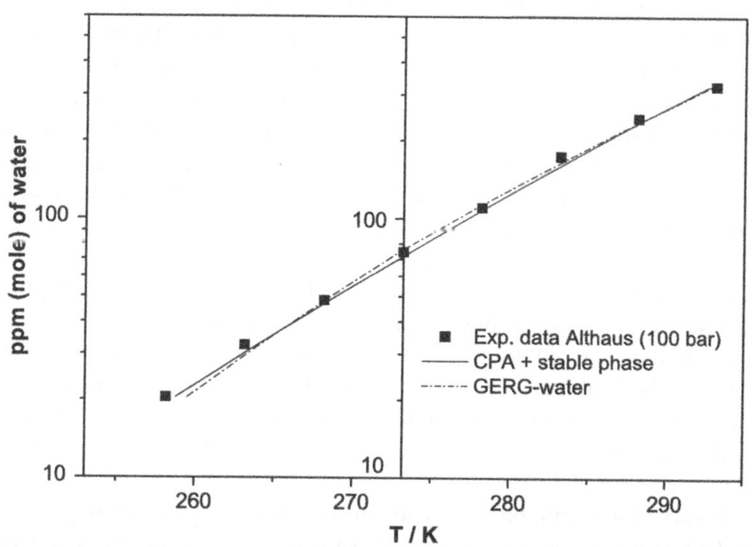

Figure 10.16 *Prediction with CPA and the GERG-water EoS of the water content in equilibrium with the most stable phase for a gas mixture (mol%: 93.22 C_1, 2.91 C_2, 0.71 C_3, 1.94 N_2, 0.85 CO_2, 0.09 i-C_4, 0.04 n-C_4, 0.11 C_{5+}) at 100 bar. Reprinted with permission from Fluid Phase Equilibria, Data and prediction of water content of high pressure nitrogen, methane and natural gas by G. K. Folas, G. M. Kontogeorgis et al., 252, 1–2, 162–174 Copyright (2007) Elsevier*

parameter containing two interaction parameters. Moreover, they used a more complex expression for the radial distribution function for mixtures compared to the average mixture radial distribution function of CPA. The results were not very satisfactory. Perakis *et al.*[35] and Voutsas *et al.*[36] have also employed a PR–CPA, which resembles the CPA EoS (as presented in Chapter 9). The PR–CPA of Perakis–Voutsas[35,36] employs the vdW1f mixing rules and the same simplified mixture radial distribution function as SRK CPA (Chapter 9). These authors have applied PR–CPA to several VLE systems containing CO_2: CO_2–water–ethanol, CO_2–water–acetic acid and CO_2 with various polar compounds (acetone, diethyl ether). They obtained good results by treating CO_2 as a self-associating molecule (with four sites). They used the geometric mean rule for both the cross-association energy and cross-association volume parameters, thus a combining rule which is different from both ECR and CR-1 rules, discussed in Chapter 9. A four-site CO_2 molecule was used by Button and Gubbins[56] in their extension of SAFT to CO_2–water–alkanolamines. The alkanolamines have either five (MEA) or six (DEA) sites. A large k_{ij} parameter ($= -0.24$) is needed for CO_2–water in the temperature range 423–523 K, but the article was more focused on the ternary mixture rather than the CO_2–water binary. Ruffine *et al.*[37] have used CPA (as presented here, see equations in Chapter 9) for VLE of binary H_2S mixtures (with water, methanol and alkanes). Satisfactory results are obtained when H_2S is considered to be a self-associating molecule (3B scheme) and using yet another choice for the combining rules for the cross-associating parameters: the geometric mean rule for the cross-association energy and the arithmetic mean rule for the cross-association volume. Folas *et al.*[38] modeled CO_2–water VLE using $k_{12} = -0.066$ optimized from data within the range 298–318 K as an intermediate step in their subsequent study of multicomponent mixtures with CO_2 (and N_2), water and dimethyl ether. Such a simplified approach provided satisfactory representation of CO_2–water VLE at all temperatures and in general, as also shown by Kontogeorgis *et al.*,[39] CPA correlates well the CO_2 solubility in water over extensive pressure ranges.

All of the above investigations for mixtures containing CO_2 and H_2S contain an inherent degree of approximation as in reality CO_2 and H_2S are non-self-associating molecules, but on the other hand they can solvate with water and the lower molecular weight alcohols, especially methanol and ethanol. Indeed such solvation (Lewis acid (LA) and Lewis base (LB) interactions) may be quite strong in some cases. According to Hyatt,[40] for example, CO_2 behaves as a Lewis base (LB) similar to ethers or ethylacetate, i.e. it is expected to have a basic (β) Kamlet–Taft parameter around 0.45–0.5. It is thus expected to interact via LA–LB interactions with hydrogen bonding compounds, especially those having a high value of the Kamlet–Taft acidic parameter (α), e.g. water (1.17, 0.18), methanol (0.93, 0.62) and ethanol (0.83, 0.77), where the two values indicate the acid–basic solvatochromic (Kamlet–Taft) parameters, respectively. From the modeling point of view, not explicitly accounting for the solvation yields a poor description of the water solubility in CO_2 as shown in Figures 10.17a and b (see also Kontogeorgis *et al.*)[39].

Thus, the solvation in mixtures between CO_2–water and H_2S–water must be explicitly accounted for. Instead of assuming association sites for these acid gases, Kontogeorgis *et al.*[41] showed that satisfactory phase equilibrium calculations are obtained with CPA for mixtures of acid gases with water, alcohols–-glycols and hydrocarbons (both binary and ternary) when the gases are modeled as non-self-associating molecules, but the solvation is explicitly accounted for via the modified CR-1 rule. A typical result is presented in Figures 10.17a and 17.17b for the water solubility in CO_2 at very high pressures as well as the CO_2 mole fraction in water.

For the CO_2–water binary system emphasis was put on simultaneously correlating both the water solubility in CO_2 as well as the CO_2 solubility in the aqueous phase while maintaining the temperature independency of the interaction parameters. In order to correlate both solubilities simultaneously, a binary interaction parameter k_{12} as well as a cross-association volume parameter ($\beta^{A_iB_j}$) should be used. As can be seen in Figure 10.17a, the correlation of the water solubility in CO_2 is very satisfactory over an extended temperature range. The performance of the model regarding the solubility of CO_2 in the aqueous phase (Figure 10.17b) is

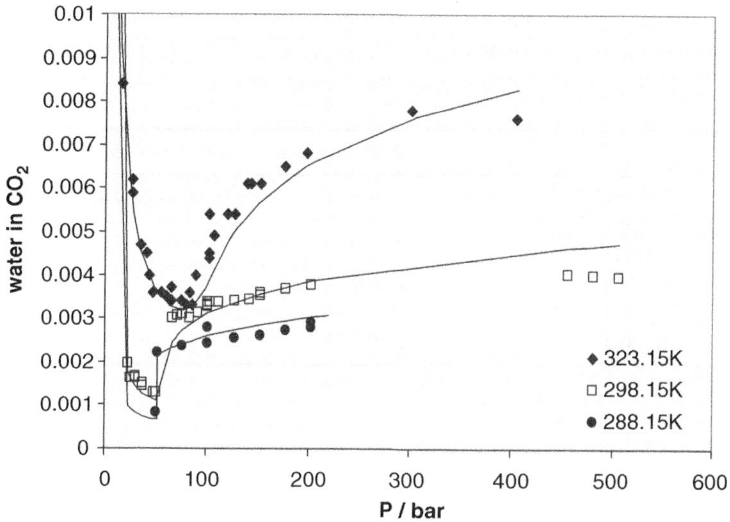

Figure 10.17a Correlation with CPA of the water mole fraction in CO_2 at three different temperatures using the 4C scheme. The binary interaction parameters are $k_{12} = 0.06$ and $\beta^{A_iB_j} = 0.075$. The minimum in the water solubility is reproduced only when solvation is explicitly accounted for (using the modified CR-1 rule). Experimental data from Gillespie and Wilson, GPA Research Report RR-48, 1982

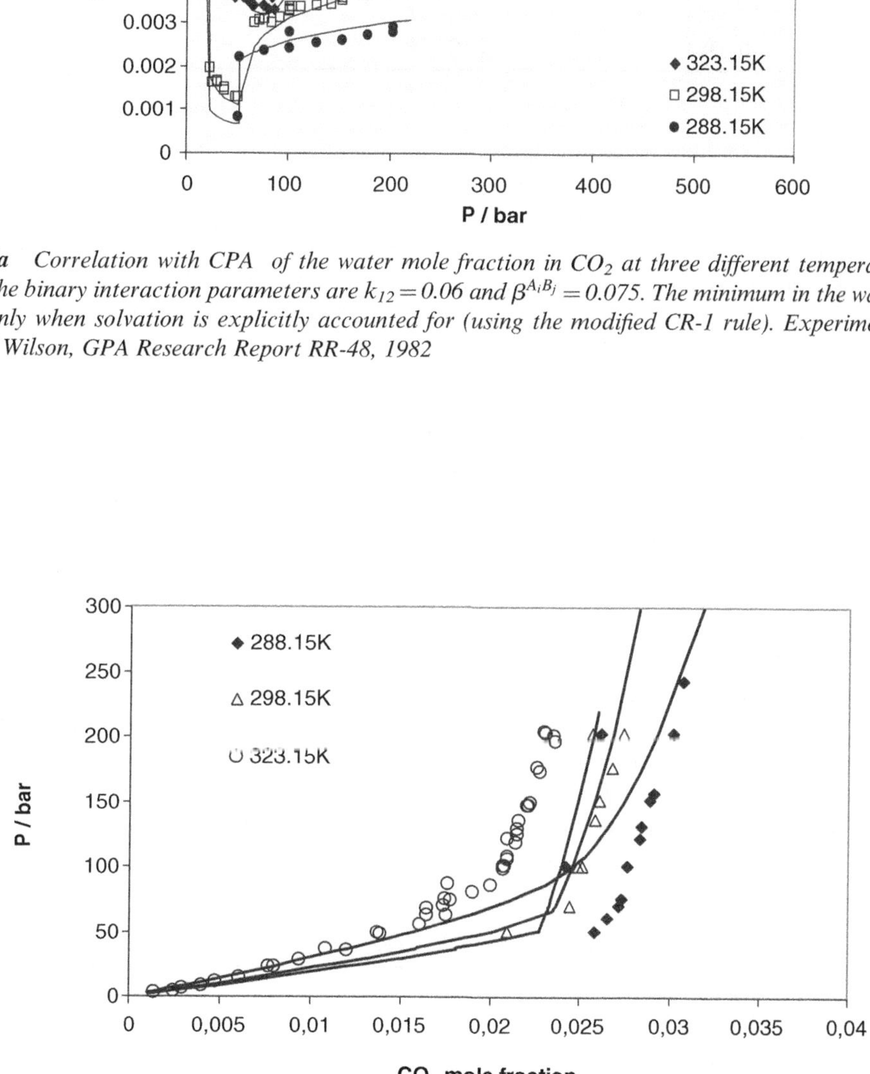

Figure 10.17b CO_2 mole fraction in water phase with the CPA EoS. The binary interaction parameters are $k_{12} = 0.06$ and $\beta^{A_iB_j} = 0.075$ (same as in Figure 10.17a). Experimental data are from King et al., *J. Supercrit. Fluids*, **1992**, 4 and Tekenouchi and Kennedy, *Am. J. Sci.*, **1964**, 262

Table 10.2 *Optimum k_{ij} values for CPA and average deviations in composition and pressure for CO_2–methanol. Experimental data are from Hong and Kobayashi, Fluid Phase Equilib., **1988**, 41, 269 Ohgaki and Katayama, J. Chem. Eng. Data, **1976**, 21, 53 and Yoon et al., J. Chem. Eng. Data, **1993**, 38, 53,* $\Delta P\% = \frac{1}{N}\sum_i \left|\frac{P_i^{exp} - P_i^{cal}}{P_i^{exp}}\right|$

T (K)	k_{ij}	$\beta^{A_iB_j}$	ΔP (%)	$100\Delta y$
230.15	−0.0051	—	4.78	0.008
298.15	−0.0051	—	5.67	0.239
313.15	−0.0051	—	3.33	0.354
230.15	0.0384	0.0199 8	1.3	0.003
298.15	0.0384	0.0199 8	3.32	0.123
313.15	0.0384	0.0199 8	2.55	0.184

a bit inferior, especially at higher pressures; however, the model still provides acceptable deviations from the experimental data for the pressure and temperature range considered.

Mixtures with alcohols have also been investigated. It can be concluded that solvation is important in CPA modeling for mixtures of CO_2 or H_2S with small alcohols as well, particularly methanol and ethanol. Solvation is of less or no significance for acid gases/heavy alcohols (propanol and beyond) systems. Tables 10.2 and 10.3 present some typical results for CO_2–methanol and H_2S–methanol binary systems respectively and the importance of explicitly accounting for the solvation. A typical result is also illustrated in Figure 10.18 for the H_2S–methanol system.

The importance and actual role of solvation in mixtures of acid gases and water/polar chemicals is not fully understood, especially for multicomponent mixtures (CO_2–water–alcohol–hydrocarbons and CO_2–water–hydrocarbons). We summarize some conclusions from recent studies. For mixtures of CO_2–water–methanol and CO_2–water–ethanol, satisfactory results are obtained when the solvation between CO_2 and alcohol is accounted for. Surprisingly, the solvation of CO_2–water does not seem to have much influence on the prediction of multicomponent phase equilibria.

Mixtures of CO_2 or H_2S and glycols can be well correlated without the need explicitly to account for the solvation. Figures 10.19a, b and 10.20 show some examples, for the CO_2–MEG and CO_2–DEG systems, as well as for the multicomponent mixture CO_2–water–MEG.

Regarding the CO_2–glycol mixtures, it is concluded that CPA using a single binary interaction parameter provides satisfactory correlation of the solubility of CO_2 in the glycol phase. The solubility of glycols in the gas phase is, however, underestimated when only k_{12} is used in the physical term. Use of an extra cross-association volume parameter ($\beta^{A_iB_j}$) enables us to correlate simultaneously both solubilities (i.e. explicitly accounting for solvation phenomena between CO_2 and glycols). However, even when emphasizing

Table 10.3 *Optimum k_{ij} values for CPA and average deviations in composition and pressure for H_2S–methanol. Experimental data are from Leu et al., Fluid Phase Equilib., **1992**, 72, 163. ΔP% as in Table 10.2*

T (K)	k_{ij}	$\beta^{A_iB_j}$	ΔP (%)	$100\Delta y$
298.15	−0.0261	—	4.32	1.63
348.15	−0.0396	—	6.54	2.57
398.15	−0.1011	—	4.24	4.12
298.15	0.0000	0.05	1.27	1.61
348.15	0.0798	0.14	1.72	2.43
398.15	−0.0500	0.09	1.26	4.56

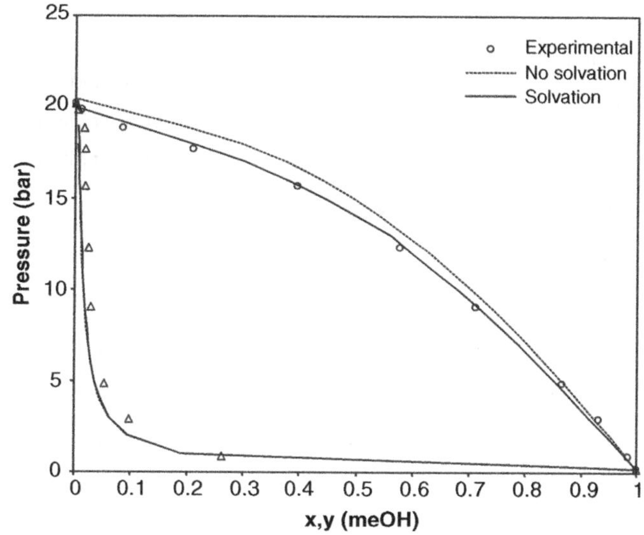

Figure 10.18 VLE with CPA for the system methanol (meOH) and H_2S at 298.15 K. Experimental data are from Leu et al., Fluid Phase Equilib., 1992, 72, 163. The calculations are without solvation using $k_{12} = -0.026$ (dashed line) and with solvation $k_{12} = 0.0$, $\beta^{A_iB_j} = 0.05$ (solid line). Reprinted with permission from Institut Francais du Petrole, Solvation Phenomena in Association Theories with applications to oil & gas and chemical industries by G. M. Kontogeorgis, G. K. Folas, N. Muro-Sune, F. Roca Leon, M. L. Michelsen, Oil & Gas Science and Technology-Rev. IFP, 63, 305–319 Copyright (2008) Institut Francais du Petrole

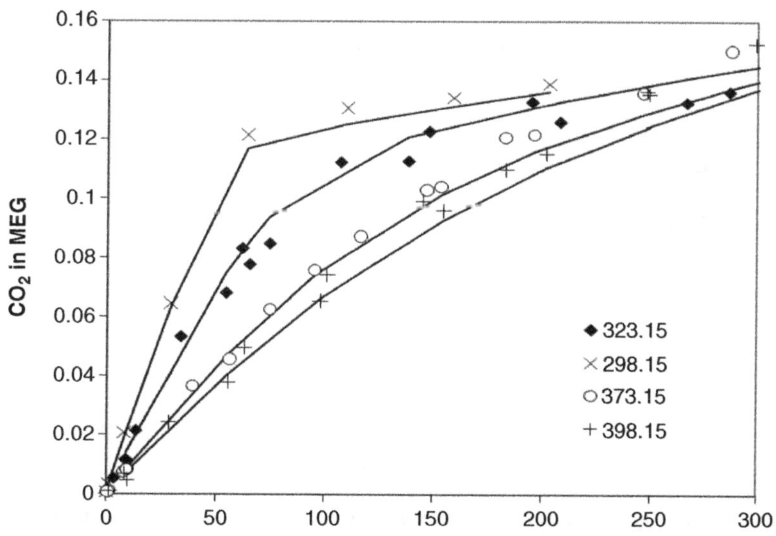

Figure 10.19a VLE of CO_2–MEG with CPA in the temperature range 323.15–398.15 K using a single interaction parameter, $k_{12} = 0.05$. Experimental data are from Jou et al., Chem. Eng. Commun., **1990**, 87, 223 and Zheng et al., Fluid Phase Equilibria, **1999**, 155, 277

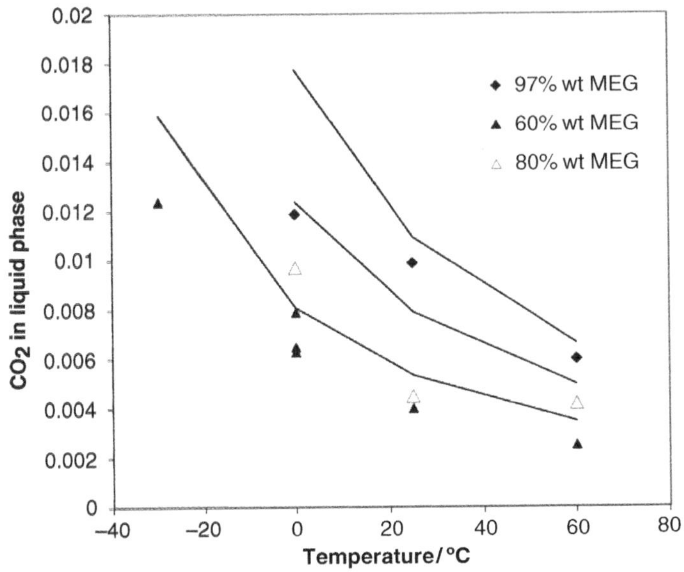

Figure 10.19b *Prediction with CPA of the CO_2 solubility in the aqueous phase for the ternary system MEG–water– CO_2 at different temperatures and MEG concentrations (expressed in wt%). The pressure is 5 bar. Experimental data are from the Gas Processors Association (GPA), Research Report 183, Tulsa, USA*

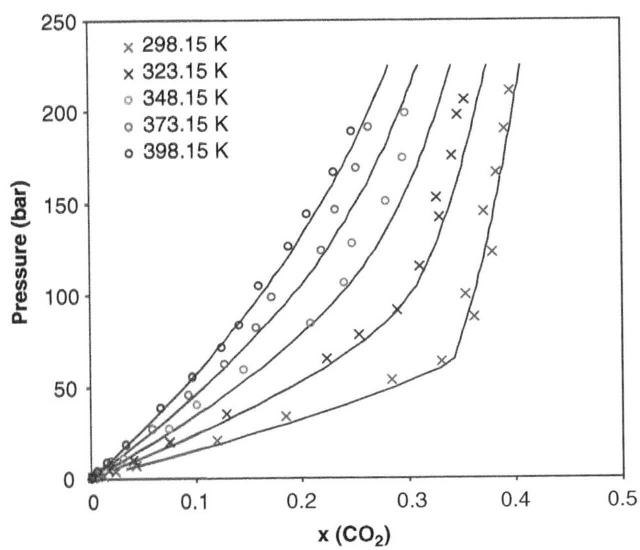

Figure 10.20 *VLE prediction ($k_{12} = 0$) of CO_2–DEG using CPA. Experimental data are from F.-Y. Jou, F.D. Otto, A.E. Mather, Fluid Phase Equilib.,* ***2000****, 175, 53. Reprinted with permission from Institut Francais du Petrole, Solvation Phenomena in Association Theories with applications to oil & gas and chemical industries by G. M. Kontogeorgis, G. K. Folas, N. Muro-Sune, F. Roca Leon, M. L. Michelsen, Oil & Gas Science and Technology-Rev. IFP, 63, 305–319 Copyright (2008) Institut Francais du Petrole*

only the CO_2 solubility in MEG (Figure 10.19a) the prediction using CPA of the ternary MEG–water–CO_2 system over an extended temperature range and for different wt% MEG is very satisfactory, as can be seen in Figure 10.19b. Equally good results are obtained for H_2S–DEG. CO_2–phenol VLE can also be correlated well with CPA using a single k_{12} (0.068 at 398.15 K) without explicitly accounting for solvation effects. The same is the case for CO_2–methylphenol ($k_{12} = 0.06$ at 328.15 K).

An alternative approach for modeling acid gas mixtures with a CPA-type EoS has been recently presented by Perfetti *et al.*[57,58] They developed a polar CPA EoS, based on a non-primitive MSA theory, with the polar term expressed via the Padé approximant and the integrals given by Gubbins and Twu[59]. The polar term is similar to the ones used in polar SAFT EoS, which are discussed in Chapter 13.

Perfetti *et al.*[57,58] applied the polar CPA (termed CPAMSA) to water–H_2S, water–methane and water–CO_2 mixtures. Water is treated as both polar and associating compound (six pure compound parameters), H_2S is treated as polar non-associating compound (four pure compound parameters), whereas methane and CO_2 are treated as 'inert' (non-polar, non-self-associating compounds) and have thus only three parameters (critical properties and acentric factors). For mixtures, mixing rules are needed for the dipole moment appearing in the MSA term. Some preliminary first results are promising (especially for the pure compounds) but improvements will be needed. For example, there is a need in the current version of CPAMSA for two different sets of interaction parameters for water–methane (for the liquid and gas phase), which are moreover temperature dependent and have been correlated using a three-parameter temperature correlation. The best results are obtained for water–H_2S for which the interaction parameters are also small, about 0.02 in the 60–180 °C range (roughly temperature independent), while higher interaction parameter values are needed for water–CO_2, which are clearly dependent on temperature as they vary from 0.05 (at 50 °C) up to −0.2 (300 °C). No results were presented with temperature-independent interaction parameters.

A common denominator in all the aforementioned discussions is that capturing the temperature dependency of the solubilities involving water and acid gases is still very much a challenging task, especially when this is attempted with temperature-independent parameters. On the other hand, the SRK EoS with the Huron–Vidal mixing rule and seven adjustable NRTL parameters can satisfactorily describe solubilities in water–CO_2–methane mixtures over extensive temperature ranges, as shown by Austegard *et al.*[60] The authors have also considered SRK with vdW1f mixing rules and Huron–Vidal/NRTL with five adjustable parameters, and it is clear that substantial improvements are obtained when the number of parameters is increased from five to seven. The error of CO_2 solubility in water is reduced to half and the representation of solubilities in water–methane is also improved. The SRK with the classical (vdW1f) mixing rules cannot be used, yielding errors over 100% in many cases.

10.6 Reservoir fluids

A characterization procedure has been developed for CPA[42] so that the model can be applied to reservoir fluids. Calculations presented for reservoir fluids/water and reservoir fluids/water/methanol glycols show promising results. However, data are available for very few systems, especially gas condensates, and more data are required for an extensive investigation. Another issue to be investigated is whether standard characterization procedures previously developed for SRK are sufficient or whether new procedures like the one developed by Yan *et al.*[42] should be preferred. The next section presents the C_{7+} characterization which has been developed for CPA[42] together with some results for gas condensates with water and water–methanol.

10.6.1 Heptanes plus characterization

General

The heptanes plus (C_{7+}) fraction in any petroleum reservoir fluid generally contains numerous heavy compounds which cannot be identified by the existing analytical technology. Even if they could be identified, their critical properties and other EoS parameters are often unknown. Compared to the compounds lighter than C_7 which are often called 'well-defined' since there is enough information for direct application of an EoS model, the ill-defined C_{7+} fraction must be characterized before applying any EoS. A C_{7+} fraction is usually analyzed either by true boiling point distillation or by gas chromatography. The results are expressed by a number of cuts defined by boiling point (T_b) range. The specific gravity (*SG*) and molecular weight (*MW*) for each cut of the whole C_{7+} fraction are typically the only additional information known. The purpose of C_{7+} characterization is to estimate the composition of the C_{7+} fraction in terms of an appropriate number of pseudo components, and the EoS model parameters associated with these components.

For C_{7+} in reservoir fluids, two characterization procedures are often used: the method by Pedersen *et al.*[43] and that by Whitson *et al.*[44] Both methods have three steps in common:

1. Determination of the detailed molar composition in the C_{7+} fraction such as the mole fraction of single carbon number (*SCN*) components.
2. Estimation of EoS model parameters for *SCN* components.
3. Lumping of *SCN* components into a few pseudo components.

In principle, both methods can be used directly with CPA for reservoir fluids applications. However, since CPA uses T_{cm}, P_{cm} and ω_m (see Chapter 9) different from the experimental T_c, P_c and ω, it is more consistent to estimate T_{cm}, P_{cm} and ω_m for *SCN* components in the C_{7+} fraction instead of using the existing correlations for T_c, P_c and ω. A modified version of the method by Pedersen *et al.*[43] has been developed.[42] The modification is made only in the second step, where a set of correlations for T_{cm}, P_{cm} and ω_m is proposed. The first and the last characterization steps are kept the same, i.e. using the exponential decay molar distribution for *SCN* components and the equal mass lumping with mass-weighted averaging. It should be noted that the correlations proposed below can also be used with other characterization methods such as the one by Whitson *et al.*[44]

The new correlations

The T_{cm}, P_{cm} and ω_m values used in developing the new correlations are based on the DIPPR database. Data for 28 *n*-alkanes and 316 other hydrocarbons are used. Numerous correlations of the critical properties have been proposed using two of the three parameters T_b, *SG* and *MW* as input information, but Yan *et al.*[42] have used the two-step perturbation method[45] in order to develop correlations for the CPA parameters. In the perturbation method, the properties of a fraction with given boiling point T_b and specific gravity *SG* are estimated in two steps: first, the properties of the normal alkane at the same T_b are calculated; second, the properties of the fraction are estimated by using the specific gravity difference $\Delta SG = SG - SG_0$ as the perturbation parameter, where SG_0 is the specific gravity of the normal alkane. The correlations constructed using the perturbation method usually give smaller overall deviations. Besides, they can guarantee very accurate description of the properties of *n*-alkane and thus those of paraffinic oils.

At first, the correlations for normal alkanes have been developed (from propane to *n*-C_{36}):

$$T_{cm0} = \frac{(1885.45947 + 0.222337924\,T_b)T_b}{950.853406 + T_b} \tag{10.25}$$

$$\ln P_{cm0} = -4.052\,825\,58 \times 10^{-12} T_b^4 + 8.761\,257\,76 \times 10^{-9} T_b^3 - 7.457\,830\,4 \times 10^{-6} T_b^2 - 1.099\,729\,89 \times 10^{-4} T_b + 4.160\,592\,95 \tag{10.26}$$

$$\omega_{m0} = \exp\left(\frac{-2553.065\,3 + 3.684\,18\,T_b}{608.722\,6 + T_b}\right) \tag{10.27}$$

In the above equations, T_b and T_{cm0} are in K, and P_{cm0} is in bar. The subscript 0 refers to the properties of n-alkanes. Soave's correlation[46] is used to calculate the specific gravity for n-alkanes:

$$SG_0 = (1.8\,T_b)^{1/3}\left(11.7372 + 3.336 \times 10^{-3} T_b - 976.3 T_b^{-1} + 3.257 \times 10^5 T_b^{-2}\right)^{-1} \tag{10.28}$$

For the perturbation step, ΔSG is used to account for the aromaticity of the fraction. Aromatic compounds generally have higher densities than normal alkanes at the same T_b. And as a general trend, the larger ΔSG is, the higher are the differences between T_{cm} and T_{cm0}, and between P_{cm} and P_{cm0}. The final equations proposed by Yan *et al.*[42] are:

$$T_{cm}/T_{cm0} = (1 - 12.069\,079\,5\Delta SG + 22.862\,656\,2\Delta SG^2 + 89.711\,581\,8\Delta SG^3)/(1 - 12.631\,138\,6\Delta SG + 30.677\,947\,2\Delta SG^2 + 62.469\,896\,5\Delta SG^3) \tag{10.29}$$

$$\ln(P_{cm}/P_{cm0}) = \Delta SG[-677.989\,269 + (76624.406 - 29811.874\,9/SG)\Delta SG]/(1 + 10949.220\,2\Delta SG + 28099.157\,3\Delta SG^2) \tag{10.30}$$

The CPA acentric factor ω_m is back-calculated by matching the T_b of the fraction. The direct vapor pressure calculation procedure proposed by Soave[47] can be used, which does not need any iteration. Equation (10.27) is used only if T_b exceeds T_c for extremely heavy compounds.

10.6.2 Applications of CPA to reservoir fluids

The characterization method presented in Section 10.6.1 has been tested for CPA against the reservoir fluids data from Pedersen *et al.*[48,49], where water and methanol concentrations are also measured. Both reservoir fluids have a high percentage of methane (69% in the 2001 article and 73% in the 1996 article).

The calculation results are shown in Figures 10.21–10.24, together with the calculation results from Pedersen *et al.*[49] using the SRK EoS with the Huron–Vidal mixing rule (SRK/HV), which uses three interaction parameters (five parameters if they are assumed to be temperature dependent). CPA captures correctly both the trend and the order of magnitude of experimental data even in the case of pure prediction, i.e. when $k_{ij} = 0$ is used. The water content in the gas phase increases significantly with temperature and decreases slightly with pressure. CPA with zero k_{ij} accurately predicts the water content and performs better than the SRK/HV with fixed temperature-independent interaction parameters (NRTL's g_{ij} parameters). The methane content in water increases with temperature and pressure. CPA with zero k_{ij} captures the trend but overpredicts the values. If, however, temperature-dependent k_{ij} parameters are used, the CPA prediction for multicomponent mixtures is in good agreement with the experimental data and comparable to the results of SRK/HV which uses more adjustable parameters.

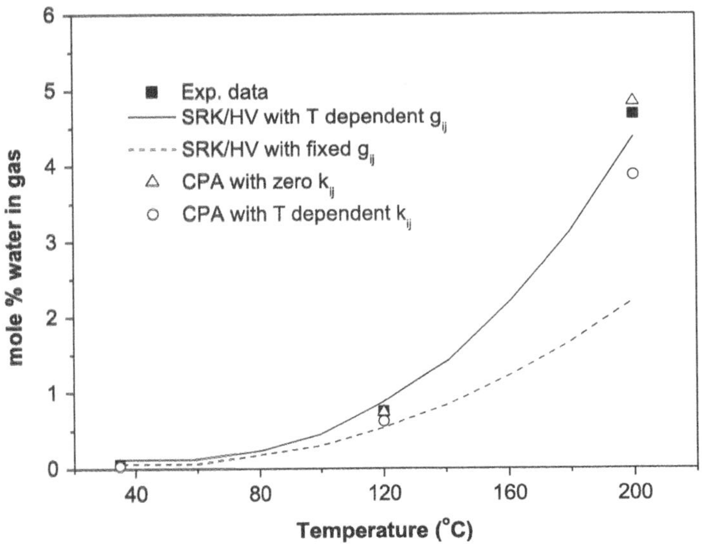

Figure 10.21 Water content in the hydrocarbon phase for a gas condensate mixture at 1000 bar with CPA and SRK/HV. Experimental data are from Pedersen et al.[49] Reprinted with permission from Fluid Phase Equilibria, Application of the CPA equation of state to reservoir fluids in presence of water and polar chemicals by G. M. Kontogeorgis, E.S. Stenby and W. Yan, 276, 1, 75–85 Copyright (2009) Elsevier

An example is shown in Table 10.4. The CPA prediction is based only on the k_{ij} between methanol and water, and between methane and water. The results by Pedersen *et al.*[48] using the SRK/HV EoS are also provided in Table 10.4. In general, both models perform well and show satisfactory agreement with the experimental data. Nevertheless, CPA seems to be better in reproducing the methanol concentration in the

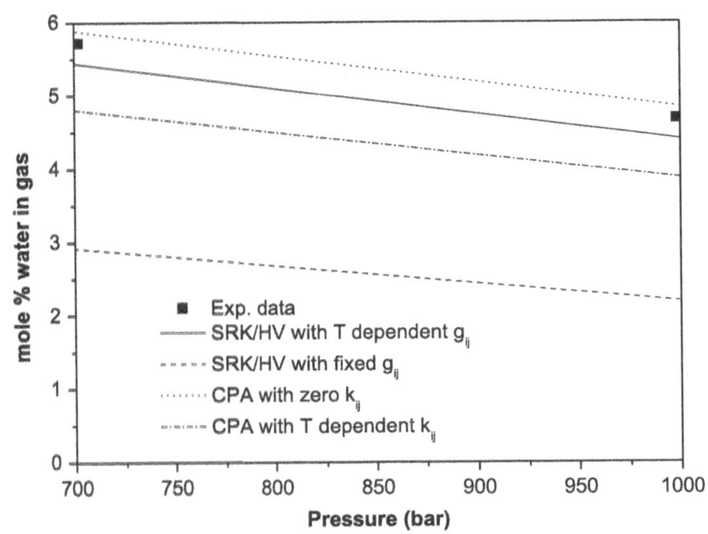

Figure 10.22 Molar concentration of water in hydrocarbon phase for a gas condensate mixture at 200 °C with CPA and SRK/HV. Experimental data are from Pedersen et al.[49] Reprinted with permission from Fluid Phase Equilibria, Application of the CPA equation of state to reservoir fluids in presence of water and polar chemicals by G. M. Kontogeorgis, E.S. Stenby and W. Yan, 276, 1, 75–85 Copyright (2009) Elsevier

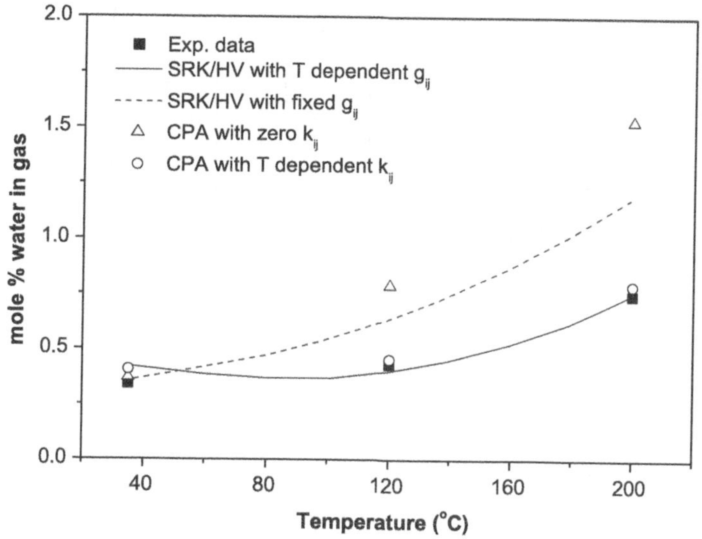

Figure 10.23 *Molar concentration of methane in water phase (gas condensate mixture) at 1000 bar with CPA and SRK/HV. Experimental data are from Pedersen et al.*[49] *Reprinted with permission from Fluid Phase Equilibria, Application of the CPA equation of state to reservoir fluids in presence of water and polar chemicals by G. M. Kontogeorgis, E.S. Stenby and W. Yan, 276, 1, 75–85 Copyright (2009) Elsevier*

hydrocarbon liquid phase. Yan *et al.*[42] present results with CPA also for two synthetic natural gases with different concentrations of water and methanol as well as for LLE of two condensate oils with MEG and water. The performance is satisfactory. Unfortunately, there is a lack of data for mixtures with heavier oils containing water and methanol or other inhibitors and it is therefore as yet not possible to evaluate broader the proposed characterization method for reservoir fluids and polar chemicals.

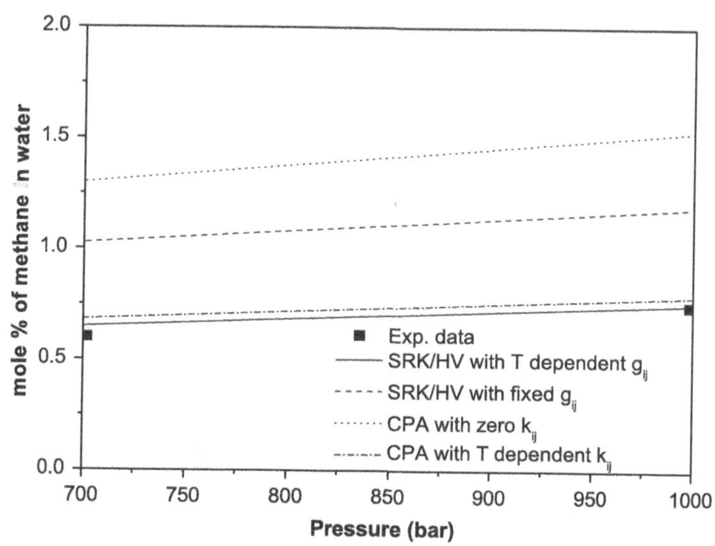

Figure 10.24 *Molar concentration of methane in water (gas condensate mixture) at 200°C with CPA and SRK/HV. Experimental data are from Pedersen et al.*[49] *Reprinted with permission from Fluid Phase Equilibria, Application of the CPA equation of state to reservoir fluids in presence of water and polar chemicals by G. M. Kontogeorgis, E.S. Stenby and W. Yan, 276, 1, 75–85 Copyright (2009) Elsevier*

Table 10.4 *Phase compositions (mol%) for mixture 2 in the article by Pedersen et al.[48] calculated with the CPA and SRK/HV EoS. Reprinted with permission from Fluid Phase Equilibria, Application of the CPA equation of state to reservior fluids in presence of water and polar chemicals by G. M. Kontogeorgis, E.S. Stenby and W. Yan, 276, 1, 75–85 Copyright (2009) Elsevier. The values in parentheses are deviations (%) from the experimental data*

Component	Feed	Hydrocarbon liquid phase			Hydrocarbon vapor phase			Aqueous phase		
		Exp.	CPA	SRK/HV	Exp.	CPA	SRK/HV	Exp.	CPA	SRK/HV
		$P = 60.3$ bar and $T = 3.6\,°C$								
Res. fluid	84.76	99.799	99.800 (0.0)	99.675 (−0.1)	99.957	99.940 (0.0)	99.936 (0.0)	—	0.602	—
Methanol	2.99	0.201	0.183 (−8.7)	0.288 (43.3)	0.0429	0.0452 (5.3)	0.0441 (2.8)	18.68	19.185 (2.7)	18.210 (−2.5)
Water	12.25	—	0.0163	—	—	0.0152	—	81.32	80.213 (−1.4)	81.400 (0.1)
		$P = 149.9$ bar and $T = 7.7\,°C$								
Res. fluid	64.04	99.812	99.780 (0.0)	99.741 (−0.1)	99.812	99.922 (0.0)	99.909 (0.0)	—	0.876	—
Methanol	6.72	0.188	0.199 (5.9)	0.214 (13.8)	0.188	0.0638 (−7.1)	0.0636 (−7.4)	18.68	18.388 (−1.6)	18.440 (−1.3)
Water	29.22	—	0.0204	—	—	0.0140	—	81.32	80.736 (−0.7)	80.930 (0.5)

10.7 Conclusions

The CPA EoS satisfactorily describes glycol–aromatic hydrocarbon phase equilibria (VLLE) using the modified CR-1 rule. Moreover, satisfactory results are obtained for the prediction of multicomponent glycol–water–BTEX (and other hydrocarbons) VLLE including high-pressure water–MEG–methane equilibria. The predictions of the multicomponent mixtures are based solely on binary interaction parameters regressed from binary data. The modified CR-1 rule is also useful for modeling acid gases (CO_2, H_2S) in mixtures with water and methanol (but its significance is smaller for mixtures of acid gases with heavier alcohols or glycols).

CPA has been extended to oil-specific applications, gas hydrate equilibria (where CPA is combined with the van der Waals–Platteeuw model) and reservoir fluids (using a recently developed characterization method). The results are satisfactory, but in the case of reservoir fluids only data for a few gas condensates in contact with water and polar chemicals were available for testing the model. In the case of gas hydrates, new parameters for the Langmuir coefficients have been optimized for use with the CPA model. Satisfactory results are obtained for gas hydrate dissociation/formation curves in all cases (single and mixed gases, condensates, and in the presence of inhibitors such as methanol and glycols).

Some first applications of CPA to electrolytes[51–54] and asphaltenes[55] have appeared recently. For extension to electrolytes, an additional electrostatic mean spherical approximation (MSA) term has been added in some studies[51–53] and the Debye–Hückel term in others.[54] In both the work of Lin *et al.*[51] and Wu and Prausnitz[52] the association term was only used for water and no mixed solvents were investigated. Lin *et al.*[51] showed that, in addition to the activities of solutes and solvent in such solutions, the density of the solution could be well represented by this electrolyte CPA. The extension of this electrolyte CPA to a wider temperature range and to electrolyte solutions containing non-electrolytes such as monoethylene glycol and methanol has not yet been attempted. Wu and Prausnitz[52] have, on the other hand, applied their PR–CPA to high-pressure methanol solubility in water–NaCl mixtures as well as water–NaCl mean ionic activity coefficients. The electrolyte extensions of CPA will be discussed together with other electrolyte equations of state in Chapter 15.

References

1. S.O. Derawi, G.M. Kontogeorgis, E.H. Stenby, T. Haugum, A.O. Fredheim, *J. Chem. Eng. Data*, **2002**, *47(2)*, 169.
2. G.K. Folas, G.M. Kontogeorgis, M.L. Michelsen, E.H. Stenby, E. Solbraa, *J. Chem. Eng. Data*, **2006**, *51(3)*, 977.
3. G.K. Folas, O.J. Berg, E. Solbraa, A.O. Fredheim, G.M. Kontogeorgis, M.L. Michelsen, E.H. Stenby, *Fluid Phase Equilib.*, **2007**, *251(1)*, 52.
4. S.O. Derawi, M.L. Michelsen, G.M. Kontogeorgis, E.H. Stenby, *Fluid Phase Equilib.*, **2003**, *209(2)*, 163.
5. G.K. Folas, G.M. Kontogeorgis, M.L. Michelsen, E.H. Stenby, *Ind. Eng. Chem. Res.*, **2006**, *45(4)*, 1527.
6. S.O. Derawi, M.L. Michelsen, G.M. Kontogeorgis, E.H. Stenby, *Ind. Eng. Chem. Res.*, **2003**, *42(7)*, 1470.
7. G.K. Folas, J. Gabrielsen, M.L. Michelsen, E.H. Stenby, G.M. Kontogeorgis, *Ind. Eng. Chem. Res.*, **2005**, *44*, 3823.
8. J.M. Prausnitz, R.N. Lichtenthaler, E.G. de Azevedo, *Molecular Thermodynamics of Fluid-Phase Equilibria* (3rd edition). Prentice Hall International, **1999**.
9. G.K. Folas, G.M. Kontogeorgis, M.L. Michelsen, E.H. Stenby, *Fluid Phase Equilib.*, **2006**, *249(1–2)*, 67.
10. D. Avlonitis, *Chem. Eng. Sci.*, **1994**, *49*, 1161.
11. J.H. van der Waals, J.C. Platteeuw, *Adv. Chem. Phys.*, **1959**, *2*, 1.
12. A.A. Shapiro, E.H. Stenby, Multicomponent adsorption: approaches to modeling adsorption equilibria. *Encyclopedia of Surface and Colloid Science* (2nd edition). Taylor & Francis, 2002, p. 4180.
13. L. Lundgaard, J.M. Mollerup, *Fluid Phase Equilib.*, **1991**, *70*, 199.

14. J. Munck, S. Skjold-Jørgensen, P. Rasmussen, *Chem. Eng. Sci.*, **1988**, *43*, 2661.
15. E.M. Hendriks, B. Edmonds, R.A.S. Moorwood, R. Szczepanski, *Fluid Phase Equilib.*, **1996**, *117*, 193.
16. G.K. Folas, E.W. Froyna, J. Lovland, G.M. Kontogeorgis, E. Solbraa, *Fluid Phase Equilib.*, **2007**, *252*, 162.
17. W.R. Parrish, J.M. Prausnitz, *Ind. Eng. Chem. Process Des. Dev.*, **1972**, *11*, 26.
18. K.S. Pedersen, P.L. Christensen, *Phase Behavior of Petroleum Reservoir Fluids*. Taylor & Francis, 2006.
19. G.D. Holder, G. Corbhand, K.D. Papadopoulos, *Ind. Eng. Chem. Fundam.*, **1980**, *79*, 282.
20. E.D. Sloan, *Clathrate Hydrates of Natural Gases* (2nd edition). Marcel Dekker, 1998.
21. V.T. John, K.D. Papadopoulos, G.D.A Holder, *AlChE J.*, **1985**, *31*(2), 252.
22. E.D. Sloan, F.M. Khoury, R. Kobayashi, *Ind. Eng. Chem. Fundam.*, **1976**, *15*, 318.
23. H.-J. Ng, D.B. Robinson, *Ind. Eng. Chem. Fundam.*, **1980**, *19*, 33.
24. J.B. Klauda, S.I. Sandler, *Ind. Eng. Chem. Res.*, **2000**, *39*, 3377.
25. J.B. Klauda, S.I. Sandler, *J. Phys. Chem. B*, **2002**, *106*, 5722.
26. G.K. Folas, Modelling of complex mixtures containing hydrogen bonding molecules. PhD Thesis, Technical University of Denmark, 2006.
27. B.Edmonds, R.A.S. Moorwood, R.A Szczepanski, *Fluid Phase Equilib.*, **1999**, *158–160*, 481.
28. J. Madsen, K.S. Pedersen, M.L. Michelsen, *Ind. Eng. Chem. Res.*, **2000**, *39*, 1111.
29. X.-S. Li, H.-J. Wu, P. Englezos, *Ind. Eng. Chem. Res.*, **2006**, *45*, 2131.
30. G.M. Kontogeorgis, G.K. Folas, N. Muro-Suñé, N. von Solms, M.L. Michelsen, E.H. Stenby, *Fluid Phase Equilib.*, **2007**, *261*(*1–2*), 205.
31. F.E. Anderson, J.M. Prausnitz, *AIChE J.*, **1986**, *32*, 1321.
32. Y. Du, T.-M. Guo, *Chem. Eng. Sci.*, **1990**, *45*(*4*), 893.
33. M.L. Michelsen, J.M. Mollerup, *Thermodynamic Models: Fundamentals and Computational Aspects*. Tie-Line Publications, 2004.
34. O. Pfhol, A. Pagel, G. Brunner, *Fluid Phase Equilib.*, **1999**, *157*, 53.
35. C. Perakis, E. Voutsas, K. Magoulas, D. Tassios, *Fluid Phase Equilib.*, **2006**, *243*(*1–2*), 142.
36. E.C. Voutsas, C. Perakis, G. Pappa, D.P. Tassios, *Fluid Phase Equilib.*, **2007**, *261*, 343.
37. L. Ruffine, P. Mougin, A. Barreau, *Ind. Eng. Chem. Res.*, **2006**, *45*, 7688.
38. G.K. Folas, G.M. Kontogeorgis, M.L. Michelsen, E.H. Stenby, *Ind. Eng. Chem. Res.*, **2006**, *45*(*4*), 1516.
39. G.M. Kontogeorgis, M.L. Michelsen, G. Folas, S. Derawi, N. von Solms, E.H. Stenby, *Ind. Eng. Chem. Res.*, **2006**, *45*(*14*), 4869.
40. J.A. Hyatt, *J. Org. Chem.*, **1984**, *49*, 5097.
41. G.M. Kontogeorgis, G.K. Folas, N. Muro-Suñé, F. Roca Leon, M.L. Michelsen, *Oil Gas Technol – Rev. IFP*, **2008**, *63*, 305.
42. W. Yan, G.M. Kontogeorgis, E.H. Stenby, *Fluid Phase Equilib.*, **2009**, *276*(*1*), 75.
43. K.S. Pedersen, P. Thomassen, Aa. Fredenslund, *Characterization of gas condensate mixtures*. In: L.G. Chorn, G.A. Mansoori, Eds, Advances in Thermodynamics, Vol. *1*, C_{7+} Fraction Characterization. Taylor & Francis, 1989.
44. C.H. Whitson, T.F. Anderson, I. Søreide, *C_{7+} characterization of related equilibrium fluids using the gamma distribution*. In: L.G. Chorn, G.A. Mansoori, Eds, Advances in Thermodynamics, Vol. *1*, C_{7+} Fraction Characterization. Taylor & Francis, 1989.
45. C.H. Twu, *Fluid Phase Equilib.*, **1984**, *16*, 137.
46. G. Soave, *Fluid Phase Equilib.*, **1998**, *143*, 29.
47. G. Soave, *Fluid Phase Equilib.*, **1986**, *31*, 203.
48. K.S. Pedersen, M. Michelsen, A. Fredheim, *Fluid Phase Equilib.*, **1996**, *126*, 13.
49. K.S. Pedersen, J. Milter, C. Rasmussen, *Fluid Phase Equilib.*, **2001**, *189*, 85.
50. D. Jerinic, J. Schmidt, K. Fischer, L. Friedel, *Fluid Phase Equilib.*, **2008**, *264*, 253.
51. Y. Lin, K. Thomsen, J.-C. de Hemptinne, *AIChE J.*, **2007**, *53*(*4*), 989.
52. J. Wu, J.M. Prausnitz, *Ind. Eng. Chem. Res.*, **1998**, *37*, 1634.
53. R. Inchekel, J.-C. De Hemptinne, W. Furst, *Fluid Phase Equilib.*, **2008**, *271*, 19.
54. H. Haghighi, A. Chapoy, B. Tohidi, *Ind. Eng. Chem. Res.*, **2008**, *47*, 3983.
55. M.A. Fahim, S.I. Andersen, *SPE 93517*, **2005**.

56. J.K. Button, K.E. Gubbins, *Fluid Phase Equilib.*, **1999**, *158–160*, 175.
57. E. Perfetti, R. Thiery, J. Dubessy, *Chem. Geol.*, **2008**, *251*, 58.
58. E. Perfetti, R. Thiery, J. Dubessy, *Chem. Geol.*, **2008**, *251*, 50.
59. K.E. Gubbins, C.H. Twu, *Chem. Eng. Sci.*, **1978**, *33*, 863.
60. A. Austegard, E. Solbraa, G. De Koeijer, M.J. Mølnvik, *Trans. IChemE, Part A, Chem. Eng. Res. Des.*, **2006**, *84*, 781.
61. H. Haghighi, A. Chapoy, R. Burgess, B. Tohidi, *Fluid Phase Equilib.*, **2009**, *276*, 24.
62. M.P. Breil, I. Tsivintzelis, G.M. Kontogeorgis, **2009**, Modelling of phase equilibria with CPA using the homomorph approach (submitted).

11

Applications of CPA to Chemical Industries

11.1 Introduction

In order to apply a thermodynamic model widely in the chemical industry, the model must be capable of being used for many different types of chemicals (e.g. organic acids, alcohols, amines), some of which are polar and associating, while others may only exhibit polar character (such as esters, ethers, ketones). Multifunctional chemicals, e.g. alkanolamines and glycolethers, as well as chemicals with ionic and oligomeric character, are also important in many applications such as removal of CO_2 from coal-fired power plants, as antifoamers, scale and wax inhibitors, corrosion inhibitors, etc. The phase equilibria of such mixtures may be complex and include both cross-association and solvation phenomena. Application of a model naturally requires that pure compound and mixture parameters are available. There are a vast number of potential chemicals and in this chapter we discuss applications of CPA which have appeared in the literature, some very recently, while Chapter 12 presents a methodology on how association models can be applied to new compounds not previously investigated. All pure compound and binary interaction parameters of CPA are available in Appendices A and B on the companion website at www.wiley.com/go/Kontogeorgis. The applications of CPA related to the chemical industry, which will be discussed in this chapter, are:

1. Aqueous mixtures of heavy alcohols.
2. Mixtures containing inert, associating and polar non-associating compounds like ethers, esters, amines and ketones.
3. Mixtures with organic acids with emphasis on acetic acid. Other organic acids will also be considered (formic, propanoic).
4. Multifunctional chemicals such as glycolethers and alkanolamines.

Other recent applications involve aromatic acids[1,2] and heavy esters of relevance to biodiesel.[3] Characteristic results from these applications will be illustrated. Comparisons of CPA to 'its' base model, the SRK EoS, will be presented for characteristic mixtures. Some comparisons of CPA with SRK using the classical and the Huron–Vidal mixing rules have been recently presented.[9,29]

Thermodynamic Models for Industrial Applications: From Classical and Advanced Mixing Rules to Association Theories
Georgios M. Kontogeorgis and Georgios K. Folas

11.2 Aqueous mixtures with heavy alcohols

Simultaneous correlation of VLE and LLE for cross-associating systems can be difficult,[4] as can be seen in Figure 11.1 for water–butanol. CPA with the CR-1 rule is used. When the interaction parameter is fitted to LLE, the prediction of VLE is not very satisfactory and vice versa. When the ECR combining rule is used instead of CR-1, the model cannot simultaneously correlate both solubilities in the LLE phase diagram. Typical behavior of the ECR rule for heavy alcohols–water mixtures is presented in Figure 11.2 for the octanol–water mixture. LLE investigations of various water and heavy alcohol systems have resulted in the same conclusion, i.e. that CPA/CR-1 performs better than CPA/ECR for such mixtures over extensive temperature ranges. This can be seen also from the results shown in Figure 11.3 for water with pentanol. Notice that the simultaneous description of VLE and LLE for water–pentanol is more satisfactory than for water–butanol. Similar results are obtained for water–pentanol with PC–SAFT.[5]

Still, CPA can describe, at least qualitatively correctly, with the same parameters VLE, LLE and SLE for the binary system water–butanol. ECR is employed in the calculations of Figure 11.4 using a binary interaction parameter obtained either from VLE (−0.096) or from SLE data (−0.115). These results clearly demonstrate not only the sensitivity of the LLE correlation, but also the challenges related to simultaneous correlation of different types of phase equilibria, especially when LLE is also included. In conclusion:

- An interaction parameter is almost always needed for satisfactory correlation of cross-associating mixtures, and typically negative interaction parameter values are required (this indicates that the degree of cross-association is somewhat underestimated).
- ECR is recommended for water with light alcohols (methanol, ethanol and propanol) and MEG, but the CR-1 rule should be preferred for aqueous mixtures with heavy alcohols and glycols (DEG, TEG, etc.). In many cases, for symmetric systems, the difference between the two combining rules is small, but as the asymmetry increases CPA with the CR-1 rule in most cases performs better than CPA with ECR, especially for LLE of water and heavy alcohol mixtures.

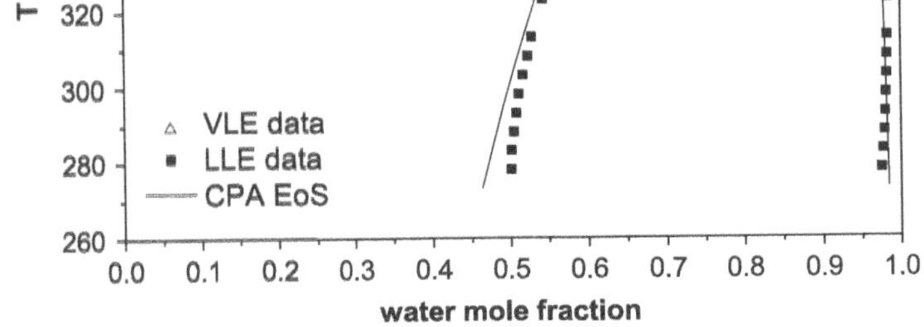

Figure 11.1 *Water–butanol VLE and LLE with CPA and CR-1 using $k_{12} = -0.065$. Experimental data are from Sørensen and Arlt, Liquid–Liquid Equilibrium Data Collection (Binary Systems), DECHEMA Chemistry Data Series, Frankfurt, 1980, Vol. 5, Part 1, Hessel et al., Z. Phys. Chem. (Leipzig),* ***1965****, 229, 199. Reprinted with permission from Ind. Eng. Chem. Res., by Folas et al., 44, 3823 Copyright (2005) American Chemical Society*

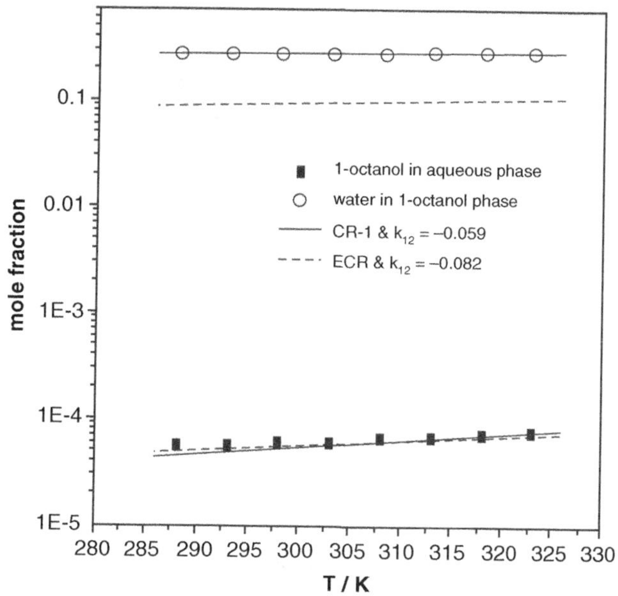

Figure 11.2 *LLE correlation for 1-octanol–water with CPA and different combining rules. Experimental data are from Dallos et al., J. Chem. Thermodyn., **1995**, 27, 447. Reprinted with permission from Ind. Eng. Chem. Res., by Folas et al., 44, 3823 Copyright (2005) American Chemical Society*

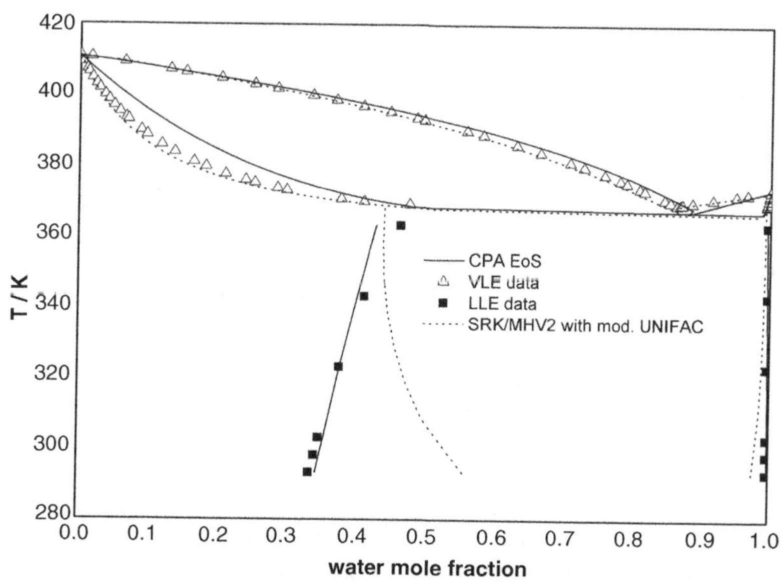

Figure 11.3 *VLE and LLE for n-pentanol–water using SRK/MHV2 with modified UNIFAC and CPA with CR-1 and $k_{12} = -0.037$ optimized from LLE data. Experimental data are from Sørensen and Arlt, Liquid–Liquid Equilibrium Data Collection (Binary Systems), DECHEMA Chemistry Data Series, Frankfurt, 1980, Vol. 5, Part 1. Reprinted with permission from Fluid Phase Equilibria, Modelling of associating mixtures for applications in the oil & gas and chemical industries by G. M. Kontogeorgis, G. Folas et al., 261, 1–2, 205–211 Copyright (2007) Elsevier*

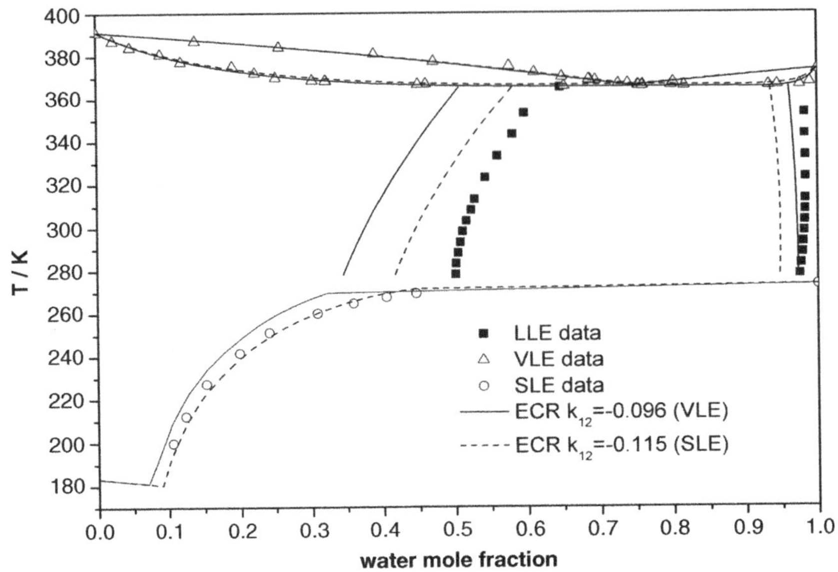

Figure 11.4 *Water–butanol VLE, LLE and SLE using CPA and ECR. Experimental data are from Sørensen and Arlt, Liquid–Liquid Equilibrium Data Collection (Binary Systems), DECHEMA Chemistry Data Series, Frankfurt, 1980, Vol. 5, Part 1, Hessel et al., Z. Phys. Chem. (Leipzig), **1965**,229, 199 and Lohman et al., J. Chem. Eng. Data, **1997**, 42, 1170. Reprinted with permission from Institut Francais du Petrole, Solvation Phenomena in Association Theories with applications to oil & gas and chemical industries by G. M Kontogeorgis, G. K. Folas, N. Muro-Sune, F. Roca Leon, M. L. Michelsen, Oil & Gas Science and Technology-Rev. IFP, 63, 305–319 Copyright (2008) Institut Francais du Petrole*

- ECR can satisfactorily correlate alcohol–alcohol VLE using a small value of the interaction parameter. In agreement with other studies, e.g. for alcohol–alkanes, small differences are observed in using the 2B or 3B schemes for alcohols.

11.3 Amines and ketones

The application of CPA to chemicals like amines and ketones has been a challenging task, especially when cross-associating mixtures of ketones or amines with water were considered. Some results are shown in Figure 11.5 and Tables 11.1–11.3 and the major conclusions of this investigation are summarized as follows:

- Ketones are, strictly speaking, non self-associating compounds. In these cases, CPA reduces to SRK and CPA can correlate (but cannot predict) ketone–alkane VLE, e.g. as shown on the left side of Figure 11.5. CPA can predict ketone–alkane VLE including the azeotropic behavior of these systems[9] if ketones (acetone here) are assumed to be self-associating compounds.
- When acetone is assumed to be self-associating, CPA can satisfactorily describe water–acetone VLE (Figure 11.5, right). The k_{12} values range between -0.17 and -0.114 with an average percentage deviation in pressure between 6 and 2% in the temperature range of 298–523 K. For comparison the correlation performance with SRK using one interaction parameter is shown in Table 11.1.
- An additional polar term is, in principle, required for modeling ketones. The assumption that acetone is a self-associating component is an engineering approach in order to maintain simplicity and avoid introducing additional terms in the CPA model.
- Amines are weakly self-associating compounds, but unlike glycols and lower alcohols, they are miscible with hydrocarbons even at very low temperatures. CPA can correlate well with a temperature-independent

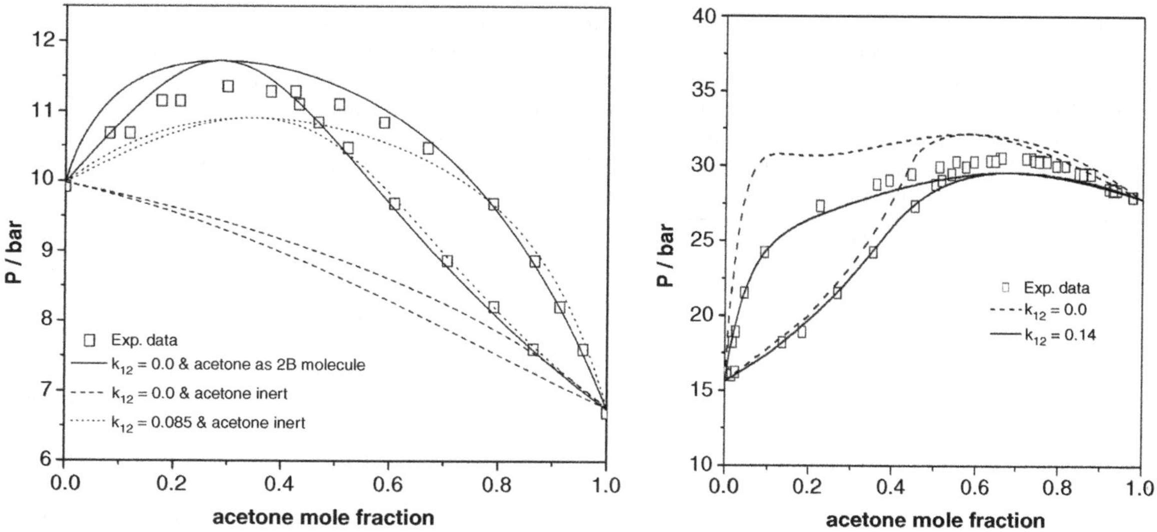

*Figure 11.5 Application of CPA to acetone-containing systems. Left: VLE prediction and correlation of acetone–pentane at 397.7 K with CPA when acetone is considered as inert or a self-associating compound. Experimental data from Campbell et al., J. Chem. Eng. Data, **1986**, 31, 424. When acetone is considered to be an inert compound, CPA can correlate the phase diagram using an interaction parameter. However, without introducing self-association for acetone, CPA cannot predict ($k_{12} = 0.0$) the azeotropic behavior of mixtures of ketones and alkanes. Right: VLE prediction ($k_{12} = 0.0$) and correlation for acetone–water at 473.15 K with ECR and $k_{12} = -0.14$. Acetone in considered to be a pseudo-associating compound, described via the 2B scheme. Experimental data from Griswold and Wong, AIChE Symp. Ser., **1952**, 48, 18. Reprinted with permission from Ind. Eng. Chem. Res., Application of the Cubic-Plus-Association Equation of State to Mixtures with Polar Chemicals and High Pressures by Georgios K. Folas, Georgios M. Kontogeorgis et al., 45, 4, 1516–1526 Copyright (2006) American Chemical Society*

k_{ij} and often predict amine–alkane VLE, using either the two- or three-site association scheme for amines, 2B and 3B.[10] It is difficult to say which of the association schemes (i.e. 2B or 3B) is superior. The results with CPA are better than with SRK, especially at the lower temperatures. Some typical results for methylamine mixtures are shown in Table 11.2, and similar results are obtained for diethylamine mixtures as well.

Table 11.1 VLE correlation results for acetone–water with CPA and ECR. Given in parentheses are the results with SRK (using the van der Waals one fluid mixing rules and one interaction parameter). Reprinted with permission from Ind. Eng. Chem. Res., Application of the Cubic-Plus-Association Equation of State to Mixtures with Polar Chemicals and High Pressures by Georgios K. Folas, Georgios M. Kontogeorgis et al., 45, 4, 1516–1526 Copyright (2006) American Chemical Society

T (K)	k_{12} with ECR	ΔP (%)	Δy 100
298.15	−0.171 (−0.283)	6.6 (25)	2.0 (5.2)
323.15	−0.14 (−0.254)	5.5 (17.6)	1.5 (3.5)
373.15	−0.14 (−0.227)	2.6 (11.7)	1.4 (5.6)
423.15	−0.14 (−0.190)	2.5 (8.6)	0.6 (3.5)
473.15	−0.14 (−0.162)	2.1 (7.3)	0.7 (2.8)
523.15	−0.14 (−0.126)	3.0 (6.8)	1.3 (3.7)
Average CPA		**3.7**	**1.3**
Average SRK		**12.8**	**4.1**

Table 11.2 Average percentage deviation between experimental and calculated pressures and vapor phase mole fractions for binary mixtures containing methylamine and alkanes using the CPA and SRK equations of state. The results with $k_{12}=0$ are predictions. Experimental data: A. Maczynski and A. Skrzecz, TRC-database for Chemistry and Engineering – Vapor-Liquid Equilibrium Data for Binary Systems, electronic version 1998. After Kaarsholm et al.[10] *Reprinted with permission from Ind. Eng. Chem. Res., Extension of the Cubic-plus-Association (CPA) Equation of State to Amines by Mads Kaarsholm, Samer O. Derawi, Michael L. Michelsen and Georgios M. Kontogeorgis, 44, 12, 4406–4413 Copyright (2005) American Chemical Society*

Associating compound	Diluent	T (K)	Model	k_{ij}	ΔP (%)	Δy
Methylamine	n-Hexane	218	CPA (2B)	0.0042	6.1	0.0057
			CPA (3B)	−0.0078	10.1	0.0105
			SRK	0.0641	29.6	0.0251
			CPA (2B)	0	8.3	0.007
			CPA (3B)	0	7.0	0.0083
			SRK	0	33.7	0.0586
Methylamine	n-Hexane	233	CPA (2B)	0.009	3.3	0.0029
			CPA (3B)	0.0005	5.2	0.0045
			SRK	0.0713	22.6	0.0145
Methylamine	n-Hexane	263	CPA (2B)	0.0123	1.6	0.0016
			CPA (3B)	0.0130	3.4	0.0055
			SRK	0.0930	16.3	0.0212
Methylamine	n-Hexane	293	CPA (2B)	0.0087	1.1	0.0022
			CPA (3B)	0.0119	2.5	0.0061
			SRK	0.0960	10.5	0.0179
			CPA (2B)	0	2.6	0.0056
			CPA (3B)	0	3.6	0.0105
			SRK	0	23.1	0.0680
Methylamine	n-Butane	218	CPA (2B)	0.026	0.9	
			CPA (3B)	0.0171	0.9	
			SRK	0.041	10.6	
			CPA (2B)	0	8.7	
			CPA (3B)	0	6.2	
			SRK	0	22.2	
Methylamine	n-Butane	233	CPA (2B)	0.0275	0.8	
			CPA (3B)	0.0192	0.8	
			SRK	0.055	7.4	
Methylamine	n-Butane	288	CPA (2B)	0.0271	0.3	
			CPA (3B)	0.0225	0.6	
			SRK	0.0873	3.2	
			CPA (2B)	0	5.1	
			CPA (3B)	0	4.4	
			SRK	0	16.1	

- The representation of VLE of cross-associating mixtures containing amines is more difficult, especially for mixtures with water. Some results are shown in Table 11.3. Relatively large negative interaction parameters are needed to obtain satisfactory correlation.

11.3.1 The case of a strongly solvating mixture: acetone–chloroform

A well-studied solvating mixture, discussed also in Chapter 2, is that of acetone and chloroform. Both are inert, i.e. non self-associating, molecules. Acetone is a highly polar molecule, but the most important characteristic

Table 11.3 *Average percentage deviation between experimental and calculated pressures and vapor phase mole fraction for binary mixtures containing cross-associating systems using the CPA and SRK equations of state. Experimental data: Srivastava et al., J. Chem. Eng. Data,* ***1985****, 30, 313, A. Maczynski and A. Skrzecz, TRC-database for Chemistry and Engineering – Vapor-Liquid Equilibrium Data for Binary Systems, electronic version 1998. The results with* $k_{12} = 0$ *are predictions. After Mads Kaarsholm et al.*[10] *Reprinted with permission from Ind. Eng. Chem. Res., Extension of the Cubic-plus-Association (CPA) Equation of State to Amines by Mads Kaarsholm, Samer O. Derawi, Michael L. Michelsen and Georgios M. Kontogeorgis, 44, 12, 4406–4413 Copyright (2005) American Chemical Society*

Associating compound 1	Associating compound 2	*T* (K)	Model	k_{ij}	ΔP (%)	Δy
Diethylamine	Methanol	398.58	CPA(2B)–CR-1	−0.0739	1.1	
			CPA(2B)–ECR	−0.1234	0.3	
Diethylamine	Methanol	297.97	CPA(2B)–CR1	−0.2006	2.1	
			CPA(2B)–ECR	−0.2361	2.8	
Diethylamine	Ethanol	313.15	CPA(2B)–CR–1	−0.1126	2.2	
			CPA(2B)–ECR	−0.1300	2.0	
Diethylamine	Water	329.95	CPA(2B)–CR–1	−0.2914	6.1	0.020
			CPA(2B)–ECR	−0.4158	16.4	0.049
			SRK	−0.3597	13.9	0.022
Ethylamine	Ethanol	293.15	CPA (2B)	−0.1587	2.2	0.003
			CPA (3B)	−0.1499	2.5	0.0034
			SRK	−0.1270	1.9	0.0034
Ethylamine	Ethanol	Isobar (102 kPa)	CPA (2B)	−0.18	0.3 (*T*)	0.038
			CPA (3B)	−0.17	0.3 (*T*)	0.039
			SRK	−0.14	0.3 (*T*)	0.0346
Ethylamine	Water	Isobar (80 kPa)	CPA (2B)	−0.35	1.5 (*T*)	0.046
			CPA (3B)	−0.33	1.4 (*T*)	0.045
			SRK	−0.33	1.0 (*T*)	

in this connection is the strong hydrogen bonding between acetone and chloroform. Both VLE and SLE data are available and the strong solvation at low temperatures is evidenced by the characteristic solid phase at intermediate concentrations, similar to the one we met previously (in Chapters 9 and 10) for MEG–water and methanol–water. Thus, a demanding test for the model is the simultaneous description of VLE at high temperatures, both freezing curves and the solid complex behavior. The acetone–chloroform mixture has been modeled with CPA without explicitly accounting for the polarity of the compounds. A systematic investigation has shown that the task of describing the acetone–chloroform phase equilibria over the whole temperature range is particularly challenging. Satisfactory results for both VLE and LLE were *not* obtained when:

1. Acetone is treated as a self-associating compound with only one negative association site (2B scheme as discussed in the previous section) which can associate with the positive site of chloroform and following the procedure illustrated in Chapter 9, i.e. the modified CR-1 rule (Equation (9.10)).
2. Acetone is treated as an inert molecule with one negative site capable of solvating with chloroform.
3. It did not help if both cross-associating parameters ε, β were fitted to the experimental data or whether k_{ij} was also included in the physical term.

However, satisfactory results are obtained if acetone is treated as an inert molecule, assuming that *both* lone electron pairs of the oxygen atom can participate in hydrogen bonds with the hydrogen of chloroform. CPA, with the 'increased solvation' incorporated in the way described, results in very good correlation of both VLE and SLE as shown in Figures 11.6 and 11.7. There are both 'strengths' and 'weaknesses' of the approach presented for modeling acetone–chloroform, which are summarized in Table 11.4.

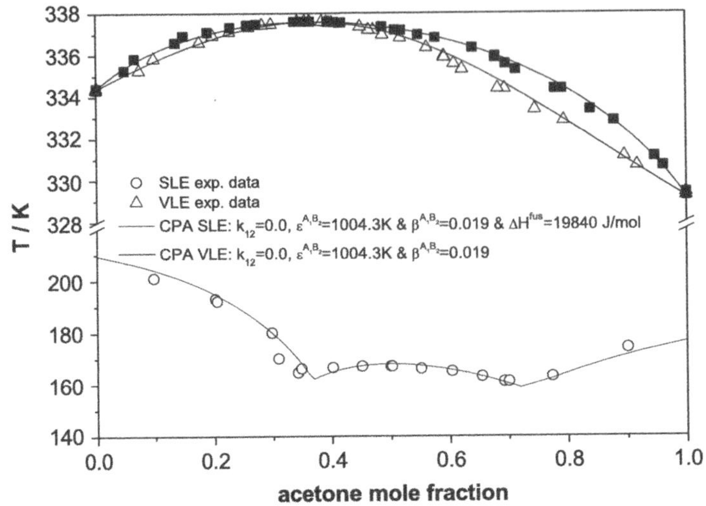

Figure 11.6 *Simultaneous VLE and SLE correlation of acetone–chloroform using the CPA EoS. The solid complex is modeled with the approach presented in Appendix 9.D. Both acetone and chloroform are treated as inert compounds. Acetone's two lone electron pairs are assumed to participate in hydrogen bonds with chloroform. The melting temperature (K) and heat of fusion (in J/mol) of acetone and chloroform are, respectively, 176.6, 5720 and 209.6, 8800. Prausnitz et al.*[12] *reported that the heat of fusion of the solid complex is approximately 11 000 J/mol, which is lower than the optimized value obtained for the solid complex, following the aforementioned methodology. Experimental data are from Campbell et al., Can. J. Chem.,* ***1960****, 38, 652, Kudryavtseva et al., Zh. Prikl. Khim. (Leningrad),* ***1963****, 36, 1231 and Kojima et al., J. Chem. Eng. Data,* ***1991****, 36, 343*

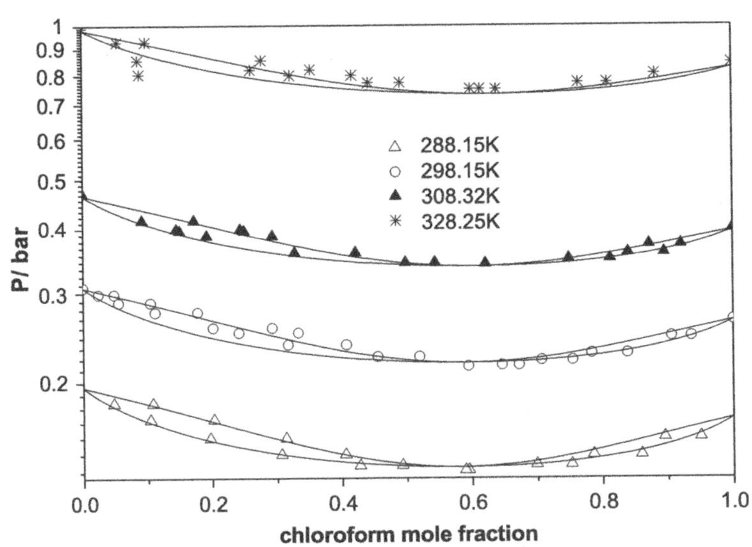

Figure 11.7 *VLE correlation of acetone–chloroform for various isotherms. Binary interaction parameters and modeling information as in Figure 11.6. Experimental data are from Temir et al., Fluid Phase Equilib.,* ***1981****, 6, 113, Roeck et al., Z. Phys. Chem. (Frankfurt),* ***1957****, 11, 41 and Beckmann et al., Z. Phys. Chem. (Leipzig),* ***1915****, 89, 235*

Table 11.4 *Modeling of acetone–chloroform: strengths and weaknesses*

Strengths	Weaknesses
Good results for VLE and SLE using only parameters fitted from SLE. VLE results are predictions $k_{ij}=0$, only solvating parameters used	Acetone is a polar compound ($\mu = 3$ D), so a more appropriate description should account for the polarity If acetone is treated as an inert compound, then acetone–hydrocarbon phase equilibria cannot be predicted with SRK (though correlation can be achieved with a k_{ij} value)
Solvation is very strong, thus it is physically correct to account explicitly for it	The approach when acetone is treated as an inert compound is different from that illustrated previously (Section 11.3) where acetone was treated as a self-associating compound in order to describe acetone–-hydrocarbon and acetone–water VLE
Acetone and chloroform are treated as inert compounds, in agreement with their physical picture, and only capable of solvating with each other No explicit accounting for the polarity is needed, i.e. no extra polar term is needed and thus no major modeling/computational changes in the model	

11.4 Mixtures with organic acids

CPA has been extensively applied to mixtures with organic acids and other chemicals (ethers, esters, alcohols, ketones). More specifically the following categories have been considered:

- organic acids (formic, acetic, propionic) + hydrocarbons or gases;
- organic acids + polar non-associating compounds (esters, ethers, ketones);
- organic acids + alcohols;
- organic acids + water;
- multicomponent mixtures: acids + water + hydrocarbons or gases.

The main conclusions from the investigation project on acids can be summarized as follows:

- Organic acids have been modeled using the one-site (1A) scheme, see Tables 8.11 and 8.12 in Chapter 8. This is the scheme which provides the best results for acid–alkane VLE and virial coefficients for organic acids.[11] The 2B scheme was tested and shown to yield better results than 1A for acetic acid–water VLE, but for consistency reasons the 1A scheme is retained in all calculations shown here.
- VLE and LLE of acids with hydrocarbons, even aromatic hydrocarbons, can be described satisfactorily with a single interaction parameter (k_{ij}). A typical result for acetic acid–octane VLE is presented in Figure 11.8. Unlike mixtures of water or glycols with aromatic hydrocarbons (Chapter 10), solvation does not need to be explicitly accounted for in mixtures of organic acids with aromatic hydrocarbons, not even in the case of LLE. This is similar behavior as in alcohol–aromatic hydrocarbons for which it was not necessary to account explicitly for the solvation effects. Only limited LLE data are available for mixtures of aromatics and organic acids. A typical example for formic acid–benzene is presented in Figure 11.9.
- Excellent representation of gas solubility data in acetic acid is obtained for various gases (CO_2, CO, H_2, N_2, O_2) using either no interaction parameters or a temperature-independent k_{ij}. An exception is

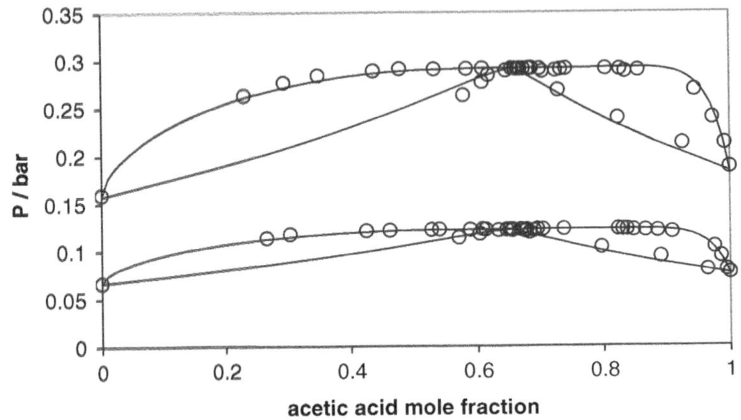

Figure 11.8 VLE for acetic acid + n-octane system with CPA at T = 323.2 K (lower curve) and T = 343.2 K (upper curve) with $k_{12} = 0.06$. CPA calculations using 1A site model for acetic acid. Experimental data from Plesnar et al., J. Chem. Eng. Data, ***1996****, 41, 799. After Derawi et al.[11] Reprinted with permission from Fluid Phase Equilibria, Application of the CPA equation of state to organic acids by G. M. Kontogeorgis, S. O. Derawi et al., 225, 1–2, 107–113 Copyright (2004) Elsevier*

methane–acetic acid where a temperature-dependent k_{ij} is required. A typical result for acetic acid–CO_2 is presented in Figure 11.10.

- Good results are generally obtained for ether + acid and ester + acid VLE (Figures 11.11 and 11.12). Accounting for solvation effects may help in the case of ethers. In the case of heavy esters like butyl and isobutyl acetate, better results are obtained when these esters are allowed to associate using the 2B scheme. As

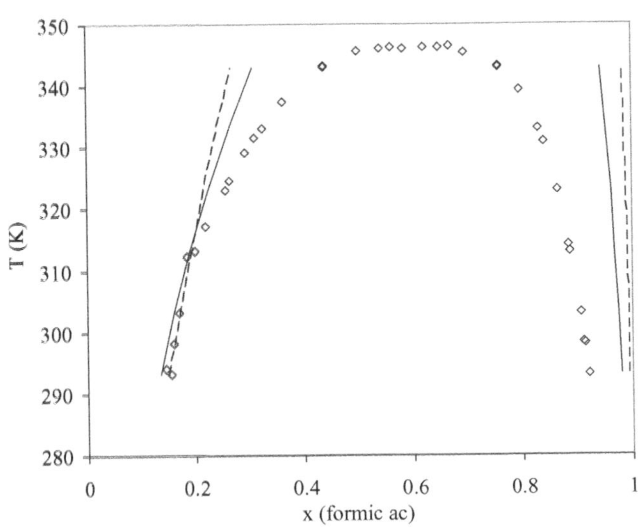

Figure 11.9 LLE correlation for the system formic acid–benzene with CPA. The points are the experimental data (Ewins, J. Chem. Soc., ***1914****, 105, 350). CPA calculations considering benzene as inert, $k_{12} = 0.133$ (——) and CPA calculations considering solvation, with $k_{12} = 0.238$ and $\beta^{A_iB_j} = 0.0958$ (– – –). Reprinted with permission from Fluid Phase Equilibria, Modelling of associating mixtures for applications in the oil & gas and chemical industries by G. M. Kontogeorgis, G. Folas et al., 261, 1–2, 205–211 Copyright (2007) Elsevier*

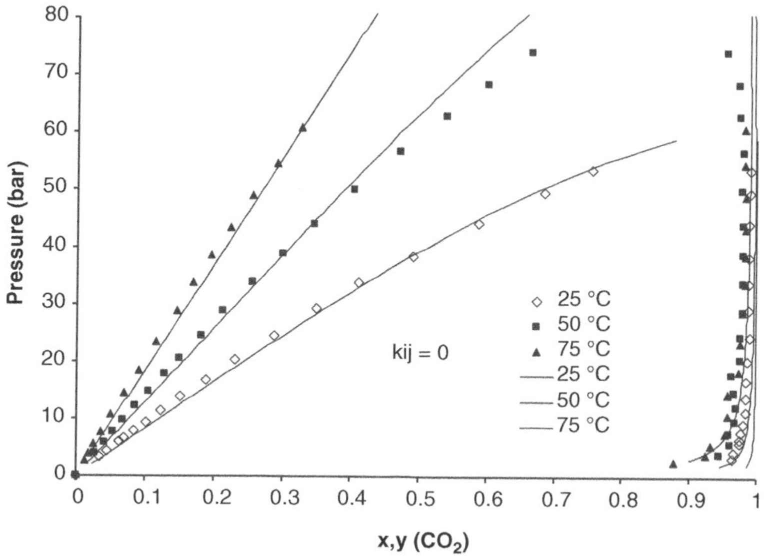

Figure 11.10 *Pxy diagram for carbon dioxide(1)–acetic acid(2) at 25, 50 and 75 °C. The lines are pure predictions. Experimental data are from Fu and Sandler, Ind. Eng. Chem. Res.,* ***1995****, 34, 1897. Reprinted with permission from Ind. Eng. Chem. Res., Phase Equilibrium Modelling for Mixtures with Acetic Acid Using an Association Equation of State by Nuria Muro-Suñé, Georgios M. Kontogeorgis et al., 47, 15, 5660–5668 Copyright (2008) American Chemical Society*

seen in Section 11.3, this 'practical trick' has been used with success for other polar non-self-associating compounds like acetone. The negative k_{ij} values for ester–acid systems may indicate that the cross-association is underestimated and that the solvation may need to be explicitly accounted for. Some results are shown Figures 11.11 and 11.12. More results are presented by Kontogeorgis *et al.*[7] and Muro-Suñé *et al.*[13]

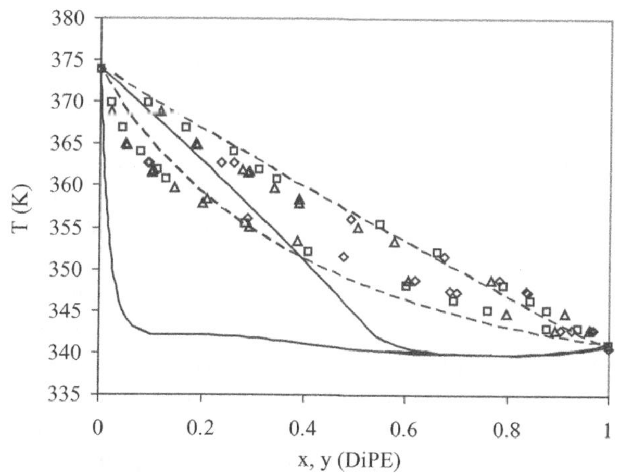

Figure 11.11 *VLE correlation with CPA for the system diisopropyl ether–formic acid at 101.32 kPa. Experimental data are from Grewer and Schmidt, Chem.-Ing.-Tech.,* ***1973****, 45, 1063, Hunsmann and Simmrock, Chem.-Ing.-Tech.,* ***1966****, 38, 1053 and Gilburd et al., Zh. Prikl. Khim. (Leningrad),* ***1983****, 56, 684. CPA calculations with* $k_{12}=0$ *(——) and* $k_{12}=-0.1414$ *(– – –)*

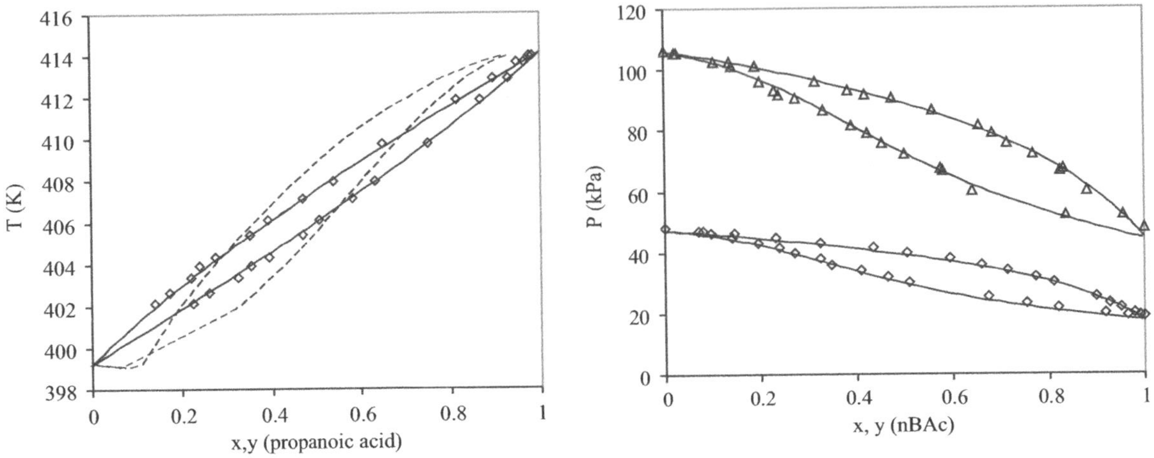

Figure 11.12 *Left: VLE correlation with CPA for the system n-butyl acetate– propionic acid at 101.32 kPa. Experimental data are from Ionescu et al., Rev. Roum. Chim.,* ***1993****, 38, 457. CPA calculations with n-butyl acetate as inert and* $k_{12} = -0.0869$ *(– – –) and with n-butyl acetate as 2B associating fluid and* $k_{12} = -0.0053$ *(——). Right: The system n-butyl acetate–n-heptane. Experimental data at 347.85 K (◇) and 373.15 K (Δ) from Scheller et al., J. Chem. Eng. Data,* ***1969****, 14, 17. CPA calculations with n-butyl acetate as inert and* $k_{12} = 0.0409$*. CPA calculations with n-butyl acetate as 2B associating fluid and* $k_{12} = 0.0314$ *at both temperatures are almost identical*

- An excellent description of alcohol–acid VLE is obtained in the majority of cases. There are extensive data for these systems. A small negative interaction parameter value is needed, which indicates a moderate underestimation of cross-association. The ECR and CR-1 rules perform similarly (see Figure 11.13).
- Satisfactory results are obtained for mixtures of acids with other polar chemicals as well, such as the binary mixture acetic acid–acetone presented in Figure 11.14 or acetic acid–acetic anhydride shown in Figure 11.15.
- The performance of the CPA model is not satisfactory for design purposes for mixtures containing water with the small organic acids (formic, acetic). CPA correlation is somewhat better for propionic acid, as shown in Figure 11.16.
- In general, for aqueous mixtures with formic and acetic acid, large negative k_{ij} are required and the relative volatilities are not represented accurately. Prediction ($k_{ij} = 0$) is not possible. The performance of CPA cannot be improved by changes in the cross-association strength. A significant change in the model, using the Huron–Vidal mixing rule for the energy term, is required in order to correct for the problems for water–acid VLE. The improved performance is achieved at the cost of extra parameters. The CPA–HV model is presented in Appendix 11.A. Table 11.5 summarizes some results and the interaction parameters with both CPA and CPA–HV.
- Representation of water and small organic acid VLE, especially water with formic and acetic acid, is a very challenging task for SAFT models in general, as will be discussed in Chapter 13. The results with other SAFT variants, e.g. CK–SAFT, simplified SAFT,[14] etc., have not been very satisfactory.[15,16] Moreover, large k_{ij} values were needed as in the case of CPA. For example, for water–acetic acid, CPA with ECR requires $k_{ij} = -0.145$, while with simplified SAFT $k_{ij} = -0.179$ is used and a value around -0.1 (depending on the pure acid parameter values) is needed in SAFT in the work of Wolbach and Sandler[16]. A typical result for the VLE of water–formic acid is presented in Figure 11.17. Perakis *et al.*[17] suggested using, in connection to their PR–CPA EoS, the three-site scheme (3B) for water to model water–acetic acid VLE. They obtained rather good results, using the geometric mean rule for both the cross-association

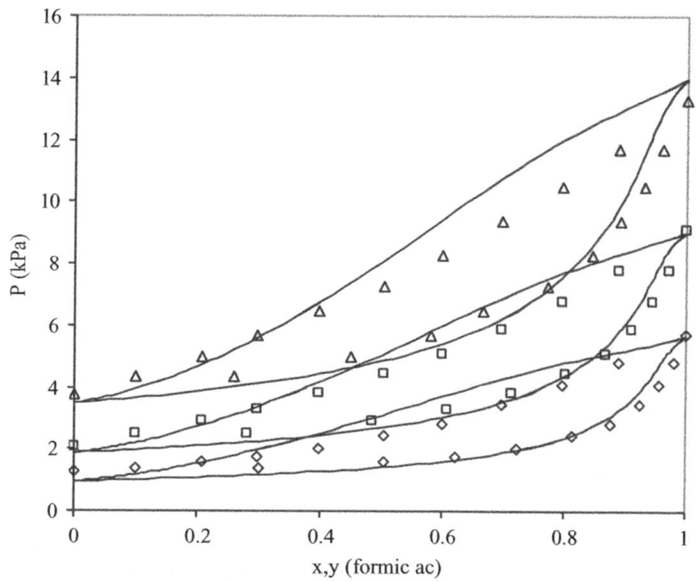

Figure 11.13 *VLE correlation with CPA for the system formic acid–1-butanol. Experimental data at 298.15 K (◇), at 308.15 K (□) and at 318.15 K (Δ), and CPA calculations with ECR and* $k_{12} = 0.0651$. *Reprinted with permission from Institut Francais du Petrole, Solvation Phenomena in Association Theories with applications to oil & gas and chemical industries by G. M Kontogeorgis, G. K. Folas, N. Muro-Suñé, F. Roca Leon, M. L. Michelsen, Oil & Gas Science and Technology-Rev. IFP, 63, 305–319 Copyright (2008)*

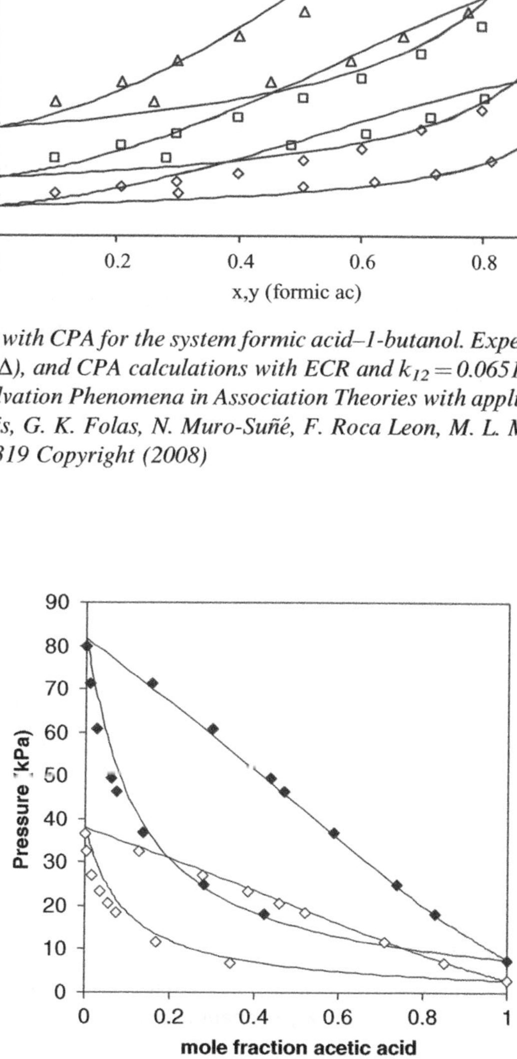

Figure 11.14 *Pxy diagram with CPA for the system acetic acid–acetone at T = 303.15 K (open symbols) and 323.15 K (filled symbols). Acetone is modeled as a 2B self-associating molecule. Experimental data are from Meehan et al., Chem. Eng. Sci., **1965**, 20, 757. The curves are predictions. Reprinted with permission from Ind. Eng. Chem. Res., Phase Equilibrium Modelling for Mixtures with Acetic Acid Using an Association Equation of State by Nuria Muro-Suñé, Georgios M. Kontogeorgis et al., 47, 15, 5660–5668 Copyright (2008) American Chemical Society*

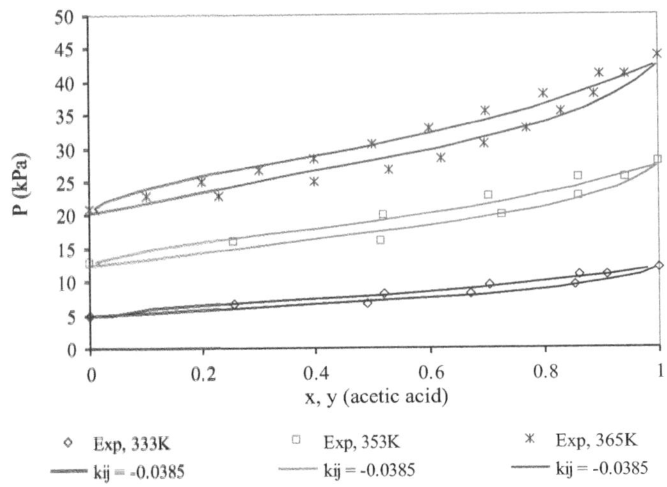

Figure 11.15 *Binary VLE for acetic acid with acetic anhydride using CPA. Experimental data are from Cherbov, Zh. Russ. Fiz. Khim. Obshch.,* ***1930****, 62, 1509 and Tatscheff et al., Z. Phys. Chem. (Leipzig),* ***1968****, 237, 53. Reprinted with permission from Ind. Eng. Chem. Res., Phase Equilibrium Modelling for Mixtures with Acetic Acid Using an Association Equation of State by Nuria Muro-Suñé, Georgios M. Kontogeorgis et al., 47, 15, 5660–5668 Copyright (2008) American Chemical Society*

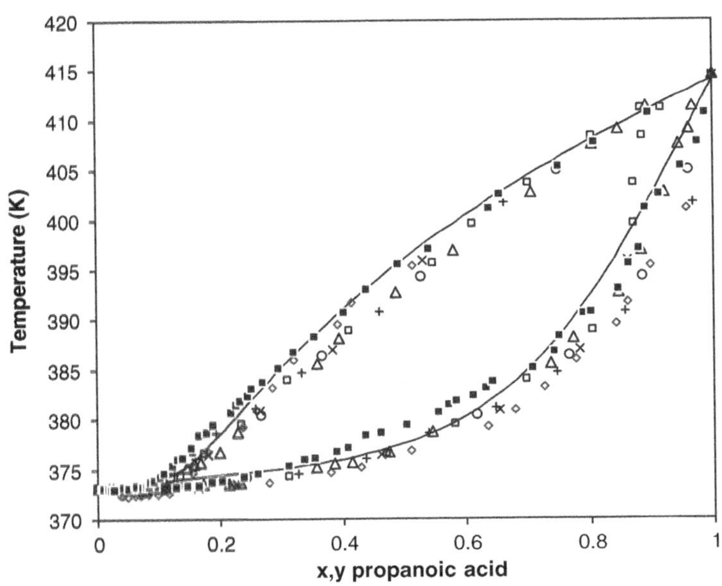

Figure 11.16 *VLE with CPA for the system propanoic acid–water at P = 1 atm. CPA calculations with $k_{12} = -0.21$ using the CR-1 rule. The (points) data are from five different sources: Ito et al., J. Chem. Eng. Data,* ***1963****, 8, 315; Dakshinamurty et al., J. Appl. Chem.,* ***1961****, 11, 226; F. Rivenq, Bull. Soc. Chim. Fr.,* ***1961****, 7, 1392; Kushner et al., Russ. J. Phys. Chem., USSR,* ***1967****, 41, 121; and Johnson et al., Can. J. Technol.,* ***1954****, 32, 179. After Kontogeorgis et al.[7] Reprinted with permission from Fluid Phase Equilibria, Modelling of associating mixtures for applications in the oil & gas and chemical industries by G. M. Kontogeorgis, G. Folas et al., 261, 1–2, 205–211 Copyright (2007) Elsevier*

Table 11.5 Acetic acid–water VLE with CPA (using ECR) and CPA/HV (Appendix 11.A). Parameter sets for CPA/HV and average deviations from experimental data with both models. Deviations are shown for the vapor phase mole fraction (Δy), pressure ($\Delta P\%$) and relative volatility ($\Delta\alpha\%$)

Set	a_{ij}	A_{12} (K)	A_{21} (K)	D_{12}	Parameters fitted to
1	0.4126	−77.17	−831.38	0.6219	372.8 K
2	0.3	$-321.4+0.6895T$	−157.6 to $1.8972T$	0.6725	313.15, 372.8 and 462.05 K

	Δy			ΔP (%)			$\Delta\alpha$ (%)		
T (K)	CPA/HV set 1	CPA	CPA/HV set 2	CPA/HV set 1	CPA	CPA/HV set 2	CPA/HV set 1	CPA	CPA/HV set 2
502.9	0.0051	0.00924	0.0044	1.15	2.24	2.11	4.66	9.5	3.79
462.1	0.0043	0.0122	0.0023	1.51	1.99	2.54	4.31	13.2	2.25
372.8	0.0029	0.0162	0.0046	4.01	3.11	3.09	2.42	14.9	3.77
313.15	0.0293	0.0216	0.0069	7.98	2.7	3.19	23.46	13.8	6.71
293.15	0.0444	0.02	0.0064	10.72	2.56	3.90	39.07	12.5	5.14
Total	**0.0172**	**0.0158**	**0.00492**	**5.07**	**2.52**	**2.97**	**14.78**	**12.8**	**4.33**

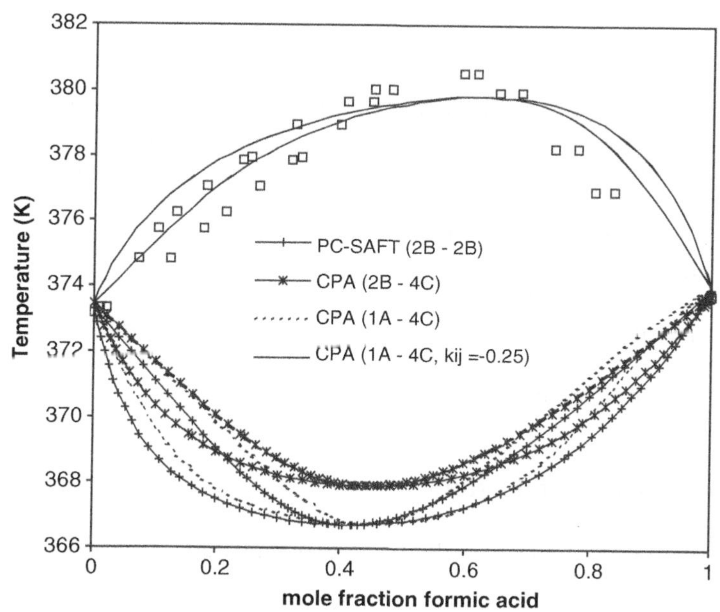

Figure 11.17 VLE for the system formic acid–water at 1 atm using CPA with various association schemes and PC–SAFT using the 2B parameters presented for water and formic acid by Gross and Sadowski.[5] All the curves which show a trend 'opposite to the experimental data' are predictions with the various models ($k_{12}=0$). Experimental data are from Chalov and Aleksandrova, Gidroliz. Lesokhim. Prom., **1959**, 10, 15. Reprinted with permission from Fluid Phase Equilibria, Modelling of associating mixtures for applications in the oil & gas and chemical industries by G. M. Kontogeorgis, G. Folas et al., 261, 1–2, 205–211 Copyright (2007) Elsevier

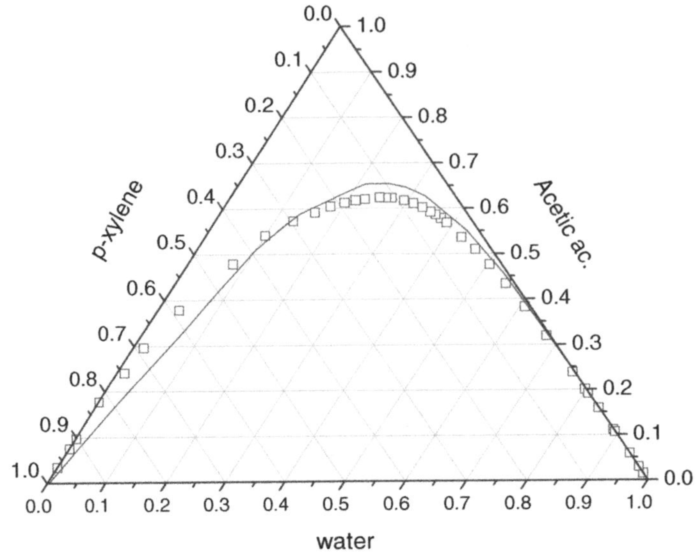

*Figure 11.18 Experimental data (□) and CPA calculations (——) for the ternary system p-xylene–acetic acid–water at 298 K. The interaction parameters are as follows: $k_{ij} = -0.223$ for water–acetic acid, for water–p-xylene $k_{ij} = -0.0133$, $\beta^{A_iB_j} = 0.0667$ and for acetic acid–p-xylene $k_{ij} = 0.0171$. Experimental data are from Suresh and Beckman, Fluid Phase Equilib., **1994**, 99, 219. Reprinted with permission from Ind. Eng. Chem. Res., Phase Equilibrium Modelling for Mixtures with Acetic Acid Using an Association Equation of State by Nuria Muro-Suñé, Georgios M. Kontogeorgis et al., 47, 15, 5660–5668 Copyright (2008) American Chemical Society*

energy and volume, and using k_{ij} and an l_{ij} parameter. These parameters are corrections to the geometric mean rules in the cross-dispersion energy and the cross-association energy.

- Despite the limitations for water–acetic acid, the performance of CPA for multicomponent systems is satisfactory, even when using k_{ij} and the classical approach for modeling cross-association. This has been shown for multicomponent VLE of water–CO_2–acetic acid and LLE for water–acetic acid with benzene and hexane[18] as well as water–acetic acid–xylene (see Figure 11.18).
- Recently, Oliveira *et al.*[35] applied CPA to LLE and SLE of water with fatty acids (C_5–C_{12}). Both CR-1 and ECR were tested and the results are satisfactory with both combining rules, especially the latter. The interaction parameter k_{ij} can be correlated well with the fatty acid's chain length. SLE calculations are performed using the k_{ij} obtained from LLE.

11.5 Mixtures with ethers and esters

Figures 11.19–11.23 show some typical results for aqueous mixtures with ethers and esters. The main conclusions are:

- Ethers and esters are treated as inert compounds without explicitly accounting for the polarity.
- LLE data are available for certain polar compounds like ethers and esters with water. Figure 11.19 shows calculations with CPA for dipropyl ether (DPE) and ethyl acetate but similar results have been obtained for other ethers and esters in water. These mixtures can be considered to exhibit 'induced association' or solvation, i.e. only one of the compounds in a binary mixture is self-associating though cross-association is present. The term 'induced association' can also be used for mixtures with glycols or water and aromatic

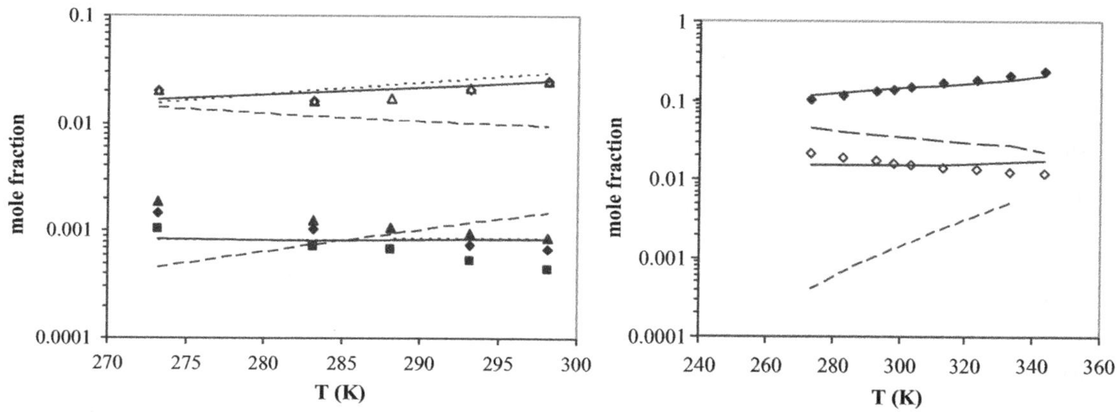

Figure 11.19 *LLE for water–ethers or esters with CPA. The importance of accounting for the solvation is illustrated. Left: LLE for the system DPE–water, mole fraction of water in organic phase (◆, Δ) and of DPE in aqueous phase (◇, ▲, ■). CPA calculations with: one-site solvation, $k_{12} = -0.1769$ and $\beta^{A_iB_j} = 0.196$ (••••••); two-site solvation, $k_{12} = -0.177$ and $\beta^{A_iB_j} = 0.0555$ (—); no solvation, $k_{12} = -0.328$ (– – –). Reprinted with permission from Institut Francais du Petrole, Solvation Phenomena in Association Theories with applications to oil & gas and chemical industries by G. M Kontogeorgis, G. K. Folas, N. Muro-Suñé, F. Roca Leon, M. L. Michelsen, Oil & Gas Science and Technology-Rev. IFP, 63, 305–319 Copyright (2008) Institut Francais du Petrole. Right: LLE for the system water–ethyl acetate, mole fraction of water in ethyl acetate (◆) and mole fraction of ethyl acetate in water (◇). CPA calculations without solvation and $k_{12} = -0.25$ (—). CPA calculations with 2B (two-site) solvation, $k_{12} = -0.111$ and $\beta^{A_iB_j} = 0.105$ (– – –). Data from Sørensen and Arlt, Liquid–Liquid Equilibrium Data Collection (Binary Systems), DECHEMA Chemistry Data Series, Frankfurt, 1980, Vol. 5*

hydrocarbons, where solvation is present due to Lewis acid–Lewis base interactions and not due to hydrogen bonding. Induced association seems as discussed, in the context of CPA, to be particularly important for aqueous or glycol mixtures with polar and aromatic compounds, especially for obtaining accurate correlation of LLE. It is less important in the case of alcohols or acids with aromatic hydrocarbons. When solvation is significant, satisfactory results are obtained with CPA when it is explicitly accounted for using the modified CR-1 rule.

- The need for explicitly accounting for the solvation is apparent as, without it, the predicted solubilities are much lower than the experimental data, and large, negative k_{ij} values are needed. Moreover, even when interaction parameters are used, LLE cannot be appropriately modeled. For example, as shown in Figure 11.19 for water–DPE, the experimental solubility of water in the organic phase is higher than the solubility of DPE in the aqueous phase but the model predicts the opposite solubility behavior if solvation is not explicitly taken into account.
- Equally satisfactory results are obtained using either the one-site or two-site type of solvation scheme (i.e. considering one or two solvating sites in the oxygen atom of ethers or esters). Somewhat better results are obtained with the two-site solvation scheme. The one-site scheme has been investigated in depth as it is consistent with the representation of alcohols, discussed previously, which are described using the 2B scheme, i.e. one site for both lone pairs of electrons in the oxygen atom.
- VLE is also improved when solvation is accounted for, as shown in Figure 11.20.
- For mixtures with gases (CO_2, N_2), water and ethers, VLLE data for two gases (CO_2, N_2), water and dimethyl ether (DME) have been measured by Laursen *et al.*[19]. Due to the complex interactions and high pressures involved, these data have been used for testing CPA and Figure 11.21 shows the result for the CO_2–water–DME, which is quite satisfactory. Surprisingly, less satisfactory results were obtained for

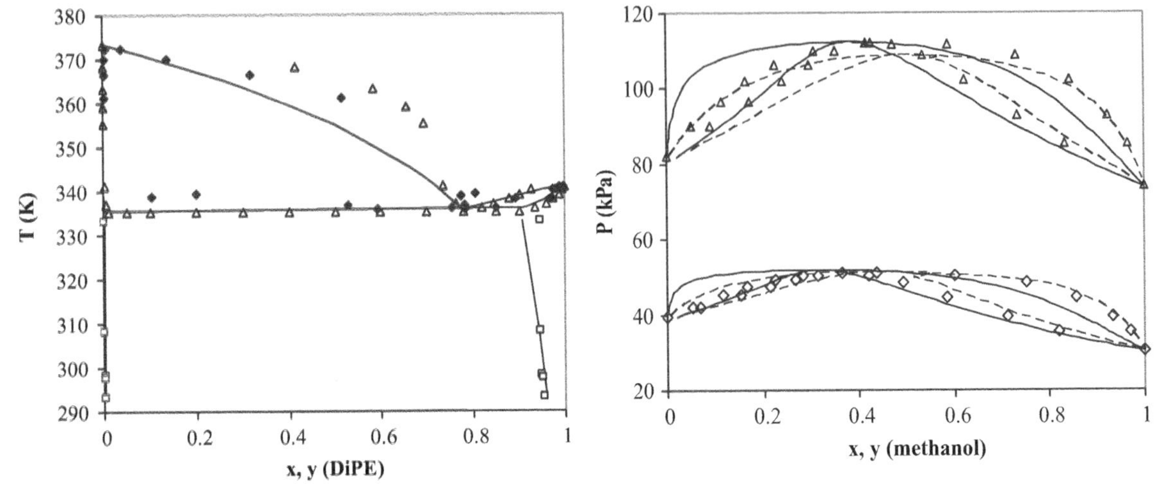

Figure 11.20 *Left: VLLE with CPA for the system DiPE–water at 101.32 kPa, from VLE data are Hunsmann et al., Chem.-Ing.-Tech., **1966**, 38, 1053 (◇) and Yorizane et al., Kagaku Kogaku, **1967**, 31, 451 (Δ). LLE data are from Sørensen and Arlt, Liquid–Liquid Equilibrium Data Collection (Binary Systems), DECHEMA Chemistry Data Series, Frankfurt, 1980, Vol. 5 (□). CPA calculations with solvation: $k_{12} = -0.2236$ and $\beta^{A_iB_j} = 0.2373$ (——). Interaction parameters were obtained from the LLE data. Right: VLE with CPA for the system ethyl propyl ether–ethanol (Farkova et al., Fluid Phase Equilib., **1995**, 109, 53) at 310 K (◇) and 330 K (Δ). CPA calculations without solvation: $k_{12} = -0.0994$ at 310 K (——). With solvation (using only one association site): $k_{12} = 0.0367$ and $\beta^{A_iB_j} = 0.177$ (– – –)*

N_2–water–DME and the reasons for this are not clear. Similar results to CPA were obtained by Laursen[20], both for binary and ternary mixtures of DME, using the SRK EoS with the MHV1 mixing rule and NRTL.

- Good results are obtained for the complex phase equilibrium of the quaternary VLLE of the mixture CO_2–DME–water–methanol, as shown in Figure 11.22. Notice the relatively high value of the binary

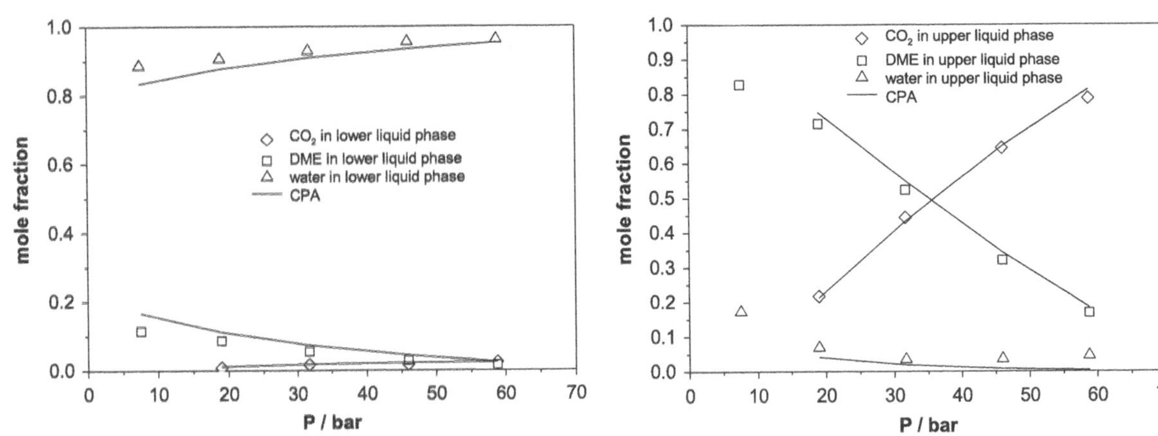

Figure 11.21 *VLLE for CO_2–DME–water with CPA. Left: Lower liquid phase (i.e. aqueous phase). Right: lower liquid phase. Experimental data are from Laursen et al.[19] Reprinted with permission from Ind. Eng. Chem. Res., Application of the Cubic-Plus-Association Equation of State to Mixtures with Polar Chemicals and High Pressures by Georgios K. Folas, Georgios M. Kontogeorgis et al., 45, 4, 1516–1526 Copyright (2006) American Chemical Society*

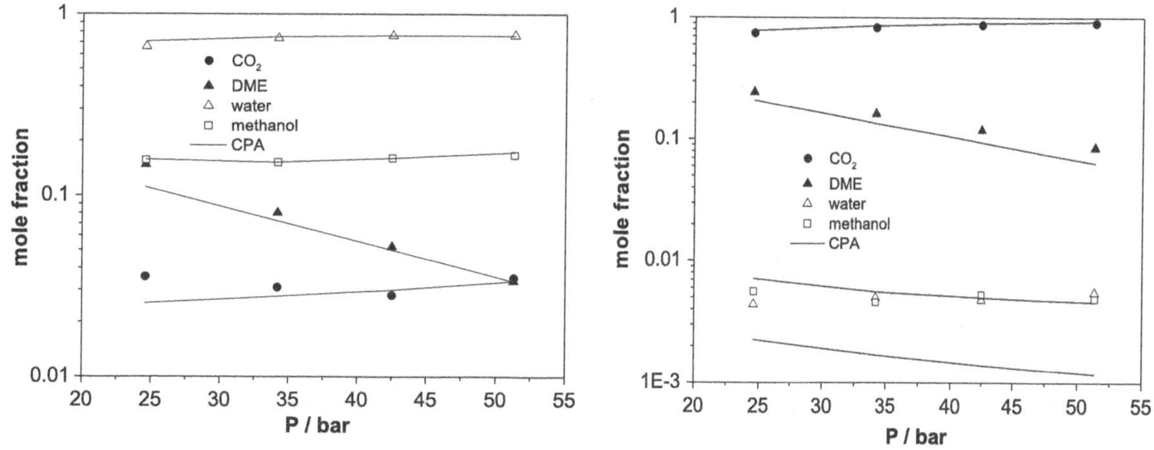

Figure 11.22 *VLLE for CO_2–DME–water–methanol with CPA. Left: Lower liquid phase (i.e. aqueous phase). Right: Vapor phase. Experimental data are from Laursen et al.*[19]

interaction parameters for DME–water (k_{12}=−0.16) and DME–methanol (k_{12}=0.221). These high values may be due to the assumption, built into the modified CR-1 rule, that the cross-association energy is equal to half of the association energy value of the hydrogen bonding compound (see Figure 11.23). Still, the multicomponent, multiphase equilibria is satisfactorily predicted.

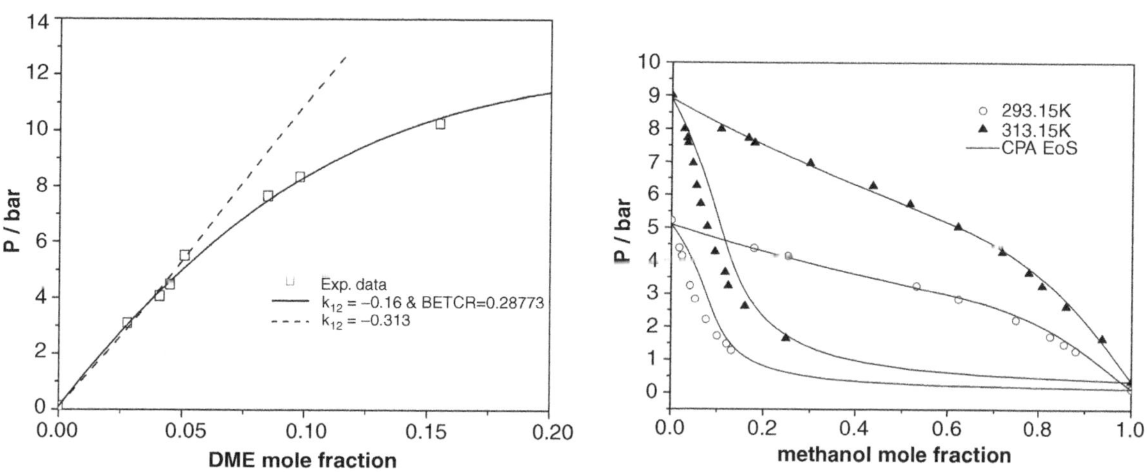

Figure 11.23 *Left: VLE correlation with CPA for the water–DME mixture at 323.15 K when accounting for the solvation (——) and without accounting for the solvation (– – –). Experimental data are from Pozo de Fernadez et al., J. Chem. Eng. Data,* ***1984****, 29, 324. Reprinted with permission from Ind. Eng. Chem. Res., Application of the Cubic-Plus-Association Equation of State to Mixtures with Polar Chemicals and High Pressures by Georgios K. Folas, Georgios M. Kontogeorgis et al., 45, 4, 1516–1526 Copyright (2006) American Chemical Society. Right: VLE correlation with CPA of methanol–DME mixture at two different temperatures using the modified CR-1 rule, k_{12}=0.221 and $\beta^{A_iB_j}$ = 2.15 at 293.15 K or $\beta^{A_iB_j}$ = 1.41 at 313.15 K. Experimental data are from Chang et al., J. Chem. Eng. Data,* ***1982****, 27, 293*

11.6 Multifunctional chemicals: glycolethers and alkanolamines

Two types of multifunctional chemicals, glycolethers and alkanolamines, have been recently studied with CPA.[21,22] In both cases a so-called first-level approach was followed, i.e. certain simplifying assumptions were made:

- No intramolecular association is considered.
- No distinction is made between the N and O atoms in alkanolamines.
- The lone oxygen atoms or number of oxygen atoms are not considered in glycolethers.
- Polarity is not considered explicitly and only the association term is used.
- Existing association schemes (2B, 3B and 4C) have been used.

A difficulty in using SAFT-type approaches in such families of multifunctional chemicals is the lack of plentiful experimental vapor pressures and liquid density data which are needed in the pure compound parameter estimation.

Selected results for glycolether mixtures are presented in Figures 11.24–11.26. The most important conclusions are:

- Excellent VLE and LLE correlation is obtained for all mixtures (except those with water) using a single k_{ij} value per binary and in many cases pure predictions ($k_{ij} = 0$) yield satisfactory results.
- There is little difference in using the 2B or the 3B association scheme for the glycolethers, except for aqueous systems.
- Very good results are obtained for mixtures of glycolethers with polar or associating compounds such as alcohols. There is almost no difference whether ECR or the CR1 rule is used.

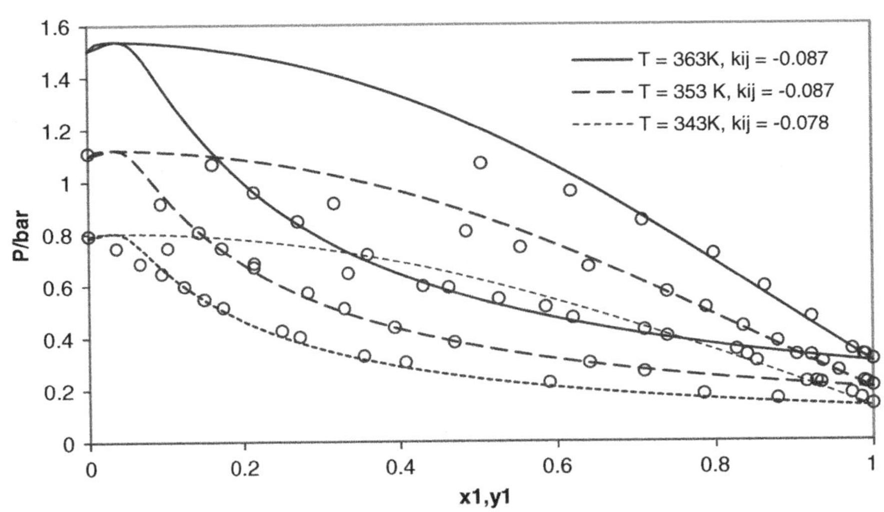

Figure 11.24 *Comparison between experimental data (Chandak et al., J. Chem. Eng. Data,* ***1977****, 22, 137) and CPA correlation (lines) for the system 2-methoxyethanol,* C_1E_1 *(2B) + ethyl acetate with the optimal* k_{ij}*at different temperatures. Similar results are obtained using the 3B scheme for* C_1E_1*. Reprinted with permission from Fluid Phase Equilibria, Modelling of phase equilibria of glycol ethers mixtures using an association model by N. M. Garrido, G. K. Folas and G. M. Kontogeorgis, 273, 1–2, 11–20 Copryight (2008) Elsevier*

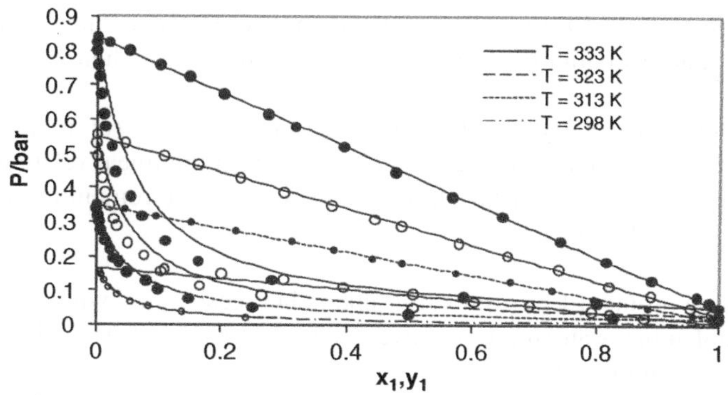

Figure 11.25 Comparison between experimental data (Antosik et al., J. Chem. Eng. Data, **1999**, 44, 368) and CPA prediction (lines) with $k_{ij}=0$ for the system 2-ethoxyethanol, C_2E_1 (2B) + methanol. Using $k_{ij}=0$, the average percentage deviation in pressure (P) 4.3% at 298.15 K and 1.0% at 313.15 K. Reprinted with permission from Fluid Phase Equilibria, Modelling of phase equilibria of glycol ethers mixtures using an association model by N. M. Garrido, G. K. Folas and G. M. Kontogeorgis, 273, 1–2, 11–20 Copryight (2008) Elsevier

- For mixtures with water, a large negative k_{ij} is required for describing the phase behavior (indicating that the degree of solvation is underestimated). In this case CPA with the 2B scheme for glycolethers performs better than CPA/3B (though with a somewhat higher k_{ij} value). Moreover, CPA with ECR performs better than CPA with CR-1 in most cases. High negative k_{ij} values were needed, as was the case for aqueous mixtures with glycols, alcohols and amines.

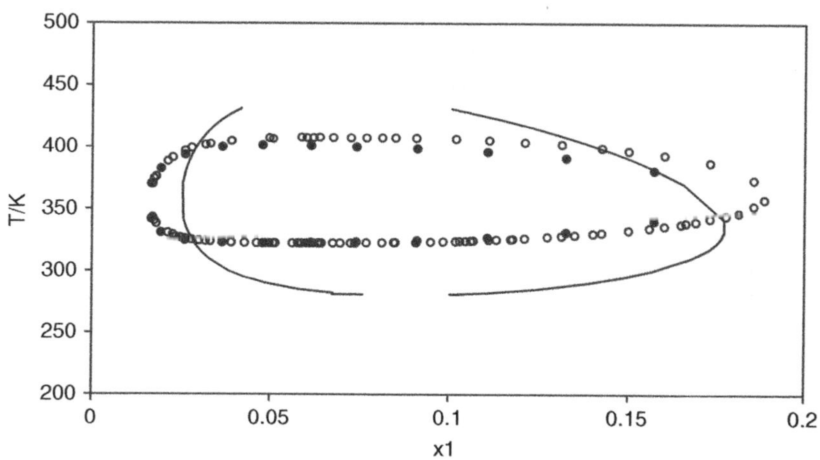

Figure 11.26 Comparison between experimental data for the LLE system: 2-butoxyethanol (C_4E_1) (3B) + water (4C), using the CR-1 rule and $k_{ij} = -0.192$; 3B is the only choice in this case. Experimental data are from: Aizpiri et al., Chem. Phys., **1992**, 165, 131; Gmehling and Onken, Vapor-liquid equilibrium data collection, DECHEMA Chemistry Data Series, Vol. V, Part 4, Frankfurt, 1977 and Schneider, G., Z. Phys. Chem. Neue Folge, **1963**, 37, 333. Reprinted with permission from Fluid Phase Equilibria, Modelling of phase equilibria of glycol ethers mixtures using an association model by N. M. Garrido, G. K. Folas and G. M. Kontogeorgis, 273, 1–2, 11–20 Copryight (2008) Elsevier

- The correlation of the closed LLE loop for butoxyethanol (C_4E_1) and water is particularly difficult to represent accurately and a qualitative correlation is possible only when the 3B scheme is used in CPA.
- Glycolether parameters particularly suitable for LLE representation (of mixtures with alkanes) are obtained when, first, the co-volume parameter is obtained from the linear relationship with the van der Waals volume ($b(\text{l/mol}) = 0.0018V_w(\text{dm}^3/\text{kmol}) - 0.0132$) and the remaining parameters are fitted to vapor pressures and liquid densities.

Figures 11.27–11.30 show some LLE results for alkanolamines with hydrocarbons and the VLE of MDEA–methane. Alkanolamine–water mixtures have also been considered. The following points summarize the most important results from this investigation:

- The best choice for MEA is 4C and this scheme is, moreover, used for all three alkanolamines studied (MEA, DEA, MDEA).
- Use of LLE data (for alkanolamines with alkanes) is important for selecting the best parameters for the alkanolamines among sets which fit vapor pressures and liquid densities of pure compounds equally well.
- LLE correlation for alkanolamine–alkane is satisfactory with a temperature-independent k_{ij} parameter.
- MEA–benzene LLE is satisfactorily represented by explicitly accounting for the solvation using the modified CR-1 rule, similar to what was done for water or glycols with aromatic hydrocarbons.

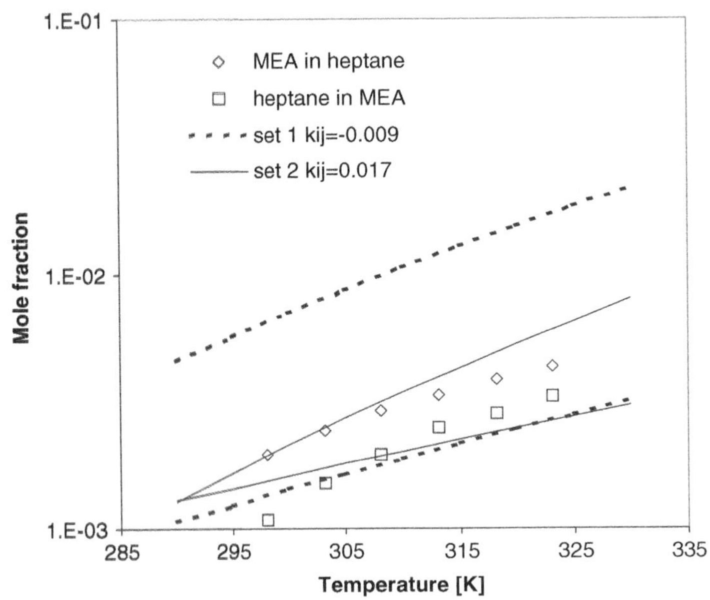

Figure 11.27 *MEA–heptane LLE with CPA using the 4C scheme for MEA and fitted k_{ij}. Set 1 is based only on vapor pressure and liquid density data, while set 2 is also based on these LLE data. The representation of MEA vapor pressure and liquid density is excellent with both parameter sets. Experimental data from Gustin and Renon, J. Chem. Eng. Data,* ***1973****, 18(2), 164. After Avlund et al.*[22]

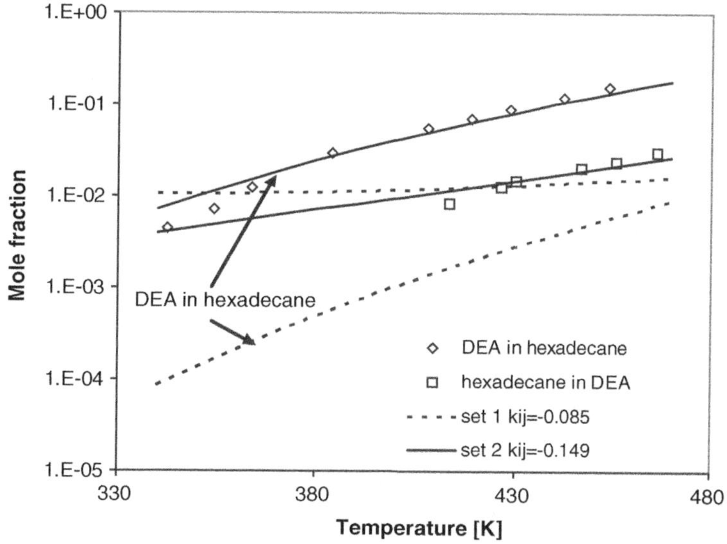

Figure 11.28 DEA–hexadecane LLE with CPA using the 4C scheme for DEA and fitted k_{ij}. Set 1 is based only on vapor pressure and liquid density data, while set 2 is also based on these LLE data. The representation of MEA vapor pressure and liquid density is excellent with both parameter sets. Experimental data are from Abdi and Meisen, J. Chem. Eng. Data, **1998**, 43(2), 138. After Avlund et al.[22]

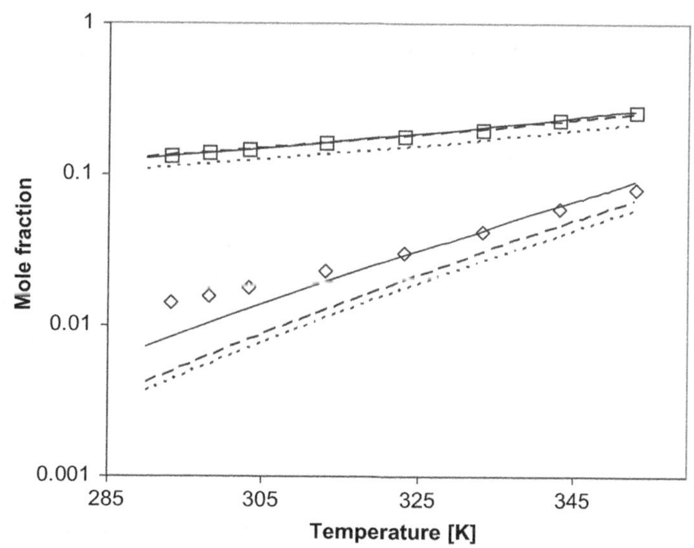

Figure 11.29 MEA–benzene LLE with CPA using the 4C scheme for MEA. The upper lines and points are the solubility of benzene in MEA and the lower ones indicate the solubility of MEA in benzene. Dotted line: $k_{ij}=0$; dashed line: $k_{ij}=-0.0065$; and solid line: $k_{ij}=0.0016$ and $\beta^{A_iB_j}=0.0067$. Experimental data are from Sørensen and Arlt, Liquid–liquid equilibrium data collection (Binary Systems), 1995. Reprinted with permission from Ind. Eng. Chem. Res., Modeling Systems Containing Alkanolamines with the CPA Equation of State by A. S. Avlund, G. M. Kontogeorgis and M. L. Michelsen, 47, 19, 7441–7446 Copyright (2008) American Chemical Society

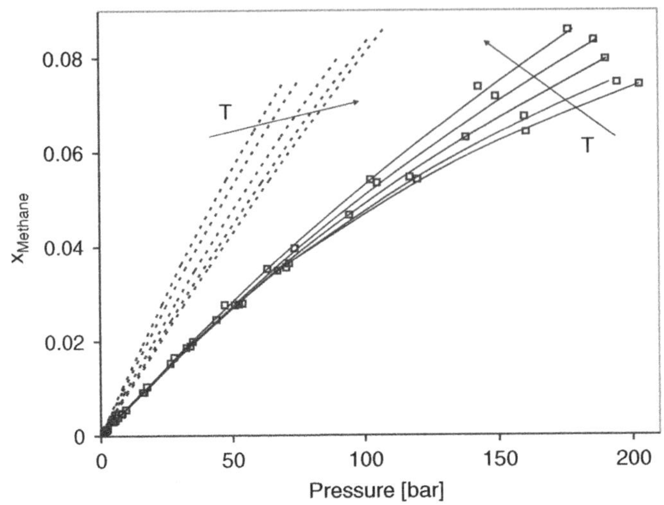

Figure 11.30 *MDEA–methane VLE at five temperatures (298, 313, 343, 373 and 403 K) with CPA. The dotted line is k_{ij} = 0 and the solid line is correlation with a single temperature-independent interaction parameter value, k_{ij} = 0.164. Reprinted with permission from Ind. Eng. Chem. Res., Modeling Systems Containing Alkanolamines with the CPA Equation of State by A. S. Avlund, G. M. Kontogeorgis and M. L. Michelsen, 47, 19, 7441–7446 Copyright (2008) American Chemical Society*

- The parameter estimation of MDEA was possible when the assumption was made that the association parameters are the same as for DEA. This assumption is justified by the similarities in the Kamlet–Taft parameters of the two compounds (Table 11.7).
- Alkanolamine–water VLE can be well correlated with both CR-1 and ECR (a bit better with the former) using rather large, negative but temperature-independent interaction parameters.

Recently, Avlund *et al.*[32] applied CPA to alkanolamine systems using a so-called 4D association scheme for MEA and a so-called 6A association scheme for DEA and MDEA. In the 4D scheme, the amine (NH_2) and hydroxyl (OH) groups have different association parameters, those of the OH group are obtained from alcohols, while the amine group association parameters are fitted to experimental vapor pressure and liquid density data (together with the three parameters of the physical term). In the 6A scheme for MDEA, the hydroxyl and amine groups are each assigned two (identical) sites. While the 4D and 6A schemes are possibly more realistic for alkanolamines compared to the 4C scheme, the improvement in the description of alkanolamine–hydrocarbon LLE and alkanolamine–water VLE is small.

New association schemes have also been used by Breil and Kontogeorgis[33] in their application of CPA to heavy glycol mixtures. They used the so-called 6D scheme for triethylene glycol (six equal sites for the four oxygen atoms and the two hydrogen atoms) and the so-called 7D scheme for tetraethylene glycol (seven equal sites for the five oxygen atoms and the two hydrogen atoms). The differences from the 4C scheme previously used for both glycols are not significant, but the new scheme represents better than the simpler (4C) scheme the glycol–alkane LLE as well as glycol–methane VLE and other properties (infinite dilution activity coefficients, excess enthalpies and glycol–water VLE). Some characteristic results are shown in Figures 11.31–11.33.

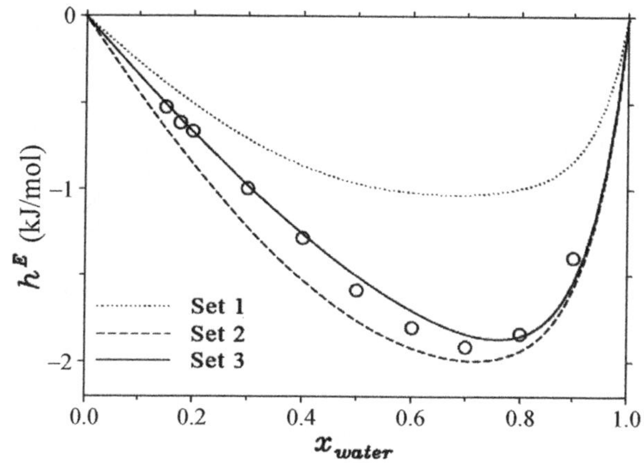

Figure 11.31 *Excess enthalpy of the TEG–water mixture at 25 °C with the CPA EoS. Sets 1 and 2 are both using the 4C scheme, while set 3 is based on the 6D scheme for TEG. Experimental data are from Hamam, S.E.M.; Benson, G.C.; Kumaran, M.K., Excess enthalpies of (water + diethylene glycol) and (water + triethylene glycol). J. Chem. Thermodyn.**1985**, 17, 973–976. Reprinted with permission from Ind. Eng. Chem. Res., Thermodynamics of Triethylene Glycol and Tetraethylene Glycol Containing Systems Described by the Cubic-Plus-Association Equation of State by Martin P. Breil and Georgios M. Kontogeorgis, 48, 11, 5472–5480 Copyright (2009) American Chemical Society*

11.7 Complex aqueous mixtures

Phase equilibria of aqueous systems are often complex. Phase equilibrium results for several binary and multicomponent mixtures containing water, alcohols or glycols and hydrocarbons were presented in this chapter and in Chapters 9 and 10. Mixtures of water with low molecular ethers and esters, organic acids as well

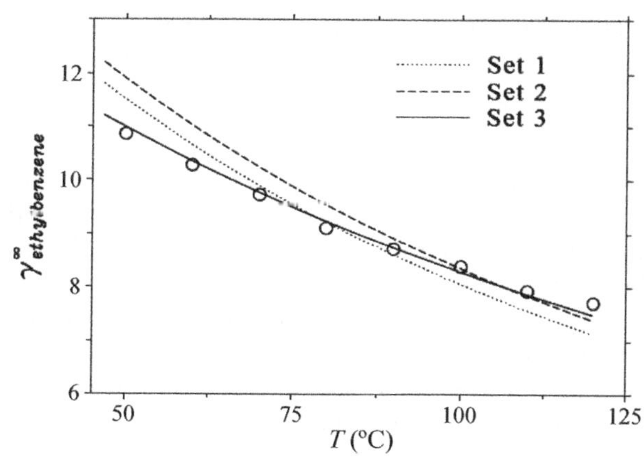

Figure 11.32 *Activity coefficient of ethylbenzene at infinite dilution in TEG calculated using the CPA EoS. Sets 1 and 2 are both using the 4C scheme, while set 3 is based on the 6D scheme for TEG. Experimental data are from Sun, P.-P.; Gao, G.-H.; Gao, H., Infinite dilution activity coefficients of hydrocarbons in triethylene glycol and tetraethylene glycol. J. Chem. Eng. Data **2003**, 48, 1109–1112. Reprinted with permission from Ind. Eng. Chem. Res., Thermodynamics of Triethylene Glycol and Tetraethylene Glycol Containing Systems Described by the Cubic-Plus-Association Equation of State by Martin P. Breil and Georgios M. Kontogeorgis, 48, 11, 5472–5480 Copyright (2009) American Chemical Society*

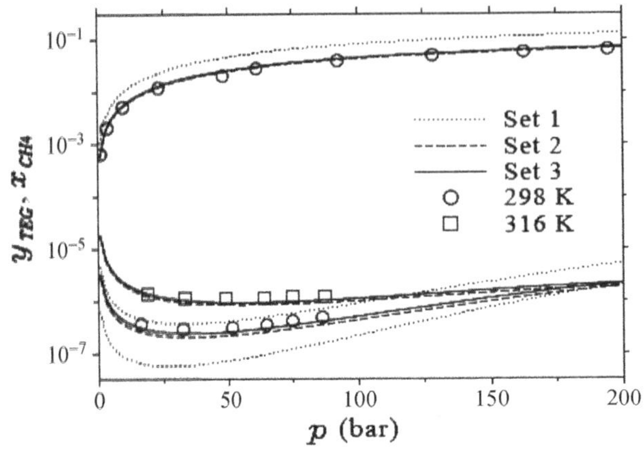

*Figure 11.33 Solubility of TEG in methane (both vapor and liquid phases) as a function of pressure at two temperatures calculated using the CPA EoS. Sets 1 and 2 are both using the 4C scheme, while set 3 is based on the 6D scheme for TEG. Experimental data are from Jerinic et al., Fluid Phase Equilib., **2008**, 264, 253 and Jou et al., Fluid Phase Equilib., **1987**, 36, 121. Reprinted with permission from Ind. Eng. Chem. Res., Thermodynamics of Triethylene Glycol and Tetraethylene Glycol Containing Systems Described by the Cubic-Plus-Association Equation of State by Martin P. Breil and Georgios M. Kontogeorgis, 48, 11, 5472–5480 Copyright (2009) American Chemical Society*

as with acetone, alkanolamines and glycolethers were presented in Sections 11.3–11.6. Water is miscible with many polar and/or hydrogen bonding compounds when cross-association is strong, e.g. with small alcohols, glycols, glycolethers, alkanolamines and acetone. For these systems only VLE and sometimes SLE data are available. On the other hand, LLE data and partial miscibility are present (at low or over extended temperature ranges) in aqueous systems if:

- cross-hydrogen bonding interactions are weak, e.g. water and heavy alcohols or esters;
- weak Lewis acid–Lewis base interactions are present, e.g. water with aromatic hydrocarbons;
- no cross-interactions are present, e.g. water–alkanes.

Accounting for the solvation is not so important in the case of VLE or SLE calculations. It is, however, crucial, to account explicitly for the solvation (Lewis acid–Lewis base or hydrogen bonding) interactions between water and aromatic-containing compounds or esters, when LLE data are available. The modified CR-1 rule is an efficient way of describing induced association (solvation) effects.

These conclusions have been further verified by recent measurements and modeling with CPA for a variety of aqueous systems; for natural phenolic compounds presented by Mota *et al.*[2], for fluorocarbons by Oliveira *et al.*[3,26] and for heavy esters of relevance to biodiesel. Figures 11.34–11.36 show typical results from these three publications. More specifically, it is observed that:

1. Fluorocarbons are compounds characterized by very weak van der Waals forces and are more hydrophobic than hydrocarbons, thus exhibiting extremely low solubilities in water. Actually their solubility in water has not been measured. As was also the case for hydrocarbons, the aromatic fluorocarbons exhibit much higher solubilities than the non-aromatic ones. In this case both solubilities have been measured, explicitly accounting for the solvation results in an excellent representation of the LLE of water with aromatic fluorocarbons with CPA (Figure 11.34, left). The same cross-association volume parameter ($\beta^{A_iB_j}$) and k_{ij} can be used for the two aromatic perfluorocarbon–water systems

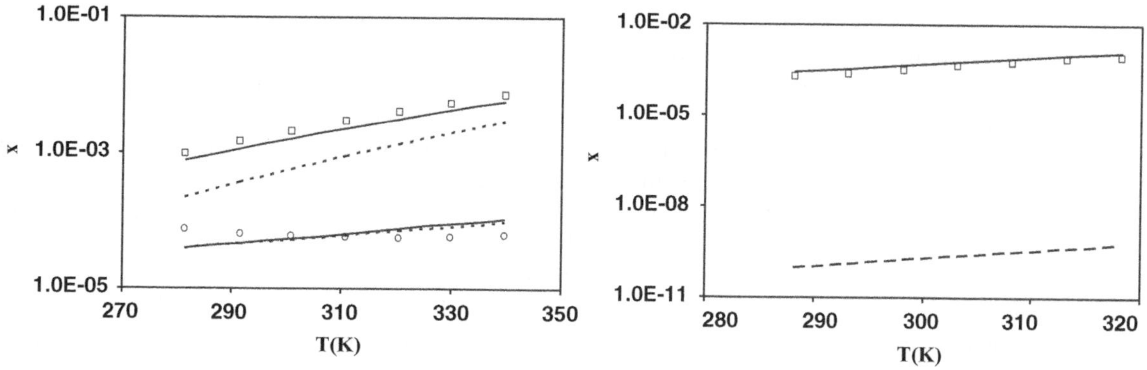

Figure 11.34 LLE for water–fluorocarbons with CPA. Symbols represent experimental data and lines are model calculations. The upper plot is the hydrocarbon phase and the lower plot is the aqueous phase. Left: Perfluorobenzene (C_6F_6) and water (dashed line: only k_{ij} is fitted; solid line: solvation is accounted for). Right: Perfluorohexane (C_6F_{14}) and water. No experimental data are available for the aqueous phase and thus only model calculations are shown. Reprinted with permission from Ind. Eng. Chem. Res. by Oliveira et al., 46, 1415. Copyright (2007) American Chemical Society

considered. There is no solvation present in the water–non-aromatic fluorocarbons and an excellent representation of the water solubility in the fluorocarbon phase is obtained with a single interaction parameter over the temperature range for which experimental data are available. CPA, in agreement with what is anticipated from experiment, predicts much lower solubilities in the water phase compared to n-alkanes (Figure 11.34, right).

2. CPA, using a single cross-association volume parameter ($\beta^{A_iB_j}$) (for all esters) and a system-specific k_{ij}, can satisfactorily correlate with the modified CR-1 rule the water solubility in several heavy esters and

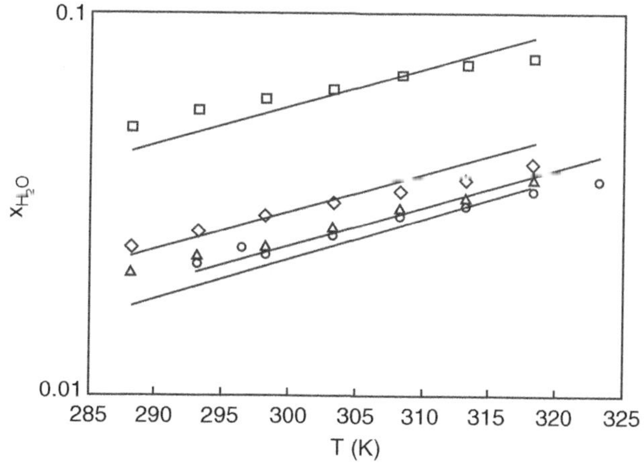

Figure 11.35 LLE for water and heavy esters and biodiesel with CPA. Symbols represent experimental data (water solubility) and lines are model calculations. In all cases the solvation with water is explicitly accounted for. From top to bottom: ethyl butanoate, ethyl decanoate, biodiesel F (triangles) and methyl tetradecanoate (circles). Reprinted with permission from Ind. Eng. Chem. Res., Prediction of Water Solubility in Biodiesel with the CPA Equation of State by M. B. Oliveira, J. A. P. Coutinho et al., 47, 12, 4278–4285 Copyright (2008) American Chemical Society

biodiesels in the 288–323 K temperature range. One solvating site is associated with each ester group. A global deviation of 7% is achieved for the 11 esters considered and 16% for the prediction of the water solubility in six commercial biodiesels (which are mixtures of four esters). The interaction parameter can be expressed by a linear relationship against the chain length. The authors estimated parameters for 46 alkyl esters, acetates, formates and unsaturated esters. In agreement with previous studies, the energy and especially the co-volume parameters of the oxygenated compounds considered are shown to follow generalized plots with the van der Waals volume. Oliveira *et al.*[31] are currently extending the CPA EoS to heavier esters using model parameters estimated by a group contribution method.

3. In addition to the alkanolamines and glycolethers discussed in Section 11.6, one more family of multifunctional chemicals has been recently modeled with CPA. Mota *et al.*[2] applied CPA to phenolic aromatic acids, most of which contain, in addition to the carboxylic group, one or two hydroxyl groups. The authors have shown (Figure 11.36) that CPA using a single system-dependent interaction parameter can correlate the solubility of eight solid compounds in water (SLE) in the temperature range 288–323 K. Several of the data against which the model has been tested were measured by the same authors. Fusion enthalpies and melting temperatures are not available and are estimated by a group contribution method. A significant difficulty in applying CPA to such compounds is their multifunctional character and the lack of experimental vapor pressure/liquid density data for some of them. A semi-predictive estimation scheme has been developed for the pure compound CPA parameters. First, the CPA parameters for seven aromatic compounds were fitted to experimental vapor pressures and liquid density data, then the parameters of the physical term were expressed as functions of the critical properties and van der Waals volume (the former estimated by a group contribution method). These generalized correlations were finally used to estimate the parameters of gallic/ferulic/caffeic acids, whereas the association parameters are the same as for the compounds in the database. Different association parameters are used for the carboxyl and hydroxyl groups. It was, thus, possible to obtain CPA parameters for three aromatic acids for which no experimental pure compound data are available. The CPA EoS with this methodology is currently being applied by the same investigators to other multifunctional phenolic compounds. The same authors applied further CPA to model the aqueous solubility of polycyclic aromatic hydrocarbons (PAHs),[34] with very good results. A

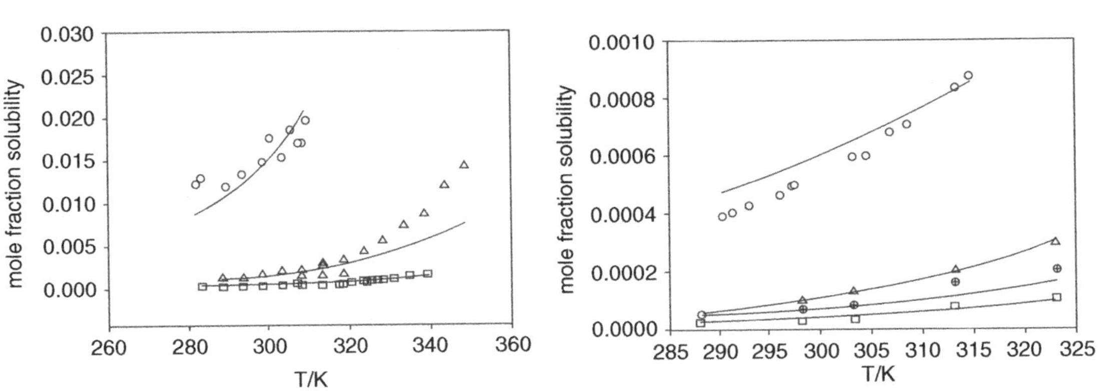

Figure 11.36 *SLE for water–phenolic acids/phenolic compounds with CPA. Symbols represent experimental data and lines are model calculations. Left: From top to bottom, phenol ($k_{ij}=-0.02$), gallic acid ($k_{ij}=-0.13$), salicylic acid ($k_{ij}=-0.06$). Right: From top to bottom, benzoic acid ($k_{ij}=-0.04$), caffeic acid ($k_{ij}=-0.10$), ferulic acid ($k_{ij}=-0.13$), trans-cinnamic acid ($k_{ij}=-0.05$). Reprinted with permission from Ind. Eng. Chem. Res., Aqueous Solubility of Some Natural Phenolic Compounds by F. L. Mota, A. J. Queimada et al.,47, 15, 5182–5189 Copyright (2008) American Chemical Society*

single, compound-specific, temperature-independent cross-association volume parameter is used, together with the modified CR-1 rule. No other interaction parameters were used (i.e. $k_{ij} = 0$). The average deviation between experimental and predicted aqueous solubilities is 6% for 11 PAHs.

11.8 Concluding remarks

Figure 11.37 summarizes the different types of cross-association (and solvation) which are possible in mixtures containing at least one self-associating or highly polar compound. We have used the term 'cross-association' for mixtures having two self-associating compounds. In this case either Elliott's (ECR) or the CR-1 rule should be used. We use the term 'solvation' or 'induced association' in the general case of Lewis acid–Lewis base (LA–LB) types of solvating mixtures as well as solvating systems with only one hydrogen bonding compound. Table 11.6 summarizes the interaction parameters for many such types of mixtures investigated with CPA. Table 11.7 provides the acid/base Kamlet–Taft[23] solvatochromic parameters for some of the compounds studied here. These parameters offer one way to quantify the expected solvation

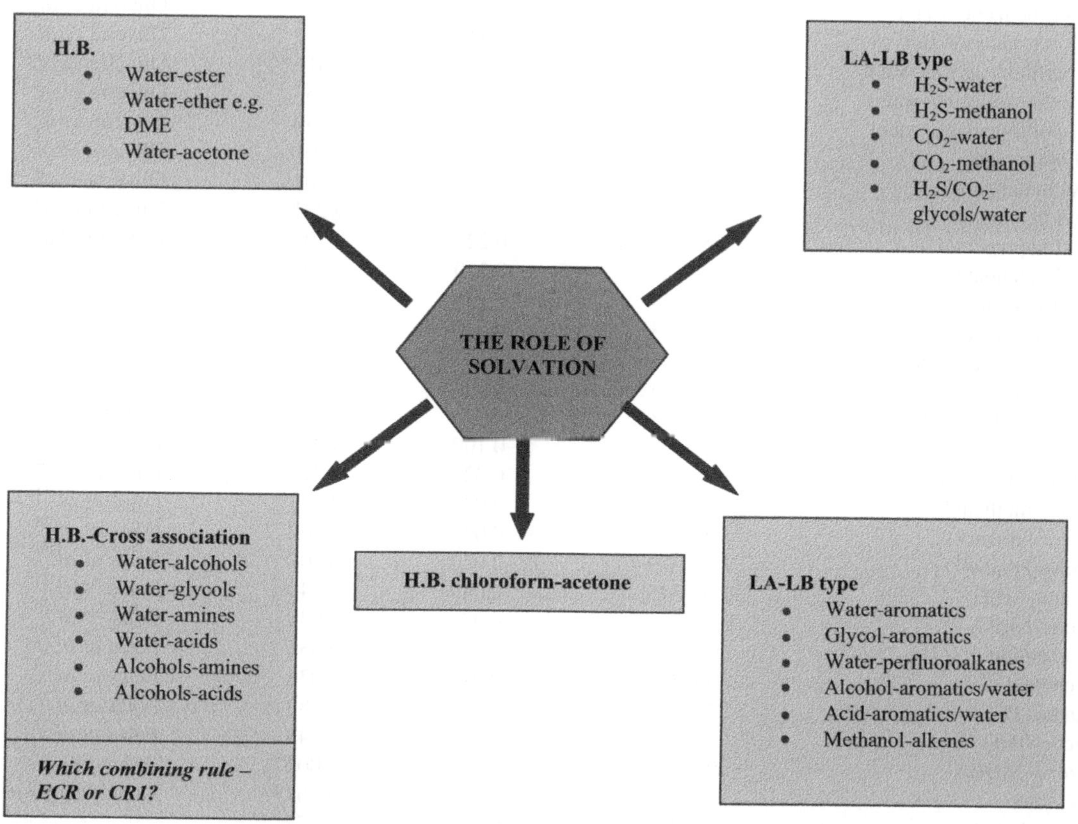

Figure 11.37 The various types of cross-association and solvation together with some examples, for mixtures containing self-associating (or polar) compounds

Table 11.6 *Binary interaction parameters for solvating systems. In all cases except for the acid gas and alcohol mixtures, they are estimated from LLE data*

System	T range (K)	k_{ij}	$\beta^{A_iB_j}$	Reference
Benzene–water	273–473	0.0355	0.079	Folas et al.[24]
Toluene–water	273–473	0.0095	0.060	Folas *et al.*[24]
Ethylbenzene–water	303–568	−0.0165	0.051	Folas *et al.*[24]
m-Xylene–water	373–473	−0.0165	0.039	Folas *et al.*[24]
o-Xylene–water		−0.022	0.051	Oliveira *et al.*[25]
p-Xylene–water	290–450	−0.0133	0.0667	Muro-Suñé *et al.*[13]
p-Xylene–water		−0.054	0.063	Oliveira *et al.*[25]
1,3,5-Trimethylbenzene–water	280–370	−0.011	0.063	Oliveira *et al.*[25]
Propylbenzene–water		0.012	0.059	Oliveira *et al.*[25]
Butylbenzene–water		−0.048	0.020	Oliveira *et al.*[25]
Pentylbenzene–water		−0.050	0.039	Oliveira *et al.*[25]
Isopropylbenzene–water	260–550	−0.057	0.046	Oliveira *et al.*[25]
Styrene–water		0.033	0.079	Oliveira *et al.*[25]
1-Hexene–water	310–496	0.0355	0.021	Folas *et al.*[24]
1-Octene–water	310–540	−0.0165	0.021	Folas *et al.*[24]
1-Decene–water	310–550	−0.0685	0.021	Folas *et al.*[24]
Perfluorobenzene–water	280–340	−0.034	0.032	Oliveira *et al.*[26]
Perfluorotoluene–water	270–340	−0.034	0.032	Oliveira *et al.*[26]
1-Br-perfluorooctane–water	290–320	0.0	0.012	Oliveira *et al.*[26]
1-H-perfluorooctane–water	290–320	0.0	0.066	Oliveira *et al.*[26]
1,8-H-perfluooctane–water	290–320	0.0	0.138	Oliveira *et al.*[26]
Ethyl butanoate	288–323	−0.254	0.201	Oliveira *et al.*[3]
Propyl butanoate	288–323	−0.238	0.201	Oliveira *et al.*[3]
Methyl hexanoate	288–323	−0.234	0.201	Oliveira *et al.*[3]
Methyl heptanoate	288–323	−0.221	0.201	Oliveira *et al.*[3]
Methyl octanoate	288–323	−0.210	0.201	Oliveira *et al.*[3]
Ethyl decanoate	288–323	−0.166	0.201	Oliveira *et al.*[3]
Methyl dodecanoate	288–323	−0.150	0.201	Oliveira *et al.*[3]
Methyl tetradecanoate	288–323	−0.123	0.201	Oliveira *et al.*[3]
Methyl hexadecanoate	288–323	−0.092	0.201	Oliveira *et al.*[3]
Methyl octadecanoate	288–323	−0.075	0.201	Oliveira *et al.*[3]
Methyl oleate	288–323	−0.100	0.201	Oliveira *et al.*[3]
Benzene–methanol	330–355	0.020	0.010	Folas *et al.*[24]
Toluene–methanol	290–380	0.034	0.029	Folas *et al.*[24]
Benzene–ethanol	290–380	0.022	0.002	Folas *et al.*[24]
Toluene–ethanol	290–380	0.020	0.003	Folas *et al.*[24]
Benzene–MEG	280–345	0.049	0.040	Folas *et al.*[27]
Toluene–MEG	280–380	0.051	0.042	Folas *et al.*[27]
Benzene–DEG	290–350	0.028	0.035	Folas *et al.*[28]
Toluene–DEG	300–380	0.046	0.033	Folas *et al.*[28]
Benzene–TEG	280–560	0.032	0.083	Folas *et al.*[27]
Toluene–TEG	280–560	0.038	0.048	Folas *et al.*[27]
Benzene–MDEA	290–350	0.0016	0.0067	Avlund *et al.*[22]
EPE–water		−0.1914	0.2617	Kontogeorgis *et al.*[8]
EPE–methanol	310–330	0.0367	0.177	
EPE–water	270–300	−0.1931	0.2917	Kontogeorgis *et al.*[8]

Table 11.6 *(Continued)*

System	*T* range (K)	k_{ij}	$\beta^{A_iB_j}$	Reference
DPE–water	270–300	−0.177	0.196	Kontogeorgis *et al.*[8]
DiPE–water		−0.2236	0.237 3	
DME–water	323	−0.16	0.287 7	
Ethylacetate–water	280–340	−0.111	0.105	
Methylacetate–water	280–360	−0.083	0.135	
Propylacetate–water	290–360	−0.119	0.086	
Butylacetate–water	290–360	−0.124	0.304	Kontogeorgis *et al.*[8]
CO_2–methanol	230–313	0.0384	0.019 98	Kontogeorgis *et al.*[8]
CO_2–ethanol	230–313	0.1158	0.0797	
CO_2–water	323–444	0.1	0.2	Kontogeorgis *et al.*[8]
H_2S–water	298–318	0.0	0.06	Kontogeorgis *et al.*[8]
H_2S–methanol	298	0.0	0.05	Kontogeorgis *et al.*[8]
	398	−0.05	0.09	

EPE = ethyl propyl ether, DPE = dipropyl ether, DiPE = diisopropyl ether, DME = dimethyl ether.

phenomena. The following points summarize the most important conclusions from the study of CPA to different chemicals including several of the various solvating systems:

- CPA has been applied to a wide variety of chemicals (alcohols, glycols, amines, ketones, organic acids, ethers, esters, etc.).
- Cross-associating mixtures are challenging, and Elliott's rule is the best choice for mixtures of water with methanol, ethanol, propanol (VLE and SLE) and other systems, e.g. alcohol–alcohols. The CR-1 rule is

Table 11.7 *Kamlet–Taft solvatochromic parameters for compounds included in Table 11.6 and for alkanolamines. Data from Kamlet et al.*[23] *and for alkanolamines from Lagalante et al.*[30]

Compound	Base parameter, β	Acid parameter, α
DiPE	0.49	0.0
DBE	0.46	0.0
DEE	0.47	0.0
DPE	0.46	0.0
Ethylacetate	0.45	0.0
Methylacetate	0.42	0.0
Toluene	0.11	0.0
Benzene	0.10	0.0
Methanol	0.62	0.93
Ethanol	0.77	0.83
MEG	0.52	0.90
Water	0.18	1.17
MEA	0.72	0.40
DEA	0.68	0.59
MDEA	0.59	0.53
Acetone	0.48	0.08
Chloroform	0.00	0.44
CO_2	0.45–0.47	0.0

recommended for mixtures of water with heavy alcohols. Correlation of LLE is satisfactory in the latter cases.

- The strongest LA–LB type of solvation is observed in mixtures of 'highly associating' compounds like water and glycols ('4C-type' molecules) with aromatic, olefinic or perfluoro-aromatic hydrocarbons, as well as in mixtures with esters and ethers. In these cases LLE data are available and the solvation should be explicitly accounted for.
- For LA–LB solvation where only VLE data are available, the solvation is often less important (compared to LLE) and can often be safely ignored. This is true for modeling mixtures of alcohols or acids with aromatic hydrocarbons.
- Very good phase equilibrium results are obtained for VLE and LLE of mixtures containing chemicals and alkanes. In the case of acetone, prediction (i.e. using $k_{ij}=0$) with CPA is possible only if acetone is considered to be a self-associating compound. Alternatively, the polarity of acetone should be explicitly accounted for, introducing a drastic change in the model (an additional polar term).
- For the solvating hydrogen bonding systems, e.g. chloroform–acetone, or for mixtures of water with ethers or esters, the solvation is always very strong and cannot be neglected.
- The conclusions in this chapter are in agreement with the Kamlet *et al.*[23] solvatochromic parameters. For example, water and glycols with a very high value for the acidic solvatochromic parameter solvate with basic compounds like aromatic hydrocarbons and also with the strong basic ethers and esters. The same is true for the interaction of a strong LA (chloroform) with a strong LB (acetone). Methanol is a strong LA but also a strong LB, thus its interaction with basic compounds is somewhat less pronounced compared to compounds which are 'mostly acidic' like water or MEG.
- Mixtures of organic acids are generally well represented except for the VLE of water with small organic acids. In this case, e.g. for water–formic acid and water–acetic acid, the CPA/HV modification is recommended (described in Appendix 11.A).

Appendix 11.A The CPA/Huron–Vidal[13] approach (CPA/HV)

This variant of CPA has been developed specifically for water–acetic acid VLE. The most important difference from the CPA equation is that, instead of the van der Waals one-fluid mixing rule for the energy parameter, the Huron–Vidal (HV) mixing rule is applied to the a parameter in SRK as follows:

$$\frac{a}{b}=\sum_i z_i\frac{a_{ii}}{b_{ii}}-\frac{g^E}{\ln 2} \tag{11.1}$$

where the modified NRTL model is used as the activity coefficient model:

$$\frac{g^E}{RT}=\sum_i z_i\frac{\sum_j z_j G_{ji} C_{ji}}{\sum_j z_j G_{ji}} \tag{11.2}$$

$$G_{ji}=b_j\exp(-a_{ij}C_{ji}) \tag{11.3}$$

$$C_{ji}=\frac{A_{ji}}{RT} \tag{11.4}$$

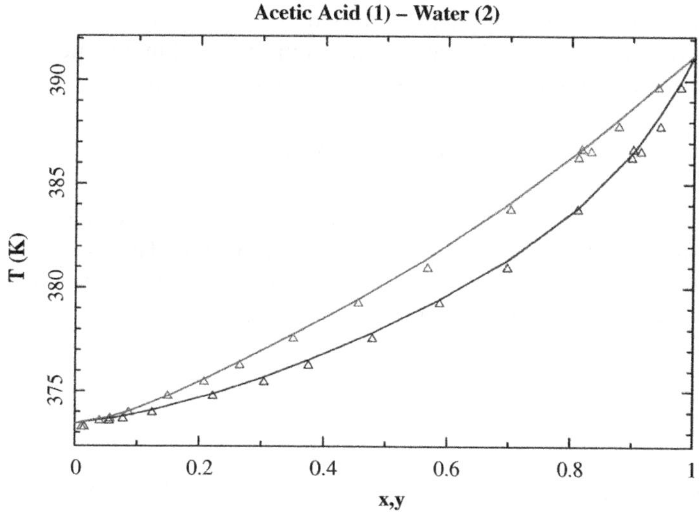

***Figure 11.38** VLE for the system acetic acid(1)–water(2) at 1 bar with the CPA/Huron–Vidal (HV) model. Experimental data are from Ito and Yoshida, J. Chem. Eng. Data, **1963**, 8, 315. The calculations are performed with CPA using the HV mixing rule and the following parameter values: $A_{12} = -8.5$ and $A_{21} = -981.6$ and $D_{12} = 0.601$. The a_{ij} value is fixed at 0.3. The 1A scheme is used for acetic acid. Reprinted with permission from Ind. Eng. Chem. Res. by Muro-Suñé et al., 47, 5660 Copyright (2008) American Chemical Society*

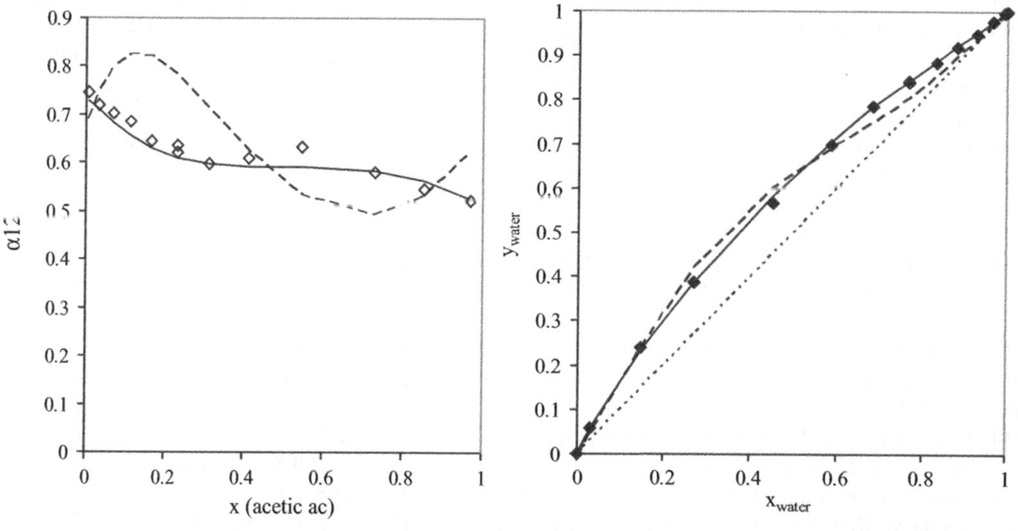

***Figure 11.39** VLE for the system acetic acid–water at 372.8 K using the CPA EoS and different mixing rules. Experimental data (◇) are from Freeman and Wilson, AIChE Symp. Ser., **1985**, 81, 14. CPA calculations with ECR and $k_{ij} = -0.223$ (– – –). CPA/Huron–Vidal calculations with parameter set 1 (——), see Table 11.5. Plots of: left, relative volatilities; right, water concentrations in liquid (x_{water}) and in vapor (y_{water})*

Thus, there are three binary parameters required (A_{ij}, A_{ji} and $\alpha_{ij} = \alpha_{ji}$). There is a fourth binary parameter (D_{ij}) which is not related to the HV mixing rule, but results from the combining rule for the cross-associating strength (the strength of association between an acid site and a water site):

$$\Delta_{ij} = D_{ij}\sqrt{\Delta_{ii}\Delta_{jj}} \tag{11.5}$$

An important advantage of the HV mixing rule is that by setting $\alpha_{ij} = 0$ and by choosing A_{ij} and A_{ji} appropriately, the classical one-fluid mixing rule with k_{ij} is recovered. This is in line with the philosophy of CPA, which itself reduces to SRK in the absence of association. Two examples of the excellent correlation performance of CPA/HV are shown in Figures 11.38 and 11.39.

References

1. J. Kofod, Modeling of phase equilibria for mixtures of relevance to the PTA process. MSc Thesis, Technical University of Denmark, 2007.
2. F.L. Mota, A.J. Queimada, S.P. Pinho, E.A. Macedo, *Ind. Eng. Chem. Res.*, **2008**, *47*, 5182.
3. M.B. Oliveira, F.R. Varanda, I.M. Marrucho, A.J. Queimada, J.A.P. Coutinho, *Ind. Eng. Chem. Res.*, **2008**, *47*, 4278.
4. G.M. Kontogeorgis, M.L. Michelsen, G. Folas, S. Derawi, N. von Solms, E.H. Stenby, *Ind. Eng. Chem. Res.*, **2006**, *45(14)*, 4869.
5. J. Gross, G. Sadowski, *Ind. Eng. Chem. Res.*, **2002**, *41*, 5510.
6. G.K. Folas, J. Gabrielsen, M.L. Michelsen, E.H. Stenby, G.M. Kontogeorgis, *Ind. Eng. Chem. Res.*, **2005**, *44*, 3823.
7. G.M. Kontogeorgis, G.K. Folas, N. Muro-Suné, N. von Solms, M.L. Michelsen, E.H. Stenby, *Fluid Phase Equilib.*, **2007**, *261*, 205.
8. G.M. Kontogeorgis, G.K. Folas, N. Muro-Suñé, F. Roca Leon, M.L. Michelsen, *Oil Gas Sci. Technol. – Rev. IFP*, **2008**, *63*, 305.
9. G.K. Folas, G.M. Kontogeorgis, M.L. Michelsen, E.H. Stenby, *Ind. Eng. Chem. Res.*, **2006**, *45(4)*, 1516.
10. M. Kaarsholm, S.O. Derawi, M.L. Michelsen, G.M. Kontogeorgis, *Ind. Eng. Chem. Res.*, **2005**, *44(12)*, 4406.
11. S.O. Derawi, J. Zeuthen, M.L. Michelsen, E.H. Stenby, G.M. Kontogeorgis, *Fluid Phase Equilib.*, **2004**, *225*, 107.
12. J.M. Prausnitz, R.N. Lichtenthaler, E.G. de Azevedo, *Molecular Thermodynamics of Fluid-Phase Equilibria* (3rd edition). Prentice Hall International, 1999.
13. N. Muro-Suñé, G.M. Kontogeorgis, N. von Solms, M.L. Michelsen, *Ind. Eng. Chem. Res.*, **2008**, *47*, 5660.
14. Y.-H. Fu, S.I. Sandler, *Ind. Eng. Chem. Res*, **1995**, *34*, 1897.
15. J.P. Wolbach, S.I. Sandler, *Ind. Eng. Chem. Res.*, **1997**, *36*, 4041.
16. J.P. Wolbach, S.I. Sandler, *Ind. Eng. Chem. Res.*, **1998**, *37*, 2917.
17. C.A. Perakis, E.C. Voutsas, K.G. Magoulas, D.P. Tassios, *Ind. Eng. Chem. Res.*, **2007**, *46(3)*, 932.
18. G.K. Folas, S.O. Derawi, M.L. Michelsen, E.H. Stenby, G.M. Kontogeorgis, *Fluid Phase Equilib.*, **2005**, *228–229*, 121.
19. T. Laursen, P. Rasmussen, S.I. Andersen, *J. Chem. Eng. Data*, **2002**, *47*, 198.
20. T. Laursen, Measurements and modeling of VLLE at elevated pressures. PhD Thesis, Technical University of Denmark, 2002.
21. N.M.F. Garrido, G.K. Folas, G.M. Kontogeorgis, *Fluid Phase Equilib.*, **2008**, *273*, 11.
22. A.S. Avlund, G.M. Kontogeorgis, M.L. Michelsen, *Ind. Eng. Chem. Res.*, **2008**, *47*, 7441.
23. M.J. Kamlet, J.M. Abboud, M.H. Abraham, R.W. Taft, *J. Org. Chem.*, **1983**, *48*, 2877.
24. G.K. Folas, G.M. Kontogeorgis, M.L. Michelsen, E.H. Stenby, *Ind. Eng. Chem. Res.*, **2006**, *45*, 1527.
25. M.B. Oliveira, J.A.P. Coutinho, A.J. Queimada, *Fluid Phase Equilib.*, **2007**, *258*, 58.
26. M.B. Oliveira, M.G. Freire, I.M. Marrucho, G.M. Kontogeorgis, A.J. Queimada, J.A.P. Coutinho, *Ind. Eng. Chem. Res.*, **2007**, *46*, 1415.
27. G.K. Folas, G.M. Kontogeorgis, M.L. Michelsen, E.H. Stenby, E. Solbraa, *J. Chem. Eng. Data*, **2006**, *51(3)*, 977.

28. G.K. Folas, G.M. Kontogeorgis, M.L. Michelsen, E.H. Stenby, *Ind. Eng. Chem. Res.*, **2006**, *45*, 1527.
29. G.K. Folas, G.M. Kontogeorgis, M.L. Michelsen, E.H. Stenby, *Fluid Phase Equilib.*, **2006**, *249*, 67.
30. A.F. Lagalante, M. Spadi, T.J. Bruno, *J. Chem. Eng. Data*, **2000**, *45*, 382.
31. M.B. Oliveira, F.R. Varanda, M.J. Melo, I.M. Marrucho, A.J. Queimada, J.A.P. Coutinho, Prediction of water solubility in biodiesel by a group contribution CPA EoS. Presented at the ESAT Conference, Cannes, France, 2008, p. 608.
32. A.S. Avlund, G.M. Kontogeorgis, M.L. Michelsen, Application of an association model to complex fluids. Presented at the AIChE Conference, Philadelphia, USA, 2008.
33. M.P. Breil, G.M. Kontogeorgis, *Ind. Eng. Chem. Res.*, **2009**, *48(11)*, 5472.
34. M.B. Oliveira, V.L. Oliveira, J.A.P. Coutinho, A.J. Queimada, *Ind. Eng. Chem. Res.*, **2009**, *48(11)*, 5530.
35. M.B. Oliveira, M.J. Pratas, I.M. Marrucho, A.J. Queimada, J.A.P. Coutinho, *AIChE J.*, **2009**, *55(6)*, 1604.

12

Extension of CPA and SAFT to New Systems: Worked Examples and Guidelines

12.1 Introduction

Application of CPA and other association equations of state (EoS) to new associating systems (for which parameters are not available in the literature) includes the following three (often interconnected) stages:

1. Choice of pure component parameters.
2. Establishment of the number of possible association sites and choice of an appropriate association scheme.
3. Cross-association and solvation schemes.

Stage 2 is at first based on Tables 8.11 and 8.12 (Chapter 8). However, various association schemes are possible for a specific compound and, moreover, various sets of parameters of CPA/SAFT models can be obtained which equally well correlate the available vapor pressure and liquid density data. Experience has shown that, when possible, it is best to select the most appropriate parameter set for an associating compound using, in addition to pure compound properties, LLE data of the compound under investigation with an inert compound, e.g. *n*-alkane. For example, water–hexane LLE data can be used to determine the optimum parameters for water, among the various sets which represent the vapor pressure and liquid density data of water equally well.

In this way, reliable parameters were estimated in the cases of water and glycols as many LLE data are available for these highly immiscible systems with alkanes. LLE data for partially miscible systems like methanol or ethanol with alkanes (for which VLE data are also available) are useful for the same purpose. There are families of compounds such as amines where LLE data are not available for mixtures with alkanes. Then, optimum sets of parameters can be estimated using, in addition to pure compound properties, other 'sensitive' data, e.g. VLE at low temperatures. Second, virial coefficients can also be used but their accuracy, especially at low temperatures, should be evaluated carefully.

In this chapter, we will illustrate how association models like CPA and SAFT can be extended to 'new' compounds and systems, not previously studied, where the association sites/schemes and types of cross-association/solvation are not known a priori. A working approach will be illustrated first for the case study of sulfolane, while other systems (aniline and phenols) will be briefly considered next. The underlying purpose of

Thermodynamic Models for Industrial Applications: From Classical and Advanced Mixing Rules to Association Theories
Georgios M. Kontogeorgis and Georgios K. Folas

the chapter is to provide 'mechanisms and working tools' for extending association EoS to new compounds and mixtures without significant changes to the framework, e.g. without introducing new terms in the models for explicit accounting of the polarity and other effects beyond those already considered in the models (physical, chain and association terms).

12.2 The case of sulfolane: CPA application

12.2.1 Introduction

This section will illustrate the application of CPA and later also the sPC–SAFT EoS to model the phase equilibria for mixtures containing sulfolane. The reason for choosing sulfolane is two-fold: (1) mixtures containing sulfolane are of industrial importance and reveal complex phase equilibria, where association models are in principle applicable; and (2) sulfolane has not been previously studied with association EoS.

Finally, this example will elucidate the applicability of association models to highly polar compounds, due to the flexibility of Wertheim's association term. In this section we will first demonstrate a methodology for obtaining optimum pure component parameters for sulfolane. The way of assigning and evaluating the association sites will also be discussed. Such a procedure is essential for the satisfactory correlation of different types of phase equilibria (VLE, LLE and SLE) of mixtures of sulfolane with both associating and non-associating components. The CPA EoS is initially used in this section, but the proposed methodology is equally well applicable to any SAFT variant, and this will be illustrated in the next section for the sPC–SAFT EoS.

Sulfolane is used as an extraction as well as a reaction solvent. It is used to separate aromatic hydrocarbons (benzene, toluene and xylenes), to separate *n*-propyl alcohol and sec-butyl alcohol and purify natural gas streams, and for fractionation of fatty acids into saturated and unsaturated components. It is also used as a reaction solvent for the preparation of aromatic sulfonic acids, pyridines, isocyanates and pharmaceuticals. It can also be involved in the halogen exchange process and polymerization process. It is used as a plasticizer and curing agent. The chemical structure of sulfolane is shown in Figure 12.1. The presence of the two lone pairs of electrons per oxygen atom as well as of the sulfur atom makes sulfolane a weak electron donor (with a donor number of 14.81), which can consequently participate in hydrogen bonding in the presence of an associating molecule. The sulfur–oxygen double bond is highly polar (with a dipole moment in the liquid phase of 4.8 D), which as expected influences the phase behavior of mixtures with sulfolane.[1]

12.2.2 Sulfolane: is it an 'inert' (non-self-associating) compound?

Based on its chemical structure alone, sulfolane is not a classical self-associating molecule. Therefore, we will first assume that it is an inert molecule; in this case CPA reduces to SRK EoS with fitted parameters to vapor pressure and liquid density data. The pure component parameters of sulfolane are given in Table 12.1, providing very satisfactory correlation of vapor pressure and liquid density data.

Having obtained a consistent set of pure component parameters for sulfolane, the VLE of binary systems of sulfolane–hydrocarbons (HC) will then be considered. If the polar nature of sulfolane is pronounced, the VLE of those systems is expected to be seriously in error. As demonstrated for example by von Solms *et al.*[2] using

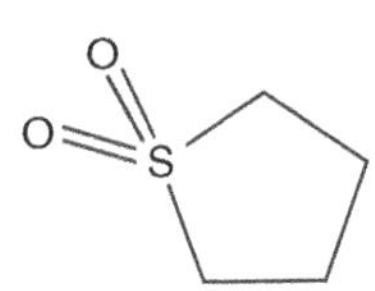

Figure 12.1 *Chemical structure of sulfolane, a very polar compound with a dipole moment equal to 4.8 D*

Table 12.1 *CPA pure component parameters for sulfolane, for $T_r = 0.55–0.90$*

	b (l/mol)	a_0 (bar l^2/mol^2)	c_1	ε^{AB}/R (K)	β^{AB}	% AAD ΔP^a	% AAD $\Delta\rho^a$
Inert	0.0939	36.115	0.878	—	—	0.3	0.9
2B	0.0952	32.674	0.692	2381.5	0.030	0.9	1.7
3B	0.0951	27.081	1.122	879.48	0.575	0.3	1.6

[a]Here:

$$\Delta P(\%) = 100 \times \frac{1}{NP}\sum_{1}^{NP}\frac{|p^{exp} - p^{calc}|}{p^{exp}} \quad \text{and} \quad \Delta\rho(\%) = 100 \times \frac{1}{NP}\sum_{1}^{NP}\frac{|\rho_{exp} - \rho_{calc}|}{\rho_{exp}}$$

with NP the number of experimental data points.

the sPC–SAFT EoS and Folas *et al.*[3] using the CPA EoS, the VLE of the acetone–pentane system is seriously in error with both models, if the polar nature of acetone is not explicitly accounted for (i.e. considering acetone to be an inert component).

However, in the case of sulfolane it is found that very satisfactory results can be obtained for the binary mixtures of sulfolane–hydrocarbons tested, as presented in Table 12.2, using in all cases a binary interaction parameter equal to zero (i.e. prediction, see also Figure 12.2). These results could be attributed to the fact that the physical term of the model accounts (implicitly) for the polar interactions between sulfolane molecules, which are rather weak at ambient conditions, as explained below. Polar effects are pronounced at lower temperatures, but when the temperature increases, such effects tend to become less important for phase equilibrium calculations. In order to investigate the polar nature of the sulfolane–hydrocarbon mixtures considered in Table 12.2, we will further investigate the SLE of the sulfolane–benzene system, where experimental data are available over a limited concentration range. For the SLE calculations the value of heat of fusion $\Delta H_{Tm}^{fus} = 1427.7\ \text{J/mol}$ is used at the melting temperature of $T_m = 301.6\ \text{K}$ assuming that the difference in the heat capacity is zero. During the phase transition between plastic phase I and crystalline phase II the value of the enthalpy of the phase transition used is $\Delta H_{Tm}^{fus} = 5353.9\ \text{J/mol}$, while the heat capacity difference used is $\Delta Cp_{Tr} = 45.51\ \text{J/mol/K}$. Those values are in accordance with the ones reported by Domańska *et al.*[4] Assuming ideal behavior in the solid phase and real solution behavior in the liquid phase, the experimental activity coefficients can be calculated; they are found to be $\sim$1 for temperatures up to 280 K. Hence, at increased temperatures, where the VLE data are available, polar effects are indeed negligible, which provides an explanation of the zero binary interaction parameter used for the correlation of the mixtures with the CPA EoS. At lower temperatures the system deviates from ideality. This explains the sensitivity of the SLE calculations with CPA (shown in Figure 12.3) to the binary interaction parameter, as well as the inferior results

Table 12.2 *VLE correlation of sulfolane(1)–hydrocarbon(2) binary systems with the CPA and sPC–SAFT EoS. Experimental data are from Benoit et al., Can. J. Chem.,* ***1969****, 47, 4195 and Karvo, J. Chem. Thermodyn.,* ***1980****, 12, 1175. Δp% as defined in Table 12.1*

Component (2)	T (K)	CPA, inert		CPA, 2B		sPC–SAFT, 2B	
		k_{12}	ΔP (%)	k_{12}	ΔP (%)	k_{12}	ΔP (%)
Benzene	303.15	0.0	4.7	−0.016	5.1	−0.02	5.1
	313.15	0.0	4.6	−0.016	4.8	−0.02	4.8
	323.15	0.0	4.2	−0.016	4.1	−0.02	4.1
	333.15	0.0	4.3	−0.016	3.7	−0.02	3.7
Toluene	303.15	0.0	6.0	−0.01	5.2	−0.01	5.3
	313.15	0.0	5.9	−0.01	4.1	−0.01	4.3
	323.15	0.0	5.6	−0.01	3.5	−0.01	3.6
	333.15	0.0	5.3	−0.01	3.4	−0.01	3.7

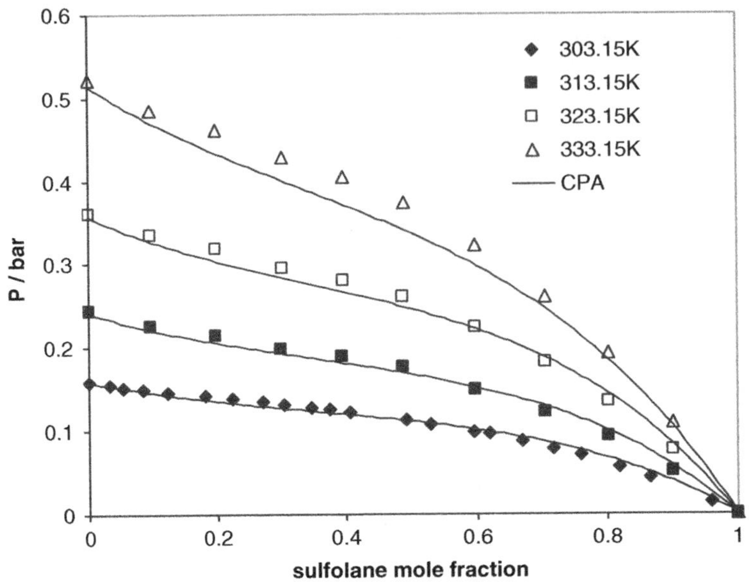

Figure 12.2 *VLE correlation with CPA for the sulfolane–benzene mixture with a binary interaction parameter equal to zero ($k_{12}=0.0$). Experimental data are from Benoit et al., Can. J. Chem.,* ***1969****, 47, 4195 and Karvo, J. Chem. Thermodyn.,* ***1980****, 12, 1175*

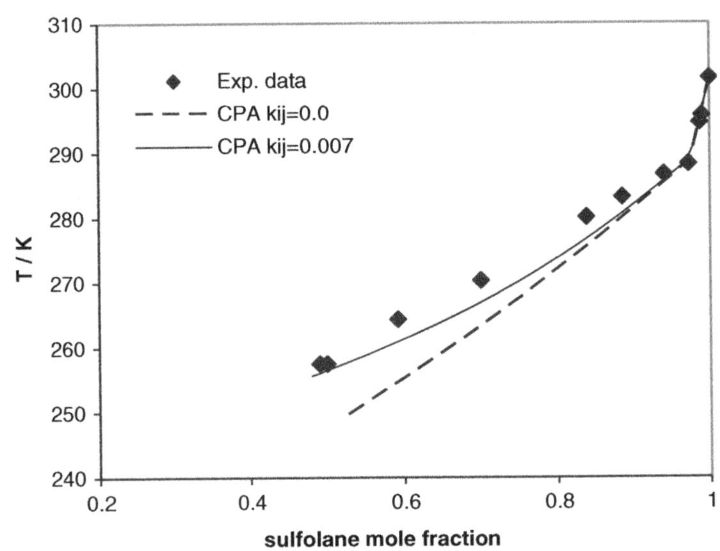

Figure 12.3 *SLE prediction ($k_{12}=0.0$) and correlation ($k_{12}=0.007$ obtained from SLE data) with CPA for the sulfolane–benzene mixture. Experimental data are from Domańska et al.*[4]

obtained when $k_{12}=0$ is used (optimum value from VLE). However, a temperature-independent binary interaction parameter ($k_{12}=0.007$) for sulfolane–benzene, obtained from SLE data, can be successfully used for the VLE region of this system.

Binary mixtures of sulfolane with associating components will be considered next. When considering these components, e.g. water or methanol, the electron donor groups of the sulfolane molecule can participate in hydrogen bonds. Indeed, when assuming that sulfolane does not interact (i.e. solvate) with water, then the VLE correlation of the sulfolane–water system is seriously in error, implying that the binary interaction parameter in the physical term of the model cannot account for the pronounced attractive forces between the molecules. The solvation between sulfolane and water molecules can, however, be satisfactorily taken into account within the framework of the Wertheim theory. As shown by Folas *et al.*[5] for the CPA EoS, which can be straightforwardly applied to SAFT variants as well, the use of the modified CR-1 rule provides satisfactory results for binary and multicomponent systems of solvating but not self-associating components, as for binary mixtures of BTEX and water or glycols, and for ternary mixtures containing BTEX, glycols and water. Recently Oliveira *et al.*[6] showed that the modified CR-1 rule can be successfully applied to binary water–aromatic fluorocarbon mixtures.

We can assume, for example, that the sulfolane molecule has four negatives sites, which is in agreement with the chemical structure of the molecule (i.e. two lone pairs of electrons per oxygen atom), which are able to form hydrogen bonds with the two hydrogen atoms of the water molecule. Excellent correlation results can be achieved in this way, using a zero binary interaction parameter in the physical term (k_{12}). The use of the modified CR-1 rule assumes that the cross-association energy equals half the energy of the associating component (water in this case). This approach would normally require both the use of a binary interaction parameter in the physical term (k_{12}) and a value for the cross-associating volume, which are regressed from experimental data. This is, however, only an approximation, since there is in principle no reason for the cross-association energy to be half that of the associating component.

In order to elucidate further the flexibility of the model, we will follow a slightly different procedure, which results in a different mathematical contribution of the physical and association terms of CPA, when compared to the modified CR-1 rule. In this case, both cross-association energy and volume will be fitted to the experimental data, while setting the binary interaction parameter in the physical term (i.e. SRK term) equal to zero. Such an alternative procedure is applied to the water–sulfolane and methanol–sulfolane binary systems, providing satisfactory results over an extended temperature range. Typical results are shown in Figures 12.4 and 12.5, demonstrating the adequacy of the in-built temperature dependency of the association strength and of the association term in general.

It is important to notice that both the SRK term and the Wertheim term contribute to the in-built temperature dependency of the model. With respect to the association term, the in-built temperature dependency is maintained by simultaneously fitting the energy and volume parameters of the cross-association strength. As Folas *et al.*[5] showed, the in-built temperature dependency of the SRK term alone does not provide satisfactory results when fitting the cross-association strength of the model, $\Delta^{A_iB_j}$, to a fixed value (i.e. eliminating the in-built temperature dependency of the association term).

The analysis so far indicates that binary mixtures containing sulfolane can be satisfactorily correlated assuming sulfolane to be an inert molecule. CPA (or actually SRK with fitted parameters to vapor pressure and liquid density data) correlates well the VLE and SLE of sulfolane–hydrocarbon systems. Moreover, accounting for the solvation in the presence of associating components, excellent VLE results are obtained for thermodynamically difficult systems such as water–sulfolane or methanol–sulfolane, where classical EoS fail. VLE correlation results for the solvating mixtures are not explicitly presented in this section; however, the errors are low and very similar to those presented in Table 12.3 when the 2B association scheme is used for sulfolane (for reasons explained later).

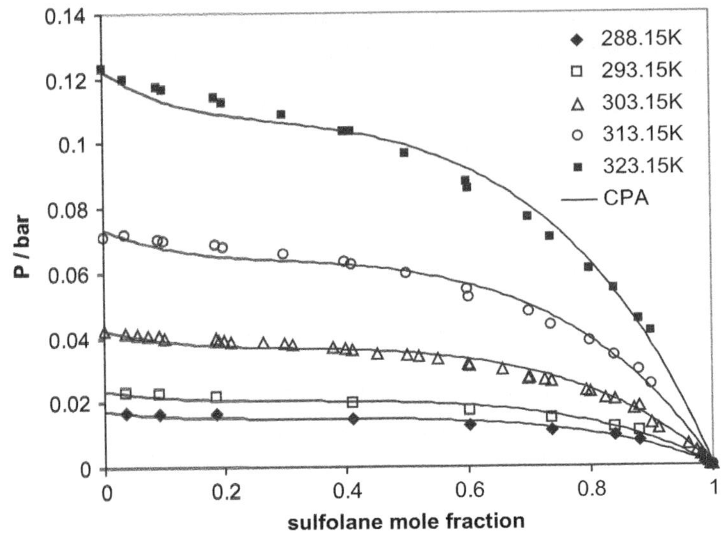

Figure 12.4 VLE correlation with CPA for sulfolane–water using $\varepsilon^{AB} = 1635.8\,K$ and $\beta^{AB} = 0.029$ ($k_{ij} = 0$). Experimental data are from Benoit et al., Can. J. Chem., **1968**, 46, 3215 and Tommila et al., Suomi Kemistil. B, **1969**, 42, 95

12.2.3 Sulfolane as a self-associating compound

In the lack of LLE data, the conclusion would have been that the thermodynamics of sulfolane mixtures can be satisfactorily represented by the CPA model assuming that sulfolane is an inert component. Some important additional factors should be considered, however:

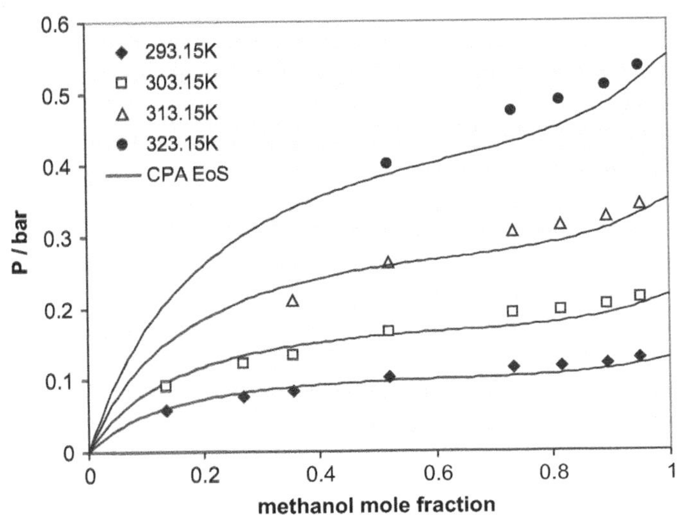

Figure 12.5 VLE correlation with CPA for sulfolane–methanol using $\varepsilon^{AB} = 2381.5\,K$ and $\beta^{AB} = 0.003$ ($k_{ij} = 0$). Experimental data are from Tommila et al., Suomi Kemistil. B, **1969**, 42, 95

*Table 12.3 VLE correlation of sulfolane(1)–associating component(2) binary systems with the CPA and sPC–SAFT EoS, using the 2B association scheme for sulfolane. Experimental data are from Benoit et al., Can. J. Chem., **1968**, 46, 3215 and Tommila et al., Suomi Kemistil. B, **1969**, 42, 95. Δp% as defined in Table 12.1*

Component (2)	T (K)	CPA		sPC–SAFT	
		k_{12}	ΔP (%)	k_{12}	ΔP (%)
Water	288.15	−0.15	7.4	−0.05	5.7
	293.15	−0.15	5.6	−0.05	4.6
	303.15	−0.15	5.6	−0.05	4.6
	313.15	−0.15	4.8	−0.05	4.2
	323.15	−0.15	6.5	−0.05	3.7
Methanol	288.15	0.06	6.3	0.04	5.4
	293.15	0.06	6.4	0.04	5.0
	303.15	0.06	6.4	0.04	4.7
	313.15	0.06	4.0	0.04	4.1
	323.15	0.06	3.5	0.04	3.3

1. Infinite dilution activity coefficient data, as presented by Moollan *et al.*[7] reveal that the activity coefficients of sulfolane in various hydrocarbons are of magnitude similar to those for methanol–hydrocarbon mixtures. For example, the infinite dilution activity coefficients are 58, 75 and 98 for sulfolane–pentane, hexane and heptane, respectively.
2. Recent LLE data for sulfolane–cycloalkane binary mixtures have been reported by Ko *et al.*[8] and Im *et al.*[9]
3. The infinite dilution activity coefficients for sulfolane–benzene and sulfolane–toluene are only around 2.4 and 4.8, respectively, a clear indication of solvation in the latter system (compare the values reported for sulfolane–hexane and sulfolane–heptane).

Considering, for example, the LLE of the sulfolane–methylcyclohexane mixture, a major limitation is revealed if sulfolane is assumed to be an inert molecule. When fitting the binary interaction parameter to the solubility of methylcyclohexane in the sulfolane phase, the solubility of sulfolane in the hydrocarbon phase is erroneously calculated to be higher than that of methylcyclohexane in the sulfolane phase. The LLE correlation of the system is shown in Figure 12.6. This behavior, which is opposite to the physical picture of the system, is characteristic of the other binary sulfolane–alkane mixtures considered. In order to overcome such a limitation we assume that sulfolane is a self-associating molecule. Table 12.1 presents CPA pure component parameters for sulfolane assuming that it behaves as either a 2B or a 3B molecule. To obtain the sulfolane pure component parameters for the 2B and 3B schemes presented in Table 12.1, LLE binary data of sulfolane with various alkanes have been considered together with vapor pressure and liquid density data, for screening purposes. This is because several sets of pure component parameters can equally well correlate vapor pressure and liquid density data within the framework of association theories. The challenge is, in this case, to obtain a set of parameters which also provides satisfactory LLE correlation results for the binary mixtures of the associating compound of interest. A procedure for obtaining a set of parameters which equally well correlates the LLE of the binary systems of interest and also provides satisfactory correlation of vapor pressure and liquid density data is the following:

1. All five parameters should initially be simultaneously regressed based on vapor pressure and liquid density ‘experimental’ data (which are, in principle, generated from correlation equations provided from DIPPR). This set of parameters should be further used for correlating the LLE of binary mixtures of the

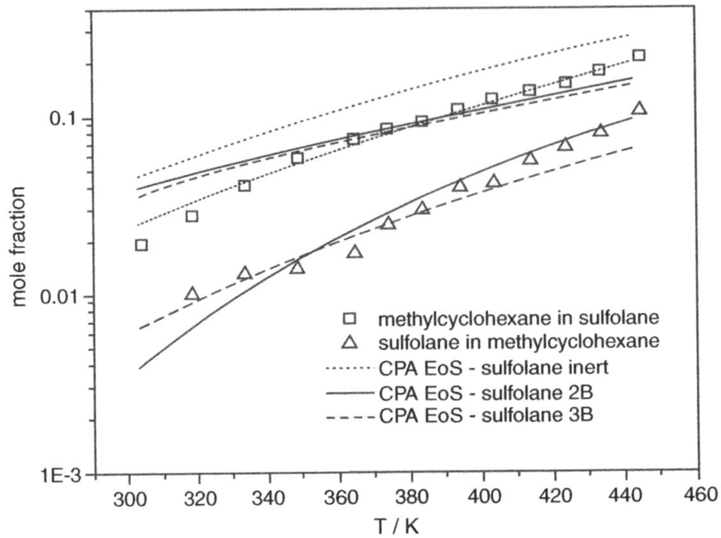

Figure 12.6 *LLE correlation with CPA for the sulfolane–methylcyclohexane mixture with a common and temperature-independent binary interaction parameter* $k_{12} = 0.043$ *for the inert, 0.035 for the 3B and 0.049 for the 2B schemes. Experimental data are from Im et al.*[9]

associating compound (sulfolane in our example), preferably with an inert component. This is because the uncertainty of the combining rule (needed for cross-associating mixtures) is eliminated. The binary interaction parameter should be normally fitted to the mutual solubility of one component in one of the two phases. Parameters which are directly related to measured quantities e.g. enthalpy or entropy of hydrogen bonding for specific compounds or similar ones, as already discussed in Chapters 9–11, should maintain values that are in reasonable agreement to experimentally determined ones. Furthermore, if such measurements are available they provide excellent initial values for the minimization procedure.

2. Further, the value of the association energy should be increased by a factor of 0.1 of the optimized value in order to investigate whether this results in correlated solubilities which are closer to the experimental ones. Before performing LLE calculations, the remaining four pure component parameters should be regressed using vapor pressure and liquid density data, whereas for LLE calculations with the new set of parameters the binary interaction parameter might be slightly different from the one obtained initially.
3. If the calculated solubilities are further removed from the experimental ones when compared to those obtained in step 1, then the value of the association energy parameter of the initial set of parameters (step 1) should be reduced by a factor of 0.1 and the aforementioned procedure should be repeated.

Such a methodology is helpful for an initial screening of an appropriate value of the association energy parameter. Tuning this parameter is preferable to the association volume one, because of the pronounced temperature dependency of the association energy in the expression for the association strength. Occasionally, while increasing or decreasing the value of the association energy, the association volume parameter should be kept constant (i.e. allowing a simultaneous regression of the remaining three parameters). This is because the influence of the contribution of the association term to the overall performance of the model becomes more pronounced in this case. Such a remark is applicable not only to CPA but to other SAFT variants as well. Following such a methodology provides a safe screening of the trend of the association parameters (i.e. increased or decreased energy and volume parameters with respect to the values initially obtained) and consequently of the overall contribution of the association term. It is important to note that such a challenge in obtaining an optimum set of parameters is related to mixtures that exhibit LLE. When only VLE or SLE data are considered, most of the sets of pure component parameters perform very similarly, when pure component vapor pressure and liquid density data are satisfactorily correlated with only slightly different values of the binary interaction parameters.

Returning to our example, the LLE correlation of the sulfolane–methylcyclohexane mixture is very similar with both the 2B and 3B association schemes, as shown in Figure 12.6. Similar conclusions can be obtained for the LLE correlation of the other mixtures where experimental LLE data are available. Furthermore, when modeling sulfolane as a self-associating compound, the model correctly calculates the solubility of sulfolane in the alkane phase to be lower than the solubility of alkane in the sulfolane phase. Figure 12.7 presents correlation results with the NRTL model as published by Ko *et al.*[8] using a temperature-dependent expression for the interaction parameter τ_{ij} for the sulfolane–cyclooctane mixture. Using an activity coefficient model with seven parameters (per binary system) explicitly fitted to the experimental data does not provide superior results compared to CPA, especially if temperature-dependent binary interaction parameters are used in CPA.

The VLE of sulfolane–methanol and sulfolane–water mixtures is considered next, to allow for further evaluation of the two different association schemes for sulfolane. ECR is used in both cases since this rule provides lower errors when compared to CR-1. The 2B scheme for sulfolane is superior for these complex mixtures when compared to the 3B association scheme. As Figure 12.8 demonstrates for the VLE of the sulfolane–water mixture, CPA using a single and temperature-independent binary interaction parameter correlates the experimental data very satisfactorily with the 2B scheme (for the VLE correlation results see Table 12.3). When the 3B scheme is used, higher errors are seen (at least 2% higher error in vapor pressure for sulfolane–water, when compared to the 2B scheme), even when the binary interaction parameter is fitted per isotherm. The 3B scheme predicts an incorrect phase split (i.e. the maximum in pressure); consequently, when forcing the model to predict the correct phase behavior by tuning the binary interaction parameter, the error in vapor pressure increases. The VLE of the methanol–sulfolane binary system is equally satisfactory using a common binary interaction parameter of 0.06, as Table 12.3 shows.

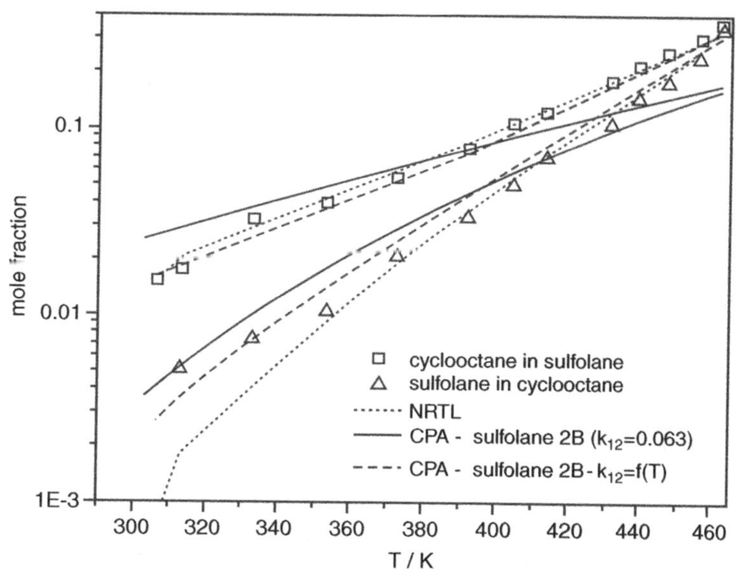

Figure 12.7 *LLE correlation with CPA and NRTL for sulfolane–cyclooctane. A common and temperature- independent binary interaction parameter $k_{12} = 0.063$ is used for CPA and the 2B scheme for sulfolane, or a temperature-dependent k_{12}. The performance of NRTL with a temperature-dependent τ_{ij} expression is also presented. Experimental data and the parameters for NRTL model are from Ko et al.*[8]

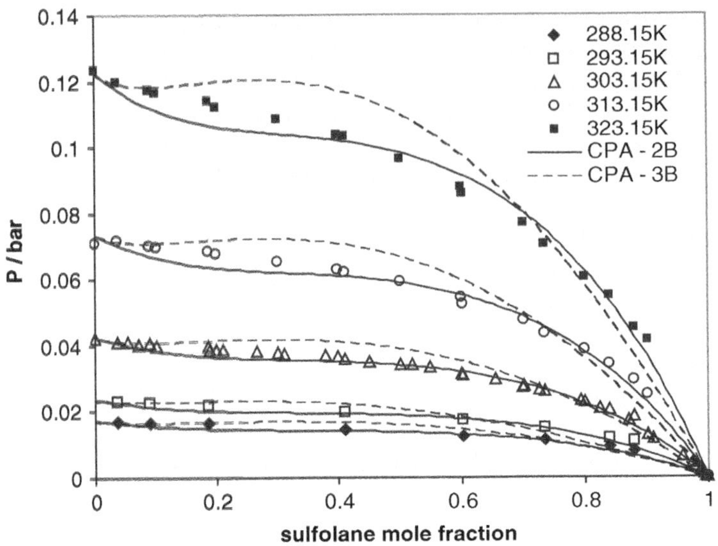

Figure 12.8 *VLE correlation for sulfolane–water with a common and temperature-independent binary interaction parameter $k_{12} = -0.15$ using the 2B scheme for sulfolane, or the 3B scheme with $k_{12} = -0.035$. Experimental data are from Benoit et al., Can. J. Chem., **1968**, 46, 3215 and Tommila et al., Suomi Kemistil. B, **1969**, 42, 95*

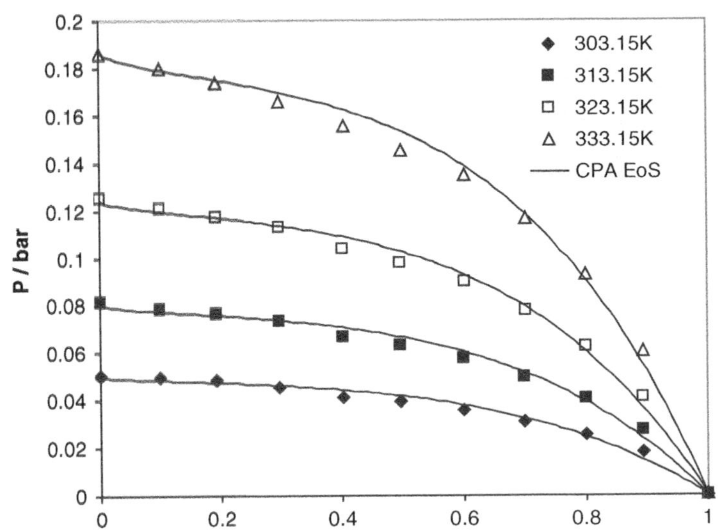

Figure 12.9 *VLE correlation with CPA for sulfolane–toluene with a common and temperature-independent binary interaction parameter $k_{12} = -0.01$ using the 2B scheme for sulfolane. Experimental data are from Karvo, J. Chem. Thermodyn., **1980**, 12, 1175*

We have further calculated with CPA the VLE of sulfolane–aromatic hydrocarbon systems in order to evaluate the 2B scheme parameters for sulfolane. An indication of solvation, between sulfolane and aromatic hydrocarbons, is that these binary systems are completely miscible even at low temperatures, whereas the aliphatic hydrocarbons with the same carbon number are partially miscible. For instance, sulfolane–heptane is partially miscible while only VLE data exist for sulfolane–toluene. Satisfactory VLE correlation results are obtained in all cases. Typical VLE results for the sulfolane–toluene mixture are presented in Figure 12.9, using a common and temperature-independent binary interaction parameter $k_{12} = -0.01$. An interesting observation is that a single and low value of the binary interaction parameter in the physical term of the model is sufficient in CPA to describe the solvating effects between the sulfolane and toluene molecules. As Folas *et al.*[5] showed for alcohol–aromatic hydrocarbon binary mixtures, in cases of weak solvating effects the use of a binary interaction parameter accounts satisfactorily for the solvation even for typical associating components such as methanol and ethanol. In these cases, the use of the modified CR-1 rule is not needed. The VLE correlation results are summarized in Table 12.2.

12.3 Application of sPC–SAFT to sulfolane-related systems

This section presents the application of sPC–SAFT to mixtures containing sulfolane. We will: (1) illustrate certain similarities with respect to the methodology used in association models when they are applied to complex phase equilibria; and (2) identify differences between CPA and PC–SAFT. The two models use similar association terms, but CPA employs a semi-empirical approach for the physical interactions (i.e. the SRK EoS) compared to a theoretically based one used in sPC–SAFT.

Table 12.4 presents pure component parameters for sulfolane with the sPC–SAFT EoS, assuming that sulfolane is either an inert molecule, or has two or three association sites (2B and 3B schemes), in a way similar to what was described in the previous section for CPA. Sulfolane–hydrocarbon VLE is satisfactorily correlated when sulfolane is assumed to be an inert molecule, with an accuracy similar to that presented in Table 12.2 for CPA. However, the performance of sPC–SAFT for LLE of sulfolane–hydrocarbon systems is poor. Actually, the solubility of sulfolane in the hydrocarbon phase is erroneously calculated to be higher than the solubility of the hydrocarbon in the sulfolane phase. This was also shown to be the case when CPA was used.

The LLE correlation with sPC–SAFT for sulfolane–hydrocarbon systems is satisfactory and equally good using either the 2B or the 3B scheme. Furthermore, the results are overall similar to CPA, as typically demonstrated in Figures 12.10 and 12.11 in the case of sulfolane–cyclohexane and sulfolane–methylcyclohexane LLE, respectively. The performance of CPA for the system sulfolane–methylcyclohexane was presented in Figure 12.6. Figure 12.11 shows the correlation results with the NRTL model as published by Im *et al.*[9] using a temperature-dependent interaction parameter τ_{ij} with seven parameters per binary mixture. Even though a single binary interaction parameter is used over the entire temperature range with CPA and sPC–SAFT, the results are satisfactory with the two association EoS. This observation by no means implies

Table 12.4 *The sPC–SAFT pure component parameters for sulfolane, for $T_r = 0.55$–0.90. Percentage deviations are as defined in Table 12.1*

	σ (Å)	ε/k (K)	m	ε^{AB}/R (K)	κ^{AB}	% AAD ΔP	% AAD $\Delta\rho$
Inert	3.5050	394.52	3.3236	—	—	0.4	1.5
2B	3.6483	365.05	3.0465	1280	0.959	0.4	2.2
3B	3.7593	384.59	2.8336	1137	0.372	0.4	2.4

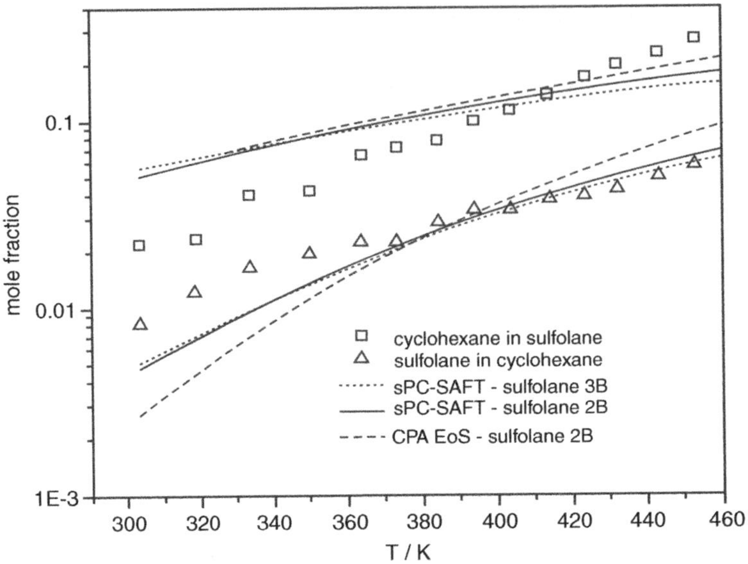

Figure 12.10 LLE correlation for the sulfolane–cyclohexane mixture with CPA and sPC–SAFT. A common and temperature-independent binary interaction parameter $k_{12} = 0.071$ is used for CPA with the 2B scheme for sulfolane. For sPC–SAFT, $k_{12} = 0.031$ (2B scheme for sulfolane) and $k_{12} = 0.029$ (3B scheme for sulfolane). Experimental data are from Ko et al.[8]

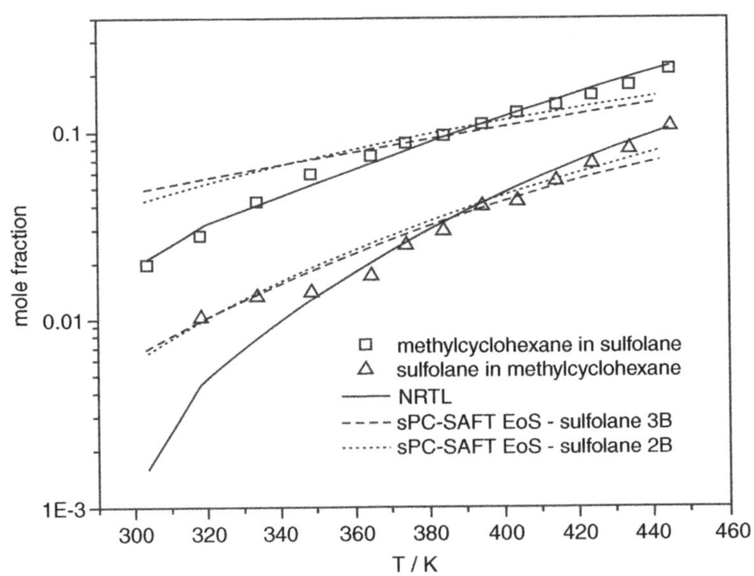

Figure 12.11 LLE correlation for the sulfolane–methylcyclohexane mixture using sPC–SAFT and NRTL. For sPC–SAFT a common and temperature-independent binary interaction parameter is used; $k_{12} = 0.018$ for the sulfolane 2B scheme and $k_{12} = 0.016$ for the sulfolane 3B scheme. The performance of NRTL with a temperature-dependent τ_{ij} expression is also presented. Experimental data and the parameters for the NRTL model are from Im et al.[9]

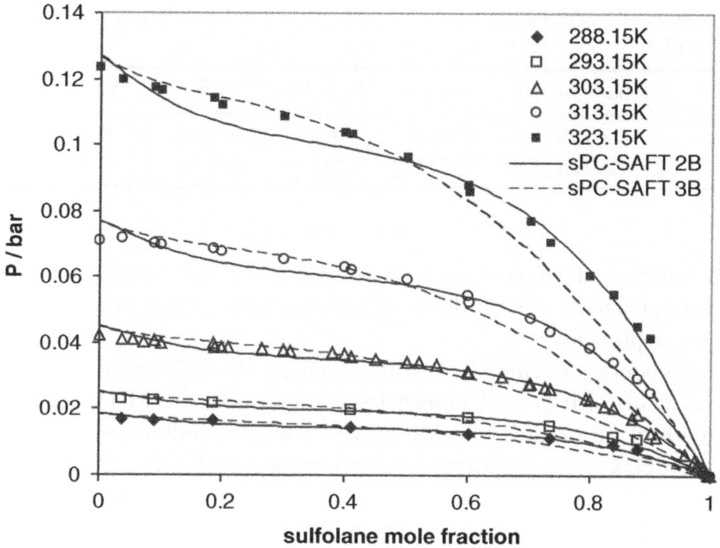

Figure 12.12 *VLE correlation with sPC–SAFT for sulfolane–water with a common and temperature-independent binary interaction parameter $k_{12} = -0.05$ with the 2B scheme for sulfolane, or the 3B scheme, and $k_{12} = 0.0$ (overall optimum). Experimental data are from Benoit et al., Can. J. Chem., **1968**, 46, 3215 and Tommila et al., Suomi Kemistil. B, **1969**, 42, 95*

that local-composition-based activity coefficient models do not provide satisfactory phase equilibria results for complex mixtures, especially when they are tailor-made for specific systems. Improved results can be obtained with CPA and PC–SAFT by using temperature-dependent binary interaction parameters. The case study of sulfolane indicates that association models are promising tools for applications of interest to the chemical industry. More applications of CPA and sPC–SAFT to systems of interest to the chemical industry will be presented in the next section, where additional features of the association models will be described.

Similar to CPA, the 3B association scheme for sulfolane combined with sPC–SAFT provides inferior VLE results for the complex phase equilibria of sulfolane–water and sulfolane–methanol systems. Typical results for the water–sulfolane mixture are presented in Figure 12.12. The correlative performance of sPC–SAFT is almost identical to CPA when the 2B scheme is used for sulfolane (see Tables 12.2 and 12.3), with the use of a lower value of the binary interaction parameter (k_{12}). Finally the VLE correlation of sulfolane–hydrocarbon mixtures is also very similar for both models, as Table 12.2 shows.

12.4 Applicability of association theories and cubic EoS with advanced mixing rules (EoS/G^E models) to polar chemicals

Association theories, especially when applied within an engineering framework where some simplifications are allowed, often provide a flexible tool for thermodynamic calculations, as shown in the previous section for sulfolane. Some recent publications present successful phase equilibria calculations with association models for complex systems of polar chemicals, such as ketones or ethers with sPC–SAFT by von Solms *et al.*[2] and CPA using SRK or PR for the physical interactions by Folas *et al.*[3] and Voutsas *et al.*[10] respectively. In these

Table 12.5 Aniline pure component parameters for the CPA EoS. Percentage deviations are as defined in Table 12.1

	b (l/mol)	a_0 (bar l^2/mol^2)	c_1	ε^{AB}/R (K)	β^{AB}	% AAD ΔP	% AAD $\Delta\rho$
2B	0.0854	23.155	0.9249	1307.4	0.061	0.1	0.4
3B	0.0856	23.326	0.9215	1203	0.053	0.1	0.5

publications the polar components were treated as pseudo-associating ones, providing very satisfactory correlations of binary mixtures but also predictions of multicomponent ones (see also Chapters 10 and 13 for further discussion of this approach).

Classical engineering models, e.g. cubic EoS with advanced (EoS/G^E) mixing rules, offer an alternative approach to association theories. It is well known for example that the MHV1 and MHV2 mixing rules, originating from the work of Mollerup[11] and Michelsen[12], when combined with a classical EoS, provide satisfactory results for complex phase equilibria calculations of associating and polar components (Dahl and Michelsen[13]; for details see Chapter 6). A limitation of this method is, however, that VLE UNIFAC tables do not provide satisfactory LLE results. Another widely used engineering model within the oil industry is the SRK EoS using the Huron–Vidal mixing rule. This flexible and rather simple model, especially when compared to association EoS, provides satisfactory results for multicomponent, multiphase equilibria of complex mixtures of hydrocarbons–water–gas hydrate inhibitors.[14,15] The applicability of association and cubic EoS with advanced mixing rules to additional mixtures of interest to the chemical industry is briefly discussed in this section, in order to compare alternative approaches for phase equilibria calculations of complex systems.

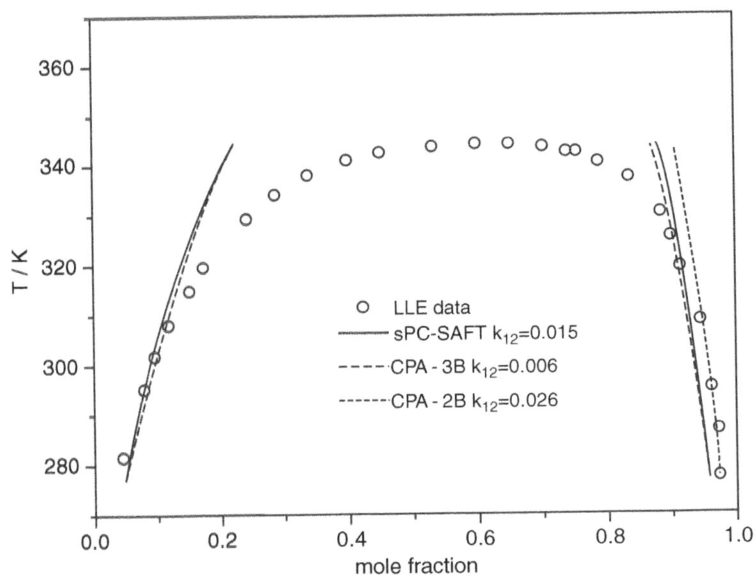

Figure 12.13 *LLE correlation for aniline–octane with the CPA (using both the 3B and 2B association schemes for aniline) and sPC–SAFT EoS with the 3B scheme. Experimental data are from Arlt and Onken, Chem. Eng. Commun.,* ***1982****, 15, 207*

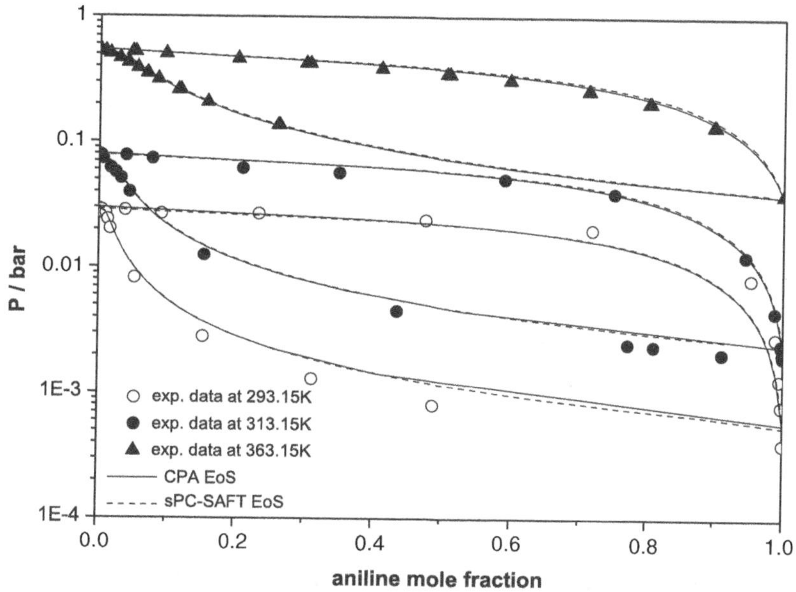

Figure 12.14 *VLE prediction ($k_{ij} = 0$) for the aniline–toluene binary mixture with the CPA and sPC–SAFT EoS, using the 2B association scheme for aniline. Experimental data are from Schneider, Z. Phys. Chem. (Frankfurt), **1960**, 24, 165 and Billes et al., Acta Chim. (Budapest), **1963**, 35, 147*

Association models and classical EoS with advanced mixing rules can satisfactorily correlate VLE of binary mixtures with aniline. ESD (Elliott–Suresh–Donohue) and sPC–SAFT were recently applied to such systems by Grenner *et al.*[16] The pure component parameters for aniline with CPA are presented in Table 12.5, using the 2B and 3B association schemes for aniline, as also done with sPC–SAFT by Grenner *et al.*[16] The LLE and VLE performance of CPA is very similar to sPC–SAFT, as Figure 12.13 shows for the LLE of aniline–octane binary mixtures (average error in vapor pressure of 3.5%) and Figure 12.14 for the VLE of aniline–toluene over an extended temperature range. Satisfactory VLE results can be obtained with SRK using the Huron–Vidal mixing rule combined with the modified NRTL, but inferior results are obtained with the MHV2 mixing rule combined with the modified UNIFAC of Larsen (see Figure 12.15). The Matthias– Copeman expression is used in SRK for both models. LLE prediction of aniline–water is also satisfactorily correlated with SRK using the Huron–Vidal mixing rule with modified NRTL, providing similar results to CPA and sPC–SAFT with ECR, as Figure 12.16 demonstrates. The aniline partition coefficient (defined as the mole fraction of aniline in toluene over the mole fraction of aniline in the aqueous phase) is predicted equally well, both with an association model (CPA) and with SRK using the Huron–Vidal mixing rule for the ternary mixture of aniline–water–toluene (see Figure 12.17), where several interactions, due to both the presence of an associating component and the aromatic ring of toluene, occur. Both models perform better compared to UNIQUAC or NRTL as presented by Sørensen and Arlt[17].

12.5 Phenols

Determining the 'optimum' five parameters for associating compounds for CPA (and other SAFT-type models) cannot often be based solely on vapor pressure and liquid density data. This is because several parameter sets can reproduce the pure compound data within experimental accuracy. As already discussed, one

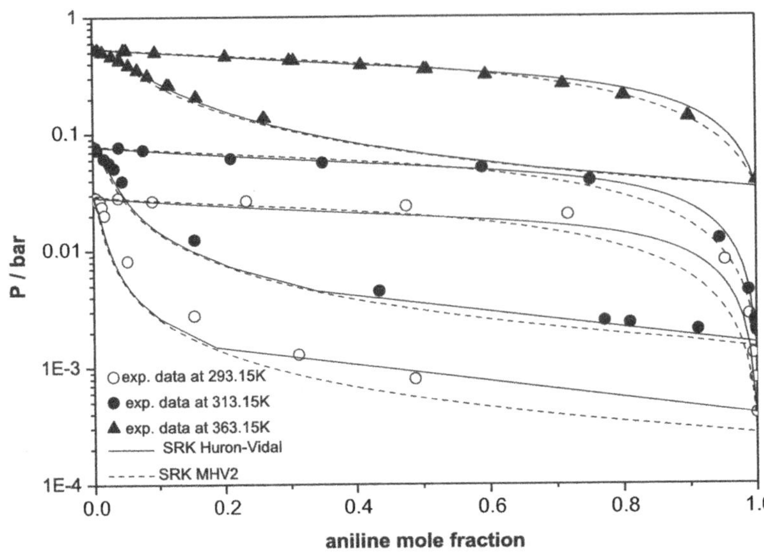

Figure 12.15 VLE correlation for aniline(1)–toluene(2) with the SRK EoS using the Huron–Vidal mixing rule combined with modified NRTL ($a_{12} = 0.2$, $g_{12}-g_{22}/R = 300$ and $g_{21}-g_{11}/R = 145$), and the SRK EoS using the MHV2 mixing rule combined with the modified UNIFAC of Larsen et al. (see Chapter 5 for details)

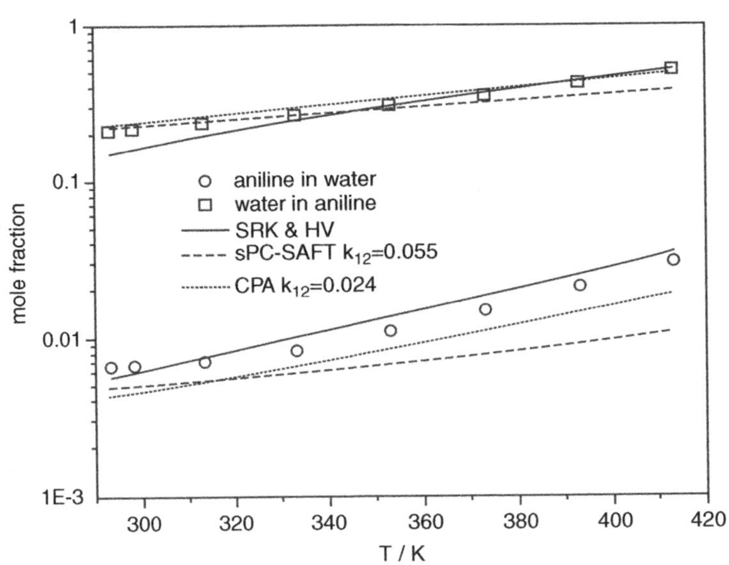

Figure 12.16 LLE correlation for the aniline(1)–water(2) binary mixture with CPA and sPC–SAFT EoS using the 3B association scheme for aniline, and SRK EoS using the Huron–Vidal (HV) mixing rule combined with modified NRTL ($a_{12} = 0.2$, $g_{12}-g_{22}/R = 8800$, $g_{21}-g_{11}/R = 111$). Experimental data are from Sørensen and Arlt[17]

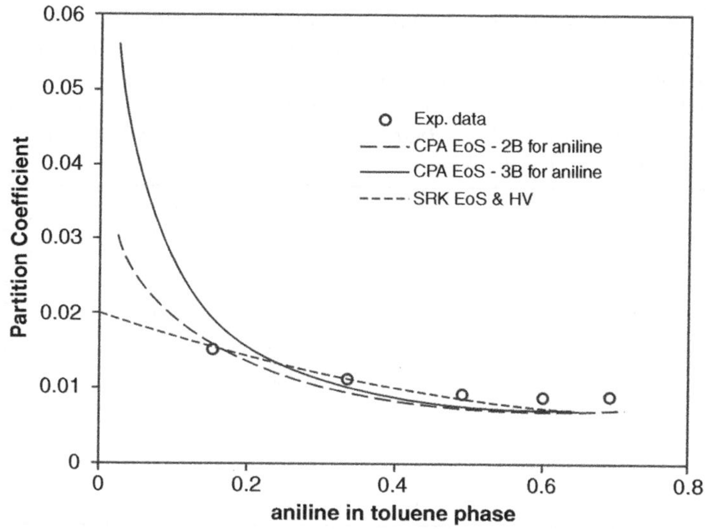

Figure 12.17 *Prediction of the aniline partition coefficient at 298.15 K for the ternary mixture aniline(1)–water (2)–toluene(3). The models used are: CPA EoS with two different association schemes for aniline; and SRK using the Huron–Vidal (HV) mixing rule combined with modified NRTL. CPA accounts for the solvation between water and toluene using the modified CR-1 rule. Experimental data are from Sørensen and Arlt.*[17] *The following interaction parameters are used for CPA: 2B aniline,* $k_{12} = -0.026$, $k_{13} = 0.0$, $k_{23} = 0.0095$, $\beta^{A_iB_j} = 0.06$; *3B aniline,* $k_{12} = 0.024$, $k_{13} = 0.0$, $k_{23} = 0.0095$, $\beta^{A_iB_j} = 0.06$

way to select an optimum set is by incorporating mixture data in the parameter selection. The best choice are data of the self-associating compound with an alkane, but as such data may not always be available, selection can sometimes be based on cross-associating mixtures. This will be illustrated in this section for *m*-cresol (assumed to be described by the 2B scheme also used for alcohols). Several parameter sets are obtained based on pure compound properties and they are shown in Table 12.6. The different sets of parameters are initially estimated from the correlation of the experimental vapor pressure and liquid density data in the reduced temperature range 0.5–0.9. Parameter set *1* is obtained by fixing the value of the co-volume parameter, b, during the estimation procedure. For parameter set 4 the co-volume and the association energy parameters (ε) are fixed. For parameter set 2 the association energy was fixed, while for parameter sets 3 and 5 none of the parameters were a priori fixed.

Table 12.6 *CPA pure compound parameter sets for m-cresol and average deviations between experimental and calculated vapor pressures and liquid densities. After Kontogeorgis et al.*[19] *Percentage deviations are as defined in Table 12.1*

Set	a_0 (bar l^2/mol^2)	b (l/mol)	c_1	ε^{AB} (bar l/mol)	β^{AB}	ΔP (%)	$\Delta\rho$ (%)
1	28.411	0.097 54	0.820 74	200.08	0.0067	0.11	0.55
2	28.821	0.097 57	0.931 41	170.00	0.0068	0.39	0.58
3	28.459	0.097 69	0.824 13	199.30	0.0067	0.13	0.52
4	28.718	0.098 84	0.923 36	170.00	0.0093	0.54	1.27
5	25.500	0.096 51	0.650 73	248.71	0.0086	1.16	0.75

Table 12.7 *Average deviations in pressure and vapor compositions with different sets of parameters for m-cresol–alkanes. Experimental data are from Simnick et al., Fluid Phase Equilib., **1979**, 3, 145; Niederbroecker and Schmelzer, Int. Electron. J. Phys.-Chem. Data, **1997**, 3, 197; Schmelzer et al., Int. Electron. J. Phys.-Chem. Data, **1996**, 2, 153. Percentage deviations are as defined in Table 12.1. After Kontogeorgis et al.*[19]

Alkane	T (K)	Set	k_{ij}	ΔP (%)	$\Delta y \cdot 100$
Methane	462, 542	1	−0.034	3.2	0.37
		3	−0.0326	3.1	0.37
		5	−0.0019	5.3	0.71
n-Nonane	373, 413	1	0.0062	2.3	0.53
		3	0.0066	2.4	0.54
		5	0.0075	2.2	0.75
n-Decane	373, 433	1	0.0017	2.2	0.80
		3	0.0021	2.3	0.80
		5	0.0059	3.5	1.53

As can be seen from Table 12.6, all sets correlate vapor pressure and liquid density very well, thus the final choice could be based on other properties and mixture data. From spectroscopic data,[18] the enthalpy of hydrogen bonding is 2096 K, which is closer to sets 2 and 4 (2% deviation), and 1 and 3 (14% deviation), but somewhat different from set 5 (40% deviation). The VLE performance with CPA for mixtures of m-cresol with alkanes (Table 12.7) using some of the sets of Table 12.6 is very good but cannot be used for selecting the optimum parameter set. Thus, LLE data for the m-cresol–water mixture have been used and the various sets have been tested using the two 'well-accepted' combining rules, ECR and CR-1. The results are presented in terms of average deviations in Table 12.8. The CR-1 rule is the best choice and, moreover, several sets (1, 3, 5) perform satisfactorily, with set 1 being the best choice overall, considering all the properties studied.

Table 12.8 *Average deviations of compositions with different sets of parameters for* m-cresol(1)–water(2) with CPA using the ECR and CR-1 rules. *Experimental data are from: Varhanickova et al., J. Chem. Eng. Data, **1996**, 40, 448; Sidgwick et al., J. Chem. Soc., **1915**, 107, 1202; Pfohlet et al., Fluid Phase Equilib., **1997**, 141, 179. Percentage deviations are as defined in Table 12.1. After Kontogeorgis et al.*[19]

Set	Rule	k_{ij}	Δx_1^{II} (%)	Δx_1^{I} (%)
1	ECR	−0.05	52.1	19.9
	CR-1	−0.032	21.9	18.6
2	ECR	−0.0488	61.4	22.4
	CR-1	−0.0324	31.2	21.6
3	ECR	−0.505	52.3	20.10
	CR-1	−0.033	22.0	18.7
4	ECR	−0.053	58.0	21.9
	CR-1	−0.0363	27.3	20.7
5	ECR	−0.064	39.5	19.5
	CR-1	−0.0446	12.1	16.9

12.6 Conclusions

Due to the complexity of the association models and the mixtures such models intend to describe, a 'pragmatic' modeling approach is often useful, especially for applications involving compounds not previously studied. To summarize the key points of such a modeling approach:

- Association schemes should at first be chosen based on the chemistry of the associating molecules, but the parameters should be checked and if necessary fine tuned to mixture data. Optimum data for such fine tuning are LLE data of the associating compound under study with an alkane (or other inert compounds) but if not available cross-associating mixtures, e.g. with water, could be used.
- All mixtures involving either two self-associating compounds or one self-associating compound and a compound capable of Lewis acid–Lewis base interactions require combining rules for the association parameters. ECR and CR-1 are good choices for cross-associating mixtures (see also Chapter 11).
- It has been previously shown that modified CR-1 is a successful combining rule for representing solvating systems. Explicitly accounting for the solvation is, in the case of CPA, of great importance for:
  - water–BTEX and water–alkenes LLE/VLE;
  - glycol–BTEX LLE/VLE;
  - water–perfluoro-aromatics LLE;
  - water–esters, ethers LLE/VLE;
  - CO_2 or H_2S–water and methanol VLE.
- Explicit account of the solvation is not important and a single interaction parameter in the cross-dispersion energy term (k_{ij}) is sufficient for:
  - CO_2 or H_2S–alcohols and glycols VLE;
  - alcohols–BTEX VLE;
  - acids–BTEX VLE (and formic acid–benzene LLE).
- Strongly polar compounds can be represented with CPA either as 'inert' compounds (allowing for solvation with self-associating compounds) or as 'pseudo-associating compounds'. It is not always easy to determine a priori which is the best modeling approach. However, if LLE data are available with alkanes (as in the case of sulfolane) then the 'pseudo-associating' approach is likely to yield the best results.
- The performance of CPA and PC–SAFT for many polar and associating mixtures is similar, provided the development of the models, e.g. parameter estimation, is carried out in the same way.

References

1. L. Jannelli, A. Lopez, R. Jalent, L. Silvestri, *J. Chem. Eng. Data*, **1982**, *27*, 28.
2. N. von Solms, M.L. Michelsen, G.M. Kontogeorgis, *Ind. Eng. Chem. Res.*, **2004**, *43*, 1803.
3. G.K. Folas, G.M. Kontogeorgis, M.L. Michelsen, E.H. Stenby, *Ind. Eng. Chem. Res.*, **2006**, *45(4)*, 1516.
4. U. Domańska, W.C. Moollan, T.M. Letcher, *J. Chem. Eng. Data*, **1996**, *41*, 261.
5. G.K. Folas, G.M. Kontogeorgis, M.L. Michelsen, E.H. Stenby, *Ind. Eng. Chem. Res.*, **2006**, *45*, 1527.
6. M.B. Oliveira, M.G. Freire, I.M. Marrucho, G.M. Kontogeorgis, A.J. Queimada, J.A.P. Coutinho, *Ind. Eng. Chem. Res.*, **2007**, *46*, 1415.
7. W.C Moollan, U.M. Domańska, T.M. Letcher, *Fluid Phase Equilib.*, **1997**, *128*, 137.
8. M. Ko, J. Im, J. Yong Sung, H. Kim, *Fluid Phase Equilib.*, **2006**, *51*, 636.
9. J. Im, H. Lee, S. Lee, H. Kim, *Fluid Phase Equilib.*, **2006**, *246*, 34.
10. E. Voutsas, C. Perakis, G. Pappa, D. Tassios, *Fluid Phase Equilib.*, **2007**, *261*, 343.
11. J. Mollerup, *Fluid Phase Equilib.*, **1986**, *25*, 323.
12. M.L. Michelsen, *Fluid Phase Equilib.*, **1990**, *60*, 213.

13. S. Dahl, M.L. Michelsen, *AIChE J.*, **1990**, *36*(*12*), 1829.
14. K.S. Pedersen, M.L. Michelsen, A.O. Fredheim, *Fluid Phase Equilib.*, **1996**, *126*, 13.
15. G.K. Folas, G.M. Kontogeorgis, M.L. Michelsen, E.H. Stenby, *Fluid Phase Equilib.*, **2006**, *249*, 67.
16. A. Grenner, J. Schmelzer, N. von Solms, G.M. Kontogeorgis, *Ind. Eng. Chem. Res.*, **2006**, *45*, 8170.
17. J.M. Sørensen, W. Arlt, *Liquid–Liquid Equilibrium Data Collection, DECHEMA Chemistry Data Series*. Frankfurt/Main, 1979.
18. A. Kziazckzac, K. Moorthi, *Fluid Phase Equilib.*, **1985**, *23*(2), 153.
19. G.M. Kontogeorgis, G.K. Folas, N. Muro-Suñé, F. Roca Leon, M.L. Michelsen, *Oil Gas Sci. Technol. – Rev. IFP*, **2008**, *63*, 305.

13

Applications of SAFT to Polar and Associating Mixtures

13.1 Introduction

It could be argued that the majority of the applications of SAFT variants are found in the field of non-associating mixtures, especially polymer solutions and blends (Chapter 14). A few applications for associating mixtures were developed in the decade 1990–2000, but since then the number of applications for associating mixtures has increased substantially and some characteristic publications are summarized in Table 13.1. Mostly, applications to aqueous mixtures as well as systems with alcohols and organic acids have been considered. Few publications deal with multifunctional chemicals such as glycols and alkanolamines, while non-associating polar chemicals (ketones, esters, nitriles, etc.) have also been considered (in most cases with polar and sometimes quadrupolar/induction terms being added). Comparisons against other models, especially the CPA, ESD and NRHB EoS, have been presented in the literature. From a scientific and development point of view, the nature of association schemes and the combining rules for solvating/cross-associating mixtures are worthy of investigation. We will illustrate the major conclusions and present some typical results for some of these applications, roughly divided according to the associating compound considered.

13.2 Water–hydrocarbons

Aqueous systems with hydrocarbons have been studied with various SAFT models with varying degrees of success. There is some disagreement in the overall assessment of the model in the various publications. The different performance of SAFT variants for water–hydrocarbons may be due to details of the specific SAFT model investigated (sites of water, mixing rules, which dispersion term is used, etc.) but they seem more to be due to the parameterization for water, i.e. determination of pure water parameters. The major findings in water–hydrocarbon studies are summarized below in chronological order:

1. The first investigations with the original SAFT[9,46] for water–alkanes indicate that SAFT has severe difficulties in representing both water and alkane solubility with the same interaction parameter (k_{ij}).

Thermodynamic Models for Industrial Applications: From Classical and Advanced Mixing Rules to Association Theories
Georgios M. Kontogeorgis and Georgios K. Folas

Table 13.1 *Some applications of SAFT for associating (and polar) mixtures. Comparisons against some other models are provided in some cases. (Applications of CPA are not shown here, as they are presented in Chapters 9–12)*

Application	SAFT variant	Reference
Acid–hydrocarbon VLE	CK–SAFT	Huang and Radosz[1]
CO_2–alcohol VLE		
Water–ethane		
Water–alcohol (acid)–hydrocarbon LLE	CK–SAFT	Suresh and Beckman[2]
Alcohols, acids, water VLE	Simplified and CK–SAFT	Fu and Sandler[3]
Especially cross-associating mixtures		
Water–alkane LLE	Hard-sphere SAFT (a modification of original SAFT)	Galindo *et al.*[4]
Water, alcohols + hydrocarbons	SAFT–LJ	Kraska and Gubbins[5,6]
Water, alcohols, acids – emphasis on cross-associating mixtures	CK–SAFT	Wolbach and Sandler[7,8]
Water–alkane LLE	CK–SAFT with two mixing rules	Economou and Tsonopoulos[9]
Methanol–pentane VLE and LLE	Original SAFT	Jog *et al.*[10]
Water–alkane (V)LLE	CK–SAFT vs. CPA	Voutsas *et al.*[11]
Henry's law constants, water–alkanes		Boulougouris *et al.*[12]
CO_2–water–ethanol, phenols (V)LLE	Original SAFT, CPA–PR, etc.– comparison against PR	Zhang *et al.*[13]
		Pfohl and Budich[14]
		Pfhol *et al.*[15]
		Li[16]
Alcohols	SAFT1	Adidharma and Radosz[17]
Alcohols, water, amines, acetic acid (VLE and LLE)	PC–SAFT	Gross and Sadowski[18]
CO_2–water–alcohols/glycols–hydrocarbons VLE (binary and ternary)	Original SAFT	Li and Englezos[19,20]
Water–alkanes	SAFT–VR	Patel *et al.*[21]
Alcohol–alkane VLE	Original SAFT	Passarello and Tobaly[22]
Methanol–hexane	PC–SAFT vs. cubic EoS	Gupta and Olson[23]
Acetone–hexane		
Formic acid–water		
Methanol–alkane VLE and high-pressure LLE	CK–SAFT and PC–SAFT	Yarrison and Chapman[24]
Polar–inert mixtures treating polar compounds as self-associating	sPC–SAFT	Von Solms *et al.*[25]
Perfluoroalkanes + water etc.	Soft-SAFT	Dias *et al.*[26,27]
		Melo *et al.*[89]
Glycols	Soft-SAFT	Pedrosa *et al.*[28]
Alcohol–hydrocarbons VLE	Original and SAFT–VR (GC versions)	Tamouza *et al.*[29]
Water–alkane LLE	sPC–SAFT, ESD	Grenner *et al.*[30]
Aniline–water, water–aniline–hydrocarbons		
CO_2–water, methanol–propane	sPC–SAFT	Von Solms *et al.*[31]
CO_2–alcohols	Crossover soft-SAFT	Llovell and Vega[32,33]

Table 13.1 *(Continued)*

Application	SAFT variant	Reference
VLE and LLE with mixtures containing alcohols and glycols	sPC–SAFT	Grenner *et al.*[34,35]
Water and water–alkanes LLE + analysis using spectroscopic data	SAFT–VR, sPC–SAFT and CPA	Clark *et al.*[36] Grenner *et al.*[37]
Water–alkanes LLE	PC–SAFT vs. ESD and CPA	Voutsas *et al.*[38]
Amine-containing mixtures		
Polar and associating mixtures		
Amino acids + alcohol/water	PC–SAFT	Fuchs *et al.*[39]
Induced association interactions (ethers, esters, nitriles, ketones, water)	Polar PC–SAFT	Kleiner and Sadowski[40]
SLE with complexes (alcohols, water, phenols, bisphenols, etc.)	PC–SAFT	Tumakaka *et al.*[41] Prikhodko *et al.*[42]
Pure compounds (alkanes etc.)	SAFT vs. NRHB	Tsivintzelis *et al.*[43]
A large variety of mixtures (Danner standard VLE database) and LLE for water–alkanes, glycol–alkanes, etc.	sPC–SAFT vs. NRHB	Grenner *et al.*[44] Tsivintzelis *et al.*[45]

When the former is correlated well, the latter is largely underpredicted and the k_{ij} values are typically positive and high, which is exactly what was seen also with cubic EoS! The results do not seem to depend much on the number of sites used for water or the mixing/combining rules used, although some effect of the mixing rules has been reported.[9] Nevertheless, there have been some successful results reported in these earlier studies, most notably by Huang and Radosz[1] for the CK–SAFT and Kraska and Gubbins[5,6] for the LJ–SAFT, especially for water–light alkanes (methane, ethane).

2. Later investigations[11,12,47] with both the SAFT and the SRK CPA and PR–CPA models seem to verify the earlier conclusions with SAFT. Actually the results with SAFT seem to be inferior to those of CPA (see Chapter 9). Problems are sometimes attributed 'loosely' to the hydrophobic effect,[48] but as mentioned previously (Chapter 9), the CPA EoS does not suffer from these limitations and water–hydrocarbons LLE are satisfactorily represented. The large differences in the performance of CPA and SAFT for aqueous hydrocarbon mixtures is difficult to explain, considering that the models have similar association terms and CPA has a semi-empirical physical term.
3. With the exception of the water 2B parameters reported for PC–SAFT by Gross and Sadowski[18], most recent investigations tend to agree that a 4C association scheme best represents water–alkane phase equilibria and other aqueous mixtures with SAFT, irrespectively of the SAFT variant employed.[30,36,49] There are recent studies which question the superiority of the 4C scheme for water with SAFT-type models,[50] but both the success of the 4C scheme for phase equilibrium and verification from molecular orbital calculations[7] leave little room for doubt: *4C is currently the best choice for SAFT models when water should be modeled!* It is worth mentioning that a recent investigation by Kleiner[111] for PC–SAFT further supports the use of the 4C scheme for water, compared to the 2B and 3B schemes.
4. Recent investigations[35,37,45] illustrate that PC–SAFT can indeed represent rather well water–alkane and water– aromatic LLE using newly estimated water parameters. These new water parameters have a

physical significance, when compared against spectroscopic and simulation data, as is explained below. Table 13.2 presents SAFT parameters from various SAFT models compared to values from molecular simulation for the dispersion energy and to spectroscopy values for the association parameters. Figures 13.1–13.3 present some LLE results, while Figure 13.4 shows the water solubility in CO_2 at various pressures. The agreement is satisfactory with sPC–SAFT using recently presented water PC–SAFT parameters.[30]

Table 13.2 *SAFT parameters for water from various SAFT models. 'Experimental' association energy: 1813 K.*[88,51] *'Simulation' dispersion energies can be as low as 60 K (range 74–160 from Errington et al.*[82]*)*

SAFT variant	Reference	m	Dispersion energy (K)	Association energy (K)	κ^{AB}
SRK–sCPA	Kontogeorgis *et al.*[52]	—	—	2003	0.069 2
CK–SAFT	Huang and Radosz[80]	1.179	528.17	1809	0.015 93
SAFT 3 site	Economou and Tsonopoulos[9] Sandler and Wolbach[53]	1.179	528.17	1809	0.015 93
SAFT 4 site	Economou and Tsonopoulos[9]	1.236	431.69	1368.1	0.036 47
Simplified SAFT	Fu and Sandler[3]	2.00	188.23	825.9	119.93×10^{-3}
CK–SAFT	Button and Gubbins[54]	1.047	504	1365	0.024 1
Original SAFT, 4 site	Li and Englezos[20]	0.98	433.91	1195.20	0.038
PSAFT	Karakatsani *et al.*[55]	1.00	52.13	1982	0.073 7
PC–PSAFT	Karakatsani *et al.*[55]	1.00	42.77	1973	0.070 6
CK–SAFT	Boulougouris *et al.*[12]	2.85	167	1634	0.337 4
PR–CPA	Wu and Prausnitz[47]	—	—	1477	0.456 7
SAFT–VR	Patel *et al.*[21]	1.00	253	1366	1.028
CK–SAFT 3 site HF	Wolbach and Sandler[7]	1.278	385.12	2286	0.030 91
CK–SAFT 4 site HF	Wolbach and Sandler[7]	1.406	212.91	1809	0.091 09
CK–SAFT 3 site DFT	Wolbach and Sandler[7]	1.00	615.94	1627	0.010 98
CK–SAFT 4 site DFT	Wolbach and Sandler[7]	1.00	546.63	1237	0.022 03
APACT	Economou and Donohue[56]			2418 (2 site) 2618 (3 site)	
Anderko EoS	Anderko[90]			3505	
Original SAFT 4C	Li and Englezos[19]	0.982	433.91	1195.2	0.038
SRK–CPA	Voutsas *et al.*[11]	—	—	1793.6	0.115 1
CK–SAFT	Voutsas *et al.*[11]	2.853	167.10	1634.7	0.337 4
PC–PSAFT	Karakatsani *et al.*[71]	1.75	169.53	1131.0	0.200 1
tPC–PSAFT	Karakatsani *et al.*[71]	1.60	58.06	1639.5	0.181 2
tPC–PSAFT 4C	Karakatsani *et al.*[72]	2.815	150.7	1575	0.351 8
PC–SAFT 2B	Gross and Sadowski[18]	1.0656	366.51	2500.7	0.034 87
sPC–SAFT 4C	Grenner *et al.*[30]	1.50	180.3	1804.22	0.094 2
sPC–SAFT 4C	Grenner *et al.*[37]	2.61	140.39	1694.77	0.587 9
PC–SAFT 3B	Kleiner[111]	3.2542	196.21	1800.6	0.046 00
PC–SAFT 4C	Kleiner[111]	3.7923	138.60	1718.2	0.029 10

HF = Hartree–Fock, DFT = density functional theory.

We will now comment on the recent studies of water–alkanes with sPC–SAFT.[30] Several water–SAFT parameter sets have been investigated with the purpose of studying the effect of parameterization in SAFT modeling of water–alkanes. In addition to earlier studies with the PC–SAFT 2B parameters[18] and the three 4C (four-site) sets recently presented by von Solms *et al.*[81] (all of which assumed segment numbers 2.00, 2.75 and 3.50 for water), Grenner *et al.*[30] proposed a new 4C PC–SAFT parameter set which is based on physically justified arguments:

1. The number of segments for water should be small, around unity, as water is a low-molecular-weight compound.
2. The dispersion energy parameter should be low, as molecular simulation studies by Errington *et al.*[82] suggest a dispersion energy for water between 74 and 160 K.
3. The association energy should be close to the value reported by Koh *et al.*[51] which is $\varepsilon^{AB} = 1813$ K. The water parameter set obtained from simultaneous fitting of vapor pressure and liquid density data results to the values of $m = 1.5$, $\varepsilon = 180.3$ K, $\varepsilon^{AB} = 1804.22$ K, which are in good agreement with the 'expected' physically reasonable values of these parameters.

Figure 13.1 shows the predicted ($k_{ij} = 0$) results for the water–octane LLE. The solubility of water in octane is predicted to be too high with all sets except for that by Grenner *et al.*[30] The 2B water parameters of Gross and

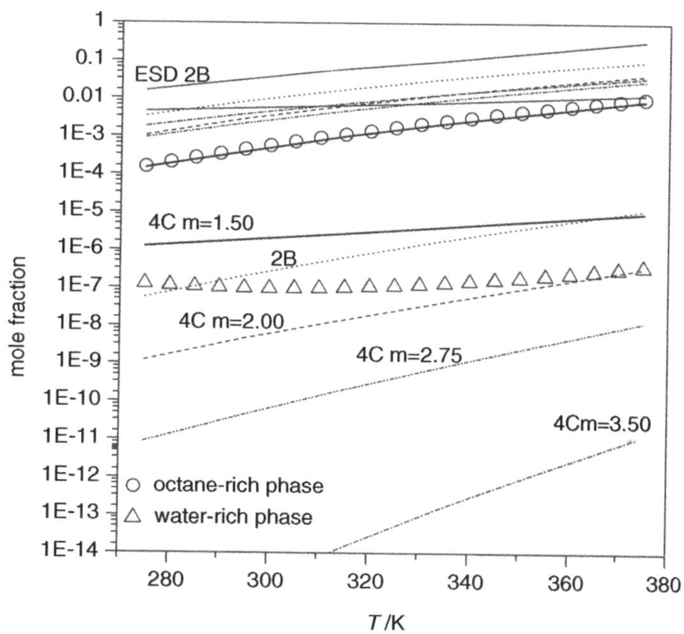

Figure 13.1 *Prediction ($k_{ij} = 0$) of LLE for octane–water with the sPC–SAFT EoS using various parameter sets for water. Experimental values are from Tsonopoulos, Fluid Phase Equilib.,* ***2001****, 186, 185. Prediction of simplified PC–SAFT with water parameters of von Solms et al.*[81] *(dashed lines) with the 4C scheme, Gross and Sadowski*[18] *with 2B water parameters (line indicated as 2B), Grenner et al.*[30] *with water 4C (solid line) and the ESD model with 2B (indicated as ESB 2B). Reprinted with permission from Ind. Eng. Chem. Res., Comparison of Two Association Models (Elliott-Suresh-Donohue and Simplified PC-SAFT) for Complex Phase Equilibria of Hydrocarbon–Water and Amine-Containing Mixtures by Andreas Grenner, Jürgen Schmelzer, Nicolas von Solms, and Georgios M. Kontogeorgis, 45, 24, 8170–8179 Copyright (2006) American Chemical Society*

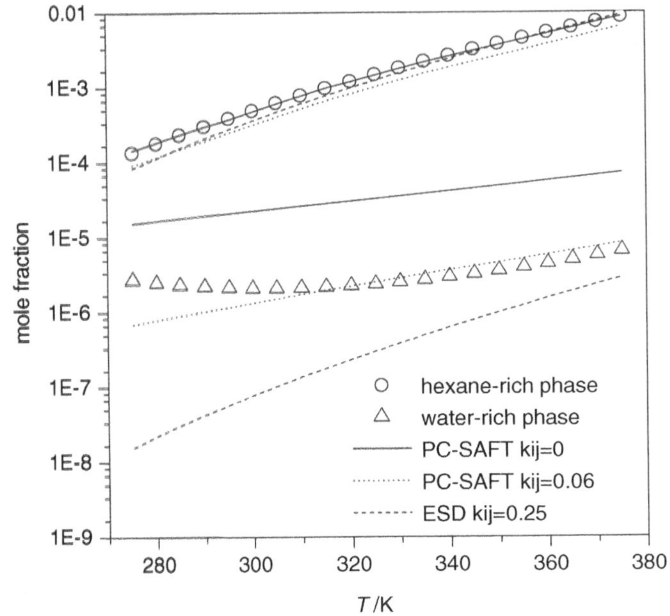

Figure 13.2 *LLE for hexane–water using sPC–SAFT (with 4C scheme for water and parameters from Grenner et al.[30]) and ESD. Experimental values are of Tsonopoulos, Fluid Phase Equilib.,* ***2001****, 186, 185. sPC–SAFT prediction ($k_{ij} = 0$) is the solid line; sPC–SAFT k_{ij} fit to the water-rich phaseand ESD fit to the hexane-rich phase*

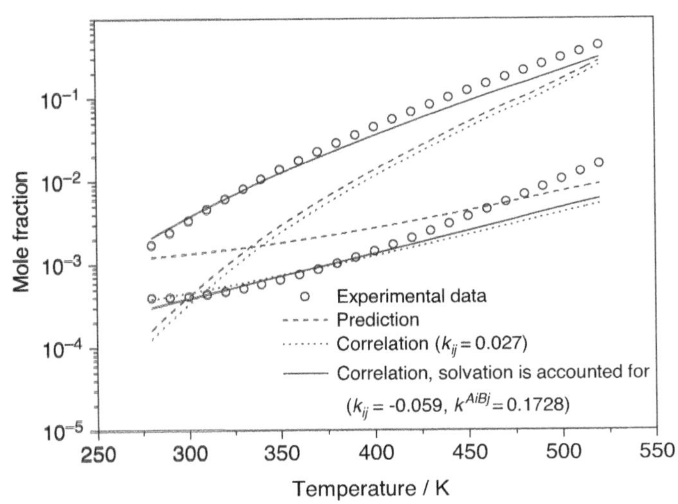

Figure 13.3 *Benzene–water LLE with sPC–SAFT. Best results are obtained when solvation is explicitly accounted for. Experimental data are from Tsonopoulos et al., AIChE J.,* ***1983****, 29, 990. Reprinted with permission from Ind. Eng. Chem. Res., Evaluation of the Nonrandom Hydrogen Bonding (NRHB) Theory and the Simplified Perturbed-Chain–Statistical Associating Fluid Theory (sPC-SAFT). 2. Liquid–Liquid Equilibria and Prediction of Monomer Fraction in Hydrogen Bonding Systems by I. Tsivintzelis, A. Grenner et al., 47, 15, 5651–5659 Copyright (2008) American Chemical Society*

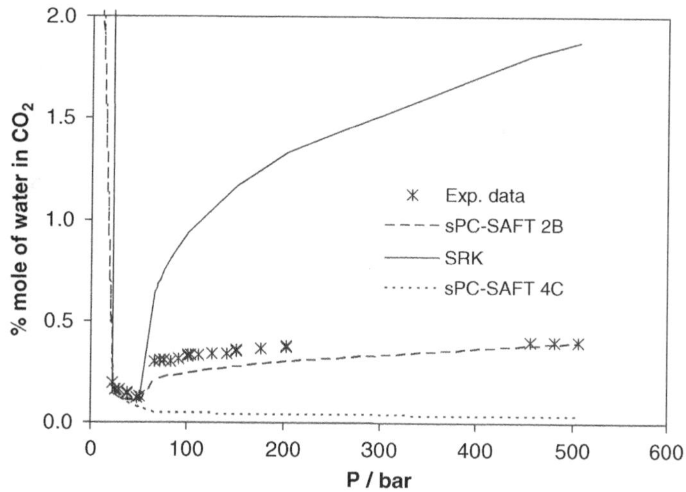

Figure 13.4 *Prediction ($k_{ij} = 0$) of the water solubility in CO_2 by SRK and the simplified PC–SAFT model at 298.15 K. Water has been assigned a 2B and 4C scheme in sPC–SAFT. In both cases, no attempt has been made to account for the solvation in CO_2–water mixtures. The sPC–SAFT with the 4C scheme can also describe the phase behavior if the solvation between water and CO_2 is explicitly accounted for*

Sadowski[18] give too high a solubility of octane in water. The prediction with sPC–SAFT using the Grenner *et al.*[30] parameters is very satisfactory. The octane-rich phase is predicted very well but the hydrocarbon solubility in the water-rich phase is predicted higher than the experimental value. From an industrial application point of view, the results of the hydrocarbon-rich phase are more important. Equally satisfactory results have been obtained for the LLE of other water–hydrocarbon systems studied (water with pentane up to octane, decane, cyclohexane, 1-hexene, 1-octene). One typical result is shown in Figure 13.2, where the effect of the interaction parameter, k_{ij}, is also seen.

13.3 Alcohols, amines and alkanolamines

13.3.1 General

Other than water, associating compounds have also received attention from SAFT models. Relatively few investigations have been reported for amines and alkanolamines,[1,18,54,94] but more studies have been reported for alcohol and phenol-containing mixtures, e.g. with alkanes, gases or water (VLE and LLE).

Figures 13.5–13.10 illustrate some representative results, while Tables 13.3 and 13.4 present a comparison of association parameters for methanol and other alcohols and glycols from several SAFT variants against spectroscopic data. Figures 13.5 and 13.6 present some results with CPA and PC–SAFT (calculations from two different sources) for methanol with two hydrocarbons. There is an emphasis on representing both VLE and LLE, and thus phase equilibria over an extensive temperature range. Figure 13.7 presents high-pressure LLE for methanol–pentane. Figure 13.8 shows the prediction of the azeotrope at low methanol concentrations for methanol–propane at 20 °C, azeotrope; there is ‘industrial’ evidence for its existence. Figure 13.9 shows the results of sPC–SAFT for two cross-associating systems, and finally Figure 13.10 illustrates the possible need for explicitly accounting for solvation for mixtures of methanol with olefinic hydrocarbons.[57]

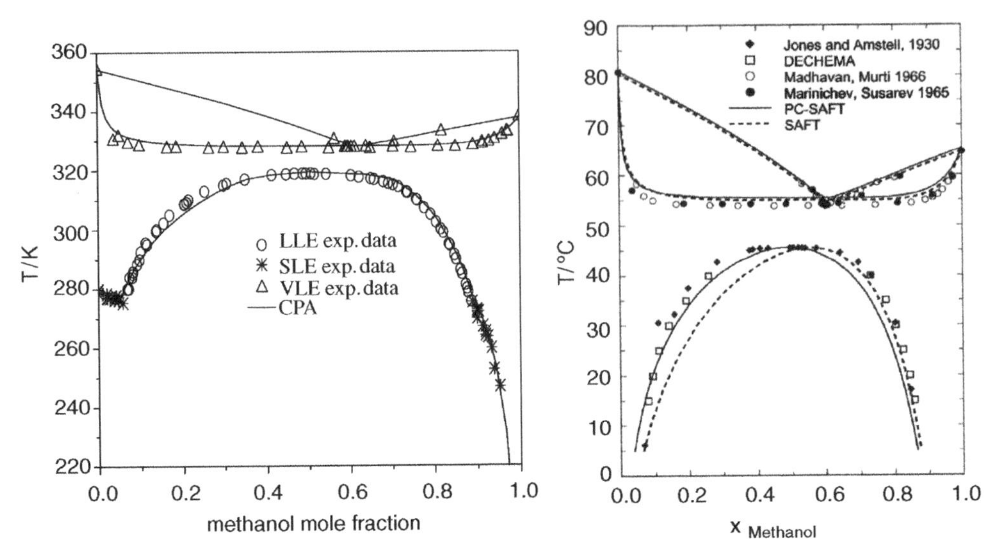

Figure 13.5 *VLE/LLE/SLE correlations for methanol–cyclohexane with CPA (left) and PC–SAFT and SAFT (right). The k_{ij} for SAFT is 0.044 and for PC–SAFT 0.051. For CPA the k_{ij} is 0.04. PC–SAFT/SAFT calculations are from J. Gross & G. Sadowski. Reprinted with permission from Ind. Eng. Chem. Res., Application of the Perturbed-Chain SAFT Equation of State to Associating Systems by Joachim Gross and Gabriele Sadowski, 41, 22, 5510–5515 Copyright (2002) American Chemical Society. CPA calculations are Reprinted from G. Folas et al. with permission from Fluid Phase Equilibria, Recent applications of the cubic-plus-association (CPA) equation of state to industrially important systems by G. Folas, G. M. Kontogeorgis et al., 228–229, PPEPPD 2004 Proceedings, 121–126 Copyright (2005) Elsevier*

13.3.2 Discussion

Besides the general studies on the performance of the models, two aspects of SAFT modeling which were specifically studied are: (1) the choice of association scheme for alcohols; and (2) the influence of combining rules for cross-associating mixtures, e.g. alcohol–water. From these investigations, the following points summarize the major conclusions on the SAFT application to alcohol mixtures.

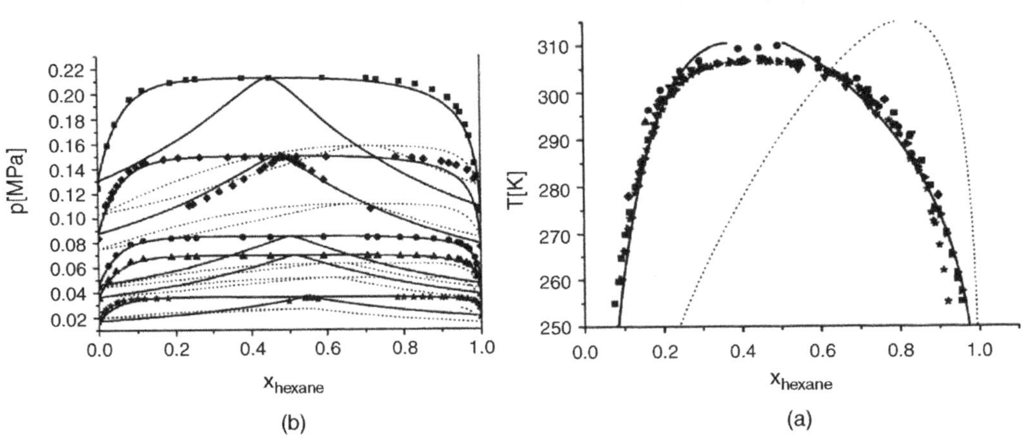

Figure 13.6 *VLE and LLE for methanol–hexane with PC–SAFT. Reprinted with permission from Ind. Eng. Chem. Res., Industrial Needs in Physical Properties by Sumnesh Gupta and James D. Olson, 42, 25, 6359–6374 Copyright (2003) American Chemical Society*

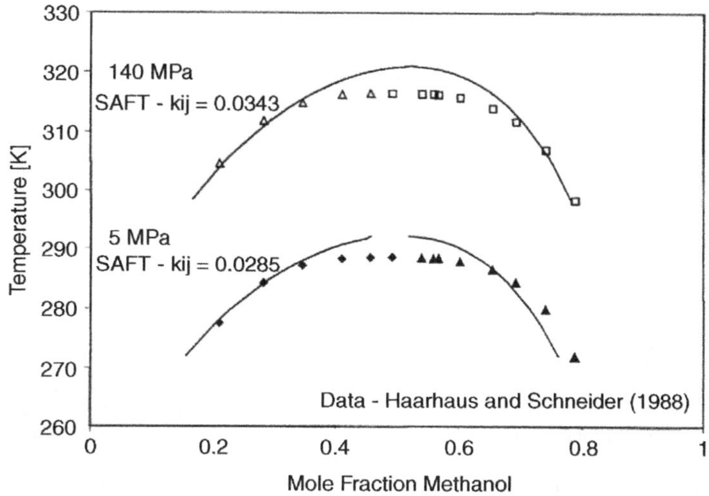

Figure 13.7 *SAFT correlation for methanol–pentane LLE at two different pressures. Experimental data are from Haarhaus and Schneider, J. Chem. Thermodyn.,* ***1988****, 20, 1121. Reprinted with permission from Fluid Phase Equilibria, Extensions and applications of the SAFT equation of state to solvents, monomers and polymers by K. Jog Prasanna, A. Garcia-Cuellar and W. G. Chapman, 158–160, 321–326 Copyright (1999) Elsevier*

General

- Excellent alcohol–hydrocarbon VLE is obtained. The results with the various SAFT variants and CPA are very similar and a small system-specific (but often temperature-independent) k_{ij} is needed.
- Very good results are obtained for alcohol–alkane LLE as well, especially for methanol–hydrocarbons. In most cases the same interaction parameter can be used for both VLE and LLE (for the few systems for which VLE and LLE data are available, e.g. methanol with various alkanes). Again the performance is

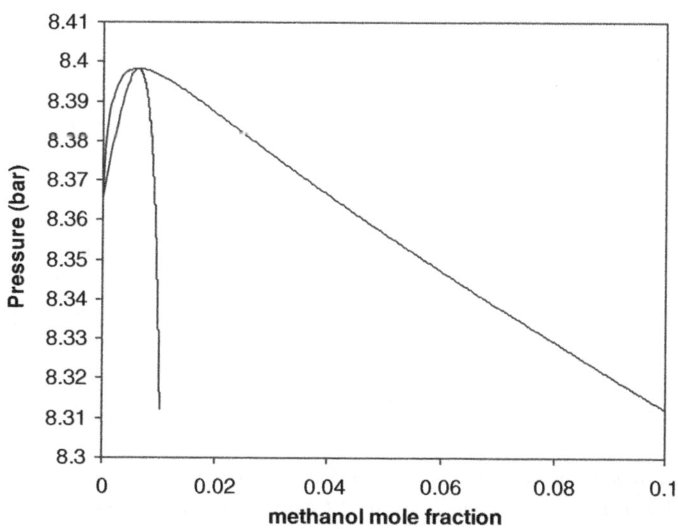

Figure 13.8 *Prediction of the azeotrope with sPC–SAFT for the system methanol–propane at low methanol concentrations. The temperature is 20 °C. There is industrial evidence for the existence of this azeotrope*

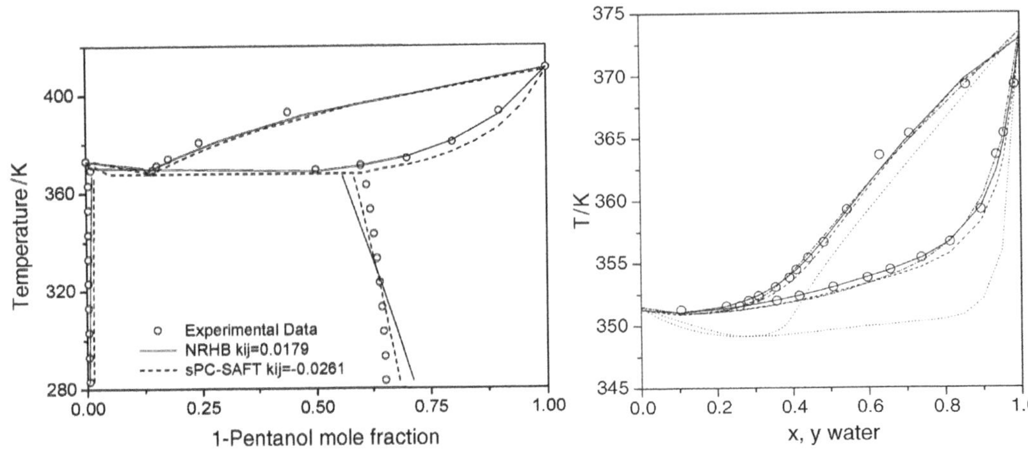

Figure 13.9 *Comparison between sPC–SAFT and NRHB for alcohol–water phase equilibria. Left: Water–1-pentanol LLE and VLE with sPC–SAFT and NRHB. Reprinted with permission from Ind. Eng. Chem. Res., Evaluation of the Nonrandom Hydrogen Bonding (NRHB) Theory and the Simplified Perturbed-Chain–Statistical Associating Fluid Theory (sPC-SAFT). 2. Liquid–Liquid Equilibria and Prediction of Monomer Fraction in Hydrogen Bonding Systems by I. Tsivintzelis, A. Grenner et al., 47, 15, 5651–5659 Copyright (2008) American Chemical Society. Right: VLE for ethanol + water at 101.32 kPa. Lines are predictions ($k_{ij} = 0$) and correlations with NRHB or sPC–SAFT, respectively. Dashed line: NRHB prediction; solid line: NRHB correlation ($k_{ij} = -0.0127$); dotted line: sPC–SAFT prediction; dashed–dotted line, sPC–SAFT correlation ($k_{ij} = -0.0414$). Reprinted with permission from Ind. Eng. Chem. Res., Evaluation of the Nonrandom Hydrogen Bonding (NRHB) Theory and the Simplified Perturbed-Chain–Statistical Associating Fluid Theory (sPC-SAFT). 1. Vapor–Liquid Equilibria by A. Grenner, I. Tsivintzelis, G. M. Kontogeorgis et al.,47, 15, 5636–5650 Copyright (2008) American Chemical Society*

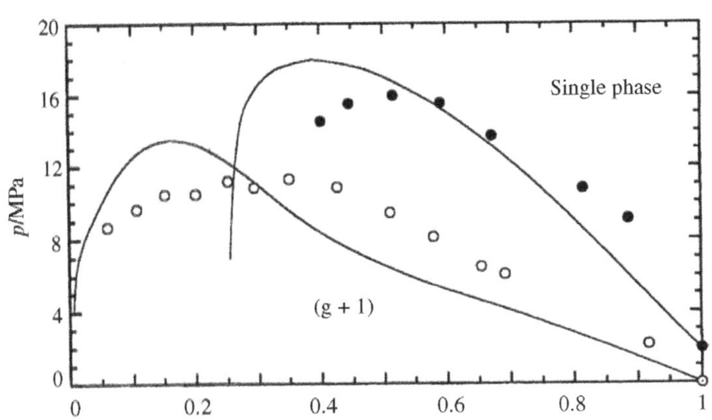

Figure 13.10 *Methanol–ethylene phase equilibria with SAFT. No solvation between methanol and ethylene is considered. Reprinted with permission from the Journal of Chemical Thermodynamics, (Methanol + ethene): phase behavior and modeling with the SAFT equation of state by B. M. Hasch, E. J. Maurer et al., 26, 6, 625–640 Copyright (1994) Elsevier*

Table 13.3 SAFT parameters for methanol. Experimental value for the enthalpy of association of methanol: 2630 K[93]

SAFT	Reference	Association energy (K)	κ^{AB}
Original	Chapman *et al.*[58]	2964	0.053
CK–SAFT	Huang and Radosz[80]	2714	4.856×10^{-2}
Simplified SAFT	Fu and Sandler[3]	2710.4	119.107×10^{-3}
CK–SAFT	Wolbach and Sandler[7]	2003	6.783×10^{-2}
CK–SAFT HF, 3B	Wolbach and Sandler[7]	1930	0.044 72
CK–SAFT DFT, 2B	Wolbach and Sandler[7]	1729	1.134×10^{-2}
CK–SAFT DFT, 3B	Wolbach and Sandler[7]	1470	0.016 41
PC–SAFT	Gross and Sadowski[18] Kleiner and Sadowski[40]	2899.5	0.035 176
sPC–SAFT	Von Solms *et al.*[59]	2090.2	0.1460
Original	Passarello and Tobaly[22]	2150.2	0.018 17
Original	Li and Englezos[19]	2320.8	0.019
PSAFT	Karakatsani *et al.*[55]	2718.3	0.0537
PC–PSAFT	Karakatsani *et al.*[55]	2766.6	0.0499
tPC–PSAFT	Karakatsani *et al.*[72]	2836.7	0.0459

similar to CPA, while various SAFT variants (CK–SAFT, PC–SAFT, simplified SAFT, etc.) perform similarly for alcohol–alkanes.

- The accurate representation of mixtures with methanol and aromatic or olefinic hydrocarbons requires the use of solvation schemes, otherwise the results may not be very satisfactory, as illustrated in Figure 13.10.

Specific

- As with CPA, the azeotropic behavior of methanol–propane at low methanol concentrations (for which there is industrial evidence) is well predicted (Figure 13.8).
- There is some controversy in the literature[13,14,16] about how well SAFT models represent CO_2–alcohol and CO_2–phenol mixtures, especially VLLE in CO_2–water–alcohols and phenols. More systematic investigations are needed in order to clarify the performance of the models in these cases. Earlier investigations[1] reveal some problems for CO_2–alcohols in the near critical area.

On the association scheme for alcohols

There are a few investigations on the choice of association scheme for alcohols. Most studies use either the 2B or the 3B scheme and Adidharma and Radosz[17] and Passarello and Tobaly[22] suggest using the 3B scheme, although both schemes in principle perform similarly for most alcohols. As discussed by Wolbach and Sandler[7], molecular orbital quantum chemistry calculations illustrate that the 3B scheme is possibly the best choice for methanol, a conclusion which is in agreement with some investigations with CPA, presented in Chapters 9 and 10.

On the combining rules for cross-associating mixtures

- Besides Equations (8.49–8.51) (Chapter 8), other combining rules have also been investigated for the cross-association parameters. In particular, the geometric mean rule for the cross-association energy

Table 13.4 *SAFT parameters for alcohols (in general) and glycols. Experimental value for the enthalpy of association of ethanol (and heavier alcohols): 2526–3007 K.[84] Many authors consider that the 'experimental' values for the heats of hydrogen bonding of alcohols heavier than ethanol are close to the value of ethanol, i.e. heats of hydrogen bonding for alcohols do not depend considerably on alcohol chain length. For phenol, experimental enthalpy of association: 2343 K[92]*

SAFT	Reference	Association energy (K)	κ^{AB}
CK–SAFT for ethanol	Huang and Radosz[80]	2759	2.920×10^{-2}
CK–SAFT for propanol	Huang and Radosz[80]	2619 (2605–2556 for butanol–hexanol)	1.968×10^{-2} 1.639×10^{-2} to 1.873×10^{-2} for butanol–hexanol
Simp. SAFT for ethanol	Fu and Sandler[3]	2802.7	5.6634×10^{-2}
tPC–PSAFT for ethanol	Karakatsani *et al.*[72]	2549.4	0.0503
PC–PSAFT for ethanol	Karakatsani *et al.*[55]	2635.8	0.0419
Simp. SAFT for propanol	Fu and Sandler[3]	2511.7 (2458.6 for butanol)	4.142×10^{-2} to 4.9971×10^{-2} for butanol
tPC–PSAFT for propanol	Karakatsani *et al.*[72]	1976.3	0.0196
PC–PSAFT for propanol	Karakatsani *et al.*[55]	2128.0	0.0171
tPC–PSAFT for butanol	Karakatsani *et al.*[72]	2565.2	0.0051
PC–PSAFT for butanol	Karakatsani *et al.*[55]	2766.5	0.0011
sPC–SAFT for 1-propanol	Kouskoumvekaki *et al.*[94]	2370	0.014 57
sPC–SAFT for 1-butanol	Kouskoumvekaki *et al.*[94]	2504	0.009 66
sPC–SAFT for 2-methyl propanol	Kouskoumvekaki *et al.*[94]	2554	0.002 93
PC–SAFT for ethanol (2B)	Gross and Sadowski[18]	2653.4	0.032 384
Original for ethanol (2B)	Li and Englezos[19]	2616.5	0.012
Original for MEG (4C)	Li and Englezos[19]	2375.26	0.020
Original for DEG (4C)	Li and Englezos[19]	1859.7	0.031
Original for TEG (4C)	Li and Englezos[19]	2470.02	0.061
sPC–SAFT for all alcohols except methanol (2B)	Grenner *et al.*[34]	2811.02	0.0033
sPC–SAFT for all glycols (4C)	Grenner *et al.*[35]	2080.03	0.0235

and the arithmetic mean rule for the cross-association volume have been investigated by several researchers.[3,19,20,60,61,87] Possibly these mixing rules are inspired by the use of the geometric mean rule for the dispersion energy and the arithmetic mean rule for cross-segment diameter. Small differences are observed between the various combining rules for SAFT (equations 8.49–8.51), as could be expected (due to similarities in the segment diameter values).

- Alcohol–alcohol VLE is well represented with small or even zero interaction parameters and possibly which combining rule is used for the cross-association parameters has little influence for these systems.
- The complexes between water and alcohols (or other compounds) can be well described using the PC–SAFT EoS and a 'reaction scheme' similar to that presented for CPA in Chapter 9, Appendix 9.D.[41,42]
- Several studies, e.g. by Fu and Sandler[3] and Englezos and co-workers,[19,20,60] illustrate the need for rather high k_{ij} values for certain cross-associating mixtures like water–alcohol mixtures, e.g. a $k_{ij} = -0.1$ value

has been reported for water–methanol. This may be due to the parameterization of SAFT and the choice of association schemes. As Figure 13.9 illustrates, rather small k_{ij} values are needed with sPC–SAFT using the 4C scheme for water and the parameters from Grenner *et al.*[30] A detailed study of water–alcohol cross-associating binary mixtures using a generalized approach for obtaining alcohol pure component parameters has recently been presented by Grenner *et al.*[34]

13.3.3 Study of alcohols with generalized associating parameters

Systematic studies of alcohol mixtures VLE, LLE and SLE have been carried out with sPC–SAFT using the generalized parameters presented in Chapter 8 (Equation (8.44b)). The new alcohol parameters give good results for VLE with alkanes and aromatic hydrocarbons as well for mixtures with water where cross-association is present. Furthermore, VLE at high pressures with CO_2 were modeled successfully. Figure 13.11 (left) shows the VLE for 2-propanol + benzene and confirms that the parameter approach also gives good results for 2-alcohol/hydrocarbon mixtures. Figure 13.11 (right) presents the successful prediction ($k_{ij}=0$) of the SLE for 1-octanol + dodecane. The prediction of the mutual solubility for alcohol + water systems yields deviations lower than 50% for both phases. If the k_{ij} is fitted to the alcohol-rich phase, the solubility of the water-rich phase will be overestimated. Predictions of VLE at 308.15–328.15 K for the ternary systems ethanol + benzene + hexane show excellent results with deviations lower than 5% in the bubble pressure. The LLE in the system 1-butanol + water + benzene at 298.15 K was also satisfactorily predicted.

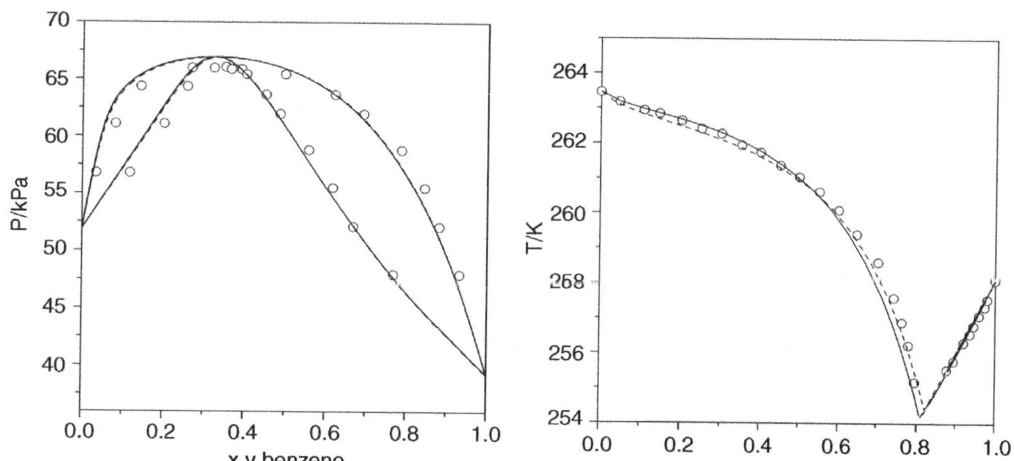

Figure 13.11 *Phase equilibria with sPC–SAFT for alcohol–hydrocarbon mixtures using various parameter sets for alcohols. Left: VLE for 2-propanol–benzene at 333.15 K. The lines are correlations with simplified PC–SAFT: dashed line ($k_{ij}=0.01349$), 2-propanol parameter;[18] solid line ($k_{ij}=0.0030$) 2-propanol parameters from Grenner et al.[34] Right: SLE for 1-octanol + dodecane. The lines are prediction ($k_{ij}=0$) with simplified PC–SAFT: dashed line, 1-octanol parameter;[18] solid line, 1-octanol parameter from Grenner et al.[34] Reprinted with permission from Fluid Phase Equilibria, Modeling phase equilibia of alkanols with the simplified PC-SAFT equation of state and generalized pure compound parameters by A. Grenner, G. M. Kontogeorgis et al., 258, 1, 83–94 Copyright (2007) Elsevier*

13.4 Glycols

Few investigations with glycols and glycolethers with SAFT have been reported, e.g. the study of glycolether– water closed loop with SAFT–HS[91] and the study of glycol-containing mixtures VLE with soft SAFT.[28] A systematic investigation of glycols with sPC–SAFT using the generalized parameters presented in Chapter 8 (Equation (8.44a)) has been reported.[35,45] The simplified PC–SAFT EoS has been applied for correlation and prediction of VLE and LLE in mixtures containing glycol oligomers + water, hydrocarbons, aromatic hydrocarbons, methane, N_2 and CO_2 with satisfactory results. The performance of simplified PC–SAFT is comparable to or slightly better than other glycol studies with different SAFT types or the CPA EoS. For some systems, constant k_{ij} values can be used, e.g. the solubility of CO_2 in diethylene glycol (DEG) shown in Figure 13.12 (left). Also the cross-associating systems like glycol oligomers + water could be modeled satisfactorily. By employing the same pure compound parameters as used for the VLE, LLE in glycol + heptane and glycol + aromatic hydrocarbon systems were correlated. Satisfactory results are obtained for three glycols (MEG, DEG, PG) with heptane, see Figures 13.12 (right) and 13.13.

For the glycol aromatic systems, solvation was explicitly accounted for with reasonably good results. Finally, prediction for VLE and LLE for two ternary systems was performed. The vapor phase compositions in the system water + propylene glycol + monoethylene glycol could be predicted successfully with very low deviations, while somewhat larger deviations for the bubble pressure occur. Also the binodal curve at 298.15 K in the system diethylene glycol + water + benzene was predicted satisfactorily. These studies have shown that it is possible to use PC–SAFT for phase equilibrium calculations for alcohols and glycols using generalized parameters for the associating compounds. The results are as successful as those presented in the literature using compound-specific parameters. Furthermore, the results show that the generalized parameter estimation procedure is capable of giving reliable parameters.

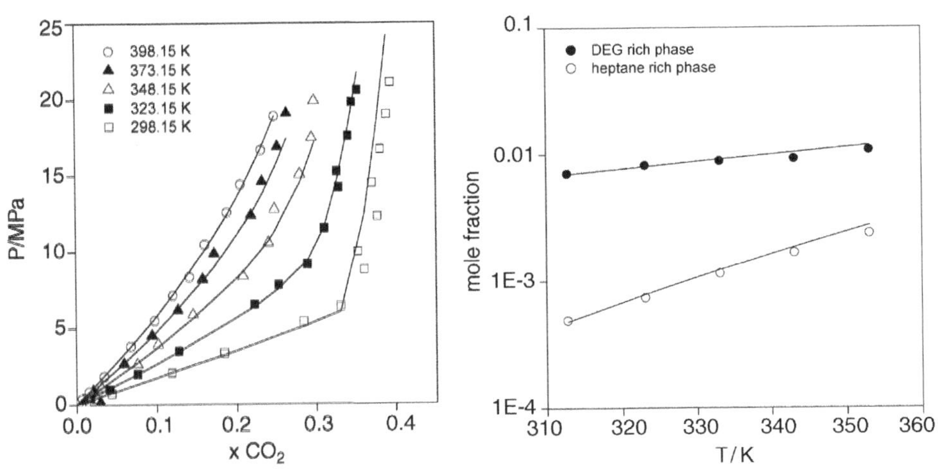

Figure 13.12 *Phase equilibria with sPC–SAFT for glycol-containing mixtures using generalized parameters for glycols. Left: CO_2 solubility in DEG. The lines are correlations with simplified PC–SAFT using $k_{12} = -0.0116$. Right: LLE for DEG–heptane. The lines are correlations with simplified PC–SAFT using $k_{12} = 0.032$. Reprinted with permission from Fluid Phase Equilibria, Application of a PC-SAFT to glycol containing systems-PC-SAFT towards a predictive approach by A. Grenner, G. M. Kontogeorgis et al., 261, 1–2, 248–257 Copyright (2007) Elsevier*

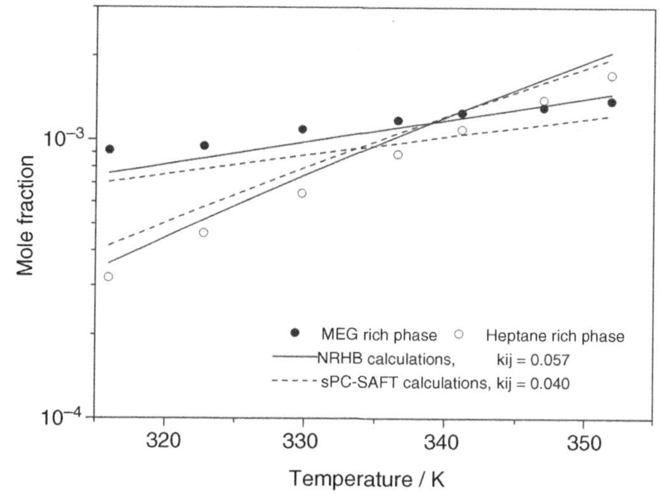

Figure 13.13 *LLE for MEG–n-heptane with sPC–SAFT and NRHB. Experimental data are taken from Derawi et al., J. Chem. Eng. Data,* ***2002,*** *47, 169. Reprinted with permission from Ind. Eng. Chem. Res., Evaluation of the Nonrandom Hydrogen Bonding (NRHB) Theory and the Simplified Perturbed-Chain–Statistical Associating Fluid Theory (sPC-SAFT). 2. Liquid–Liquid Equilibria and Prediction of Monomer Fraction in Hydrogen Bonding Systems by I. Tsivintzelis, A. Grenner et al., 47, 15, 5651–5659 Copyright (2008) American Chemical Society*

13.5 Organic acids

Few investigations have been reported for mixtures with organic acids: Fu and Sandler[3] used simplified SAFT (for formic acid or acetic acid + water), while Wolbach and Sandler[8] used SAFT for some acid–water mixtures. Gupta and Olson[23] applied SAFT to the formic acid–water system and Kouskoumvekaki *et al.*[62] used sPC–SAFT for water–acetic acid VLE. Parameters for some acids have been reported also in one of the early SAFT articles.[58] Table 13.5 presents a comparative overview for the association parameters of acetic acid

Table 13.5 *Association parameters for acetic acid with various SAFT models. Experimental values for the enthalpy of association are in the range 6949–8266 K. (References presented by Derawi et al., Fluid Phase Equilib.,* ***2004****, 225, 107)*

SAFT	Reference	Association energy (K)	κ^{AB}
Original	Chapman *et al.*[58]	7200	0.000 53
CK–SAFT	Huang and Radosz[80]	3941	39.26×10^{-3}
	Wolbach and Sandler[8] (standard estimation procedure)		
CK–SAFT	Wolbach and Sandler[8] (HF quantum chemistry method)	5751	3.465×10^{-3}
CK–SAFT	Wolbach and Sandler[8] (DFT/B3LYP quantum chemistry method)	5490	4.157×10^{-3}
Simplified SAFT	Fu and Sandler[3]	5810.5	28.840×10^{-3}
sPC–SAFT	Kouskoumvekaki *et al.*[62]	2756.7	0.2599
PC–SAFT	Gross and Sadowski[18]	3044.4	0.075 55

Table 13.6 Comparison of the performance of various SAFT-type models for water–acetic acid VLE at two temperatures. Average deviations for the pressure and relative volatility are shown. The subscripts in each model indicate the association schemes of water and acetic acid, e.g. $SAFT_{4C-2B}$ indicates SAFT using the 4C scheme for water and the 2B scheme for acetic acid

Model	T (K)	k_{ij}	% AAD in pressure	% AAD in relative volatility
$SAFT_{4C-2B}$	372.8	0.2335	2.9	106
Or. $PC\text{–}SAFT_{2B-2B}$		−0.1221	3.7	5.9
$PC\text{–}PSAFT_{4C-2B}$		−0.0228	4.4	24.6
ESD_{2B-2B}		0.008	2.2	10.0
Sim. $PC\text{–}SAFT_{4C-2B}$		−0.15	2.1	21.8
Sim. $PC\text{–}SAFT_{2B-2B}$		−0.08	2.9	16.5
Or. $PC\text{–}SAFT_{4C-2B}$	462.1	−0.0896	1.4	3.3
$PC\text{–}PSAFT_{4C-2B}$		0.0091	2.5	17.1
ESD_{2B-2B}		0.027	1.1	11.3
Sim. $PC\text{–}SAFT_{4C-2B}$		−0.15	1.0	9.8
Sim. $PC\text{–}SAFT_{2B-2B}$		−0.08	2.4	4.0

Sim. = simplified, Or. = original (for PC–SAFT).

from various SAFT models. As previously seen in the case of CPA, the performance of most SAFT models for water–acetic acid is not satisfactory for industrial applications. An illustration of the results obtained with different SAFT models is shown in Table 13.6, where the percentage deviation in both pressure and relative volatility is presented with several SAFT variants for water–acetic acid VLE at two temperatures of practical interest. The best results, among the models shown in Table 13.6, are obtained with the original PC–SAFT using the 2B scheme for both acetic acid and water, an assumption which seems doubtful for both compounds, especially for water.

13.6 Polar non-associating compounds

Unfortunately, for highly polar but not self-associating compounds, e.g. ketones, nitriles and esters, poor results are obtained with the various SAFT variants. For example, the azeotropic behavior of acetone–alkanes cannot be predicted and a k_{ij} value is required. When $k_{ij} = 0$ a near-ideal liquid behavior is obtained which is in contradiction to the experimental data, see Figures 13.14 and 13.15a. Similar problems have been observed for other highly polar mixtures. Two approaches proposed to correct for this problem are:

1. A further development of the models by inclusion of a separate dipolar term; a number of approaches are discussed in Section 13.6.1.
2. Assuming that the polar compounds can be treated as 'pseudo' self-associating compounds, thus estimating five parameters,[25] an approach which was illustrated also for CPA in Chapter 11.

In both cases, satisfactory results are obtained, as illustrated in Figures 13.14–13.16, for ketone-alkanes, even in the absence of any interaction parameter. Each approach has shortcomings and advantages, as will be illustrated in Section 13.6.1. Notice, however, as shown in Figure 13.15b, the difficulties of the PC–SAFT EoS in satisfactorily representing both VLE and LLE for acetone–hexane with a single interaction parameter.

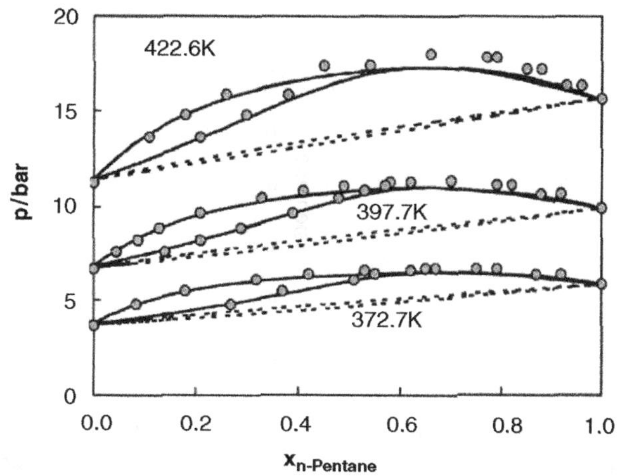

Figure 13.14 VLE predictions ($k_{ij} = 0$) for acetone–pentane with PC–SAFT (dashed line) and with polar PC–SAFT (solid line). Reprinted with permission from Fluid Phase Equilibria, Application of the Perturbed-Chain SAFT Equation of State to Polar Systems by Feelly Tumakaka and Gabriele Sadowski, 217, 2, 233–239 Copyright (2004) Elsevier

The importance of explicitly accounting for polar effects may be even more pronounced when extended temperature ranges and different types of phase equilibria (VLE and LLE) are considered.

In the cases of induced association when we have mixtures of self-associating and polar compounds, the approaches presented in Chapter 8 can be used. Satisfactory VLE results are obtained.[40,44] A typical example is shown in Figure 13.16. The induced association is further discussed in Section 13.6.4.

13.6.1 Theories for extension of SAFT to polar fluids

A number of approaches have been developed recently for extending SAFT to polar fluids by explicitly accounting for dipolar and other interactions. Three of the most well-known popular methods are:

A. The polar (dipolar/quadrupolar) terms proposed by Chapman's group.[64,65]
B. The polar (dipolar/quadrupolar) terms proposed by Gross[66], Gross and Vrabec[67].
C. The polar (dipolar/quadrupolar/cross-polar) terms proposed by Economou's group.[55]

In Economou's approach, there are two versions of the model, one using an additional pure compound parameter (indicated as C1 in this discussion) and a version which takes into account the whole polynomial series that approximate the integrals of Stell theory, so that no additional pure component parameter is needed (denoted as C2 here). The Economou method is presented in detail in Section 13.6.2. Special features of the Chapman and Gross–Vrabec models are discussed in Section 13.6.3.

There are important differences and similarities between the three approaches (A, B and C), which are hereafter briefly presented.

Some common features in the A, B and C approaches are the following:

1. All three theories are developed based on third-order perturbation theories, therefore the polar terms are given in a Padé approximation form for the Helmholtz free energy.
2. None of the three theories takes into account the first-order perturbation terms, $A_1/(NkT)$, which are zero within the numerical uncertainties for spheres and non-zero but small for the hard dumbbell and hard spherocylinder fluids, as shown by Vega[68].

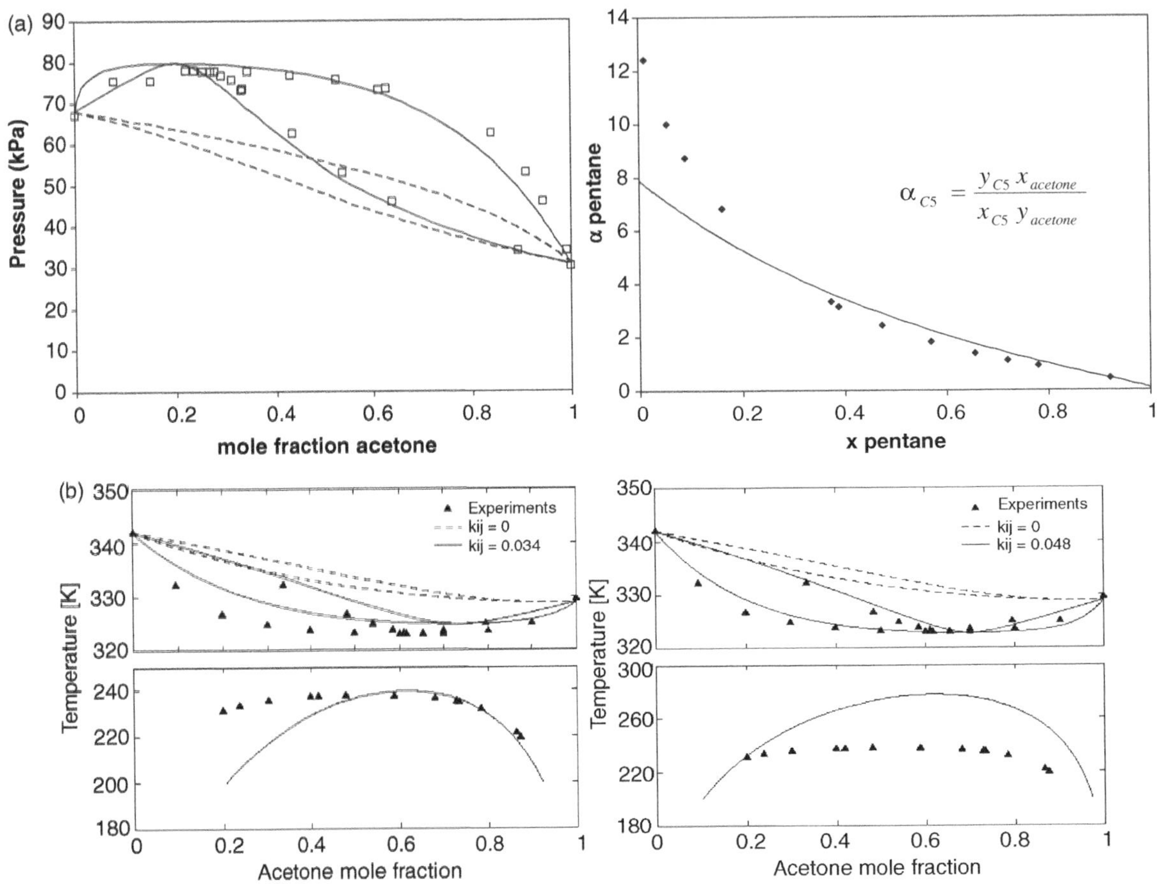

Figure 13.15 (a) Phase equilibria for acetone–pentane with sPC–SAFT using various assumptions for acetone. All calculations are predictions ($k_{ij}=0$). Left: VLE prediction for acetone–pentane with sPC–SAFT at 25 °C. The dashed line corresponds to the model when acetone is treated as an 'inert' compound (non-self-associating), while the solid line is based on acetone being considered as an associating compound. Right: Prediction of relative volatility of pentane in the pentane–acetone mixture with sPC–SAFT, by treating acetone as a self-associating compound. Reprinted with permission from Ind. Eng. Chem. Res., Applying Association Theories to Polar Fluids by Nicolas von Solms, Michael L. Michelsen, and Georgios M. Kontogeorgis, 43, 7, 1803–1806 Copyright (2004) American Chemical Society (2004); (b) VVLE and LLE for acetone–n-hexane with sPC–SAFT. Left: The interaction parameter is fitted to LLE data. Right: The interaction parameter is fitted to VLE data. Acetone is considered an 'inert' (non-self-associating) compound and no polar terms are included in the sPC–SAFT calculations. It can be seen that simultaneous accurate representation of VLE and LLE with a single value of the interaction parameter is not possible. From A. Tihic, Group contribution sPC–SAFT equation of state. PhD Thesis, The Technical University of Denmark, 2008

3. In the case of mixture calculations, no additional binary interaction parameter, k_{ij}, is introduced (when non-self-associating molecules are included).

Some common features in B and C2 are the following:

1. No additional pure component parameter is introduced.
2. It is possible to use tabulated values of multipolar moments (in C1 as well).

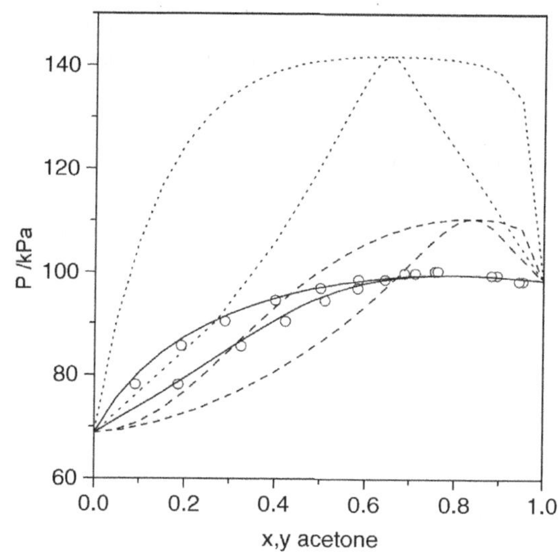

Figure 13.16 *VLE for acetone–methanol at 328.15 K with sPC–SAFT (lines): dotted line, standard approach prediction ($k_{ij} = 0$); dashed line, standard approach k_{ij} fit ($k_{ij} = -0.0772$); solid line, correlation ($k_{ij} = 0$) using the combining rule, of Equation (8.49) but using $\kappa^{A_iB_j} = 3.063$ (fitted to mixture data). The standard approach implies no explicit treatment of the solvation between acetone and methanol. Reprinted with permission from Ind. Eng. Chem. Res., Evaluation of the Nonrandom Hydrogen Bonding (NRHB) Theory and the Simplified Perturbed-Chain–Statistical Associating Fluid Theory (sPC-SAFT). 1. Vapor–Liquid Equilibria by A. Grenner, I. Tsivintzelis, G. M. Kontogeorgis et al., 47, 15, 5636–5650 Copyright (2008) American Chemical Society*

3. The triple integrals are replaced by simple power series in density.
4. A systematic improvement in the vapor pressure and prediction of saturated liquid density has been obtained for strongly polar substances (in C1 as well).
5. It has been found with both theories that the reference fluid EoS do not perfectly agree with results from molecular simulations for non-polar fluids.

Some common features in A and C1 and C2 are the following:

1. They all use some kind of 'segment approach'. In A the dipoles are assumed to be located on certain segments of a chain molecule. In C1 and C2 dipoles are assumed to be uniformly distributed on all segments, i.e. each segment possesses an average dipole moment.
2. The coefficients of the second- and third-order Helmholtz free energy terms are identical.

There are, however, a number of differences between the various methods.

Some differences between B vs. A and C are the following:

1. Only in B are the coefficients of the polynomials that approximate the integrals of the theory adjusted to molecular simulation data.
2. The coefficients of the Helmholtz energy terms of B do not match the coefficients of the other models, including the coefficients of the solution theory of Gubbins and Twu[69] and the perturbation theory for multipolar and ionic liquids of Larsen *et al.*[70]
3. Only in B is the third-order two-body term ($A_{3,2}/(NkT)$) completely neglected. This may be considered a serious simplification, especially for the treatment of mixtures.

4. Only in B are the double and triple integrals defined in a different way: the double integral is temperature and density dependent, while the triple integral is only density independent.

Some differences between C vs. A and B are the following:

1. The reduced polar parameters have a different chain length dependence:

$$\tilde{\mu} = 85.12 \frac{\mu/m}{\sqrt{(u/k)\sigma^3}} \quad \text{and} \quad \tilde{Q} = 85.12 \frac{Q/m}{\sqrt{(u/k)\sigma^5}} \quad \text{in C}$$

versus:

$$\tilde{\mu} = \frac{\mu}{\sqrt{u/km\sigma^3}} \quad \text{and} \quad \tilde{Q} = \frac{Q}{\sqrt{u/km\sigma^5}} \quad \text{in A and B}$$

2. Only in C is the assumption of additivity of interaction energies used; that is, the residual Helmholtz energy per mole of segments due to polar interactions is written as $a_{mol} = m a_{seg}$. Thus, there appears to be a kind of reverse way of thinking in developing model C in contrast to models A and B. In model C a segmental expression is used for the residual Helmholtz energy due to polar interactions and then, using the assumption of additivity, an expression is obtained for the residual Helmholtz energy per mole. In approaches A and B the starting point is an expression for a polar molecular sphere and then, by using the fact that $d_{molecule} = (m d^3_{segment})^{1/3}$, an expression for the contribution per segment is derived. It is perhaps noteworthy that, following this way of thinking, Chapman and co-workers[64,65] found a different relationship between a_{mol} and a_{seg}, that is:

$$a_{mol} = \frac{1}{m} a_{seg}$$

Table 13.7 provides a partial list of mixtures for which the models have been tested.

Table 13.7 *Mixtures to which approaches A, B and C have been applied*[40,55,63–67,71,72,95–103]

A (Chapman)	B (Gross)	C (Economou)
Acetone + alkanes	CO_2 + alkanes, aromatic hydrocarbons (VLE, LLE) CO_2 + nonanol	Acetone + alkanes
Various ketones (with 5–10 carbon atoms) + alkanes	Aldehydes, ketones + alkanes	CO_2 + alkanes
Ethers, esters, epoxies, cyclic ethers	Acetonitrile, various nitriles + alkanes, chloro-alkanes	CO_2 + aromatic hydrocarbons
Polar (co-)polymers + alkanes	DMSO, DMF + alkane	Alcohol + alkanes, alcohols + water
Ethylacetate + CC_6 VLE and H^E	Chloro- and nitro-alkanes	Water + alkanes
CO_2 + CC_6	Methanol + decane VLE/LLE	Ionic liquids (also with CO_2)
Polyethers + alkanes LLE	Ethers, THF	Alcohol + aromatic hydrocarbons
	Oxides, ethoxyalkanes	Mixtures with aromatic hydrocarbons and polar compounds
	Water–alcohols–esters	Water–ethanol–cyclohexane VLLE

13.6.2 Application of the tPC–PSAFT EoS to complex polar fluid mixtures

The model

Economou and co-workers[55,71,72] extended PC–SAFT to polar fluids, namely dipolar, quadrupolar and polarizable fluids, using the theory of Larsen *et al.*[70] The engineering version of the model is called truncated PC–PSAFT (tPC–PSAFT) and uses a truncated version of the multipolar perturbation expansion. Two extra terms are added in PC–SAFT, due to polar and induction effects, and the EoS in terms of the residual Helmholtz free energy is:

$$\frac{a^{res}(T,\rho)}{RT} = \frac{a^{hs}(T,\rho)}{RT} + \frac{a^{chain}(T,\rho)}{RT} + \frac{a^{assoc}(T,\rho)}{RT} + \frac{a^{disp}(T,\rho)}{RT} + \frac{a^{polar}(T,\rho)}{RT} + \frac{a^{ind}(T,\rho)}{RT} \quad (13.1)$$

where T and ρ are the temperature and the molar density of the system, respectively, and R is the universal gas constant. The polar and induced polar terms are expressed via the simple Padé approximants proposed by Larsen *et al.*[70]:

$$\frac{a^{polar}}{RT} = m\frac{a_2^{polar}}{1-a_3^{polar}/a_2^{polar}} \qquad \frac{a^{ind}}{RT} = m\frac{a_2^{ind}}{1-a_3^{ind}/a_2^{ind}} \quad (13.2)$$

The subscripts 2 and 3 in Equation (13.2) denote the second- and third-order terms in the perturbation expansion for polar and induced polar interactions, respectively. The third-order term for polar interactions consists of a two-body and a three-body term so that:

$$\frac{a_3^{polar}}{RT} = \frac{a_{3,2}^{polar}}{RT} + \frac{a_{3,3}^{polar}}{RT} \quad (13.3)$$

The tPC–PSAFT model uses a simplified version of the perturbation expansion for polar interactions. In order to account for the higher order terms that are omitted, a new pure component parameter was introduced to account for the spatial extent of polar interactions compared to hard-core repulsive interactions. As a result, the polar terms in tPC–PSAFT assume the following form:

$$\frac{a_2^{polar}}{RT} = -\left(\frac{1}{\tilde{T}}\right)^2 \frac{\eta}{K^3}\left[\frac{4}{3}\tilde{\mu}^4 + \frac{12}{5}\frac{\tilde{\mu}^2\tilde{Q}^2}{K^2} + \frac{12}{5}\frac{\tilde{Q}^4}{K^4}\right] \quad (13.4)$$

$$\frac{a_{3,2}^{polar}}{RT} = \left(\frac{1}{\tilde{T}}\right)^3 \frac{\eta}{K^8}\left[\frac{6}{5}\tilde{\mu}^4\tilde{Q}^2 + \frac{144}{175}\frac{\tilde{\mu}^2\tilde{Q}^4}{K^2} + \frac{72}{245}\frac{\tilde{Q}^6}{K^4}\right] \quad (13.5)$$

$$\frac{a_{3,3}^{polar}}{RT} = \left(\frac{1}{\tilde{T}}\right)^3 \frac{\eta^2}{K^3}\left[\frac{10}{9}\tilde{\mu}^6 + \frac{159}{125}\frac{\tilde{\mu}^4\tilde{Q}^2}{K^2} + \frac{689}{1000}\frac{\tilde{\mu}^2\tilde{Q}^4}{K^4} + \frac{243}{800}\frac{\tilde{Q}^6}{K^6}\right] \quad (13.6)$$

where the reduced dipole and quadrupole moments are expressed as:

$$\tilde{\mu} = 85.12 \frac{\mu/m}{\sqrt{(u/k)\sigma^3}} \tag{13.7}$$

$$\tilde{Q} = 85.12 \frac{Q/m}{\sqrt{(u/k)\sigma^5}} \tag{13.8}$$

where μ is the dipole moment of the fluid (in D) and Q is its quadrupole moment (in D Å). K is a dimensionless quantity that accounts for the spatial range of polar interactions compared to hard-sphere interactions, $K = \sigma_p/\sigma$. Consequently, σ_p (or equivalently v^p) is the adjustable effective polar interactions' segment diameter. The expressions for the dipole-induced two-body and three-body terms of tPC–PSAFT are:

$$\frac{a_2^{ind}}{RT} = -\frac{8}{\tilde{T}} \frac{\eta}{K^3} \tilde{\mu}^2 \tilde{a} \tag{13.9}$$

$$\frac{a_3^{ind}}{RT} = 10 \left(\frac{1}{\tilde{T}}\right)^2 \frac{\eta^2}{K^3} \tilde{\mu}^4 \tilde{a} \tag{13.10}$$

where:

$$\tilde{\alpha} = \frac{\alpha/m}{\sigma^3}$$

and α is the polarizability of the fluid (in Å^3).

The extension of the polar terms requires mixing rules and the following ones are used for the reduced multipole moments and polarizability as a function of the multipole moments and polarizability of the pure components and the composition.

For the case of two-body terms, this is:

$$\tilde{\mu} = \frac{\sum_i \sum_j x_i x_j m_i m_j \sqrt{\tilde{\mu}_i \tilde{\mu}_j}}{\left(\sum_i x_i m_i\right)^2} \tag{13.11}$$

$$\tilde{Q} = \frac{\sum_i \sum_j x_i x_j m_i m_j \sqrt{\tilde{Q}_i \tilde{Q}_j}}{\left(\sum_i x_i m_i\right)^2} \tag{13.12}$$

and:

$$\tilde{\alpha} = \frac{\sum_i \sum_j x_i x_j m_i m_j \sqrt{\tilde{\alpha}_i \tilde{\alpha}_j}}{\left(\sum_i x_i m_i\right)^2} \tag{13.13}$$

As for the three-body terms, it is:

$$\tilde{\mu} = \frac{\sum_i \sum_j \sum_k x_i x_j x_k m_i m_j m_k (\tilde{\mu}_i \tilde{\mu}_j \tilde{\mu}_k)^{1/3}}{\left(\sum_i x_i m_i\right)^3} \tag{13.14}$$

$$\tilde{Q} = \frac{\sum_i \sum_j \sum_k x_i x_j x_k m_i m_j m_k (\tilde{Q}_i \tilde{Q}_j \tilde{Q}_k)^{1/3}}{\left(\sum_i x_i m_i\right)^3} \tag{13.15}$$

and:

$$\tilde{\alpha} = \frac{\sum_i \sum_j \sum_k x_i x_j x_k m_i m_j m_k (\tilde{\alpha}_i \tilde{\alpha}_j \tilde{\alpha}_k)^{1/3}}{\left(\sum_i x_i m_i\right)^3} \tag{13.16}$$

Note that both three-body and two-body cross-interactions are expressed via the geometric mean rule, which is rigorously true only for two-body interactions.

Finally, additional mixing rules are necessary for the cross three-body dipole–dipole–quadrupole and dipole–quadrupole–quadrupole terms:

$$(\tilde{\mu}^2 \tilde{Q})^{1/3} = \frac{\sum_i \sum_j \sum_k x_i x_j x_k m_i m_j m_k (\tilde{\mu}_i \tilde{\mu}_j \tilde{Q}_k)^{1/3}}{\left(\sum_i x_i m_i\right)^3} \tag{13.17}$$

$$(\tilde{\mu} \tilde{Q}^2)^{1/3} = \frac{\sum_i \sum_j \sum_k x_i x_j x_k m_i m_j m_k (\tilde{\mu}_i \tilde{Q}_j \tilde{Q}_k)^{1/3}}{\left(\sum_i x_i m_i\right)^3} \tag{13.18}$$

Results

tPC–PSAFT contains six pure compound parameters fitted to vapor pressures and liquid densities and typically only one k_{ij} parameter fitted to mixture data. The model has been applied with success to polar/quadrupolar non-associating molecules and mixtures (ketones, aromatic hydrocarbons, etc.) and hydrogen bonding fluids including several mixtures that exhibit cross-associating effects, e.g. water–alcohols as well as ionic liquids. Some typical results are shown in Figures 13.17 and 13.18, where the 2B scheme is used for alcohols and the 4C scheme for water.

From these and pertinent investigations we conclude that:

- Excellent prediction of various 'sensitive' properties of polar and quadrupolar fluids, e.g. second virial coefficients, isochoric specific heat capacities and Joule–Thomson coefficients are obtained.
- Excellent prediction is achieved for ketone–alkane VLE, systems which cannot be predicted well without explicitly accounting for polarity.
- PC–SAFT (with or without accounting for polarity) can predict the azeotropic behavior in methanol–propane at low methanol concentrations, for which there is industrial evidence.
- Excellent VLE and LLE results are obtained for methanol–alkane VLE and LLE with a single k_{ij} and over extensive temperature and pressure ranges.
- Especially good results are obtained for water–alkanes (both solubilities) except for the minimum in the hydrocarbon solubility. The results are similar to CPA, although with a little higher but constant k_{ij}, around 0.2 for various water–alkanes, ranging from water–pentane up to water–decane. Equally good results were obtained for water–alkenes and water–cycloalkanes LLE.

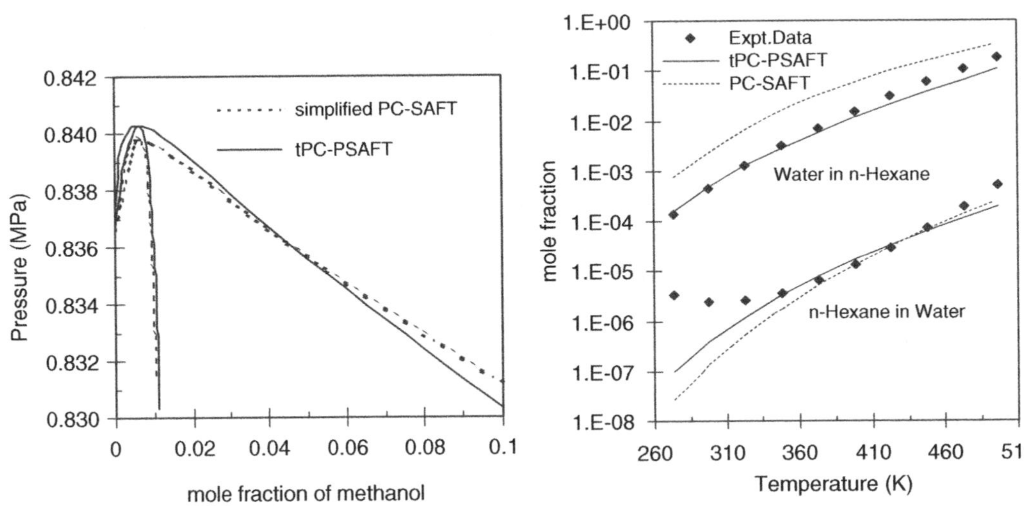

Figure 13.17 *Polar PC–SAFT (in the form of tPC–PSAFT) against non-polar PC–SAFT versions for two associating mixtures. Left: Prediction of the azeotrope formed in the methanol–propane mixture at 298.15 K with tPC–PSAFT and simplified PC–SAFT EoS. Only dipolar and association contributions are used for methanol. Right: Water– n-hexane LLE at three-phase equilibrium pressure. Experimental data (points) and model (tPC–PSAFT and PC–PSAFT) correlation (lines). PC–SAFT results are based on a 2B scheme for water with parameters from Gross and Sadowski.[18] Reprinted with permission from Ind. Eng. Chem. Res., Evaluation of the Truncated Perturbed Chain-Polar Statistical Associating Fluid Theory for Complex Mixture Fluid Phase Equilibria by Eirini K. Karakatsani, Georgios M. Kontogeorgis, and Ioannis G. Economou, 45, 17, 6063–6074 Copyright (2006) American Chemical Society*

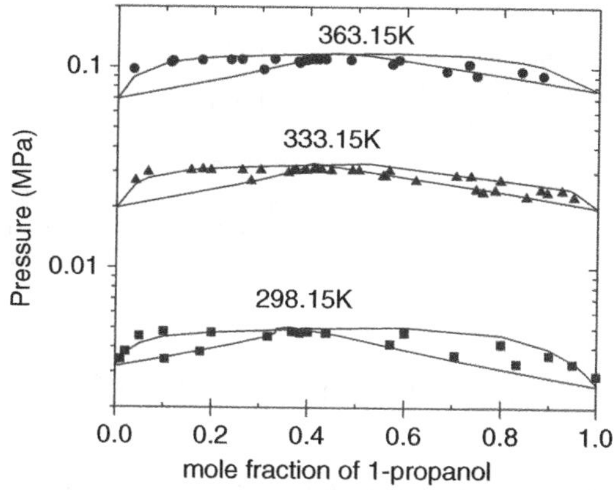

Figure 13.18 *VLE for 1-propanol–water, experimental data (points) and tPC–PSAFT correlations (lines). Reprinted with permission from Ind. Eng. Chem. Res., Evaluation of the Truncated Perturbed Chain-Polar Statistical Associating Fluid Theory for Complex Mixture Fluid Phase Equilibria by Eirini K. Karakatsani, Georgios M. Kontogeorgis, and Ioannis G. Economou, 45, 17, 6063–6074 Copyright (2006) American Chemical Society*

- VLE of aqueous mixtures with methanol, ethanol, propanol and butanol are predicted very well using the combining rules presented in Chapter 8, and rather low k_{ij} values are required (around 0.03, depending somewhat on the alcohol considered). Simultaneous description of VLE and LLE of water–butanol is still a challenging problem and the results are similar to those for CPA shown in Chapter 11. Due to lack of data, the quadrupole of propanol and butanol is assumed to be the same as for ethanol.
- Very good results are obtained for a few ternary polar mixtures, e.g. VLE of ethanol–acetone–benzene and acetone–chloroform–benzene, water–ethanol–cyclohexane VLLE and the quaternary ethanol–acetone–benzene–hexane system.
- Recently successful results have been obtained for CO_2 solubility in imidazolium-based ionic liquids.

13.6.3 Discussion: comparisons between various polar SAFT EoS

The similarities and differences of the various polar terms from a theoretical development point of view were presented in Section 13.6.1. All terms considered are formulated in the form of a Padé approximant, but the details of the second- and third-order perturbations differ. The polar model by Economou and co-workers and typical results were presented in Section 13.6.2. Some key observations from the other polar terms as well as a comparative evaluation of the models are presented in this section.

Chapman's model

This model contains, as mentioned above, four pure compound parameters for polar non-associating fluids, the extra parameter being the fraction of dipolar segments. The (functional) dipole moment could be taken from vacuum values, but sometimes this parameter has been adjusted to experimental data as well. The model has been combined with both CK–SAFT and PC–SAFT and small differences are observed. The most extensive application has been for mixtures of ketones and alkanes. The VLE including the azeotropic behavior are predicted very well without interaction parameters, whereas SAFT (without the polar term) predicts ideal solution behavior in these cases (Figure 13.14). Even when correlation is considered, polar PC–SAFT performs better than PC–SAFT.

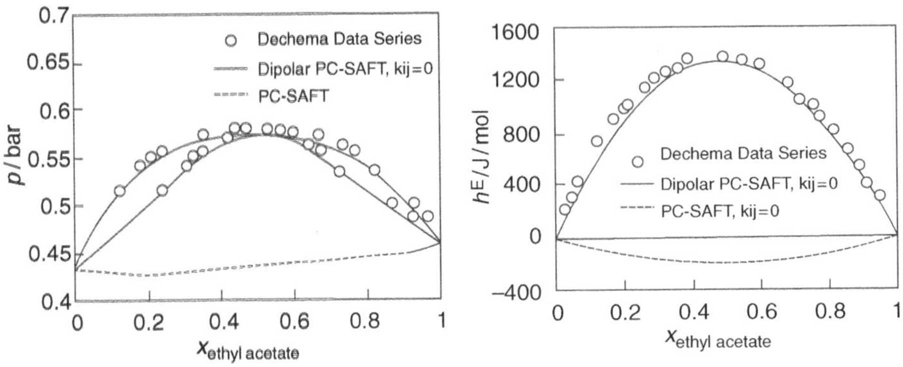

Figure 13.19 *The system ethylacetate–cyclohexane with PC–SAFT (dashed lined) and dipolar PC–SAFT (Chapman model). The calculations are made with $k_{ij}=0$. Right: VLE at 55 °C. Left: Excess enthalpy at 25 °C. Reprinted with permission from Fluid Phase Equilibria, Thermodynamic modeling of complex systems using PC-SAFT by F. Tumakaka, J. Gross and G. Sadowski, 228–229, 2, PPEPPD 2004 Proceedings, 89–98 Copyright (2005) Elsevier*

Other success stories involve mixtures with esters and CO_2. As can be seen in Figure 13.19, excellent VLE and excess enthalpies are obtained with polar PC–SAFT for ethylacetate–cyclohexane (with $k_{ij}=0$), whereas PC–SAFT predicts ideal solution behavior and the opposite trend for the enthalpy concentration dependency.

In the case of CO_2 mixtures with alkanes, again excellent VLE prediction ($k_{ij}=0$) is obtained, which ensures that the quadrupolar effects are properly accounted for. PC–SAFT can correlate CO_2–alkane VLE behavior but at a cost of interaction parameters typically higher than 0.1.

Gross–Vrabec model

This model has been applied to a large variety of polar mixtures, including some highly polar compounds, with dipole moments higher than 3 D. The most important results obtained are:

- Very good CO_2–alkane VLE is achieved, with k_{ij} values much smaller than when quadrupolar forces are not explicitly accounted for (in PC–SAFT). The simultaneous VLE/LLE of CO_2–C_{16} is described very well.
- In all cases, the k_{ij} values with polar PC–SAFT are much smaller than those with PC–SAFT, typically reduced to half.
- Highly polar systems such as DMFO and especially nitriles (dipole moments around 4 D) with alkanes cannot be correlated well with PC–SAFT even with large k_{ij} values, whereas polar PC–SAFT describes these systems very well and with small k_{ij} values. A typical example is shown in Figure 13.20.
- Representation of LLE is still difficult even for polar SAFT, e.g. for DMFO + CC6 or CO_2–C_{12}. Simultaneous correlation of VLE/LLE or of both branches of LLE is not always possible.

Kleiner and Sadowski[40] showed recently that the Gross–Vrabec polar PC–SAFT, with experimental dipole moments, performs very well for polar cross-associating mixtures exhibiting induced association, such as alcohols or water with esters or ethers, acetone–water, methanol–butyronitrile and THF–water. In all cases, they showed that the best performance is obtained when the induced association is explicitly accounted for (using the combining rules discussed in Chapter 8, text following Equation (8.49)), i.e. assuming that the cross-association energy is equal to half the value of the association energy for the hydrogen bonding compound. Inclusion of the polarity does not help when phase equilibria of associating compounds were considered and was therefore not included for water and alcohols. An example is shown in Figure 13.21 for the complex water–acetone system. Induced association is discussed in more detail in Section 13.6.4.

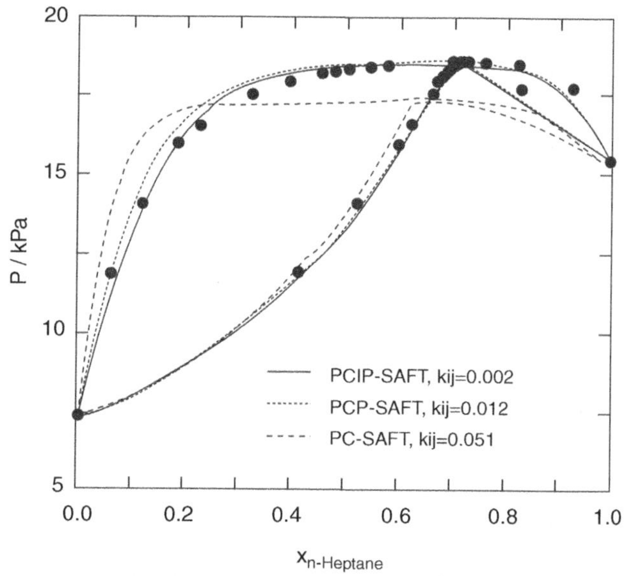

Figure 13.20 *VLE for the mixture butyronitrile–heptane at 45 °C. Butyronitrile has a dipole moment equal to 4.1 D. Comparison of experimental data (points) to correlation results of PC–SAFT, polar PC–SAFT (PCP–SAFT) and the polar PC–SAFT which accounts for polarizability (PCIP–SAFT). The polar model from Gross is used. Reprinted with permission from AICHE, An Equation of State Contribution for Polar Components:Polarizable Dipoles by Matthias Kleiner and Joachim Gross,42, 11, 3170–3180 Copyright (2006) Wiley-VCH*

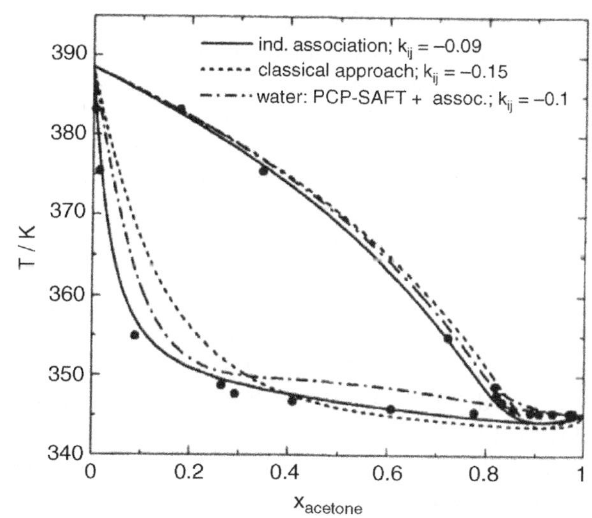

Figure 13.21 *VLE of the acetone–water mixture at 1.7 bar. Points are experimental data and lines are correlations with various versions of polar PC–SAFT (Gross model): solid line, induced association; dashed line, classical approach (i.e. no induced association); dashed–dotted line, both association and polarity of water are explicitly accounted for. Reprinted with permission from Journal of Physical Chemistry C, Modeling of Polar Systems Using PCP-SAFT: An Approach to Account for Induced-Association Interactions by Matthias Kleiner and Gabriele Sadowski, 111, 43, 15533–15544 Copyright (2007) American Chemical Society*

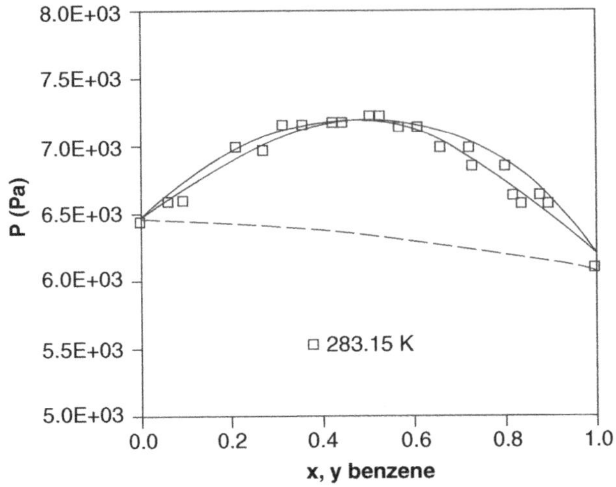

Figure 13.22 *Prediction of benzene–cyclohexane VLE using the GC–VR–SAFT. The full line is when the quadrupole moment of benzene is included and the dashed line is when it is ignored. All calculations are with $k_{ij} = 0$. Reprinted with permission from Fluid Phase Equilibria, Application of GC-SAFT EOS to polar systems using a segment approach by D. NguyenHuynh, J.-P. Passarello, P. Tobaly and J.-C. de Hemptinne, 264, 1–2, 62–75 Copyright (2008) Elsevier*

GC polar SAFT model

Finally, it is worth discussing the polar GC–SAFT method developed by the research group of de Hemptinne and co-workers.[104–107] This and the other GC–SAFT methods were presented in Chapter 8 (Section 8.5) but recently the model has been further extended using polar and quadrupolar terms. The polar GC–SAFT has been applied to polar (esters) and associating compounds (alcohols) as well as mixtures of those compounds with various types of hydrocarbons, including alkylbenzenes and polyaromatic hydrocarbons. The polar term of Gubbins and Twu[69] has been used and in the most recent publications the segment polar approach from Chapman's group has been adopted. Very satisfactory results are obtained and the importance of polar/quadrupolar effects is evident as no interaction parameters are needed (except for ester–alcohols, where results are improved when cross-association is accounted for).

One example illustrating the importance of the quadrupolar effects is shown in Figure 13.22.

The polar/quadrupolar GC–SAFT has been applied, and group parameter tables are available, with various SAFT variants (original SAFT, SAFT–VR and PC–SAFT) and with the differences being rather small, especially for the latter two SAFT models (SAFT–VR and PC–SAFT). Cross-association effects are important for alcohol–ester mixtures (Figure 13.23).

The polar GC–SAFT is under continuous development and recent applications include water–alkanes[108] and a variety of polar/associating mixtures containing alcohols, esters, ketones, ethers, acid gases and hydrocarbons.[109]

Comparison between polar terms

Table 13.8 gives an overview of some of the most characteristic comparisons that have been reported and which include more than one polar terms. The major conclusions from these comparisons are also included

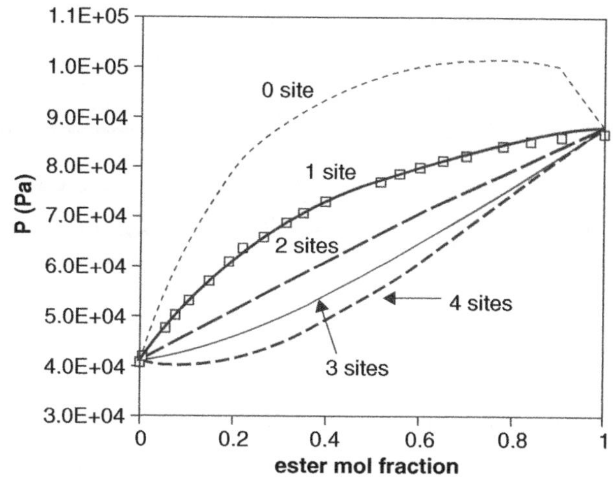

Figure 13.23 *Influence of the cross-association site number on the VLE phase diagram of methyl propanoate–propanol at 348.15 K. The calculations are made with the polar GC–PC–SAFT. It is of paramount importance to include the cross-association effects. The one-site cross-association model provides the best results. The cross-association parameters are those of 1-alcohols. Reprinted with permission from Fluid Phase Equilibria, Application of GC-SAFT EoS to polar systems using a segment approach by D. NguyenHuynh, J.-P. Passarello, P. Tobaly and J.-C. de Hemptinne, 264, 1–2, 184–200 Copyright (2008) Elsevier*

in the table. In many cases the differences between various polar terms are very small, despite their differences in derivation and background. One example is shown in Figure 13.24.

A recent investigation considering all three major polar terms (Chapman, Gross and Economou) has been published by Al-Saifi *et al.*[96] They compared the predictive capabilities of PC–SAFT to all three different polar terms, using pure compound parameters estimated solely from pure compound data and $k_{ij} = 0$ for the binary systems. They considered only VLE for alcohol–alkanes, alcohol–water, glycol–water and water–propane.

Table 13.8 *Comparison of the performance between various polar SAFT EoS*

Reference	EoS compared	Conclusions
Kleiner and Gross[95]	Polar PC–SAFT Induced polar PC–SAFT	Very similar results The dipole moment effect is much greater than the polarizability term
Sauer and Chapman[98]	Polar CK–SAFT Polar PC–SAFT (using Chapman's dipolar term)	Very similar results for ketone–alkanes, in all cases better than when CK– or PC–SAFT are used without the polar terms
Dominik *et al.*[99]	Polar PC–SAFT using Chapman's and Fischer's polar terms	Very similar results for various mixtures of ethers, ketones and esters with alkanes (VLE, LLE, excess enthalpies)
Al-Saifi *et al.*[96]	Polar PC–SAFT using the Chapman, Gross and Economou terms	Chapman's term best overall for alcohol–alkanes Problems with false phase splits for methanol–alkanes at low temperatures All terms tested for pure predictions ($k_{ij} = 0$)

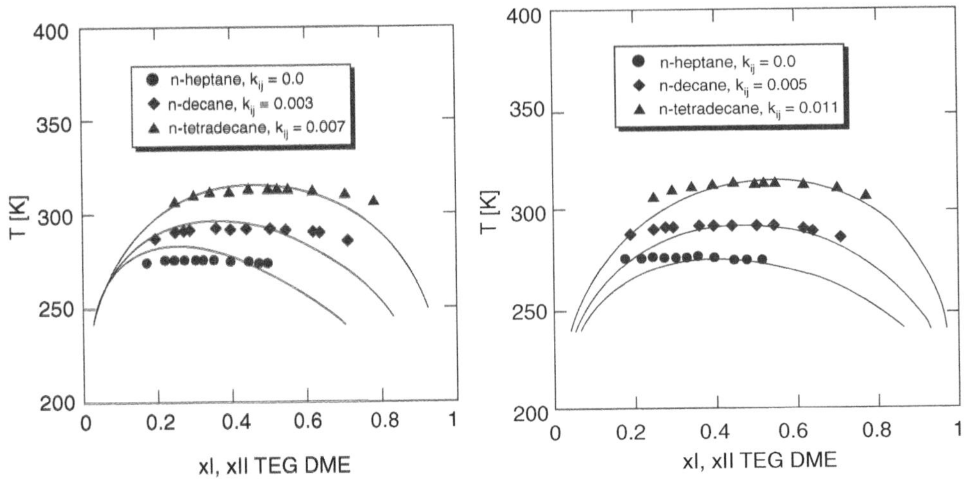

Figure 13.24 *LLE of TEG DME–alkanes with polar PC–SAFT. Left: Chapman's polar model is used. Right: Fischer's model is used for the polarity. Similar results with the two versions of polar PC–SAFT have been presented also for excess enthalpies for the same system (at 323 K). Reprinted with permission from Ind. Eng. Chem. Res., by Dominik et al., 44, 6928 Copyright (2005) American Chemical Society*

They concluded that overall Chapman's polar term performs best, although a single 'winner model' for all cases could not be established. Chapman's polar term combined with PC–SAFT resulted in false splits for methanol–alkanes at low temperatures. It also resulted in the highest contributions from polarity to the chemical potential values, as shown in Figure 13.25.

While this investigation[96] certainly provided insights on the capabilities and limitations of the three polar terms combined with PC–SAFT, from an engineering point of view a number of cases for the same mixtures which were not considered but are of interest are: LLE for methanol–alkanes and simultaneous VLE/LLE; and especially the prediction of multicomponent LLE and VLLE of water–alcohol–alkanes. In addition, the water 2B model is used, which is in disagreement with water's physical picture (the three- or four-site association models would have been more realistic).

In brief, we can conclude that:

- Polar non-associating mixtures can be predicted very well with polar SAFT EoS, and in many cases it is safe simply to set $k_{ij}=0$.
- Polarizability effects are less important than dipolar interactions, but the quadrupolar terms improve the performance of SAFT EoS for CO_2-containing mixtures as well as for systems with (poly)aromatic hydrocarbons.
- The effect of quadrupolar/dipolar and other multipolar interactions has not been systematically evaluated but it may be small in many practical applications.
- It is not always clear a priori whether the dipole or the quadrupole moment can be taken from the 'vacuum' experimental values or whether they should be considered adjustable parameters fitted to experimental data.
- In the presence of polar self-associating compounds, the effect of polarity may be small and in some cases overshadowed by the most important hydrogen bonding effects (see also Section 13.6.4).

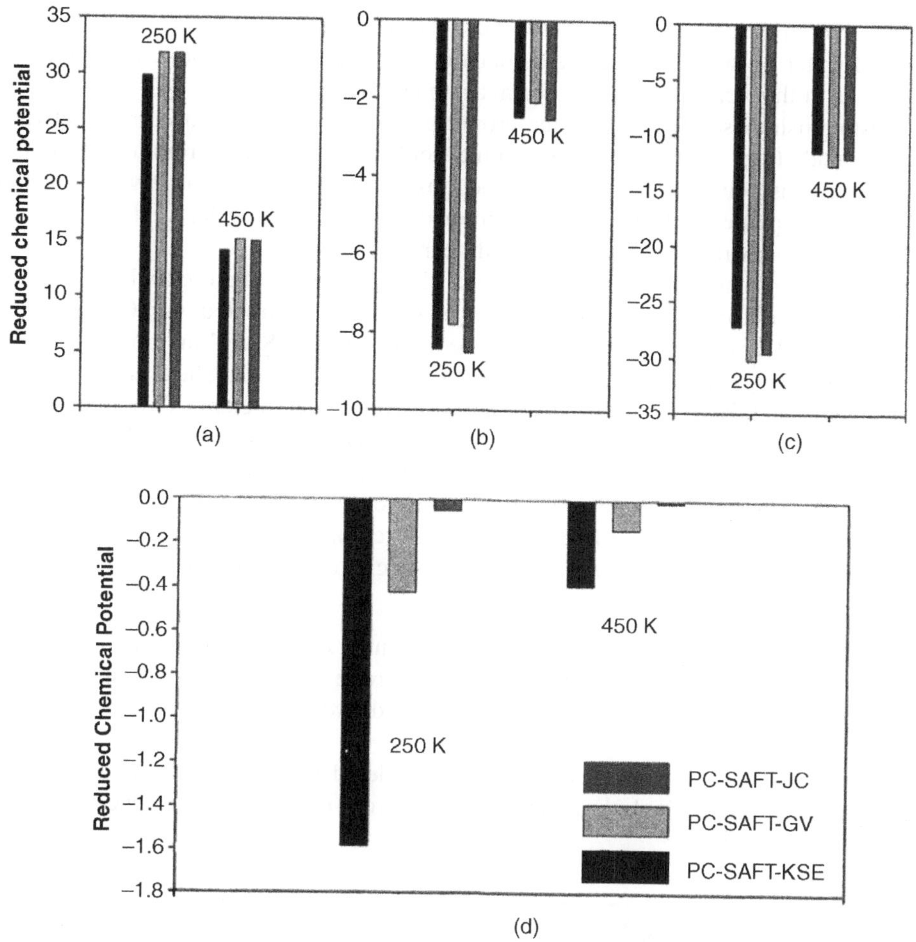

Figure 13.25 *Reduced chemical potential of 1-propanol at 250 and 450 K from the dipolar PC–SAFT using the Chapman (JC), Gross (GV) and Economou (KSE) polar terms. The various contributions are as follows: (a) repulsion; (b) hydrogen bonding; (c) dispersion; and (d) dipole–dipole. Reprinted with permission from Fluid Phase Equilibria, Prediction of vapor-liquid equilibrium in water-alcohol-hydrocarbon systems with the dipolar perturbed-chain SAFT equation of state by Nayef M. Al-Saifi, Esam Z. Hamad and Peter Englezos, 281, 1–2, 82–93 Copyright (2008) Elsevier*

- For mixtures of polar and hydrogen bonding compounds, it is often far more important to consider explicitly the cross-association (induced association) between the two compounds rather than the polar character of the hydrogen bonding compound (see also Section 13.6.4).
- Future studies should definitely attempt to clarify among others, also the relative importance of the various polar interactions as compared to hydrogen bonding effects (self- and cross-association ones).

13.6.4 The importance of solvation (induced association)

Many mixtures containing only one self-associating compound (e.g. water or alcohols) and a polar compound (e.g. ketones, ethers or esters) can exhibit cross-association, i.e. creation of hydrogen bonding between different compounds in the solution. Weaker cross-association interactions can also exist between aromatic hydrocarbons and water. In some cases cross-association is present even if both compounds are non self-associating, such as in the well-known chloroform–acetone case.

In order to distinguish these effects from the cross-association interaction occurring between two self-associating compounds (e.g. water–alcohols or amine–acids), we call these cross-association interactions 'solvation' or 'induced association'. Solvation interactions can be very strong, as in water–acetone, and can markedly change the properties of the mixture in solution often by increasing the solubility, e.g. the increased solubility of aromatic compared to the aliphatic hydrocarbons in water.

Because of their importance, solvation effects can almost never be ignored when such mixtures are modeled with association EoS. We saw how solvating systems can be treated with CPA in Chapters 9–12, using the modified CR-1 rule. Similar concepts can be applied to PC–SAFT and other SAFT models. The CR-1 and Elliott combining rules as shown in Equations (8.49)–(8.51) cannot be used directly, as one of the two compounds does not have, strictly speaking, association parameters. Two possible approaches, discussed in Chapter 8 (text following Equation (8.51)), is to treat $\kappa^{A_iB_j}$ as an adjustable parameter (Lyngby approach)[44,45] or set it equal to the value of the self-associating compound (Dortmund approach).[40] In both cases, the cross-association energy is simply taken by Equation (8.49) or (8.50), i.e. it is set equal to half the value of the association energy of the hydrogen bonding compound ($\varepsilon^{A_iB_j} = \varepsilon^{assoc}/2$). Both investigations,[40,44,45] despite the difference in the way induced association is accounted for, lead to the following conclusions:

- The solvation phenomena should almost always be accounted for, especially for aqueous mixtures. They are crucial for aqueous systems, and they can rarely be ignored even for relatively simpler cross-associating mixtures, such as those between alcohols and chlorinated hydrocarbons, as shown for one example in Figure 13.26.
- The performance of PC–SAFT when solvation is considered is always better than when the classical approach is used, i.e. using a single k_{ij} parameter in the cross-dispersion term. One example is shown in

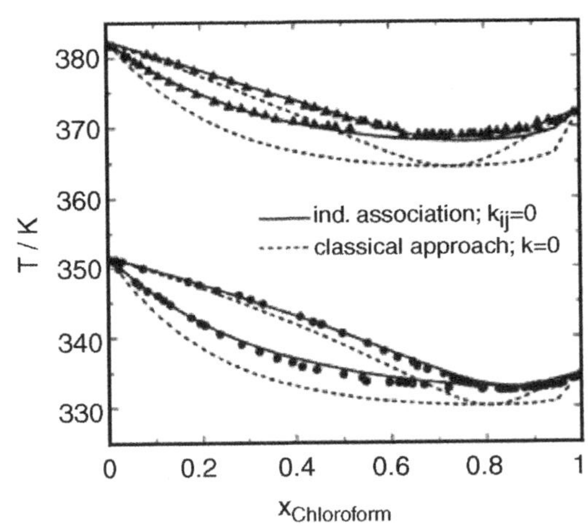

Figure 13.26 *VLE of chloroform–ethanol at 1 bar using the polar PC–SAFT with or without accounting for the solvation. No parameters were fitted to experimental data. Reprinted with permission from Journal of Physical Chemistry C, Modeling of Polar Systems Using PCP-SAFT: An Approach to Account for Induced-Association Interactions by Matthias Kleiner and Gabriele Sadowski, 111, 43, 15533–15544 Copyright (2007) American Chemical Society*

Figure 13.16. In many cases a single interaction parameter is needed, i.e. when $\kappa^{A_iB_j}$ is adjusted to experimental data, then k_{ij} can be set to zero.

- Even for highly polar compounds, e.g. butyronitrile, and even if the polar interactions are explicitly accounted for via a polar PC–SAFT, phase equilibria can still be described successfully only when induced association is accounted for, as shown by Kleiner and Sadowski[40]. One example is shown in Figure 13.27. Simultaneous representation of VLE and LLE for butanone–water with polar PC–SAFT is possible only if the induced association is accounted for. The polar term is used for the ketone but not for water. Similar conclusions have been obtained for other aqueous systems as well, e.g. with methyl methacrylate (VLE and LLE), ethylacetate (VLE and LLE), furfural (VLE and LLE) and butyronitrile (LLE). The results are successful despite the fact that water is treated (in polar PC–SAFT) with the simple and somewhat unrealistic 2B association scheme (two-site molecule).
- The system water–butanone LLE with sPC–SAFT was shown in Figure 2.2 (over a wider temperature range compared to Figure 13.27). It can be concluded that sPC–SAFT cannot, without accounting explicitly for polarity, describe phase equilibria for this highly polar mixture, even when the polar compounds are assumed to be self-associating.
- LLE of water–aromatic hydrocarbons can only be represented when solvation is accounted for and k_{ij} is also included (Figure 13.3).
- For highly polar systems, accounting for polarity, hydrogen bonding and solvation phenomena may be of importance but care should be exercised. As shown in Figures 13.3, 13.21, and 13.27. it is best to 'ignore' polar effects in water and assume that they are absorbed in the association term of polar PC–SAFT. Nonetheless, solvation should be explicitly accounted for.

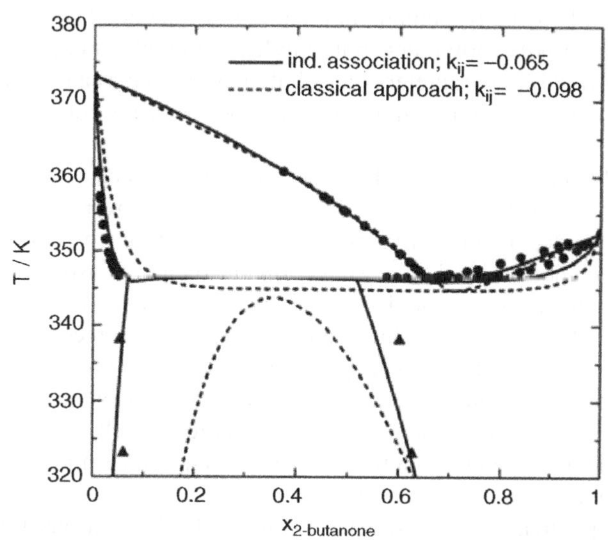

Figure 13.27 *VLE (high temperatures) and LLE (low temperatures) for 2-butanone–water at 1 bar with polar PC–SAFT. The correlation results are shown when solvation is considered (induced association) and when solvation is not accounted for (classical approach). The polar term is not used for water. Reprinted with permission from Journal of Physical Chemistry C, Modeling of Polar Systems Using PCP-SAFT: An Approach to Account for Induced-Association Interactions by Matthias Kleiner and Gabriele Sadowski, 111, 43, 15533–15544 Copyright (2007) American Chemical Society*

- Similar conclusions on the importance of solvation have been obtained for other EoS which account explicitly for hydrogen bonding phenomena such as the CPA and the NRHB EoS.

13.7 Flow assurance (asphaltenes and gas hydrate inhibitors)

Solid compounds such as gas hydrates, waxes and asphaltenes may create significant problems during oil production and processing. Thermodynamic models are useful in determining conditions for creating such 'solid nuisances' as well as predicting, e.g. for gas hydrates, what type and how many (thermodynamic) inhibitors are needed for a specific application.

The performance of an EoS in gas hydrate formation is closely linked to water–hydrocarbon–inhibitor (alcohols, glycols) phase equilibria. Li *et al.*[60,61] showed that (original) SAFT coupled with the van der Waals–Platteeuw model (mentioned in Chapter 10) provides excellent prediction of the inhibition effect of several polar inhibitors on single gas hydrate formation (even up to concentrations of inhibitors as high as 50 wt%). They showed that SAFT can be used for mixed gas hydrate systems as well.

Other relevant investigations are the early study by Suresh and Beckman[2], who reported water–alcohol– hydrocarbon LLE with SAFT, and the recent ones by the Aveiro group[28] with soft SAFT for glycols.

Of the three types of solid compounds mentioned, the most complex ones (and difficult to inhibit) are the asphaltenes, sometimes also called 'the cholesterol of petroleum'. Asphaltenes are solid, polar, polyaromatic compounds which can plug reservoir wells and can precipitate during production, transportation, refining and processing of crude oil. Asphaltenes have a high molecular weight (around 1000–4000 g/mol) and they also contain some heteroatoms (especially S, O, N), but there is great uncertainty and many debates about their actual structure. They are therefore often 'simply' defined as a 'solubility class', i.e. asphaltenes are insoluble in heptane or pentane but soluble in toluene. They are the 'most polar' part of petroleum, as polar as certain polar polymers (polyethylene terephthalate, polyvinyl chloride, polyvinyl acetate). For example, the solubility parameters of asphaltenes are in the same range as those of these polar polymers. Asphaltenes are stabilized in oil in the presence of (the equally polar but smaller) resin molecules.

The precipitation of asphaltenes depends on the various conditions such as temperature, pressure, but also injection of gases, e.g. CO_2 or methane injected for enhancing oil recovery.

As efforts to increase oil production intensify with deeper reservoirs and production systems using a variety of enhanced oil recovery techniques based on gas injection, the danger of asphaltene precipitation increases. Asphaltenes can easily precipitate as pressure is reduced but also if the oil is diluted by a light hydrocarbon, e.g. a gas such as methane, CO_2 or nitrogen which can be injected for oil recovery purposes. In this case, the resin concentration decreases and asphaltenes may precipitate and form a deposit. The temperature has an interesting effect in asphaltenes: at high temperatures, the polar resin–asphaltenes interactions are decreased and flocculation may occur. At even higher temperatures, asphaltenes melt and redissolve in oil, producing in this case a stable system. Thus, depending on the temperature and oil composition, flocculation may either decrease or increase with increasing temperature.

It is difficult to predict asphaltene precipitation based on simple observations. For example, data show that the precipitation is not directly related to either asphaltene content in the oil or the molecular weight of the oil.

Thus, predicting the asphaltene precipitation is immensely important for economical and environmental reasons, although there are significant difficulties due to lack of reliable data and limited understanding of the chemical nature of the association with resins, the polar compounds which stabilize asphaltenes in oil.

Table 13.9 *Application of SAFT (and CPA) for flow assurance (asphaltenes and gas hydrates)*

SAFT variant	Reference	Application
Original	Wu *et al.*[73,74]	Asphaltenes
SAFT–VR	Buenrostro-Gonzalez *et al.*[75]	Asphaltenes
PC–SAFT	Ting *et al.*[76]	Asphaltenes
	Gonzalez *et al.*[77,85,86]	
	Pedersen and Sørensen[110]	
CPA	Fahim and Andersen[78]	Asphaltenes
	Edmonds *et al.*[79]	
Original SAFT	Li *et al.*[60]	Gas hydrates (single gas + inhibitors: methanol, MEG, TEG and glycerol)
Original SAFT	Li *et al.*[61]	Gas hydrates (C1, C2, C3, mixed gases – no inhibitors)

We saw in Chapter 4 some simple 'asphaltene' models based on the regular solution theory and the Flory–Huggins model. Such simple models are merely of qualitative value. Due to the extensive association–solvation phenomena, it is tempting to develop SAFT to asphaltene mixtures.

Table 13.9 presents some applications of various SAFT variants (for completion, we have also included CPA applications) for asphaltenes (and gas hydrates).

Figures 13.28 and 13.29 illustrate two recent results from the application of PC–SAFT to asphaltene precipitation. In these applications the association term of PC–SAFT is not used. In another recent publication

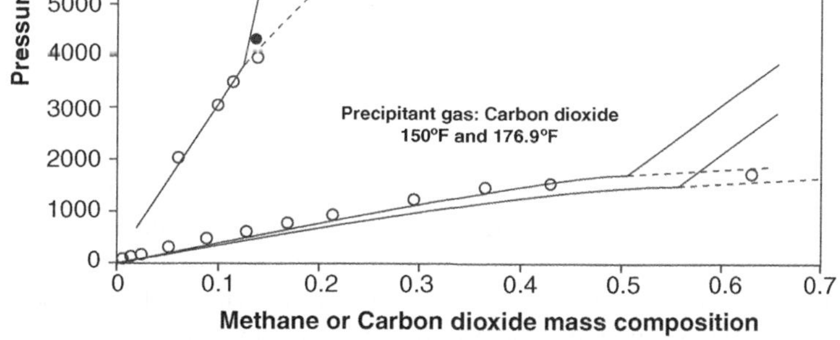

Figure 13.28 *The effect of CO_2 injection against methane addition on asphaltene precipitation and SAFT calculations. The points are the experimental data (open circles, bubble points; filled circles, asphaltene stability onset). The lines are predictions with PC–SAFT without accounting for association. Reprinted with permission from Energy & Fuels, Prediction of Asphaltene Instability under Gas Injection with the PC-SAFT Equation of State by Doris L. Gonzalez, P. David Ting, George J. Hirasaki, and Walter G. Chapman, 19, 4, 1230–1234 Copyright (2005) American Chemical Society*

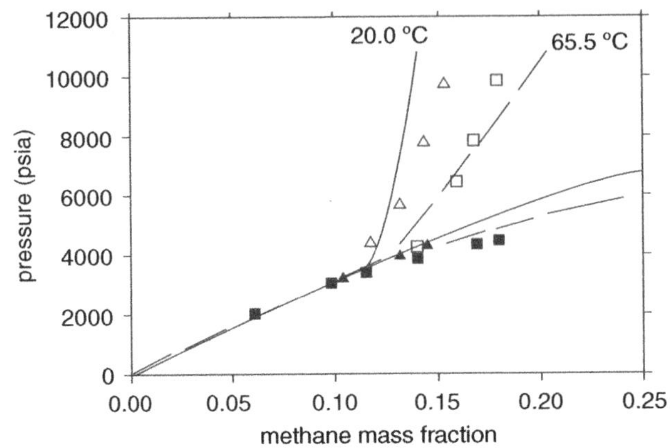

Figure 13.29 *Asphaltene instability onsets (open symbols) and bubble points (filled symbols) for a model live oil at 20 and 65.5 °C. The PC–SAFT predictions are drawn as lines. The EoS parameters of asphaltenes were fitted to precipitation data from oil titrations with n-alkanes at ambient conditions. Six pseudo-components were used. Reprinted with permission from Petroleum Science and Technology, Modeling of Asphaltene Phase Behavior with the SAFT Equation of State by P. David Ting, George J. Hirasaki and Walter G. Chapman, 21, 3, 647–661 Copyright (2003) Taylor and Francis*

related to PC–SAFT for asphaltenes,[110] the association term was not used either. However, in other works on SAFT for asphaltenes,[73–75] the 'full' version of the model, i.e. including the association term, has been used (with different approximations of the number of association sites per asphaltene molecule).

The preliminary results which have been presented so far are promising, but there are significant difficulties in both interpretation and generalization of these results:

- The role of SAFT's association term in asphaltene calculations has not yet been established.
- The models have so far been used mostly for correlation as some asphaltene precipitation data are needed for fitting the model's parameters.
- The models have been tested for few systems, partly due to the lack of (publicly available) asphaltene precipitation data. Thus, there is need for more measurements.

13.8 Concluding remarks

A number of successful applications of SAFT (especially PC–SAFT) for polar and associating mixtures have been illustrated in this chapter. These applications include aqueous mixtures (emphasis on water–alkane LLE), as well as mixtures with alcohols and glycols. Generalized parameters can be used for alcohols and glycols. For polar non-associating mixtures an additional term which accounts explicitly for polar effects could be added. Various approaches have been proposed for this, such as the successful approach with the truncated polar PC–SAFT. Future challenges which remain to be addressed include:

- Validating independent methods for obtaining the association parameters, e.g. via quantum chemistry or spectroscopy.
- Multifunctional associating molecules such as glycolethers and alkanolamines.
- Assessing the balance of or need for additional terms, especially for describing polar and quadrupolar effects. A comparative analysis of terms, e.g. via simulation studies, would be useful.

- A description of the critical area.
- Addressing the limitations of Wertheim's theory, e.g. cooperativity, intramolecular association and steric effects.

References

1. S.H. Huang, M. Radosz, *Ind. Eng. Chem. Res.*, **1991**, *30*, 1994.
2. J. Suresh, E.J. Beckman, *Fluid Phase Equilib.*, **1994**, *99*, 219.
3. Y.-H. Fu, S.I. Sandler, *Ind. Eng. Chem. Res.*, **1995**, *34*, 1897.
4. P.A. Galindo, J. Whitehead, G. Jackson, A.N. Burgess, *J. Phys. Chem.*, **1996**, *100*, 6781.
5. T. Kraska, K.E. Gubbins, *Ind. Eng. Chem. Res.*, **1996**, *35*, 4727.
6. T. Kraska, K.E. Gubbins, *Ind. Eng. Chem. Res.*, **1996**, *35*, 4738.
7. J.P. Wolbach, S.I. Sandler, *Ind. Eng. Chem. Res.*, **1997**, *36*, 4041.
8. J.P. Wolbach, S.I. Sandler, *Ind. Eng. Chem. Res.*, **1998**, *37*, 2917.
9. I.G. Economou, C. Tsonopoulos, *Chem. Eng. Sci.*, **1997**, *52(4)*,511.
10. P.K. Jog, A. Garcia-Cuellar, W.G. Chapman, *Fluid Phase Equilib.*, **1999**, *158–160*, 321.
11. E.C. Voutsas, G.C. Boulougouris, I.G. Economou, D.P. Tassios, *Ind. Eng. Chem. Res.*, **2000**, *39*, 797.
12. G.C. Boulougouris, E.C. Voutsas, I.G. Economou, D.N. Theodorou, D.P. Tassios, *J. Phys. Chem. B*, **2001**, *105*, 7792.
13. Z.-Y. Zhang, J.-C. Yang, Y.-G. Li, *Fluid Phase Equilib.*, **2000**, *169*, 1.
14. O. Pfohl, M. Buddich, *Fluid Phase Equilib.*, **2001**, *189*, 179.
15. O. Pfohl, A. Pagel, G. Brunner, *Fluid Phase Equilib.*, **1999**, *157*, 53.
16. Y.-G. Li, *Fluid Phase Equilib.*, **2001**, *189*, 187.
17. H. Adidharma, M. Radosz, *J. Phys. Chem. B*, **2001**, *105*, 9822.
18. J. Gross, G. Sadowski, *Ind. Eng. Chem. Res.*, **2002**, *41*, 5510.
19. X.-S. Li, P. Englezos, *Ind. Eng. Chem. Res.*, **2003**, *42*, 4953.
20. X.-S. Li, P. Englezos, *Fluid Phase Equilib.*, **2004**, *224*, 111.
21. B.H. Patel, P. Paricaud, A. Galindo, G.C. Maitland, *Ind. Eng. Chem. Res.*, **2003**, *42*, 3809.
22. J.-P. Passarello, P. Tobaly, *Ind. Eng. Chem. Res.*, **2003**, *42*, 5383.
23. S. Gupta, J.D. Olson, *Ind. Eng. Chem. Res.*, **2003**, *42*, 6359.
24. M. Yarrison, W.G. Chapman, *Fluid Phase Equilib.*, **2004**, *226*, 195.
25. N. von Solms, M.L. Michelsen, G.M. Kontogeorgis, *Ind. Eng. Chem. Res.*, **2004**, *43(7)*,1803.
26. A.M.A. Dias, J.C. Pamies, J.A.P. Coutinho, I.M. Marrucho, L.F. Vega, *J. Phys. Chem. B*, **2004**, *108*, 1450.
27. A.M.A. Dias, A.I. Caco, J.A.P. Coutinho, L.M.N.B.F. Santos, M.M. Pineiro, L.F. Vega, M.F.C. Gomes, I.M. Marrucho, *Fluid Phase Equilib.*, **2004**, *225(1–2)*,29.
28. N. Pedrosa, J.C. Pamies, J.A.P. Coutinho, I.M. Marrucho, L.F. Vega, *Ind. Eng. Chem. Res.*, **2005**, *44*, 7027.
29. S. Tamouza, J.-Ph. Passarello, P. Tobaly, J.-Ch. de Hemptinne, *Fluid Phase Equilib.*, **2005**, *228–229*, 409.
30. A. Grenner, J. Schmelzer, N. von Solms, G.M. Kontogeorgis, *Ind. Eng. Chem. Res.*, **2006**, *45*, 8170.
31. N. von Solms, I. Kouskoumvekaki, M.L. Michelsen, G.M. Kontogeorgis, *Fluid Phase Equilib.*, **2006**, *241*, 344.
32. F. Llovell, L.F. Vega, *J. Phys. Chem. B*, **2006**, *110*, 1427.
33. F. Llovell, L.F. Vega, *J. Phys. Chem. B*, **2006**, *110*, 1350.
34. A. Grenner, G.M. Kontogeorgis, N. von Solms, M.L. Michelsen, *Fluid Phase Equilib.*, **2007**, *258(1)*,83.
35. A. Grenner, G.M. Kontogeorgis, N. von Solms, M.L. Michelsen, *Fluid Phase Equilib.*, **2007**, *261(1–2)*,248.
36. G.N.I. Clark, A.J. Haslam, A. Galindo, G. Jackson, *Mol. Phys.*, **2006**, *104*, 3561.
37. A. Grenner, G.M. Kontogeorgis, M.L. Michelsen, G.K. Folas, *Mol. Phys.*, **2007**, *105(13–14)*,1797.
38. E. Voutsas, C. Perakis, G. Pappa, D.P. Tassios, *Fluid Phase Equilib.*, **2007**, *261*, 343.
39. D. Fuchs, J. Fischer, F. Tumakaka, G. Sadowski, *Ind. Eng. Chem. Res.*, **2006**, *45*, 6578.
40. M. Kleiner, G. Sadowski, *J. Phys. Chem. C*, **2007**, *111*, 15544.
41. F. Tumakaka, I.V. Prikhodko, G. Sadowski, *Fluid Phase Equilib.*, **2007**, *260*, 98.
42. I.V. Prikhodko, F. Tumakaka, G. Sadowski, *Russ. J. Appl. Chem.*, **2007**, *80(4)*,542.

43. I. Tsivintzelis, T. Spyriouni, I.G. Economou, *Fluid Phase Equilib.*, **2007**, *253(1)*,19.
44. A. Grenner, I. Tsivintzelis, G.M. Kontogeorgis, I.G. Economou, C. Panayiotou, *Ind. Eng. Chem. Res.*, **2008**, *47 (15)*,5636.
45. I. Tsivintzelis, A. Grenner, G.M. Kontogeorgis, I.G. Economou, C. Panayiotou, *Ind. Eng. Chem. Res.*, **2008**, *47 (15)*,5651.
46. M.-L. Yu, Y.-P. Chen, *Fluid Phase Equilib.*, **1994**, *94*, 149.
47. J. Wu, J.M. Prausnitz, *Ind. Eng. Chem. Res.*, **1998**, *37*, 1634.
48. J.M. Prausnitz, R.N. Lichtenthaler, E.G. de Azevedo, *Molecular Thermodynamics of Fluid-Phase Equilibria* (3rd edition). Prentice Hall International, **1999**.
49. I.G. Economou, *Ind. Eng. Chem. Res.*, **2002**, *41*, 953.
50. S. Aparicio-Martinez, K.R. Hall, *Fluid Phase Equilib.*, **2007**, *254*, 112.
51. C.A. Koh, H. Tanaka, J.M. Walsh, K.E. Gubbins, J.A. Zollweg, *Fluid Phase Equilib.*, **1993**, *83*, 51.
52. G.M. Kontogeorgis, I.V. Yakoumis, H. Meijer, E.M. Hendriks, T. Moorwood, *Fluid Phase Equilib.*, **1999**, *158–160*, 201.
53. S.I. Sandler, J.P. Wolbach, M. Castier, G. Escobedo-Alvarado, *Fluid Phase Equilib.*, **1997**, *136*, 15.
54. J.K. Button, K.E. Gubbins, *Fluid Phase Equilib.*, **1999**, *158–160*, 175.
55. E.K. Karakatsani, T. Spyriouni, I.G. Economou, *AIChE J.*, **2005**, *51*, 2328.
56. I.G. Economou, M.D. Donohue, *Ind. Eng. Chem. Res*, **1992**, *31*, 2388.
57. B.M. Hasch, E.J. Maurer, L.F. Ansanelli, M.A. McHugh, *J. Chem. Thermodyn.*, **1994**, *26*, 625.
58. W.G. Chapman, K.E. Gubbins, G. Jackson, M. Radosz, *Ind. Eng. Chem. Res.*, **1990**, *29*, 1709.
59. N. von Solms, M.L. Michelsen, G.M. Kontogeorgis, *Ind. Eng. Chem. Res.*, **2003**, *42*, 1098.
60. X.-S. Li, H.-J. Wu, P. Englezos, *Ind. Eng. Chem. Res.*, **2006**, *45*, 2131.
61. X.-S. Li, H.-J. Wu, Y.-G. Li, Z.-P. Feng, L.-G. Tang, S.-S. Fan, *J. Chem. Thermodyn.*, **2007**, *39*, 417.
62. I.A. Kouskoumvekaki, G. Krooshof, M.L. Michelsen, G.M. Kontogeorgis, *Ind. Eng. Chem. Res.*, **2004**, *43(3)*,826.
63. F. Tumakaka, G. Sadowski, *Fluid Phase Equilib.*, **2004**, *217*, 233.
64. P.K. Jog, S.G. Sauer, J. Blaesing, W.G. Chapman, *Ind. Eng. Chem. Res.*, **2001**, *40*, 4641.
65. P.K. Jog, W.G. Chapman, *Mol. Phys.*, **1999**, *97*, 307.
66. J. Gross, *AIChE J.*, **2005**, *51*, 2556.
67. J. Gross, J. Vrabec, *AIChE J.*, **2006**, *52*, 1194.
68. C. Vega, *Mol. Phys.*, **1992**, *75*, 427.
69. K.E. Gubbins, C.H. Twu, *Chem. Eng. Sci.*, **1978**, *33*, 863.
70. B. Larsen, J.C. Rasaiah, G. Stell, *Mol. Phys.*, **1977**, *33*, 987.
71. E.K. Karakatsani, I.G. Economou, *J. Phys. Chem. B*, **2006**, *110*, 9252.
72. E.K. Karakatsani, G.M. Kontogeorgis, I.G. Economou, *Ind. Eng. Chem. Res.*, **2006**, *45(17)*,6063.
73. J. Wu, J.M. Prausnitz, A. Firoozabadi, *AIChE J.*, **2000**, *46(1)*,197.
74. J. Wu, J.M. Prausnitz, A. Firoozabadi, *AIChE J.*, **1998**, *44(5)*,1188.
75. E. Buenrostro-Gonzalez, C. Lira-Galeana, A. Gil-Villegas, J. Wu, *AIChE J.*, **2004**, *50(10)*,2552.
76. P.D. Ting, G.J. Hirasaki, W.G. Chapman, *Pet. Sci. Technol.*, **2003**, *21(3–4)*,647.
77. D.L. Gonzalez, P.D. Ting, G.J. Hirasaki, W.G. Chapman, *Energy Fuels*, **2005**, *19*, 1230.
78. M.A. Fahim, S.I. Andersen, *SPE*, 93517, **2005**.
79. B. Edmonds, R.A.S. Moorwood, R. Szczepanski, X. Zhang, M. Heyward, R. Hurle, Measurement and prediction of asphaltene precipitation from live oils. Third International Symposium on Colloid Chemistry in Oil Production, Asphaltenes and Waxes Deposition (ISCOP'99), Mexico, November 1999.
80. S.H. Huang, M. Radosz, *Ind. Eng. Chem. Res.*, **1990**, *29*, 2284.
81. N. von Solms, M.L. Michelsen, C.P. Passos, S.O. Derawi, G.M. Kontogeorgis, *Ind. Eng. Chem. Res.*, **2006**, *45*, 5368.
82. J.R. Errington, G.C. Boulougouris, I.G. Economou, A.Z. Panagiotopoulos, D.N. Theodorou, *J. Phys. Chem. B*, **1998**, *102*, 8865.
83. G.K. Folas, S.O. Derawi, M.L. Michelsen, E.H. Stenby, G.M. Kontogeorgis, *Fluid Phase Equilib.*, **2005**, *228–229*, 121.
84. G.C. Pimentel, A.L. McClellan, *The Hydrogen Bond*. W.H. Freeman, 1960.
85. D.L. Gonzalez, G.J. Hirasaki, J. Creek, W.G. Chapman, *Energy Fuels*, **2007**, *21*, 1231.

86. D.L. Gonzalez, F.M. Vargas, G.J. Hirasaki, W.G. Chapman, *Energy Fuels*, **2008**, *22*, 757.
87. Z.-Y. Zhang, J.-C. Yang, Y.-G. Li. *Fluid Phase Equilib.*, **2000**, *169*, 1.
88. W.A.P. Luck, *Angew. Chem. Int. Ed. Engl.*, **1980**, *19*, 28.
89. N.J.P. de Melo, A.M.A. Dias, M. Blesic, L.P.N. Rebelo, L.F. Vega, J.A.P. Coutinho, I.M. Marrucho, *Fluid Phase Equilib.*, **2006**, *2*, 210.
90. A. Anderko, *Fluid Phase Equilib.*, **1991**, *65*, 89.
91. M.N. García-Lisbona, A. Galindo, G. Jackson, *Mol. Phys.*, **1998**, *93*, 57.
92. A. Kziazckzac, K. Moorthi, *Fluid Phase Equilib.*, **1985**, *23(2–3)*,153.
93. A. Nath, E. Bender, *Fluid Phase Equilib.*, **1981**, *7*, 275.
94. I.A. Kouskoumvekaki, N. von Solms, M.L. Michelsen, G.M. Kontogeorgis, *Fluid Phase Equilib.*, **2004**, *215(1)*,71.
95. M. Kleiner, J. Gross, *AIChE J.*, **2006**, *52*, 1951.
96. N.M. Al-Saifi, E.Z. Hamad, P. Englezos, *Fluid Phase Equilib.*, **2008**, *271*, 82.
97. F. Tumakaka, J. Gross, G. Sadowski, *Fluid Phase Equilib.*, **2005**, *228–229*, 89.
98. S.G. Sauer, W.G. Chapman, *Ind. Eng. Chem. Res.*, **2003**, *42*, 5687.
99. A. Dominik, W.G. Chapman, M. Kleiner, G. Sadowski, *Ind. Eng. Chem. Res.*, **2005**, *44*, 6928.
100. J. Gross, O. Spuhl, F. Tumakaka, G. Sadowski, *Fluid Phase Equilib.*, **2003**, *42*, 1266.
101. E.K. Karakatsani, I.G. Economou, *Fluid Phase Equilib.*, **2007**, *261*, 265.
102. E.K. Karakatsani, I.G. Economou, M.C. Kroon, C.J. Peters, G.-J. Witkamp, *J. Phys. Chem. C*, **2007**, *111*, 15487.
103. M.C. Kroon, E.K. Karakatsani, I.G. Economou, G.-J. Witkamp, C.J. Peters, *J. Phys. Chem. B*, **2006**, *110*, 9262.
104. T.X.N. Thi, S. Tamouza, P. Tobaly, J.-Ph. Passarello, J.-Ch. de Hemptinne, *Fluid Phase Equilib.*, **2005**, *238*, 254.
105. D.N. Huynh, M. Benamira, J.-Ph. Passarello, P. Tobaly, J.-Ch. de Hemptinne, *Fluid Phase Equilib.*, **2007**, *254*, 60.
106. D.N. Huynh, J.-Ph. Passarello, P. Tobaly, J.-Ch. de Hemptinne, *Fluid Phase Equilib.*, **2008**, *264*, 62.
107. D.N. Huynh, A. Falaix, J.-Ph. Passarello, P. Tobaly, J.-Ch. de Hemptinne, *Fluid Phase Equilib.*, 2008, *264*, 184.
108. M. Mourah, J.-Ch. de Hemptinne, J.-Ph. Passarello, P. Tobaly, N. Ferrando, Water solubility in n-alkanes for LLE conditions using GC-PC–SAFT EoS. Presented at the ESAT Conference, France 2008, pp. 343–346.
109. D.N. Huynh, J.-Ph. Passarello, P. Tobaly, J.-Ch. de Hemptinne, GC-SAFT as a predictive tool for computing VLE and LLE of systems involved in oil and gas industry. Presented at the ESAT Conference, France 2008, pp. 203–209.
110. K.S. Pedersen, C.H. Sørensen, *SPE*, 110483 **2008**.
111. M. Kleiner, Thermodynamic modeling of complex systems: polar and associating fluids and mixtures. PhD Thesis, Technical University of Dortmund, 2008.

14

Application of SAFT to Polymers

14.1 Overview

One of the most important first applications of SAFT was the description of phase equilibria of polymer mixtures and thus many of the applications of the original SAFT (or CK–SAFT)[1] included high-pressure phase equilibria for polyethylene and other polyolefins. Of the newest SAFT variants, PC–SAFT especially has been extensively applied to polymers, while applications of SAFT–VR and soft-SAFT to polymers have been reported as well. A variety of polymers and co-polymers have been considered, as well as different types of phase equilibria (VLE, LLE, SLLE, gas solubilities, etc.) and conditions (low and high pressures). Table 14.1 summarizes several applications of SAFT variants and Sections 14.3 and 14.4 will illustrate some applications, especially of simplified PC–SAFT. In Section 14.2 we will address an important issue when extending SAFT models (and other equations of state (EoS)) to polymers, i.e. how the parameters of polymers are obtained. Computational aspects are discussed in Appendix 14.B.

14.2 Estimation of polymer parameters for SAFT-type EoS

14.2.1 Estimation of polymer parameters for EoS: general

Obtaining reliable SAFT parameters for polymers (and for other EoS as well) is a difficult task and currently an active research area. This is because methods employed for low-molecular-weight compounds cannot be used since critical properties and vapor pressure data are not available (or have no meaning) for polymers. Thus, indirect methods are needed and various approaches have been tested:

1. Use of polymer density data alone (sometimes over extensive pressure and temperature conditions) may not always result in EoS parameters which are suitable for phase equilibria, as shown for one example in Figure 14.1. Figure 14.2 shows results for the same system with sPC–SAFT using other PDMS parameters (see discussion later).
2. Use of polymer volumetric data together with phase equilibria data (VLE or LLE) for mixtures (of the polymer under investigation with one or more solvents) is a useful approach for obtaining reliable parameters for SAFT, suitable for phase equilibrium calculations. This method has been investigated extensively with PC–SAFT.[14,15,32]

Thermodynamic Models for Industrial Applications: From Classical and Advanced Mixing Rules to Association Theories
Georgios M. Kontogeorgis and Georgios K. Folas

Table 14.1 *Some applications of SAFT models to polymers*

Polymer application	SAFT variant	Reference
Poly(E-*co*-P) + olefins	CK–SAFT	Chen *et al.*[2,41,42]
Poly(E-*co*-P) + ethylene, propylene	CK–SAFT	Gregg *et al.*[3]
Polymer–solvent VLE	CK–SAFT vs. Holten-n Andersen, UNIFAC–FV, GC–Flory	Wu and Chen[4]
Polyolefins, co-polymers + ethylene	CK–SAFT	Folie and Radosz[5]
Poly(E-*co*-VAC) + ethylene, ethylene/VAC	CK–SAFT	Folie *et al.*[6]
Various co-polymers: Poly(E-*co*-acrylic acid), Poly(E-*co*-methyl acrylate) + alkanes, DME	CK–SAFT	Lee *et al.*[7]
PE–ethylene	CK–SAFT vs. SL and polymer–SRK	Orbey *et al.*[8]
PE + ethylene, PB–butene	CK–SAFT, SLE	Koak *et al.*[9]
PE/ethylene, Cp, densities	SAFT vs. polymer–SRK, SL	Bokis *et al.*[10]
Poly(E-*co*-VAC), Poly(E-co-hexene) + solvents, gases	CK–SAFT	Kinzl *et al.*[11]
EMA co-polymers	CK–SAFT	Jog *et al.*[12]
PE fractionation	PC–SAFT	Cheluget *et al.*[13]
High-pressure PE–pentane	PC–SAFT	Gross and Sadowski[14]
PP, PE + solvents, PP–pentane–CO_2, PEP–butene	PC–SAFT	Tumakaka *et al.*[15]
Polymer–solvent VLE	SAFT vs. PR–WS and NLF–HB	Kang *et al.*[16]
Co-polymers	PC–SAFT	Gross *et al.*[17]
P(E-MA) co-polymers	PC–SAFT	Becker *et al.*[18]
Polar co-polymer	PC–SAFT/CK–SAFT	Chapman *et al.*[19]
LDPE–solvent PS–CO_2–CC_6 Poly(ethylene-*co*-butene)/C_3 Poly(ethylene-*co*-alkylacrylate)/ethylene	PC–SAFT	Sadowski[20]
Polymer–solvent low-pressure VLE	sPC–SAFT	Kouskoumvekaki *et al.*[21]
Nylon mixtures	sPC–SAFT	Kouskoumvekaki *et al.*[22]
Poly(E-*co*-VAC)/ethylene Poly(E-*co*-MA)/propylene, butane	Polar PC–SAFT	Tumakaka and Sadowski[23]
Polymer–solvent low-pressure LLE	sPC–SAFT	Von Solms *et al.*[24]
Ternary polymer–solvent VLE	sPC–SAFT vs. entropic–FV, U–FV, GCLF, FH/Hansen, GC–Flory and PV EoS	Lindvig *et al.*[25]
Binary and ternary polymer–solvent LLE	sPC–SAFT	Lindvig *et al.*[26]
Polar co-polymers	PC–SAFT	Tumakaka *et al.*[27]
Gas solubilities in polymers	sPC–SAFT	von Solms *et al.*[28]
Poly(E-*co*-MA) and PE/ethylene, co-polymers	PC–SAFT and CK–SAFT	Spyriouni and Economou[29]
Polycarbonates	PC–SAFT	Van Schilt *et al.*[30]
PDMS–pentane	PC–SAFT	Kruger *et al.*[31]
Polymer–solvent LLE Method for estimating polymer parameters	sPC–SAFT	Kouskoumvekaki *et al.*[32]
Gas–silicon polymers	sPC–SAFT	Economou *et al.*[33]

Table 14.1 *(Continued)*

Polymer application	SAFT variant	Reference
Blends, PVAC–water	sPC–SAFT	Von Solms *et al.*[34]
Polymer–solvent LLE		
Poly(E-*co*-methacrylic acid) co-polymers	PC–SAFT	Kleiner *et al.*[35]
PE–solvents	sPC–SAFT vs. SAFT–VR	Haslam *et al.*[36]
PEG mixtures	Soft-SAFT	Pedrosa *et al.*[37]

3. Extrapolation methods based on the trends of the SAFT parameters with the molecular weight,[32] see Section 14.2.2.
4. Group contribution methods.[38,39] The method is discussed in Chapter 8 and some examples for parameter estimation for polymers are shown in Appendix 14.A.

Table 14.2 presents the polymer parameters for PC–SAFT estimated using various methods. Method 2 is useful but the parameters of pure polymers are often sensitive to the type of information employed, as illustrated in Tables 14.3–14.5 for three polymers. Thus, the polymer parameters are not unique and may describe well some systems but fail for others, without being able to determine this beforehand.

14.2.2 The Kouskoumvekaki *et al.* method

The method of Kouskoumvekaki *et al.*[32] is based on the extrapolation of PC–SAFT parameters without including mixture data in the parameter estimation. The method is briefly presented here together with an illustrative example for PMMA. The starting point of the method is the assumption that the linear trends for the

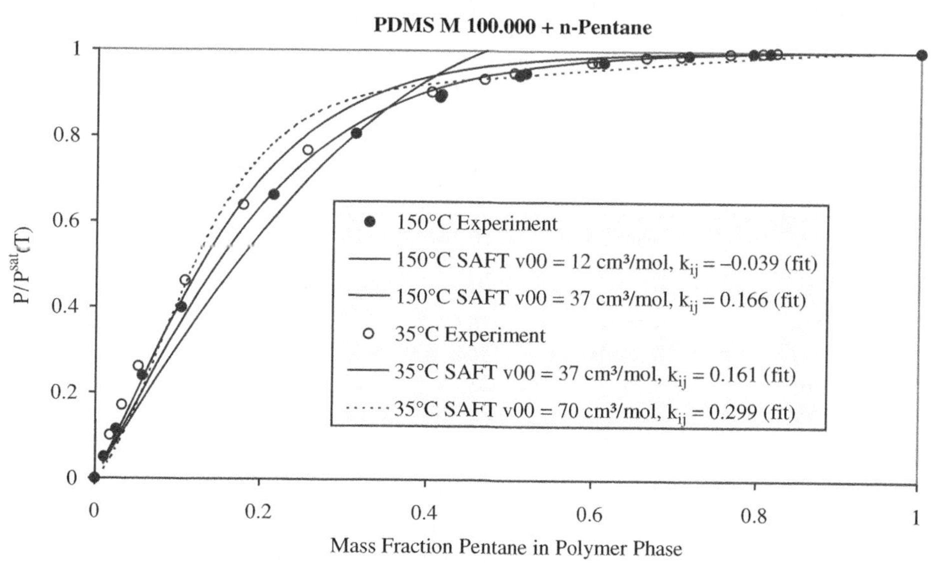

Figure 14.1 *Pressure–composition phase diagram for the system polydimethylsiloxane (PDMS)–n-pentane with the SAFT EoS. Using PDMS pure component parameter sets that reproduce the PDMS densities well, no acceptable reproduction of the binary equilibria is possible. From Pfohl et al.*[40] *(supplementary material from R. Dohrn, personal communication)*

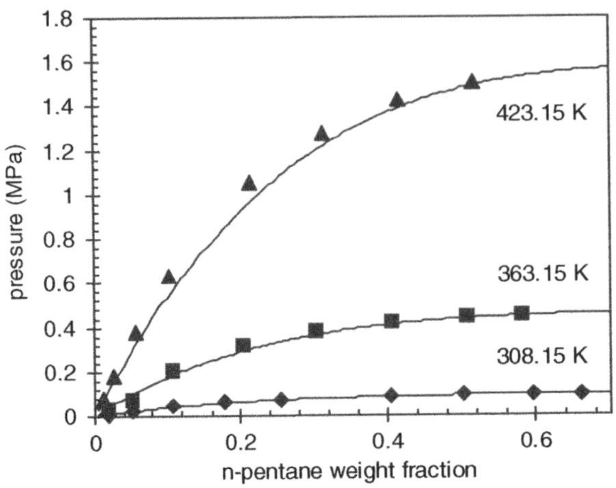

***Figure 14.2** PDMS–n-pentane phase equilibria at 308.15, 363.15 and 423.15 K. Experimental data (points) and sPC–SAFT prediction (solid curves: $k_{ij} = 0.0$). Equally successful results have been presented by Kruger et al.[31] Reprinted with permission from Molecular Simulation, Solubility of gases and solvents in silicon polymers: molecular simulation and equation of state modeling by Ioannis G. Economou, Zoi A. Makrodimitri and Georgios M. Kontogeorgis, 33, 9, 851–860 Copyright (2007) Taylor and Francis*

***Table 14.2** PC–SAFT parameters for polymers obtained using the group contribution (GC) scheme of Tihic et al.[39] and from other methods (types 2 and 3, see discussion above). The percentage average absolute deviations (AAD %) between calculated and experimental liquid densities (ρ) are also reported. The temperature range for the melt polymer density calculations is, for PMMA and PVAc, from 305 to 385 K, and for the other polymers 390–470 K. The methods of Becker et al.[18], Gross et al.[14,17] and Arce and Aznar[43] belong to type 2 of the estimation methods discussed above. Reprinted with permission from Ind. Eng. Chem. Res., A Predictive Group-Contribution Simplified PC-SAFT Equation of State: Application to Polymer Systems by Amra Tihic, Georgios M. Kontogeorgis et al., 47, 15, 5092–5101 Copyright (2008) American Chemical Society*

Polymer	m/MW (–)	σ (Å)	ε/k (K)	Reference	AAD % ρ
PMMA	0.0262	3.6511	267.64	GC[39]	4.1
	0.027	3.5530	264.60	Kouskoumvekaki *et al.*[32]	1.0
	0.0262	3.60	245.0	Becker *et al.*[18]	1.8
PBMA	0.0269	3.7598	267.21	GC[39]	1.9
	0.0241	3.8840	264.70	Kouskoumvekaki *et al.*[32]	1.0
	0.0268	3.75	233.8	Becker *et al.*[18]	3.4
PVAc	0.0292	3.3372	227.68	GC[39]	2.0
	0.0299	3.4630	261.60	Kouskoumvekaki *et al.*[32]	0.4
	0.0321	3.3970	204.60	Gross *et al.*[17]	6.0
PS	0.0202	4.1482	367.17	GC[39]	2.9
	0.0205	4.1520	348.20	Kouskoumvekaki *et al.*[32]	0.6
	0.1900	4.1071	267.00	Gross and Sadowski[14]	1.2
PP	0.0255	4.0729	279.64	GC[39]	2.7
	0.0248	4.1320	264.60	Kouskoumvekaki *et al.*[32]	1.1
	0.0231	4.1000	217.00	Gross and Sadowski[14]	5.5
	0.0270	4.0215	289.33	Arce and Aznar[43]	1.4

Table 14.2 (*Continued*)

Polymer	m/MW (–)	σ (Å)	ε/k (K)	Reference	AAD % ρ
PIB	0.0210	4.3169	296.46	GC[39]	3.4
	0.0233	4.1170	267.60	Kouskoumvekaki *et al.*[32]	1.4
	0.0235	4.1000	265.50	Gross and Sadowski[14]	1.0
PMA	0.0292	3.4704	259.21	GC[39]	1.3
	0.0292	3.5110	268.30	Kouskoumvekaki *et al.*[32]	0.5
	0.0292	3.50	243	Becker *et al.*[18]	5.2
	0.0309	3.5000	275.00	Gross *et al.*[17]	3.3
BR	0.0293	3.8413	284.70	GC[39]	4.9
	0.0245	4.1440	275.50	Kouskoumvekaki *et al.*[32]	0.9
	0.0140	4.2000	230.00	Gross *et al.*[17]	7.0
PBA	0.0286	3.6486	261.43	GC[39]	2.2
	0.0259	3.9500	224.00	Becker *et al.*[18]	4.5
PEA	0.0289	3.5478	260.15	GC[39]	1.9
	0.0271	3.6500	229.00	Becker *et al.*[18]	3.1
PPA	0.0287	3.6048	260.86	GC[39]	1.5
	0.0262	3.8000	225.00	Becker *et al.*[18]	2.4

segment number, m, and the product (segment number multiplied by segment energy, $m\varepsilon/k$) against molecular weight (MW), which are valid for the alkane series (as presented in Figure 14.3):

$$\begin{aligned} m &= 0.025\,37\,MW + 0.9081 \\ m\varepsilon/k &= 6.918\,MW + 127.3 \end{aligned} \tag{14.1}$$

are also valid for any polymer series. Thus the following equations can be written:

$$\begin{aligned} m &= A_m MW + B_m \\ m\varepsilon/k &= A_\varepsilon MW + B_\varepsilon \end{aligned} \tag{14.2}$$

Table 14.3 *Estimation of PS parameters for sPC–SAFT using various methods. The five methods are as follows: (1) a single binary LLE data set for the system PS–cyclohexane (Gross and Sadowski method/parameters); (2) a single binary VLE data set for PS–cyclohexane; (3) a single binary VLE data set of PS–cyclohexane, with the binary interaction parameter excluded from the regression; (4) a large number of binary VLE data sets of PS with acetone, benzene, toluene, carbon tetrachloride, chloroform, MEK, propylacetate, nonane and cyclohexane; (5) the same number of binary VLE data sets as in (4), with the binary interaction parameter excluded from the regression. The percentage average absolute deviations (AAD %) between calculated and experimental liquid densities (ρ) are also reported*

Method of pure polymer parameters regression	m/MW	σ (Å)	ε/k (K)	AAD % ρ
(1) *PVT* + single binary LLE	0.0190	4.107	267.00	5.1
(2) *PVT* + single binary VLE	0.0242	3.939	366.43	0.3
(3) *PVT* + single binary VLE excl. k_{ij}	0.0214	4.061	312.06	0.9
(4) *PVT* + all binary VLE	0.0364	3.354	277.92	0.4
(5) *PVT* + all binary VLE excl. k_{ij}	0.0390	3.243	243.81	0.8

Table 14.4 *Estimation of PVC parameters for sPC–SAFT using various methods. Single VLE is the binary system PVC/1,4-dioxane, all (binary) VLE are the following: PVC with dibutyl ether, 1,4 dioxane, CCl_4, toluene, tetrahydrofurane and vinyl chloride. The percentage average absolute deviations (AAD %) between calculated and experimental liquid densities (ρ) are also reported*

Method	m/MW	σ (Å)	ε/k (K)	AAD % ρ
Kouskoumvekaki *et al.*[32]	0.021 01	3.724	315.93	0.5
PVT + single VLE incl. k_{ij}	0.012 08	4.726	541.56	2.7
PVT + single VLE excl. k_{ij}	0.029 83	3.213	221.19	1.1
PVT + all VLE incl. k_{ij}	0.009 71	5.030	495.86	1.4
PVT + all VLE excl. k_{ij}	0.014 20	4.298	360.92	0.8
GC[39]	0.019 10	3.8358	314.56	0.8
Arce and Aznar[43]	0.021 36	3.7486	370.42	0.2

Table 14.5 *Estimation of PDMS parameters for sPC–SAFT using various methods. The percentage average absolute deviations (AAD %) between calculated and experimental liquid densities (ρ) are also reported*

Method	m/MW	σ (Å)	ε/k (K)	AAD% ρ
PVT only	0.032 45	3.531	204.95	0.1
PVT + single VLE excl. k_{ij}	0.022 40	4.070	248.77	0.2
PVT + single VLE incl. k_{ij}	0.022 29	4.076	248.49	0.2
PVT + 10 VLE excl. k_{ij}	0.039 98	3.225	159.65	2.7
PVT + 7 binary VLE excl. k_{ij}	0.022 64	4.055	248.36	0.2

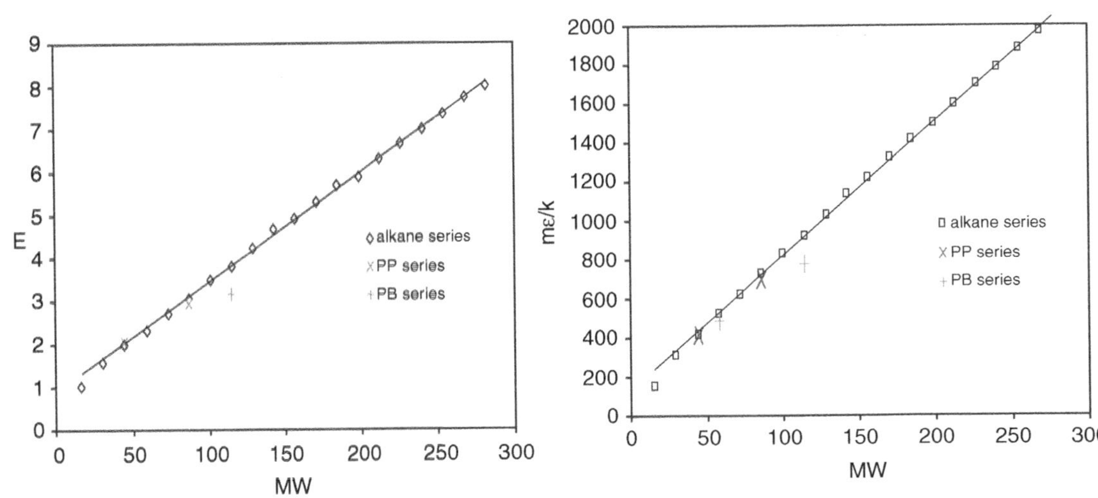

Figure 14.3 *Left: The segment m of PC–SAFT against molecular weight (MW) for the alkane, isobutylene (PIB) and propylene (PP) series. Right: The parameter mε/k of PC–SAFT against MW for the alkane, PIB and PP series. Reprinted with permission from Ind. Eng. Chem. Res., Novel Method for Estimating Pure-Component Parameters for Polymers: Application to the PC-SAFT Equation of State by Irene A. Kouskoumvekaki, Nicolas von Solms, Georgios M. Kontogeorgis et al., 43, 11, 2830–2838 Copyright (2004) American Chemical Society*

There are four unknowns, but it is further assumed that B_m and B_ε are universal and can be determined from any homologous series, thus the values from the alkane series are used:

$$\begin{aligned} B_m &= 0.9081 \\ B_\varepsilon &= 127.3 \end{aligned} \tag{14.3}$$

Then, Equation (14.2) can be written as:

$$\begin{aligned} A_m &= (m - 0.9081)/MW \\ A_\varepsilon &= (m\varepsilon/k - 127.3)/MW \end{aligned} \tag{14.4}$$

There are only two unknowns, m, ε/k taken from the monomer, and then for high-molecular-weight polymers, the segment number and energy parameters are estimated as:

$$\begin{aligned} \frac{m}{MW} &= A_m \\ \frac{\varepsilon}{k} &= \frac{A_\varepsilon}{A_m} \end{aligned} \tag{14.5}$$

The final parameter of the model, σ, is fitted to PVT data of the polymer.

The method will be illustrated with one example.

For PMMA, the corresponding monomer is methyl-isobutyrate and its parameters (obtained in the usual way from vapor pressure and liquid density data) are $m = 3.660\,55$, $\varepsilon/k = 234.109$ K. As the molecular weight is 102.133 kg/mol, the A_m and A_ε parameters can be easily calculated from Equations (14.4) and are found to be 0.0270 and 7.144, respectively. Finally, from Equations (14.5), we calculate the segment number and energy parameter of the polymer:

$$\frac{m}{MW} = A_m = 0.0270 \tag{14.6}$$

$$\frac{\varepsilon}{k} = \frac{A_\varepsilon}{A_m} = 264.6 \tag{14.7}$$

The segment diameter is then the only parameter fitted to polymer PVT data and is found equal to 3.553 Å with a percentage error in density of only 1%. Excellent results are obtained for other polymers as well, as shown in Table 14.6, while Figure 14.4 shows a typical LLE result for PMMA in two solvents.

14.2.3 Polar and associating polymers

Compared to the extensive use of SAFT for polyolefins and other non-polar polymers, relatively few investigations have been reported for polar and especially for associating polymers,[22,37,45] but current research is devoted to the subject, also in relation to certain polar/associating co-polymers. Some recent results for a few associating polymers will be discussed here.

Polyethylene glycol

Pedrosa *et al.*[37] have applied soft-SAFT to polyethylene glycol (PEG) mixtures. Satisfactory results can in some (but not all) cases be obtained for such polymers using parameters entirely based on low-molecular-weight homologues.

Table 14.6 PC–SAFT parameters calculated using the method of Kouskoumvekaki et al.[32] *and average absolute deviations between calculated and experimental liquid density of several polymers*

Polymer	m/MW	σ (Å)	ε/k (K)	T range (K)	P range (bar)	AAD% ρ
PS	0.0205	4.152	348.2	390–470	1–1000	0.6
PIB	0.0233	4.117	267.6	325–385	1–1000	1.4
PP	0.0248	4.132	264.6	445–565	1–1000	1.1
PVAc	0.0299	3.463	261.6	310–370	1–800	0.4
PE	0.0254	4.107	272.4	410–470	1–1000	1.0
PMMA	0.027	3.553	264.6	390–430	1–1000	1.0
PBMA	0.0241	3.884	264.7	310–470	1–1000	1.0
BR	0.0263	4.008	279.4	275–325	1–1000	1.0
PMA	0.0292	3.511	268.3	310–490	1–1000	0.5
PB	0.0245	4.144	275.5	410–510	1–1000	0.9

We present hereafter an application of sPC–SAFT for PEG, shown in Figure 14.5. The pure compound parameters can be obtained by extrapolation of the ethylene glycol oligomers parameters (from MEG up to tetra EG):

$$\begin{aligned} m &= 0.0192\,M + 0.7924 \\ m\sigma^3 &= 1.3121\,M + 8.5441 \\ m\varepsilon/k &= 6.1866\,M + 224.49 \end{aligned} \tag{14.8}$$

while the association parameters (association volume and energy) are assumed to be constant for all PEG: $\varepsilon^{HB} = 2080.03$ K, $\kappa^{HB} = 0.0235$. Similar to glycols, PEG is assumed to be represented as a 4C-type molecule.

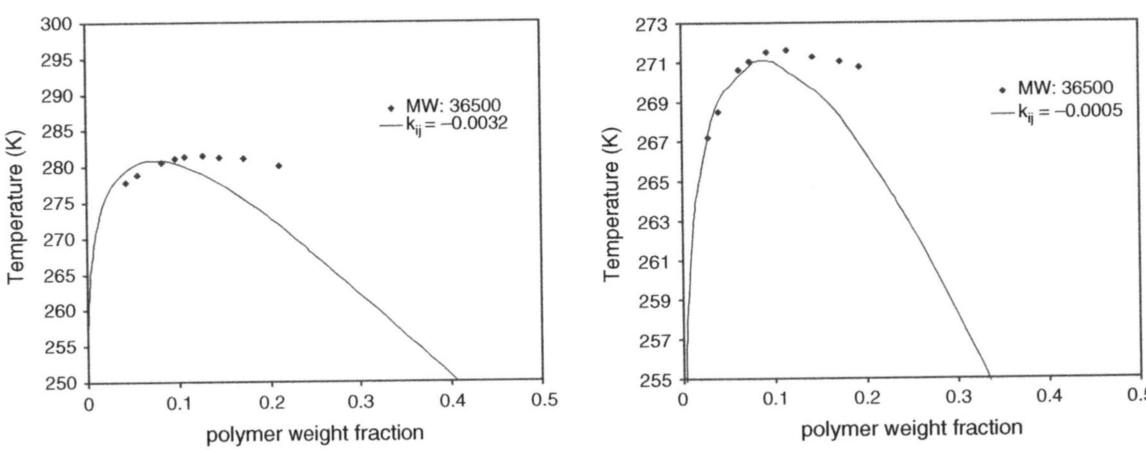

Figure 14.4 LLE with sPC–SAFT for PMMA with chlorobutane (left) and heptanone (right). A small k_{ij} is used in both cases. The PMMA parameters are estimated using the method of Kouskoumvekaki et al.[32] *Reprinted with permission from Ind. Eng. Chem. Res., Novel Method for Estimating Pure-Component Parameters for Polymers: Application to the PC-SAFT Equation of State by Irene A. Kouskoumvekaki, Nicolas von Solms, Georgios M. Kontogeorgis et al., 43, 11, 2830–2838 Copyright (2004) American Chemical Society*

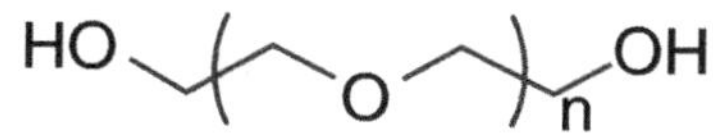

Figure 14.5 *Chemical structure of PEG.*

However, as the PEG structure (see Figure 14.5) indicates, the intermolecular interactions are also influenced by association through the ether groups. In order to account for this, the ether groups are assumed to have one association site and a linear correlation with respect to the molar mass (M in eq. (14.9)) is used to calculate the number of those sites. Since diethylene glycol has one ether group, triethylene glycol has two ether groups and tetraethylene glycol has four ether groups, the number of additional association sites in the molecule attributed to the ether groups can be approximated by:

$$N_{assoc\text{-}sites} = 0.022\,M - 1.409 \tag{14.9}$$

Using these assumptions, satisfactory PEG–water VLE correlations are obtained with sPC–SAFT at various temperatures and molecular weights of PEG, as Figure 14.6 illustrates. Equally good results are obtained at other conditions as well, but the performance is less satisfactory when the association sites of the ether groups are not accounted for. A simpler modeling procedure using the above-mentioned parameters (Equation (14.8)) but without considering the ether group sites results in satisfactory modeling for VLE of PEG with hydrocarbons, e.g. propane and benzene.

Nylon parameters

Kouskoumvekaki *et al.*[22] proposed an approach for obtaining PC–SAFT parameters for another family of associating polymers, the polyamides or nylon polymers. At first m, ε and the two association parameters of nylon-6 were assumed to be equal to the parameter values for ε-caprolactam, whereas the segment diameter

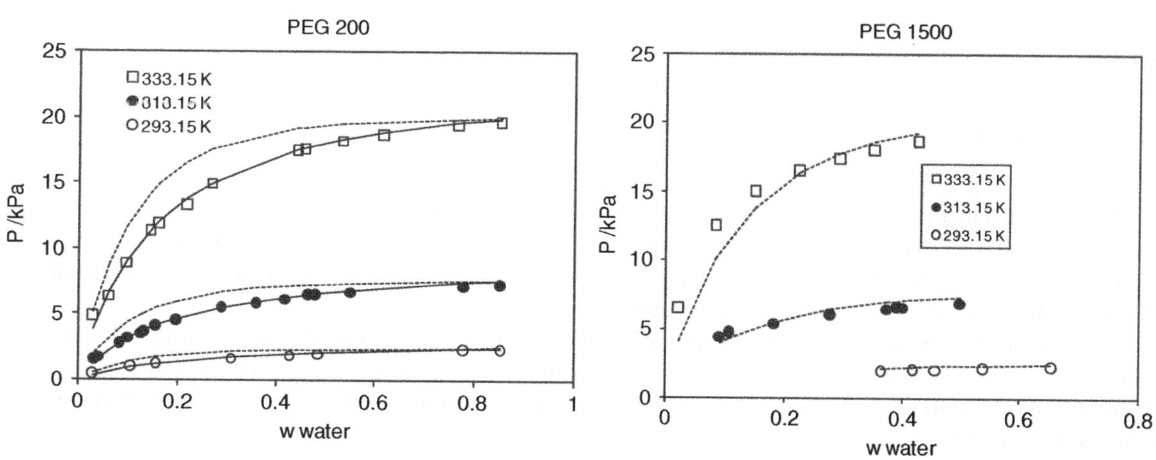

Figure 14.6 *Left: VLE for PEG 200 + water. Lines are correlations with simplified PC–SAFT: dashed line, $k_{ij} = 0$; solid line, optimized k_{ij}. Right: VLE for PEG 1500 + water. Lines are predictions with simplified PC–SAFT, $k_{ij} = 0$. Experimental data are from Herskowitz and Gottlieb, J. Chem. Eng. Data,* ***1985****, 30, 233*

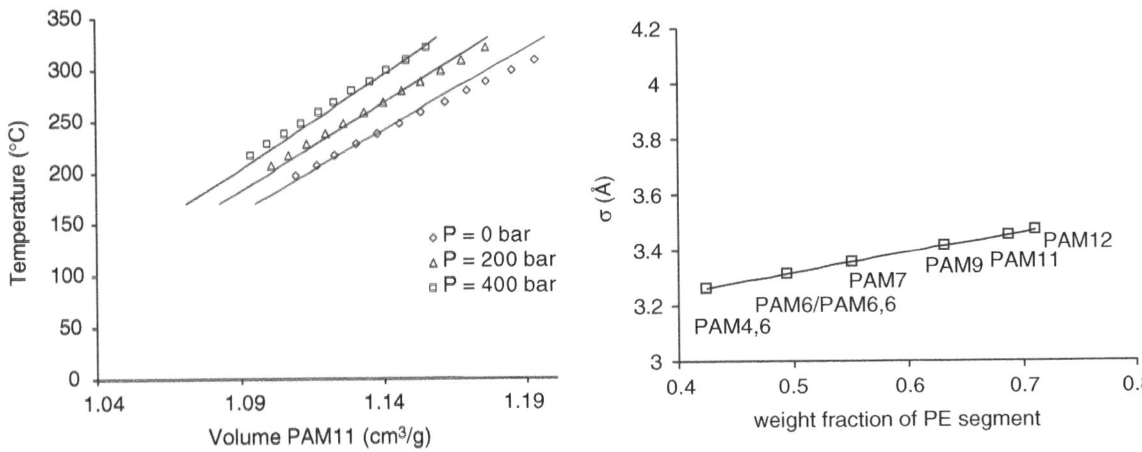

Figure 14.7 *Left: Liquid volume of polyamide 11 with sPC–SAFT against the experimental data of Zoller and Walsh, Standard PVT Data for Polymers, Technonomic Publishing Company, Lancaster, PA, 1995. The PC–SAFT parameters are: m/MW = 0.036,* $\varepsilon/k = 335.374$ K, $\sigma = 3.51$ Å, $\varepsilon^{AB} = 1623$ K, $\kappa^{AB} = 0.003\,995$. *Right: Optimum segment diameter* (σ) *values of sPC–SAFT for various polyamides against the weight fraction of the linear polyethylene segments, w(PE), in the polyamide chain for seven polyamide types* ($\sigma = 0.7127w(PE) + 3.0245$). *Reprinted with permission from Ind. Eng. Chem. Res., Application of the Simplified PC-SAFT Equation of State to the Vapor– Liquid Equilibria of Binary and Ternary Mixtures of Polyamide 6 with Several Solvents by Irene A. Kouskoumvekaki, Georgios M. Kontogeorgis et al., 43, 3, 826–834 Copyright (2004) American Chemical Society*

has been fitted to the liquid density of nylon-6. It has been further assumed that all the associating sites of the polymer chain are active in forming hydrogen bonds, i.e. the number of active sites of each site type per polymer molecule is equal to the number of monomer units in the chain. All polyamides studied ((4,6), 6, (6,6), 7, 9, 11 and 12) are assumed to have the same segment number, dispersion and association energies and association volume (see caption of Figure 14.7) and only the segment diameter is optimized to the polymer density. The agreement is excellent between experimental density data and sPC–SAFT results and, moreover (see Figure 14.7, right), the optimum segment diameter values depend linearly on the percentage of the linear polyethylene segment in the polyamides.

14.2.4 Parameters for co-polymers

Co-polymers are polymers composed of different types of segments, e.g. denoted as α, β, which, moreover, can be connected in different ways as random, alternating, graft or block. SAFT models can be extended to co-polymers in different ways, but here we will limit the discussion to the method presented by Gross *et al.*[17] for the extension of PC–SAFT to co-polymers. The dispersion, hard-chain and association contributions of PC–SAFT for co-polymers are presented by Gross *et al.*[17]

The sequence of segments along the co-polymer chain needs not to be known exactly in PC–SAFT. The number of binary segment–segment contacts is defined by two quantities, the segment fraction $z_{i\alpha}$, which represents the co-polymer composition, and the bonding fraction $B_{i\alpha i\beta}$, both of which are given by the following equations:

$$z_{i\alpha} = \frac{m_{i\alpha}}{m_i} \tag{14.10}$$

$$m_{i\alpha} = w_{i\alpha} M_i \left(\frac{m}{M}\right)_\alpha \tag{14.11}$$

where:

$m_{i\alpha}$ = number of segments of type α in the co-polymer chain i which is composed of total m_i segments
$w_{i\alpha}$ = mass fraction of monomers of type α in the co-polymer chain i
M_i = molecular mass of the co-polymer i

The bonding fraction $B_{i\alpha i\beta}$ denotes the number of binary segment–segment contacts within polymer chain i that are of type α and β (and thus includes information on the monomer sequence along the polymer backbone). Under certain assumptions, the bonding fraction for a random co-polymer with $z_{i\beta} < z_{i\alpha}$ is calculated as:

$$\begin{aligned} B_{i\alpha i\beta} &= 2\frac{m_{i\beta}}{(m_{i\alpha}-1)} \\ B_{i\alpha i\alpha} &= 1-B_{i\alpha i\beta} \\ B_{i\beta i\beta} &= 0 \end{aligned} \tag{14.12}$$

The combining rules of Equations (8.47b) can be used for calculating the cross-diameter and cross-dispersion energy between the different segments of the co-polymer. The interaction parameters between each one of the various co-polymer segments and the solvent can be obtained from independent experiments on binary homopolymer–solvent mixtures. The interaction parameter between the two co-polymer segments inside the co-polymer must be estimated from co-polymer data.

14.3 Low-pressure phase equilibria (VLE and LLE) using simplified PC–SAFT

Several SAFT variants have been applied to low-pressure phase equilibria calculations for mixtures including polymers. VLE and LLE for binary and multicomponent (especially ternary) mixtures have been considered. We will illustrate the potential of the SAFT theory using the simplified PC–SAFT which has been applied to:

- VLE for binary polymer solvent mixtures;[21]
- LLE for binary polymer–solvents;[24]
- ternary mixed solvent–polymer VLE;[25]
- ternary mixed solvent–polymer LLE;[26]
- absorption of light gases in polyethylene;[36]
- aqueous polymer–solvent VLE and polymer blends.[34]

Recent applications[33,48] of PC–SAFT include silicon polymer phase equilibria, high-pressure polymer–solvent VLE, as well as a study of phase equilibria of mixtures containing polar polymers and polar or associating solvents.

Figures 14.8–14.18 present some typical results, while the most important observations can be summarized as follows:

1. Both the original and sPC–SAFT EoS perform equally well for polymer–solvent VLE. The performance is satisfactory, as shown for two examples in Figures 14.8a and 14.8b. In the second case, the polymer

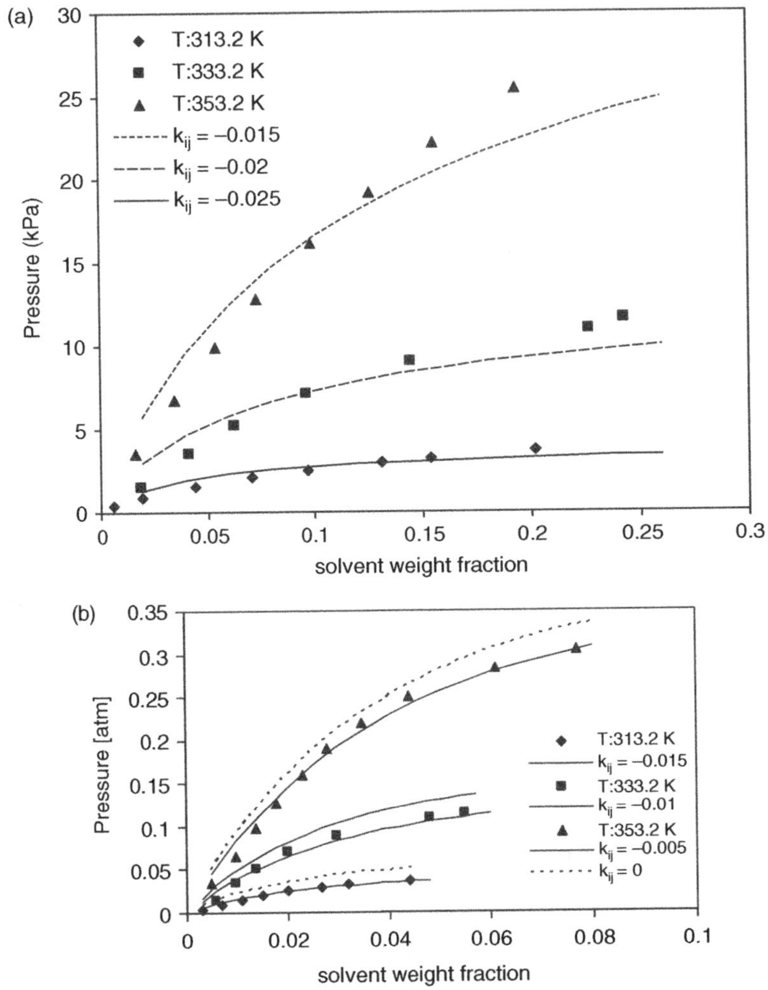

Figure 14.8 *(a) VLE correlation with the simplified PC–SAFT EoS for PVAC(110 000)–2-methyl-1-propanol at three temperatures. Reprinted with permission from Ind. Eng. Chem. Res., Novel Method for Estimating Pure-Component Parameters for Polymers: Application to the PC-SAFT Equation of State by Irene A. Kouskoumvekaki, Nicolas von Solms, Georgios M. Kontogeorgis et al., 43, 11, 2830–2838 Copyright (2004) American Chemical Society (b) VLE correlation with the simplified PC–SAFT EoS for PIB–2-methyl-1-propanol at three temperatures. Lines are simplified PC–SAFT correlations with PIB parameters obtained from the GC method.*[38,39] *After Tihic et al.*[48]

parameters have been estimated from the group contribution method of Tihic *et al.*[38,39] In some cases the performance is somewhat less satisfactory in the presence of associating solvents, especially water (Figure 14.17). An interaction parameter is almost always needed in order to obtain satisfactory correlation of polymer–solvent VLE. It is not always possible to establish trends of the k_{ij} values with characteristics of the mixtures, e.g. of the molecular weight or the van der Waals volume of solvent for a polymer–solvent series, as shown by Tihic *et al.*[48] for PVC mixtures.

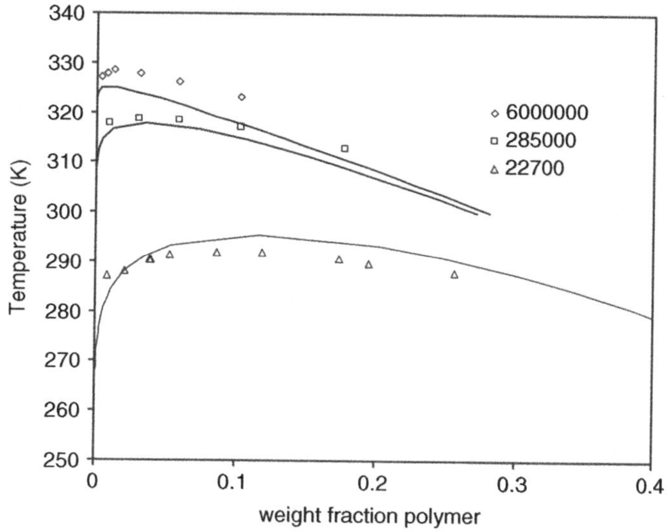

*Figure 14.9 LLE for polyisobutylene–diisobutyl ketone with sPC–SAFT. The sPC–SAFT parameters for diisobutyl ketone were obtained by fitting to experimental liquid density and vapor pressure data in the temperature range 260–600 K. These data were from the DIPPR database. The parameters were: $m = 4.6179$, $\varepsilon/k = 243.72$ K and $\sigma = 3.7032$ Å. Average percent deviations were 1.03% for vapor pressure and 0.64% for liquid density. Lines are simplified PC–SAFT correlations with $k_{ij} = 0.0053$, the same at all three molecular weights. Experimental data from Shultz and Flory (1952, J. Am. Chem. Soc., **74**, 4760). Reprinted with permission from Fluid Phase Equilibria, A novel approach to liquid-liquid equilibrium in polymer systems with application to simplified PC-SAFT by Nicolas van Solms, Georgios M. Kontogeorgis et al., 222–223, 1, 87–93 Copyright (2004) Elsevier*

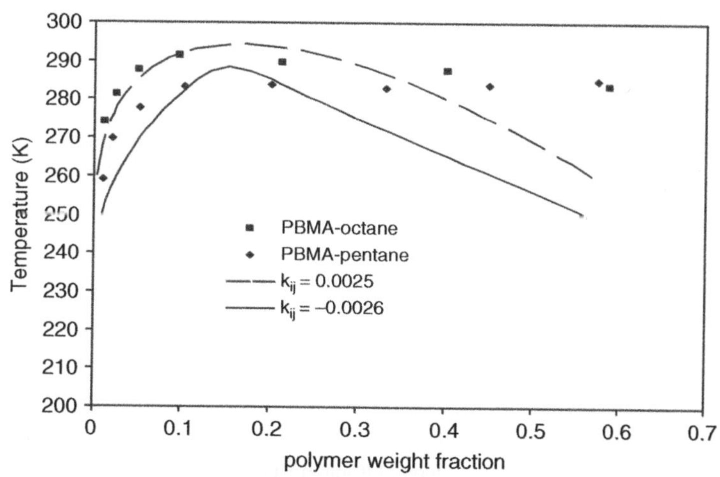

*Figure 14.10 LLE for PBMA–alkanes with sPC–SAFT. The molecular weight of PBMA is 11 600 g/mol. The experimental data are from Saraiva et al. (**1994**, SEP 9412, Internal Report, Department of Chemical and Biochemical Engineering, Technical University of Denmark). Reprinted with permission from Ind. Eng. Chem. Res., Novel Method for Estimating Pure-Component Parameters for Polymers: Application to the PC-SAFT Equation of State by Irene A. Kouskoumvekaki, Nicolas von Solms, Georgios M. Kontogeorgis et al., 43, 11, 2830–2838 Copyright (2004) American Chemical Society*

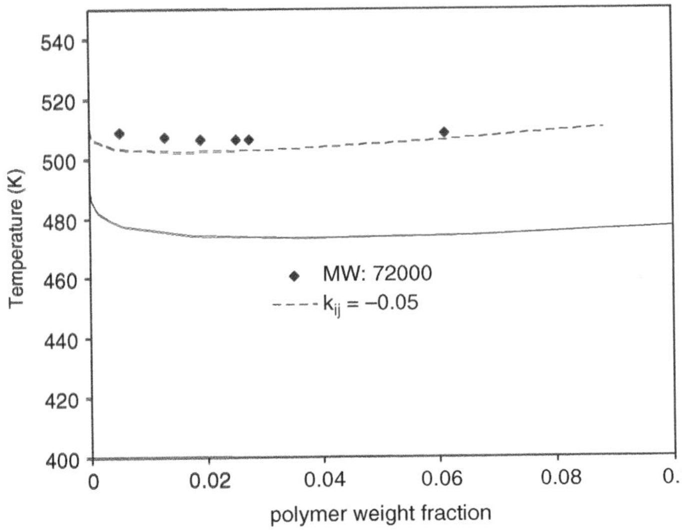

Figure 14.11 Prediction and correlation of LCST phase behavior with sPC–SAFT for PIB–octane. The molecular weight of PIB is 72 000 g/mol. The experimental data are from Liddell and Swinton (**1970**, Discuss. Faraday Soc., **49**, 115). The solid line is prediction with $k_{ij} = 0$. Reprinted with permission from Ind. Eng. Chem. Res., Novel Method for Estimating Pure-Component Parameters for Polymers: Application to the PC-SAFT Equation of State by Irene A. Kouskoumvekaki, Nicolas von Solms, Georgios M. Kontogeorgis et al., 43, 11, 2830–2838 Copyright (2004) American Chemical Society

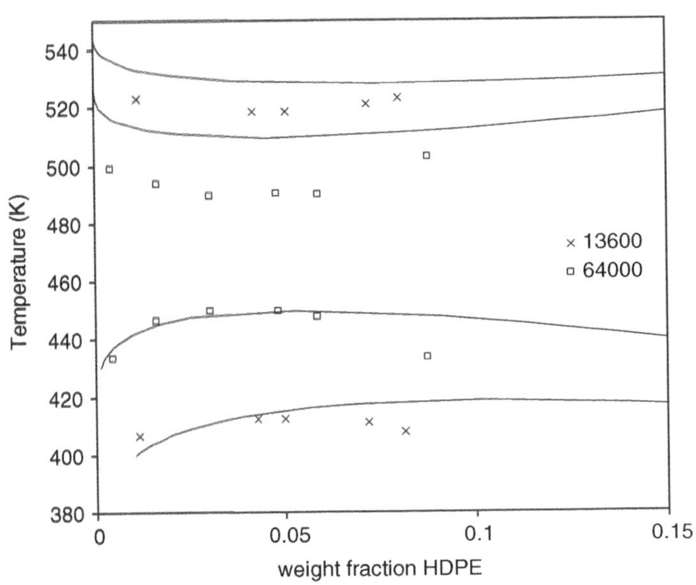

Figure 14.12 LLE for HDPE–butyl acetate. The system displays both upper and lower critical solution temperature behavior. The experimental data are for molecular weights of 13 600 and 64 000 g/mol (Kuwahara et al., Polymer, **1974**, 15, 777). Lines are simplified PC–SAFT correlations with $k_{ij} = 0.0156$ for both molecular weights. Reprinted with permission from Fluid Phase Equilibria, A novel approach to liquid-liquid equilibrium in polymer systems with application to simplified PC-SAFT by Nicolas van Solms, Georgios M. Kontogeorgis et al., 222–223, 1, 87–93 Copyright (2004) Elsevier

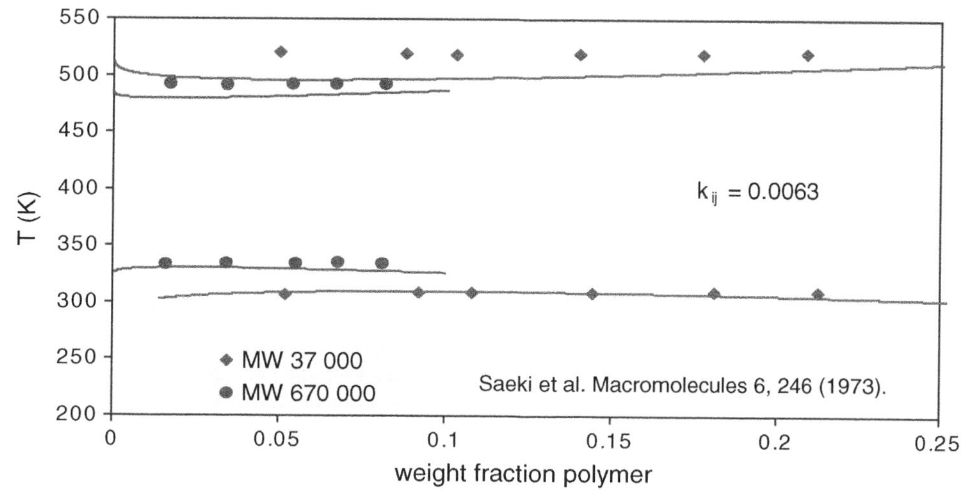

Figure 14.13 *LLE with sPC–SAFT for PS–methyl cyclohexane. One k_{ij} is used for both UCST and LCST*

2. UCST-type LLE phase behavior is correlated well using a single temperature-independent k_{ij} (adjusted to the UCST value), also independent of the polymer molecular weight. Small k_{ij} values are sufficient, but the results can be quite sensitive to the k_{ij} value used. Moreover, the correct upper critical solution concentration is not always obtained, neither is the rather flat shape of the coexistence curves always well represented. Calculated curves are sometimes sharper than the experimental data. Figures 14.9 and 14.10 illustrate two examples.

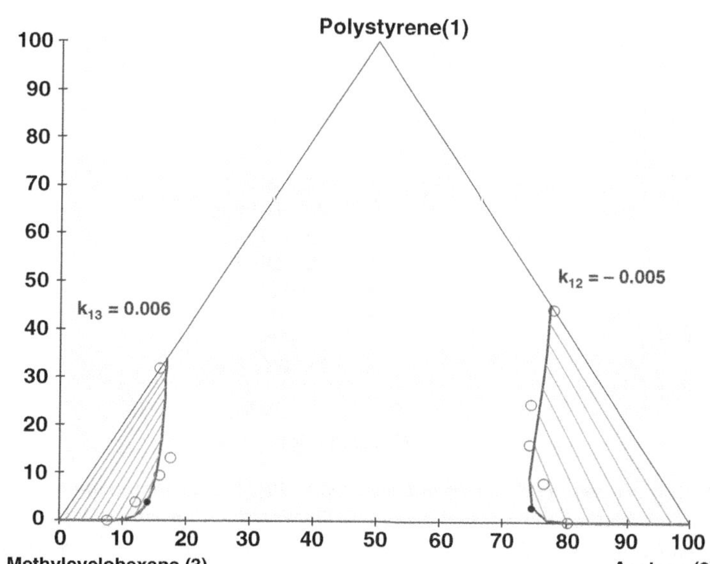

Figure 14.14 *Ternary LLE with sPC–SAFT for PS(300 000)–acetone–methylcyclohexane. The interaction parameter between the two solvents is zero. After Lindvig et al.*[26]

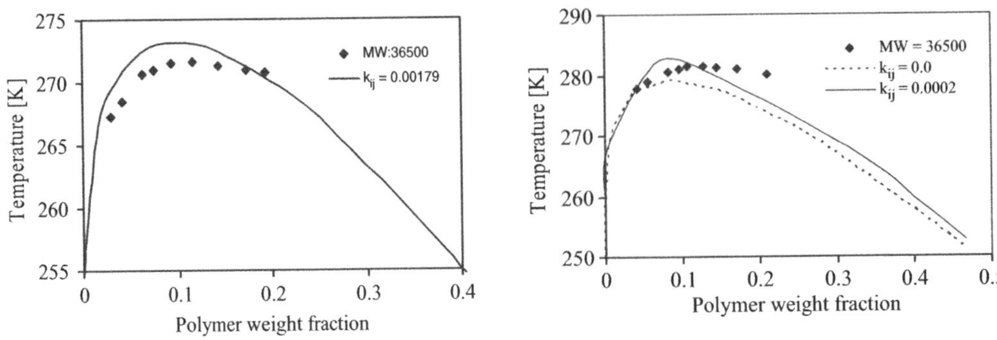

Figure 14.15 *LLE for PMMA–solvent with sPC–SAFT. The lines are simplified PC–SAFT correlations with PMMA parameters obtained from the GC method. Left: LLE for PMMA– 4-heptanone. Experimental data are from Wolf and Blaum, J. Polym. Sci.,* ***1975****, 13, 1115. Right: LLE for PMMA–chlorobutane. Reprinted with permission from Ind. Eng. Chem. Res., A Predictive Group-Contribution Simplified PC-SAFT Equation of State: Application to Polymer Systems by Amra Tihic, Georgios M. Kontogeorgis et al., 47, 15, 5092–5101 Copyright (2008) Elsevier*

3. Some polymer–solvent systems, especially nearly athermal mixtures, e.g. alkanes with polyolefins, only exhibit LCST. As Figure 14.11 shows, sPC–SAFT can describe such behavior well and even prediction (with $k_{ij} = 0$) is satisfactory.
4. Unlike the mixtures shown in Figures 14.9–14.11, many polymer–solvent systems exhibit both a UCST-type behavior at low temperatures and a LCST behavior at high temperatures. Two examples are shown in

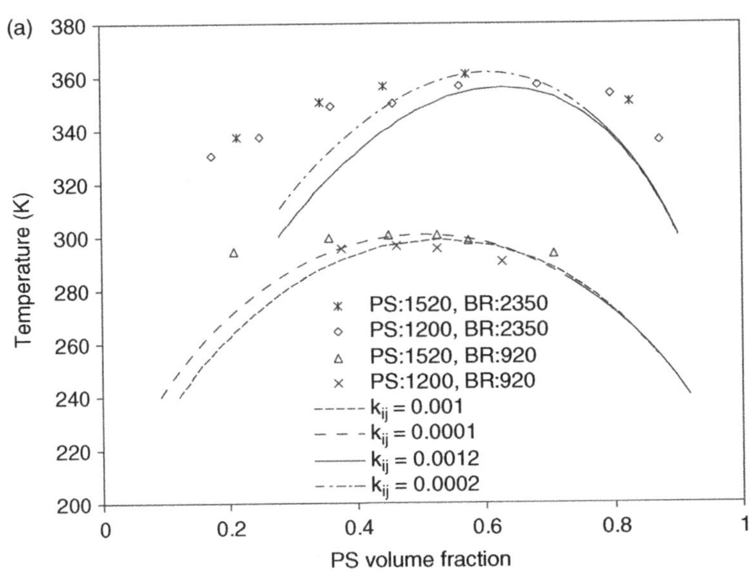

Figure 14.16 *(a) LLE for a polystyrene (PS) and butadiene rubber (BR) blend with the simplified PC–SAFT EoS. Results are shown with various values for the interaction parameter. Reprinted with permission from Fluid Phase Equilibria, Capabilities, limitations and challenges of a simplified PC-SAFT equation of state by Nicolas van Solms, Georgios M. Kontogeorgis et al., 241, 1–2, 344–353 Copyright (2006) Elsevier (b) The interaction parameter of sPC–SAFT estimated from LLE data from various polystyrene (PS) blends against the PS molecular weight, MW. The interaction parameters can be approximated by the equation:* $k_{ij} = -9.07 \times 10^{-7} MW(\mathrm{PS}) + 1.61 \times 10^{-3}$

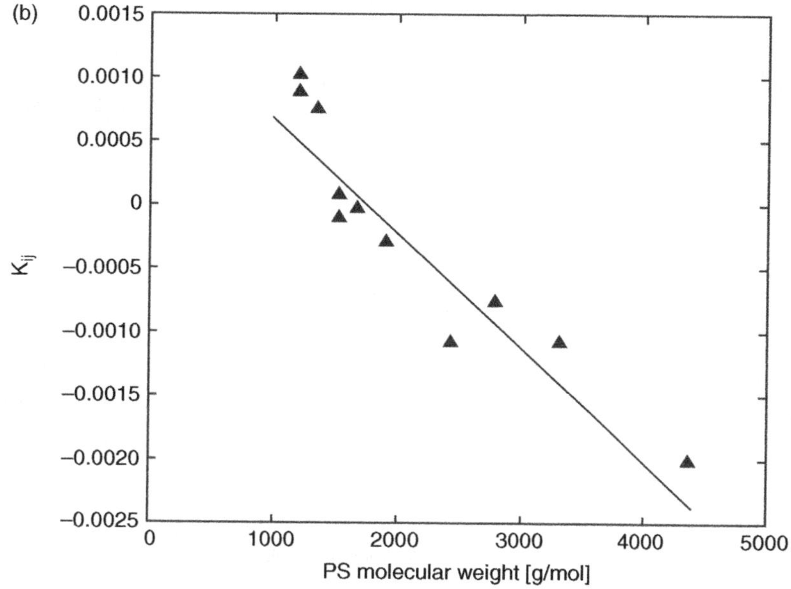

Figure 14.16 (*Continued*)

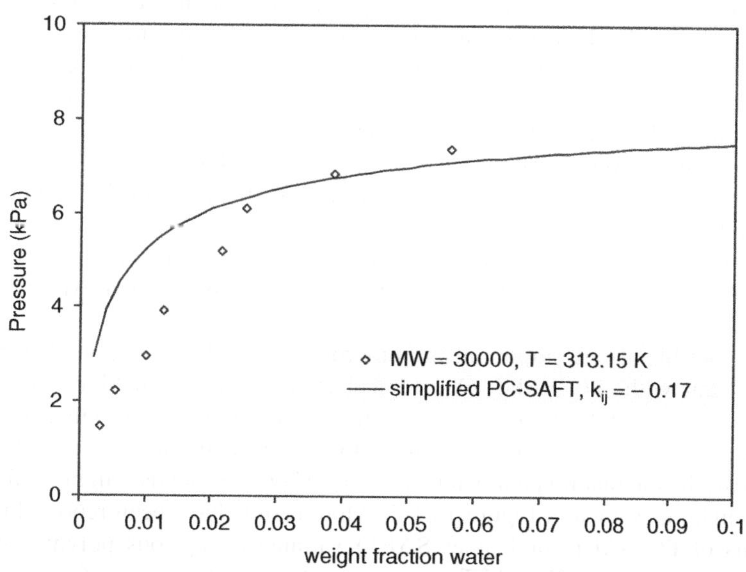

Figure 14.17 *VLE calculation for poly(vinylacetate)–water with the simplified PC–SAFT. Reprinted with permission from Fluid Phase Equilibria, Capabilities, limitations and challenges of a simplified PC-SAFT equation of state by Nicolas van Solms, Georgios M. Kontogeorgis et al., 241, 1–2, 344–353 Copyright (2006) Elsevier*

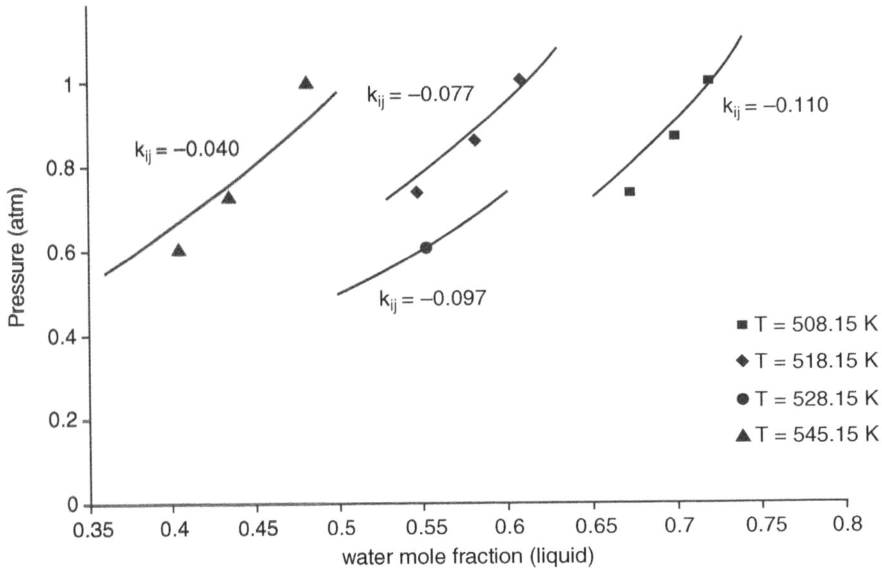

Figure 14.18 *VLE correlation for nylon-6 water at four temperatures with the simplified PC–SAFT EoS. Reprinted with permission from Ind. Eng. Chem. Res., Application of the Simplified PC-SAFT Equation of State to the Vapor–Liquid Equilibria of Binary and Ternary Mixtures of Polyamide 6 with Several Solvents by Irene A. Kouskoumvekaki, Georgios M. Kontogeorgis et al., 43, 3, 826–834 Copyright (2004) American Chemical Society*

Figures 14.12 and 14.13. In both cases, a single binary interaction parameter can correlate the data for both molecular weights shown. The binary interaction parameter was adjusted to give a good correlation for the UCST. The LCST curve is rather insensitive to k_{ij}. Nevertheless, the LCST curve is reasonably well correlated using this value. The prediction ($k_{ij} = 0$) is also good for the LCST curve, but the UCST curve will be substantially underpredicted if $k_{ij} = 0$ is used.

5. Ternary polymer–mixed solvent VLE and LLE have been studied with sPC–SAFT[25,26] and one example of ternary LLE is shown in Figure 14.14. The results are satisfactory with k_{ij} parameters obtained from fitting binary mixture data alone.
6. Figure 14.15 shows two LLE results with sPC–SAFT with the polymer parameters estimated from the GC method of Tihic *et al.*[39] Satisfactory agreement is obtained between experimental data and the model.
7. Mixtures of polymer blends are important in numerous practical applications but they have not been investigated in detail with sPC–SAFT. One typical result is shown in Figure 14.16a. Simultaneous correlation of the variation of the binodal curves with respect to molecular weight and the concentration is difficult. Moreover, there is significant sensitivity of the results to the values of the interaction parameters used. Nevertheless, the interaction parameters fitted to various PS blends roughly follow a trend with the PS molecular weight, as shown in Figure 14.16b. More results have been reported recently.[48]
8. Few applications of PC–SAFT (and other SAFT) variants to aqueous polymer systems have been presented. Two examples with sPC–SAFT (using the water parameters from Gross and Sadowski[14]) are shown in Figures 14.17–14.18 for the water activity in PVAc and the nylon–water mixture. In the first case, the agreement is rather poor; in the second it is satisfactory, although in both cases rather large interaction parameter values are required. Despite the large k_{ij}, sPC–SAFT has been applied with success to the ternary water–nylon–caprolactam mixture.

14.4 High-pressure phase equilibria

Many of the first applications of SAFT (and of PC–SAFT) are related to high-pressure polyethylene technology (including co-polymers of various polyolefins). Figures 14.19–14.21 show some earlier and more recent results with PC–SAFT illustrating that the model can:

- describe very well high-pressure VLE and LLE of systems containing polyethylene;
- correlate high-pressure UCST and LCST for PS-containing systems;
- describe gas solubilities in polymers, often with temperature-independent interaction parameters.

CO_2 and other gases decrease the polymer solubility in solvents, i.e. they act as anti-solvents. Figure 14.21 illustrates that PC–SAFT captures the effect of compressed gases on polymer–solvent LLE at high temperatures and pressures. Similarly good results are obtained for the system PS–cyclehexane–CO_2. Equally successful P–T plots have been presented for polyethylene in a variety of solvents.[17] PC–SAFT captures the changing slope of the cloud-point curves from the lighter (e.g. ethylene) solvents towards heavier ones (e.g. butane), using solvent-specific but temperature-independent interaction parameters.

Silicon polymers

Accurate knowledge of the solubility of gases and solvents in polymers is crucial for the efficient design of industrial processes, such as for a polymer separation process following a polymerization reaction, and of novel polymer-based materials, e.g. for highly pure polymers. Silicon polymers are used widely in many industrial applications including adhesives, coatings, elastomeric seals and membrane materials for gas and liquid separations. In all cases, polymers exist in mixtures with one or more solvents. The solubility of various organic compounds and gases in different silicon-containing polymers is calculated using atomistic simulation and tested against the sPC–SAFT predictions. Selected results are presented in Figures 14.22 and 14.23.[33]

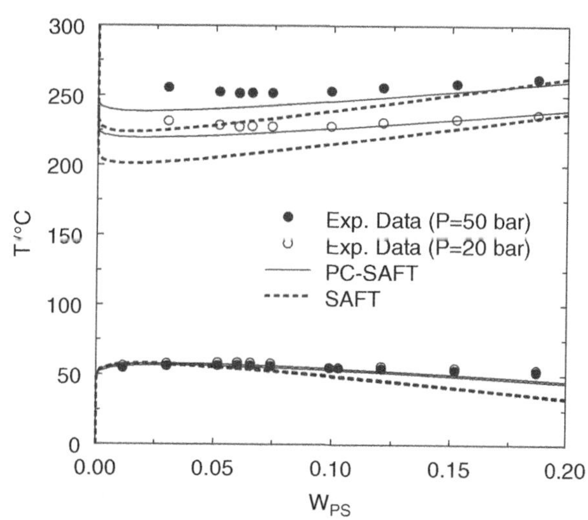

Figure 14.19 *Liquid–liquid demixing of PS–methylcyclohexane at pressures of 20 and 50 bar. (PS: $M_w = 405$ kg/mol, $M_w/M_n = 1.832$.) Comparison of PC–SAFT ($k_{ij} = 0.007$) and SAFT ($k_{ij} = 0.023$) correlation results to experimental data (Enders and de Loos, Fluid Phase Equilib., **1997**, 139, 335). The polymer is modeled as monodisperse. Reprinted with permission from Ind. Eng. Chem. Res., Modeling Polymer Systems Using the Perturbed-Chain Statistical Associating Fluid Theory Equation of State by Joachim Gross and Gabriele Sadowski, 41, 5, 1084–1093 Copyright (2002) American Chemical Society*

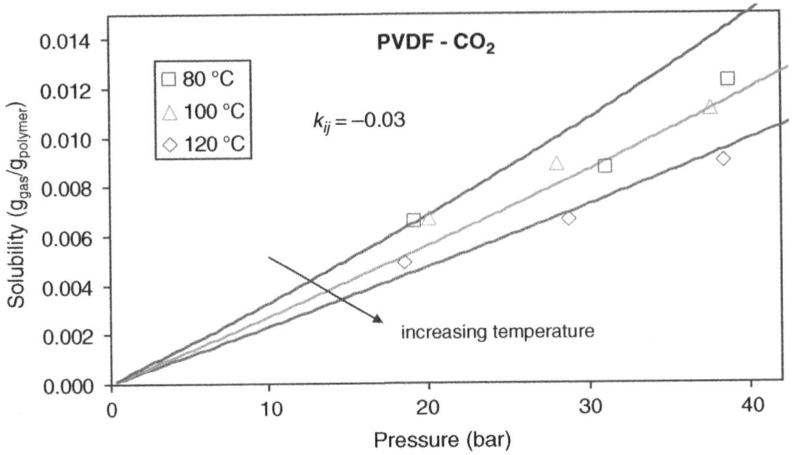

Figure 14.20 CO_2 *solubilities in PVDF with the simplified PC–SAFT. Experimental data are from von Solms et al., Eur. Polym. J.,* ***2004****, 41, 341. Equally satisfactory results are obtained for methane and* CO_2 *in HDPE and for methane and* CO_2 *in PA-11. Reprinted with permission from Ind. Eng. Chem. Res., Prediction and Correlation of High-Pressure Gas Solubility in Polymers with Simplified PC-SAFT by Nicolas von Solms, Michael L. Michelsen, and Georgios M. Kontogeorgis, 44, 9, 3330–3335 Copyright (2005) American Chemical Society*

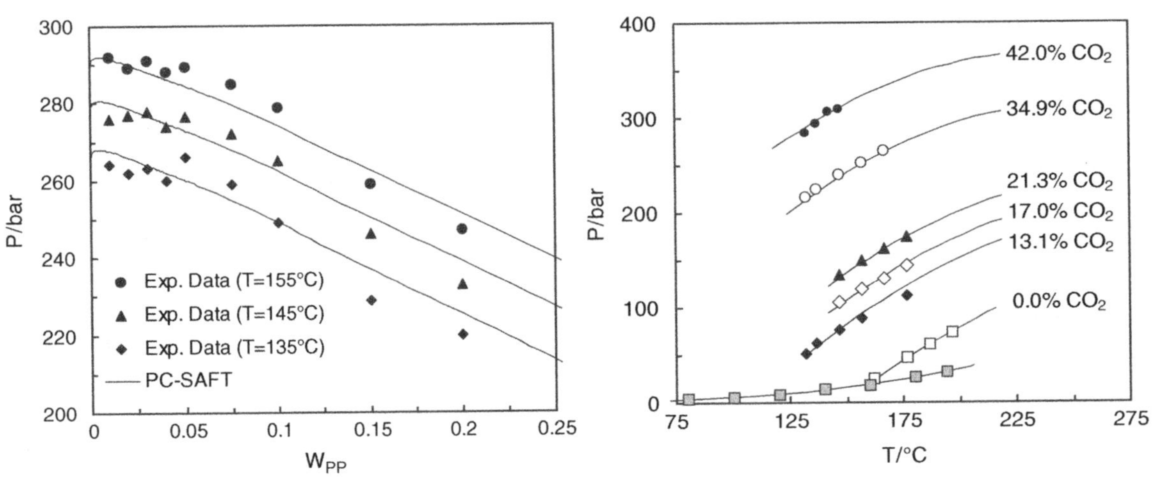

Figure 14.21 *Left: LLE of polypropylene–propane at three temperatures in a pressure–weight fraction plot. (PP:* $M_w = 290$ *kg/mol,* $M_w/M_n = 4.4$*.) Comparison of experimental cloud points (Whaley et al., Macromolecules,* ***1997****, 30, 4882) to PC–SAFT calculations (*$k_{ij} = 0.0242$*). The polymer was modeled using three pseudo-components. Reprinted with permission from Ind. Eng. Chem. Res., Modeling Polymer Systems Using the Perturbed-Chain Statistical Associating Fluid Theory Equation of State by Joachim Gross and Gabriele Sadowski, 41, 5, 1084–1093 Copyright (2002) American Chemical Society. Right: Cloud-point curve of PP–n-pentane–*CO_2 *for various* CO_2 *contents. Initial polymer weight fraction* $w_{PP} = 0.03$ *(before the addition of* CO_2*). Comparison of experimental cloud points (Martin et al., Fluid Phase Equilib.,* ***1999****, 154, 241) to PC–SAFT calculations (PP–n-pentane,* $k_{ij} = 0.0137$*; PP–*CO_2*,* $k_{ij} = 0.177$*; n-pentane–*CO_2*,* $k_{ij} = 0.143$*).Reprinted with permission from Fluid Phase Equilibria, Modeling of polymer phase equilibria using Perturbed-Chain SAFT by F. Tumakaka, J. Gross and G. Sadowski, 194–197, 1–2, 541–551 Copyright (2002) Elsevier*

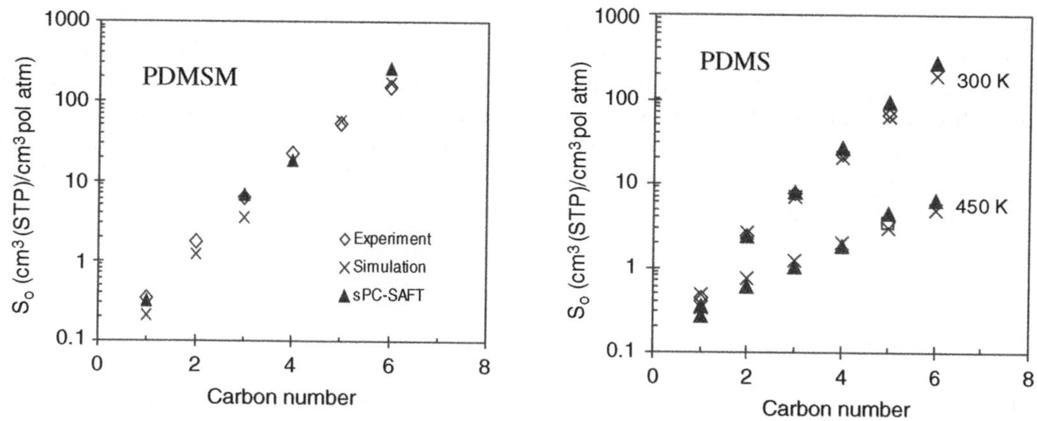

Figure 14.22 *Experimental data (open points), molecular simulation data (×) and sPC–SAFT predictions (▲) for the infinite dilution solubility coefficient of n-alkanes in PDMSM and PDMS at 300 K and 0.1 MPa. For PDMS, an experimental point at 423 K and simulation (×) and sPC–SAFT (▲) predictions at 450 K are also shown. PDMSM is poly(dimethylsilamethylene) and PDMS is poly(dimethylsiloxane). After Economou et al.*[33] *Reprinted with permission from Molecular Simulation, Solubility of gases and solvents in silicon polymers: molecular simulation and equation of state modeling by Ioannis G. Economou, Zoi A. Makrodimitri and Georgios M. Kontogeorgis, 33, 9, 851–860 Copyright (2007) Taylor and Francis*

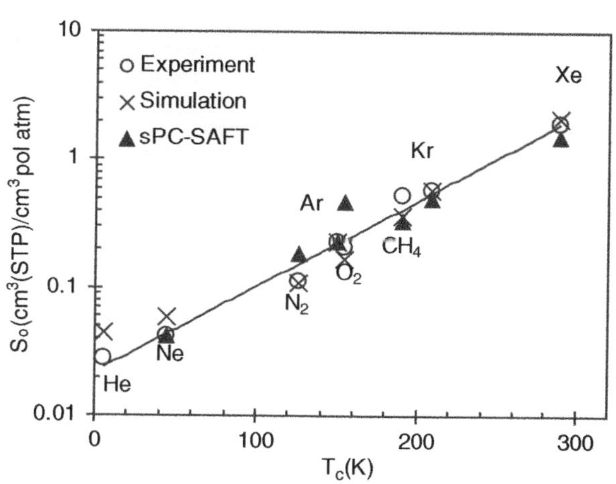

Figure 14.23 *Experimental data (circles), molecular simulation data (×) and simplified PC–SAFT (▲) predictions for infinite dilution solubility coefficient S_o of gases in PDMS at 300 K. The solid curve is a fit to experimental data. After Economou et al.*[33] *Reprinted with permission from Molecular Simulation, Solubility of gases and solvents in silicon polymers: molecular simulation and equation of state modeling by Ioannis G. Economou, Zoi A. Makrodimitri and Georgios M. Kontogeorgis, 33, 9, 851–860 Copyright (2007) Taylor and Francis*

As can be seen in Figure 14.22, solubility increases with the *n*-alkane carbon number. Simulation results are in very good agreement with experiment, when available. Furthermore, for a given *n*-alkane, S_o values are very similar for the various polymers. In other words, the chemical structure for these polymers has very little effect on the solubility of the solutes. The S_o of various solutes in a polymer correlate very nicely with the solute's experimental critical temperature. In Figure 14.23, experimental data, molecular simulation calculations and simplified PC–SAFT predictions are shown for eight different gases in PDMS. Predictions from both models are in very good agreement with the experimental data in all cases.

14.5 Co-polymers

Figure 14.24 shows a typical example from one of the earlier investigations with co-polymers using CK–SAFT. More results have been reported from Chen *et al.*[2,41,42], Gregg *et al.*[3], Lee *et al.*[7], Folie *et al.*[6], Kinzl *et al.*[11] and Jog *et al.*[12] Many of these earlier applications are for polyolefin co-polymers, e.g. poly(ethylene-*co*-propylene) in olefins, but some applications for polar co-polymers have also been reported. Moreover, most of these earlier investigations were with the 'original' SAFT (CK–SAFT), while many of the recent investigations for co-polymer mixtures have been done with PC–SAFT (from the research groups of G. Sadowski and W. Chapman). Typical results are shown in Figures 14.25–14.28 and these are briefly discussed hereafter.

Figure 14.25 from Gross *et al.*[17] shows high-pressure equilibrium for mixtures of EMA and propylene for different repeat-unit compositions of EMA. As repeat units of methyl acrylate (MA) are added to the polyethylene chain, the demixing pressure at first declines, but then increases as the composition of MA

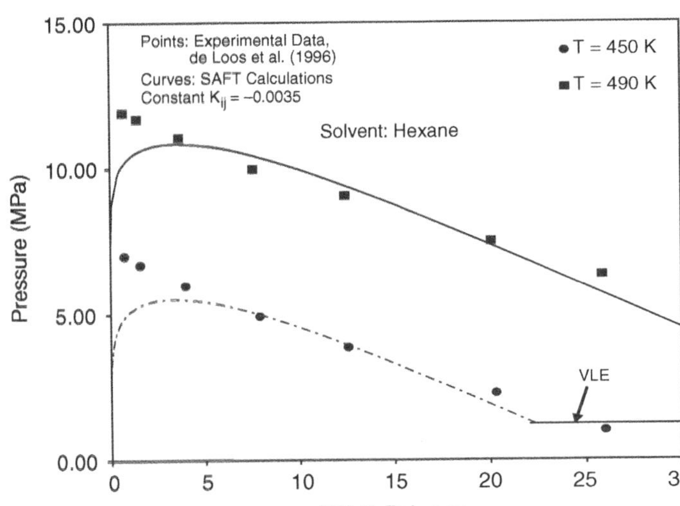

Figure 14.24 *Cloud-point isobars for the system poly(ethylene-octene) ($M_n = 33\,000$) and hexane from experimental data points and the SAFT EoS (curves). Reprinted with permission from Ind. Eng. Chem. Res., Modeling of Liquid–Liquid-Phase Separation in Linear Low-Density Polyethylene–Solvent Systems Using the Statistical Associating Fluid Theory Equation of State by Prasanna K. Jog, Walter G. Chapman et al., 41, 5, 887–891 Copyright (2002) American Chemical Society*

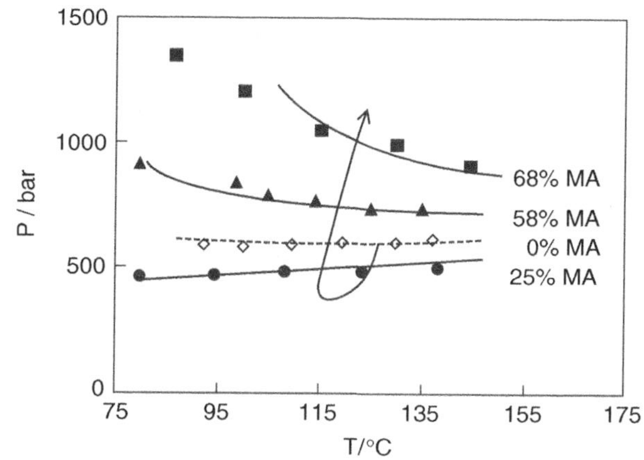

Figure 14.25 *High-pressure equilibrium for mixtures of poly-(ethylene-co-methyl acrylate) (EMA) and propylene for different repeat-unit compositions of the EMA. Comparison of experimental cloud-point measurements to calculation results of the PC–SAFT EoS. (EMA (0% MA) is equal to LDPE: open diamonds and dashed line.) Reprinted with permission from Ind. Eng. Chem. Res., Modeling Copolymer Systems Using the Perturbed-Chain SAFT Equation of State by Joachim Gross, Oliver Spuhl, et al., 42, 6, 1266–1274 Copyright (2003) American Chemical Society*

increases. This effect is correctly predicted by PC–SAFT. As with all co-polymer studies so far with PC–SAFT, only one k_{ij} is typically fitted to co-polymer data, i.e. k_{ij} between the two different co-monomers (the unlike co-polymer segments) or between E and MA in the example of Figure 14.25. The k_{ij} between the different segments and the solvents are obtained from the corresponding homopolymers. The 'special' effect illustrated in Figure 14.25, i.e. first decreasing demixing pressure with added polar units and after a certain point increasing demixing pressure, is not unique for the PEMA–propylene mixtures. Actually the behavior illustrated in Figure 14.25 is true of many more polar co-polymer mixtures, e.g. poly(ethylene-*co*-methylacrylate) or poly (ethylene-*co*-ethyl acrylate) with ethylene or poly(ethylene-*co*-vinyl acetate) with various solvents. But there are also several co-polymers, especially non-polar ones, e.g. poly(ethylene-*co*-butene)/propane, where the trend is a monotonic increase of the demixing pressure with increasing butane percentage. As Figure 14.26 shows for two mixtures, PC–SAFT can capture both effects in all cases studied.

Finally, Figures 14.27 and 14.28 show the results for associating co-polymer mixtures. PC–SAFT captures the large changes in phase behavior even when few acid groups are added. Equally satisfactory results are obtained in the case of solvent/co-solvent mixtures as well. It is worth mentioning that in the results shown in Figures 14.27 and 14.28, the polydispersity of the co-polymer is accounted for. Investigations for accounting for polydispersity in SAFT have been presented by Chapman *et al.*[19], Spyriouni and Economou[29] and Kleiner *et al.*[35]

14.6 Concluding remarks

SAFT models are standard approaches today for polymer calculations and especially the 'original' (CK–SAFT) and PC–SAFT have found many applications for a wide variety of polymer mixtures and types of phase equilibria. Other SAFT variants (soft-SAFT, SAFT–VR) are also currently being applied to polymers. Many successful applications have been presented in this chapter including VLE and LLE as well as

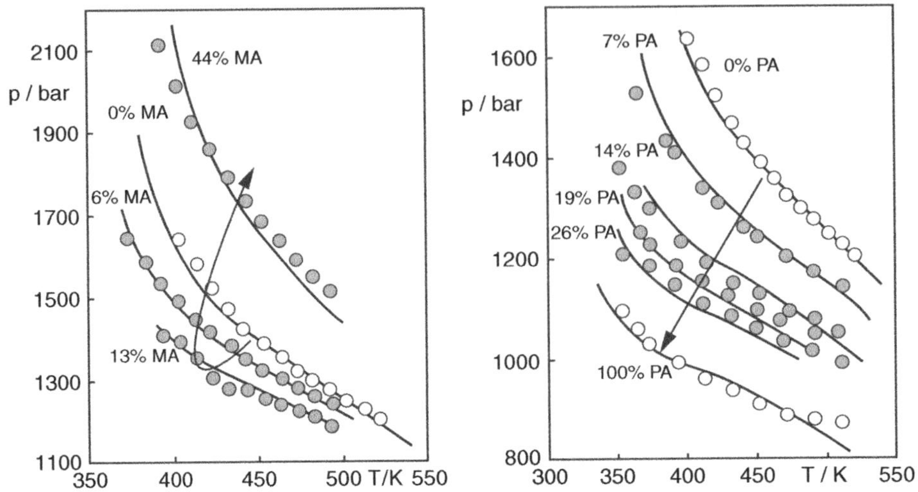

Figure 14.26 *Cloud-point pressures for poly(ethylene-co-alkylacrylate)–ethylene systems at different co-polymer compositions (mole percent of acrylate monomer in the backbone is indicated). Left: Solubility of poly(ethylene-co-methylacrylate) with 0% MA = LDPE. Right: Solubility of poly(ethylene-co-propylacrylate) with 0% PA = LDPE. The polymer content is about 5 wt%. Symbols are the experimental data and lines are the PC–SAFT calculations. Reprinted with permission from Macromol. Symp., Thermodynamics of Polymer Systems by G. Sadowski, 333–346, Copyright (2006) Wiley-VCH*

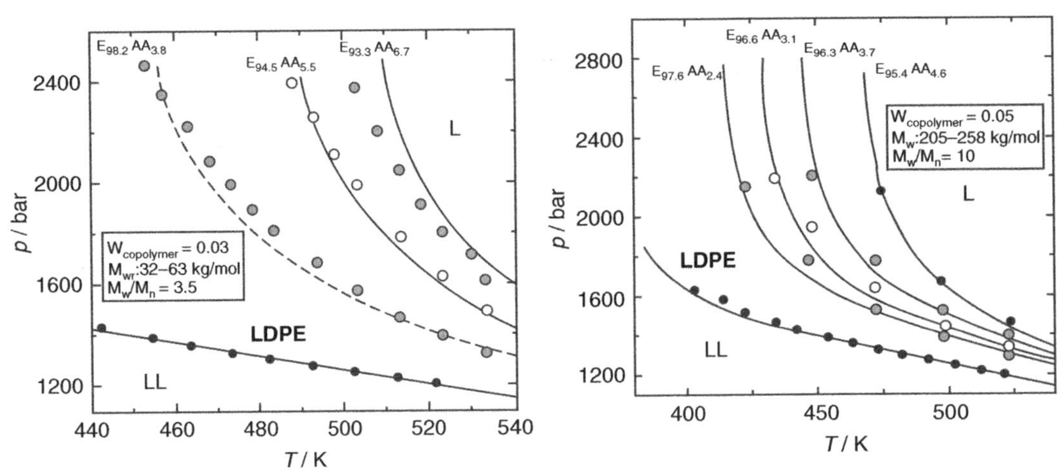

Figure 14.27 *Cloud-point pressure curves for poly(ethylene-co-acrylic)–ethene at different ethylene (E) and acrylic acid (AA) concentrations (as mole percent) of the monomers in the co-polymer backbone. These are indicated as subscripts. Results are shown for two different polymer concentrations indicated in the figures. The molecular weights and polydispersity indices are also given. Points are the experimental data (Buback and Latz, Macromol. Chem. Phys.,* ***2003****, 638; Beyer and Oeellrich, Helv. Chim. Acta,* ***2002****, 659) and lines are the PC–SAFT calculations. Reprinted with permission from Fluid Phase Equilibria, Phase equilibria in polydisperse and associating copolymer solutions: Poly (ethene-co-(meth)acrylic acid)–monomer mixtures by Matthias Kleiner, Feelly Tumakaka et al., 241, 1–2, 113–123 Copyright (2006) Elsevier*

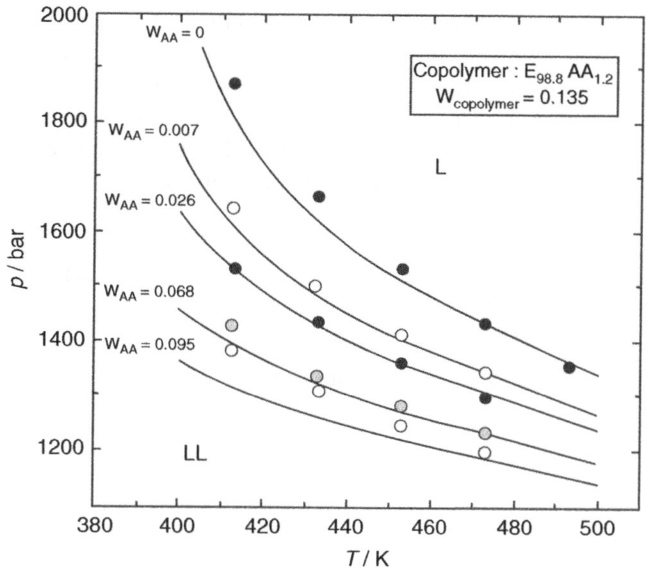

Figure 14.28 *Cloud-point pressure curves for the ternary mixture of poly(ethylene-co-acrylic)–ethene–acrylic acid for different acrylic acid (AA) percentages in the monomer mixture. Points are the experimental data (Wind, PhD Thesis, Technical University of Darmstadt, 1992) and lines are the PC–SAFT calculations. Reprinted with permission from Fluid Phase Equilibria, Phase equilibria in polydisperse and associating copolymer solutions: Poly(ethene-co-(meth)acrylic acid)–monomer mixtures by Matthias Kleiner, Feelly Tumakaka et al., 241, 1–2, 113–123 Copyright (2006) Elsevier*

co-polymer mixtures and at both low and high pressures. Few comparisons of SAFT variants with each other or with other models have been reported, even fewer for polymer mixtures. Earlier comparative evaluations include those by Wu and Chen[4] for polymer–solvent VLE (SAFT vs. UNIFAC–FV and GC–Flory EoS), Bokis *et al.*[10] for polymer–solvent VLE (SAFT vs. cubic EoS) and Orbey *et al.*[44] for polyethylene systems (SAFT vs. SL and polymer SRK). More recent investigations include that by Kang *et al.*[16] for polymer–solvent VLE (SAFT vs. PR for polymers) and by Lindvig *et al.*[25] for ternary polymer–mixed solvent VLE (SAFT vs. various free-volume models and EoS for polymers). Some comparisons between SAFT variants for polymers are included in the work of Spyriouni and Economou[29] (SAFT vs. PC–SAFT for homo- and co-polymer phase equilibria) and Haslam *et al.*[36] (sPC–SAFT vs. SAFT–VR for the absorption of light gases in polyethylene). In many cases, when the parameterization is performed in the same way, especially the way in which the polymer parameters are estimated, the performance of the various SAFT variants is very similar.[36] This conclusion is further confirmed for non-polymeric mixtures when SAFT variants are compared to certain other advanced models which also account explicitly for association effects. One extensive such comparison presented recently[46,47] illustrates that sPC–SAFT and NRHB perform satisfactorily and overall similarly for a large database of VLE (104 mixtures) and LLE systems as well as in predicting the monomer fraction of hydrogen bonding mixtures.

SAFT is available today in commercial simulators, as CAPE-Open modules and numerous stand-alone packages available worldwide, e.g. from IVC-SEP, Department of Chemical and Biochemical Engineering, Technical University of Denmark (http://www.ivc-sep.kt.dtu.dk/), The University of Dortmund (http://thwww.chemietechnik.uni-dortmund.de/de/textonly/content/PC-Saft/PC-Saft.html),

Infochem (www.infochemuk.com), VLXE (www.vlxe.com) and PE Software (O. Pfohl, S. Petkov, G. Brunner, *Ind. Eng. Chem. Res.*, 2000, 39*(11)*, 4439). Important future developments which will secure the widespread applications of the SAFT approach for polymeric systems will include methods for estimating the parameters for polymers, including associating polymers, as well as improving the performance of the approach for aqueous solutions and mixtures with polar and associating solvents. It is also important to extend the SAFT approach to multifunctional polymers, e.g. dendrimers and hyperbranched polymers.

Appendix 14.A Examples of parameter estimation for polymers using the GC simplified PC–SAFT[39]

14.A.1 Example: poly(methyl methacrylate) (PMMA)

The repeating unit of the polymer is:

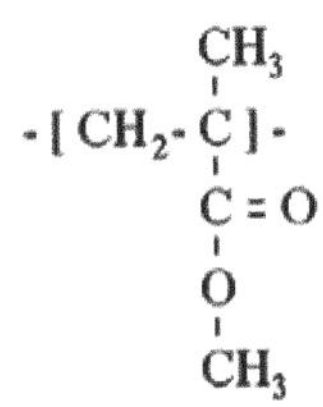

The molecular weight of the repeating unit is 100.12 g/mol. There are only first-order groups (FOG) present. Using the values of the FOG contributions, the following results are obtained:

FOG (i)	Occurrences n_i	m_i	$m_i\sigma_i^3$	$m_i\varepsilon_i/k$
CH_3	2	1.288 724	68.339 10	258.7732
CH_2	1	0.384 329	24.339 81	102.3238
C	1	−0.492 08	2.3254 15	−10.9830
COO	1	1.439 11	32.513 28	351.1344
		$\sum n_i \cdot m_i =$ 2.620 08	$\sum n_i \cdot m_i\sigma_i^3 =$ 127.5176	$\sum n_i \cdot m_i\varepsilon_i/k =$ 701.2483

From these values, the following PC–SAFT parameters are calculated:

$$m/MW = 2.620\,08/100.12 = \underline{0.0262}$$

$$\sigma = \sqrt[3]{(m\sigma^3)/m} \quad = \sqrt[3]{127.517\,60/2.620\,08} = \underline{3.6511}\ \text{Å}$$

$$\varepsilon/k = (m\varepsilon/k)/m = 701.248\,32/2.620\,08 = \underline{267.64\text{K}}$$

With this parameter set calculated from the group contribution (GC) scheme, the percentage average absolute deviation (AAD %) between simplified PC–SAFT calculations and experimental liquid density is 4.1%, in the temperature range 300–400 K and pressures up to 1000 bar.

14.A.2 Example: poly(isopropyl methacrylate) (PIPMA)

The repeating unit of the polymer is:

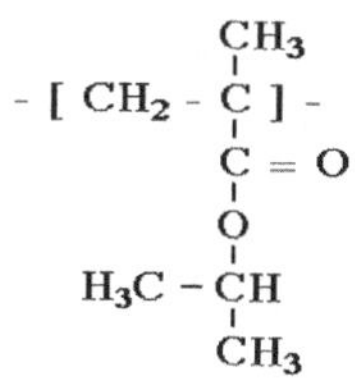

The molecular weight of the repeating unit is 128.17 g/mol.

There are both first-order groups (FOG) and second-order groups (SOG) present. Using the values of the FOG and SOG contributions, the following results are obtained:

FOG (i)	Occurrences n_i	m_i	$m_i\sigma_i^3$	$m_i\varepsilon_i/k$
CH_3	3	1.933 09	102.508 60	388.159 8
CH_2	1	0.384 329	24.339 81	102.323 8
CH	1	0.043 834	13.953 91	68.208 42
C	1	−0.492 080	2.3254 15	−10.983 0
COO	1	1.439 110	32.513 28	351.134 4
		$\sum n_i \cdot m_i =$ 3.308 28	$\sum n_i \cdot m_i\sigma_i^3 =$ 175.641 07	$\sum n_i \cdot m_i\varepsilon_i/k =$ 898.8434

SOG (j)	Occurrences n_j	m_j	$m_j\sigma_j^3$	$m_j\varepsilon_j/k$
$(CH_3)_2CH$	1	0.016 26	0.280 87	−9.836 15
		$\sum n_j \cdot m_j =$ 0.016 26	$\sum n_j \cdot m_j\sigma_j^3 =$ 0.280 87	$\sum n_j \cdot m_j\varepsilon_j/k =$ −9.836 15

From these values, the following PC–SAFT parameters are calculated:

$$m/MW = (3.308\,28 + 0.016\,26)/128.17 = \underline{0.02594}$$

$$\sigma = \sqrt[3]{(m\sigma^3)/m} = \sqrt[3]{(175.922/3.3245)} = \underline{3.7543\text{Å}}$$

$$\varepsilon/k = (m\varepsilon/k)/m = 889.0073/3.3245 = \underline{267.41\text{K}}$$

With this parameter set calculated from the group contribution (GC) scheme, the percentage average absolute deviation (AAD %) between simplified PC–SAFT calculated and experimental liquid density is 2.1%, in the temperature range 300–400 K and pressures up to 1000 bar.

Appendix 14.B Computational aspects: the 'alternating tangents method' for LLE calculations

Phase equilibrium calculations in asymmetric mixtures such as those containing polymers can frequently cause computational difficulties, especially in cases where two highly non-ideal liquid phases must be modeled with the same equation of state. Von Solms *et al.*[24] proposed a novel method for finding liquid–liquid coexistence (binodal) compositions in binary polymer–solvent mixtures, the so-called 'method of alternating tangents'. The algorithm is robust and traces the full temperature composition curve for both UCST- and LCST-type systems through the critical solution temperature. The algorithm has been successfully applied for all polymer molecular weights encountered in the experimental literature, and also extended to ternary LLE in systems containing polymers by Lindvig *et al.*[26] The equation of state employed for the calculations is the sPC–SAFT, although in principle the method should be applicable to any equation of state capable of predicting phase equilibrium in polymer systems.

The method finds the spinodal compositions as a starting point for finding the binodals, as well as upper or lower, or both, critical solution temperatures, and is best illustrated by reference to Figure 14.29. The first step in the procedure is to determine whether a spinodal point exists (this is a necessary condition for phase separation). The spinodal condition is:

$$\frac{\partial^2 g^{res}/RT}{\partial x} = 0$$

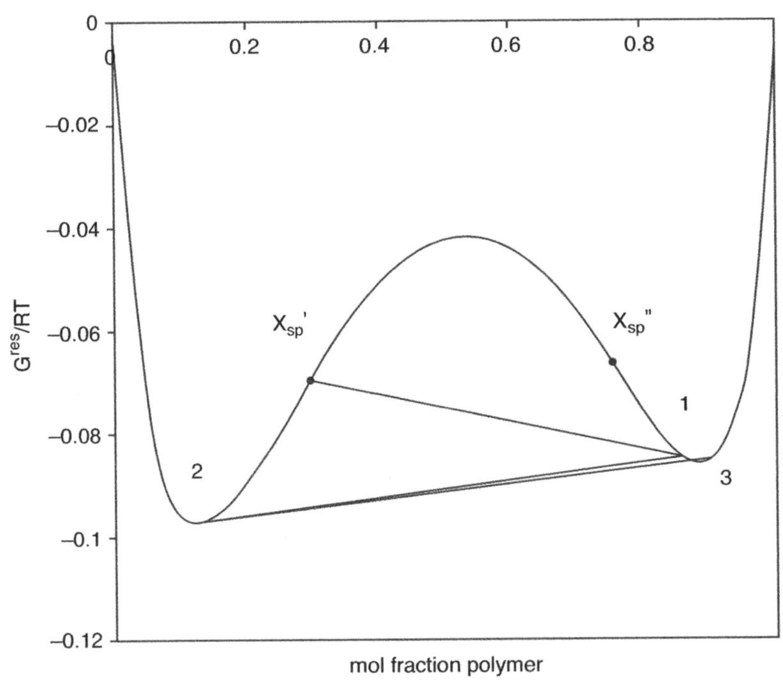

Figure 14.29 *Residual Gibbs energy for a typical binary mixture exhibiting liquid–liquid phase equilibrium. Reprinted with permission from Fluid Phase Equilibria, A novel approach to liquid-liquid equilibrium in polymer systems with application to simplified PC-SAFT by Nicolas van Solms, Georgios M. Kontogeorgis et al., 222–223, 1, 87–93 Copyright (2004) Elsevier*

and the corresponding compositions are shown in Figure 14.29 as x'_{sp} and x''_{sp}, where g^{res} is the residual Gibbs energy. The criterion for the location of the consolute temperature (UCST or LCST) is that the two spinodal points (x'_{sp}, x''_{sp}) have equal values.

The equilibrium compositions (binodal points) are then found by the method of alternating tangents using the spinodal as the starting point. Thus, from point x'_{sp} a line is found tangent to the curve (but at a composition greater than x''_{sp}) at point 1. From this point a tangent is found at composition less than x'_{sp} at point 2. A tangent is then found at point 3 and so on. This procedure is repeated until the change in the compositions at the tangent points is within a certain tolerance. In this way the double tangent is located and the compositions are the equilibrium compositions. For this procedure, Newton's method can be implemented as follows: the tangent is found when the slope of the chord connecting the reference point to the curve is equal to the slope of the curve at the point of tangent. Referring to Figure 14.29, point 1 should satisfy the following condition:

$$f(x_1) = \frac{g(x_1) - g(x'_{sp})}{x_1 - x'_{sp}} - \frac{dg}{dx}(x_1) = 0 \tag{14.13}$$

where $f(x_1)$ is the Newton target function to be solved and $g(x_1)$ is shorthand for:

$$\frac{g^{RES}}{RT}(x_1)$$

while:

$$\frac{dg}{dx}(x_1) = \ln x_1 + \ln \hat{\phi}_1 - \ln(1 - x_1) - \ln \hat{\phi}_2 \tag{14.14}$$

To implement Newton's routine, the derivative of Equation (14.13) is required, which is given by:

$$\frac{df}{dx}(x_1) = \frac{(x_1 - x'_{sp})\frac{dg}{dx}(x_1) - (g(x_1) - g(x'_{sp}))}{(x_1 - x'_{sp})^2} - \frac{d^2g}{dx^2}(x_1) \tag{14.15}$$

Naturally, composition derivatives of the fugacity coefficients are required, since the value of:

$$\frac{d^2g}{dx^2}(x_1)$$

is required. Note, however, that since only a single equation has to be solved, the secant method, which does not require additional derivatives, would be an excellent alternative.

References

1. S.H. Huang, M. Radosz, *Ind. Eng. Chem. Res.*, **1990**, *29*, 2284.
2. S.-J. Chen, I.G. Economou, M. Radosz, *Macromolecules*, **1992**, *25*, 4987.
3. C.J. Gregg, S.J. Chen, F.P. Stein, M. Radosz, *Fluid Phase Equilib.*, **1993**, *83*, 375.
4. C.-S. Wu, Y.-P. Chen, *Fluid Phase Equilib.*, **1994**, *100*, 103.
5. B. Folie, M. Radosz, *Ind. Eng. Chem. Res.*, **1995**, *34*, 1501.
6. B. Folie, C. Gregg, G. Luft, M. Radosz, *Fluid Phase Equilib.*, **1996**, *120*, 11.
7. S.-H. Lee, B.M. Hasch, M.A. McHugh, *Fluid Phase Equilib.*, **1996**, *117*, 61.
8. H. Orbey, C.P. Bokis, C.-C. Chen, *Ind. Eng. Chem. Res.*, **1998**, *37*, 1567.
9. N. Koak, R.M. Visser, T.W. de Loos, *Fluid Phase Equilib.*, **1999**, *158–160*, 835.
10. C. Bokis, H. Orbey, C.-C. Chen, *Chem. Eng. Prog.*, **1999**, *April*, 39.
11. M. Kinzl, G. Luft, H. Adidharma, M. Radosz, *Ind. Eng. Chem. Res.*, **2000**, *39*, 541.
12. P.K. Jog, W.G. Chapman, S.K. Gupta, R.D. Swindoll, *Ind. Eng. Chem. Res.*, **2001**, *41(5)*, 887.
13. E.L. Cheluget, C.P. Bokis, L. Wardhaugh, C.-C. Chen, J. Fisher, *Ind. Eng. Chem. Res.*, **2002**, *41*(5), 968.
14. J. Gross, G. Sadowski, *Ind. Eng. Chem. Res.*, **2002**, *41*, 1084.
15. F. Tumakaka, J. Gross, G. Sadowski, *Fluid Phase Equilib.*, **2002**, *194–197*, 541.
16. J.W. Kang, J.H. Lee, K.-P. Yoo, C.S. Lee, *Fluid Phase Equilib.*, **2001**, *194–197*, 77.
17. J. Gross, O. Spuhl, F. Tumakaka, G. Sadowski, *Ind. Eng. Chem. Res.*, **2003**, *42*, 1266.
18. F. Becker, M. Buback, H. Latz, G. Sadowski, F. Tumakaka, *Fluid Phase Equilib.*, **2004**, *215*, 263.
19. W.G. Chapman, S.G. Sauer, D. Ting, A. Ghosh, *Fluid Phase Equilib.*, **2004**, *217*(2), 137.
20. G. Sadowski, *Macromol. Symp.*, **2004**, *206*, 333.
21. I.A. Kouskoumvekaki, N. von Solms, M.L. Michelsen, G.M. Kontogeorgis, *Fluid Phase Equilib.*, **2004**, *215*, 71.
22. I.A. Kouskoumvekaki, G. Krooshof, M.L. Michelsen, G.M. Kontogeorgis, *Ind. Eng. Chem. Res.*, **2004**, *43(3)*, 826.
23. F. Tumakaka, G. Sadowski, *Fluid Phase Equilib.*, **2004**, *217*, 233.
24. N. von Solms, I.A. Kouskoumvekaki, T. Lindvig, M.L. Michelsen, G.M. Kontogeorgis, *Fluid Phase Equilib.*, **2004**, *222–223*, 87.
25. T. Lindvig, I.G. Economou, R.P. Danner, M.L. Michelsen, G.M. Kontogeorgis, *Fluid Phase Equilib.*, **2004**, *220(1)*, 11.
26. T. Lindvig, M.L. Michelsen, G.M. Kontogeorgis, *Ind. Eng. Chem. Res.*, **2004**, *43*, 1125.
27. F. Tumakaka, J. Gross, G. Sadowski, *Fluid Phase Equilib.*, **2005**, *228–229*, 89.
28. N. von Solms, M.L. Michelsen, G.M. Kontogeorgis, *Ind. Eng. Chem. Res.*, **2005**, *44*, 3330.
29. T. Spyriouni, I.G. Economou, *Polymer*, **2005**, *46*, 10772.
30. M.A. Van Schilt, R.M. Wering, W.J. van Meerendonk, M.F. Kemmere, W.J. Keurentjes, M. Kleiner, G. Sadowski, T.W. de Loss, *Ind. Eng. Chem. Res.*, **2005**, *44*, 3363.
31. K.-M. Kruger, O. Pfhol, R. Dohrn, G. Sadowski, *Fluid Phase Equilib.*, **2006**, *241*, 138.
32. I.A. Kouskoumvekaki, N. von Solms, T. Lindvig, M.L. Michelsen, G.M. Kontogeorgis, *Ind. Eng. Chem. Res.*, **2004**, *43*, 2830.
33. I.G. Economou, Z.A. Makrodimitri, G.M. Kontogeorgis, A. Tihic, *Mol. Simulation*, **2007**, *33(9–10)*, 851.
34. N. von Solms, I. Kouskoumvekaki, M.L. Michelsen, G.M. Kontogeorgis, *Fluid Phase Equilib.*, **2006**, *241*, 344.
35. M. Kleiner, F. Tumakaka, G. Sadowski, H. Latz, M. Buback, *Fluid Phase Equilib.*, **2006**, *241*, 113.
36. A.J. Haslam, N. von Solms, C.S. Adjiman, A. Galindo, G. Jackson, P. Paricaud, M.L. Michelsen, G.M. Kontogeorgis, *Fluid Phase Equilib.*, **2006**, *243*, 74.
37. N. Pedrosa, L.F. Vega, J.A.P. Coutinho, I.M. Marrucho, *Ind. Eng. Chem. Res.*, **2007**, *46*, 4678.
38. A. Tihic, G.M. Kontogeorgis, N. von Solms, M.L. Michelsen, *Fluid Phase Equilib.*, **2006**, *248*, 29.
39. A. Tihic, G.M. Kontogeorgis, N. von Solms, M.L. Michelsen, L. Constantinou, *Ind. Eng. Chem. Res.*, **2008**, *47*, 5092.
40. O. Pfohl, C. Riebesell, R. Dohrn, *Fluid Phase Equilib.*, **2002**, *202*(2), 289.
41. S.-J. Chen, M. Radosz, *Macromolecules*, **1992**, *25*, 3089.
42. S.-J. Chen, I.G. Economou, M. Radosz, *Fluid Phase Equilib.*, **1993**, *83*, 391.

43. P.F. Arce, M. Aznar, *J. Supercrit. Fluids*, **2005**, *34*, 177.
44. H. Orbey, C.P. Bokis, C.-C. Chen, *Ind. Eng. Chem. Res.*, **1998**, *37*, 4481.
45. O. Pfohl, R. Dohrn, *Fluid Phase Equilib.*, **2004**, *217*, 189.
46. A. Grenner, I. Tsivintzelis, G.M. Kontogeorgis, I.G. Economou, C. Panayiotou, *Ind. Eng. Chem. Res.*, **2008**, *47(15)*, 5636.
47. I. Tsivintzelis, A. Grenner, G.M. Kontogeorgis, I.G. Economou, C. Panayiotou, *Ind. Eng. Chem. Res.*, **2008**, *47(15)*, 5651.
48. A. Tihic, N. von Solms, M.L. Michelsen, G.M. Kontogeorgis, L. Constantinou, *Fluid Phase Equilib.*, **2009**, *281(1)*, 70.

Part D

Thermodynamics and Other Disciplines

15

Models for Electrolyte Systems

15.1 Introduction: importance of electrolyte mixtures and modeling challenges

15.1.1 Importance of electrolyte systems and coulombic forces

Mixtures containing strong electrolytes, e.g. salts, or systems with weak electrolytes, e.g. acid gases, are present in many applications and therefore electrolyte thermodynamics is of importance in many contexts and diverse fields such as:

- CO_2 and H_2S removal by absorption using aqueous solutions of alkanolamines and ammonia.
- Novel solvents such as ionic liquids.
- Pharmaceuticals, amino acids, proteins and other applications in biotechnology.
- Corrosion in wet gas pipelines.
- Scale formation in oil production and in heat exchangers.
- Production of fertilizers using ion exchange and salts.

Despite its importance and even the fact that certain aspects are somewhat controversial (e.g. discussion about standard states, measurements of single ion activity coefficients, Lewis–Randal vs. McMillan–Mayer framework), relatively less attention has been given to modeling electrolyte mixtures compared to non-electrolyte thermodynamics.

A review of electrolyte thermodynamics has been presented by Loehe and Donohue[1], covering theories and models in the period 1985–1997. Prausnitz *et al.*[2] discussed the fundamentals of electrolyte thermodynamics as well as local composition models also with applications to gas solubility and biotechnology. Local composition models are presented also in the short review by Pinsky and Takano[3], which includes computational details of a few key activity coefficient models. The recent reviews by Lin *et al.*[4] and Tan *et al.*[5] include discussions on electrolyte equations of state, with special emphasis on those combining SAFT theory and electrolyte theories. Finally, the electrolyte chapter in Michelsen and Mollerup[6] contains a full derivation and discussion from a modern perspective of the Debye–Hückel theory, as well as standard states and theories for dipolar ions (Kirkwood theory). Full derivatives and computational aspects are also presented.

Thermodynamic Models for Industrial Applications: From Classical and Advanced Mixing Rules to Association Theories
Georgios M. Kontogeorgis and Georgios K. Folas

Electrolytes, e.g. salts, dissociate into ions when dissolved in polar solvents, especially water or water–alcohol mixtures. Strong electrolytes dissociate completely. Due to the much longer range of Coulombic forces compared to the van der Waals and related forces, solutions containing charged ions are much more non-ideal than solutions containing only neutral molecules. The Coulombic force is expressed as:

$$F_c = \frac{e^2 Z_i^2}{4\pi\varepsilon_0 D r^2} \tag{15.1}$$

where e is the electronic charge and Z is the ion valency; ε_0 is the permittivity of vacuum and D is the dielectric constant (or relative permittivity). The total permittivity is thus $\varepsilon = \varepsilon_0 D$. For a vacuum $D=1$ (per default) but $D=78.4$ for water at 25 °C. D is reduced with concentration for water–solvent mixtures (see Section 15.2.3). Thus, Coulombic forces are decreased in a medium with high dielectric constant (relative permittivity), D.

Ions also interact with water or the solvents, and thus a proper description of electrolyte solutions requires consideration of both the long-range electrostatic interactions and the short-range energetic interactions. Before discussing the models which have been proposed for electrolyte solutions, however, it is necessary to present some of the peculiarities of electrolytes which differentiate them from non-electrolyte solutions, the concepts of electroneutrality, the various standard states used and finally the mean ionic activity coefficients used for salts. These are discussed next.

15.1.2 Electroneutrality

It is always important to account for the constraint of the electroneutrality of electrolyte solutions. We discuss this by considering a salt dissociating into ν_C cations C (or $+$) and ν_A anions A (or $-$), having positive charges Z_C (or $+Z$) and negative charges Z_A (or $-Z$), respectively. We use, thus, either the symbols C or $+$ for cations and A or – for anions. The dissociation of a salt can in general be expressed as:

$$M_{\nu_C} X_{\nu_A} = \nu_C M^{+Z} + \nu_A X^{-Z} \tag{15.2}$$

e.g. $H_2SO_4 \leftrightarrow 2H^{+1} + 1SO_4^{2-}$. The electroneutrality requirement is:

$$\nu_C Z_C + \nu_A Z_A = 0 \tag{15.3}$$

with Z being the valency.

15.1.3 Standard states

Mole fraction is normally used as the concentration unit in non-electrolyte thermodynamics. When electrolytes are considered, however, the molality, m, is often used. It is defined as moles of salt (solute) per kg of solvent (s), e.g. water ($m_j = n_j / \sum_s n_s M_s$; $j =$ ions or salt, $s =$ solvent). Although, in principle, molality can go from zero to infinity, most applications handle molalities up to moderately concentrated solutions (molality around 20).

Other units such as molarity (= moles of salt per liter of solution) are less frequently used (at least in the literature related to electrolyte activity coefficient models).

The combined effect of molality and valency is often represented via the so-called (molal) ionic strength defined as:

$$I = \frac{1}{2}\sum_i m_i Z_i^2 \tag{15.4}$$

I has the same units as molality, i.e. mol/kg solvent.

Due to the introduction of the 'molality concept', different standard states are often used for water (the typical solvent in electrolyte applications) and ions/salts. Justifiable objections have been reported to the need for introducing the molality concept in electrolyte thermodynamics,[6] but it is still in widespread use because it facilitates calculations and the link to the experimental data for electrolyte-containing systems.

The various standard states are summarized hereafter. For water, the well-known standard symmetric activity coefficients (where $x_i \to 1 \Rightarrow \gamma_i \to 1$) are used:

$$\mu_i = \mu_i^o + RT\ln(x_i\gamma_i) \tag{15.5}$$

As the properties of ions cannot be measured/are not reported independently of other ions in the solutions, the usual standard states are the rational asymmetric activity coefficients and the molal-based activity coefficients, defined as follows.

The rational asymmetric activity coefficients ($\sum_i x_i \to 0 \Rightarrow \gamma_i^* \to 1$): Normalized so that the activity coefficients are one at infinite dilution:

$$\mu_i = \mu_i^* + RT\ln(x_i\gamma_i^*) \quad \text{where} \quad \gamma_i^* = \frac{\gamma_i}{\gamma_i^\infty} \tag{15.6}$$

The molal-based activity coefficients (hypothetical ideal solution at unit molality, m_0*:* $\sum_i m_i \to 0 \Rightarrow \gamma_i^m \to 1$*):*

$$\mu_i = \mu_i^m + RT\ln\left(\frac{m_i\gamma_i^m}{m_0}\right) \quad \text{where} \quad x_i = m_i x_w M_w \quad \text{and} \quad \gamma_i^m = \gamma_i^* x_w \tag{15.7}$$

All three standard states are related to each other as:

$$\mu_i^* = \mu_i^o + RT\ln\gamma_i^\infty \tag{15.8a}$$

$$\mu_i^m = \mu_i^* + RT\ln(M_w m_0) = \mu_i^o + RT\ln(\gamma_i^\infty M_w m_0) \tag{15.8b}$$

The product of the activity coefficient and the mole fraction defines the activity in the various concentrations/standard states, for both water (the solvent) and the ions:

$$a_w = x_w\gamma_w \tag{15.9a}$$

$$a_i = x_i \gamma_i \tag{15.9b}$$

$$a_i^* = x_i \gamma_i^* \tag{15.9c}$$

$$a_i^m = x_i \gamma_i^m \tag{15.9d}$$

Relationships can be easily derived for converting activity coefficients from one concentration scale to another. For example, for a binary mixture of a solute (2) in a solvent (s), the relationship between the rational asymmetric and the molal-based activity coefficients is:

$$\gamma_2^m = \frac{\gamma_2^*}{1 + 0.001 M_s m_2} \tag{15.10a}$$

and similarly for the mean ionic activity coefficients (see Section 15.1.4):

$$\gamma_\pm^m = \frac{\gamma_\pm^*}{1 + 0.001 \nu M_s m_\pm} \tag{15.10b}$$

When an equation of state is used, the asymmetric activity coefficient based on mole fraction is calculated from the fugacities as:

$$\gamma_i^* = \frac{\varphi_i(T, P, x_i)}{\varphi_i^\infty(T, P, x_i \to 0)} \tag{15.11}$$

15.1.4 Mean ionic activity coefficients (of salts)

For a salt dissociating in cations and anions (see also Equation (15.2)), where:

$$\nu = \nu_C + \nu_A \tag{15.12}$$

we can define the mean ionic concentrations for a salt solution, i.e. the mean ionic molality:

$$m_\pm = \left(m_+^{\nu_C} m_-^{\nu_A}\right)^{1/\nu} \tag{15.13a}$$

and the mean ionic mole fraction:

$$x_\pm = \left(x_+^{\nu_C} x_-^{\nu_A}\right)^{1/\nu} \tag{15.13b}$$

Similar relationships can define the mean ionic activity coefficients – again similar expressions for the various concentrations and standard states:

Mean rational ionic activity coefficients:

$$\gamma_\pm = \left(\gamma_+^{\nu_C} \gamma_-^{\nu_A}\right)^{1/\nu} \tag{15.14a}$$

Mean rational asymmetric activity coefficients:

$$\gamma_{\pm}^{*} = \left(\gamma_{+,*}^{\nu_C}\gamma_{-,*}^{\nu_A}\right)^{1/\nu} \tag{15.14b}$$

Mean molal ionic activity coefficients:

$$\gamma_{\pm}^{m} = \left(\gamma_{+,m}^{\nu_C}\gamma_{-,m}^{\nu_A}\right)^{1/\nu} \tag{15.14c}$$

Mean molal ionic activity:

$$a_{\pm}^{m} = \left(a_{+}^{\nu_C}a_{-}^{\nu_A}\right)^{1/\nu} = m_{\pm}\gamma_{\pm}^{m} \tag{15.14d}$$

For strong and fully dissociating electrolytes, e.g. salts, the above expressions for the mean ionic molality and mean ionic activity coefficients can be simplified (from Equation (15.13a)):

$$m_{\pm} = \left(\nu_C^{\nu_C}\nu_A^{\nu_A}\right)^{1/\nu} m \tag{15.15}$$

(where m is the salt's molality). This is because the molality of ions is given as ν_{ion} multiplied by the salt's molality (non-dissociated form) ($m_{+} = \nu_{+}m$, $m_{-} = \nu_{-}m$).

For example, for 1 : 1 electrolytes such as NaCl ($\nu_C = \nu_A = 1$, $\nu = 2$):

$$m_{\pm} = m_{NaCl}$$

$$\gamma_{\pm}^{m} = \sqrt{\gamma_{Na^+}^{m}\gamma_{Cl^-}^{m}}$$

Finally, similarly to Equation (15.10b), mole-fraction-based mean ionic activity coefficients can be converted to mean molal activity coefficients via:

$$\gamma_{\pm}^{m} = \frac{\gamma_{\pm,*}}{1 + 0.001\nu M_s m} \tag{15.16}$$

where ν is defined in Equation (15.12) and m is the salt's molality.

15.1.5 Osmotic activity coefficients

Osmotic (instead of 'ordinary') activity coefficients are often used for representing the activity of solvents in electrolyte solutions.

The water (solvent) activity can be measured by isopiestic measurements and is related to the so-called osmotic coefficient Φ:

$$\Phi = \frac{-\ln a_w}{M_w \nu m_s} = -\frac{n_w}{\nu n_s}\ln a_w \tag{15.17}$$

Osmotic coefficients are more convenient ways to represent the water activity as in dilute solutions the water activity is close to unity, e.g. it has the value 0.999 965 in 0.001 M NaCl at 25 °C.

Using the Gibbs–Duhem equation, the relationship between osmotic and mean ionic molal activity coefficients can be derived:

$$\ln \gamma_{\pm}^{m} = \Phi - 1 + \int_{0}^{m_s} (\Phi - 1) \frac{dm_s}{m_s} \tag{15.18a}$$

$$\Phi = 1 + \frac{1}{m_s} \int_{0}^{m_s} m_s d \ln \gamma_{\pm}^{m} \tag{15.18b}$$

where m_s is the salt molality.

15.1.6 Salt solubility

The equilibrium between an aqueous phase and a solid salt ($C_{\nu_C} A_{\nu_A} \cdot n\mathrm{H_2O}$) consisting of ν_C cations, ν_A anions and n water molecules is expressed as:

$$C_{\nu_C} A_{\nu_A} \cdot n\mathrm{H_2O} \Leftrightarrow \nu_C C_{aq} + \nu_A A_{aq} + n\mathrm{H_2O} \tag{15.19}$$

where the solubility product of the salt is given as function of the activities:

$$K_{salt} = \alpha_C^{\nu_C} \alpha_A^{\nu_A} \alpha_w^{n} \tag{15.20}$$

As previously, the symmetric mole fraction activity coefficient is used for water and the asymmetric ones for ions.

The ratio of the right-hand to the left-hand side of Equation (15.20) (i.e. the product of activities over the solubility product) is called the solubility index (SI). It gives the degree of saturation. Values of SI < 1 imply unsaturated solutions, values of SI > 1 are for supersaturated solutions (and SI = 1 for saturation).

Alternatively, this equation can be written as:

$$\ln K_{salt} = \sum_{w+i} \nu_i \ln(x_i \gamma_i) - (\nu_C + \nu_A) \ln M_w \tag{15.21}$$

Summation is over water and all ions and γ_i is the mole-fraction-based activity coefficients (symmetric for water and asymmetric for ions).

15.2 Theories of ionic (long-range) interactions

15.2.1 Debye–Hückel vs. mean spherical approximation

Most theories for long-range ionic interactions typically fall under what is called ‘primitive models’ (PMs). In these primitive models, ions are considered to be point charges while the solvent is not treated ‘molecularly’, but as a dielectric continuum characterized by its dielectric constant. In the general PM theories, ions are considered to have different dimensions (e.g. ionic diameters), while in the so-called restrictive primitive model (RPM) all ions have the same (average) diameter.

The ionic long-range forces dominate, especially in dilute solutions. The oldest theory which offers a framework for modeling electrostatic interactions between ions in dilute solutions is the Debye–Hückel (DH) theory. The DH theory only accounts for the energy due to charging up molecules in a system. It should, thus, be emphasized that the DH theory does not contain a description of the interactions between ions and water, or the interaction between solvents (water etc.), thus a full model should include contributions for the short-range interactions as well.

The DH theory is derived from electrostatics coupled with some assumptions:

- ions are not distributed randomly;
- the solution is electrically neutral;
- ions are regarded as a sphere of radius α_i with a point charge in the centre of the sphere;
- around a central ion there will be an overweight of ions with the opposite charge.

It should be emphasized that the radius α_i is not the ionic radius but rather the closest distance that any other ion can approach the ion i, i.e. the so-called 'distance of closest approach'. Nevertheless, we expect that α_i reflects the size of the ion.

The expression for the excess Helmholtz energy from the DH theory is derived upon solving the Poisson equation which gives the electrical potential ψ_i (in J/C = V) as a function of the distance r from the center of a central ion i (for spherical coordinates):

$$\frac{1}{r^2}\frac{d}{dr}\left(r^2\frac{d\psi_i}{dr}\right) = -\frac{\rho_i}{\varepsilon_0 D} \tag{15.22}$$

coupled with the Boltzmann distribution for the volumetric charge density around ion i, ρ_i (C/m^3):

$$\rho_i = eN_A \sum_{all\ ions} \frac{n_j z_j}{nV} \exp\left(-\frac{z_j e \psi_j}{kT}\right) \tag{15.23}$$

where e is the electronic charge. The Boltzmann distribution describes how the distribution of the ions differs from the average distribution due to electrostatic interactions.

Combining and rearranging Equations (15.22) and (15.23), and expressing the result in a linearized form, we get the Poisson–Boltzmann (PB) equation, which can be solved analytically:

$$\frac{1}{r^2}\frac{d}{dr}\left(r^2\frac{d\psi_i}{dr}\right) = \kappa^2 \psi_i \tag{15.24}$$

where the key property, the inverse Debye screening length κ, is defined as:

$$\kappa^2 = \frac{N_A}{kT\varepsilon}\sum_{ions} c_i q_i^2 = \frac{e^2 N_A^2}{RTV\varepsilon}\sum_{ions} n_i Z_i^2 = \frac{e^2 N_A}{kTV\varepsilon}\sum_{ions} n_i Z_i^2 = \frac{F^2}{\varepsilon RT}\sum_{ions} c_i Z_i^2 \tag{15.25}$$

where $c_i = n_i/V$ (concentration), $F = eN_A$ (Faraday's constant) and $\varepsilon = \varepsilon_0 D$.

Equation (15.25) illustrates a key property in electrolyte theories, κ^{-1} (length expressed, for example, in nm). This is the so-called screening or Debye length, which determines the range of ionic interactions (or the effective thickness of an ion cloud). Beyond this length the Coulombic interactions can be ignored. Debye and Hückel[7] solved Equation (15.24) and their result, which is an expression for the excess Helmholtz energy, is given in Table 15.1. There are no other assumptions involved except those implicit in the PB equation, and in

Table 15.1 *A^E/kT (excess Helmholtz energy) equations for describing strong ionic interactions. As discussed in the text, these equations represent in reality the electrostatic contributions to the Helmholtz energy. The inverse screening Debye length κ is taken from Equation (15.25)*

Theory	Equation for A^E/kT
Debye–Hückel (complete)	$-\frac{1}{4\pi\varepsilon kT}\sum_i \frac{x_i q_i^2}{(\kappa\alpha_i)^3}\kappa\left[\frac{3}{2}+\ln(1+\kappa\alpha_i)-2(1+\kappa\alpha_i)+\frac{1}{2}(1+\kappa\alpha_i)^2\right]$ or $-\frac{1}{4\pi\varepsilon kT}\sum_i \frac{x_i q_i^2}{3}\kappa\chi_i$ $\chi_i=\frac{3}{(\kappa\alpha_i)^3}\left[\frac{3}{2}+\ln(1+\kappa\alpha_i)-2(1+\kappa\alpha_i)+\frac{1}{2}(1+\kappa\alpha_i)^2\right]$ α_i = distance of closest approach (in general different from the sum of the two radii of ions due to the presence of water molecules). It is close to the hydrated ion diameter $q_i = Z_i e$ $\varepsilon = \varepsilon_0 D$
Debye–Hückel (simplified) If all $\alpha_i = \alpha$, for all ions	$-\frac{4VA_{DH}}{b^3}\left[\frac{3}{2}+\ln\left(1+b\sqrt{I}\right)-2\left(1+b\sqrt{I}\right)+\frac{1}{2}\left(1+b\sqrt{I}\right)^2\right]$ $A_{DH}=\frac{F^2}{8\pi\varepsilon RT}\sqrt{\frac{2F^2}{\varepsilon RT}}=\frac{1}{16\pi N_A}\left(\frac{b}{\alpha}\right)^3$ $b=\alpha\sqrt{\frac{2F^2}{\varepsilon RT}}$ $\kappa\alpha=b\sqrt{I}$ $I=\frac{1}{2}\sum_i c_i Z_i^2$
MSA (complete implicit, CI) – no mixing rules needed $\rho_i=\frac{n_i N_A}{V}=\frac{N_i}{V}$ (number density) Γ, Ω, P_n are the three MSA parameters that can be solved by iteration methods	$-\frac{Ve^2}{4\pi\varepsilon kT}\left[\Gamma\sum_i\frac{\rho_i Z_i^2}{1+\Gamma\sigma_i}+\frac{\pi}{2\Theta}\Omega P_n^2\right]+\frac{V\Gamma^3}{3\pi}$ $\Theta=1-\frac{\pi}{6}\sum_i\rho_i\sigma_i^3$ $\Omega=1+\frac{\pi}{2\Theta}\sum_i\frac{\rho_i\sigma_i^3}{1+\Gamma\sigma_i}$ $P_n=\frac{1}{\Omega}\sum_i\frac{\rho_i\sigma_i z_i}{1+\Gamma\sigma_i}$ $\Gamma=\frac{1}{2}\sqrt{\frac{e^2}{\varepsilon kT}}\sqrt{\sum_i\rho_i\left(\frac{Z_i-\frac{\pi}{2\Theta}\sigma_i^2 P_n}{1+\Gamma\sigma_i}\right)^2}$ Note: $\frac{e^2}{\varepsilon kT}=\frac{\kappa^2}{\sum_i\rho_i Z_i^2}$

Table 15.1 *(Continued)*

Theory	Equation for A^E/kT
MSA (simplified implicit, SI) – derived from the general expression (CI) on assuming that $P_n = 0$	$\frac{V\Gamma^3}{3\pi} - \frac{\alpha_{LR}^2}{4\pi}\Gamma\sum_i \frac{n_i Z_i^2}{1+\Gamma\sigma_i}$ The shielding parameter Γ: $4\Gamma^2 = \alpha_{LR}^2 \sum_i \frac{n_i}{V}\left[\frac{Z_i}{1+\Gamma\sigma_i}\right]^2$ $\alpha_{LR}^2 = \frac{e^2 N_A}{\varepsilon_0 DkT}$
MSA (simplified explicit, SE) – all ions have the same effective average diameter	$-\frac{2\Gamma^3 V}{3\pi}\left(1+\frac{3}{2}\sigma\Gamma\right)$ $\sigma = \frac{\sum_i n_i\sigma_i}{\sum_i n_i} = \frac{\sum_i \rho_i\sigma_i}{\rho}$ $\Gamma = \frac{1}{2\sigma}(\sqrt{1+2\sigma\kappa}-1)$
Born term (change in free energy of ion transfer from vacuum or gas phase to a medium with dielectric constant D)	$-\frac{N_A e^2}{4\pi\varepsilon_0 kT}\left(1-\frac{1}{\varepsilon}\right)\sum_{ions}\frac{n_i Z_i^2}{\sigma_i}$
SR2 (short-range ionic interactions)	$\sum_i\sum_j \frac{n_i n_j W_{ij}}{V(1-\xi_3)}$ The summation is over all molecular and ionic species with at least one species being an ion W_{ij} is an ion–ion or ion–molecule interaction parameter ξ_3 (packing fraction of ions) is calculated as: $\xi_3 = \frac{\pi N_A}{6}\sum_i \frac{n_i\sigma_i^3}{V}$ where the summation is over all species
Ion–dipole term (from *Henderson et al.*)[20]	$\frac{A_2^{id}/kT}{1-A_3^{idd}/A_2^{id}}$ *id* and *idd* represent two-particle and three-particle ion–dipole interactions (two dipoles with an ion; two ion–dipole interactions are ignored); A_2 and A_3 terms are expressed as functions of concentration, solvent dipole moment, density and hard-sphere diameters

principle the A^E of Table 15.1 can be implemented in engineering equations of state so that the 'other' interactions such as repulsions and attractions are also included. The A^E is actually the electrostatic contribution to the Helmholtz energy according to the McMillan–Mayer framework (derivation assuming the solvent is a dielectric continuum), but we will use the superscript E (in A^E) for both excess and electrostatic contributions. In principle, corrections are needed when such terms are combined with equations of state based on the well-known Lewis–Randall framework, but these corrections are often small, as discussed in Section 15.4.

An alternative to the DH theory way of representing electrostatic interactions is given by the more recent MSA theory[15–17] (mean spherical approximation theory), which belongs to the so-called 'integral methods'. It is a more complex theory than DH, and it is derived from statistical mechanics, namely the Ornstein–Zernike equation at a specific closure. There are both implicit and explicit versions of the MSA theory and the simplest (explicit) version is when a common ion size diameter is used. The implicit versions of MSA (see Table 15.1) should be solved iteratively. Especially, the complete implicit form of the MSA term (CI in Table 15.1) requires the solution of the implicit screening parameter Γ before it can be used to calculate thermodynamic properties.

It is often stated[2,8] that MSA should be superior compared to DH for concentrated solutions, as ion–ion and ion–solvent interactions become more important. Few investigations, e.g. Galindo *et al.*[9], have provided comparisons between MSA and DH (using the same terms for the short-range interactions). Very similar results are obtained in the case of vapor pressures, e.g. for NaCl–water, but densities are better represented with MSA, particularly at high concentrations. Lin[10] showed via a Taylor series expansion that the differences between MSA and DH are very small also in mathematical form, when the assumption of the same ion diameters is made (comparison between the restrictive primitive models). Table 15.1 shows the equations for the excess Helmholtz energy from MSA including several simplifications.

15.2.2 Other ionic contributions

The DH or MSA theories indeed capture the basic characteristics of (dilute) electrolyte solutions and many engineering models (activity coefficients and equations of state) use solely either the DH or the MSA terms for describing the interactions involving ions. Nevertheless, both the DH and MSA theories ignore many important features of electrolyte solutions, such as:

- charge–dipole or quadrupole interactions;
- ion hydration;
- effects induced by ions.

In addition, for dipolar ions (zwitterions) such as amino acids, peptides and proteins (sometimes called 'complex ions'), dipolar ion–ion interactions may play a significant role, as these complex ions have a very large dipole moment (often above 10 D (Debye)). Such interactions could be accounted for using, for example, the Kirkwood theory, as will be discussed in Chapter 19.

Due to the limitations of the DH or MSA theories to account for 'all' the ion-related interactions, many engineering models have been developed which contain additional terms associated with ionic interactions especially: (1) a Born term;[18,19] and (2) a short-range ionic term. These terms are also included in Table 15.1.

The Born term describes the effect of solvation (or hydration) of ions and calculates the energy to transfer an ion from vacuum to a solvent with a dielectric constant D (or in general from a medium with a lower relative permittivity ε_1 to a medium having a higher relative permittivity equal to ε_2). It thus accounts for the redirection of polar molecules around ions. The Born or solvation energy, in other words, represents the extent ions will solvate (ion–solvent interactions) or the energy of single ions even when they do not interact at all. Interestingly enough, the solution of the PB equation *does* contain both the Born term and what was afterwards known as the Debye and Hückel term (shown in Table 15.1). The Born term was not used, however, by Debye and Hückel and it has also been ignored in several of the subsequent investigations.

The short-range SR2 term was first presented by Planche and Renon[11] and is derived from a non-primitive model for electrolyte systems. It was then used in the Fürst–Renon electrolyte EoS[12] and other electrolyte EoS more recently. Unlike the other electrolyte terms, SR2 contains binary interaction parameters which account for solvation and can be correlated to the Stokes 'solvated' diameter (see Section 15.5.1). The SR2 term accounts for the short-range interactions between ions and polar molecules.

15.2.3 The role of the dielectric constant

All the expressions in Table 15.1 contain the dielectric constant. For dilute solutions it can be assumed that the dielectric constant is identical to that of a pure solvent, but for concentrated solutions, it decreases with concentration and the effect of salts should, in principle, be included. Especially for mixed solvents, the dielectric constant will be a function of concentration as well. Correlations are, thus, needed for the dielectric constant vs. concentration relationships. These correlations must be used when calculating the concentration derivatives of the Helmholtz energy with respect to the dielectric constant.

For mixed solvent systems, it can be shown[6] that the dielectric constant of liquid mixtures can be approximated as the volume fraction average of the dielectric constant of the pure compounds:

$$D = \frac{1}{V}\sum_{i=1}^{NC} V_i D_i \qquad (15.26)$$

where V_i and D_i are the pure component volume and dielectric constants of pure solvents, e.g. of water and a alcohol.

The decrease of the dielectric constant (relative permittivity) of water at 25 °C with salt concentration can be described for chloride salt, bromide salt and divalent ion aqueous solutions using the correlation presented by Michelsen and Mollerup[6] (2007, p. 174). Two other well-known expressions are those by Pottel[13] and Simonin *et al.*[14]:

Pottel[13]:

$$\varepsilon - 1 = (\varepsilon_s - 1)\frac{2(1-\xi_3)}{2+\xi_3} \qquad (15.27)$$

The parameter ξ_3 is defined as in Table 15.1 but the summation in ξ_3 for use in Equation (15.27) is over ions alone. The solvent relative permittivity (ε_s or D) can be calculated from Equation (15.26).

Simonin et al.[14] *(containing an adjustable parameter* α*):*

$$\varepsilon - \frac{\varepsilon_s}{\left(1+\alpha\sum_{ions} x_i\right)} \qquad (15.28)$$

In equations (15.27) and (15.28) ε_s is the relative permittivity of the solvent, e.g. water. This is usually taken from expressions in the literature, as a function of the temperature and density; see also Michelsen and Mollerup.[6] For an extensive discussion on equations for the dielectric constant see Michelsen and Mollerup[6].

15.3 Electrolyte models: activity coefficients

15.3.1 Introduction

While the expressions shown in Table 15.1 have been extensively used, especially over the past 15–20 years, in EoS suitable for electrolyte solutions (see Sections 15.4 and 15.5), the most widely used approach for electrolytes for engineering applications is possibly still the activity coefficient models. Activity coefficient models are often preferred in engineering practice as several of them have been systematically developed over many years and extensive parameter tables are available. Most activity coefficient models contain an electrolyte term based on some form of the DH equation and a local composition model, e.g. NRTL or UNIQUAC, to describe the short-range interactions. Table 15.2 shows various simplifications or truncated forms of the DH equation, while Table 15.3 summarizes the basic characteristics of three of the most well-known electrolyte activity coefficients (see also Chapter 5 for extended UNIQUAC).

It is important to emphasize that whereas the expression from Table 15.1 of DH corresponds to A^E (actually to the electrostatic contribution to A) and not G^E (i.e. the electrostatic G), the electrostatic (excess) Gibbs energy equation from the DH theory can be derived.[21a,b] Then, the electrostatic contribution to the activity coefficient from DH can be calculated as (assuming that the partial molar volume of ions is zero):

$$\ln \gamma_j^{elec} = -\frac{F^2}{4\pi\varepsilon RT}\frac{\kappa Z_j^2}{6N_A}\left[2\chi_j + \frac{\sum_k n_k\sigma_k Z_k^2}{\sum_k n_k Z_k^2}\right] \tag{15.29a}$$

where:

$$\sigma_k = \left(\frac{d(\kappa\chi_k)}{d\kappa}\right)_{T,P,n} = \frac{3}{(\kappa\alpha_k)^3}\left[1+\kappa\alpha_k - \frac{1}{1+\kappa\alpha_k} - 2\ln(1+\kappa\alpha_k)\right] \tag{15.29b}$$

Equations (15.29a) and (15.29b), without the 'RT' factor, rigorously define the contribution of the chemical potential due to the charging process, rather than the activity coefficient.

Several of the expressions in Table 15.2 are derived from Equation (15.29a) when simplifying assumptions are made, as follows:

1. For very dilute solutions where both $\chi_i = 1$ and all 'closest approach distances' (similar to the hydrated ionic diameters) are equal to each other ($\alpha_i = \alpha$), then the 'Debye–Hückel limiting law' is obtained:

$$\ln \gamma_i^* = -Z_i^2 A_{DH}\sqrt{I} \tag{15.30}$$

2. For dilute solutions where only the assumption of equal 'closest approach distances' is made ($\alpha_i = \alpha$), then the 'extended Debye–Hückel limiting law' is obtained (if molarity is substituted by molality for dilute solutions):

$$\ln \gamma_i^* = -Z_i^2\frac{A_{DH}\sqrt{I}}{I + B\alpha\sqrt{I}} \tag{15.31}$$

In the derivation of Equations (15.30) and (15.31), the assumption that the electrical chemical potential is equivalent to the corresponding contribution to the activity coefficient is adopted.

The data for the experimental mean ionic activity coefficients for aqueous salt solutions show that in some cases the activity coefficients decrease continuously with increasing concentration, while in other cases they pass through a minimum. The DH-based expressions of Table 15.2, especially the first two, are valid at rather low molalities (below 0.1 M) and fail at higher salt concentrations, where the short-range interactions dominate. Prausnitz *et al.*[2] showed that the Hückel equation provides a good correlation of binary water–salt data up to high molality values. However, a more general appropriate method to extend the DH approach to high concentrations is by combining the equations of Table 15.2 with a local composition approach (such as the

Table 15.2 *Simplifications or modifications of the DH equation*

Name	Equation for $\ln \gamma_\pm^*$	Applicability range
Debye–Hückel 'limiting law'	$-\frac{1}{\nu}\sum_i \nu_i Z_i^2 A\sqrt{I}$ $\nu = \sum_i \nu_i$ where: $A = \sqrt{2\pi N_A d_0}\left(\frac{e}{\sqrt{4\pi\varepsilon_0 DkT}}\right)^3$ $I = 0.5\sum_i m_i Z_i^2$ (ionic strength) Special case: salts with one type of cations (C) and anions (A): $-A\lvert z_C z_A\rvert\sqrt{I}$	Up to 0.001 molal (very dilute solutions)
'Extended' Debye–Hückel 'law', also used in extended UNIQUAC[22,23]	$-\frac{A\lvert z_C z_A\rvert\sqrt{I}}{1+B\alpha\sqrt{I}}$ or generally: $-\frac{1}{\nu}\sum_i \nu_i Z_i^2 \frac{A\sqrt{I}}{1+B\alpha\sqrt{I}}$ $B = \sqrt{\frac{2e^2 N_A d_0}{\varepsilon_0 DkT}}$ α = distance of closest approach ($3.5\text{–}6.2\times10^{-10}$ m, same for all ions) $\ln\gamma_w^{DH} = M_w \frac{2A}{b^3}\left[1+b\sqrt{I}-\frac{1}{1+b\sqrt{I}}-2\ln(1+b\sqrt{I})\right]$ where: $b = B\alpha$	Up to 0.1 molal
Hückel,[24] 1925	$-\frac{A\lvert z_C z_A\rvert\sqrt{I}}{1+B\alpha\sqrt{I}} + CI$ A and B defined as in the extended DH equation above, while C is an adjustable parameter. It accounts indirectly for the composition dependency of the dielectric constant	Up to high molalities
Bromley,[25] 1973	$\ln\gamma_\pm^m = -\frac{A\lvert z_C z_A\rvert\sqrt{I}}{1+\sqrt{I}} + CI$ $C/\ln 10 = \frac{(0.06+0.6B_{CA})\lvert z_C z_A\rvert}{\left(1+\frac{1.5}{\lvert z_C z_A\rvert}I\right)^2} + B_{CA}$ $B_{CA} = B_C + B_A + \delta_A\delta_C$	Up to ionic strengths of 6 mol/kg Can be applied to multicomponent solutions

Table 15.3 *Engineering-oriented activity coefficient models for concentrated electrolyte solutions*

Model	Reference	Comments
Pitzer (DH + virial-type equation with series in increasing power of molality). A Bromley term may be included for short-range ion–ion interactions	Pitzer[26,27]	Widely used but complex Many modifications and parameter tables available Need for ternary parameters for ternary systems No built-in temperature dependency Up to 6 molal
e-NRTL (NRTL + Pitzer and DH + Born term for mixed solvents)	Chen and co-workers[28–33]	Salt-specific parameters Developed by ASPEN Technology and available in ASPEN's simulator Limited number of parameters have been published – fewer parameters available compared to extended UNIQUAC Physical principles: • like ion repulsion: the local composition of cations around cations is zero and the same for anions (i.e. large repulsive forces between charged ions) • local electroneutrality: around a central solvent molecule, the net ionic charge is zero, i.e. the distribution of ions is such that this requirement is fulfilled
Extended UNIQUAC (UNIQUAC + extended DH, see Table 15.2 and Chapter 5).	Thomsen and co-workers[22,23] and references 112–117 in Chapter 5	Ion-specific parameters Satisfactory reproduction of activity coefficient/ osmotic coefficient, excess enthalpy and excess heat capacity
SRK for the gas phase	Takano[3,34]	Thermal properties yield better T-dependent interaction parameters The Takano version contains an additional Born term for mixed solvents

models presented in Chapter 5) or similar activity coefficient models. The basic features of three widely used and successful models are summarized in Table 15.3, illustrating the similarities and differences. Some of their key features are discussed hereafter.

15.3.2 Comparison of models

Possibly the first major engineering (activity coefficient) model for electrolytes is the one developed by Pitzer, by combining the DH equation with a virial expansion (including up to three virial coefficients) to represent short-range interactions. The mean activity coefficient on a molality basis can be expressed by the complex but analytical equation:

$$\ln \gamma_{\pm}^{m} = |z_+ z_-| f^{\gamma} + m\left(\frac{2\nu_+ \nu_-}{\nu}\right) B_{MX}^{\gamma} + m\left[\frac{2(\nu_+ \nu_-)^{3/2}}{\nu}\right] C_{MX}^{\gamma} \tag{15.32}$$

Detailed values and expressions (as functions of ionic strength and temperature) for all the parameters f, B, C, etc., are presented elsewhere, e.g. see Prausnitz *et al.*[2] The Pitzer model is used extensively and provides very accurate results for the mean ionic activity coefficients up to very high molalities as well as SLE (salt solubilities) in aqueous mixed salt systems. However, a very large number of parameters are typically needed and mixed salt systems require additional parameters.

The e-NRTL model uses both a DH formula (the so-called Pitzer–Debye–Hückel) and a Born term for the long-range interactions:

$$\frac{g_{PDH}^{E,*}}{RT} = -\left(\sum_k x_k\right)(1000/M_s)^{1/2}(4A_\varphi I_x/\rho)\ln(1+\rho I_x^{1/2}) \tag{15.33}$$

$$\frac{g_{Born}^{E}}{RT} = \left(\frac{e^2}{2kT}\right)\left(\frac{1}{D_s} - \frac{1}{D_w}\right)\left(\sum_i \frac{x_i z_i^2}{r_i}\right)10^{-2} \tag{15.34}$$

where: A_ϕ is the DH parameter ($=\frac{1}{3}A, A$ from Table 15.2), which is a function of the solvent dielectric constant and density; I_x is the mole-fraction-based ionic strength ($I_x = \frac{1}{2}\sum_i x_i Z_i^2$); D_s and D_w are the dielectric constants of mixed solvents and pure water.

In general, we can comment for all these local composition and virial-type electrolyte activity coefficient models (Table 15.3):

- They are not predictive. They can at best be considered semi-empirical and they typically require a large number of interaction parameters which increase as the number of species grows. Essentially all the adjustable parameters are present in the short-range terms (not in the long-range DH part).
- They require an extensive number of experimental data for the parameter estimation (VLE, LLE, SLE, osmotic activity coefficients, heat of mixing and heat capacities). The thermal properties are useful for evaluating the range of applicability of models over an extended temperature range and also in model development, especially for obtaining parameters suitable over broad temperature ranges.
- They have been applied to a wide variety of systems and applications including salt solubilities (SLE) in aqueous mixed salt mixtures.

The e-NRTL model has been applied to several industrial strong acid processes such as sulfuric and nitric acid plants,[29,35] see Figure 15.1. Extended UNIQUAC has been extensively applied to both single and mixed solvents (water–alcohols) and salts (including a few quaternary systems), heavy metal ions and various types of phase equilibria (VLE, LLE, SLE), high-pressure SLE and vapor–liquid–solid equilibria of scaling minerals, high-pressure CO_2–water–NaCl VLE, mean ionic and osmotic activity coefficients as well as thermal properties. The model has an extensive parameter table, see references 110–117 in Chapter 5 as well as the recent review by Thomsen[23]. Figures 15.2–15.4 show a few characteristic results[36–38].

Electrolyte activity coefficient models contain a large number of adjustable parameters which are best obtained by simultaneous fitting of experimental data for many aqueous salt (or other electrolyte) solutions. Thus, extensive databases are very useful for that purpose. One of these electrolyte databases, developed by the IVC-SEP group at the Department of Chemical and Biochemical Engineering, Technical University of Denmark (www.ivc-sep.kt.dtu.dk), contains more than 100 000 experimental data points (activity coefficients, enthalpy of mixing and heat capacities, degree of dissociation, gas solubility, salt solubility, VLE, LLE, density, gas hydrate formation) for binary, ternary and quaternary solutions. The database is continuously being updated.

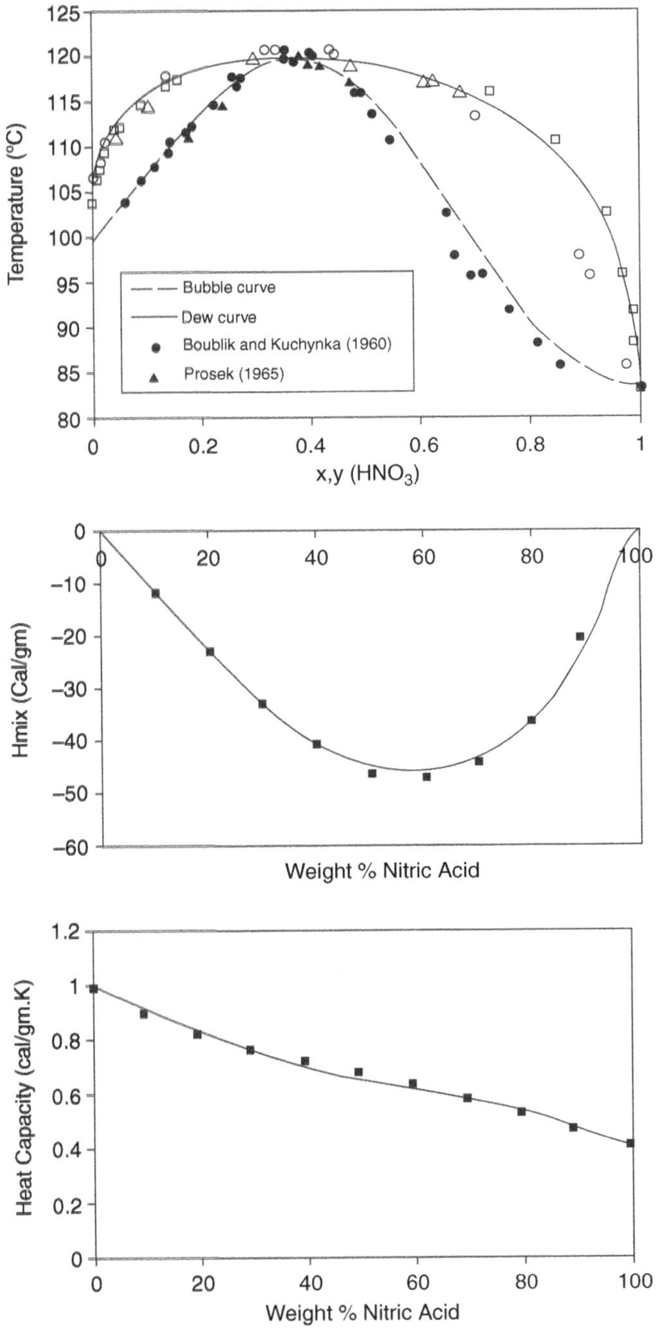

Figure 15.1 *Nitric acid–water calculations with the e-NRTL EoS. From top to bottom: VLE at 1 atm, heat of mixing at 25 °C and heat capacity at 20 °C. Reproduced with permission from Fluid Phase Equilibria, Applied thermodynamics in chemical technology: current practice and future challenges by Paul M. Mathias, PPEPPD 2004 Proceedings 228–229, 49–57 Copyright (2005) Elsevier*

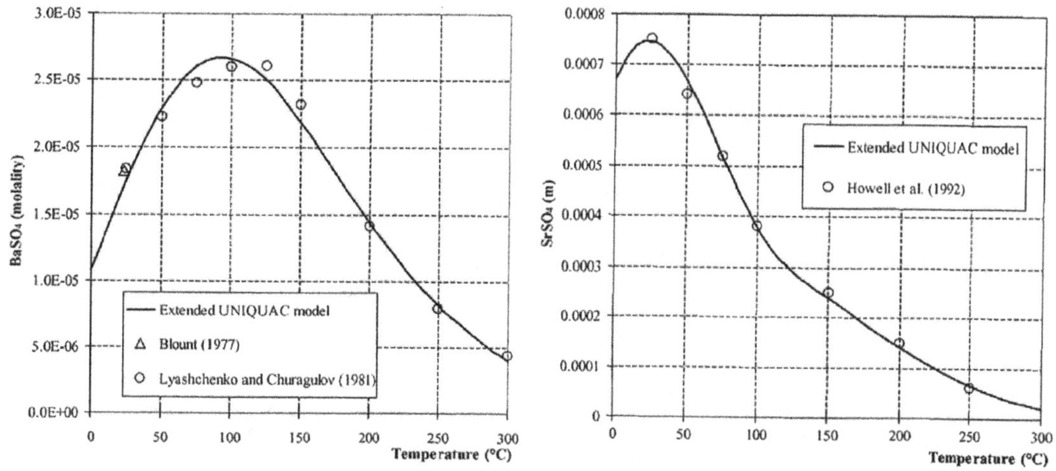

Figure 15.2 *Experimental and calculated results with extended UNIQUAC solid–liquid phase diagram of aqueous salt solutions at high pressures: left, $BaSO_4$–water at 500 bar; right, $SrSO_4$–water at 200 bar. Reprinted with permission from Geothermics, Prediction of mineral scale formation in geothermal and oilfield operations using the extended UNIQUAC model: Part I. Sulfate scaling minerals by A.V. Garcia, K. Thomsen, E.H. Stenby, 34, 1, 61–97 Copyright (2005) Elsevier*

15.3.3 Application of the extended UNIQUAC approach to ionic surfactants

Only a few group contribution models have been developed for electrolytes, e.g. the e-UNIFAC variants by Kikic *et al.*[39], Yan *et al.*[40] and Achard *et al.*[41] The last model has been modified and used by Cheng[42] to estimate the critical micelle concentration (CMC) of aqueous surfactant solutions. Cheng

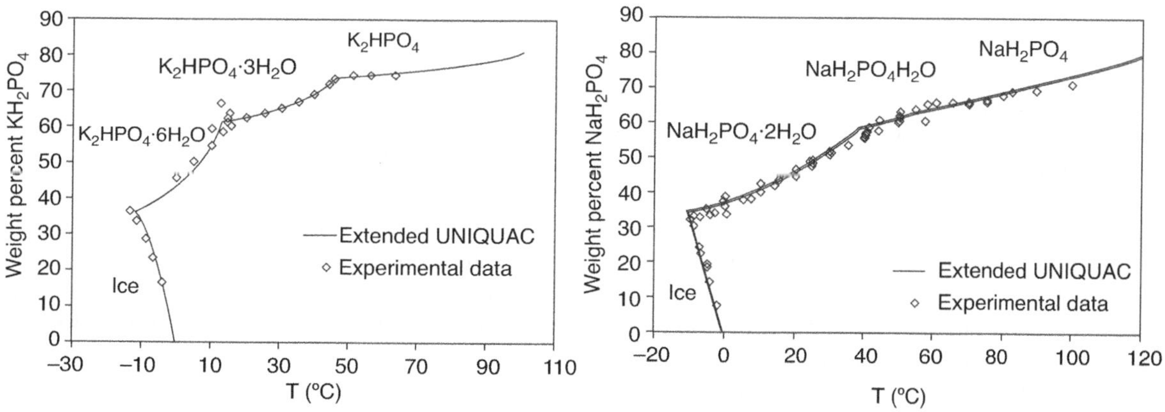

Figure 15.3 *Solubility of K_2HPO_4 in water (left) and solubility of NaH_2PO_4 in water (right). The calculations are with extended UNIQUAC. Reprinted with permission from Ind. Eng. Chem. Res., Modeling of Vapor–Liquid–Solid Equilibria in Acidic Aqueous Solutions by Søren Gregers Christensen and Kaj Thomsen, 42, 18, 4260–4268 Copyright (2003) American Chemical Society*

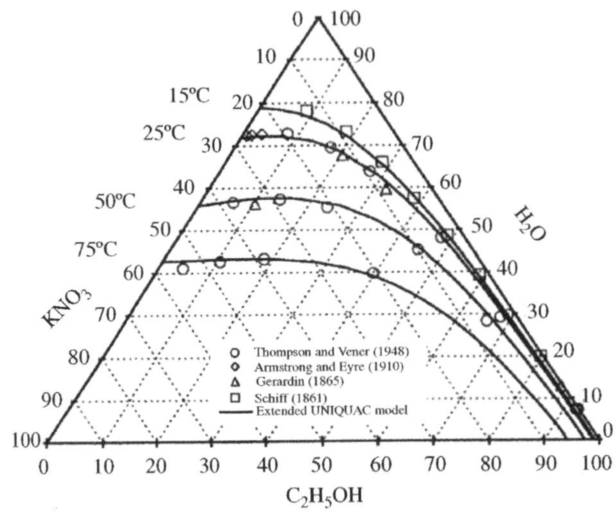

Figure 15.4 *Solubility (in mass percent) of potassium nitrate in aqueous ethanol solutions at four temperatures (15–75 °C). The calculations are with extended UNIQUAC. Reproduced with permission from Chemical Engineering Science, Extended UNIQUAC model for correlation and prediction of vapor–liquid–liquid–solid equilibria in aqueous salt systems containing non-electrolytes. Part B. Alcohol (ethanol, propanols, butanols)–water–salt systems by Kaj Thomsen, Maria C. Iliuta and Peter Rasmussen, 59, 17, 3631–3647 Copyright (2004) Elsevier*

employed the approach of Chen *et al.*[30] for ionic surfactants, according to which CMC can be estimated assuming LLE between surfactants in the aqueous solutions and the micellar phase (see also Section 18.7.3 for non-ionic surfactants with the UNIFAC approach). The LLE equation for surfactants can be expressed as:

$$x_w^{CMC}\gamma^*_{\pm,w} = x_{micelle}\gamma^*_{\pm,micelle} \tag{15.35}$$

where w indicates the aqueous phase and *micelle* the micellar phase.

At and above the CMC, the micelle particles are composed only of surfactant molecules, i.e. the mole fraction of surfactant molecules in the micelle is unity ($x_{micelle} = 1$). The CMC, i.e. the x_w^{CMC}, can be estimated using suitable activity coefficient models for the estimation of mean ionic activity coefficients. In Cheng's model for ionic surfactants, the molar excess Gibbs energy of the electrolyte solution can be expressed as the sum of two contributions: one from the long-range ion–ion interaction that exists beyond the immediate neighborhood of ionic species; and the other from short-range interactions that exist in the immediate neighborhood of any species. The asymmetric Pitzer–Debye–Hückel (PDH) term is used to account for long-range ion–ion interactions and the UNIFAC model is used for short-range (SR) interactions. The asymmetric activity coefficients are expressed as following:

$$\ln \gamma^*_{\pm,aq} = \ln\gamma^{*,PDH}_{\pm,aq} + \ln\gamma^{*,SR}_{\pm,aq} \tag{15.36a}$$

$$\ln \gamma^*_{\pm,micelle} = \ln\gamma^{*,PDH}_{\pm,micelle} + \ln\gamma^{*,SR}_{\pm,micelle} \tag{15.36b}$$

The PDH formula is normalized to mole fractions of unity for solvents and zero for electrolytes. The expression for the activity coefficient of any species is:

$$\ln\gamma_i^{*,PDH} = -\left(\frac{1000}{M_s}\right)^{1/2} A_\phi \left[\left(\frac{2Z_i^2}{\rho}\right)\ln(1+\rho I_x^{1/2}) + \frac{Z_i^2 I_x^{1/2} - 2I_x^{3/2}}{1+\rho I_x^{1/2}}\right] \tag{15.37}$$

$$I_x = \frac{1}{2}\sum Z_i^2 x_i \tag{15.38}$$

where I_x is the ionic strength on a mole fraction basis, M_s is the molecular weight of the solvent, A_ϕ is the usual DH parameter and ρ is the 'closest approach' parameter. The value of ρ depends on the electrolyte as well as the expression used to represent the short-range interactions. The DH parameter A_ϕ is temperature dependent:

$$\begin{aligned} A_\phi = & -61.4453\exp\left(\frac{T-273.15}{273.15}\right) + 2.864\,468\left[\exp\left(\frac{T-273.15}{273.15}\right)\right]^2 \\ & + 183.5379\ln\left(\frac{T}{273.15}\right) - 0.682\,022\,3(T-273.15) \\ & + 7.875\,695\times10^{-4}\left[T^2-(273.15)^2\right] + 58.957\,88\left(\frac{273.15}{T}\right) \end{aligned} \tag{15.39}$$

Cheng's modifications compared to Achard *et al.*'s[41] model are as follows:

- No hydration equations are used in the estimation of ionic radii.
- The UNIFAC R and Q values are calculated using the ionic radii.
- A new normalization is used for R and Q so that the condition $Q/R < 1$ is fulfilled for ions (unlike the Achard *et al.* and previous UNIQUAC versions for electrolytes).
- The number of adjustable parameters is decreased to only two per single salt solution (those between cation–water and anion–water, both of which are temperature dependent). If improved results are required, the cation–anion interaction can be also adjusted to experimental data.
- The linear temperature-dependent UNIFAC is used for the short-range interactions.

Cheng[42] showed that the results of his e-UNIFAC model are comparable to those of extended UNIQUAC and e-NRTL in the correlation of mean ionic and osmotic activity coefficients for aqueous salt solutions at 25 °C and higher temperatures. The proposed electrolyte UNIFAC approach can be used with success for organic electrolyte and ionic surfactant aqueous solutions as well, as shown in Figures 15.5 and 15.6. Equally satisfactory results were obtained for the CMC dependency with temperature and chain length for sodium alkyl sulfates, sodium alkyl sulfonates and potassium carboxylates, but the model performed less satisfactorily for sodium carboxylates.

More electrolyte activity coefficient models have been applied to ionic surfactant solutions. Satisfactory estimations of the CMC for sodium carboxylates were presented by Chen *et al.*[30] using their segment-based e-NRTL model. In this model (which is a variation of the polymer-NRTL and e-NRTL models, previously

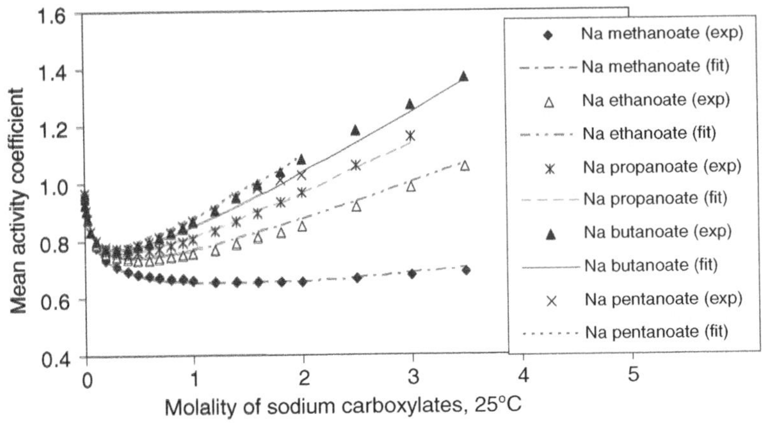

Figure 15.5 Correlation results for organic electrolyte solutions (sodium carboxylates) with Cheng's e-UNIFAC model. Four interaction parameters were fitted to these data (CH_2/COO^-, COO^-/water, CH_2/Na^+, CH_2/water). The water/CH_2 parameter is taken from UNIFAC and the Na^+/water one is fitted to NaCl–water data. From Cheng[42]

presented by the same authors), the excess Gibbs energy has three terms: the NRTL, the PDH and the Flory–Huggins term. The organic ions are treated as oligomers that consist of hydrocarbon and ionic segments, each having different interactions with the surrounding species. Based on Equation (15.35) and with six adjustable ion pair-specific parameters (all in the NRTL term), Chen *et al.*[30] satisfactorily correlated mean ionic activity coefficients and water activity for lower sodium carboxylates as well as CMC for the heavier sodium carboxylates (with $n > 5$). The e-NRTL can be readily extended to other families of ionic surfactants.

Another segment-based UNIQUAC model which has been applied to CMC has been presented by Li *et al.*[43,68] These investigators emphasized the development of a SAFT theory for surfactants and thus the results will be discussed in Section 15.5.3.

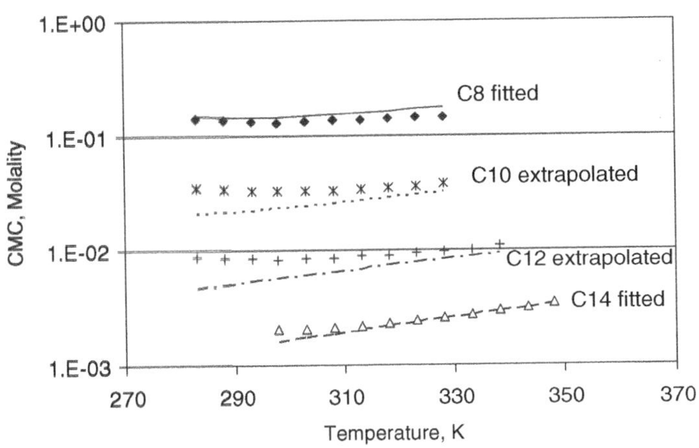

Figure 15.6 Correlation and extrapolation results for four sodium alkyl sulfates C_nOSO_3Na with Cheng's e-UNIFAC model ($n = 8$, 10, 12 and 14). Only C8 and C14 systems were used in the parameter estimation and three interaction parameters were fitted to the data (CH_2/OSO_3^- temperature dependent and water/OSO_3^- temperature dependent). With the same number of parameters, an even better description of all four systems can be achieved if all data are included in the parameter estimation. From Cheng[42]

15.4 Electrolyte models: Equation of State

15.4.1 General

In principle, any of the electrolyte theories (Table 15.1) based on (one of the various forms of) DH or MSA equations can be used in combination with any 'ordinary' EoS in order to develop an EoS for electrolyte solutions which can be used over extensive temperature, pressure and concentration ranges. Indeed, especially over the past 10–15 years, the literature has been rich in such electrolyte EoS. As association models (CPA, SAFT, etc.) are also becoming increasingly popular, several electrolyte CPA and SAFT variants have been presented. Table 15.4 summarizes some of the most recent approaches. An alternative model to these 'direct' electrolyte EoS is to use an EoS/G^E approach (see Chapter 6) and incorporate an electrolyte activity coefficient model in, for example, a cubic EoS. This has been done by Gmehling and co-workers[44,45] or via the e-LCVM model, which will be discussed in the section presenting models for CO_2–water–alkanolamine solutions (Section 15.6.4). Earlier EoS for electrolytes mostly focused on high-pressure water–salt–gas systems, e.g. the models by Harvey and Prausnitz[46] and Aasberg-Petersen *et al.*[47] The emphasis with these early approaches was on describing the salting-out effects of gases (as well as natural gas mixtures) induced by salts. A comparison of the various electrolyte EoS approaches is difficult. As observed from the overview in Table 15.4, these electrolyte EoS differ in many respects:

- The choice of EoS for representing physical (non-electrolyte) interactions.
- Inclusion or not of an association term, but also whether association is considered only for the solvent(s) or also between solvent (water) and ions.
- Which theory (DH or MSA) is used for the long-range ionic interactions.
- Whether 'auxiliary' ionic terms are used, such as the Born term and the short-range ionic term (SR2).

But the differences between these electrolyte EoS also include:

- Application ranges (temperature, pressure, VLE and also LLE/SLE).
- Which salts they have been applied to.
- Whether they are applied only to binary water–salt systems or also to multicomponent mixtures (mixed solvents and/or mixed salts).

There are also significant differences in the parameterization and details in the model development especially related to:

- Whether the parameters are ion or salt specific.
- How many adjustable parameters are included and how the accuracy of the models is affected by simplifications carried out for reducing the number of parameters.
- How water is described.
- How many experimental data are used in the parameter estimation and whether the parameters are estimated simultaneously for many salts or not.
- How many simplifications are made overall in the model and which special effects have been studied, e.g. expressions for the concentration dependency of the dielectric constants.

Of special interest is the broad applicability of these electrolyte EoS so that they can be used beyond 'ordinary' salt–solvent mixtures, e.g. to acid gas–aqueous alkanolamine mixtures (see Section 15.6) and applications containing biomolecules (see Chapter 19). Due to the difficulties mentioned above, instead of a 'direct' comparison of the models and claims on better accuracy of one approach over the others, we outline in the next section the special characteristics of some of the most interesting approaches, together with some remarks on

Table 15.4 Electrolyte EoS

EoS	Reference	Non-electrolyte term (SR part)	Electrolyte term (LR part)	Application range
e-PACT	Jin and Donohue[48–50]	PACT	MSA + corrections	$\gamma_{\pm}$, densities and SLE for 50 water–single and mixed salts, including one triple salt system Weak electrolytes (water with ammonia, CO_2, SO_2, H_2S)
e-SRK (ion specific)	Fürst and Renon[12] Zuo *et al.*[51,52] Vu *et al.*[53]	Modified SRK	MSA (SI) + SR2 Pottel expression for dielectric constant	(1) 25 °C, binary and ternary osmotic coefficients (28 halide, 35 non-halide systems) (2) VLE and LLE for mixed solvent–water (3) Water–methanol –salts gas hydrate formation
e-SRK (ion specific)	Lin *et al.*[4]	SRK	Truncated DH (as activity coefficient)	25 °C: $\gamma_{\pm}$, Φ, apparent molar volumes and SLE for multicomponent systems containing water, Na^+, H^+, Ca^{2+}, Cl^-, OH^- and $SO_4{}^{2-}$
e-PR (MSW; salt specific)	Myers *et al.*[54]	PR with volume translation	MSA (SE) + Born	$\gamma_{\pm}$, Φ, densities for 138 water–salts at 25 °C and higher temperatures (for seven salts) Φ for two ternary mixed salt–water
e-PR (ion specific)	Lin *et al.*[4]	PR with volume translation	MSA (SE and SI) + Born	25 °C: $\gamma_{\pm}$, Φ, apparent molar volumes and SLE for multicomponent systems containing water, Na^+, H^+, Ca^{2+}, Cl^-, OH^- and $SO_4{}^{2-}$
e-CPA (ion specific)	Wu and Prausnitz[55]	PR + association from SAFT (Wertheim)–CPA variant	MSA (CI) + Born	VLE for water–HC–salts $\gamma_{\pm}$ water–salts Water activity Only NaCl considered
e-CPA (ion specific)	Lin *et al.*[4]	SRK + association from SAFT (Wertheim)–CPA variant	MSA (SI) + Born	25 °C: $\gamma_{\pm}$, Φ, apparent molar volumes and SLE for multicomponent systems containing water, Na^+, H^+, Ca^{2+}, Cl^-, OH^- and $SO_4{}^{2-}$
e-CPA (ion specific)	Inchekel *et al.*[56]	SRK + association from SAFT (Wertheim)–CPA variant	MSA (SI) + Born + SR2	$\gamma_{\pm}$, osmotic coefficients at 25 °C and volumes for 10 aqueous salt solutions (up to 6 M)
e-CPA (salt specific)	Haghighi *et al.*[112]	sCPA	DH (in activity coefficient form)	Freezing point depressions of eight water–salt systems as well as some mixed salt–water

e-SAFT-VR (ion specific)	Galindo *et al.*[9] Gil-Villegas *et al.*[57] Paricaud *et al.*[8] Patel *et al.*[58]	SAFT–VR	MSA (SE)	VLE, densities for nine water–salts (0–100 °C) and one aqueous ternary (mixed salt) Water–methane–salts Water–alkane–salts
e-PC SAFT (ion specific)	Cameretti *et al.*[59] Held *et al.*[60,61]	PC–SAFT	Complete DH	115 salt–water systems VLE, densities, $\gamma_{\pm}$
e-SAFT (SAFT1–RPM; ion specific, one salt-specific parameter)	Tan *et al.*[62] Ji *et al.*[63,64]	SAFT No association between ions	MSA (SE)	$\gamma_{\pm}$, Φ, VLE and densities for water–single and mixed salt for six water–alkali solutions (mostly at 25 °C) VLE and $\gamma_{\pm}$, Φ, for eight mixed salt–water systems SLE for two ternary and one quaternary (mixed salt) systems CO_2–water–NaCl (salting out)
e-SAFT LJ (salt specific)	Liu *et al.*[65–67]	SAFT–LJ Ion–solvent association + dipole–dipole	MSA (low-density expansion for non- primitive model; CI) + ion–dipole	Densities, Φ and $\gamma_{\pm}$ for 30 water– electrolytes (25 °C) $\gamma_{\pm}$ for 13 mixed electrolyte aqueous solutions at 25 °C
e-SAFT LJ	Li *et al.*[43,68]	SAFT–LJ + dipole–dipole No association term is included	MSA	CMC for ionic and non-ionic aqueous surfactant solutions

the possible applicability of the methods. As the approaches of Table 15.4 can be roughly divided into (1) cubic EoS + electrolyte, (2) e-CPA and (3) e-SAFT, we will accordingly group the discussion of the various electrolyte EoS.

Before proceeding with the presentation of the electrolyte EoS, the 'special' McMillan–Mayer framework used in electrolyte thermodynamics will be discussed, compared to the 'classical' Lewis–Randall framework.

15.4.2 Lewis–Randall vs. McMillan–Mayer framework

Activity coefficient models and EoS for electrolyte systems combine short-range terms (known from the previous chapters of the book, e.g. UNIQUAC, NRTL, CPA and SAFT) with long-range terms based on the primitive theory (DH or MSA), where the solvent is not a component but rather a continuous medium characterized by its dielectric constant. Thus, when we combine ('add') these contributions, we combine two different frameworks with different independent variables:

- the Lewis–Randall (LR) framework (independent variables = T, P, mole numbers of all species) for the short-range interactions;
- the McMillan–Mayer (MM) framework (independent variables = T, V, mole number of solute, chemical potential of solvent) for the long-range interactions.

Thus, it appears that some inconsistency exists when adding terms based on different frameworks, and in principle thermodynamic properties from MM should be converted before being added to short-range terms, expressed on the basis of the 'ordinary' LR framework. The literature is rich with extensive (and sometimes animated) discussions[117] on whether such conversions should be considered or not and also on what are the practical implications of this 'apparent inconsistency' by using two different frameworks. Prausnitz *et al.*[2] listed several references mentioning that for single solvent–salt mixtures, the use of the different frameworks does not significantly affect the calculations, but care should be exercised for mixed solvents. Most of the e-EoS presented in Table 15.4, e.g. the MSW[54] and e-SAFT by Radosz and co-workers[62–64], do not include any corrections and comment on the small significance of the MM vs. LR frameworks for calculations and the fact that small inconsistencies may well be absorbed in the adjustable parameters that all models have. One research group (e-CPA by Wu and Prausnitz)[55] did include corrections due to the two different frameworks (as they considered the mixed water–methane solvent), but no comparisons of the results are shown with and without these corrections.

15.5 Comparison of electrolyte EoS: capabilities and limitations

15.5.1 Cubic EoS + electrolyte terms

The Fürst–Renon EoS

The Fürst-Renon[12] EoS is of interest not only because it is one of the first electrolyte EoS to have been extensively used for electrolytes, but also because the authors made several plausible assumptions for reducing the number of parameters. Fürst and Renon considered osmotic activity coefficients for binary and ternary (mixed salt)–water mixtures. The SR2 term has a central role in this model. The water diameter is set to 2.52 Å in both MSA and SR2 terms. The model has, in principle, three types of parameters: the ionic co-volumes and the ionic diameters as well as the W_{ik} parameters in the SR2 term (see Table 15.1).

Ionic co-volumes and ionic diameters are correlated:

$$\sigma_i = \left(\frac{6b_i}{\pi N_A}\right)^{1/3} \tag{15.40}$$

The interaction parameters (W) between anions (aa), between cations (cc) as well as between anions and water (aw) are set to zero (due to the lower solvation of anions), leaving only the cation–anion (W_{ca}) and cation–water (W_{cw}) interaction parameters to be determined from experimental data.

Using experimental osmotic coefficients for numerous halide solutions (Cl^-, Br^-, I^-), Fürst and Renon showed that the co-volumes and interaction parameters in their EoS are strongly correlated and can be related to the Stokes diameter for cations (σ_c^S), the Pauling diameters for (the less solvated) anions (σ_c^P), or their combinations:

$$b_c = \lambda_1(\sigma_c^S)^3 + \lambda_2 \tag{15.41a}$$

$$b_a = \lambda_1(\sigma_a^P)^3 + \lambda_2 \tag{15.41b}$$

$$W_{cw} = \lambda_3(\sigma_c^S) + \lambda_4 \tag{15.41c}$$

$$W_{ca} = \lambda_5(\sigma_c^S + \sigma_a^P)^4 + \lambda_6 \tag{15.41d}$$

When Equations (15.41a–d) are used, the model is essentially predictive (of course the λ_i parameters are fitted to experimental data). For 28 aqueous halide solutions, the percentage deviation is 2.6% for the osmotic coefficients, when Equations (15.41a–d) are used, but it is reduced to 1% when a single parameter (W_{ca}) is adjusted, whereas all other parameters are given by Equations (15.41a–c).

With few exceptions, equally satisfactory results are obtained for other salts (containing anions other than Cl^-, Br^- and I^-), but improved results can be achieved if special parameters λ_5, λ_6 are fitted for each anion family together with the W_{ca} parameter. Even better results can be obtained when the anionic diameter is adjusted as well. Using parameters solely from the binary water–salt systems, the Fürst–Renon EoS predicts very satisfactorily the osmotic activity coefficients for 30 mixed salt–water systems. Typically the deviations in osmotic coefficients range between 2 and 5%.

In later publications,[51,52] the Fürst–Renon e-EoS has been applied with success to VLE, activity coefficients and LLE for various mixed solvent–water mixtures. VLE for 31 mixed solvent–water mixtures and LLE for 13 mixed solvent (water–butanols, water–ketones)–salts (NaCl, KBr, KCl) have been considered.

The Myers et al.[54] *EoS*

In contrast to Fürst and Renon, Myers *et al.*[54] developed a salt-specific e-EoS (abbreviated here as MSW) which they applied to 138 water–salt systems (and a few mixed salt–water ones). They used the restrictive primitive model version of MSA where all ion diameters are assumed to have the same values. Pure water is described with three temperature-dependent parameters to ensure good representation of vapor pressures and liquid densities. Unlike Fürst and Renon, Myers *et al.*[54] used the pure water dielectric constant, ignoring the effect of salt concentration in the dielectric constant. They considered first the mean ionic and osmotic

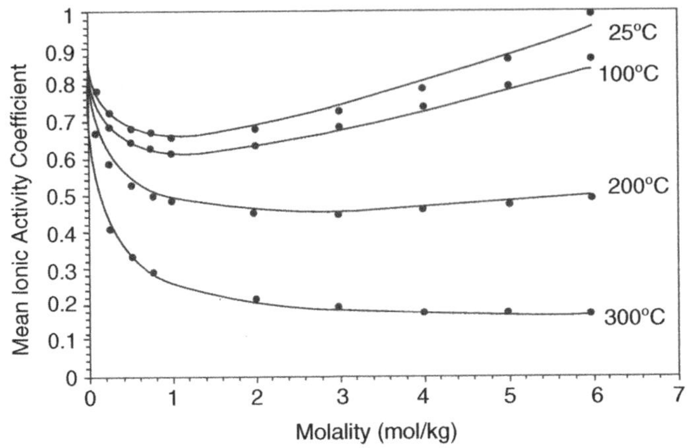

Figure 15.7 *Mean ionic activity coefficients for water–NaCl at various temperatures using the Myers et al.*[54] *EoS. Similar results are obtained for other aqueous salt solutions. Reprinted with permission from Ind. Eng. Chem. Res., An Equation of State for Electrolyte Solutions Covering Wide Ranges of Temperature, Pressure and Composition by Jason A. Myers, Robert H. Wood and Stanley I. Sandler, 41, 13, 3282–3297 Copyright (2002) American Chemical Society*

activity coefficients at room temperature. The equation also has a volume translation but this is not used for salts and, moreover, the interaction parameter k_{ij} between water–ions is set to zero. Then, the MSW EoS has three salt-specific parameters (the energy and co-volumes as well as the diameter) which are fitted to activity coefficient data at 25 °C. With three adjustable parameters per salt, a very good representation is achieved (deviations around 2%), even at much higher molalities ($M > 6$) than those used in the parameter estimation. The MSW EoS was also used for modeling densities, osmotic and mean ionic activity coefficients at higher temperatures for seven water–salt systems, but then both the energy parameters and the diameters are made temperature dependent and a non-zero interaction parameter between water and salt is required. The representation is satisfactory in the 0–300 °C and 1–120 bar temperature and pressure ranges by employing six adjustable salt-specific parameters. A typical example is shown in Figure 15.7. It is possible to reduce somewhat the number of adjustable parameters by utilizing a correlation between the salt co-volume and diameters.

Lin *et al.*[4] further tested the Myers *et al.*[54] EoS using either the simplified explicit or implicit versions of MSA. They used an ion-specific version of MSW, with three adjustable parameters for ions (the energy and co-volume parameter of Peng–Robinson as well as the ion diameter). The volume translation is zero for ions. They also employed k_{ij} for interactions between ions. The Lin *et al.*[4] version of the MSW EoS was tested for multicomponent systems containing water, Na^+, H^+, Ca^{2+}, Cl^-, OH^- and SO_4^{2-} and a variety of properties (mean ionic and osmotic activity coefficients, apparent molar volumes and SLE). The discussion of the results is presented in the next subsection, as they compared this EoS to an e-CPA they had also developed.

15.5.2 e-CPA EoS

Three different extensions of CPA to electrolytes have been presented by the group of Prausnitz[55] and recently by Thomsen and co-workers[4] and by the group of de Hemptinne.[56] There are notable similarities between all three models (besides using the CPA as the 'base' model for describing attractive, repulsive and association contributions):

1. They all use both MSA and Born terms.
2. All three models employ implicit forms of MSA, which do not assume the same ionic diameter.
3. They are all ion-specific models.

There are also some important differences:

1. CPA by Wu and Prausnitz[55] is the only one of the three models to employ the complete version of MSA. To our knowledge, it is one of the few known electrolyte EoS which uses the complete version of MSA.
2. Only the French group's e-CPA[56] includes an additional SR2 term for short-range ionic interactions.
3. All models use association for water (the 4C scheme in the two most recent versions, the 3B scheme by Wu and Prausnitz)[55] but only the e-CPA by Wu and Prausnitz employs association between water and ions (assigning 10 sites for cations and 14 for anions). This water–ion association approximates the electrostatic interactions between water and ions.
4. Only Lin *et al.*[4] have considered mixed salts and SLE and, moreover, they have compared various electrolyte EoS.
5. The e-CPA by Lin *et al.*[4] contains three ionic parameters (the energy and co-volume as well as the ionic diameter) and an interaction parameter, k_{ij} (between water–ions and in some cases also between ions).
6. In the e-CPA by the group of de Hemptinne[56] and by Wu and Prausnitz[55], the ionic co-volumes, b_i, are not separate parameters but are linked to ionic diameters, σ_i ($\sigma_i = (6b_i/\pi N_A)^{1/3}$). This equation (Equation (15.40)) was also used in the Fürst–Renon EoS.
7. The e-CPA by Wu and Prausnitz[55] contains three adjustable parameters for ions (the acentric factor is zero; the association volume is 0.001 for all ions). The energy and co-volumes are calculated from correlations with the (Lennard-Jones) energy and diameter (Equation (15.44)), but the association energies of Na^+ and Cl^- are fitted to water activities and mean ionic activity coefficients. Mixed solvents require an additional fitted parameter (Equation (15.42)).
8. The e-CPA from the group of de Hemptinne[56] contains two ionic parameters (the SRK energy parameter and the ionic diameter) as well as binary interaction parameters (k_{ij}) between water–ions (the other k_{ij} are zero) or alternatively the W_{ij} interaction parameters of the SR2 term.

Some specific applications of the electrolyte CPA EoS are presented next.

The e-CPA by Wu and Prausnitz[55] has been applied to water–NaCl (water activity and mean ionic activity coefficients) as well as the water–methane–NaCl system (Figure 15.8). In the case of the mixed solvent, the concentration dependency of the dielectric constant is considered, so here the Born term influences the phase equilibria calculations:

$$\frac{1}{D} = \frac{1}{D_w}(1 + k_D x_{HC}) \qquad (15.42)$$

where w refers to water, HC to hydrocarbon, and k_D is a dimensionless constant that depends on the hydrocarbon (and is a parameter fitted to solubility data).

The correlation results are satisfactory up to high temperatures (300 °C) even though the descriptions of the alkane solubility in water and water–methane VLE are not very satisfactory. Moreover, the values chosen for the number of sites associated with ions seem rather high when compared, for example, to the hydration numbers of Na^+ (=6) and Cl^- (=7). In general the approach is promising but the applications so far have been limited to a single solvent and VLE/activity coefficients. It must be established whether the developed correlations can be used for estimating the parameters of other ions, including the association parameters.

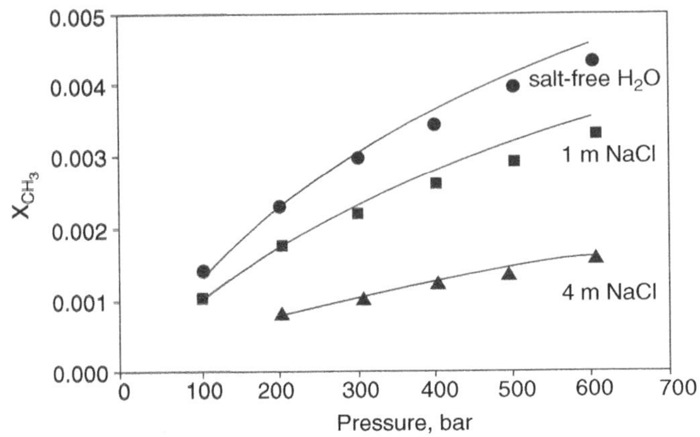

Figure 15.8 *Solubility of methane in pure water and in 1m and 4m aqueous NaCl solutions at 125 °C using the e-CPA EoS by Wu and Prausnitz.[55] Reprinted with permission from Ind. Eng. Chem. Res., Phase Equilibria for Systems Containing Hydrocarbons, Water, and Salt: An Extended Peng–Robinson Equation of State by Jianzhong Wu and John M. Prausnitz, 37, 5, 1634–1643 Copyright (1998) American Chemical Society*

The de Hemptinne group's e-CPA[56] focuses on the activity coefficients and apparent molar volumes of 10 aqueous salt solutions in terms of applications. Most importantly this e-CPA includes an investigation of SR2 and Born terms to account for short-range and solvation effects, as well as the Pottel and Simonin (Equations (15.27) and (15.28)) expressions for the dielectric constant and concentration dependency. Moreover, for two salt–water mixtures, a comparative analysis of the various electrolyte terms is performed. The authors have concluded that:

- The preferred approach is to use the Born term with the Simonin dielectric constant expression, which includes an additional adjustable parameter.
- The optimized ionic diameters are larger than the Pauling diameters but they have reasonable values and follow the expected trends for the cations and anions considered (see Table 15.5 where ionic diameters from other models are also presented).

Table 15.5 *Optimized ionic diameters (in Å) from various e-CPA and e-SAFT EoS. The values are compared to Pauling values and other 'experimental' ionic diameters (accounting for hydration)*

Ion	$\sigma_{Pauling}$ (σ_{exp})	$\sigma_{e\text{-}CPA}$ (Hemptinne)[56] 10 salts	$\sigma_{e\text{-}CPA}$ (Hemptinne)[56] 2 salts	$\sigma_{e\text{-}CPA}$ (Wu)[55]	$\sigma_{e\text{-}CPA}$ (Lin)[4] Full	$\sigma_{e\text{-}CPA}$ (Lin)[4] No SLE data	$\sigma_{e\text{-}PCSAFT}$ (Sadowski)[60]
Li^+	1.2 (1.86)	3.487	—	—	—	—	1.82
Na^+	1.95 (2.40)	3.172	3.104	1.9	2.59	16.05	2.41
K^+	1.98 (3.10)	3.315	—	—	—	—	2.97
Mg^+	1.3	1.770	—	—	—		2.32
Ca^{2+}	2.66	2.614	2.350	—	3.26	6.14	2.88
Cl^-	3.6	4.096	4.095	3.82	4.83	0.88	3.06
Br^-	3.9	4.373	—	—	—	—	3.46
H^+	—	—	—	—	2.76	8.73	2.27
SO_4^{2-}	—	—	—	—	5.11	3.02	2.45
OH^-	—	—	—	—	7.15	0.66	1.64

- The Born contribution is positive and nearly counterbalances the (negative) MSA contribution, thus appearing to be significant in the analysis. If the dielectric constant is assumed to be (salt) concentration independent, then the Born term does not contribute to the activity coefficients. There is hardly any contribution from the association term.
- Excellent results are obtained by using the same ionic diameter values in all expressions where the diameter enters (MSA, SR2, Born).
- The best results are obtained when solvent–ion k_{ij} parameters are used (in the attractive part of SRK), rather than with the W_{ij} parameters, especially for the apparent molar volumes (osmotic activity coefficients are less influenced by the approach used). Except for K^+ and Mg^{2+}, all the other ion–water k_{ij} are negative, of the order of 0.4.

The e-CPA by Lin *et al.*[4] has been tested for the systems containing water and six ions (Na^+, H^+, Ca^{2+}, Cl^-, OH^- and $SO_4{}^{2-}$). All the ionic parameters (and k_{ij}) were regressed simultaneously to aqueous electrolyte solutions at 25 °C (VLE, apparent molar volumes, osmotic coefficients and SLE data). Over 1300 experimental data points were used. SLE phase diagrams have been presented for 10 ternary mixed salt–water systems, e.g. Na_2SO_4 + NaCl + water, Na_2SO_4 + NaOH + water and $CaCl_2$ + NaCl + water (and others containing HCl, $CaSO_4$ and $Ca(OH)_2$) as well as apparent molar volumes for various salts and mean ionic and osmotic activity coefficients for various aqueous salt solutions. The performance of this e-CPA is satisfactory but similar to the e-PR tested by the same authors. It appears that the association term has a small effect. The worst model among those compared by the authors was the e-SRK (especially for the apparent molar volumes) but it must be emphasized that in this e-SRK only an 'activity-coefficient-type' version has been used with a simplified DH equation. The authors concluded that simultaneous representation of activity coefficients, apparent molar volumes and SLE requires that pure ionic parameters and interaction parameters between all ions–water as well as certain ion–ion parameters are simultaneously optimized. If SLE data are excluded, the number of interaction parameters is reduced, e.g. cation–water $k_{ij}=0$ and all ion–ion $k_{ij}=0$. Moreover, if only activity coefficients or apparent molar volumes are considered, then the number of interaction parameters can be further reduced, e.g. all $k_{ij}=0$. The authors found small differences between the various e-EoS they tested for the systems they considered, even when different versions of MSA were employed. They found essentially no reason for including the association term, as e-CPA and the versions of MSW EoS perform very similarly. The results are overall encouraging and for a wide range of properties. Unfortunately, the values of the water–ion interaction parameters are rather high and, moreover, two size parameters are used for ions (co-volume of the cubic EoS, b, and ionic diameters, σ), which must be closely intercorrelated. As can be seen in Table 15.5, the σ values are often not in good agreement with experimental values.

Besides these ion-specific e-CPA EoS, a somewhat different approach based on CPA which uses salt-specific parameters has been recently presented by Haghighi *et al.*[112] These researchers have developed an electrolyte CPA by combining simplified CPA (as described in Chapter 9) with a modified DH term expressed in activity coefficient form. They have essentially used the approach previously described by Aasberg-Petersen *et al.*[47] ($\ln \gamma_i = \ln \varphi_i^{EoS} + \ln \gamma_i^{Electrostatic}$; EoS = CPA and $Electrostatic$ = modified DH). There are no mixture adjustable parameters in the CPA term, but there is a water–salt interaction parameter, which is temperature dependent and salt concentration dependent, in the modified DH term. In total, five parameters per salt are regressed based on the freezing point depression of aqueous solutions in the presence of salt. This e-CPA correlates well the data used in parameter estimation and predicts also the freezing depression curves for some aqueous mixed salt systems. Water vapor pressures in the presence of salts are also shown to be in good agreement with experimental data.

15.5.3 e-SAFT EoS

Various variants of SAFT have been extended to electrolytes. SAFT–VR and PC–SAFT especially have been thoroughly applied to electrolytic mixtures, though in different ways. In the case of SAFT–VR, the MSA is used, while Sadowski and co-workers[59–61] have used the full DH equation (first row in Table 15.1).

e-SAFT–VR (SAFT–VRE)

The electrolyte version of SAFT–VR[8,9,58] (abbreviated SAFT–VRE) combines SAFT–VR with the restrictive MSA term, and no additional terms are used (polar interactions, Born, ion–dipole, etc.). The only hydrogen bonding interactions allowed are those between water molecules, which are modeled using the 4C scheme. The experimental Pauling ionic radii are used, and thus the only adjustable parameters of the model are the (per ion) dispersive ion–water interaction energy and the range of ion–water interaction. These parameters have been fitted to VLE and densities of nine water–salts and excellent representation of these data is obtained over extensive molality and temperature ranges. However, the activity coefficients (mean ionic or osmotic) and SLE for mixed salt–water systems have not been systematically considered.

In the 2003 publication,[58] the salting out (i.e. decrease of solubility) of methane and heavy n-alkanes in water by strong electrolytes as modeled. Experimental data were only available for water–methane–NaCl, but calculations for other salts and heavy alkanes were also performed, as can be seen in Figures 15.9

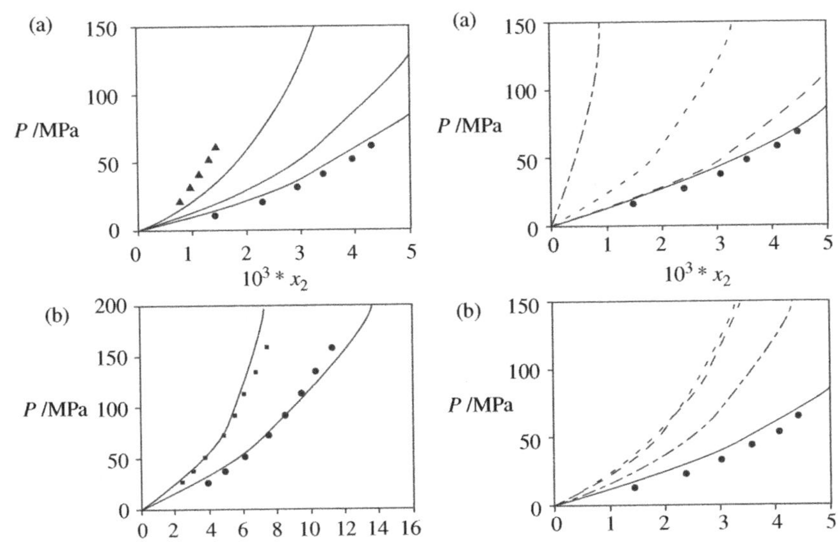

Figure 15.9 Left: Salting out of methane (methane solubilities in water-rich phase) by NaCl for water(1)–methane(2) with the SAFT–VRE EoS at 398.15 K (a) (from bottom to upper curves: no salt, 1 M, 4 M) and 443.15 K (b) (from bottom to upper curves: no salt, 5.2 M). Right: (a) Salting out of methane (methane solubilities in water-rich phase) by LiCl (dashed–dotted curve), NaCl (short-dashed curve) and KCl (long-dashed curve) for water(1)–methane(2) with the SAFT–VRE EoS. The temperature is 398.15 K, 4 M salt concentration; (b) salting out of methane (methane solubilities in water-rich phase) by NaI (dashed–dotted curve), NaCl (short-dashed curve) and NaBr (long-dashed curve) for water(1)–methane(2) with the SAFT–VRE EoS. The temperature is 398.15 K, 4 M salt concentration. The points and solid lines correspond to data and calculations for water + methane without salt. Reprinted with permission from Ind. Eng. Chem. Res., Prediction of the Salting-Out Effect of Strong Electrolytes on Water + Alkane Solutions by B. H. Patel, P. Paricaud, A. Galindo, and G. C. Maitland, 42, 16, 3809–3823 Copyright (2003) American Chemical Society

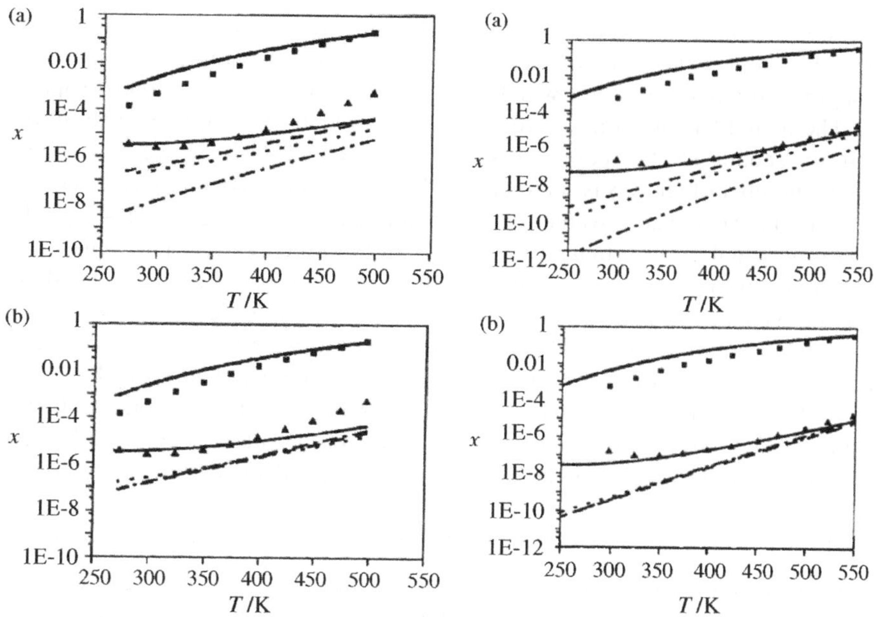

Figure 15.10 *Left: LLE calculations with the SAFT–VRE EoS for water–hexane with and without salt present (upper curves/data: water concentrations in hexane; lower curves/data: hexane solubilities in water): (a) dashed–dotted line, 2 M LiCl; short-dashed line, 2 M NaCl; long-dashed line, 2 M KCl; (b) dashed–dotted line, 2 M NaI; short-dashed line, 2 M NaCl; long-dashed line, 2 M NaBr. The points and solid lines correspond to data and calculations for water + hexane without salt. Right: LLE calculations with SAFT–VRE for water–decane with and without salt present (upper curves/data: water concentrations in hexane; lower curves/data: hexane solubilities in water). Lines as for the left side. Reprinted with permission from Ind. Eng. Chem. Res., Prediction of the Salting-Out Effect of Strong Electrolytes on Water + Alkane Solutions by B. H. Patel, P. Paricaud, A. Galindo, and G. C. Maitland, 42, 16, 3809–3823 Copyright (2003) American Chemical Society*

and 15.10. Two interaction parameters were used per system for adjusting the water–methane/alkane phase behavior (to correct for the cross-dispersion energy and the cross-square-well length terms), while the previously obtained[9] water–ion parameters were used. No alkane–ion interactions are considered. Even with these few adjustable parameters, SAFT–VRE is shown to predict correctly the salting out of methane from water due to NaCl and interesting conclusions are extracted on the salting-out effects for other alkanes and from various salts. There are not much experimental data for water–alkane–salt mixtures, but SAFT–VRE predicts that the compositions of the alkanes in the water phase are seen to be reduced by orders of magnitude upon salt addition. This large salting-out effect may find several applications, e.g. in water purification of oil traces.

e-PC–SAFT

e-PC–SAFT is a recent extension of PC–SAFT to strong electrolytes (salt–solvent including mixed salts) which is based on the full DH equation. The authors intend to develop the model further for aqueous protein solutions containing electrolytes, and the first results have been recently presented.[59–61] It is the only well-known e-EoS which employs the DH rather than the MSA equation. No additional terms (Born, SR2) are used. The association term of PC–SAFT is used only for water, which is assumed to have only two association sites (2B scheme). In the most recent publication,[60] a four-parameter temperature dependency is used for the segment diameter of water which ensures excellent representation of vapor pressures and liquid densities. The model has two ion-specific parameters which are simultaneously fitted to numerous experimental salt–water data. The two parameters are the ionic ('hydrated') diameter and the ion dispersion energy (the segment number of all ions is set to unity). The well-known Berthelot rule is used for the cross-dispersion interaction between water–water and water–ion pairs ($\varepsilon_{ij} = (\varepsilon_i \varepsilon_j)^{1/2}(1 - k_{ij})$), but all interaction parameters are set equal to zero; that is, it is sufficient to fit the ion dispersion energy alone. Ion–ion dispersion interactions are neglected. In the first publication[59] ion parameters for 12 alkali halides and sulfates were fitted to vapor pressures and liquid densities. These properties as well as vapor pressures for two aqueous mixed salt solutions are well represented. As the mean ionic activity coefficients are not represented satisfactorily in this way, the (ionic) parameters for these and additional salts were refitted in the most recent publication[60] using both densities and activity coefficients. In total e-PC–SAFT has been used for 115 single salt systems with very good results for densities (0.8%) and VLE (3.3%) and satisfactory mean ionic activity coefficients (9.2%). Some typical results are shown in Figures 15.11 and 15.12.

In addition to the satisfactory correlation results, e-PC–SAFT shows further positive features:

- Good predictive results for RbCl, without using data for this salt in the parameter estimation.
- A description of the reversed mean ionic activity coefficients series for alkali hydroxides and fluorides.
- Physically meaningful values for the ion diameters (close to experimental hydrated diameter values, see Table 15.5) and ion dispersion energies, e.g. when presented for various salt families or ion series.
- Physically correct trends of mean ionic activity coefficients, e.g. smallest alkali cations yield higher activity coefficients due to the stronger hydration.

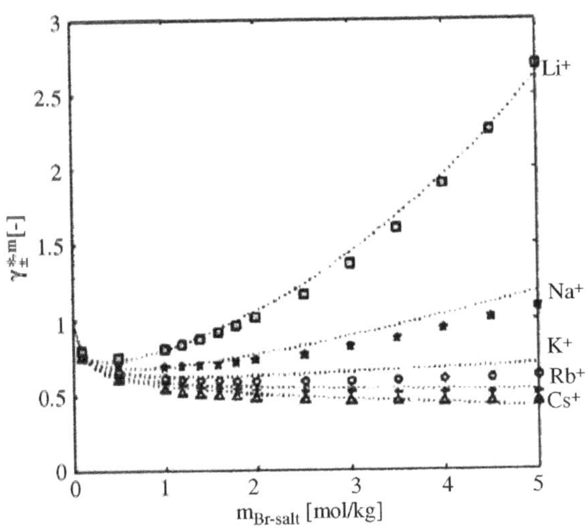

Figure 15.11 *Mean ionic activity coefficients of five bromide salts in aqueous solutions at 25 °C using e-PC–SAFT (lines) as a function of salt molality: squares, LiBr; stars, NaBr; circles, KBr; crosses, RbBr; triangles, CsBr. Activity coefficients decrease with increasing size of the cation: $Li^+ > Na^+ > K^+ > Rb^+ > Cs^+$. Reprinted with permission from Fluid Phase Equilibria, Modeling aqueous electrolyte solutions: Part 1. Fully dissociated electrolytes by Christoph Held, Luca F. Cameretti, Gabriele Sadowski, 270, 1–2, 87–96 Copyright (2008) Elsevier*

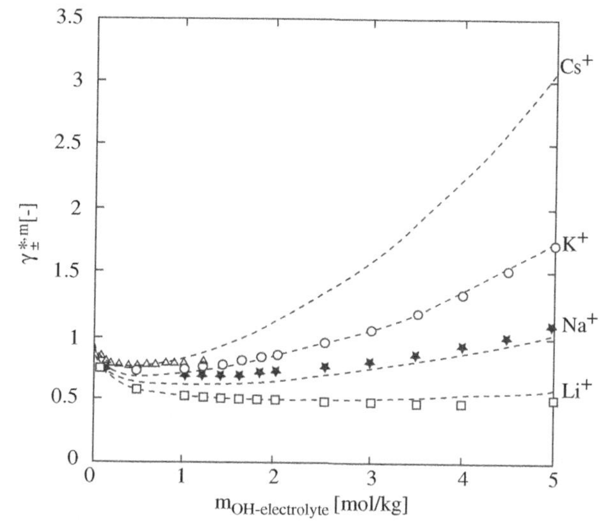

Figure 15.12 *Mean ionic activity coefficients of aqueous solutions of four hydroxides at 25 °C using e-PC–SAFT (lines) as a function of salt molality: squares, LiOH; stars, NaOH; circles, KOH; crosses, RbOH; triangles, CsOH. Activity coefficients increase with increasing size of the cation: $Cs^+ > K^+ > Na^+ > Li^+$. Reprinted with permission from Fluid Phase Equilibria, Modeling aqueous electrolyte solutions: Part 1. Fully dissociated electrolytes by Christoph Held, Luca F. Cameretti, Gabriele Sadowski, 270, 1–2, 87–96 Copyright (2008) Elsevier*

Other e-SAFT EoS

Besides the electrolyte versions of SAFT–VR and PC–SAFT, other SAFT variants have also been extended to electrolytes (see Table 15.4) and some of them present special characteristics and/or applications that deserve discussion.

The group of Radosz[62–64] have developed an e-SAFT extension of their own SAFT variant, called SAFT1. They used the common-ion diameter version of MSA, i.e. the so-called restrictive primitive model. The association term of SAFT is also employed in this e-SAFT, though only for water, which is assigned the 4C scheme. The dispersion term accounts solely for water–ion and water–water dispersion interactions (but no binary interaction parameters are used). The novelty of the approach lies mostly in the parameter estimation. These researchers have developed a hybrid approach using: (1) ion-specific parameters (three: the ion segment volume and ion segment dispersion energy, which are functions of temperature, as well as the range λ of the square-well potential); and (2) one salt-specific parameter (the salt hydrated diameter, which may also absorb the water–ion cross-association).

The authors comment that this approach is a compromise combining the positive features of both ion-specific (transferable parameters) and salt-specific (often better fit) approaches. Very good results are obtained for densities, mean ionic and osmotic activity coefficients for several salt–water systems at 25 °C, but the authors find that it is crucial to use the 'more sensitive' mean ionic activity coefficients for the parameter estimation and, moreover, that simultaneous parameter estimation is needed for the various salt–water mixtures in order to obtain reliable universal ionic diameters and satisfactory results. The hydrated diameters obtained are of the order of 3.6–5.8 Å (similar to those reported by Myers *et al.*[54]). The authors also carried out a brief analysis of the effect of the various terms of the EoS; they concluded that the ionic term has little effect on vapor pressure and density calculations (which are determined mostly by the short-range interactions), but it is of crucial importance for the mean ionic activity coefficients. They presented one example (NaCl–water at 3.26 M and 25 °C) where the contribution of the dispersion and association terms in the residual Helmholtz energy is almost equally large and in total of a value about 12, whereas the ionic term's contribution is only 0.064. Using the following mixing rule (based on salt mole fractions on a solvent-free basis) for the salt hydrated diameter, the authors have extended the e-SAFT1 EoS to eight mixed salt–solvent mixtures:

$$d = \sum_j \sum_i x'_i x'_j d_{ij}$$
$$d_{ij} = \frac{d_i + d_j}{2}(1 - l_{ij}) \qquad (15.43a)$$

The mixture interaction parameter l_{ij} corresponds to the interaction between two salts and has been fitted to experimental data (osmotic activity coefficients). The values vary quite a lot, between 0.2 and 0.9, and there is also one negative l_{ij} value (−0.6 for water–NaCl–LiCl). The representation of osmotic activity coefficients and densities is satisfactory, including a single quaternary system (NaCl–KCl–LiCl–water) which was used for predictions. As something new compared to most other investigations with e-association EoS, the authors also considered salt solubility (SLE) for two of the mixed salt systems (NaCl–KCl–water and NaBr–KBr–water). The solubility product, K_{sp}, is taken from experimental data. Excellent results are obtained for the NaCl–KCl system (Figure 15.13), but the solubility is overestimated for the NaBr–KBr system, especially at high ionic strengths (where experimental solubility product data are not available).

e-SAFT1 has been also shown[64] to represent well CO_2–water–NaCl systems (including accurate representation of CO_2–water; both CO_2 and water solubilities). For the ternary mixture a single salt–CO_2 parameter is needed. To achieve these successful results, Ji *et al.*[64] assumed that CO_2 is self-associating with three association sites (3B). Moreover, the cross-interaction association energy and volume parameters (between

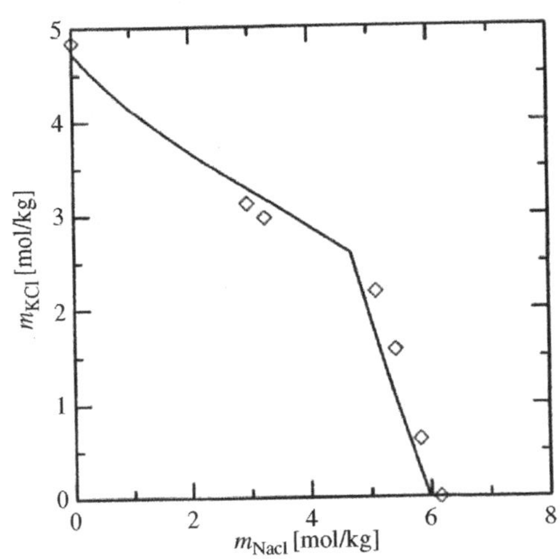

Figure 15.13 *Experimental (points) and calculated (line) solubility (molality) for the NaCl–KCl–water system using the e-SAFT1 EoS. Reprinted with permission from Ind. Eng. Chem. Res. X. Ji, S.P. Tan, H. Adidharma, M. Radosz, 44, 7584. Copyright (2005) American Chemical Society*

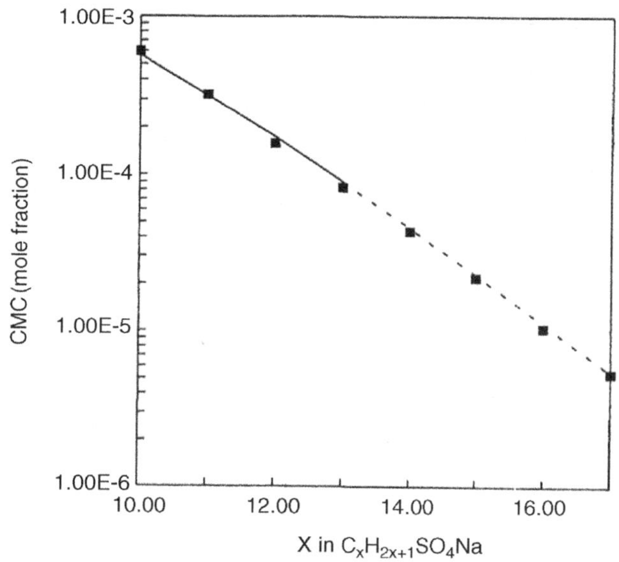

Figure 15.14 *Calculated (solid line) and extrapolated (dashed line) predicted CMC values for aqueous surfactant solutions at 40 °C using the SAFT–MSA method by Li and co-workers. Reprinted with permission from Fluid Phase Equilibria, Study on ionic surfactant solutions by SAFT equation incorporated with MSA by Xiao-Sen Li, Jiu-Fang Lu and Yi-Gui Li, 168, 1, 107–123 Copyright (2000) Elsevier*

Table 15.6 *Percentage average absolute deviations in mean ionic and osmotic activity coefficients from various electrolyte EoS. MSW is the Myers et al.*[54] *EoS*

Reference	Jin $\gamma_\pm$	MSW $\gamma_\pm$	Fürst Φ	Liu *et al.*[66] $\gamma_\pm$	Liu *et al.*[65]	Comparison basis
Liu *et al.*[67]	2.3	0.26	0.71	1.75	0.44 ($\gamma_\pm$) 0.34 (Φ)	15 water–salt at 25 °C Typical M: 0.1–6
Reference	e-SAFT1 $\gamma_\pm$ (one salt parameter, three ion parameters)		MSW γ_+ (three salt parameters, no ion parameters)		Liu *et al.*[66] $\gamma_\pm$ (two salt parameters, one ion parameter)	Comparison basis
Ji *et al.*[64]	0.6		0.12		0.23	Six water–salt at 25 °C Typical M: 0.1–6
Reference	MSW $\gamma_\pm/\Phi$	mMSW $\gamma_\pm/\Phi$	e-CPA $\gamma_\pm/\Phi$		SRK + DH $\gamma_\pm/\Phi$	Comparison basis
Lin *et al.*[4] (parameters based on activity coefficients, densities and SLE data)	2.0/1.8	1.2/1.4	1.2/1.8		1.7/2.3	Salts containing six different ions

15.5.4 Ionic liquids

Ionic liquids (ILs) have received much attention over the past 15 years for use as environmentally benign reaction and separation media. ILs are molten salts with melting temperatures close to room temperature. They have very small (almost negligible) vapor pressure and are non-flammable and odorless. Being salts, it would be natural that e-SAFT EoS would have been applied to describe mixtures with ionic liquids. So far, one SAFT variant has been applied systematically to gas solubilities and water–IL mixtures, but without consideration of the electrostatic effects (no ionic term is used).[108–111] The EoS used is the tPC–PSAFT EoS described previously, in Section 13.6.2.

Economou and co-workers have made a number of assumptions in order to estimate the parameters of ILs, since pure vapor pressure data are essentially non-existent. All ILs were assumed to have the same dipole moment as methanol (= 1.7 D), and cross-association (Lewis acid–Lewis base) interactions are considered between ILs and CO_2. In the first publication,[108] the authors estimated the IL parameters from thermodynamic data (density, enthalpy, entropy of dissolution of CO_2) as well as physicochemical data (size, polarizability, number of electrons), using Equation (15.44). First the cation and anion values are estimated and then the IL segment energy and volume parameters are calculated from combining rules. The CO_2–IL association interaction parameters are calculated from the enthalpy and entropy of dissolution of CO_2 in IL and are found to be 2200 K and 0.0017, respectively, close to the values of 1-alcohols. Satisfactory correlation of CO_2 + IL VLE is achieved over extensive T, P ranges using large values (around 0.2) of temperature-dependent interaction parameters.

For this reason, the authors have subsequently modified[109,110] the IL parameters by requiring the model to predict very low vapor pressure values over the wide temperature range considered. This approach appears to result in substantially better IL parameters. The gas solubility in imidazolium-based ILs is correlated well with much lower values of the interaction parameter (around 0.03, depending on the IL considered). The new IL parameters obtain such values that make the model assume chain-like or polymer-like characteristics for the IL molecules, which may be justified by the similarities of polymer-containing and IL-containing phase diagrams. Equally satisfactory results have been obtained for other gas + IL systems as well as the ternary system CO_2–acetone–[bmim][PF_6], with the IL–acetone parameter adjusted to the ternary data. Recently,[111] tPC–PSAFT has been applied with success to VLE of ILs with both non-polar and polar solvents as well as the ternary CO_2–water–1-butyl-3-methylimidazolium nitrate. The two nitrate-based ILs studied were considered to be self-associating molecules. The association parameters of the ILs were obtained from experimental data for the water–IL association energy or assuming that they are equal to the association parameter of water.

The CO_2–IL parameters were fitted to ternary data and the authors concluded that explicit account of IL dissociation in water via inclusion of an electrostatic term in the model appears necessary for obtaining improved and especially predictive results.

15.6 Thermodynamic models for CO_2–water–alkanolamines

15.6.1 Introduction

Alkanolamines for CO_2 and H_2S removal from flue gases and natural gas

Aqueous alkanolamine solutions are expected to represent for several years to come a mature and accepted technique for post-combustion CO_2 capture, but neither the selection of optimum solvents, especially when it comes to mixed solvents, nor the overall design of absorber–desorber units for all conditions have been optimized.[96] Data are lacking, especially pilot plant data and all types of data for mixed alkanolamines.

Table 15.7 Overview of the capabilities of alkanolamines

Amine	Absorption capacity	Reaction rate	Side reactions	Energy demand (MJ/kmol CO_2)	Solution loading (mol/mol)
MEA	Highest	Highest and formation of carbamates	SO_2, NO_2 react and form heat-stable salts	210 (140 in inhibited version) – highest! – due to high heat of reaction	0.25–0.45
DEA		High and formation of carbamates			0.4–0.8
MDEA	Lowest	Lowest (no carbamate formation, no direct reaction to CO_2)		Lower regeneration costs compared to primary amines (lower bicarbonate heat of formation than carbamate)	0.8

Table 15.7 gives an overview of the capabilities of various alkanolamines, illustrating that a compromise is needed in many cases. The structures of many important alkanolamines are shown in Figure 15.15.

MEA is the industrial standard today (used in most commercial plants), being the most stable alkanolamine with the highest absorption capacity and reaction rate. It has found widespread use despite its high heat of reaction (which means high stripper energy consumption), the fact that it can form degradation products with COS and CS_2 and that the CO_2 absorption capacity is reduced by the presence of SO_2 and O_2 in the flue gas (which should therefore be separated first). On the other hand, the most serious problem in this alkanolamine use is the high cost of the process, both the high capital costs and operating costs. The former are due to the absorption and desorption towers (actually many studies indicate that the capital cost of capture is of a similar level as the power plant itself!), and the latter are related to the energy requirement for the regeneration of the solvent, corresponding to 80% of the total cost. There is a paramount need for developing optimum alkanolamines or alkanolamine blends which can substitute MEA in an economical way. Thus, the search for new solvents implies a thorough understanding of alkanolamine (and of other chemicals) thermodynamic behavior over extensive temperature and pressure ranges. Indeed, for the design of efficient processes for the separation of CO_2 from flue gases in coal-fired power plants it is important to know the partial pressure of CO_2 over aqueous solutions containing single or mixed alkanolamines used for the capture. This is important, for example, for the quantification of the energy required for the regeneration of the alkanolamine and the design of absorber–desorber units in general. Compounds such as ammonia, hydrogen sulfide, carbon dioxide are classified as weak electrolytes and their thermodynamic behavior in water and other chemicals is affected by knowledge of both the physical and chemical equilibria involved.

The absorption of CO_2 in aqueous solutions of alkanolamines is described via a series of chemical reactions which greatly favor the solubility of the acid gas. Therefore, a thorough understanding of the chemical reactions in the system is of great importance in the thermodynamic modeling.

Irrespective of the alkanolamine present, the following reactions in the solution must be accounted for:

Ionization of water:

$$2H_2O \overset{K_1}{\longleftrightarrow} H_3O^+ + OH^- \tag{15.45a}$$

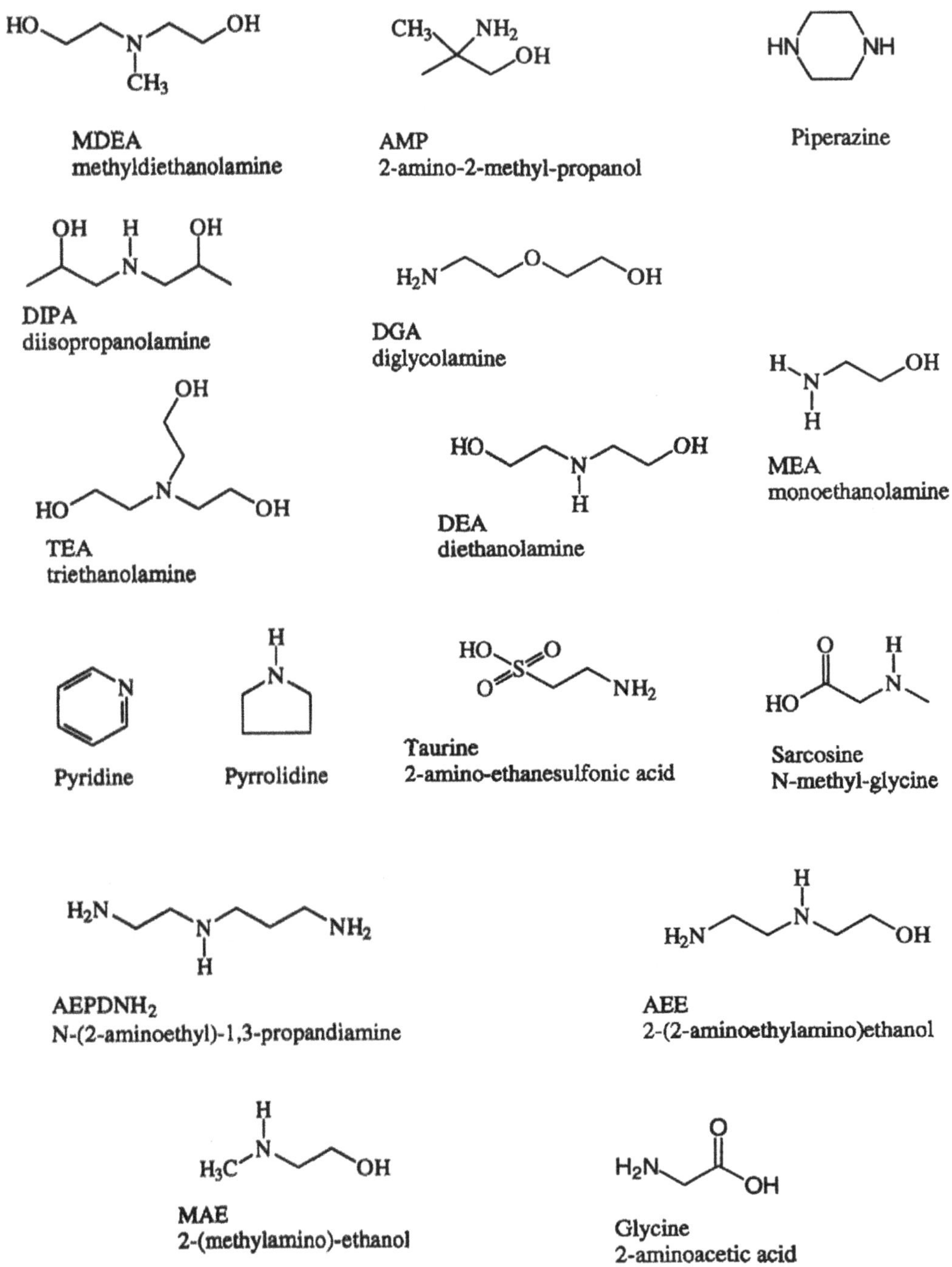

Figure 15.15 *Structure of some important alkanolamines. From Gabrielsen*[95a]

Hydrolysis and ionization of dissolved CO_2:

$$CO_2 + 2H_2O \overset{K_2}{\longleftrightarrow} H_3O^+ + HCO_3^- \qquad (15.45b)$$

(a similar equation can be written for the dissociation of hydrogen sulfide).

Dissociation of bicarbonate:

$$HCO_3^- + H_2O \overset{K_3}{\longleftrightarrow} H_3O^+ + CO_3^{2-} \qquad (15.45c)$$

(a similar reaction can be written for the dissociation of bisulfide).

Dissociation of a protonated alkanolamine (e.g. for the dissociation of protonated MDEA):

$$MDEAH^+ + H_2O \overset{K_4}{\longleftrightarrow} H_3O^+ + \mathrm{MDEA} \qquad (15.45d)$$

Similarly for MEA:

$$MEAH^+ + H_2O \leftrightarrow \mathrm{MEA} + H_2O^+ \qquad (15.45e)$$

For several alkanolamines there are no additional reactions. For example, there is no direct N–H bond in the MDEA molecule; hence, no carbamate ion is formed in the solution due to the absorption of CO_2. Other alkanolamines, e.g. AMP, are also assumed not to form stable carbamates.

But for MEA and DEA, an important additional reaction is the reversion of carbamate to bicarbonate, e.g. for MEA:

$$MEACOO^- + H_2O \leftrightarrow \mathrm{MEA} + HCO_3^- \qquad (15.45f)$$

Overview of thermodynamic models

Mostly, activity coefficient models and EoS suitable for electrolytes are used for CO_2 or H_2S–water–alkanolamines and several important approaches are summarized in Tables 15.8 and 15.9. Except for the SAFT model of Button and Gubbins[92], where chemical reactions are not considered, all the others are electrolyte models. Instead of using a 'full model' it is possible, as shown by Danckwerts and McNeil[105] and later by Kent and Eisenberg[95b] (KE) to represent approximately the CO_2 and H_2S partial pressures over aqueous solutions of some alkanolamines using a very simple methodology, which is essentially the first widely used approach for mixtures containing acid gases and aqueous alkanolamines. According to this approach, all activity coefficients and vapor phase fugacity coefficients are assumed equal to one, i.e. both ideal gas and liquid are assumed and two chemical equilibrium constants are fitted to experimental data. The KE approach has been applied to MEA, DEA and MDEA but not to mixed alkanolamines.

Most of the electrolyte models in Tables 15.8 and 15.9 are rather complex, in the sense that they require a rather large number of adjustable parameters which must be fitted to experimental data. Chemical and phase equilibria are described simultaneously, with the liquid phase being a solution containing weak electrolytes, with a substantial ionic strength. Table 15.10 presents some models used in commercial process simulators. Some of the most promising approaches will be briefly described later in the chapter (Sections 15.6.3

***Table 15.8** Activity coefficient models applied to CO_2–water–alkanolamines. e-NRTL and e-UNIQUAC are typically combined with a cubic EoS like SRK for the vapor phase. T = temperature. The abbreviations of alkanolamines are given in Figure 15.15*

Model	Reference	Application	Comments – accuracy
Pitzer	Silkenbaumer *et al.*[69]	AMP, MDEA, AMP/MDEA blend	Satisfactory results for mixed alkanolamines based on results for single alkanolamines
	Kamps *et al.*[70]	MDEA (also H_2S)	
Li–Mather (Clegg–Pitzer for liquid phase–virial or ideal gas vapor phase) 11 T-dependent, 27 T-independent parameters → 130 parameters if all seven species are included	Deshmukh *et al.*[71] Xu *et al.*[72] Weiland *et al.*[73] Li and Mather[74] Teng and Mather[75]	MEA, MDEA, DEA, AMP, mixed (MEA/MDEA, MDEA/AMP)	7% for MDEA 30% for mixed alkanolamines
e-NRTL 11 T-dependent parameters	Chen *et al.*[30]		
	Chen and Evans[28]		
	Posey and Rochelle[76]	MDEA (also H_2S)	
	Cullinane *et al.*[77]	PZ	
	Austgen *et al.*[78]	MEA, DEA	25–120 °C
	Austgen *et al.*[79]	MDEA/MEA MDEA/DEA PZ/MDEA	Loading 0–1
		MEA, DEA, MDEA, AMP, PZ, MEA/MDEA, PZ/MDEA, AMP/MDEA	30% for mixed amines
	Liu *et al.*[66]	MEA	
	Bishnoi and Rochelle[80]	MDEA	
	Bishnoi and Rochelle[80]	PZ/MDEA	30% in P
	Aroua *et al.*[81]	AMP, MDEA and blends	
	Addicks[82]	MDEA	25–200 °C Loadings up to 8 M
	Barreau *et al.*[83]	DEA	310–394 K Loadings up to 1.2
e-UNIQUAC	Kaewsichan *et al.*[84]	MEA, MDEA, MEA/MDEA blend (also H_2S)	25–120 °C Loading up to 0.1
	Addicks[82]	MDEA	25–200 °C Loadings up to 8 M
	Thomsen and Rasmussen[22]	Ammonia–water	20 °C, 0.1–2 M
	Thomsen[23]		80 °C, 0.6–12 M
	Garcia *et al.*[85]	CO_2–water	
	Faramarzi *et al.*[86]	MDEA, MEA and MEA/MDEA blends	

Table 15.9 *EoS applied to CO_2–water–alkanolamines*

Model	Reference	Application	Comments – application range
Fürst and Renon Wong–Sandler mixing rules (+ NRTL)	Vallee *et al.*[87]	DEA (also for H_2S)	37–107 °C 0.5–3.5 M, up to loading 2.34
	Chunxi and Fürst[88]	MDEA (also for H_2S)	25–120 °C Loadings up to 2 10–40% deviation
Based on SRK using Huron–Vidal mixing rules Also used a Born term (in total three ionic terms)	Solbraa[89] Huttenhuis *et al.*[118]	MDEA	10% in pressure
e-LCVM	Vrachnos *et al.*[90–91]	MDEA MEA, MDEA, MDEA/MEA	298–393 K Loading up to 1.4
SAFT	Button and Gubbins[92]	MEA, DEA	No electrolyte term is used! 40% deviation in CO_2 loading

and 15.6.4), but first the Gabrielsen model,[93–95a] which is a recent extension of the KE approach, will be presented in Section 15.6.2.

15.6.2 The Gabrielsen model

Gabrielsen *et al.*[93,94] have developed a simple model of the Kent–Eisenberg[95b] type for estimating the partial pressure of CO_2 for aqueous solutions of MEA, DEA, MDEA, PZ and AMP. In this work we will summarize the major results but the improved correlations from Gabrielsen[95a] will be presented, which also include correlations for PZ/MDEA blends. This rather simple approach assumes that only one chemical reaction equilibrium is taken into account and Henry's law and chemical reaction equilibrium constants are combined into a single constant. The major assumption of this approach is that for aqueous solutions of primary and secondary alkanolamines (MEA and DEA) the dominant reaction is that forming carbamate, while for tertiary and sterically hindered alkanolamine (MDEA, AMP), as well as for the diamine PZ, the main reaction product

Table 15.10 *Models used in commercial process simulators*

Reference	Model	Simulator	Applications
	Li–Mather	ANSIM in Pro/II 6.00 and HYSYS	MEA, DEA, TEA, MDEA, DGA + mixed alkanolamines
	Kent–Eisenberg	TSWEET – available also as add-on in ASPEN+	MEA, DEA, TEA, MDEA, DGA, DIPA, AMP
	e-NRTL		
Aroua *et al.*[81]	e-NRTL	ASPEN+	Design of CO_2 absorption using AMP, MDEA and AMP/MDEA blend
Liu *et al.*[66]	e-NRTL	ASPEN+	CO_2 removal using MEA

is bicarbonate. Other by-products are neglected and the final reactions can be written as:[95a]

$$\text{MEA}: \; R_2NH_2^+ + R_2NCOO^- \rightleftharpoons 2R_2NH + CO_2(aq.) \tag{15.46a}$$

$$\text{MDEA}: \; R_3NH^+ + HCO_3^- \rightleftharpoons R_3N + CO_2(aq.) + H_2O \tag{15.46b}$$

Based on these equations and plausible assumptions, Gabrielsen[95a] arrived at the following explicit expressions for the CO_2 pressure in the two cases:

$$\text{MEA, DEA}: \quad p_{CO_2} = K_{CO_2}\frac{X_{CO_2}{}^2}{(a_0(1-2\theta))^2} \tag{15.47a}$$

$$\text{MDEA, AMP, PZ}: \quad p_{CO_2} = K_{CO_2}X_{CO_2}\frac{\theta}{(1-\theta)} \tag{15.47b}$$

where:

$$\theta = \text{loading} = \frac{\text{mole } CO_2}{\text{mole amine}}$$

$$a_0 = \text{initial concentration of amine} = \frac{\text{amine}}{\text{amine} + H_2O}$$

K_{CO_2} = combined Henry's law and equilibrium constant for partial pressure of CO_2 over the solution
X_{CO_2} = mole fraction of chemically bound CO_2 in the solution
T = temperature (K)

The key parameter of the model is K_{CO_2}, the combined Henry's law and chemical equilibrium constant which is given for all alkanolamines by the equation:

$$\ln K_{CO_2} = A + \frac{B}{T} + C\frac{\theta}{T^2} + DX_{CO_2} + E\sqrt{X_{CO_2}} \tag{15.48}$$

The parameters have been fitted to experimental CO_2–water–alkanolamine data and are presented in Table 15.11. Details of the regression and the experimental data used are given by Gabrielsen[95a].

Notice that the first two adjustable parameters, A and B, represent the standard temperature dependence of the chemical equilibrium constant. They are used for all alkanolamine systems. The last three adjustable

Table 15.11 *Regressed parameters for the equilibrium constant used in Equation (15.48)*

System	A	B	C	D	E
MEA–CO_2	26.97 ± 1.37	-8639 ± 546	$-695\,540 \pm 112\,000$	0	0
DEA–CO_2	30.15 ± 2.20	-8839 ± 663	0	-126.2 ± 22.2	0
MDEA–CO_2	28.44 ± 1.59	-5864 ± 500	0	51.11 ± 19.6	-25.41 ± 6.46
AMP–CO_2	29.99 ± 1.29	-7985 ± 425	0	0	0
PZ–CO_2	28.78 ± 3.03	-8323 ± 1030	0	212.0 ± 59.1	0
PZ–MDEA–CO_2	24.13 ± 1.19	-5462 ± 397	0	0	0

parameters, C, D and E, involving the total loading, approximate an ionic strength dependence as suggested by Astarita *et al.*[97] to account for non-idealities in the system. The different systems correlated have different functionalities of the parameters. In general no more than three parameters are needed to correlate the partial pressure of CO_2 over the solution, with the exception of MDEA, which required four parameters.

Selected results are presented in Figures 15.16–15.18 for MEA, MDEA and AMP but similar performance is obtained for the other alkanolamines as well.[95a] It can be concluded that the concentration dependence of the model is well accounted for. Satisfactory results are obtained over an extensive temperature range, although there are uncertainties between experimental data, especially for AMP.

15.6.3 Activity coefficient models ($\gamma - \varphi$ approaches)

The electrolyte NRTL

The model and its parameters

The e-NRTL model is an extension of NRTL (Chapter 5) to electrolytes,[76,78] using NRTL for the short-range (SR) interactions and the PDH equation for representing the long-range (LR) ion–ion interactions. The equations were shown previously (Equations (15.33), (15.34) and (15.37)). The e-NRTL model has (per binary mixture) three parameters in the SR (NRTL) term (two interaction parameters and the non-randomness factor) and five parameters in the LR term (in total seven parameters). When applied to CO_2–water–alkanolamines, many of these parameters are estimated from diverse sources or are given default values. For example:

- distance of the closest approach, ρ (= 14.9, according to Pitzer);
- pure component dielectric constants, D, for alkanolamines (assumed temperature dependent, taken from Ikada *et al.*);[98]

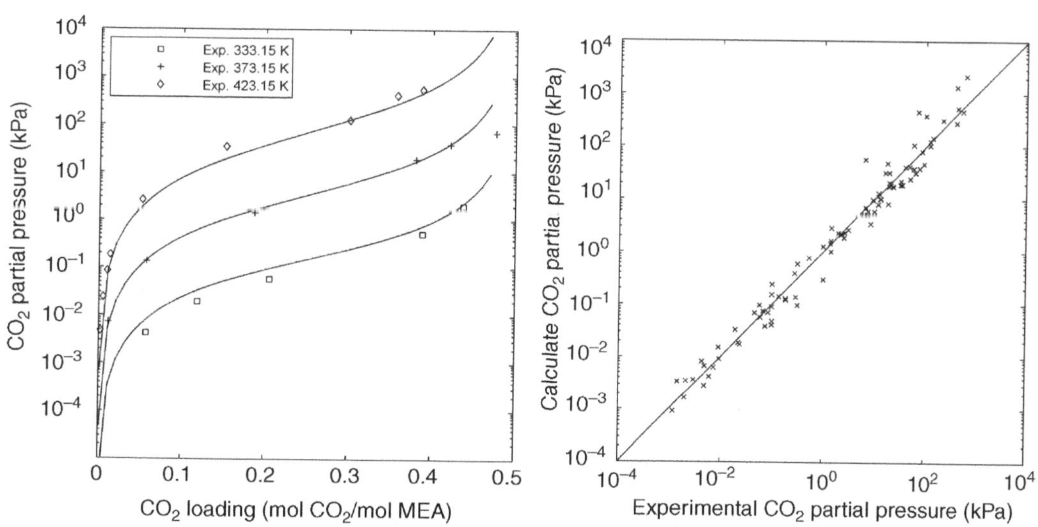

Figure 15.16 Left: Comparison of the Gabrielsen model correlation results (solid lines) with experimental data for CO_2 equilibrium partial pressures over an aqueous 30 wt% MEA solution. Right: Comparison of model correlation with all experimental data for the partial pressure of CO_2 used in the parameter regression for aqueous MEA solutions. From Gabrielsen[95a]

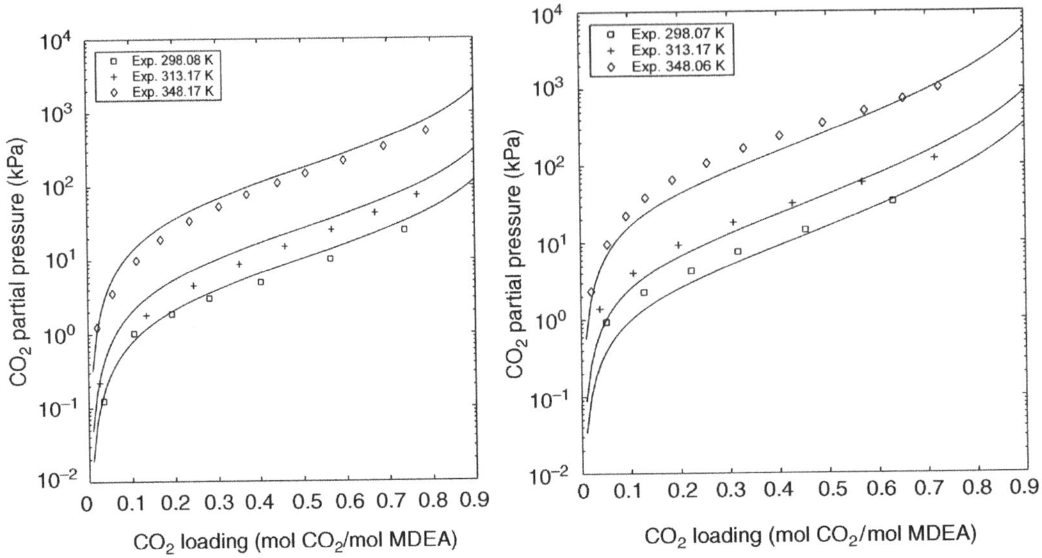

Figure 15.17 *Left: Comparison of the Gabrielsen model correlation results (solid lines) with experimental data for CO_2 equilibrium partial pressures over an aqueous 25.73 wt% MDEA solution. Right: Comparison of model correlation results (solid lines) with experimental data for CO_2 equilibrium partial pressures over an aqueous 46.88 wt% MDEA solution. From Gabrielsen*[95a]

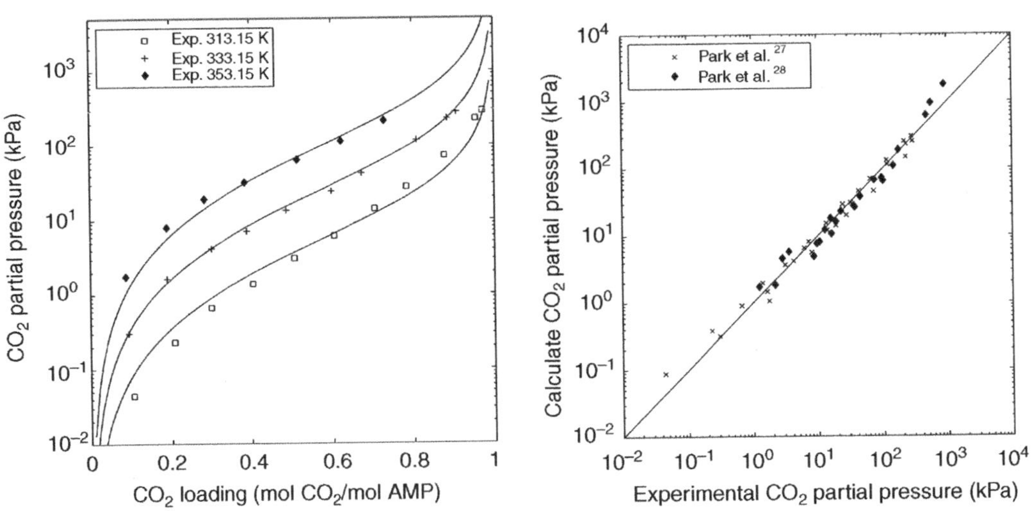

Figure 15.18 *Left: Comparison of the Gabrielsen model correlation results (solid lines) with experimental data for CO_2 equilibrium partial pressures over an aqueous 30 wt% AMP solution. Right: Comparison of model correlation with all experimental data for the partial pressure of CO_2 used in the parameter regression for aqueous AMP solutions. From Gabrielsen*[95a]

- ionic radii, r, in the Born term (assigned to 3 Å);
- NRTL's non-randomness factor, which is set to 0.2 for molecule–molecule and water–ion interactions and 0.1 for alkanolamine–ion and acid gas–ion pair interactions;[28,99]
- all ion pair–ion pair interaction parameters set to zero.

Moreover, acid gas–water and alkanolamine–water molecule–molecule parameters are typically estimated from binary experimental VLE (and other) data.[28] The MDEA–water parameters were set equal to zero in the first applications of the model (by Austgen *et al.*[78]) but they were refitted in the work of Posey and Rochelle.[76,100] Thus, the only fitted parameters to ternary data for CO_2–water–alkanolamines are the NRTL parameters between molecules and ion pairs (and not all of them, because, for species present in small amounts, they can be safely ignored). These molecule–ion pair interaction parameters are assumed in most investigations to be temperature dependent.

Applications

The e-NRTL model has been extensively applied to CO_2 and H_2S–water–alkanolamines, including mixed acid gases and several mixed alkanolamines, and only some of the applications will be highlighted here together with some illustrative examples. In one of the first applications by Austgen *et al.*[78,79], e-NRTL was applied to CO_2, H_2S–water–MEA, DEA, MDEA as well as the blends MEA/MDEA and DEA/MDEA over the 40–80 °C temperature range. Satisfactory results are obtained using 14 temperature-dependent NRTL interaction parameters (for all gases and amines). Mixed amines require one more parameter fitted to blends (MDEA protonated-carbamate MEA or DEA), but there is substantial uncertainty due to the limited data available for mixed alkanolamines.

While Austgen *et al.*[78,79] considered VLE data for various alkanolamines, Posey and Rochelle[76] focused only on CO_2 and H_2S–water–MDEA, but they considered, besides VLE data, pH, conductivity and heat of absorption data in the parameter fitting and model validation. Low acid gas loading predictions may be improved in this way. Another difference from the work of Austgen *et al.* is that Posey and Rochelle[76] fitted MDEA–water interaction parameters to VLE as well as to SLE and heat of mixing data. The model performs well (30% deviation in VLE for CO_2 and 20% in VLE for H_2S) using five temperature-dependent parameters (per gas) and the predictions for mixed gases are satisfactory without additional adjustable parameters (beyond those obtained from single gas systems). The heats of absorption are predicted well up to a certain loading, e.g. 0.7 for CO_2, but e-NRTL values do not approach the physical heat of absorption for H_2S (which is −20 kJ/mol; e-NRTL predicts about −37 kJ/mol).

Liu *et al.*[66] only considered CO_2–water–MEA, but their emphasis was the satisfactory representation of VLE data over the whole temperature range, up to 120 °C, in order to cover both absorber and desorber conditions. They emphasized that this was necessary as Austgen *et al.*'s earlier work revealed a large error in reboiler duties in the desorber, which can be attributed to high VLE errors with e-NRTL at high temperatures. They employed a two-stage estimation procedure, as previously recommended by Weiland *et al.*[73] for other models. According to this procedure, molecule–ion pair parameters were fitted once the molecule–molecule parameters were estimated (temperature dependent). The MEA–water parameters were estimated based on VLE, heat of mixing and heat capacity (but not based on SLE) data; 16 molecule–ion pair interactions are used but in this case they are left temperature independent. Speciation results were similar to those of Austgen *et al.*[78,79] but VLE predictions were better at all temperatures considered.

Recent applications with e-NRTL included other alkanolamines as well, e.g. PZ, AMP and some blend combinations. Moreover, several authors emphasized the use of speciation data for either model development or validation. Rochelle's group[80] have considered CO_2–water–PZ–MDEA and used also NMR to determine speciation. They concluded that at low loadings, the PZ carbamate is the major product, but at higher loadings, the protonated PZ carbamate dominates. NRTL parameters were fitted to both VLE and speciation data.

Aroua *et al.*[81] considered CO_2–water–MDEA, AMP and a mixture of the two. They only fitted NRTL parameters and used in total 16 adjustable parameters, i.e. 8 per amine. The deviations were 15% for MDEA and 20% for both AMP and the blend; no additional adjustable parameters were employed for the mixed alkanoalamines. Finally, a French group[83,101] studied CO_2, H_2S + water + DEA including the mixed gas. Following the usual approach for e-NRTL, they employed many default values, as discussed above, and only fitted the molecule–ion pair temperature-dependent parameters to ternary and quaternary (mixed gas) data. They considered speciation as well and percentage (%) carbamate-loading plots were also presented. The accuracy of the description is good for the single gas systems (between 20 and 25%), but it becomes progressively worse for mixed gases (40–50%), especially at lower loadings.

The extended UNIQUAC

The model was presented previously (Chapter 5, Appendix 5.C), but in the application of Kaewsichan *et al.*[84] ion–pair (salt) specific parameters are used instead of the ion-specific parameters in the e-UNIQUAC by Thomsen and co-workers,[22,23], thus reducing the number of adjustable parameters. Only the system CO_2–H_2S–MEA–MDEA has been considered and the equilibrium constants of the eight reactions involved as well as dielectric constants were obtained (as functions of temperature) from the literature.[78,79] SRK is used for estimating the vapor phase fugacities. A total of 27 temperature-dependent interaction parameters (in the residual part of UNIQUAC) have been fitted to experimental data in a three-step procedure. The parameters between molecules were first adjusted to binary data, and then the molecule–ion pair interactions were fitted to ternary data. Quaternary data were used for estimating the interactions between ion pairs (bisulfide and bicarbonate or carbamate). The performance of the model is satisfactory for all systems considered and over a temperature range of 25–120 °C, and for both single and mixed gases. The highest deviations are in the very loading region. One typical result is shown in Figure 15.19.

Recently, Faramarzi *et al.*[86] applied the extended UNIQUAC (in the ion-specific form presented by Thomsen and co-workers)[22,23] to CO_2–water–alkanolamines (MEA, MDEA and their blend). Particular emphasis has been put on developing the model in order to cover both the lower temperature range of the adsorber and the higher temperatures encountered in the stripper. An extensive database has been compiled which includes over 460 VLE data points for CO_2–water–MEA (in the loading range 0.03–2.15 and $T = 25$–140 °C) as well as 180 data points for water–MEA (VLE, SLE, excess enthalpies in the temperature range −20 to 120 °C). The database, in the case of MDEA systems, included over 710 VLE data points for CO_2–water–MDEA (loading 0.005–1.83, $T = 25$–200 °C) as well as 140 data points for water–MDEA binary (VLE, SLE, excess enthalpies in the range $T = -14$ to 175 °C).

In the case of MEA the adjustable parameters are:

- nine interaction parameters (of which seven are temperature dependent);
- the r and q of MEA, MEAH + and the MEA carbamate;
- the Gibbs energy and enthalpy of formation for MEA, $MEAH^+$ and MEA carbamate

In the case of MDEA the adjustable parameters are:

- four temperature-dependent interaction parameters;
- the r and q of MDEA and $MDEAH^+$;
- The Gibbs energy and enthalpy of formation for MDEA and $MDEAH^+$.

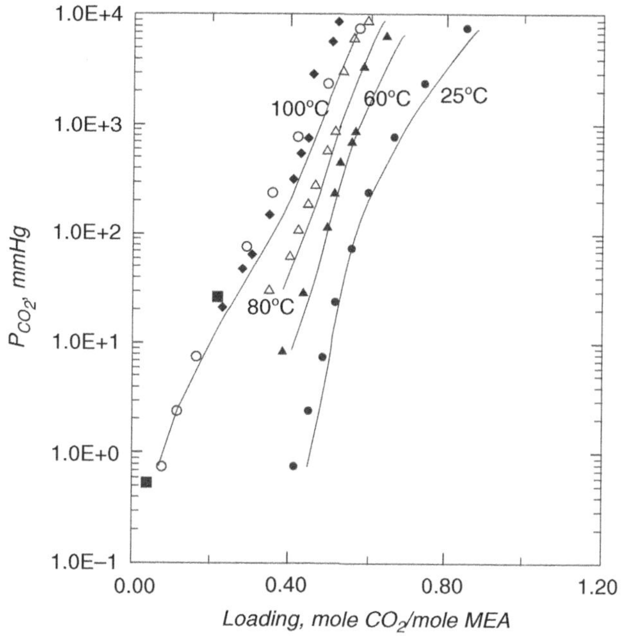

Figure 15.19 *Comparison between model predictions (solid lines) and data (points) for the CO_2 equilibrium partial pressure over a 5.0 M MEA solution with the e-UNIQUAC model. Data are from three different experimental sources are considered over the range 25–100 °C. Reprinted with permission from Fluid Phase Equilibria, Predictions of the solubility of acid gases in monoethanolamine (MEA) and methyldiethanolamine (MDEA) solutions using the electrolyte-UNIQUAC model by Lupong Kaewsichan, Osama Al-Bofersen, et al., 183–184, 1, 159–171 Copyright (2001) Elsevier*

Excellent correlation of the equilibrium pressure is achieved over the entire loading and temperature range. Equally satisfactory results were obtained for the MEA and MDEA loss as well as for SLE and excess enthalpies of the water–MEA and water–MDEA binaries. Finally, using the parameters obtained from the two alkanolamines, satisfactory correlation has been achieved also for the mixed alkanolamine (MEA/MDEA) system in the temperature range 40–180 °C where experimental data are available. Two characteristic results are shown in Figures 15.20 and 15.21.

Electrolyte-UNIFAC versions have also been applied to CO_2–water–alkanolamines mixtures. One such recent application has been presented by Jakobsen[102] and Jakobsen *et al.*[103] These authors were particularly interested in investigating how well the model can represent not only pressures but also concentrations of the various species in the solution, which were measured by the same (and other) authors based on NMR data. Various alkanolamines were considered (butyl-ethanolamines, MEA and MDEA). In general, the model predicted well the concentration of the various species at different CO_2 loadings. It was concluded that the distribution of alkanolamine species was predicted rather well, but the free CO_2 concentrations were overestimated. Moreover, at high loadings the carbamate concentration is underpredicted by the model, but the average values of the carbamate equilibrium constant obtained from the calculated concentrations are in agreement with the values obtained from previous NMR-based techniques.

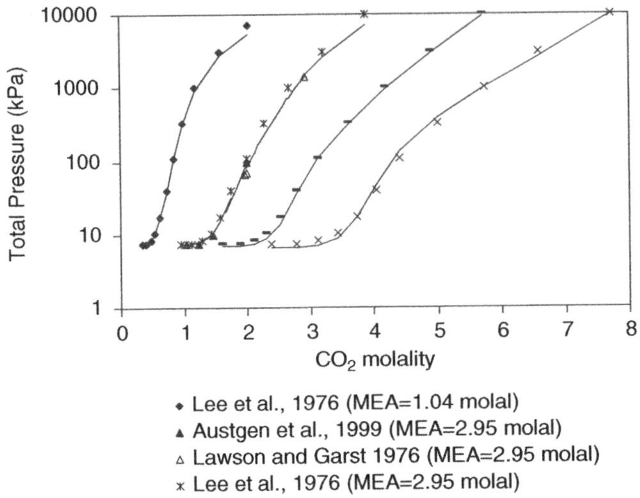

Figure 15.20 *Model results with extended UNIQUAC for the total pressure of the MEA + water + CO_2 system at 40 °C. Reprinted with permission from Fluid Phase Equilibria, Extended UNIQUAC Model for thermodynamic modeling of CO_2 absorption in aqueous alkanolamine solutions by L. Faramarzi, G. M. Kontogeorgis, K. Thomsen and E. H. Stenby, 282, 2, 121–132 Copyright (2009) Elsevier*

15.6.4 Equations of State

The e-LCVM model

LCVM (an EoS/G^E model, a combination of Peng–Robinson and an activity coefficient model) has been described in Chapter 6. Vrachnos *et al.*[90,91] have presented an extension of the LCVM model to electrolytes

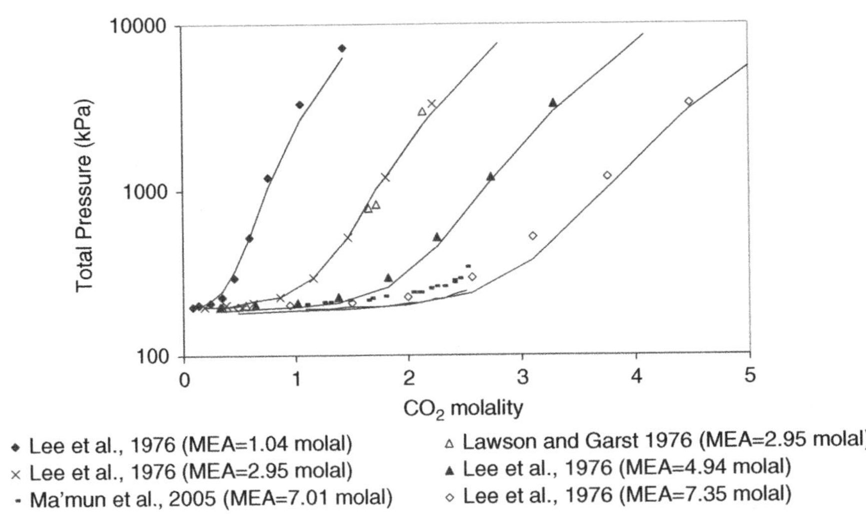

Figure 15.21 *Model results with extended UNIQUAC for the total pressure of the MEA + water + CO_2 system at 120 °C. Reprinted with permission from Fluid Phase Equilibria, Extended UNIQUAC M. model for thermodynamic modeling of CO_2 absorption in aqueous alkanolamine solutions by L. Faramarzi, G. M. Kontogeorgis, K. Thomsen and E. H. Stenby, 282, 2, 121–132 Copyright (2009) Elsevier*

which they have applied to CO_2 (as well as H_2S)–water–MEA, MDEA and their mixtures. Mixed gases and mixed alkanolamines were considered over extensive temperature and loading regions. The authors considered all eight usual chemical reactions when both gases are present and all equilibrium constants were obtained from the literature (Austgen *et al.*[78], except for the carbamate reversion which is taken from Liu *et al.*[66]). In the electrolyte version of LCVM (e-LCVM), the activity coefficient model used is the electrolyte (extended)-UNIQUAC (presented in Appendix 5.C of Chapter 5; see also Section 15.3) and the activity coefficient of a species n in solution is expressed by the equation:

$$\ln \gamma_n = \ln \gamma_n^{DH} + \ln \gamma_n^C + \gamma_n^R \tag{15.49}$$

where γ_n^C and γ_n^R are respectively the combinatorial and the residual activity coefficients of the species n in the UNIQUAC model. The DH contribution to the activity coefficient of solvent n is:

$$\ln \gamma_n^{DH} = \frac{2AM_n d_s}{b^3 d_n}\left[1 + b\sqrt{I} - \frac{1}{1 + b\sqrt{I}} - 2\ln(1 + b\sqrt{I})\right] \tag{15.50}$$

where $A = 1.327\,757 \times 10^5 d_s^{1/2}/(DT)^{3/2}$, $b = 6.359\,696 d_s^{1/2}/(DT)^{1/2}$, M_n is the molecular weight of pure solvent n (kg/mol), d_n is the density (kg/m^3) of pure solvent n, and d_s is the density of the solvent mixture (kg/m^3). The molar density of the solvent mixture, d'_s (kmol/m^3), is calculated from the following equation:

$$d'_s = \frac{1}{\left(\sum_n \frac{x'_n}{d'_n}\right)} \tag{15.51}$$

where d'_s and d'_n are the molar densities of the solvent and the mixture given in kmol/m^3 and x'_n is the salt-free mole fraction of solvent n. The dielectric constant D of the solvent mixture (water–MDEA) is obtained from the values for the pure solvents using Oster's mixing rule. For a binary solvent mixture, this rule can be approximated as:

$$D = D_1 + [(D_2 - 1)(2D_2 + 1)/2D_2 - (D_1 - 1)]x'_2 \frac{V_2}{V_1} \tag{15.52}$$

where D_1 and D_2 are the dielectric constants of water and MDEA, respectively, x'_2 is the salt-free mole fraction of MDEA and V_2 is the molar volume of MDEA. The temperature dependency of each component's dielectric constant can be represented as:

$$D_n = d^0 + \frac{d^{(1)}}{T} + d^{(2)}T + d^{(3)}T^2 + d^{(4)}T^3 \tag{15.53}$$

The UNIQUAC interaction parameters have a linear temperature dependency.

Each pure compound requires values of the critical properties and the acentric factor or the Mathias–Copeman parameters in the energy term and the five parameters in Equation (15.53) for the temperature dependency of the dielectric constant. These parameters, CO_2–water interactions as well as R and Q values for CO_2 and alkanolamines, were taken from previous publications. The remaining parameters (R and Q for ions) as well as the interaction parameters were estimated from binary (alkanolamine–water) and ternary

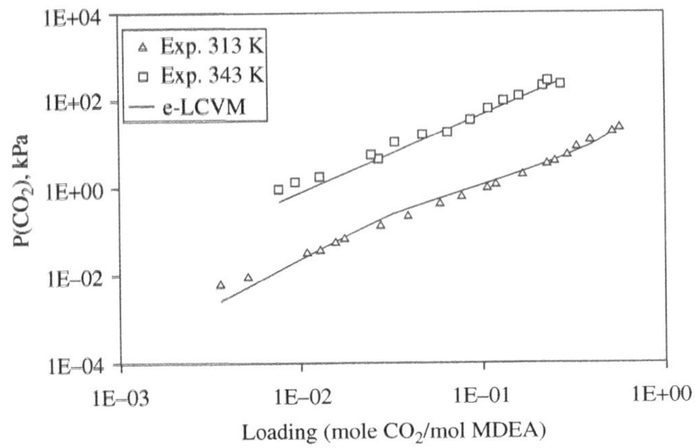

Figure 15.22 *Comparison of e-LCVM correlation results with experimental data for CO_2 equilibrium partial pressures over 35 wt% MDEA solution. The experimental data were used in parameter estimation*

data (CO_2 or H_2S–water–alkanolamines). The e-LCVM is shown to represent the experimental data very satisfactorily for both acid gases (as well as acid gas mixtures) in aqueous mixtures with MEA and MDEA as well as their blends over an extensive temperature range (40–120 °C). Two typical results using the simpler version of the model where concentration-independent interaction parameters were used are shown in Figures 15.22 and 15.23. The model predicts satisfactorily both the partial and total pressures, thus also the vapor phase composition (amine loss).

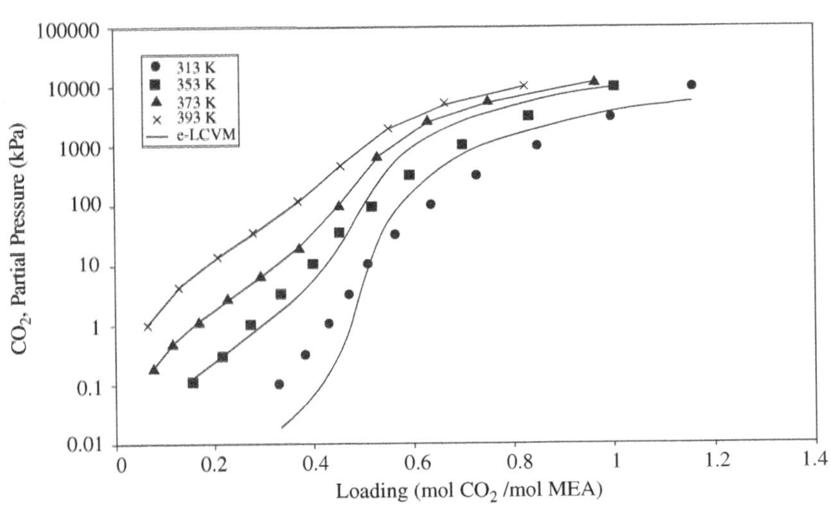

Figure 15.23 *Comparison of the e-LCVM predictions (data not used in the parameter estimation) with the experimental data for CO_2 equilibrium pressures over 5 M (30.2 wt%) MEA solutions. Reprinted with permission from Ind. Eng. Chem. Res., Thermodynamic Modeling of Acidic Gas Solubility in Aqueous Solutions of MEA, MDEA and MEA–MDEA Blends by Athanassios Vrachnos, Georgios M. Kontogeorgis and Epaminondas Voutsas,45, 14, 5148–5154 Copyright (2006) American Chemical Society*

The Fürst and Renon EoS

Overview

The model has already been presented above (Table 15.4 and Section 15.5). It has been used, in several variations, for modeling CO_2 and H_2S–water–alkanolamine mixtures. The alkanolamines considered are DEA[87] and DEA–methanol mixtures,[104] as well as MDEA with both CO_2 and H_2S.[88,89] Some investigators have used the same modified SRK EoS as Fürst and Renon did, while others[89] adopted the simpler SRK EoS. All authors employ the MSA equation for the LR interactions and the SR2 term (short-range interactions among ions), while Solbraa[89] also included a Born term. The Vallee *et al.*[87] version employs the Wong–Sandler mixing rule for the energy parameter, combined with NRTL and temperature-dependent parameters for both k_{ij} and NRTL's interaction parameters. Solbraa[89] used the Huron–Vidal mixing rules instead of the Wong–Sandler ones. The model includes:

- parameters for the pure compounds (critical properties, polar parameters in the pure energy parameter equation, ionic co-volumes, b_a and b_c);
- mixture interaction parameters (in the NRTL equation) and the symmetrical interaction parameters, W_{cs} between cation and solvent and W_{ca} between cation and anion.

The diameters are related to co-volumes via Equation (15.40), which are estimated from solvated diameters, and the anion–molecule interactions are neglected.

Characteristic applications

In the work by Vallee *et al.*[87] the e-SRK was applied to CO_2 and H_2S–water–DEA systems. Some parameters are taken from the literature (temperature dependence of the pure compound dielectric constants, molecular diameters), while certain interaction parameters (H_2O–CO_2 and H_2O–DEA) are fitted to binary data.

The equilibrium constant for carbamate formation and the temperature dependence of the W interaction parameters (in the SR2 term), the CO_2–DEA NRTL parameters and the Wong–Sandler k_{ij} are fitted to ternary VLE data:

$$\ln K = K^{(0)} + \frac{K^{(1)}}{T} + K^{(2)} \ln T + K^{(3)} T \tag{15.54}$$

$$W_{kl} = W_{kl}^{(0)} + W_{kl}^{(1)} \left(\frac{1}{T} - \frac{1}{298.15} \right) + W_{kl}^{(2)} \left(\frac{298.15 - T}{T} + \ln \frac{T}{298.15} \right) \tag{15.55}$$

In total nine parameters were fitted to the ternary data (containing CO_2 or H_2S): seven temperature-dependent ionic interaction parameters, the CO_2–water NRTL parameter and the carbamate formation equilibrium constant. The model represents the CO_2 partial pressure with an error of 21% and 15% for the H_2S partial pressure, for data included in the parameter estimation. For data which were not included in the parameter estimation, the errors varied between 15 and 35%, depending on the gas, data source and loading region.

A typical calculation is shown in Figure 15.24, where it can be seen that the highest deviations are at low loadings.

Chunxi and Fürst[88] applied the Fürst–Renon EoS to CO_2 and H_2S–water–MDEA systems in the 25–200 °C temperature range. All molecular diameters were taken from the literature except for MDEA, for which they

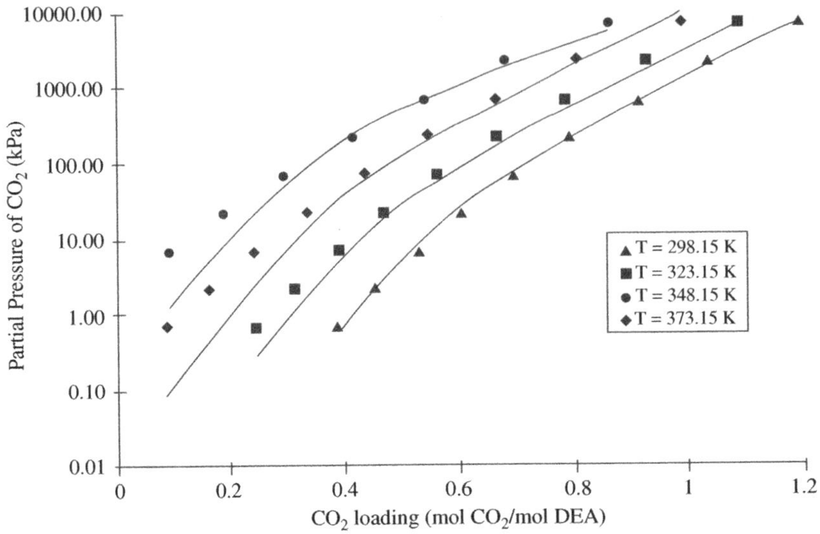

Figure 15.24 *Solubility of CO_2 in a 3.5 M DEA aqueous solution at 25, 50, 75 and 100 °C with the Fürst–Renon e-EoS (parameters used in parameter estimation). Reprinted with permission from Ind. Eng. Chem. Res., Representation of CO_2 and H_2S Absorption by Aqueous Solutions of Diethanolamine Using an Electrolyte Equation of State by Géraldine Vallée, Pascal Mougin, Sophie Jullian, and Walter Fürst, 38, 9, 3473–3480 Copyright (1999) American Chemical Society*

are estimated from the equation:

$$\sigma_{MDEA} = \sigma_{H_2O}\left(\frac{b_{MDEA}}{b_{H_2O}}\right)^{1/3} = 4.5 \times 10^{-10}\ \mathrm{m} \tag{15.56}$$

where b is the co-volume.

The Wong–Sandler/NRTL interaction parameters are temperature dependent and are fitted for mixtures of CO_2 or H_2S with water to gas solubility data. The water–MDEA parameters, however, are fitted to both VLE and SLE data, thus covering an extensive temperature range. The parameters between CO_2 or H_2S with MDEA are, due to lack of data, fitted together with the W_{ij} parameters between ions to ternary data. Only cation–water and cation–anion W parameters were considered. In total nine ionic parameters are included in the model (for both CO_2 and H_2S systems) but several of them can be estimated from relationships between the co-volume and W parameters and diameters (Equations 15.41a–d). An assumption is needed for the diameter of $MDEAH^+$, which is not available in the literature. The authors tried four different 'regression' models, with three, four, five or six adjustable parameters (diameter and W parameters). Small differences are seen between the various approaches with average deviations around 22–24%. Even three adjustable parameters (W via Equations 15.41a–d) are sufficient for satisfactory results over an extensive loading range at 25 °C. However, in order to represent data over an extensive temperature range, these three W parameters should be temperature dependent (Equation (15.55)). The model performs very well, with the highest deviations observed at low loadings, as can be seen in Figure 15.25.

In the Solbraa[89] version of the Fürst–Renon EoS, two new aspects introduced are the inclusion of the Born equation and the use of the Huron–Vidal mixing rules in the cubic EoS energy parameter. Solbraa[89] applied the

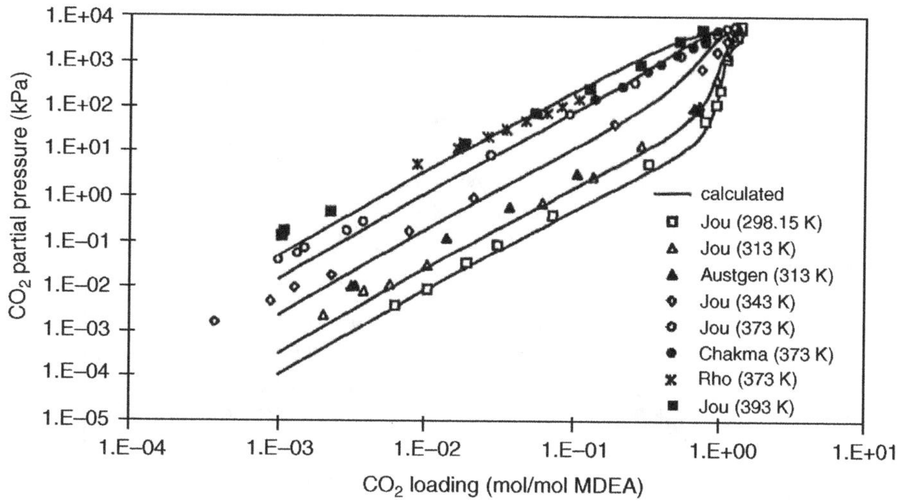

Figure 15.25 *Representation of CO_2 solubility at various temperatures with the Fürst–Renon e-EoS for CO_2–water–MDEA. The solutions have 48.9 wt% MDEA and special attention is given to the low loading range. Reprinted with permission from Chemical Engineering Science, Representation of CO_2 and H_2S solubility in aqueous MDEA soultions using an electrolyte equation of state by Li Chunxi and Walter Fürst, 55, 15, 2975–2988 Copyright (2000) Elsevier*

modified Fürst–Renon e-EoS to CO_2–water–MDEA, CO_2–water–NaCl and CO_2–methane–water–MDEA. The Huron–Vidal mixing rule offers a satisfactory representation of CO_2 (and methane) solubility in water (over an extensive temperature range) and the other molecular interactions. The model parameters, ionic co-volumes and the interaction coefficients W_{ij} are fitted to the cationic Stokes and the anionic Pauling diameter (Equations 15.41a–d) and the parameters in these relationships (in total six parameters) are fitted to experimental osmotic coefficients of aqueous salt solutions. For 28 halide solutions, the electrolyte EoS reproduces the data with an error of 2% for osmotic activity coefficients and 6% for the mean ionic activity coefficients. With this small number of parameters, Solbraa[89] could illustrate that the e-SRK EoS could predict (with 8% average deviation in the range 40–80 °C) the salting-out effect of CO_2 in aqueous NaCl and KCl solutions, i.e. the decrease of CO_2 solubility in water due to the presence of the salt, as shown in Figure 15.26 for one system. Different W interaction parameters were needed at 40 and 80 °C, however.

The application to aqueous alkanolamines requires values of the ionic diameters also for the new species and interaction parameters as well for the equilibrium constants. The diameters used for molecular components are the Lennard-Jones diameters reported by Reid *et al.*[106], except for MDEA (estimated from Equation (15.56)). The bicarbonate equilibrium constant is fitted, while the Posey values are used for the other equilibrium constants. The four ionic interaction coefficients involving MDEA are assumed to be temperature independent ($W_{MDEA^+-Water}$, W_{MDEA^+-MDEA}, $W_{MDEA^+-HCO_3^-}$, $W_{MDEA^+-CO_2}$) and they are fitted to experimental ternary data (CO_2–water–MDEA). A good (10–30% average deviation) representation of CO_2–water–MDEA is obtained in the 25–120 °C temperature interval as well as for CO_2–methane–water–MDEA. Figure 15.27 shows one calculation example.

In a recent work, Archane *et al.*[104] used the Fürst–Renon electrolyte EoS to study the effect of methanol on the CO_2–water–DEA system. Besides the inclusion of a physical solvent together with an alkanolamine, a novelty in their work was the consideration of both CO_2 solubility and liquid phase speciation data (from IR measurements) in the parameter estimation. The SRK EoS was used for the physical interactions together with Wong–Sandler mixing rule and the NRTL activity coefficient model. All molecular interactions were

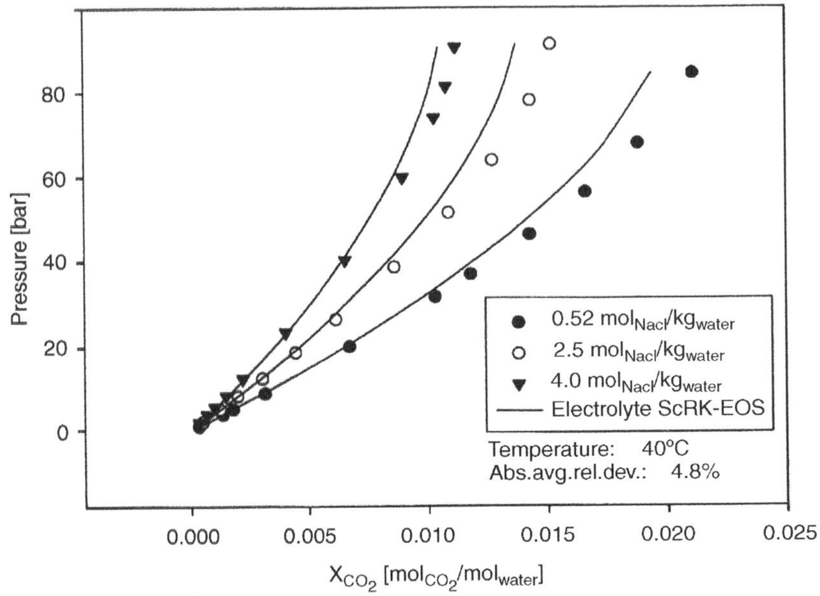

Figure 15.26 *Solubility of CO_2 in an aqueous NaCl solution at various salt concentrations using the e-SRK EoS by Solbraa. From Solbraa*[89]

estimated from binary VLE systems (except those for DEA–methanol and CO_2–DEA, due to lack of data). These adjustable parameters include both the k_{ij} in the Wong–Sandler rule and the interaction parameters in NRTL. Experimental ternary data (577 points for CO_2 or H_2S–water–DEA) were used to fit 11 parameters (4 solvated diameters, 3 W_{ij} parameters, 2 correlation parameters in W ion diameter correlations and 2 parameters in the carbamate formation constant), with excellent representation. All the remaining parameters

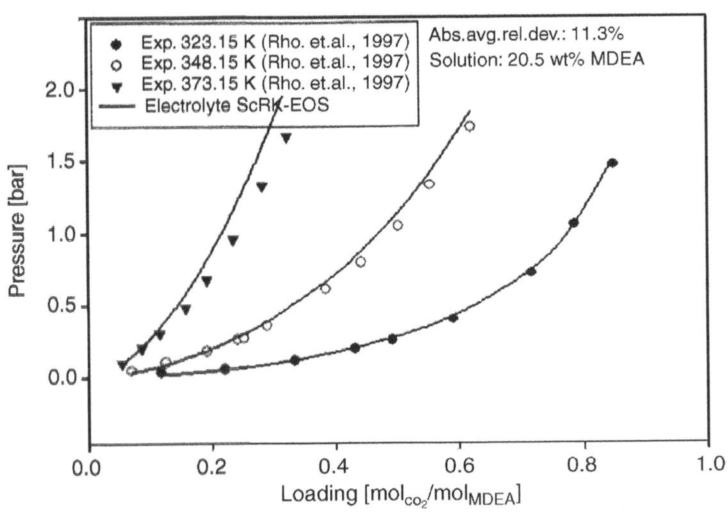

Figure 15.27 *Experimental (points) and predicted (lines) of partial pressures of CO_2 over aqueous MDEA (20.5 wt%) solutions at three temperatures with the e-SRK EoS by Solbraa. From Solbraa*[89]

were obtained from previously developed correlations (Equations 15.41a–d) or assumed to be zero, e.g. the anion–water interactions. Archane *et al.*[104] showed, moreover, that the carbamate concentrations for the water–DEA–CO_2–methanol system are properly represented only when both equilibrium pressure and CO_2 peak data are used in the parameter estimation. In addition to the above-mentioned parameters, the solvated $DEAH^+$ diameter, the W interactions between methanol and HCO_3^- and carbamate as well as the NRTL CO_2–DEA parameter had to be fitted simultaneously.

15.7 Concluding remarks

- Electrolyte thermodynamics differs greatly from the thermodynamics of non-electrolyte solutions. Many thermodynamic data are expressed in molality terms and typically mean ionic activity coefficients are used as single ion activity coefficients are not available.
- Coulombic forces are very long range. Most electrolyte models combine the Debye–Hückel (DH) theory or the mean spherical approximation (MSA) for the electrostatic interactions with an equation (activity coefficient or EoS) representing the short-range interactions. Both DH and MSA are based on the McMillan–Mayer framework (water is a dielectric continuum). While conversions may be necessary (in order to combine them with short-range terms which are based on the Lewis–Randall framework), these are often ignored in engineering electrolyte models. They may be of more importance for mixed solvent compared to single solvent or single salt solutions.
- Partly because of the LR/MM framework differences, opinions differ on whether composition-dependent dielectric constant expressions should be used in DH or MSA theories. Equations and mixing rules for the diaelectric constant are available for both water–salt and mixed solvents. Experimental data clearly show that the dielectric constants of solvents, e.g. of water, are not constant and are affected by the salts present or other molecules (mixed solvents).
- Besides the DH and MSA theories, which represent the major electrostatic interactions, many engineering models include the Born term, which accounts for solvation (hydration) effects and is important in the case of mixed solvents. Some models also have terms for short-range ionic or dipole–ion contributions.
- The most well-known activity coefficient models for electrolyte solutions are those of Pitzer, the e-NRTL and the e-UNIQUAC. Only the last has ion-specific parameters. All three models have been applied to numerous aqueous and mixed solvent salt solutions, and various properties have been considered (activity coefficients, solid–liquid equilibria, critical micelle concentration of aqueous ionic surfactants). The models are not predictive in any way and a large number of parameters are required, typically fitted to all available data. All three electrolyte activity coefficient models have been used also for weak electrolytes, e.g. CO_2 or H_2S–water–alkanolamines. In these cases some of the equilibrium constants, e.g. the carbamate formation equilibrium constant, may additionally be included in the parameter estimation.
- Over the past 10–15 years, numerous equations of state have been developed for electrolyte systems, including several electrolyte variations of CPA and SAFT. Most of them employ the MSA term, while an e-PC–SAFT based on DH theory has been developed. In several of these EoS, the effects of Born and short-range ionic effects have been evaluated as well. With association electrolyte-EoS, water is described using in most cases a four-site (4C) association scheme, but in some models only two- or three-site models are used for water. In a few of these e-SAFT EoS, association has been allowed between water and ions, but most approaches only consider the self-association of water.
- The electrolyte EoS have not been systematically applied to mixed salts and SLE, in contradiction to electrolyte activity coefficient models. Electrolyte EoS have been used for water–single salt (mean ionic, osmotic) activity coefficients, densities as well as salting out of gases in water due to the presence of salts.

Weak electrolytes, especially CO_2 and H_2S–water–alkanolamines phase equilibria, have also been modeled with the electrolyte-SRK EoS by Fürst and Renon but not by any of the electrolyte SAFT models.

- The performance of many e-EoS is very good, but it largely depends on the focus given, e.g. not all models have been developed for a large number of salts or applied to mixed salts. But there are positive features and it appears that they require fewer parameters than the electrolyte activity coefficient models. Moreover, several of these parameters, e.g. ionic diameters and segment energies, have well-defined meanings and their values, which are estimated from experimental data, can be tested independently. Several ways have been developed for reducing the number of adjustable parameters by utilizing semi-theoretical correlations or trends using the Pauling or Stokes ionic diameters.
- The comparison and evaluation of electrolyte models for weak electrolyte systems of the type CO_2 and H_2S–water–alkanolamines is difficult, as the various models have been developed and tested against different experimental data and do not include the same number of adjustable parameters. In most cases one or two equilibrium constants are also fitted together with the interaction parameters. Mixed alkanolamine and mixed acid gas data provide a way of testing the predictive capabilities of models. It seems that speciation data, which can be obtained from NMR and other techniques, may be useful in model development and testing, because using solely equilibrium pressure data is not always considered to be an optimum way to estimate the many adjustable parameters that these models contain.
- Future developments with electrolyte EoS will be facilitated by clarifying the following, still largely unresolved, points:
  - Comparison between DH and MSA theories: in which practical applications will MSA perform better?
  - What is the role, importance and significance of including or omitting the composition dependency of the dielectric constant?
  - For association models, should the association between water (solvent) and ions be included?
  - Should the ion-specific parameter be used exclusively or should some hybrid approach be preferred (including also salt-based parameters)?
  - Can an analysis of the magnitude of the various terms of the EoS (attractive, repulsive, association, ionic contributions) lead to useful information about their role and significance, maybe in combination with molecular simulation data?

True developments will emerge if electrolyte EoS can be developed from a limited amount of data, e.g. only mean ionic activity coefficients, using parameters with well-defined meaning, and then used to predict, for instance, salting out of gases, gas hydrate formation curves for mixed salt inhibitors or other applications with mixed solvents as well as SLE for aqueous systems with mixed salts.

References

1. J.R. Loehe, M.D. Donohue, *AIChE J.*, **1997**, *41(1)*, 180.
2. J.M. Prausnitz, R.N. Lichtenthaler, E.G. de Azevedo, *Molecular Thermodynamics of Fluid-Phase Equilibria* (3rd edition,). Prentice Hall International, 1999.
3. M.L. Pinsky, K. Takano, Property estimation for electrolyte systems. In: G.M. Kontogeorgis, R. Gani, *Computer-Aided Property Estimation for Process and Product Design*. Elsevier, 2004, Chapter 8.
4. Y. Lin, K. Thomsen, J.-C. de Hemptinne, *AIChE J.*, **2007**, *53(4)*, 989.
5. S.T. Tan, H. Adidharma, M. Radosz, *Ind. Eng. Chem. Res.*, **2008**, *47*, 8063.
6. M.L. Michelsen, J.M. Mollerup, *Thermodynamic Models: Fundamentals & Computational Aspects* (2nd edition,). Tie-Line Publications, 2007.
7. P. Debye, E. Hückel, *Phys. Z.*, **1923**, *24*, 185.

8. P. Paricaud, A. Galindo, G. Jackson, *Fluid Phase Equilib.*, **2002**, *194–197*, 87.
9. A. Galindo, A. Gil-Villegas, G. Jackson, A.N. Burgess, *J. Phys. Chem. B*, **1999**, *103*, 10272.
10. Y. Lin, Development of an equation of state for solutions containing electrolytes. PhD Thesis, Technical University of Denmark, 2007.
11. H. Planche, H. Renon, *J. Chem. Phys.*, **1981**, *85*, 3924.
12. W. Fürst, H. Renon, *AIChE J.*, **1993**, *39(2)*, 335.
13. R. Pottel, Dielectric properties. In: *Water: A Comprehensive Treatise*, Vol. *3.*, Plenum Press, 1973.
14. J.P. Simonin, L. Blum, P. Turq, *J. Phys. Chem.*, **1996**, *100*, 7704.
15. J.P. Simonin, O. Bernard, L. Blum, *J. Phys. Chem. B*, **1999**, *103*, 699.
16. L. Blum, *Mol. Phys.*, **1975**, *30(5)*, 1529.
17. L. Blum, J.S. Høye, *J. Phys. Chem.*, **1977**, *81(13)*, 1311.
18. M. Born, *Z. Phys.*, **1920**, *1*, 45.
19. J. Israelachvili, *Intermolecular and Surface Forces*. Academic Press 1985.
20. D. Henderson, L. Blum, A. Tani, *ACS Symp. Ser.*, **1986**, *300*, 281.
21a. M.P. Breil, Thermodynamics, experimental and modelling of aqueous electrolyte and amino acid solutions. PhD Thesis, Technical University of Denmark, 2001.
21b. M.P. Breil, J.M. Mollerup, *Fluid Phase Equilib.*, *242*, **2006**, 129.
22. K. Thomsen, P. Rasmussen, *Chem. Eng. Sci.*, **1999**, *54*, 1787.
23. K. Thomsen, *Pure Appl. Chem.*, **2005**, *77*, 531.
24. E. Hückel, *Phys. Z.*, **1925**, *26*, 93.
25. L.A. Bromley, *AIChE J.*, **1973**, *19*, 313.
26. K.S. Pitzer, *J. Phys. Chem.*, **1973**, *77*, 268.
27. K.S. Pitzer, *Activity Coefficients in Electrolyte Solutions* (2nd edition). CRC Press, 1991.
28. C.C. Chen, L.B. Evans, *AIChE J.*, **1986**, *32*, 444.
29. C.C. Chen, P.M. Mathias, *AIChE J.*, **2002**, *48(2)*, 194.
30. C.C. Chen, C.P. Bokis, P. Mathias, *AIChE J.*, **2001**, *47*, 2593.
31. C.C. Chen, Y. Song, *AIChE J.*, **2004**, *50*, 1928.
32. C.C. Chen, *AIChE J.*, **1996**, *41(11)*, 3231.
33. G.M. Bollas, C.C. Chen, P.I. Barton, *AIChE J.*, **2008**, *54(6)*, 1608.
34. K. Takano, Computer-aided design and analysis of separation processes with electrolyte systems. PhD Thesis, Technical University of Denmark, 2000.
35. P.M. Mathias, *Fluid Phase Equilib.*, **2005**, *228*, 49.
36. A.V. Garcia, K. Thomsen, E.H. Stenby, *Geothermics*, **2005**, *34*, 61.
37. S.G. Christensen, K. Thomsen, *Ind. Eng. Chem. Res.*, **2003**, *42*, 4260.
38. K. Thomsen, M. Iliuta, P. Rasmussen, *Chem. Eng. Sci.*, **2004**, *59*, 3631.
39. I. Kikic, M. Fermeglia, P. Rasmussen, *Chem. Eng. Sci.*, **1991**, *46*, 2775.
40. W. Yan, M. Topphoff, C. Rose, J. Gmehling, *Fluid Phase Equilib.*, **1999**, *162*, 97.
41. C. Achard, C.G. Dussap, J.B. Gros, *Fluid Phase Equilib.*, **1994**, *98*, 71.
42. H. Cheng, Thermodynamic Modelling of surfactants solutions. PhD Thesis, Technical University of Denmark, 2003.
43. X.-S. Li, J.-F. Lu, Y.-G. Li, *Fluid Phase Equilib.*, **1998**, *153*, 215.
44. I. Kikic, M. Fermeglia, P. Rasmussen, *Chem. Eng. Sci.*, **1991**, *46*, 2775.
45. E. Collinet, J. Gmehling, *Fluid Phase Equilib.*, **2006**, *246*, 111.
46. A.H. Harvey, J.M. Prausnitz, *AIChE J.*, **1989**, *35(4)*, 635.
47. K. Aasberg-Petersen, E. Stenby, Aa. Fredenslund, *Ind. Eng. Chem. Res.*, **1991**, *30*, 2180.
48. G. Jin, M.D. Donohue, *Ind. Eng. Chem. Res.*, **1988**, *27*, 1073.
49. G. Jin, M.D. Donohue, *Ind. Eng. Chem. Res.*, **1988**, *27*, 1737.
50. G. Jin, M.D. Donohue, *Ind. Eng. Chem. Res.*, **1991**, *30*, 240.
51. J.Y. Zuo, D. Zhang, W. Fürst, *AIChE J.*, **2000**, *46(11)*, 2318.
52. J.Y. Zuo, D. Zhang, W. Fürst, *Fluid Phase Equilib.*, **2000**, *175*, 285.
53. V.Q. Vu, P.D. Suchaux, W. Fürst, *Fluid Phase Equilib.*, **2002**, *194–197*, 361.

54. J.A. Myers, S.I. Sandler, R.H. Wood, *Ind. Eng. Chem. Res.*, **2002**, *41*, 3282.
55. J. Wu, J.M. Prausnitz, *Ind. Eng. Chem. Res.*, **1998**, *37*, 1634.
56. R. Inchekel, J.-C. de Hemptinne, W. Fürst, *Fluid Phase Equilib.*, **2008**, *271*, 19.
57. A. Gil-Villegas, A. Galindo, G. Jackson, *Mol. Phys.*, **2001**, *99(6)*, 531.
58. B.H. Patel, P. Paricaud, A. Galindo, G.C. Maitland, *Ind. Eng. Chem. Res.*, **2003**, *42*, 3809.
59. L.F. Cameretti, G. Sadowski, J.M. Mollerup, *Ind. Eng. Chem. Res.*, **2005**, *44*, 3355. And errata, same volume, p. 8944.
60. Ch. Held, L.F. Cameretti, G. Sadowski, *Fluid Phase Equilib.*, **2008**, *270*, 87.
61. Ch. Held, L.F. Cameretti, G. Sadowski, Thermodynamic properties of aqueous electrolyte/amino acid solutions. ESAT Conference Proceedings 2008, pp 70–76.
62. S.P. Tan, H. Adidharma, M. Radosz, *Ind. Eng. Chem. Res.*, **2005**, *44*, 4442.
63. X. Ji, S.P. Tan, H. Adidharma, M. Radosz, *Ind. Eng. Chem. Res.*, **2005**, *44*, 7584.
64. X. Ji, S.P. Tan, H. Adidharma, M. Radosz, *Ind. Eng. Chem. Res.*, **2005**, *44*, 8427.
65. Y. Liu, L. Zheng, S. Watanasiri, *Ind. Eng. Chem. Res.*, **1999**, *38*, 2080.
66. W.B. Liu, Y.G. Li, J.F. Lu, *Fluid Phase Equilib.*, **1999**, *160*, 595.
67. Z. Liu, W. Wang, Y. Li, *Fluid Phase Equilib.*, **2005**, *227*, 147.
68. X.-S. Li, J.-F. Lu, Y.-G. Li, *Fluid Phase Equilib.*, **2000**, *168*, 107.
69. D. Silkenbaumer, B. Rumpf, R.N. Lichtenthaler, *Ind. Eng. Chem. Res.*, **1998**, *37*, 3133.
70. A.P.-S. Kamps, A. Balaban, M. Jodecke, G. Kuranov, N.A. Smirnova, G. Maurer, *Ind. Eng. Chem. Res.*, **2001**, *40*, 696.
71. R.D. Deshmukh, A.E. Mather, *Chem. Eng. Sci.*, **1981**, *36*, 355.
72. S. Xu, Y.-W. Wang, F.D. Toot, A.E. Mather, *Chem. Eng. Process.*, **1992**, *31*, 7.
73. R.H. Weiland, T. Chakravarty, A.E. Mather, *Ind. Eng. Chem. Res.*, **1993**, *32*, 1419.
74. Y.-G. Li, A.E. Mather, *Ind. Eng. Chem. Res.*, **1994**, *33*, 2006.
75. T.T. Teng, A.E. Mather, *Can. J. Chem. Eng.*, **1989**, *67*, 846.
76. M.L. Posey, G.T. Rochelle, *Ind. Eng. Chem. Res.*, **1997**, *36*, 3944.
77. J.T. Cullinane, G.T. Rochelle, *Fluid Phase Equilib.*, **2005**, *227*, 197.
78. D.M. Austgen, G.T. Rochelle, X. Peng, C.C. Chen, *Ind. Eng. Chem. Res.*, **1989**, *28*, 1060.
79. D.M. Austgen, G.T. Rochelle, C.C. Chen, *Ind. Eng. Chem. Res.*, **1991**, *30*, 543.
80. S. Bishnoi, G.T. Rochelle, *Ind. Eng. Chem. Res.*, **2002**, *41*, 604.
81. M.K. Aroua, M.Z. Haji-Sulaiman, K. Ramasamy, *Sep. Purif. Technol.*, **2002**, *29*, 153.
82. J. Addicks, Solubility of carbon dioxide and methane in aqueous N-methydiethanolamine solutions at pressures between 100 and 200 bar. PhD Thesis, Norwegian University of Science and Technology, 2002.
83. A. Barreau, E. Blanchon le Bouhelec, K.N. Habchi Tounsi, P. Mougin, F. Lecomte, *Oil Gas Sci. Technol. – Rev. IFP*, **2006**, *61*, 345.
84. L. Kaewsichan, O. Al-Bofersen, V.F. Yesavage, M.S. Selim, *Fluid Phase Equilib.*, **2001**, *183–184*, 159.
85. A.V. Garcia, K. Thomsen, E.H. Stenby, *Geothermics*, **2006**, *35*, 239.
86. L. Faramarzi, G.M. Kontogeorgis, K. Thomsen, E.H. Stenby, *Fluid Phase Equilib.*, **2009**, *282*(2), 65.
87. G. Vallee, P. Mougin, S. Jullian, W. Fürst, *Ind. Eng. Chem. Res.*, **1999**, *38*, 3473.
88. L. Chunxi, W. Fürst, *Chem. Eng. Sci.*, **2000**, *55*, 2975.
89. E. Solbraa, Measurement and modelling of absorption of carbon dioxide into methyldiethanolamine solutions at high pressure. PhD Thesis, Norwegian University of Science and Technology, 2002.
90. A. Vrachnos, E. Voutsas, K. Magoulas, A. Lygeros, *Ind. Eng. Chem. Res.*, *43*, **2004**, 2798.
91. A. Vrachnos, G.M. Kontogeorgis, E.C. Voutsas, *Ind. Eng. Chem. Res.*, **2006**, *45(14)*, 5148.
92. J.K. Button, K.E. Gubbins, *Fluid Phase Equilib.*, **1999**, *158–160*, 175.
93. J. Gabrielsen, M.L. Michelsen, E.H. Stenby, G.M. Kontogeorgis, *Ind. Eng. Chem. Res.*, **2005**, *44*, 3348.
94. J. Gabrielsen, M.L. Michelsen, E.H. Stenby, G.M. Kontogeorgis, *AIChE J.*, **2006**, *52(10)*, 3443.

95a. J. Gabrielsen, CO_2 capture from coal fired power plants. PhD Thesis, Technical University of Denmark, 2007.

95b. R.L. Kent, B. Eisenberg, *Hydrocarbon Process.*, **1976**, *55*, 87.

95c. J. Kiepe, O. Noll, J. Gmehling, *Ind. Eng. Chem. Res.*, **2006**, *45*, 2361.

96. G. Sartori, W.S. Ho, D.W. Savage, G.R. Chludzinski, S. Wiechert, *Sep. Purif. Methods*, **1987**, *16*, 171.

97. G. Astarita, D.W. Savage, A.L. Bisio, *Gas Treating with Chemical Solvents*. John Wiley & Sons Inc., 1983.
98. E. Ikada, Y. Hida, H. Okamoto, J. Hagino, N. Koizumi, *Bull. Inst. Chem. Res. Kyoto Univ.*, **1968**, *46*, 239.
99. B.L. Mock, L.B. Evans, C.C. Chen, *AIChE J.*, **1986**, *32*, 1655.
100. M.L. Posey, K.G. Tapperson, G.T. Rochelle, *Gas Sep. Purif.*, **1996**, *10(3)*, 181.
101. E.B. Le Bouhelec-Tribouillois, P. Mougin, A. Barreau, I. Brunella, D. Le Roux, R. Solimando, *Oil Gas Sci. Technol. – Rev. IFP*, **2008**, *63(3)*, 363.
102. J.P. Jakobsen, Absorption of CO_2-modeling and experimental characterization. PhD Thesis, Norwegian University of Science and Technology, 2004.
103. J.P. Jakobsen, J. Krane, H.F. Svendsen, *Ind. Eng. Chem. Res.*, **2005**, *44*, 9894.
104. A. Archane, L. Gicquel, E. Provost, W. Fürst, *Chem. Eng. Res. Des.*, **2008**, *86*, 592.
105. P.V. Danckwerts, K.M. McNeil, *Trans. Inst. Chem. Eng.*, **1967**, *45*, 32.
106. R.C. Reid, J.M. Prausnitz, B.E. Poling, *The Properties of Gases and Liquids* (4th edition). McGraw-Hill, 1988.
107. M.S. Wertheim, *J. Chem. Phys.*, **1971**, *55*, 4291.
108. M.C. Kroon, E.K. Karakatsani, I.G. Economou, G.-J. Witkamp, C.J. Peters, *J. Phys. Chem. B*, **2006**, *110*, 9262.
109. E.K. Karakatsani, I.G. Economou, M.C. Kroon, C.J. Peters, G.-J. Witkamp, *J. Phys. Chem. C*, **2007**, *111*, 15487.
110. E.K. Karakatsani, I.G. Economou, *Fluid Phase Equilib.*, **2007**, *261*, 265.
111. E.K. Karakatsani, I.G. Economou, M.C. Kroon, M.D. Bermejo, C.J. Peters, G.-J. Witkamp, *Phys. Chem. Chem. Phys.*, **2008**, *10*, 6160.
112. H. Haghighi, A. Chapoy, B. Tohidi, *Ind. Eng. Chem. Res.*, **2008**, *47*, 3983.
113. S.P. Tan, X. Ji, H. Adidharma, M. Radosz, *J. Phys. Chem. B*, **2006**, *110*, 16694.
114. X. Ji, S.P. Tan, H. Adidharma, M. Radosz, *J. Phys. Chem. B*, **2006**, *110*, 16700.
115. X. Ji, H. Adidharma, *Ind. Eng. Chem. Res.*, **2007**, *46*, 4667.
116. X. Ji, H. Adidharma, *Chem. Eng. Sci.*, **2008**, *63*, 131.
117. J.M. Mollerup, M.P. Breil, *Fluid Phase Equilib.*, **2009**, *276*, 18.
118. P.J.G. Huttenhuis, N.J. Agrawal, E. Solbraa, G.F. Versteeg, *Fluid Phase Equilib.*, **2008**, *264*, 99.

16

Quantum Chemistry in Engineering Thermodynamics

16.1 Introduction

Quantum mechanics (QM) or quantum chemistry (QC) calculations have been extensively used in the past for the calculation of heats of formation and reaction, heat capacities, reaction pathways and transition states.[1] They are currently becoming increasingly popular for estimating phase equilibria and other properties. The starting point in QM, the Schrödinger equation, cannot be solved exactly for multielectron systems, and thus approximations are necessary. Deciding upon the method and level that the computations should be made is not a trivial task. Therefore, there are today a number of software packages available for QM calculations, e.g. Gaussian (www.gaussian.com), Schrödinger (www.schrodinger.com, distributing the software Jaguar), Turbomole (www.turbomole.com) and Gamess[2]. The packages include various calculation procedures, such as *ab initio* and density functional methods.

The purpose of this chapter is not to review these computational methods, but rather to provide a short presentation and evaluation of certain QM methods which have already found use in engineering applications. For further information on the computational methods, the reader is referred to the references of this chapter, especially the reviews by Sandler and co-workers.[1,3,4] Before presenting the methods that we will highlight in this chapter, it is important to emphasize that QM calculations can be used in different semi-direct or indirect approaches in engineering thermodynamics:

1. Calculation of the intermolecular potential from QM and then phase equilibrium calculations using molecular simulation.
2. QM calculations to determine parameters in existing thermodynamic models.
3. The continuum solvation (polarizable) models such as COSMO–RS.

The first method will not be discussed in detail in this chapter. It is currently computationally very intensive and limited to rather simple molecules. It has been used to calculate the second virial coefficients and VLE of a few molecules and systems such as methyl fluoride, methanol, acetonitrile and methanol–hydrogen fluoride. The agreement is best for the simpler molecules but for hydrogen bonding compounds such as methanol and hydrogen fluoride, corrections are required. This may be due to the fact that for hydrogen bonding molecules, the assumption of pairwise additivity is not valid due to their close proximity and the multibody effects which must be considered.

The second and third methods have been widely used over the past 10 years or so and will be discussed in this chapter. It has already been briefly mentioned (Chapter 5, Section 5.5.3) that QM calculations can provide values for the energetic interactions (parameters) of local composition models such as Wilson and UNIQUAC, indicating that UNIQUAC is possibly the theoretically more sound of the two models. QM computations can be used to improve group contribution methods by introducing corrections based on the charge and dipole moment of the functional group that is unique to the molecule in which it appears. In this way, one of the most important limitations of the group contribution methods, that of the so-called 'proximity effect', is in a way corrected. In this direction, Sandler has showed that improved octanol–water partition coefficient (K_{OW}) calculations can be obtained for both mono- and multifunctional chemicals (of environmental concern) with the so-called GCSKOW model.[5–8] This is essentially a QC solvation model for the calculation of the solvation free energy, which provides activity coefficients and partition coefficients (K_{OW}):

$$\log K_{OW,i} = \log\left(\frac{C_O^O}{C_W^W}\right) + \frac{1}{\log 10}\left[\log\left(\frac{\gamma_{i,W}^{SG,\infty}}{\gamma_{i,O}^{SG,\infty}}\right) + \frac{1}{RT}\sum_k^{N_i^k}\left(\Delta G_{k,W}^{*,chg} - \Delta G_{k,O}^{*,chg}\right)\right] \tag{16.1}$$

where the summation is over all functional groups N_i^k contained in a molecule i, SG is the Staverman–Guggenheim model, W is the water-rich and O is the octanol-rich phase. The quantity in the last parentheses is the difference in charging free energies between bringing functional group k into an aqueous and a water-saturated octanol solution.

The difference in charging free energies should be corrected for the change of electric charge and dipole moment of functional groups depending on the location of such a group in a multifunctional group, i.e. the proximity effect. This was done by Sandler and co-workers[5–8] in an effective way using the theory of Kirkwood and quick ideal gas phase quantum calculations for the charge and dipole moments for each molecule, done once for each molecule. The revised GCSKOW method provides much improved results for multifunctional chemicals, as illustrated in Figure 16.1.

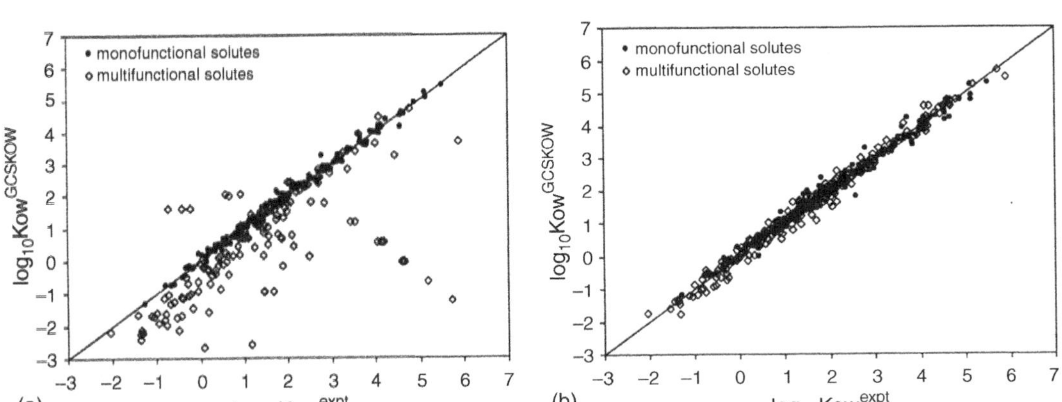

Figure 16.1 *Predictions from the GCSKOW model with (right) and without (left) multipole corrections for 204 monofunctional and 246 multifunctional compounds. Reprinted with permission from Fluid Phase Equilibria, Quantum mechanics: A new tool for engineering thermodynamics by Stanley I. Sandler, 210, 2, 147–160 Copyright (2003) Elsevier*

Evidently GCSKOW is a hybrid method using QM concepts to improve existing engineering models. In this chapter we will discuss and evaluate how QM concepts can be utilized to estimate the parameters of association (SAFT-type) models, how interaction parameters can be estimated using QM-based combining rules and finally the performance of SAFT approaches using parameters (partially) obtained from QM. The capabilities and current limitations will be elucidated. Possibly the most successful and widespread applied method based on QM which is suitable for engineering applications is the so-called continuum solvation (or polarizable) family of models such as COSMO–RS and COSMO–SAC. These models are based on the three-step process, unlike the two-step process of approach 2 above. The key concept pioneered by Klamt and co-workers[9–11] is that a molecule is deconstructed into a collection of very small surface elements and the charge density on each surface element is computed using a quantum electrostatic calculation. The unique characteristic of each molecule is its sigma (σ) profile that is a representation of charge density versus likelihood of occurrence. The sigma profiles are then used together with a statistical mechanics analysis and the excess Gibbs energy (and activity coefficients) are computed at any composition. The COSMO–RS and related models will be discussed in the next section. The range of applicability and comments on model accuracy will be discussed.

16.2 The COSMO–RS method

16.2.1 Introduction

The name COSMO–RS stands for a 'COnductor-like Screening MOdel', which is an efficient variant of dielectric continuum solvation methods in QC programs, and its extension to 'Real Solvents', which is a statistical thermodynamics approach based on the results of QC COSMO calculations.[9,10] COSMO–RS is available commercially today from the company Cosmologic (www.cosmologic.de). COSMO–RS is a two-level approach, which is illustrated in Figure 16.2.

In the first level, QC calculations have to be performed for all the compounds of interest. In these calculations, the continuum solvation model COSMO is applied in order to simulate a virtual conductor environment for the molecule. In this environment, the solute molecule induces a polarization charge density, σ, on the interface of the molecule to the conductor, i.e. on the molecular surface. These charges act back to the solute, generating a more polarized electron density than in a vacuum. The polarization charge density, σ, is a good local descriptor of the molecular surface polarity. In the second level, the statistical thermodynamics of the molecular interactions, this polarization charge density is used for quantification of the interaction energy of pairwise interacting surface segments. Computational details and implementation in TURBOMOLE can be found in Klamt[9], Klamt and Schuurmann[10] and Schafer *et al.*[11] Panayiotou[12,13] discusses the links of COSMO to the lattice theory.

16.2.2 Range of applicability

Despite the fact that COSMO–RS is a relatively new method, it has already been established as an efficient alternative to the group contribution approaches, such as UNIFAC. As mentioned, it starts directly from QC calculations for the individual molecules and expresses the intermolecular interactions based on this QC information. It does not suffer, therefore, from the 'proximity effect' and other limitations of the group contribution methods. Due to the generic functional form of the interaction energies, the method is applicable to almost the entire organic chemistry. The interaction energies are based on information from QC calculations and they only require a very limited number of element-specific parameters. An important advantage of the COSMO–RS method is its ability to handle isomers, as well as intramolecular interactions of functional

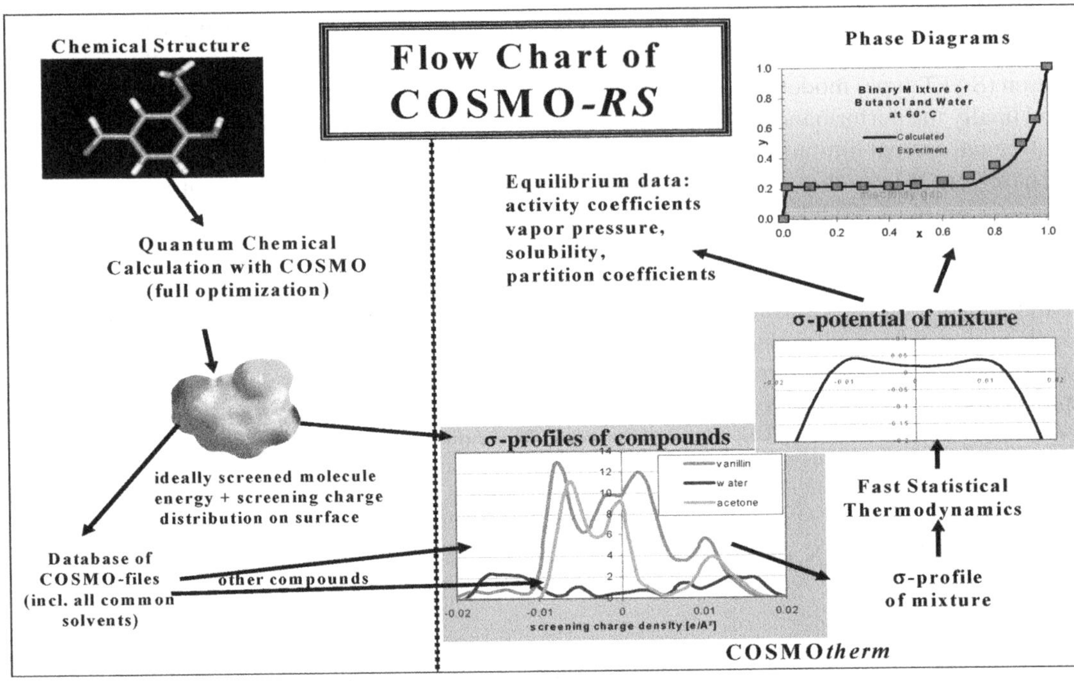

Figure 16.2 *Flow chart of COSMO–RS. Reproduced with kind permission from COSMOlogic GmbH & Co. KG. Look online for more information at www.cosmologic.de*

groups. It 'sees', for example, the difference between primary, secondary and tertiary hydroxyl groups in alcohols, because these groups differ in the underlying QC calculations. The exact thermodynamics is another merit of COSMO–RS. It gives rather homogeneous and reliable quality of results in the entire concentration range. Over the past 10 years, COSMO–RS has been successfully applied in a variety of systems, ranging from mixtures with polar and associating compounds to aqueous solutions and complex molecules for drugs and pesticides, including blood–brain partition coefficients of drugs. Several applications are listed in Table 16.1. Incidentally, it has been recently shown that COSMO–RS can be used to improve group contribution methods, by minimizing the error caused by the basic assumption that a functional group interacts independently of the molecule on which it is located. Investigations presenting comparisons against UNIFAC and other models are listed in Table 16.2. Some characteristic results illustrating the capabilities of the COSMO-related approaches as well as range of applicability are shown in Figures 16.3–16.8.

16.2.3 Limitations

Despite the fact that many exciting and successful applications have been presented with COSMO–RS since its appearance over 15 years ago,[9,10] the actual level of applicability and success is still under debate. For example:

1. Various research groups[18–20,72–74] have provided their own implementations of COSMO–RS and their conclusions are not always in agreement with those of Klamt and co-workers. This has resulted in several discussions in the literature.[19,48–51]

Table 16.1 *Applications of COSMO–RS and related models (COSMO–SAC etc.)*

Application	Reference
VLE for mixtures with polar chemicals	Banerjee *et al.*[14] Spuhl and Arlt[15] Klamt and Eckert[16,17] Lin and Sandler[6,7] (COSMO–SAC) Grensemann and Ghemling[18] Mu *et al.*[19,20]
Solvent screening	Eckert and Klamt[21,22]
Infinite dilution activity coefficients	Putnam *et al.*[23] Mu *et al.*[20]
Halocarbon mixtures (VLE, LLE, gas solubilities)	Eckert and Klamt[24]
Water–hydrocarbon LLE	Klamt[25]
Octanol–water partition coefficients	Klamt[9]
Pharmaceuticals + Pesticides, physiological solvation	Kolar *et al.*[26] Modaressi *et al.*[27] Tung *et al.*[28] Klamt *et al.*[29,30]
Ionic liquids	Diedenhofen *et al.*[31] (γ^∞) Banerjee *et al.*[14] Palomar *et al.*[32] (densities) Kato and Gmehling[33] Zhang *et al.*[34] (screening for CO_2 capture) Freire *et al.*[35] (VLE and LLE for ionic liquid + water systems)
Adsorption equilibria	Mehler *et al.*[36]
Membrane–water partition coefficients	Klamt and Eckert[37]
Blood–brain partitioning and serum albumin binding	Wichmann *et al.*[38]
Soil sorption coefficients	Klamt *et al.*[39]
Partition coefficients and solvent screening in biocatalytic two-phase reactions	Spieß *et al.*[40]
Partition in water–surfactant systems	Buggert *et al.*[41]
Solubilities in polymer systems	Tukur *et al.*[42]
Aqueous p*Ka* values for acids	Klamt *et al.*[43]
Membrane–water partition coefficients	Klamt *et al.*[70]
Vapor pressures, enthalpies of vaporization	Lin *et al.*[71]

2. COSMO–RS and related models are essentially activity coefficient models and are thus limited to low-pressure applications (like UNIFAC).
3. While the range of applicability of COSMO–RS is great and the method is predictive, its accuracy can be limited for many *standard mixtures*, e.g. alcohols or glycols with alkanes. Thus, for these mixtures, SAFT-type models often perform better. This is illustrated by some examples in Figures 16.9–16.11. As COSMO–RS is a predictive model, a fair comparison with equations of state is made if all interaction parameters are set to zero. COSMO–RS calculates qualitatively correct results and can also give the correct composition of azeotropic points. However, the bubble pressures are in general underestimated. Again the COSMO–RS results are of comparable accuracy to predictions ($k_{ij} = 0$) of PC–SAFT. Equally satisfactory results are obtained for cross-associating alcohol–water systems. A systematic analysis shows

Table 16.2 *COSMO–RS (and COSMO–SAC) against other models or in combination with other models*

Reference	Model(s)
Leonhard *et al.*[44]	COSMO–RS in PR via EoS/G^E rules
Shimoyama *et al.*[45]	COSMO–RS in SRK via EoS/G^E rules, application to water–hydrocarbon–alcohol mixtures
Constantinescu *et al.*[46]	COSMO–RS in SRK via Huron–Vidal mixing rule
Kolar *et al.*[26]	COSMO–RS vs. UNIFAC for pharmaceuticals
Tung *et al.*[28]	COSMO–SAC vs. NRTL–SAC for pharmaceuticals
Banerjee *et al.*[14]	COSMO–RS vs. Wilson, NRTL and UNIQUAC for ionic liquids
Kato and Gmehling[33]	COSMO–RS vs. UNIFAC and modified UNIFAC for ionic liquids (VLE + γ^∞)
Arlt *et al.*[47]	COSMO–RS vs. PR and modified UNIFAC for VLE/*K* factors
Eckert and Klamt[21,22]	COSMO–RS vs. UNIFAC for solvent screening (VLE, LLE, SLE)
Putnam *et al.*[23]	COSMO–RS vs. various UNIFAC variants for γ^∞
Grensemann and Gmehling[18]	COSMO–RS vs. various UNIFAC variants, UNIQUAC, ASOG for VLE
Buggert *et al.*[41]	COSMO–RS vs. UNIFAC for organic solute–water–surfactant mixtures

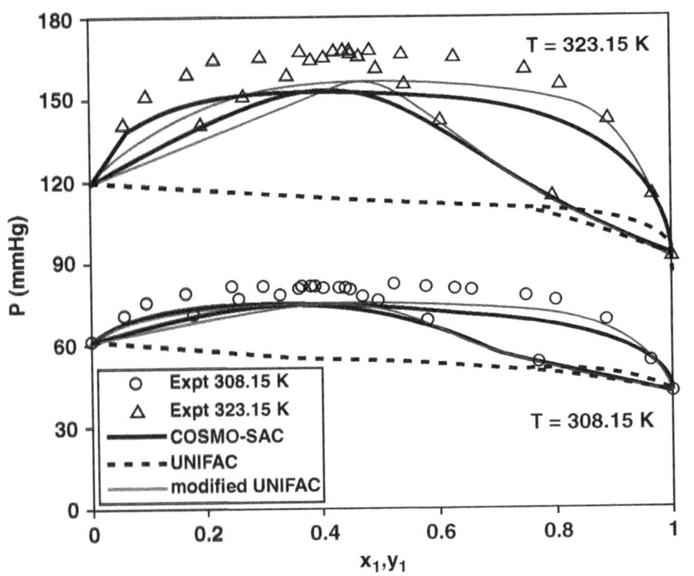

Figure 16.3 Comparison of VLE predictions from COSMO–SAC, UNIFAC and modified UNIFAC models for water (1)–1,4-dioxane(2) at 308.15 K and 323.15 K. Reprinted with permission from Fluid Phase Equilibria, Quantum mechanics: A new tool for engineering thermodynamics by Stanley I. Sandler, 210, 2, 147–160 Copyright (2003) Elsevier

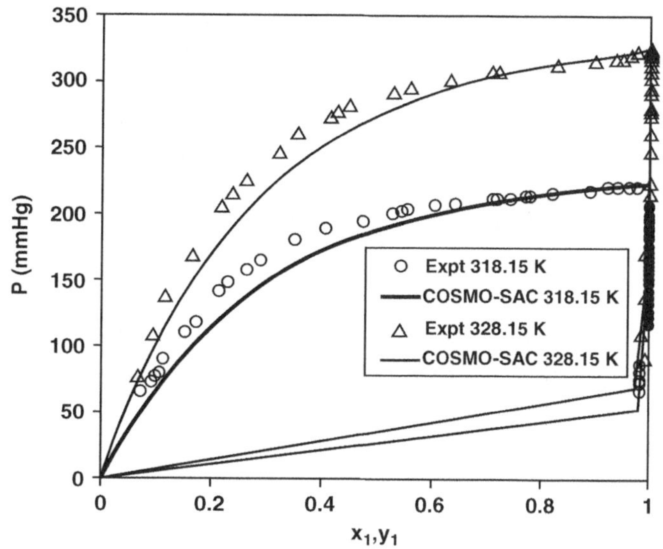

Figure 16.4 *VLE predictions from COSMO–SAC for benzene(1)–n-methylformamide(2) at temperatures 308.15 K and 323.15 K. Reprinted with permission from Fluid Phase Equilibria, Quantum mechanics: A new tool for engineering thermodynamics by Stanley I. Sandler, 210, 2, 147–160 Copyright (2003) Elsevier*

that COSMO–RS calculates VLE with low deviations and performs better than the PC–SAFT predictions ($k_{ij}=0$) for alcohol–alkane mixtures but worse compared to PC–SAFT with fitted k_{ij}.

4. Figures 16.12 and 16.13 show LLE results for a glycol–alkane and a glycol–aromatic hydrocarbon system. For both mixtures the solubility in both phases is overestimated with COSMO–RS. The performance of COSMO–RS is not satisfactory in these cases.
5. Unlike the glycol–alkane LLE, LLE in alcohol–water systems can be described satisfactorily with COSMO–RS, especially the alcohol-rich phase. The alcohol content in the water-rich phase is somewhat underestimated. A typical result is shown in Figure 16.14.
6. Limitations and the need for improved accuracy of COSMO–RS are pointed out by many authors, e.g. Modarresi *et al.*[27] Moreover, there is a need for extensive databases – libraries of sigma profiles – to become available. An example for pharmaceuticals is shown in Figure 16.15 illustrating that the performance of COSMO–RS depends on the solvent chosen.

16.3 Estimation of association model parameters using QC

Sandler and co-workers[52–57] have presented a methodology to estimate, using QC (molecular orbital) techniques, the enthalpy and entropy changes for the dimerization reactions in the vapor phase for various associating compounds (alcohols, acids, water). They compared results with two QC methods, the computationally less demanding HF (Hartree–Fock) and the more rigorous DFT (Density Functional Theory). The QC-estimated parameters were used in different ways:

1. Estimation of the association energy and volume parameters of SAFT-type models, thus reducing the number of parameters which are fitted to vapor pressures and liquid densities.

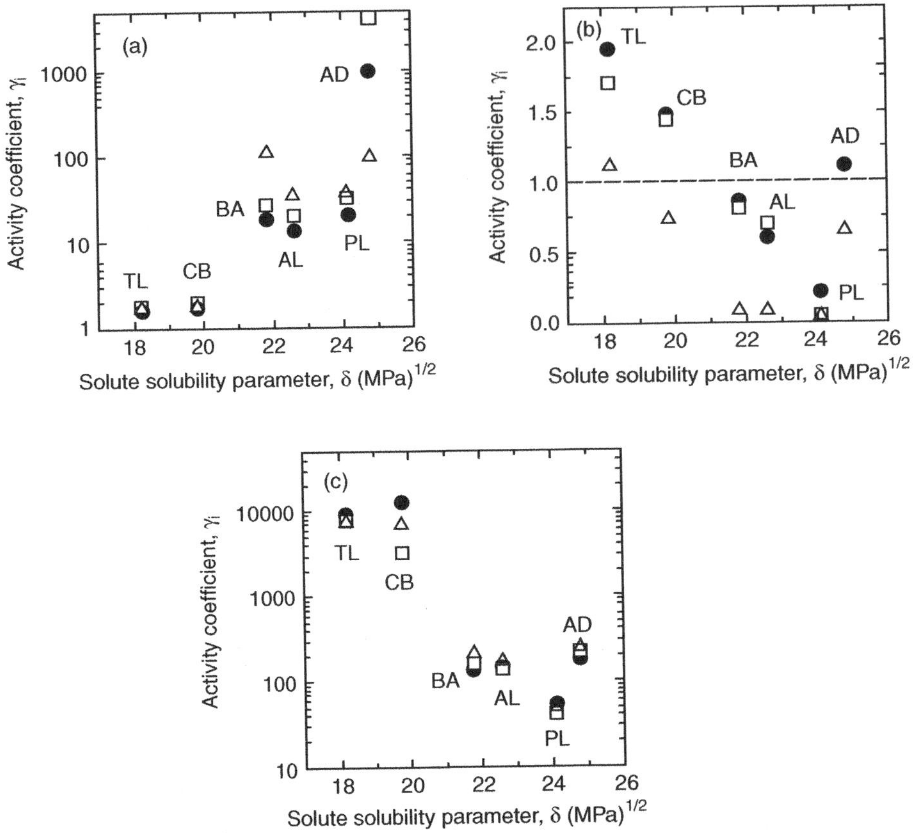

Figure 16.5 Prediction of activity coefficients of six monofunctional solutes in (a) hexane, (b) acetone and (c) water. Filled circles are the experimental data, squares indicate UNIFAC predictions and the triangles are the COSMO-RS predictions. Solute abbreviations: TL = toluene, CB = chlorobenzene, BA = benzoic acid, AL = aniline, PL = phenol and AD = acetanilide. Reprinted with permission from Fluid Phase Equilibria, Solvent selection for pharmaceuticals by P. Kolar, J.-W. Shen A. Tsuboi and T. Ishikawa; 194–197, 2, 771 Copyright (2002) Elsevier

2. Phase equilibrium calculations with SAFT-type models using association parameters from QC calculations.
3. Investigating the validity of the various association schemes, e.g. whether two or three sites should be used for alcohols and what is the correct association scheme for water.
4. Evaluating the combining rules for the association parameters, either separate combining rules for the association energy and volume or combining rules for the association strength or the cross-equilibrium constant. Examples of combining rules evaluated are the CR-1 and Elliott's rules, discussed in Chapters 7 and 8 (Equations (7.12) and (8.50)).

Details of the two molecular orbital methods (HF and DFT) can be found in Wolbach and Sandler[56,57]. Water and alcohols were assumed to form linear dimers, while acids form cyclic dimers. Table 16.3 summarizes the enthalpy and entropy change values for some of the compounds considered. Some assumptions are needed in order to use the QC-estimated values for obtaining SAFT parameters. For this purpose

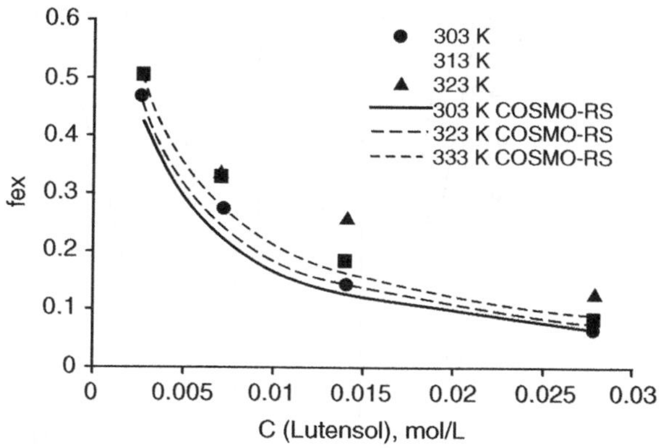

Figure 16.6 *Partitioning of p-xylene in the system p-xylene–Lutensol FSA10–water. Lines are COSMO–RS predictions and points are the experimental data. Reprinted with permission from Chem. Eng. Technol., Prediction of Equilibrium Partitioning of Nonpolar Organic Solutes in Water- Surfactant Systems by UNIFAC and COSMO-RS Models by M. Buggert, L. Mokrushina, I. Smirnova, R. Schomäcker and W. Arlt, 29, 5, 567–573 Copyright (2006) Wiley-VCH*

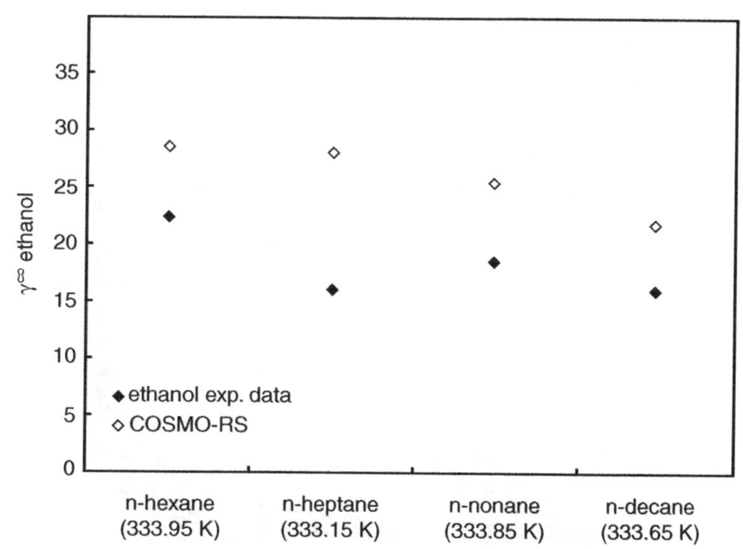

Figure 16.7 *Experimental and COSMO–RS predicted activity coefficients of ethanol at infinite dilution in different alkanes. Reprinted with permission from Ind. Eng. Chem. Res., COSMO-RS Predictions in Chemical Engineering: A Study of the Applicability to Binary VLE by Oliver Spuhl and Wolfgang Arlt, 43, 4, 852–861 Copyright (2004) American Chemical Society*

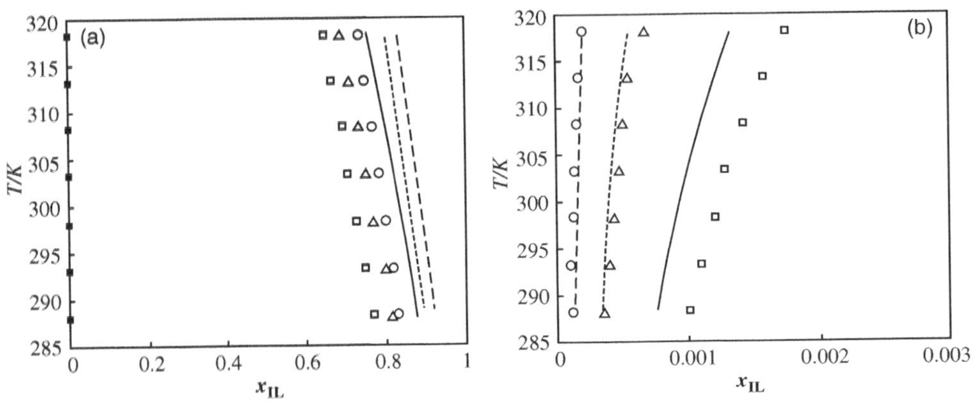

Figure 16.8 *(a) Complete liquid–liquid phase diagram for water and ionic liquids; (b) water-rich phase. Squares and full lines, [C_4mim][PF_6]; triangles and short-dashed lines, [C_6mim][PF_6]; circles and long-dashed lines, [C_8mim][PF_6]. Experimental data (points) and COSMO–RS predictions (lines). Reprinted with permission from Fluid Phase Equilibria, Evaluation of COSMO-RS for the prediction of LLE and VLE of water and ionic liquids binary systems by M. G. Freire, S. P. M. Ventura et al., 268, 1–2, 74–84 Copyright (2008) Elsevier*

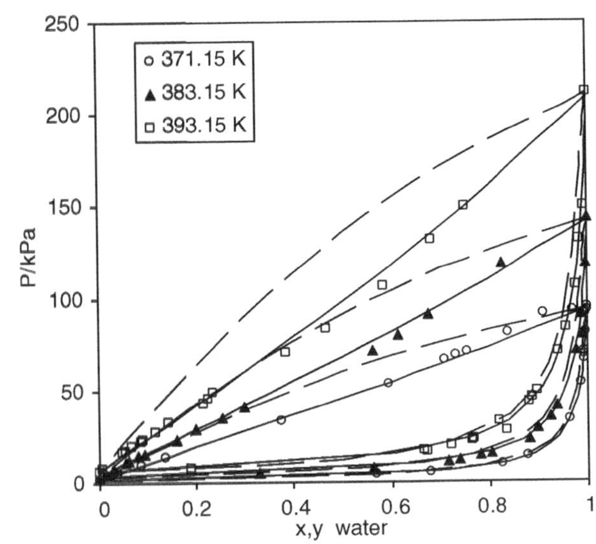

Figure 16.9 *VLE for MEG–water with COSMO–RS and sPC–SAFT. Solid lines are calculations with simplified PC–SAFT ($k_{ij} = -0.046$) and dashed lines are COSMO–RS calculations*

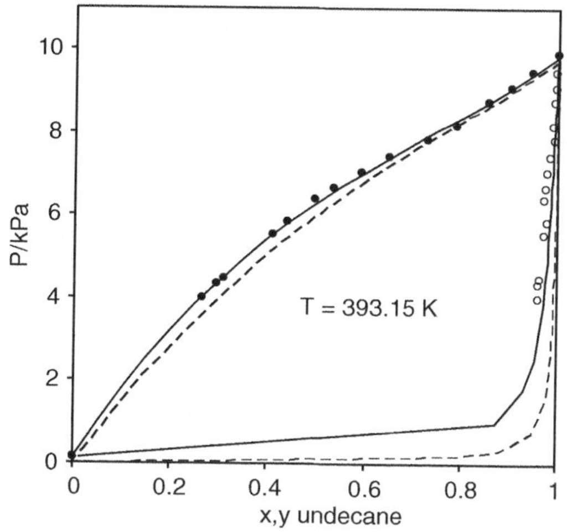

Figure 16.10 VLE for 1-tetradecanol–undecane with COSMO–RS and sPC–SAFT. Solid lines are calculations with simplified PC–SAFT ($k_{ij} = 0.0041$) and dashed lines are COSMO–RS calculations

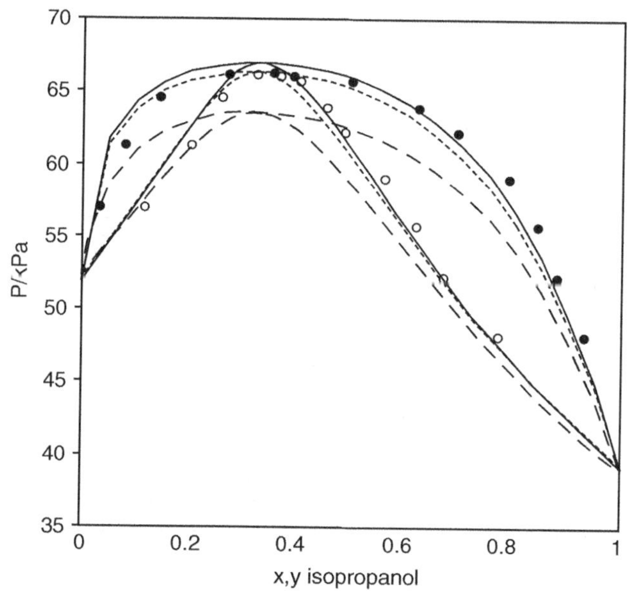

Figure 16.11 VLE for isopropanol–benzene at 333.15 K with COSMO–RS (dashed line) and sPC–SAFT. The solid lines are simplified PC–SAFT correlations ($k_{ij} = 0.002\,96$) and the short dashed lines are simplified PC-SAFT ($k_{ij} = 0$) predictions

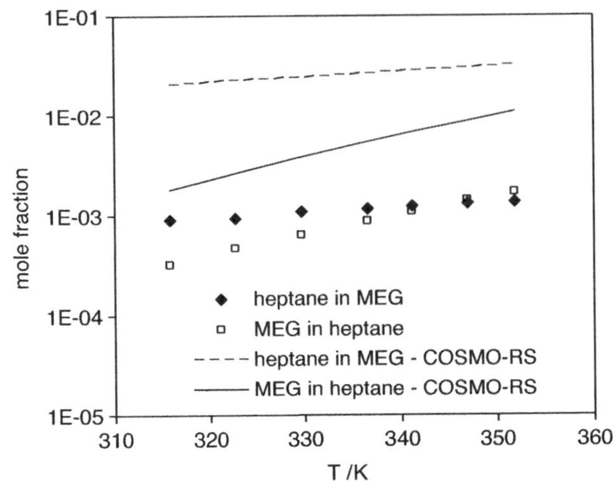

Figure 16.12 *LLE for MEG–heptane. The lines are COSMO–RS calculations. Similar results are obtained for other glycol–alkanes systems. COSMO–RS for these systems overestimates the solubility of the compounds in the two liquid phases*

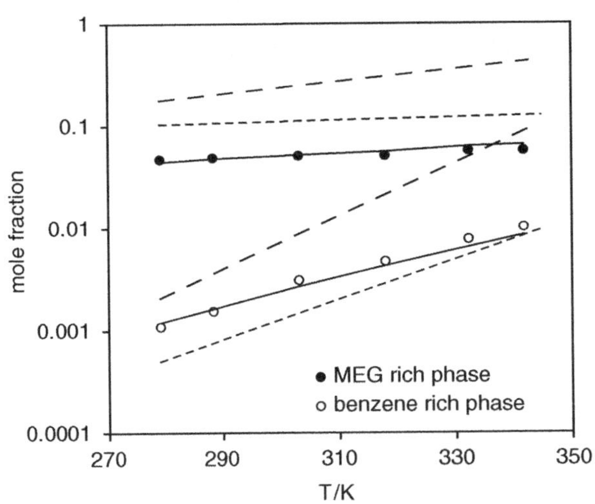

Figure 16.13 *LLE for MEG–benzene. Lines are correlations with sPC–SAFT and COSMO–RS. Short-dashed line, sPC–SAFT without accounting for solvation ($k_{ij} = -0.02$); solid line, sPC–SAFT accounting for solvation ($k_{ij} = 0.019\,9$, $\kappa_{fitted}^{A_iB_j} = 0.043$); dashed line, COSMO–RS*

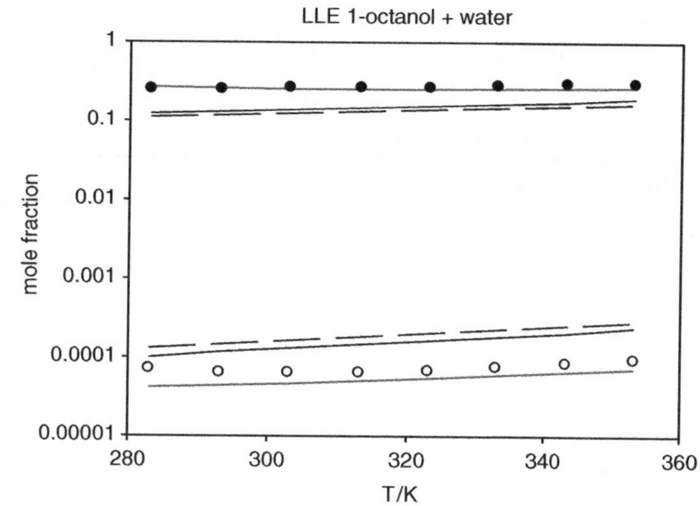

Figure 16.14 *LLE for 1-octanol–water. Lines are correlations with various models. Black lines, PC–SAFT $k_{ij} = 0$ (solid line, alcohol parameters from Grenner et al., Fluid Phase Equilib.,* ***2007****, 258(1), 83; dashed line, alcohol parameters from Gross and Sadowski, Ind. Eng. Chem. Res.,* ***2002****, 41, 5510); grey line, COSMO–RS*

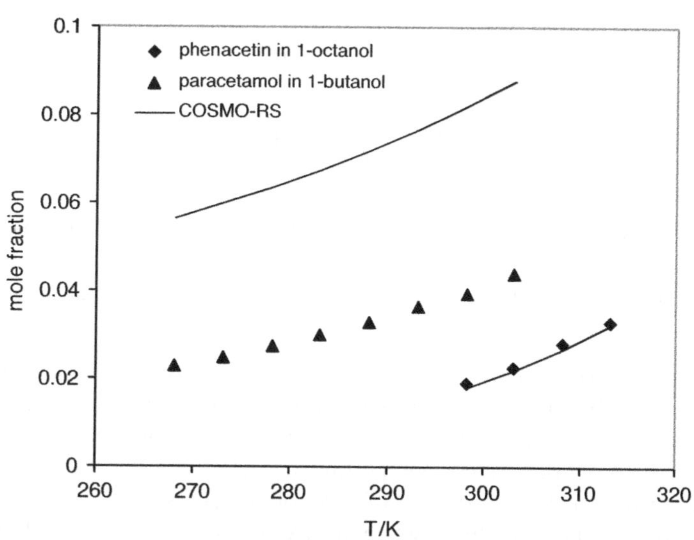

Figure 16.15 *Solubility of two pharmaceuticals in two solvents predicted by COSMO–RS*

Table 16.3 Enthalpies and entropies of hydrogen bonding for various molecules. The MO calculations are from the HF and DFT methods (first and second value, respectively, in the 'MO' column) from Wolbach and Sandler,[53–55] and the chemical theory values are from Anderko[59–62]. The various values presented with the chemical theory are based on different parameter regressions and chemical theories used. All enthalpy values are in kJ/mol and the entropy values are in J/mol K. The references for the experimental ΔH values are presented in Tables 13.2–13.5

Compound	$-\Delta H$ (MO)	$-\Delta H$ (Exp)	$-\Delta H$ (Chemical)	$-\Delta S$ (MO)	$-\Delta S$ (Chemical)
Water	16.4 12.5	15.1	29	77 88	106–108
Methanol	16.2 13.1	25 ± 3	19–26	83 94	85–91
Ethanol		21–2	22–26		89–95
Propanol		21–25	23		91
Formic acid	56.7 56.5			144 156	
Acetic acid	58.3 58.4	58–61	52	142 151	122
Propionic acid	57.4 57.5			144 156	

Sandler and co-workers[52–55] used the following equation (which depends on the association scheme assumed):

$$\begin{aligned} \Delta^{AB} &= C^{/} RTK \\ C^{/} &= 1 \qquad (2B) \\ C^{/} &= 2 \qquad (1A) \\ C^{/} &= \frac{1}{2} \qquad (3B) \\ C^{/} &= \frac{1}{4} \qquad (4C) \end{aligned} \tag{16.2}$$

together with the definitions of the association strength (Δ^{AB}) and the equilibrium constant (K) given in Equations (7.14) and (7.15) (Chapter 7). The most important conclusions from the investigations by Sandler and co-workers[52–57] and others[58] who have used QC-based association parameters in SAFT are the following:

- It is possible to eliminate one of the two association parameters of SAFT, but typically not both. The remaining parameters must be fitted to vapor pressure and liquid density data. Nevertheless, in this way the deviations in vapor pressures and liquid densities are similar to those obtained when all five parameters of association models (e.g. SAFT) are fitted to experimental data.
- Good VLE results are obtained for mixtures of alcohols or acids with alkanes as well as for cross-associating systems, often better than those when all SAFT parameters are fitted to vapor pressures and liquid densities.
- The three-site (3B) scheme is the best choice for methanol and the 4C scheme for water. Acids are modeled as one-site molecules.

- It makes little difference to use the more advanced DFT method over the simpler HF. Actually, possibly due to the cancellation of errors, the results with the HF method are more satisfactory in most cases.
- The QC-estimated enthalpy and entropy changes agree well and validate the arithmetic mean rules for the cross-enthalpy/cross-entropy terms, thus also the geometric mean rule for the equilibrium constants (Equations (7.13) in Chapter 7). Based on these rules, Wolbach and Sandler[56,57] proposed the combining rule given by Equation (8.50) (Chapter 8), which is essentially Elliott's rule for the association strength.

Despite these encouraging results, certain deficiencies or difficulties when using QC-based parameters should be mentioned. First, DFT results appear inferior in most cases compared to HF ones (with the differences between the two methods being especially pronounced for methanol). Moreover, it was not typically possible to estimate reliably both association parameters from QC values and sometimes higher interaction parameters were required for good VLE results with SAFT using QC-based parameters. Finally, in certain cases, different association schemes were used for the same molecule (for different systems). For example, water performs better using the 3B scheme in mixtures with acetone, while for all other systems water performs best when a 4C scheme is used. It should also be kept in mind that the QC values from Sandler and co-workers correspond to dimerization of isolated molecules in the vapor phase, thus they may differ from association values in the liquid state.

The values of the enthalpy and entropy changes obtained from molecular orbital (MO) calculations are in reasonably good agreement with experimental values (from spectroscopy) and with the values obtained from the chemical theory of Anderko[59–62], as shown in Table 16.3. This is especially true when the scatter in both the experimental and chemical theory values is taken into account.

It was mentioned that QM values can be used to check combining rules in association theories. Such investigations have resulted in the validation of Elliott's rule for the association strength or its equivalent geometric mean rule for the equilibrium constant. A similar investigation is shown in Table 16.4 for the modified CR-1 rule often used in SAFT-type models for systems showing induced solvation, e.g. methanol–chloroform and ethanol–acetone. Association energies (in K) estimated with the modified CR-1 rule are compared to the few experimental values available and the estimations with the method of Drago and Wayland.[63] There are only a few experimental and MO values and thus definite conclusions cannot be drawn,

Table 16.4 *Calculation of cross-association energy (enthalpy), ΔH_{12} or ε_{12} (all values are in K) with various equations of state and the Drago-Wayland method. The experimental values correspond to the enthalpy of hydrogen bond formation (in kJ/mol of reactant), while the values with PC–SAFT, CPA and NRHB are estimated with the so-called modified CR-1 rule, i.e. $\Delta H_{12} = \Delta H_{assoc}/2$ or $\varepsilon_{12} = \varepsilon_{assoc}/2$ (except for chloroform–acetone, where the value is fitted to experimental data). The sPC–SAFT and NRHB values are from Grenner et al., Ind. Eng. Chem. Res., **2008**, 47(15), 5636. The polar PC–SAFT values are from Kleiner and Sadowski, J. Phys. Chem. C, **2007**, 111, 15544. The cross-association energy values from the equations of state and the Drago-Wayland method are expressed per mole of reactant. (Note that the values reported in Table 16.4 for ABPACT are half those given by Economou et al., AIChE J., **1990**, 36, 851, which are given per mole of product)*

System	Exp.	Drago–Wayland	Polar PC–SAFT	NRHB	CPA	sPC–SAFT	ABPACT
Methanol–chloroform		2064	1250	1509	1479	1450	1848
Methanol–acetone	1270	1259	1250	1509	1479	1450	1436
Ethanol–chloroform		1989	1327	1443	1295	1406	1712
Ethanol–acetone	1743	1401	1327	1443	1295	1406	1436
Chloroform–acetone	1367	1850		1277	1004	1400	1848

but the following can be observed:

1. For mixtures of methanol or ethanol with chloroform and acetone: all models provide values rather close to those obtained by the Drago-Wayland[63] method and are also close to the two experimental values available. These results indicate the validity of the modified CR-1 rule for these induced solvating systems.
2. For chloroform–acetone: Again the models provide rather good agreement with the Drago-Wayland method and with the experimental value. In this case the cross-association energy of the models is fitted directly to the experimental data, as both molecules are not self-associating.

16.4 Estimation of size parameters of SAFT-type models from QC

16.4.1 The approach of Imperial College

Recent methods have been proposed not only for the association parameters of SAFT, but also for estimating the size parameters (segment number and segment diameter) from QM calculations. Sheldon *et al.*[64] and Pollock[65] have proposed a methodology based on QM for obtaining the σ and m parameters of SAFT–VR and then the remaining parameters are estimated in the usual way, i.e. by fitting to vapor pressure and liquid density data. The molecule is treated as a spherocylinder of diameter σ and aspect ratio m (Figure 16.16). The strategy followed consists of the following steps:

1. QM (the restricted HF formalism) is used to calculate the volume V_{QM} of a molecule and the dimensions of the smallest box containing the molecule $L_1 \leq L_2 \leq L_3$.
2. The dimensions σ and m are calculated for a spherocylinder (Figure 16.16) of the same volume and aspect ratio as the QM molecule. Then, the m and σ parameters are given by:

$$m = \frac{L_{longest}}{L_{shortest}} \tag{16.3}$$

$$\sigma = \left[\frac{4V_{QM}}{\pi(m-1/3)}\right]^{1/3} \tag{16.4}$$

3. The remaining SAFT–VR parameters (dispersion energy and square-well λ as well as association energy and volume) are fitted to experimental vapor pressures and liquid densities (up to temperatures below 90% of the critical temperature).

Parameters for several hydrocarbons, gases (N_2, CO_2, CO), water and refrigerants have been determined quantum-mechanically and the parameters are rather close to those estimated directly from fitting experimental data, as can be seen in Figures 16.17–16.19. The QM-based parameter results agree well, when used in

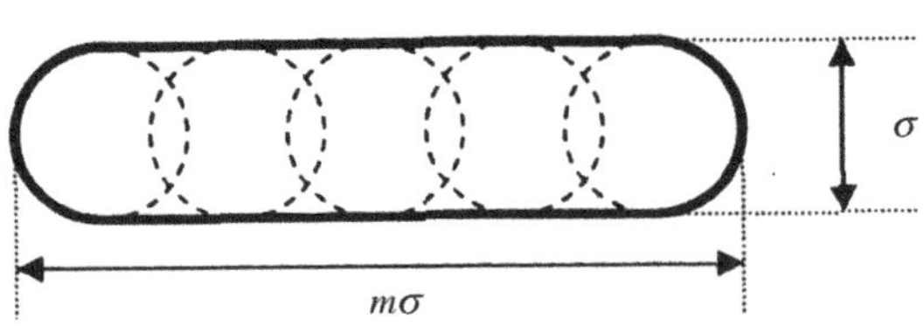

Figure 16.16 *A spherocylinder with diameter σ and aspect ratio or number of spheres $m = 4.24$. From Sheldon et al.*[64]

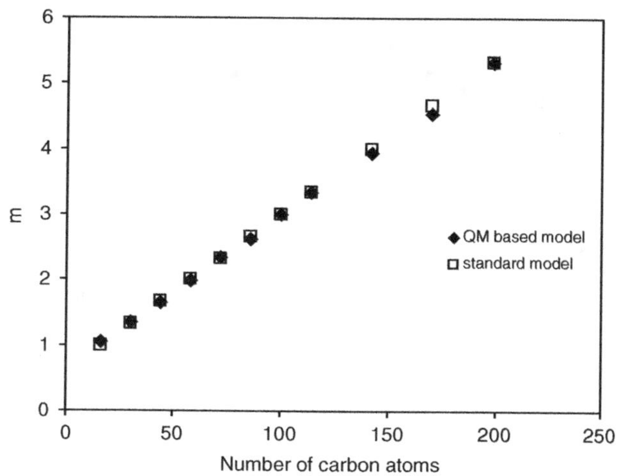

Figure 16.17 *Behavior of the SAFT–VR segment number parameter as a function of the carbon number for n-alkanes. The open symbols indicate the values obtained with the standard method (parameters fitted to vapor pressure and liquid densities) and the filled symbols are the values obtained from QC calculations (QM-based model) . Based on the values presented by Sheldon et al.[64]*

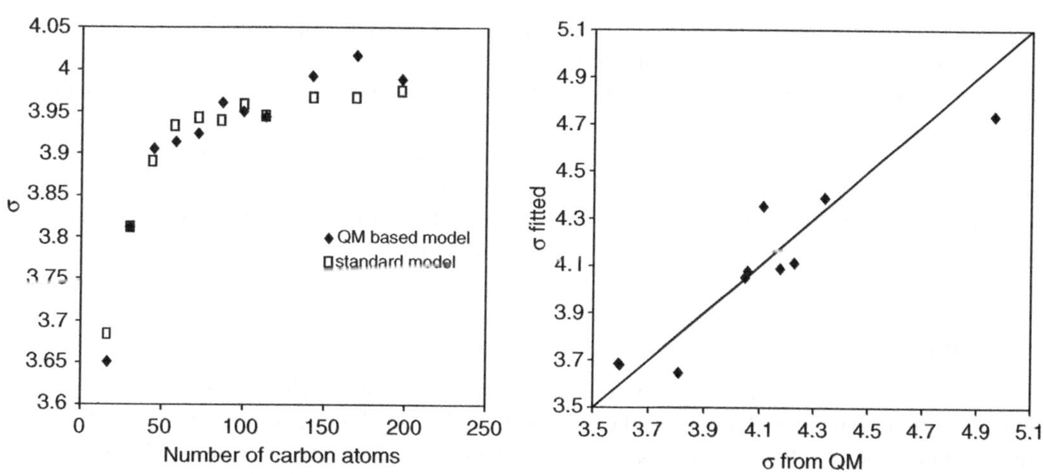

Figure 16.18 *Behavior of the SAFT–VR segment diameter parameter for n-alkanes (left) and refrigerants (right). Left: The open symbols indicate the values obtained with the standard method (parameters fitted to vapor pressure and liquid densities) and the filled symbols are the values obtained from QC calculations (QM-based model). Based on the values presented by Sheldon et al.[64] Right: Comparison of values from the standard method against QM-estimated values for refrigerants. Based on the values presented by Pollock[65]*

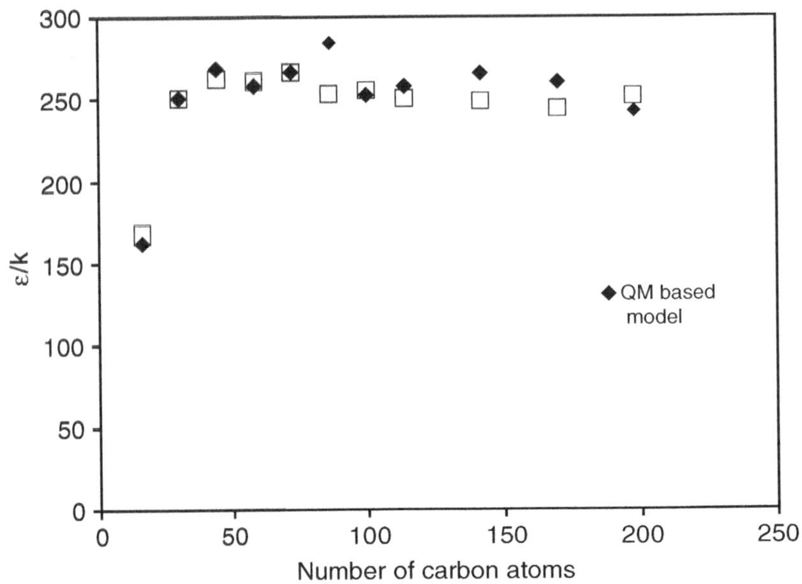

Figure 16.19 *Behavior of the SAFT–VR segment energy parameter as a function of the carbon number for n-alkanes. The open symbols indicate the values obtained with the standard method (parameters fitted to vapor pressure and liquid densities) and the filled symbols are the values obtained from QC (QM-based model). Based on the values presented by Sheldon et al.*[64]

SAFT–VR, with vapor pressure (0.7–7%) and liquid density (1–15%) data. The largest deviations are observed for density, indicating that the density calculations are very sensitive to the size parameters. The errors are more pronounced for cyclic and aromatic compounds such as benzene and cyclohexane, possibly due to limitations of the spherocylindrical model adopted. Thus, the approach recommended by Sheldon *et al.*[64] and Pollock[65] for engineering calculations is to determine only m from QM and obtain the remaining parameters (including σ) from vapor pressure and liquid density data if such accurate data are available for both properties. Despite the limitations mentioned above, in the future the methodologies based on QM may offer broadly applicable routes useful in estimating equation of state parameters.

16.4.2 The approach of Aachen

Leonhard and co-workers[66–69] have proposed in a series of recent articles an alternative approach for utilizing QC computations in SAFT-type equations of state. The final aim is to arrive at partially predictive equations of state. These authors performed QC calculations to determine various molecular properties, namely dipole moments, quadrupole moments, polarizabilities and the dipole–dipole dispersion coefficient, C_{ij}:

$$E_{disp} = -\frac{C_{ij}}{r_{ij}^6} \tag{16.5}$$

where E_{disp} is the dispersion energy and r_{ij} is the distance between the molecules.

These properties have been determined for several small to medium-sized molecules including hydrocarbons, gases, ethers, alcohols, water, nitriles and refrigerants. The QC method yields dipole moments with an accuracy of about 3%, polarizabilities and C_{ij} dispersion coefficients with an accuracy of about 5%, and quadrupole moments with an accuracy higher than that of the available few experimental data (which can be around 10%). Moreover, using QC the deviation of the unlike–dispersion interaction coefficient from the geometric mean combining rule is also calculated:

$$\left(\frac{C_{ij}}{\sqrt{C_i C_j}}\right)$$

The accuracy for the ratios of dispersion coefficients is around 0.3%, thus higher than the accuracy of the individual dispersion coefficients. The Leonhard method has been mentioned in Chapter 3 (Appendix 3. B) and as was shown there the ratio above achieves values close to the function containing ionization potentials:

$$\left(\frac{2\sqrt{I_i I_j}}{I_i + I_j}\right)$$

as expected from the Houdson–McCoubrey (*Trans. Faraday Soc.*, **1960**, *56*, 761) equation (see Equations ((3.26) and (3.31)).

By introducing certain simplifying assumptions in the QC calculations, different combining rules can be obtained, e.g. the well-known Berthelot combining rule:

$$C_{ij} = \sqrt{C_i C_j} \tag{16.6}$$

Alternatively, using a more refined approximation (assuming a certain frequency dependency of the polarizability), the so-called *London rule* is derived (also known as the Moelwyn–Hughes rule):

$$C_{ij} = 2\frac{C_i C_j}{\alpha_j^2 C_i + \alpha_i^2 C_j}\alpha_i \alpha_j \tag{16.7}$$

or equivalently written as:

$$\begin{aligned} \varepsilon_{ij} &= \frac{2}{\sigma_{ij}^6}\frac{\varepsilon_i \sigma_i^6 \varepsilon_j \sigma_j^6 \alpha_{i,s}\alpha_{j,s}}{\alpha_{j,s}^2 \varepsilon_i \sigma_i^2 + \alpha_{i,s}^2 \varepsilon_j \sigma_j^2} \\ \alpha_{i,s} &= \frac{\alpha_i}{m_i} \end{aligned} \tag{16.8}$$

The latter equation for $\alpha_{i,s}$ corresponds to the static polarizability of a molecule i divided by the number of segments in the molecule i, m.

Finally, Leonhard *et al.*[66] derived the following combining rule from QM considerations, which, as mentioned previously, resembles the combining rule first proposed by Houdson and McCoubrey:

$$\varepsilon_{ij} = \sqrt{\varepsilon_i \varepsilon_j}\left(\frac{\sqrt{\sigma_i \sigma_j}}{\sigma_{ij}}\right)^6 \left(\frac{C_{ij}}{\sqrt{C_i C_j}}\right)\left(\frac{f_{ij}}{\sqrt{f_i f_j}}\right) \tag{16.9}$$

f is a conversion factor and the term with the f factors can be approximately ignored.

Using the above combining rules, e.g. Equations (16.8) and (16.9), the interaction parameter k_{ij} often used to correct for the GM rule can be predicted solely from parameters obtained from QM calculations. These properties were then used in the polar PC–SAFT EoS by Gross and co-workers (PCP–SAFT, see Chapter 13) to perform calculations for phase equilibrium (VLE, LLE) and excess enthalpies. The remaining parameters of PCP–SAFT (segment number, diameter and energy) are fitted to vapor pressures and liquid densities, while dipole and quadrupole moments are obtained from QM. The predictions based on (partially) quantum-mechanically determined pure compound properties together with the QM-based combining rules (Equations (16.8) and (16.9)) show promising rules compared to the original PCP–SAFT (with the geometric mean combining rule), e.g. the average deviation is decreased from 10% (geometric mean rule) to 4% (London rule).

There are, however, also some limitations. Many of the problems are not necessarily associated to the QM approach of estimating the parameters but can be attributed to assumptions in the EoS itself. These problems occur especially when the model is used for mixtures for which certain of its underlying assumptions are not fully applicable, e.g. CO_2–neopentane, cyclohexane with benzene or DMF and refrigerant–alkane mixtures. The version of PCP–SAFT used did not include any dipole–quadrupole interactions. Figures 16.20 and 16.21

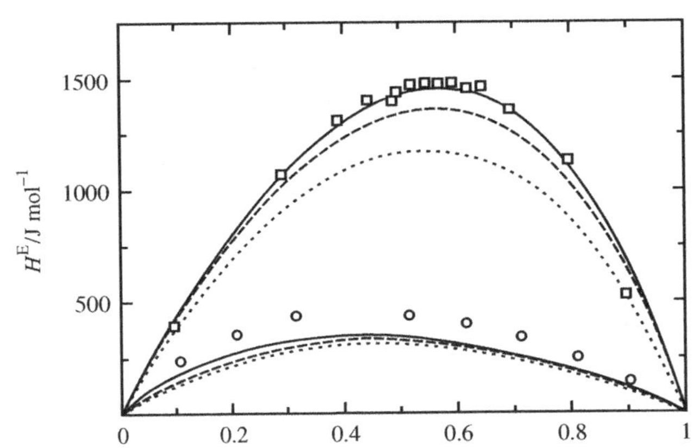

Figure 16.20 *Excess enthalpy (H^E) for the system ethane(1)–carbon dioxide(2) at 308.4 K and 4 MPa (circles) and at 308.4 K and 11 MPa (squares) with the PCP–SAFT EoS. The dotted line shows the prediction obtained with the geometric mean rule for ε (= $(\varepsilon_1\varepsilon_2)^{1/2}$), the dashed line shows the results for the London combining rule, Equation (16.8), and the solid line stands for the combining rule using QM-obtained C_6 coefficients, Equation (16.9). Reprinted with permission from Fluid Phase Equilibria, Making equation of state models predictive: Part 2: An improved PCP-SAFT equation of state by Kai Leonhard, Nguyen Van Nhu and Klaus Lucas, 258, 1, 41–50 Copyright (2007) Elsevier*

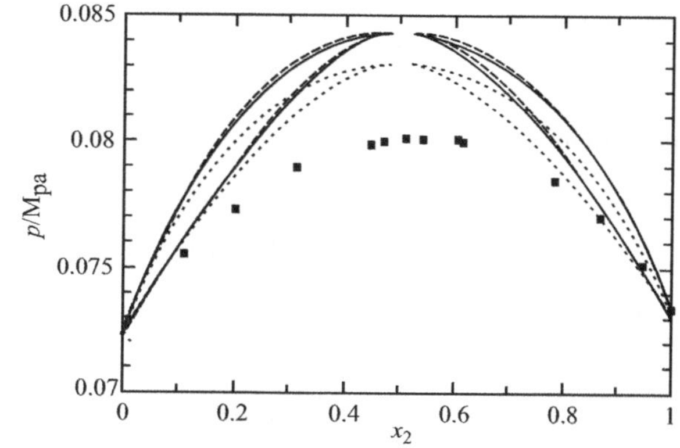

Figure 16.21 *VLE of the system cyclohexane(1)–benzene(2) at 343.15 K with the PCP–SAFT EoS. The dotted line shows the prediction obtained with the geometric mean rule for* $\varepsilon\,(= (\varepsilon_1\varepsilon_2)^{1/2})$*, the dashed line shows the results for the London combining rule, Equation (16.8), and the solid line stands for the combining rule using QM-obtained* C_6 *coefficients, Equation (16.9). Reprinted with permission from Fluid Phase Equilibria, Making equation of state models predictive: Part 2: An improved PCP-SAFT equation of state by Kai Leonhard, Nguyen Van Nhu and Klaus Lucas, 258, 1, 41–50 Copyright (2007) Elsevier*

illustrate two characteristic results. Even for these systems, however, there is an improvement when k_{ij} is obtained from any of the QM-based combining rules (Equations (16.8) and (16.9)) compared to the geometric mean rule, though the improvement is smaller compared to other systems where the EoS assumptions are valid (8% vs. 12%).

In their more recent work, Leonhard and co-workers[69] have computed molecular descriptors for the size, shape, charge distribution and dispersion interactions for 67 compounds using QC *ab initio* and DFT methods. The purpose was to develop predictive methods to estimate the parameters of the physical term of a SAFT-type model (PCP–SAFT), which also contains special polar and quadrupolar terms. Their approach can be briefly described as follows:

1. They first fitted the m, ε and σ parameters for 67 compounds based on vapor pressure and liquid density data and retained parameters for 49 compounds, for which the deviations in either of the properties is below 1.5%.
2. A fused hard-sphere model is chosen and the fitted parameters are used to estimate the volume and shape factor of a molecule:

$$V_{fit} = \frac{4}{3}\pi m_{fit}\left(\frac{\sigma_{fit}}{2}\right)^3$$
$$\alpha_{fit} = \frac{m_{fit}+1}{2} \qquad (16.10)$$

   It was difficult to find a single optimum set of element radii which fits equally well the volume and shape factor and the best set of radii (for H, C and O atoms) which could represent V within 4.6% and the shape factor with a deviation of 6%.
3. On the basis of the molecular volume, the shape factor, the dipole and quadrupole moments, the C_6 dispersion coefficients and various preliminary calculations, the authors defined the following five molecular descriptors, in order to regress EoS parameters:

(a) the geometrical segment number

$$m_0 = 2\alpha_0 - 1 \tag{16.11}$$

(b) the geometrical segment diameter

$$\sigma_0 = 2\sqrt[3]{\frac{3}{4\pi}\frac{V_0}{m_0}} \tag{16.12}$$

(c) the reduced segment energy parameter

$$\varepsilon_0 = \frac{C_6}{m_0^6 \sigma_0^6} \tag{16.13}$$

(d) the reduced dipole moment

$$\mu_0^* = \frac{\mu}{m_0 \varepsilon_0 \sigma_0^3} \tag{16.14}$$

(e) the reduced quadrupole moment

$$\Theta_0^{*4} = \frac{\Theta^4}{m_0 \varepsilon_0 \sigma_0^5} \tag{16.15}$$

4. The final step is to perform multilinear regressions between the molecular descriptors and the three parameters fitted to the 49 compounds. The equation used is:

$$P_{prediction} = \varepsilon_0 C_{\varepsilon_0}^P + m_0 C_{m_0}^P + \sigma_0 C_{\sigma_0}^P + \Theta_0^{*4} C_{\Theta_0^{*4}}^P + \mu_0^* C_{\mu_0^*}^P + C_0^P \tag{16.16}$$

where:

$P = m, \varepsilon$ or σ
C_D^P = the coefficient associated with parameter P and descriptor D
C_0^P = the y axis intercept

Dipole and quadrupole moments as well as C_6 coefficients were obtained from QM calculations, as described above. The analysis showed that four descriptors are needed for m and only three for the segment diameter, with m_0 and σ_0 being the most significant parameters for these properties, as can be expected. All five descriptors are necessary for the segment energy parameter, although ε_0 is evidently quite significant. Using these predicted parameters from Equation (16.16), PCP–SAFT provides a reasonably good prediction of vapor pressure (42%), liquid density (6%), enthalpy of vaporization (11%) and boiling point (10%) for the 49 compounds. CO_2, CO and certain triple bond molecules are not represented very well. When the energy parameter is fitted to the boiling point, and the two size parameters are taken from Equation (16.16), the results are much improved (3% in vapor pressure and enthalpy of vaporization and 4% in liquid density). Good VLE and excess enthalpy predictions for a few mixtures were obtained, with k_{ij} estimated from the London combining rule (Equation (16.8)).

16.5 Conclusions

Quantum chemistry (QC) is receiving increased interest in engineering thermodynamics and already many results can be used for practical applications. This chapter is a modest introduction focusing on the applications rather than the computational details of the different approaches. The *most complete* method today is possibly the solvation method COSMO–RS and its variations such as COSMO–SAC. These methods provide activity coefficient calculations a priori, without using any mixture data. There are still limitations and unresolved questions, but the COSMO–RS approach has already been shown to be a qualitatively very successful method, especially for solvent screening and novel applications where experimental data are scarce. An alternative to the COSMO–RS approach of using QC results, which is suitable also for high-pressure calculations, is to estimate via quantum mechanics (QM) the parameters of theoretically based models such as those of the SAFT family equations of state. This has been attempted recently for both the association parameters and the parameters of the physical terms (number of segments, segment energy and segment diameter) and the first results are promising, although the QM-estimated parameters cannot be expected to resolve whatever problems the equations of state may have.

References

1. S.I. Sandler, M. Castier, *Pure Appl. Chem.*, **2007**, *79*, 1345.
2. M.S. Gordon, M.W. Schmidt, Advances in electronic structure theory: GAMESS a decade later. In: C.E. Dykstra *et al.*, Eds, *Theory and Applications of Computational Chemistry*. Elsevier, 2005, Chapter 41.
3. S.I. Sandler, *Fluid Phase Equilib.*, **2003**, *210*, 147.
4. S.I. Sandler, Thermodynamic properties from quantum chemistry. In: T. Letcher, Ed., *Chemical Thermodynamics in Industry*. Royal Society of Chemistry, 2004, pp. 43–56.
5. S.I. Sandler, S.-T. Lin, A.K. Sum, *Fluid Phase Equilib.*, **2002**, *194–197*, 61.
6. S.-T. Lin, S.I. Sandler, *Chem. Eng. Sci.*, **2002**, *57*, 2727.
7. S.-T. Lin, S.I. Sandler, *Ind. Eng. Chem. Res.*, **2002**, *41*, 899.
8. S.-T. Lin, S.I. Sandler, *AIChE J.*, **1999**, *45*, 2606.
9. A. Klamt, *J. Phys. Chem.*, **1995**, *99*, 2224.
10. A. Klamt, G. Schuurmann, *J. Chem. Soc. Perkin Trans.*, **1993**, *2*, 799.

11a. A. Klamt, F. Eckert, *Fluid Phase Equilib.*, **2000**, *172*, 43.

11b. A. Schafer, A. Klamt, D. Sattel, J.C.W. Lohrenz, F. Eckert, *Phys. Chem. Chem. Phys.*, **2000**, *2*, 2187.

12. C. Panayiotou, *Ind. Eng. Chem. Res.*, **2003**, *42*, 1495.
13. C. Panayiotou, *J. Chem. Thermodyn.*, **2003**, *35*, 349.
14. T. Banerjee, M.K. Singh, A. Khanna, *Ind. Eng. Chem. Res.*, **2006**, *45*, 3207.
15. O. Spuhl, W. Arlt, *Ind. Eng. Chem. Res.*, **2004**, *43*, 852.
16. A. Klamt, F. Eckert, *Fluid Phase Equilib.*, **2004**, *217*, 53.
17. A. Klamt, F. Eckert, *Fluid Phase Equilib.*, **2007**, *260*, 183.
18. H. Grensemann, J. Gmehling, *Ind. Eng. Chem. Res.*, **2005**, *44*, 1610.
19. T. Mu, J. Rarey, J. Gmehling, *Ind. Eng. Chem. Res.*, **2008**, *47*, 989.
20. T. Mu, J. Rarey, J. Gmehling, *Ind. Eng. Chem. Res.*, **2007**, *46*, 6612.
21. F. Eckert, A. Klamt, *Ind. Eng. Chem. Res.*, **2001**, *40*, 2371.
22. F. Eckert, A. Klamt, *AIChE J.*, **2002**, *48*, 369.
23. R. Putnam, R. Taylor, A. Klamt, F. Eckert, M. Schiller, *Ind. Eng. Chem. Res.*, **2003**, *42*, 3635.
24. F. Eckert, A. Klamt, *Fluid Phase Equilib.*, **2003**, *210*, 117.
25. A. Klamt, *Fluid Phase Equilib.*, **2003**, *206*, 223.
26. P. Kolar, J.-W. Shen, A. Tsuboi, T. Ishikawa, *Fluid Phase Equilib.*, **2002**, *194–197*, 771.

27. H. Modaressi, E. Conte, J. Abildskov, R. Gani, P. Crafts, *Ind. Eng. Chem. Res.*, **2008**, *47*, 5234.
28. H.-H. Tung, J. Tabora, N. Variankaval, D. Bakken, C.-C. Chen, *J. Pharm. Sci.*, **2008**, *97(5)*, 1813.
29. A. Klamt, F. Eckert, M. Hornig, *J. Comput.-Aided Mol. Des.*, **2001**, *15*, 355.
30. A. Klamt, F. Eckert, M. Hornig, M.E. Beck, T. Burger, *J. Comput. Chem.*, **2002**, *23*, 275.
31. M. Diedenhofen, F. Eckert, A. Klamt, *J. Chem. Eng. Data*, **2003**, *48*, 475.
32. J. Palomar, V.R. Ferro, J.S. Torrecilla, F. Rodrigues, *Ind. Eng. Chem. Res.*, **2007**, *46*, 6041.
33. R. Kato, J. Gmehling, *J. Chem. Thermodyn.*, **2005**, *37*, 603.
34. X. Zhang, Z. Liu, W. Wang, *AIChE J.*, **2008**, *54*, 2717.
35. M.G. Freire, S.P.M. Ventura, L.M.N.B.F. Santos, I.M. Marrucho, J.A.P. Coutinho, *Fluid Phase Equilib.*, **2008**, *268*, 74.
36. C. Mehler, A. Klamt, W. Peukert, *AIChE J.*, **2002**, *48*, 1093.
37. A. Klamt, F. Eckert, *Ind. Eng. Chem. Res.*, **2008**, *47*, 1351.
38. K. Wichmann, M. Diedenhofen, A. Klamt, *J. Chem. Inf. Model.*, **2007**, *47*, 228.
39. A. Klamt, F. Eckert, M. Diedenhofen, *Environ. Toxicol. Chem.*, **2002**, *21*, 2562.
40. A.C. Spieß, W. Eberhand, M. Peters, M.F. Eckstein, L. Greiner, J. Buchs, *Chem. Eng. Process.*, **2008**, *47*, 1034.
41. M. Buggert, L. Mokrushina, I. Smirnova, R. Schomacker, W. Arlt, *Chem. Eng. Technol.*, **2006**, *29*, 567.
42. N.M. Tukur, S.M. Waziri, E.Z. Hamad, Predicting solubilities in polymer systems with COSMO–RS. AIChE Annual Meeting, Cincinnati, USA, 2005.
43. A. Klamt, F. Eckert, M. Diedenhofen, M.E. Beck, *J. Phys. Chem. A*, **2003**, *107*, 9380.
44. K. Leonhard, J. Vererka, K. Lucas, *Fluid Phase Equilib.*, **2009**, *275*(2), 105.
45. Y. Shimoyama, Y. Iwai, S. Takada, Y. Arai, T. Tsuji, T. Hiaki, *Fluid Phase Equilib.*, **2006**, *243*, 183.
46. D. Constaninescu, A. Klamt, D. Geana, *Fluid Phase Equilib.*, **2005**, *231*, 231.
47. W. Arlt, O. Spuhl, A. Klamt, *Chem. Eng. Process.*, **2004**, *43*, 221.
48. A. Klamt, F. Eckert, *Ind. Eng. Chem. Res.*, **2006**, *45*, 3766.
49. S. Wang, S.-T. Lin, J. Chang, W.A. Goddard III, S.I. Sandler, *Ind. Eng. Chem. Res.*, **2006**, *45*, 5426.
50. A. Klamt, *Ind. Eng. Chem. Res.*, **2002**, *41*, 2330.
51. A. Klamt, *Ind. Eng. Chem. Res.*, **2008**, *47*, 987.
52. S.I. Sandler, J.P. Wolbach, M. Castier, G. Escobedo-Alvarado, *Fluid Phase Equilib.*, **1997**, *136*, 15.
53. J.P. Wolbach, S.I. Sandler, *Ind. Eng. Chem. Res.*, **1997**, *36*, 4041.
54. J.P. Wolbach, S.I. Sandler, *AIChE J.*, **1997**, *43*, 1589.
55. J.P. Wolbach, S.I. Sandler, *Int. J. Thermophys.*, **1997**, *18*, 1001.
56. J.P. Wolbach, S.I. Sandler, *AIChE J.*, **1997**, *43*, 1589.
57. J.P. Wolbach, S.I. Sandler, *AIChE J.*, **1997**, *43*, 1597.
58. M. Yarrison, W.G. Chapman, *Fluid Phase Equilib.*, **2004**, *226*, 195.
59. A. Anderko, *Fluid Phase Equilib.*, **1989**, *45*, 39.
60. A. Anderko, *Fluid Phase Equilib.*, **1989**, *50*, 21.
61. A. Anderko, *Fluid Phase Equilib.*, **1991**, *65*, 89.
62. A. Anderko, *Fluid Phase Equilib.*, **1992**, *74*, 89.
63. R.S. Drago, B.B. Wayland, *J. Am. Chem. Soc.*, **1965**, *87*, 3571.
64. T.J. Sheldon, B. Ginera, C.S. Adjiman, A. Galindo, G. Jackson, D. Jacquemin, V. Wathelet, E.A. Perpète, The derivation of size parameters for the SAFT-VR equation of state from quantum mechanical calculations. In: M. Laso and E.A. Perpète, Eds, *Computer-Aided Chemical Engineering 22: Multiscale Modelling of Polymer Properties*, Part I, Chapter 7. Elsevier, 2006.
65. M. Pollock, Systematic approaches to the development of thermodynamic models for associating fluids and their mixtures. PhD Thesis, Department of Chemical Engineering, Imperial College London, 2008.
66. K. Leonhard, V.N. Nguyen, K. Lucas, *Fluid Phase Equilib.*, **2007**, *258(1)*, 41.
67. K. Leonhard, V.N. Nguyen, K. Lucas, *J. Phys. Chem. C*, **2007**, *111*, 15533.
68. M. Singh, K. Leonhard, K. Lucas, *Fluid Phase Equilib.*, **2007**, *258(1)*, 16.
69. N. van Nhu, M. Singh, K. Leonhard, *J. Phys. Chem. B*, **2008**, *112*, 5693.

70. A. Klamt, U. Huniar, S. Spycher, J. Keldenich, *J. Phys. Chem. B*, **2008**, *112*, 12148.
71. S.-T. Lin, J. Chang, S. Wang, W.A. IIIGoddard, S.I. Sandler, *J. Phys. Chem. A*, **2004**, *108*, 7429.
72. E. Mullins, R. Oldland, Y.A. Liu, S. Wang, S.I. Sandler, C.-C. Chen, M. Zwolak, K.C. Seavey, *Ind. Eng. Chem. Res.*, **2006**, *45*, 4389.
73. S. Wang, S.I. Sandler, C.-C. Chen, *Ind. Eng. Chem. Res.*, **2007**, *46*, 7275.
74. S. Wang, J.M. Stubbs, J.I. Siepmann, S.I. Sandler, *J. Phys. Chem. A*, **2005**, *109*, 11285.

17
Environmental Thermodynamics

17.1 Introduction

Prausnitz[1] wrote:

> *For about 100 years, chemical engineering has had the loyal support and applause of the public. As the public becomes increasingly critical, that support is eroding. Encouraged by the ideas of postmodernism, the applause is fading, replaced by mistrust and accusations.*

According to Prausnitz[1], postmodernism, as a tendency/philosophical movement, does not reject science, but denies that science offers the only path to the truth. Postmodernism stresses the multidimensionality of phenomena and their interaction with society but, in addition, demands that those who practice science and engineering accept the responsibilities that such interaction implies. As a part of the society it serves, the chemical engineering curriculum needs in future to be 'more human' and more 'society/human/environment oriented'. In brief, the essential messages of postmodernism to chemical engineering (and consequently also to thermodynamics and property modeling, as part of the latter) are:

- Better products, not just in terms of properties, but also healthier and with respect to the environment.
- Cleaner, safe processes, with respect to the environment and employees.
- Clean environment, minimum pollution, sustainable development

For 'environment and safety', the challenges involve the development of healthier products and cleaner, safer processes, with respect to the environment and occupational health. Thermodynamics can contribute to the following:

- Estimation of the distribution of long-lived chemicals in environmental ecosystems.[2,3] The concentration of chemicals in biota (fish, humans) can be assessed via the bioconcentration factor (BCF), which is related to thermodynamic properties such as the octanol–water partition coefficients.

Thermodynamic Models for Industrial Applications: From Classical and Advanced Mixing Rules to Association Theories
Georgios M. Kontogeorgis and Georgios K. Folas

- Design of environmentally friendly separations such as membranes, supercritical fluid extraction, adsorption, supercritical water oxidation or use of ionic liquids.
- Assessment of safety parameters such as flammability limits, flash points and the control of volatile organic compounds (VOCs) from various productions, e.g. of paints.
- Special focus on the removal (and sequestration) of CO_2 from the flue gases of power plants.
- Control of the compatibility of additives in products, e.g. of plasticizers in PVC.
- The understanding of and protection against corrosion in the chemical industry.

The focus in this chapter will be on the distribution of chemicals in the various ecosystems (air–water–soil/sediment–biota) and the role of the octanol–water partition coefficient; the chapter will close with a short discussion on supercritical fluids, an important class of environmentally friendly solvents.

17.2 Distribution of chemicals in environmental ecosystems

17.2.1 Scope and importance of thermodynamics in environmental calculations

Thermodynamics (phase equilibria) plays an important role in several applications related to the environment and environmental protection. One important application is the distribution of the so-called 'long-lived' chemicals, which are potential pollutants, in the various environmental ecosystems (compartments). The chemicals are produced in large amounts in our society (Figure 17.1).

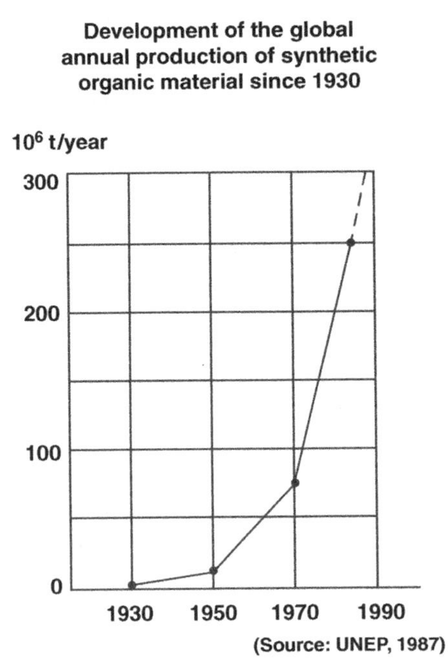

Figure 17.1 The global annual production of synthetic organic chemical materials since 1930. Reproduced with permission from Environmental Organic Chemistry by R. P. Schwarzenbach, Ph. M. Gschwend and D.M. Imboden. Copyright (1993) John Wiley and Sons, Inc.

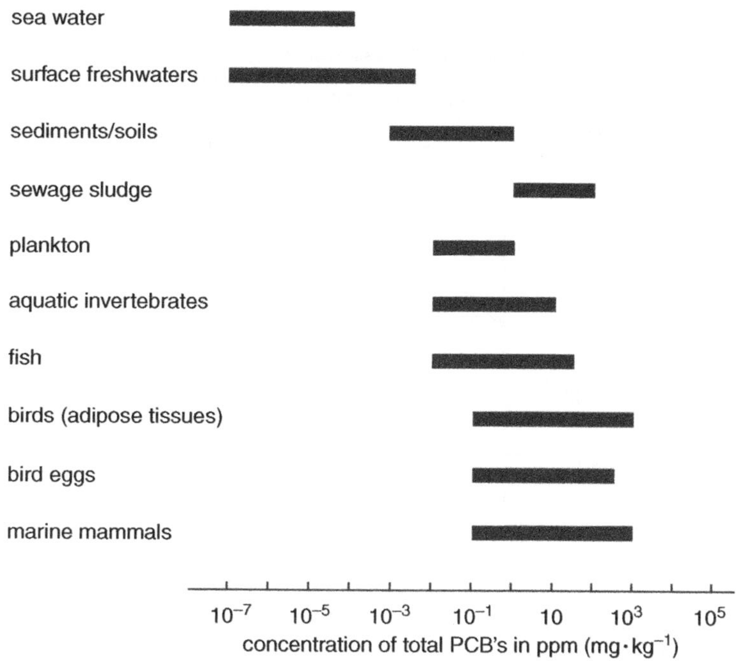

***Figure 17.2** Ranges of PCB concentrations detected in various environmental samples. The bioaccumulation and the effect of biomagnification are evident. PCBs have the general structure $C_{12}H_{10-n}Cl_m$, with $1 < n < 10$. Reproduced with permission from Environmental Organic Chemistry by R. P. Schwarzenbach, Ph. M. Gschwend and D.M. Imboden. Copyright (1993) John Wiley and Sons, Inc.*

The concentration of a pollutant in an aquatic organism or in soil can be assessed through thermodynamics, while in conjunction with other information we can estimate the concentration of chemicals in higher organisms (mammals) and finally in humans. This transfer of chemicals through the food chain (biomagnification) is a complex phenomenon but thermodynamics offers a first approximation. Figures 17.2 and 17.3 provide some indications of the dangers of such chemicals, namely the well-known DDT, previously used extensively against malaria, and PCBs (polychlorinated biphenyls).

In another example from the Great Lakes aquatic ecosystem,[6] the concentration of PCBs in Herring Gull eggs is almost 50 000 times higher than in phytoplankton. The biomagnification can be enormous in certain cases. This is due to metabolism and other phenomena. For some chemicals, like phenol, which have a half-life of a few hours, kinetic phenomena dominate. However, for slowly biodegrading chemicals like PCBs and DDT, with half-lives of the order of 3–4 years, phase equilibria dominate, and the principles discussed in this chapter are relevant. There is, of course, a third class of chemicals, in between the previous two families, where equilibrium and kinetic/mass transfer concepts are of almost equal importance. Slowly biodegrading (long-lived) chemicals such as those from the waste of the pharmaceutical, agricultural, cosmetics and related industries tend to bioaccumulate at different percentages in aquatic organisms and in several cases are then transferred via the food chain to higher organisms (mammals, humans). Several types of chemicals, e.g. pesticides of which almost 3 billion kilograms are produced annually, are known in certain cases to have severe effects on aquatic and animal populations. Examples of the great variety of the chemicals discussed are shown in Figure 17.4.

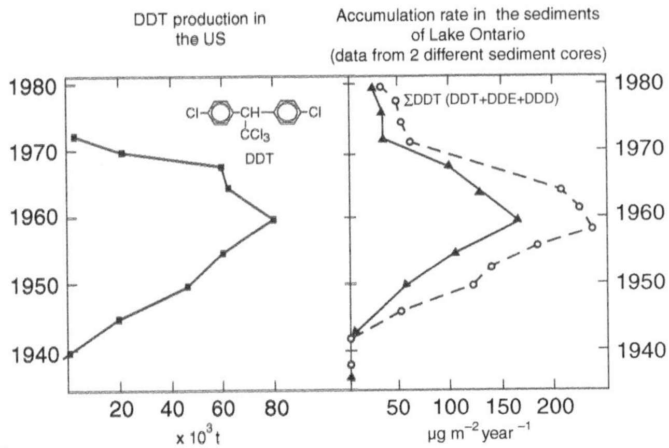
DDT production in
the US
Accumulation rate in the sediments
of Lake Ontario
(data from 2 different sediment cores)
ΣDDT (DDT+DDE+DDD)
DDT
1980
1970
1960
1950
1940
20 40 60 80
x 10³ t
50 100 150 200
μg m⁻² year⁻¹

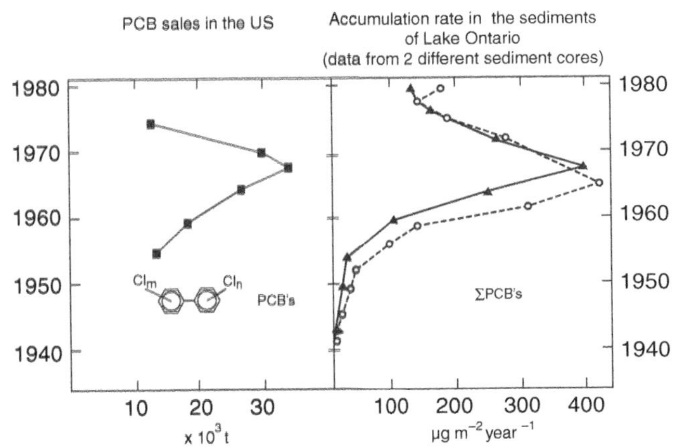
PCB sales in the US
Accumulation rate in the sediments
of Lake Ontario
(data from 2 different sediment cores)
PCB's
ΣPCB's
1980
1970
1960
1950
1940
10 20 30
x 10³ t
100 200 300 400
μg m⁻² year⁻¹

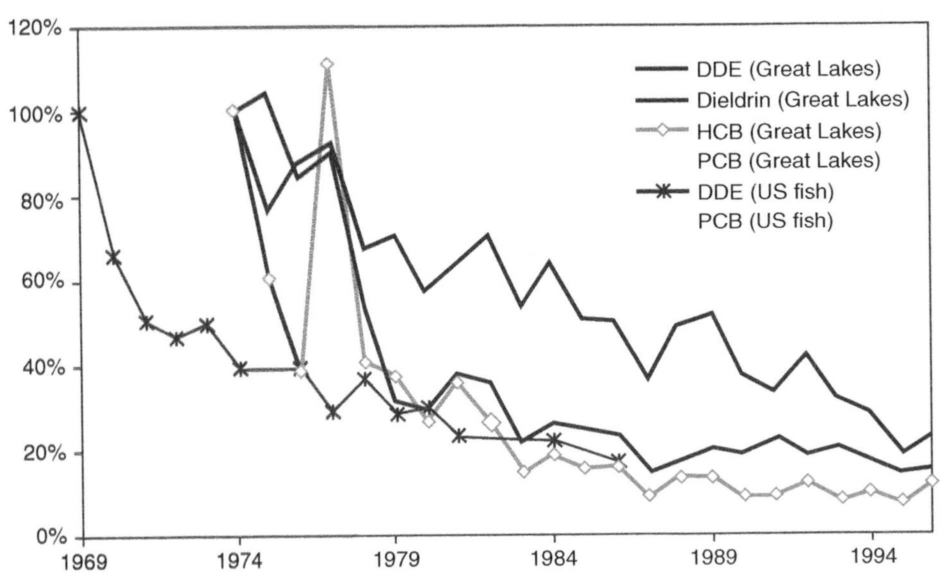
DDE (Great Lakes)
Dieldrin (Great Lakes)
HCB (Great Lakes)
PCB (Great Lakes)
DDE (US fish)
PCB (US fish)
120%
100%
80%
60%
40%
20%
0%
1969
1974
1979
1984
1989
1994

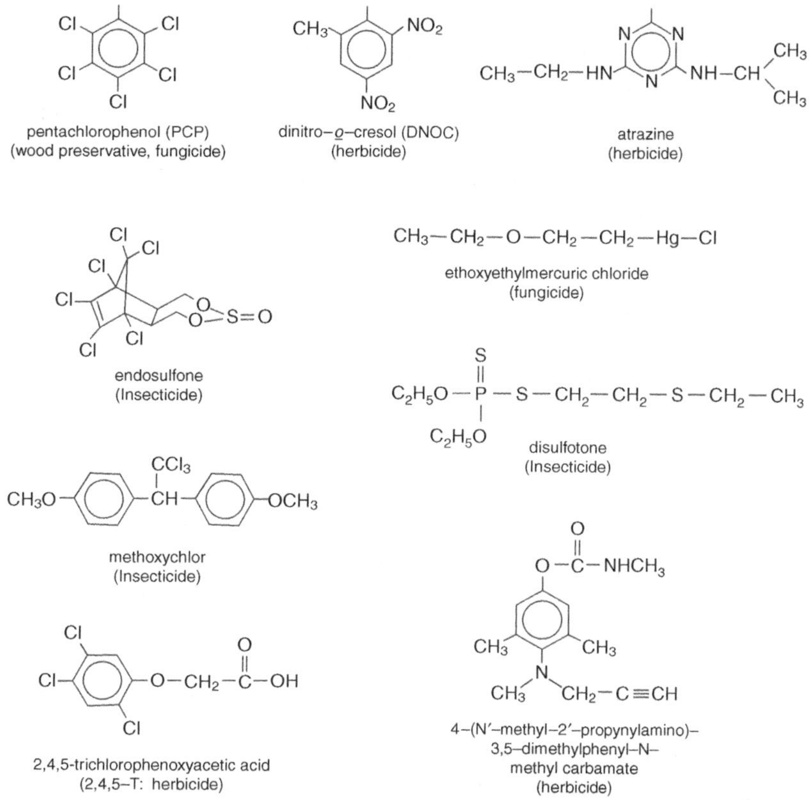

Figure 17.4 *Examples of pesticides illustrating the large structural diversity found in this group of environmental chemicals. Reproduced with permission from Environmental Organic Chemistry by R. P. Schwarzenbach, Ph. M. Gschwend and D.M. Imboden. Copyright (1993) John Wiley and Sons, Inc.*

Pesticides protect us against a number of potential enemies (fungi, plants, insects, etc.). There are numerous categories with various names: fungicides, herbicides, insecticides, bactericides, piscicides, algicides, avicides, acaricides, larvicides, etc. Many are aromatic and/or chlorinated compounds. Most pesticides have a rather complex chemical structure. The effect of DDT and other chemicals (pesticides etc.) on birds and other organisms is well known (Figure 17.2). Thus, in addition to the physical, chemical and microbiological

Figure 17.3 *Top: Historical records of the sales/production volumes of DDT and PCBs and the similarity of these trends in the accumulation rates of these chemicals in the sediments of Lake Ontario. Reproduced with permission from Environmental Organic Chemistry by R. P. Schwarzenbach, Ph. M. Gschwend and D.M. Imboden. Copyright (1993) John Wiley and Sons, Inc. Bottom: Levels of persistent pollutants in US freshwater fish (1969–1986) and Herring Gull eggs in the US/Canada Great Lakes (1974–1996), as indexed from the first year. Reprinted with permission from The Skeptical Environmnentalist: Measuring the Real State of the World by Bjørn Lomborg, Copyright (2001) Cambridge University Press. Similar results have been reported by others, e.g. Baird[6] shows the decreasing trend (since 1974) of the total PCB concentrations in Herring Gull eggs in the Lake Ontario area* analyses, proper assessment of the toxicological effects of chemicals in water and aquatic organisms gains increased interest. Among others, toxicological tests are related to European Union directives and policies for risk assessment and environmental protection.

Among the principal objectives of environmental thermodynamics is the aim to provide data and methods for predicting the risks to the environment, in terms of bioaccumulation and toxicity, caused by the various families of important and commonly encountered pollutants, as well as to monitor the behavior of such pollutants in ecosystems. We summarize the important issues:

1. What is the estimated toxicity of an organic chemical?
2. How is the chemical distributed, e.g. between water and the living population of a certain (aquatic) ecosystem?
3. Provided that the various distribution factors are known, how accurate may a risk analysis be in terms of chemical impact on a specific type of ecosystem (e.g. water or biota) and the environment as a whole?

Environmental policies in the EU and USA

Studies on the toxicity and accumulation of organic compounds and novel models for estimating octanol–water partition coefficients (K_{ow}) and the bioconcentration factors (BCFs) are of major importance: for example, in health protection via proper assessment of the effect of existing and new chemicals, design and manufacturing of 'better' ('environmentally more suitable') products, substitution of existing hazardous chemicals and designing waste water treatment processes for complex, dangerous and slowly biodegrading chemicals.

EU policies emphasize environmental protection and more specifically water quality standards and determination of the effect (fate, transport, distribution) of toxic and hazardous pollutants in aqueous ecosystems (both water and water-living organisms) (*Official Journal of the EU*, No. L 129/23, 18 May 1976; 1984; 1986; 1988). Monitoring the effect of the pollutants in ecosystems and especially water and its living organisms is extremely important for those EU countries for which fishing, agriculture and tourism play significant roles in the national economies. In addition, for certain countries, especially in the Mediterranean region where rainfall has decreased significantly over the past few years (also due to climate changes), water has essentially become a priority and vital element both for drinking and for industrial and agricultural use. In this context, the protection of water resources and determination of maximum permissible concentration of chemicals in it (necessary for establishing protection laws) are of the utmost importance. The protection of water, via the study of the distribution of organic pollutants, will assist also in dealing with the desertification problem. This problem has been more pronounced in Southern Europe over the last few years (this region is classified as of high danger with 15–25% being desert, while in the whole of Europe 10% is considered to be desert and 40% to be dry). This is partly due to overexploitation of land and to climate change. One way to determine the major pollutants and their effect on the environment, without resorting to many experiments, is via thermodynamics, as will be discussed in this chapter.

Other countries, like the USA, via its Environmental Protection Agency (EPA), have already established criteria for maximum permissible concentrations of pollutants in various ecosystems based on octanol–water and other partition coefficients. Due to the lack of experimental data and/or reliable models, these criteria cover for the time being a small part of potential pollutants. For example, only a small part of the 219 different chemical compounds in the class of PCBs is included, but in time criteria for the most important pollutants will be established.

Table 17.1 K_{ow} *values for three chemicals illustrating the different behavior which may occur*

Compound	K_{ow}	Log K_{ow}	Tendency
TEG	0.0083	−2.08	Hydrophilic
n-Butane	776	2.89	Hydrophobic
DDT	1 555 000	6.19	Strongly hydrophobic (dangerous pollutant)

TEG = triethylene glycol.

17.2.2 Introduction to the key concepts of environmental thermodynamics

The prediction of bioaccumulation as well as of the toxicity effects of chemicals on living organisms is often linked to a property which will be discussed extensively in this chapter: namely, the octanol–water partition coefficient (K_{ow}), which is defined for a chemical i as:

$$K_{ow} = \frac{C_i^o}{C_i^w} = \frac{x_i^o C^o}{x_i^w C^w} \tag{17.1}$$

where C_i^o and C_i^w are the concentrations of the chemical i in the octanol-rich phase (o) and in the water-rich phase (w), respectively, and C^o and C^w are the total molar concentrations of the octanol-rich and water-rich phases, respectively. The mole fractions can be substituted by the activity coefficients based on the assumption of LLE in the system (see later, Equation (17.21)).

K_{ow} is a thermodynamic quantity which can be measured experimentally and also estimated from thermodynamic models, as will be discussed in Section 17.2.4. When *n*-octanol is mixed with water, two liquid phases are formed, one almost pure water and the other approximately 73% mole fraction *n*-octanol. If a very small amount of a third chemical is added and allowed to reach equilibrium, its concentration in both phases can be measured and will generally be very different in these two liquid phases (octanol-rich, o, and water-rich, w, phases). A hydrophilic compound will have a higher concentration in the water phase, while a hydrophobic compound will mostly be concentrated in the octanol-rich phase.

Some examples for K_{ow} are given in Tables 17.1 and 17.2. Those chemicals that tend to go to water are called 'hydrophilic' (low K_{ow} or negative $\log(K_{ow})$), but by far most chemicals 'hate' water and prefer the 'oily'

Table 17.2 K_{ow} *and water solubility (in ppm). After Baird*[6]

Pesticide	Solubility in water	Log K_{ow}
HCB	0.0062	5.5–6.2
DDT	0.0034	6.2
Toxaphene	3	5.3
Dieldrin	0.10	6.2
Mirex	0.20	6.9–7.5
Malathion	145	2.9
Atrazine	35–70	2.2–2.7

HCB = hexachlorobenzene.

octanol; they are called hydrophobic, and if K_{ow} is very high they are dangerous pollutants. High K_{ow} values imply high biomagnification values, and this means that the chemical can move up the food chain, all the way to higher mammals and maybe humans. Notice the very high value of DDT; essentially values of $\log(K_{ow}) > 4$ imply very dangerous pollutants (high biomagnification).

K_{ow} and water solubility are, as can be understood from Equation (17.1), inversely proportional. High K_{ow} values imply low solubility in water, as can be seen in Table 17.2 for some known pesticides. Since K_{ow} and BCF are generally proportional, the same is true for the bioconcentration. Pesticides do not like water. This is why the largest amount of pesticides comes from the food we eat (over 99%) and less than 1% comes from the water we drink.

Why is K_{ow} so important in the study of ecosystems?

Octanol is an organic solvent that is considered to be a surrogate for natural organic matter. This parameter is used in many environmental studies to help determine the fate of chemicals in the environment. The environment is a complex system to model and consists of many compartments, sometimes called 'ecosystems':

- air (A);
- water (W);
- soil and sediment (S);
- biota, which are living beings, including humans (B).

Some of the compartments, e.g. air and water, are well defined, others (biota and soil) are not. More 'compartments' exist, e.g. atmospheric particulates and suspended sediment. The study of the distribution and environmental fate of long-lived (persistent) chemical pollutants in the various compartments (air, water, soil, sediment) can be accomplished using the so-called FUGACITY models by Mackay[7], which are based on the equilibrium concept. These models typically require Henry's law constant and the soil/water and K_{ow} (or BCF) coefficients as input parameters. The Mackay statement is summarized as:

Long-lived chemicals reach a state of environmental phase equilibrium, among the various environmental ecosystems; air, water, soil–sediment, biota including humans.

Thus, the concentrations of chemicals in the various ecosystems can be estimated, to some extent, if the fugacities of the chemicals in the various ecosystems are known (see Section 17.2.3). Due to the equilibrium assumption, the fugacities of the chemicals in the various ecosystems are equal:

$$f^{air} = f^{water} = f^{soil} = f^{sediment} = f^{biota} \tag{17.2}$$

Given the complexity of the environment, we realize that any model will only be very approximate. However, even the very simple approach, which is presented here, often yields reasonable results. This pioneering work is based on the principles of phase equilibria and is known as the (first-level) Mackay model.[7] The basic assumption of the model is that a chemical that has a rather high half-life in the environment, e.g. the pesticide DDT, PCBs or dioxins, will achieve a state of environmental phase equilibrium. No doubt other factors should have an effect, e.g. mass transport limitations since the diffusion of chemicals into or out of soil is very slow. Nonetheless, the Mackay model provides a first approximation to the problem of the distribution of chemicals.

The octanol–water partition coefficient, K_{ow}, is related to several other important partition coefficients, which determine the concentration of long-lived chemicals in the various environmental compartments, air (A), water (W), soil and sediment (S) and biota (B), the last also including living organisms and humans. These calculations are possible for long-lived chemicals, where the assumption of phase equilibria, Equation (17.2), is possible. The basic relationships of environmental engineering giving the fugacities of the chemicals in the various ecosystems and the role of the octanol–water partition coefficient will be presented in Section 17.2.3. K_{ow} is also related to the BCF and the toxicity indices LC_{50}. K_{ow} is also used in the medical and pharmaceutical sciences for the design of drugs and pharmaceuticals, toxicology studies and research on medicinal chemicals.

The importance of K_{ow} can be also manifested by the large data compilations available.[8,9] Data are, however, not available for all chemicals. Estimation methods are thus required for new chemicals and many models have been developed. We will see in this chapter (Section 17.2.4) both specific models for K_{ow} and how classical models such as UNIFAC perform for such applications. Risk analysis and hazard identification typically require knowledge of both the toxicity and the BCF. In certain cases additional information, mostly kinetic data such as the half-life of the pollutants, are also required for estimating the concentration of pollutants in aquatic organisms and via this estimate the chronic daily intake (CDI) parameter, which ultimately yields the risk from a certain pollutant.[10]

Biomagnification of pollutants

It was mentioned that the Mackay model is based on the assumption that a chemical is in phase equilibrium throughout the food chain. This is a reasonable assumption for many long-lived chemicals, but chemicals with high values of K_{ow} accumulate up the food chain. An example of a biomagnification effect can be seen by re-examining biotic PCB concentrations in Lake Michigan in the USA. The equilibrium concentration of PCBs in aquatic biota is expected to be 16 000 times greater than the concentration of PCBs in Lake Michigan water based on a phase equilibrium calculation only. However, biomagnification results in an additional factor of about 14 in PCB accumulation, leading to a predicted total PCB concentration in Lake Michigan trout of about 225 000 times greater than the concentration in lake water.[3] Thomann[11] has described the food chain bioaccumulation via a simple steady-state model, which was further discussed by Sandler.[3] The most important conclusions from this analysis are:

- There is no significant biomagnification of a chemical up the food chain unless K_{ow} is above 10^4, which corresponds to a value of the infinite dilution activity coefficient in water of about 3×10^5.
- There is significant biomagnification for K_{ow} values of 10^6 and larger.

Thus, insecticides such as malathion ($\log K_{ow} = 2.9$) and lindane ($\log K_{ow} = 3.85$) will not biomagnify in the food chain, but, on the other hand, dieldrin ($\log K_{ow} = 5.48$), DDT ($\log K_{ow} = 6.19$) and mirex ($\log K_{ow} = 7.5$) are likely to increase significantly up the food chain.

17.2.3 Basic relationships of environmental thermodynamics

In brief

Many thermodynamic calculations in environmental engineering can be performed using a few basic thermodynamic quantities, which are:

- the vapor pressures of chemicals, P_i^{sat};
- the water solubility (or infinite dilution activity coefficients), x_i^w, γ_i^∞;

- Henry's law constant (vapor pressure × infinite dilution activity coefficient), $H_i(\gamma_i^\infty P_i^{sat})$;
- the octanol–water partition coefficient K_{ow}.

All these properties vary over several orders of magnitude for the various chemicals. Since the phase equilibrium relationships require knowledge of the fugacities, we start by showing how we can express the fugacity of a chemical in the various environmental ecosystems. The Mackay *fugacity model*[12] uses as its key concept the property $Z = C/f$, which is named *fugacity capacity* and has units, for example, of mol/m^3 Pa. C is the concentration (mol/m^3) and f is the fugacity (e.g. in Pa). Knowing the values of the fugacity capacity Z_i of each environmental compartment, its molar concentration C_i and its volume fraction V_i, it is possible to calculate the total molar concentration and the mean fugacity capacity and fugacity values:

$$\begin{aligned} C &= \sum_i C_i \\ Z^{mean} &= \sum_i V_i Z_i \\ f^{mean} &= \frac{C}{Z^{mean}} \\ C_i &= f^{mean} Z_i \end{aligned} \tag{17.3}$$

where the summations are over all the environmental compartments. Furthermore, we demonstrate how we can estimate the fugacity capacities and the fugacities for the different environmental compartments.

Fugacity of a chemical in water

The fugacity of a chemical in water is:

$$f_i^w = x_i^w \gamma_i f_i^0 = x_i^w \gamma_i P_i^{sat} = x_i^w \gamma_i^\infty P_i^{sat} = x_i^w H_i = C_i^w H_i^c \tag{17.4}$$

where x_i^w is the mole fraction of the chemical i in water, f_i^0 is the pure component fugacity, which can be replaced by the pure compound vapor pressure P_i^{sat}, and γ_i is the activity coefficient, which is the key property in this calculation. Since in most cases of environmental interest the concentration of the chemical (pollutant) in the aqueous phase is quite low, the activity coefficient can be replaced by its value at infinite dilution γ_i^∞ as shown in Equation (17.4).

As can be seen in Equation (17.4), if the infinite dilution activity coefficient and the vapor pressure of a compound in water are known, the fugacity in rivers, lakes and oceans can be estimated. For many chemicals, Henry's law constant (H_i or H_i^c) is known instead of the activity coefficient at infinite dilution and the vapor pressure. In this case, as understood from Equation (17.4), no other piece of information is needed. Mackay's fugacity capacity is in this case:

$$Z_i^w = \frac{C_i^w}{f_i^w} = \frac{1}{H_i^c} \tag{17.5}$$

The infinite dilution activity coefficients of many chemicals in water can have very high values, which indicates that they are almost insoluble in water; or, in other words, that they are highly hydrophobic. Other chemicals with lower values of water activity coefficients are less hydrophobic (more hydrophilic). Vapor pressure values of several chemicals can be found in several handbooks. For environmental chemicals, which are often heavy complex compounds, experimental data are not available and vapor pressures need to be estimated via predictive methods. A compilation of vapor pressures for environmentally important compounds is provided by Site.[13] Vapor pressures for solid compounds can be estimated, when not available experimentally, from group contribution methods, e.g. Coutsikos *et al.*[14]

Fugacity of a chemical in air

This is the most well-established fugacity in environmental engineering. Under environmental conditions, air is an ideal gas, thus the fugacity of a chemical is equal to the partial pressure:

$$f_i^A = x_i^A P = \left(\frac{C_i^A}{C^A}\right) P = C_i^A RT \tag{17.6}$$

where x_i^A is the mole fraction of chemical i in air, C_i^A is the molar concentration of the chemical and C^A is the molar concentration of the air. R is the ideal gas constant and T is the absolute temperature. Thus, Mackay's fugacity capacity is:

$$Z_i^A = \frac{C_i^A}{f_i^A} = \frac{1}{RT} \tag{17.7}$$

The distribution of a chemical between air and water can be obtained by starting from the equilibrium equation:

$$f_i^w = f_i^A \Rightarrow C_i^w H_i^c = C_i^A RT \Rightarrow K_{AW,i} = \frac{C_i^A}{C_i^w} = \frac{H_i^c}{RT} = \frac{0.2165}{T} (\gamma_i^{w,\infty} P_i^{sat}) \tag{17.8}$$

$K_{AW,i}$ is the well-defined air-water partition coefficient, T is expressed in K and the vapor pressure is given in atm.

Fugacity of a chemical in soil and sediment

The fugacity of a chemical in soil and sediment as well as in biota (see later) cannot be estimated directly. It is instead calculated via knowledge of the distribution of the chemical between soil and water and between biota and water. We first define the distribution (ratio) of a chemical between soil or sediment and water:

$$\begin{aligned} K_d &= K_{s/w} = \frac{C_i^s}{C_i^w} \\ K_d^/ &= \frac{K_d}{\rho_s} \end{aligned} \tag{17.9}$$

where C_i^w and C_i^s are the concentrations of the chemical in water and soil or sediment, and ρ_s is the density of soil or sediment (e.g. in kg/dm^3). From Equation (17.9) it can be seen that the distribution ratio K_d is dimensionless since both concentrations are expressed in the same units (e.g. mg/dm^3), while $K_d^/$ has dimensions for example, dm^3/kg. Assuming that equilibrium is reached in the distribution of the chemical between soil or sediment and water, we have:

$$f_i^w = f_i^s \Rightarrow C_i^w H_i^c = x_i^s \gamma_i^s \tag{17.10}$$

Combining Equations (17.9) and (17.10) we have:

$$f_i^s = C_i^w H_i^c = \left(\frac{H_i^c}{K_d^/ \rho_s} \right) C_i^s \tag{17.11}$$

and Mackay's fugacity capacity is:

$$Z_i^s = \frac{C_i^s}{f_i^s} = \frac{K_d^/ \rho_s}{H_i^c} \tag{17.12}$$

Equations (17.9)–(17.12) are valid for both soil and sediment. The densities of soil and sediment are approximately 1.3 and 1.5 kg/l.

Fugacity of a chemical in biota

Similarly, as in the case of the fugacity of a chemical in soil and in sediment, we first define the distribution (ratio) of a chemical between biota and water. This defines the important concept of the BCF:

$$BCF = K_b = K_{b/w} = \frac{C_i^b}{C_i^w}$$
$$K_b^/ = \frac{K_b}{\rho_b} \tag{17.13}$$

where C_i^w and C_i^b are the concentrations of the chemical in water and biota, respectively, and ρ_b is the density of biota (e.g. in kg/dm^3). From Equation (17.13) we understand that the distribution ratio K_b is dimensionless since both concentrations are expressed in the same units (e.g. in mg/dm^3), while $K_b^/$ has dimensions for example dm^3/kg. All these are equivalent to what was previously mentioned for the soil–water and sediment–water partition coefficients (compare equation 17.13 with equation 17.9).

Assuming that equilibrium is reached in the distribution of the chemical between biota and water, we have:

$$f_i^w = f_i^b \Rightarrow C_i^w H_i^c = x_i^b \gamma_i^b \tag{17.14}$$

Combining Equations (17.13) and (17.14) we have:

$$f_i^b = C_i^w H_i^c = \left(\frac{H_i^c}{K_b^/ \rho_b} \right) C_i^b \tag{17.15}$$

and Mackay's fugacity capacity is:

$$Z_i^b = \frac{C_i^b}{f_i^b} = \frac{K_b^/ \rho_b}{H_i^c} \tag{17.16}$$

The octanol–water and other partition coefficients

The fugacities of chemicals in soil or sediment and biota can be estimated via Equations (17.11) and (17.15) provided that the distribution ratios K_d and BCF ($= K_b$) are known. These distribution ratios are difficult to assess quantitatively but to a first approximation they are related to the so-called octanol–water partition coefficient K_{ow}, which, as stated, is a measure of the hydrophobicity of a chemical. The reason that a correlation between K_d and BCF with K_{ow} exists is due to the fact that *n*-octanol is considered to be a good measure for the lipid–organic part (phase) of living organisms (biota) and for the organic part of soil and sediment. It is assumed that the (hydrophobic) organic pollutants (chemicals) preferentially partition only into the organic matter of soil or sediment and the lipid part (fatty tissue) of the biota. We assume that these chemicals enter mainly the fatty tissues of fish and humans rather than muscles, other tissues or skeletal structures. Octanol–water partition coefficients (K_{ow}) are widely considered today to be a useful measure of the accumulation of pollutants and chemicals by aquatic organisms. K_{ow} values range from below unity, in the case of hydrophilic compounds (e.g. triethylene glycol), up to several million, in the case of very dangerous compounds.

For chemicals with values of $\log K_{ow} > 4$, it is believed that they constitute dangerous pollutants which can be transferred through the food chain to higher organisms.[15] In this case, these chemicals 'biomagnify' or 'bioaccumulate', i.e. their concentration is increased as they are transferred via the food chain. For example, the well-known insecticide DDT with $\log K_{ow}$ equal to almost 6.2 has been found in fish, higher birds such as eagles and even humans. While the mechanism for food bioaccumulation is not completely understood, the important point is that its extent can be directly correlated with the value of the octanol–water partition coefficient. Table 17.3 gives an overview of the certainly non-trivial story of DDT.

Examples of K_{ow} values for typical chemicals are: K_{ow}(pyrene) = 135 000, K_{ow}(Aldrin) = 3 160 000, K_{ow}(biphenyl) = 12 300 and K_{ow}(ppDDT) = 2 344 000. The important chemicals or pollutants of interest to environmental science are PAHs (polynuclear aromatic hydrocarbons), PCBs (polychlorinated biphenyls), heavy metals (Cr, Ni, Fe, Cu, Zn, Cd, etc.) and inorganic salts. Due to this significantly large range of values (over 10 orders of magnitude: 10^{-3} to 10^{7}) and other factors (e.g. the important influence of isomerism on K_{ow}), knowledge of K_{ow} is a non-trivial problem, especially for new, structurally complex chemicals, and either experimental measurements or reliable estimation methods are required. Both are considered in Section 17.2.4.

Returning to the relationship between K_{ow} and the various distribution coefficients, the following empirical equations have been proposed:

$$\begin{aligned} BCF &= f_b K_{ow} = 0.05 K_{ow} \\ K_{sw} &= f_{oc} K_{oc} \\ K_{sdw} &= f_{oc}^/ K_{oc} \end{aligned} \tag{17.17}$$

where f_b is the lipid fraction in biota. For aquatic organisms, e.g. fish, it is often assumed that f_b is approximately 5%, but other correlations have been proposed of the type:[12]

$$\log BCF = a + b \log(K_{ow}) \tag{17.18}$$

Table 17.3 *The story of DDT*

Year	Event	Comments
<1939		Before DDT, only a very few toxic pesticides existed
1939	Discovered by Paul Muller (of Geigy Pharmaceuticals) – an excellent weapon against malaria	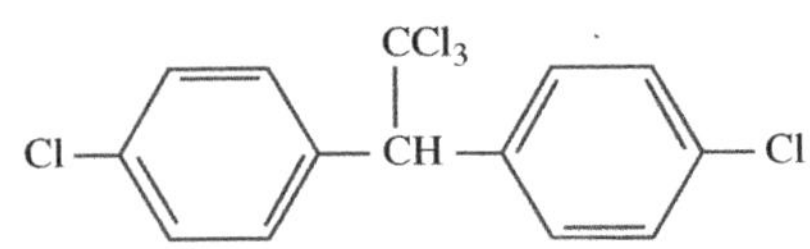 DDT The structure of DDT
1941	First DDT products Switzerland informs both Allies and Axis Powers about DDT	
1948	Paul Muller wins Nobel Prize	Paul Hermann Muller (1899–1965): Nobel Prize in Physiology or Medicine 1948 'for his discovery of the high efficiency of DDT as a poison against several arthropods'
1940–1970	Extensive use of DDT against insects responsible for typhus, malaria and in agriculture	More than 8000 tons per year of DDT used on USA farms DDT is cheap and not toxic (to humans) WHO estimates '25 million lives were saved'. DDT helped Europe and the USA make malaria vanish for ever!
1962	Rachael Carson publishes *The Silent Spring* resulting in huge anti-DDT hysteria	Eagles and pelicans almost driven to extinction Many insects develop resistance to DDT DDT accumulated in fish, birds, etc., not easily metabolized, transferred far away via the food chain
1969	Sweden and other countries ban DDT	
1970	National Academy of Sciences: 'to few chemicals does man owe as great a debt as to DDT'	
1973	EPA (USA) bans DDT for all uses 'except for public health reasons'	Many lakes, rivers, etc., are now clean, see e.g. Figure 15.3
1996	South Africa bans DDT and malaria jumps from a few thousand to more than 50 000 cases per year	USA, Institute for Environmental Health (2001): 'DDT use in USA linked to premature births in the 1960's'
December 2000	120 countries agree to phase out many chemicals including DDT	DDT banned in most of the world Still legal in 25 countries (India, China, African) against malaria – illegal in agriculture Many organizations (WWF) press for a complete phasing out of DDT
16 June 2002 (*Washington Post*)	400 scientists say: 'Bring back DDT' – but exclusively for the treatment of malaria'	'How do you think we made malaria vanish from USA? DDT!'

Table 17.3 (Continued)

Year	Event	Comments
		There is only one thing cheaper than DDT when dealing with malaria – and that is to do nothing! Rich countries: no money to poor countries if they reintroduce DDT
Today	Every year more than 2 million people die from malaria – there is no better cure than DDT, situation worse than with AIDS! WHO: 'malaria kills one child under the age of 5 every 30 seconds'	

EPA = Environmental Protection Agency, WHO = World Health Organization, WWF = World Wide Fund for Nature.

where a and b are constants depending on the specific biota system considered. The K_{sw} and K_{sdw} distribution ratios (for soil and sediment), Equation (17.17), are functions of K_{oc}, the organic carbon partition coefficient. K_{oc} is also considered to be a function of K_{ow}, e.g. $K_{oc} = 0.63K_{ow}$ (for soil) and $K_{oc} = 0.53K_{ow}$ (for sediment). Other correlations using slightly different constants have been proposed, e.g. by La Grega *et al.*[16] The fractions of organic carbon in soil and in sediment are also typically different, e.g. f_{oc} is assumed to vary between 0.2 and 3% with the most typical value equal to 0.02 and $f_{oc}^{/}$ is typically around 5% (= 0.05). Using these assumptions, the distribution ratios involving soil and sediment can be given also as functions of K_{ow}:

$$\begin{aligned} K_{sw} &= f_{oc}0.63K_{ow} = 0.02 \times 0.63 \times K_{ow} \\ K_{sdw} &= f_{oc}^{/}0.53K_{ow} = 0.05 \times 0.53 \times K_{ow} \end{aligned} \tag{17.19}$$

Similar relationships have been proposed for the distribution coefficients of chemicals between water and suspended matter (K_{mw}) and water and animal biota (K_{vw}):

$$\begin{aligned} K_{vw} &= f_v 0.41K_{ow} = 0.2 \times 0.63 \times K_{ow} \\ K_{mw} &= f_m 0.41K_{ow} = 0.17 \times 0.41 \times K_{ow} \end{aligned} \tag{17.20}$$

From Equations (17.17)–(17.20) we understand that all partition coefficients can be expressed – to some approximation – as a function of the octanol–water partition coefficient. Thus, K_{ow} is the key property in environmental engineering. Several approximate relationships between K_{oc} and K_{ow} have been reported in the literature. Sandler[3] indicates that all these simple relationships should be used with care since soil is an extremely complex heterogeneous material where, under the single heading of soils, one can find materials ranging from clays to sand. Moreover, some soils, e.g. clays, have quite different physical and chemical properties depending on whether they are wet or dry. Apparently, all these important complexities are ignored when the soils are characterized simply by their organic carbon content.

17.2.4 The octanol–water partition coefficient

Definition

The octanol–water partition coefficient was defined in Equation (17.1) but it can be rewritten using activity coefficients as:

$$K_{ow} = \frac{C_i^o}{C_i^w} = \frac{x_i^o C^o}{x_i^w C^w} = \frac{\gamma_i^w C^o}{\gamma_i^o C^w} = \frac{\gamma_i^{w,\infty} C^o}{\gamma_i^{o,\infty} C^w} \tag{17.21}$$

The variables were defined in the text following Equation (17.1).

The mole fractions are substituted by the activity coefficients based on the assumption of LLE in the system. Finally, $\gamma_i^{w,\infty}$ and $\gamma_i^{o,\infty}$ are the infinite dilution activity coefficients in the water-rich and octanol-rich phases, respectively. Use of the infinite dilution activity coefficients is justified by the fact that the distribution chemical is inserted only in very small amounts. It can be shown (see Problem 1 on the companion website at www.wiley.com/go/Kontogeorgis) that:

$$K_{ow} = 0.151 \frac{\gamma_i^{w,\infty}}{\gamma_i^{o,\infty}} \tag{17.22}$$

Equation (17.22) has been derived by considering that the water-rich phase is essentially pure water (99.99% water) while in the octanol-rich phase the solubility of water in octanol is 27.5 mol%, a result confirmed by most experimental measurements reported for the binary octanol–water system. Generally, there is some disagreement in the literature, and water solubilities in octanol between 20.7 and 27.5% have been reported.

Experimental determination of K_{ow}

Up to 15 000 experimental K_{ow} data classes are available.[9] The reason for such an abundance of K_{ow} data is due to its extensive use and its adoption by many international and governmental agencies as a physical property of organic pollutants, which is directly related to the uptake in tissues and fat of living species and thus potentially to toxicity. Various methods are available for the experimental determination of the octanol–water partition coefficients. Some of them are briefly described here.

Shake-flask method

This is the most widely used method. This classic extraction procedure is, despite being very time consuming, widely used and produces reliable results. A small amount of solute is dissolved in either aqueous or organic phase, equilibrium partition is obtained by agitation, the phases are separated and then analyzed for the solute.

Static cell measurements with gas solubility apparatus (GLE)

A carefully degassed solvent is poured into an equilibrium cell, which is submerged in a liquid water bath thermostat. The gas (solute) flows into a gas system located in an air thermostat. When the gas is admitted into the cell it is absorbed until equilibrium is reached while the solvent is continuously stirred. Using the isochoric method, the pressure in the exactly calibrated volume of the gas system, piping and the vapor phase of the cell is measured at the beginning and at the end of the experiment. Using the isobaric method, mercury is poured into the gas system reducing the gas volume and thus the pressure until it reaches the condition at the beginning of the experiment. Knowing the temperature of the gas system and the cell, the absorbed amount of gas can be calculated from the difference in pressure (isochoric) or the difference in the gas volume (isobaric). Measurements can be made in the temperature range from −10 to 80 °C and in the pressure range from 0 to 5 bar.

Gas–liquid chromatography (GLC)

The equipment is placed in a thermostat oven. A solid support is coated with the solvent and placed into the chromatographic column. The amount of liquid in the column is determined by weighting. The solute is injected through a septum of the injection block into the flow of the carrier gas (helium) using a microliter syringe. The helium has been presaturated with the solvent. In the column, the solute is retained by the adsorbed solvent. Henry's law constants and limiting activity coefficients can be calculated using the net retention time, the flow rate in the column and the mass of the solvent. The apparatus works in a temperature range from 20 to 300 °C. Using the gas solubility apparatus, a gaseous solute is dissolved either in water or in *n*-octanol and Henry's law constants determined. K_{ow} is calculated from the quotient of Henry's law constants:[17]

$$K_{i,ow} = \frac{x_i^o}{x_i^w} = \frac{H_{i,w}}{H_{i,o}} \qquad (17.23)$$

Here i is the solute, w is the water phase and o is the octanol phase. The field of application for the GLC technique covers an even wider range. It is not restricted to just gaseous solutes. Solvents, which are to some extent volatile, can also be handled by relative liquid chromatography if the limiting activity coefficient of one of the two simultaneously injected solutes is known. K_{ow} is determined from the quotient of the limiting activity coefficients in water and *n*-octanol:

$$K_{i,ow} = \frac{x_i^o}{x_i^w} = \frac{\gamma_{i,w}^\infty}{\gamma_{i,o}^\infty} \qquad (17.24)$$

Reliable data have been reported for a number of alkanes and constituent chloro-fluoro derivatives using both techniques.[18] The maximum error in Henry coefficient and infinite dilution activity coefficients is reported as between 1 and 4% in *n*-octanol. In water the error, for very poorly soluble organics, is in the region of 7% and typically lower than 4%. Determination of Henry's law constants and/or limiting activity coefficients in solvents other than *n*-octanol, e.g. in nonane and tricapryline, leads to similar liquid–liquid partition coefficients. These partition coefficients can also be used, as an alternative to K_{ow}, to describe the hydrophobicity of organic pollutants and can give an idea of the magnitude of the BCF.

Generator column method

The solid support is usually a silanized diatomaceous silica. The liquid chromatographic column is loaded by pulling through the column an unsaturated solution of the solute of interest at a known concentration in water-saturated octanol. The solute is eluted with octanol-saturated water, and the effluent is analyzed by either high-pressure liquid chromatography or GLC. For each solute a new column is needed.

Reversed-phase high-pressure liquid chromatography (RP-HPLC)

K_{ow} is deduced from aqueous solubility and Henry's law constant. This is probably the most widely used method after using solute retention volumes or chromatographic capacity factors. Unfortunately this correlation method does not give reliable results because a single correlation plot is not accurate.

Reversed-phase thin-layer chromatography (RP-TLC)

This method may be considered as the two-dimensional analogue of RP-HPLC with a precision generally inferior to RP-HPLC.

Estimation methods

The estimation methods for K_{ow} can be roughly divided into: (1) empirical direct correlation methods; (2) higher order group contribution methods; and (3) thermodynamic models. All are summarized in Table 17.4 and a short discussion follows.
Among the empirical correlations, often based on the fragment or group contribution concepts, the most widely used are:

1. The ClogP developed by Leo and co-workers.[19,20]
2. The AFC correlation model, sometimes also abbreviated as KOWWIN in its computerized form.[21]
3. The ACD method (which has not been described in the scientific literature; the reader is referred to http://www.acdlabs.com).

A common feature of these methods is that they have been developed using large databases of K_{ow} data, and they are often (in their most general formulations) rather complex. These specific correlations are often capable of distinguishing between isomers and also can take into account, to some extent, the proximity and the intermolecular effects. They are described in a monograph[22] and many of the methods also exist in computer programs readily available on the Internet. For example, Environmental Science Centre (ESC) of

Table 17.4 *Estimation methods for octanol–water partition coefficients*

Method	Type	Reference	Comments
CLogP (Leo group)	Special correlation	Leo *et al.*[19] Hansch and Leo[20]	http://www.daylight.com
AFC (Kowwin)	Special correlation	Meylan and Howard[21]	http://srcinc.com/what-we-do/free-demos.aspx
ACD	Special correlation	Sangster[22]	http://www.acdlabs.com
Gani group	Third-order GC	Marrero and Gani[23]	
Constantinou	Second-order GC	Stefanis *et al.*[24]	
K_{ow}–UNIFAC	Activity coefficient thermodynamic model	Wienke and Gmehling[25]	
Water–UNIFAC	Activity coefficient thermodynamic model	Chen *et al.*[26]	

GC = group contribution methods.

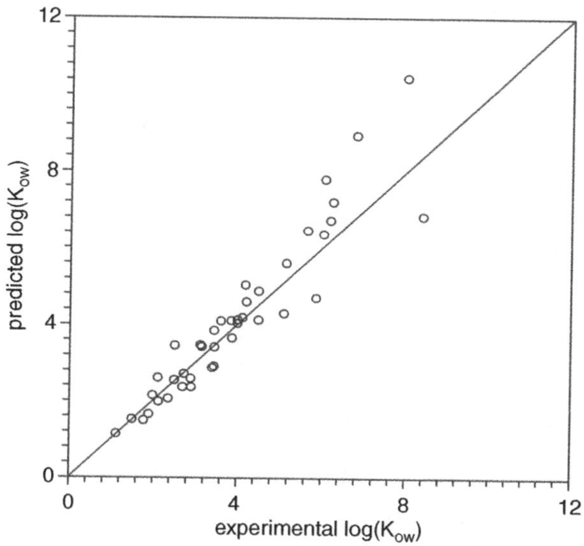

Figure 17.5 Predicted against experimental octanol–water partition coefficients with CPA. From Polyzou et al.[27]

Syracuse Research Corporation (SCR) has a computerized form of the AFC method available on its homepage (http://srcinc.com/what-we-do/free-demos.aspx). Similarly, Daylight Chemical Information Systems, Inc. (http://www.daylight.com) provides computerized versions of the Leo–Hansch ClogP. Two group contribution approaches, using second- and second/third-order groups, have been presented fairly recently by Stefanis *et al.*[24] and Marrero and Gani[23]. Alternatively, thermodynamic models can be used, including the advanced ones like SAFT and CPA. A preliminary calculation with CPA is shown in Figure 17.5.

However, it is traditional to use UNIFAC for K_{ow} calculations and both the standard methods (described in Chapter 5, e.g. UNIFAC–VLE, UNIFAC–LLE, modified UNIFAC) and the specially designed UNIFAC methods have been presented. Comparisons of standard UNIFAC methods against experimental K_{ow} data have been presented by Arbuckle[28], Banerjee and Howard[29], Li *et al.*[30], Campbell and Luthy[31], Kuramochi *et al.*[32], Lyman *et al.*,[33] Kan and Tomson,[34] and indirectly for activity coefficients at infinite dilution in aqueous systems by Voutsas and Tassios[35] and Zhang *et al.*[36]

The conclusions vary somewhat depending on the database considered, but the modified UNIFAC (of Larsen *et al.* and of Gmehling's group, see Chapter 5 for details) as well as UNIFAC–LLE perform better than the VLE UNIFACs with temperature-independent or linearly temperature-dependent parameters. The results are generally much better for monofunctional chemicals (e.g. 1-alcohols, chloroalkanes and alkylbenzenes) than for multifunctional chemicals such as polycyclic aromatic hydrocarbons. Two specially designed UNIFAC variants for K_{ow} calculations have been presented by Chen *et al.*[26] (water–UNIFAC) and Wienke and Gmehling[25]. The former uses Equation (17.22) in its development, but the factor 0.151 is ignored in the method of Gmehling[25] and is thus incorporated in the interaction parameters. The special methods due to specific parameter tables empirically correct some of UNIFAC's problems, e.g. proximity effects and polar chemicals. A systematic comparison of various UNIFAC models and two specific models for K_{ow} calculations (the AFC model mentioned earlier and the theoretically based method by Lin and Sandler[37], the GCS method) has been presented Derawi *et al.*[38] Figure 17.6 shows one result from this investigation for complex chemicals.

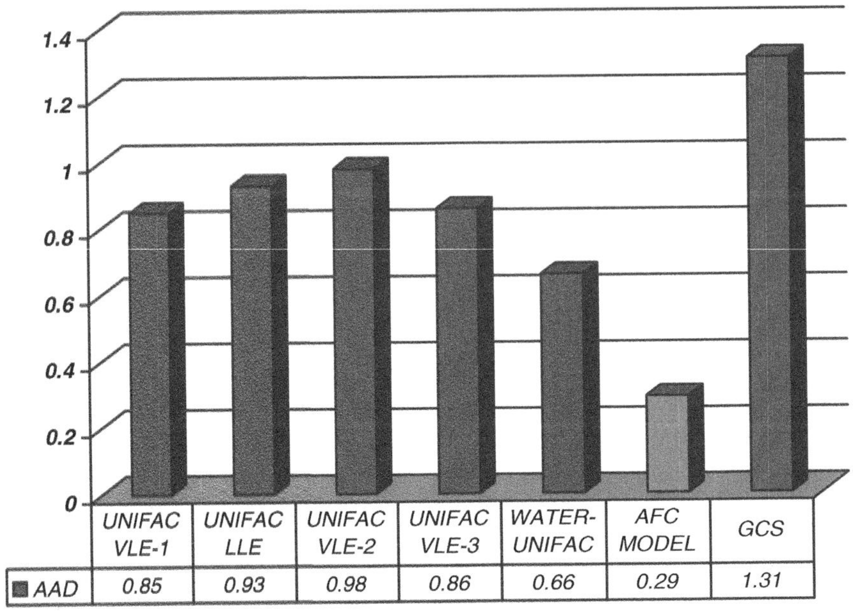

Figure 17.6 *Average absolute deviation (AAD) between experimental and predicted log K_{ow} values from various models for complex chemicals (glycols and alkanolamines). After Derawi et al.*[38]

The abbreviations used in Figure 17.6 and Tables 17.5–17.7 hereafter are explained:

- *UNIFAC VLE-1:* Original UNIFAC with parameters by Hansen *et al., Ind. Eng. Chem. Res.*, **1991**, *30*, 2352.
- *UNIFAC LLE:* Interaction parameters have been determined by fitting LLE experimental data; the LLE table was developed by Magnussen *et al., Ind. Eng. Chem. Process Des. Dev.*, **1981**, *20*, 133.

Table 17.5 *Percentage average absolute deviation (AAD %) between experimental and predicted log K_{ow} values at 298.15 K from the various UNIFAC models and the AFC correlation model:*[39]

$$AAD(\%) = \frac{1}{N}\sum_{i=1}^{N}\left|\log K_{ow}^{\exp,i} - \log K_{ow}^{cal,i}\right|$$

Solute	AAD (%)				
	UNIFAC VLE-1	UNIFAC LLE	UNIFAC VLE-2	UNIFAC VLE-3	AFC Model
N-octane	1.56	1.20	1.34	2.02	0.88
N-tetradecane	1.80	1.20	1.44	2.36	0.78
Pyrene	0.38	0.78	3.63	0.50	0.17
Benzo(a)pyrene	0.78	1.27	4.11	0.29	0.05
Octachloronaphthalene	6.84	3.72	4.12	4.52	0.08
DDT	0.24	1.01	3.87	0.78	0.60
Pentachlorobenzene	4.24	2.26	2.75	2.70	0.05
Decachlorobiphenyl	6.45	2.54	3.00	3.27	1.94
Lindane	0.50	0.66	0.32	0.22	0.41
	2.53	**1.63**	**2.73**	**1.85**	**0.55**

Table 17.6 Log K_{ow} prediction results for phthalates with various estimation methods. GCS is the 'solvation' method by Lin and Sandler[37]

Phthalates	Exp.[a]	UNIFAC VLE-1	ACD	CLogP	AFC	GCS
Dimethyl	1.6	1.74	1.62	1.56	1.66	−0.08
Diethyl	2.42	2.64	2.69	2.62	2.65	0.97
Dipropyl	3.64	3.53	3.75	3.68	3.63	2.03
Diisopropyl	2.83	3.53	3.38	3.24	3.48	1.73
Dibutyl	4.50	4.43	4.81	4.73	4.61	3.09
Diisobutyl	4.48	4.43	4.44	4.47	4.46	2.79
Dipentyl	5.62	5.32	5.87	5.79	5.59	4.15
Dihexyl	6.82	6.21	6.94	6.85	6.57	5.21
Dioctyl	8.18	8.00	9.06	8.97	8.54	7.34
Didecyl	8.83	9.79	11.19	11.08	10.5	9.46
Di-*sec*-octyl	7.06	8.00	8.69	8.71	8.39	7.03
Ditridecyl	8.4	12.47	14.38	14.26	13.45	12.6
Diallyl	2.98	3.39	3.28	3.11	3.36	0.93
Dutylbenzyl	4.73	4.79	4.99	4.98	4.84	2.37
Dicyclohexyl	4.9	6.02	5.74	5.62	6.20	5.0
Average deviation %		**12**	**14**	**13**	**13**	**37**

[a]All names for phthalates are given in substitute group.

- *UNIFAC VLE-2:* Developed by Hansen *et al.*, *SEP* 9212 (Internal Report), Department of Chemical and Biochemical Engineering, DTU, Lyngby, 1992. The equations are identical to VLE-1 and LLE except that the group interaction parameters are linearly temperature dependent.
- *UNIFAC VLE-3:* Modified version of the original UNIFAC by Larsen *et al.*, *Ind. Eng. Chem. Res.*, **1987**, 26, 2274; developed at the Technical University of Denmark.
- *WATER-UNIFAC model:* Developed by Chen *et al.*, *Chemosphere*, **1993**, *26(7)*, 1325. This model is similar to the original UNIFAC (VLE-1), but it is specifically designed for aqueous systems.

The following conclusions summarize the basic findings:

- The AFC correlation performs best for both the monofunctional and the multifunctional compounds.
- The GC solvation model by Lin and Sandler[37] fails for multifunctional compounds such as glycols and alkanolamines, although it performs very well for monofunctional compounds.

Table 17.7 Log K_{ow} prediction results for alcohol ethoxylates with various estimation methods. After Cheng et al.[40]

Compounds	Abb.	Exp.	ClogP	AFC	ACD	GCS	UNIFAC VLE-1
2-Methoxyethanol	C1E1	−0.77	−0.75	−0.91	−0.80	−1.56	−0.83
2-Ethoxyethanol	C2E1	−0.28	−0.22	−0.42	−0.27	−1.02	−0.38
3,6-Dioxa-1-octanol	C2E2	−0.54	−0.15	−0.69	−0.26	−1.89	−0.75
Isopropoxyethanol	C3E1	0.05	0.09	0.00	0.08	−0.65	0.07
2-Butoxyethanol	C4E1	0.8	0.84	0.57	0.80	0.03	0.51
3,6-Dioxadecanol	C4E2	0.56	0.91	0.29	0.81	−0.83	0.15
2-(Hexyloxy) ethanol	C6E1	1.86	1.90	1.55	1.86	1.09	1.41
3,6-Dioxa-1-dodecanol	C6E2	1.7	1.96	1.28	1.87	0.22	1.04
Average deviation			**33**	**39**	**22**	**311**	**36**

- The performance of the various UNIFAC versions varies a bit, but the water–UNIFAC version by Chen *et al.*[26], which has special water-based parameters, performs best, especially for multifunctional compounds. Among the other UNIFAC methods, the linear UNIFAC (VLE-2) is a bit worse than the others, while the UNIFAC VLE-1, LLE, VLE-3 methods perform similarly. These conclusions are valid for both mono- and multifunctional compounds.
- The results are not very satisfactory for environmentally interesting compounds (potential pollutants), as demonstrated by the results shown in Table 17.5. Again the AFC model performs best. The UNIFAC–LLE is the second-best model for these complex chemicals, while the linear UNIFAC (VLE-2) is again the worse among these models.

Tables 17.6 and 17.7 present K_{ow} results from another investigation[40] for complex chemicals and phthalates. This investigation has considered all three 'direct' computerized methods discussed above (AFC, CLogP, ACD), the UNIFAC VLE-1 and the method of Lin and Sandler.[37] From this analysis, it can be concluded that:

- The three commercial 'direct methods' for octanol–water estimations (ACD, CLogP and AFC) yield reliable results, similar to each other.
- The GCS method does not perform very well, especially, for alcohol ethoxylates.
- Of the various UNIFAC models, the UNIFAC VLE-1 performs best.
- The UNIFAC VLE-1 performs similar to the 'direct' methods for both families of compounds considered (phthalates and alcohol ethoxylates).
- Only the UNIFAC VLE-1 and the AFC methods predict the correct trend of K_{ow} with increasing hydrophilic part for the alcohol ethoxylate surfactants.

17.3 Environmentally friendly solvents: supercritical fluids

Supercritical fluid extraction (SCFE) is an environmentally friendly separation technique, which is useful in cases where traditional separation methods such as distillation are expensive or for compounds such as pharmaceuticals, enzymes, food products, etc., which cannot be produced and/or separated with traditional methods because they are thermally labile. SCFE can work at relatively low temperatures. Moreover, CO_2 is a non-toxic 'green' solvent with high solvency capacity.[41] Another positive feature of SCFE is its selectivity, of importance in the case of very similar substances, e.g. CO_2 dissolves benzoic acid in concentrations 1000 times higher than *p*-hydroxybenzoic acid. The removal of caffeine from coffee, the extraction of nicotine from tobacco as well as the extraction of high-value substances in the cosmetics and pharmaceutical industries and the extraction of flavors and fragrances are some of the processes for which SCFE has been used commercially.

Some of the shortcomings of SCFE compared to other conventional separation methods are the high operation pressures (up to 300 atm) and the associated high capital (equipment) cost, the need for batch operation, especially for solids, the use of flammable solvents (ethane, ethylene) and the difficulty in achieving optimum design due to the complexity of thermodynamic modeling and the need to select useful co-solvents. The co-solvents are low-molecular-weight volatile compounds, which are inserted in small amounts (1–5%) in the supercritical (SC) solvent and considerably modify its critical properties and density. Since, in many cases, the solubilities of solids (often heavy and complex compounds) in pure SC fluids are very low, co-solvents are required to increase the solubility and make the application commercially tractable. The co-solvents may significantly increase the solubility of the solid compound in CO_2 (up to two orders of magnitude) and often selectively. Their action is based on the strong interactions between CO_2 and the solid compound. Since in this way less amount of solvent is required and less energy as well, the SCFE method becomes more attractive from a commercial point of view. For example, it has been reported that adding 5% methanol in CO_2 increases the solubility of acridine in CO_2 almost five times at 50 °C and 200 bar. The use of co-solvents has another role as well, i.e. to increase the selectivity, which is accomplished when the molecules of the co-solvent interact selectively with only one of the solids in the mixture. For example, adding only 1% methanol in CO_2 increases the selectivity of CO_2 for acridine in a solid mixture with anthracene almost four times. This apparently makes the separation much easier using SCFE. The phenomenon is due to the hydrogen bonds that are formed between the methanol and acridine molecules.

Use of cubic EoS with advanced EoS/G^E mixing rules for SGE calculations has been shown in Chapter 6. There are no widely reliable models which can be used for solid–gas calculations for a variety of solids (polar, non-polar) and different types of co-solvents. We saw in Chapter 6 that the LCVM model yields quantitatively satisfactory predictions for binary solid–gas systems involving aromatic hydrocarbons, aliphatic acids and some alcohols, but poor results are obtained for complex solids, e.g. naproxen and cholesterol. Prediction of the co-solvent effect using the LCVM model should be restricted to relatively non-polar systems, while only a few qualitative guidelines can be given for polar ones. To choose the best co-solvent for hydrogen bonding systems, the ideas of Walsh *et al.*,[42] which are supported by recent experimental data, are recommended. These ideas can be summarized in the following statement:

> *Significant solubility enhancement in solid-SCF systems can be achieved only if solvation (cross-association) between the co-solvent and the solid takes place.*

The solvation can be either due to hydrogen bonding or due to charge-transfer complexes. Both types can be classified as Lewis acid–Lewis base (LA–LB) interactions. Thus, if the solid is a LA, a suitable co-solvent should be a strong LB, and vice versa. Use of Kamlet *et al.*'s[43] solvatochromic parameters α and β is possibly the best way to assess the acidity (LA) and basicity (LB). Coutsikos *et al.*[44] showed that the use of Walsh *et al.*'s[42] ideas can explain the significant enhancement in the solubility of hydroquinone in CO_2 using tributylphosphate (TBP) and that of propanols for naproxen's (a pharmaceutical) solubility in CO_2.

17.4 Conclusions

- Long-lived chemicals may attain a state of equilibrium in the environment. Thermodynamic relationships can then be used to estimate the amount of chemicals distributed in the various environmental ecosystems such as soil, sediment, water, air and biota.
- Air–water partitioning depends exclusively on temperature, vapor pressures and infinite dilution activity coefficients. For calculations involving the other ecosystems, the distribution coefficients must be known to estimate the fugacities in the 'less well-defined' systems such as soil, sediment and biota.
- Organic carbon–water and bioconcentration factors can be estimated by knowing the octanol–water partition coefficient of the chemicals. High values of the octanol–water partition coefficient, especially values of $\log K_{ow} > 4$, indicate very dangerous pollutants and possible biomagnification, i.e. accumulation of the chemicals in the higher members of the food chain (possibly up to humans).
- Octanol–water partition coefficient data have been collected for thousands of chemicals. When not available, the octanol–water partition coefficients can be estimated either from computerized methods (commercially available; demos also freely available on the Internet) or from thermodynamic methods, e.g. UNIFAC variants.

- UNIFAC–LLE and UNIFAC variants developed specifically for octanol–water partition coefficient calculations are the preferred approaches for octanol–water partition coefficient estimations. Many methods often fail, however, for multifunctional chemicals.

References

1. J.M. Prausnitz, *Chem. Eng. Sci.*, **2001**, *56*, 3627.
2. S.I. Sandler, *Fluid Phase Equilib.*, **1996**, *116*, 343.
3. S.I. Sandler, *J. Chem. Thermodyn.*, **1999**, *31*, 3.
4. R.P. Schwarzenbach, Ph.M. Gschwend, D.M. Imboden, *Environmental Organic Chemistry*. John Wiley & Sons, Inc., 1993.
5. B. Lomborg, *The Skeptical Environmentalist: Measuring the real state of the world*. Cambridge University Press, 2002.
6. C. Baird, *Environmental Chemistry* (2nd edition). W.H. Freeman, 1999.
7. D. Mackay, *Multimedia Environmental Models: The Fugacity Approach*. Lewis, 1991.
8. R.S. Pearlman, S.H. Yalkowsky, S. Banerjee, *J. Phys. Chem. Ref. Data*, **1984**, *13*(2), 555.
9. J. Sangster, *J. Phys. Chem. Ref. Data*, **1989**, *18*(*3*), 1111.
10. A.P. Sincero, G.A. Sincero, *Environmental Engineering: A design approach*. Prentice Hall, 1996.
11. R.V. Thomann, *Environ. Sci. Technol.*, **1989**, *23*, 699.
12. K.T. Valsaraj, *Elements of Environmental Engineering: Thermodynamics and Kinetics*. CRC Press, 1995.
13. A. Site, *J. Phys. Chem. Ref. Data*, **1997**, *26*(*1*), 157.
14. Ph. Coutsikos, E. Voutsas, K. Magoulas, D.P. Tassios, *Fluid Phase Equilib.*, **2003**, *207*(*1–2*), 263.
15. S.I. Sandler, H. Orbey, *Fluid Phase Equilib.*, **1993**, *82*, 63.
16. M.D. La Grega, Ph.L. Buckingham, J.C. Evans, *Hazardous Waste Management*. McGraw-Hill, 1994.
17. S. Maassen, A. Reichl, H. Knapp, *Fluid Phase Equilib.*, **1993**, *82*, 71.
18. S. Maassen, H. Knapp, W. Arlt, *Fluid Phase Equilib.*, **1996**, *116*, 354.
19. A. Leo, C. Hansch, D. Elkins, *Chem. Rev.*, **1971**, *71*, 525.
20. C. Hansch, A. Leo, *Substituent Constants for Correlation Analysis in Chemistry and Biology*. John Wiley & Sons, Inc., 1979.
21. W.M. Meylan, P.H. Howard, *J. Pharm. Sci.*, **1995**, *84*, 83.
22. J. Sangster, *Octanol-Water Partition Coefficients: Fundamentals and Physical Chemistry*. John Wiley & Sons, Ltd, 1997.
23. J. Marrero, R. Gani, *Fluid Phase Equilib.*, **2001**, *183–184*, 183.
24. E. Stefanis, L. Constantinou, C. Panayiotou, *Ind. Eng. Chem. Res.*, **2004**, *43*, 6253.
25. G. Wienke, J. Gmehling, *Toxicol. Environ. Chem.*, **1998**, *65*, 57.
26. F. Chen, J. Holten-Andersen, H. Tyle, *Chemosphere*, **1993**, *26*, 1325.
27. E.N. Polyzou, A.E. Louloudi, G.M. Kontogeorgis, I.V. Yakoumis, Prediction of octanol-water partition coefficients. 2nd Greek Chemical Engineering Conference, University of Thessaloniki, 16–29 May 1999, Book of Abstracts, pp. 101–104 (in Greek).
28. W.B. Arbuckle, *Environ. Sci. Technol.*, **1983**, *17*, 537.
29. S. Banerjee, P.H. Howard, *Environ. Sci. Technol.*, **1988**, *22*, 839.
30. A. Li, W.J. Doucette, A.W. Andren, *Chemosphere*, **1994**, *29*, 657.
31. J.R. Campbell, R.G. Luthy, *Environ. Sci. Technol.*, **1985**, *19*, 980.
32. H. Kuramochi, H. Noritomi, D. Hoshino, S. Kato, K. Nagahama, *Fluid Phase Equilib.*, **1998**, *144*, 87.
33. W.J. Lyman, W.F. Reehl, D.H. Rosenblatt, *Handbook of Chemical Property Estimation Methods: Environmental Behavior of Organic Compounds*. American Chemical Society, 1990.
34. A.T. Kan, M.B. Tomson, *Environ. Sci. Technol.*, **1996**, *30*(*4*), 1369.
35. E.C. Voutsas, D.P. Tassios, *Ind. Eng. Chem. Res.*, **1996**, *35*, 1438.
36. S. Zhang, T. Hiaki, K. Kojima, *Fluid Phase Equilib.*, **2002**, *198*, 15.
37. S.-T. Lin, S.I. Sandler, *Ind. Eng. Chem. Res.*, **1999**, *38*, 4081.

38. S.O. Derawi, G.M. Kontogeorgis, E.H. Stenby, *Ind. Eng. Chem. Res.*, **2001**, *40(1)*, 434.
39. J.-S. Cheng,Comparison of UNIFAC and the AFC model for the prediction of octanol-water partition coefficients of mono- and multi-functional chemicals. Internal report, Institut for Kemiteknik, Technical University of Denmark, 2001.
40. H.Y. Cheng, G.M. Kontogeorgis, E.H. Stenby, *Ind. Eng. Chem. Res.*, **2005**, *44*, 7255.
41. W. Leitner, *Nature*, **2000**, *405*, 130.
42. J.M. Walsh, G.D. Ikonomou, M.D. Donohue, *Fluid Phase Equilib.*, **1987**, *33*, 295.
43. M.J. Kamlet, J.M. Abboud, M.H. Abraham, R.W. Taft, *J. Org. Chem.*, **1983**, *48*, 2877.
44. Ph. Coutsikos, K. Magoulas, G.M. Kontogeorgis, *J. Supercrit. Fluids*, **2003**, *25(3)*, 197.

18

Thermodynamics and Colloid and Surface Chemistry

18.1 General

Colloid and surface chemistry – or in other words 'soft nanotechnology' – is a field of great importance, related to special processes and product technology, e.g. paints and coatings, surfactants and cleaning, wetting and adhesion, as well as dispersions such as those found in food, cosmetics and pharmaceutical formulations. As a subject, colloid and surface chemistry is typically taught separately from thermodynamics. However, there are many links to thermodynamics which can illustrate the underlying concepts in a clearer way (than can each science alone!) or inspire alternative ways of testing thermodynamic models. We have selected four topics in this chapter to illustrate certain interrelations between thermodynamics and colloid–surface chemistry:

1. intermolecular vs. interparticle forces and their importance in adhesion and colloid stability studies;
2. approaches for estimating hydrophilicity;
3. micellization and phase behavior of surfactant solutions;
4. adsorption.

18.2 Intermolecular vs. interparticle forces

18.2.1 Intermolecular forces and theories for interfacial tension

Intermolecular and interparticle forces were discussed in Chapter 2. Surface and interfacial tensions are manifestations of intermolecular forces. The very high surface tension of mercury (485 mN/m) is due to metallic forces ('metallic bonding', mobile electrons shared by atoms of a metal), while most liquids have surface tensions between 20 and 50 mN/m. However, strongly hydrogen bonding liquids like water and glycols have high surface tensions (72.8 mN/m for water and 66.0 mN/m for glycerol at 25 °C). They decrease with temperature, though, as the hydrogen bonding forces become less significant at higher temperatures. As can be

Thermodynamic Models for Industrial Applications: From Classical and Advanced Mixing Rules to Association Theories
Georgios M. Kontogeorgis and Georgios K. Folas

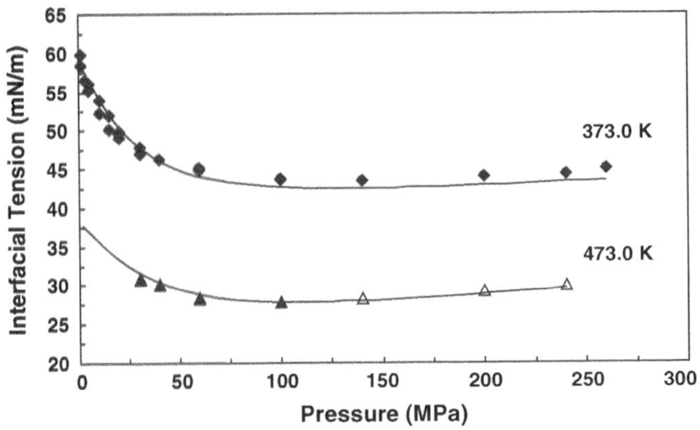

Figure 18.1 *Interfacial tension of the methane–water system at 373.0 K and 473.0 K. The line is calculation with LGT–SRK (see Section 18.5). Reprinted with permission from Fluid Phase Equilibria, Calculation of the interfacial tension of the methane-water system with the linear gradient theory by Kurt A. G. Schmidt, Georgios K. Folas and Bjørn Kvamme, 261, 1–2, 230–237 Copyright (2007) Elsevier*

seen in Figure 18.1, for example, the interfacial tension of the methane–water system goes through a minimum with increasing pressure along the higher temperature (373 and 473 K) isotherms. This minimum with increasing pressure has been exhibited in other experimental investigations for the interfacial tension of hydrocarbon–water systems.[1]

Molecular dynamics simulations[3] have shown that, at increased temperatures, the reduction of the hydrogen bonds results in (1) increased mutual penetration of methane and water, (2) a stronger 'roughening' of the surface and even periodic 'throwing' of water clusters into the methane. Both of these effects become more pronounced at higher pressures.

Differences in miscibility in, for example, aqueous solutions with hydrocarbons, fluorocarbons or alcohols can be elucidated from the values of the interfacial tensions, given for some compounds in Table 18.1. The higher the interfacial tensions, the lower the miscibility, e.g. water–alkanes have higher interfacial tensions compared to water with the corresponding aromatic hydrocarbons. On the other hand, the interfacial tensions

Table 18.1 *Liquid–liquid interfacial tensions (and liquid surface tensions) for some liquids. All values are given at room temperature*

System	Interfacial tension (mN/m)	Liquid	Surface tension (mN/m)
Water–hexane	51	Water	72.8
Water–benzene	35	*n*-Hexane	18
Water–C_6F_{14}	57.2	Benzene	28.9
MEG–hexane	16	C_6F_{14}	11.5
Water–butanol	1.8	MEG	47.7
Water–pentanol	6.0	Pentanol	25.2
Water–octanol	8.5	Octanol	27.5

in water–fluorocarbons are higher than those between water and the corresponding alkanes, as fluorocarbons are even more hydrophobic compounds compared to alkanes.

Another important application of intermolecular forces is in the development of theories for estimating interfacial tensions. Such theories are important not just for liquid–liquid interfaces (Table 18.1), for which the interfacial tensions can be measured, but especially for describing and understanding solid–liquid and solid–solid interfaces. Interfacial tensions for interfaces with solids cannot be measured directly but are of great importance for understanding wetting, adhesion, cleaning, lubrication and many other processes. Early theories for interfacial tensions are based on the extension of the Fowkes equation (often mentioned in colloid and surface science textbooks, e.g. Myers[4], Shaw[5], Hiemenz and Rajagopalan[6], Goodwin[7]):

$$\gamma_{ij} = \gamma_i + \gamma_j - 2\sqrt{\gamma_i^d \gamma_j^d} \tag{18.1a}$$

where the surface tension is divided into a dispersion (d) and specific ($spec$) component:

$$\gamma = \gamma_i^d + \gamma_i^{spec} \tag{18.1b}$$

For example, alkanes exhibit (mostly) dispersion forces (d), but for other molecules additional forces may be present. These additional forces are termed specific ($spec$). Examples of specific forces are polar, metallic, m (e.g. for mercury, Hg) and hydrogen bonding. The surface tension of n-hexane (n-C_6), mercury (Hg) and n-octanol (n-C_8OH) can be written for example as:

$$\begin{aligned} \gamma_{n\text{-C}_6} &= \gamma_{n\text{-C}_6}^d \\ \gamma_{\text{Hg}} &= \gamma_{\text{Hg}}^d + \gamma_{\text{Hg}}^m \\ \gamma_{n\text{-C}_8\text{OH}} &= \gamma_{n\text{-C}_8\text{OH}}^d + \gamma_{n\text{-C}_8\text{OH}}^{spec} \end{aligned} \tag{18.2}$$

In the case of mercury (Hg) the specific contribution includes only the metallic part, while in the case of alcohols, the specific contribution includes both polar and hydrogen bonding contributions.

An important property tightly linked to surface and interfacial tensions is the so-called 'work of adhesion'. The (thermodynamic or) ideal work of adhesion, i.e. the work to separate an interface ij into two separate surfaces i and j (in contact with air), is given by the Dupre equation:

$$W_{adh} = \gamma_i + \gamma_j - \gamma_{ij} \tag{18.3}$$

The work of adhesion can be calculated if an equation for the interfacial theory is available. For example, W_{adh} is equal to $2(\gamma_i^d \gamma_j^d)^{1/2}$ according to the Fowkes equation (Equation (18.1a)). Similarly the so-called work of cohesion can be defined for a single surface (work to break the surface), W_{coh}, which is equal to two times the value of its surface tension.

The Fowkes theory, like the other theories discussed later, has strengths and weaknesses. It is possibly the simplest method for estimating interfacial tensions but success stories have been reported, e.g. modeling the interfacial tension for mercury–water and mercury–alkanes mixtures, as well as for interfaces between water or glycol and *n*-alkanes (see Problems 2 and 3 on the companion website at www.wiley.com/go/Kontogeorgis). For other interfaces, e.g. those between water and aromatic hydrocarbons, the Fowkes equation has problems. However, the dispersion value for the surface tension of water (γ^d_{water}) can be calculated with the Fowkes equation using, for instance, water–hydrocarbon liquid–liquid interfacial tension data (Problem 1 on the companion website at www.wiley.com/go/Kontogeorgis). The obtained value ($\gamma^d_{water} = 21.8$ mN/m) is generally well accepted and in agreement with other considerations and molecular calculations (see Chapter 2), illustrating that about 30% of water's surface tension is due to dispersion forces. Extensions of the Fowkes equation have been proposed, which account explicitly for polar and hydrogen bonding effects in the expression for the interfacial tension using geometric mean rules for all terms. Two well-known theories are those by Owens and Wendt and by Hansen and Skaarup[8,9]:

Owens–Wendt:

$$\gamma = \gamma^d + \gamma^{spec} \tag{18.4}$$

$$\gamma_{ij} = \gamma_i + \gamma_j - 2\sqrt{\gamma_i^d \gamma_j^d} - 2\sqrt{\gamma_i^{spec} \gamma_j^{spec}} = \left(\sqrt{\gamma_i^d} - \sqrt{\gamma_j^d}\right)^2 + \left(\sqrt{\gamma_i^{spec}} - \sqrt{\gamma_j^{spec}}\right)^2 \tag{18.5}$$

Hansen–Skaarup (or Hansen–Beerbower):

$$\gamma_{ij} = \left(\sqrt{\gamma_i^d} - \sqrt{\gamma_j^d}\right)^2 + \left(\sqrt{\gamma_i^p} - \sqrt{\gamma_j^p}\right)^2 + \left(\sqrt{\gamma_i^h} - \sqrt{\gamma_j^h}\right)^2 \tag{18.6}$$

or alternatively:

$$\gamma_{ij} = \gamma_i + \gamma_j - 2\sqrt{\gamma_i^d \gamma_j^d} - 2\sqrt{\gamma_i^p \gamma_j^p} - 2\sqrt{\gamma_i^h \gamma_j^h} \tag{18.7}$$

where the individual surface tensions (dispersion, polar and hydrogen bonding parts) are estimated from the total value of surface tension and the Hansen solubility parameters using the equation:

$$\gamma = 0.715 V^{1/3} \left[\delta_d^2 + l(\delta_p^2 + \delta_h^2)\right] \tag{18.8}$$

Here δ_d is the dispersion part of the solubility parameter, δ_p is the polar part of the solubility parameter and δ_h is the hydrogen bonding part of the solubility parameter. In Equation (18.8), V is the molar volume (of the compound or of the repeating unit for polymers) in cm^3/mol, all solubility parameters are in $(cal/cm^3)^{1/2}$ and the surface tension is expressed in mN/m (or dyn/cm). The parameter l should generally be estimated directly from Equation (18.8) using experimental values for the surface tension. The dispersion, polar and hydrogen bonding contributions to the surface tension can then be calculated as:

$$\gamma = \gamma^d + \gamma^p + \gamma^h \tag{18.9}$$

where:

$$\gamma^d = 0.0715\ V^{1/3}\delta_d^2$$

$$\gamma^p = 0.0715\ V^{1/3} l\delta_p^2$$

$$\gamma^h = 0.0715\ V^{1/3} l\delta_h^2$$

Despite the success of the Fowkes/Owens–Wendt/Hansen–Beerbower methods for many practical applications such as for polymer surfaces, they are gradually being abandoned. This may be attributed to the doubtful use of the geometric mean rule for the polar and especially for the hydrogen bonding interactions. As discussed in Chapter 2, this geometric mean rule is rigorously valid only for the dispersion forces. One of the most successful and widely used recent methods is the van Oss *et al.*[10] theory or acid–base theory, where the hydrogen bonding (in general 'Lewis acid–Lewis base' interactions) is expressed via asymmetric combining rules:

van Oss et al.:

$$\gamma = \gamma^{LW} + \gamma^{AB} = \gamma^{LW} + 2\sqrt{\gamma^+\gamma^-} \tag{18.10a}$$

LW are the London/van der Waals forces and *AB* are the acid(+)/base(−) forces.

Furthermore:

$$\gamma_{ij} = \left(\sqrt{\gamma_i^{LW}} - \sqrt{\gamma_j^{LW}}\right)^2 + 2\left(\sqrt{\gamma_i^+} - \sqrt{\gamma_j^+}\right)\left(\sqrt{\gamma_i^-} - \sqrt{\gamma_j^-}\right) \tag{18.10b}$$

Equations (18.10a) and (18.10b) are generally valid for any interface.

There are also theories for the interfacial tension which are *not* based on the principle of dividing the surface tension into contributions due to intermolecular forces. One of the best known such equations is that proposed by Girifalco and Good[4,6,7]:

$$\gamma_{ij} = \gamma_i + \gamma_j - 2\varphi\sqrt{\gamma_i\gamma_j} \tag{18.11}$$

The correction parameter φ is available for many liquid–liquid interfaces, estimated from experimental data.

18.2.2 Characterization of solid interfaces with interfacial tension theories

Irrespective of the theory used, solid interfaces can be characterized by using interfacial tension theories in combination with the Young equation for contact angle, Equations (18.12). Characterization in this context means estimating the solid surface tension and its 'components' or parts (dispersion, specific, etc.). Knowledge of these can assist in determining which impurities are present on the surface and which modifications are required in the surface in order to improve, for example, wetting or adhesion for a certain application.

The solid characterization procedure is illustrated in Figure 18.2.

The Young equation for the contact angle and the Young–Dupre equation for the work of adhesion:

$$\cos\vartheta = (\gamma_s - \gamma_{sl})/\gamma_l \tag{18.12a}$$

$$W_{adh} = W_{sl} = \gamma_s + \gamma_l - \gamma_{sl} = \gamma_l(1 + \cos\vartheta) \tag{18.12b}$$

In Equations (18.12a) and (18.12b) the subscripts *s*, *l* and *sl* indicate solid, liquid and solid–liquid interfaces. These equations are true under the assumption that the spreading pressure is zero. This is a reasonable assumption, especially for low-surface-energy surfaces like polymers (less so for metals). The spreading pressure is the difference between the solid surface tension (in contact with air) and the solid–vapor surface tension, where the vapor is due to the presence of the liquid.

Combining Equations (18.12) with any of the theories mentioned for estimating interfacial tension (Equations (18.1), (18.5), (18.7), (18.11)) can help to estimate the surface tensions of the solid and its

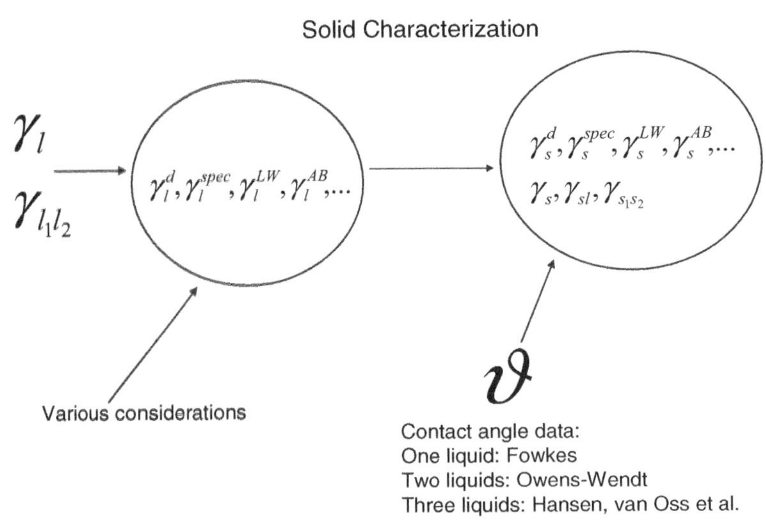

Figure 18.2 *Procedure for characterization of a solid surface. The work of adhesion and the characterization/profile of the solid surface can be obtained as well as information about the type of contaminants which may be present. Such studies are helpful in surface modification and adhesion. Data from studies of liquid–liquid interfaces are required*

'components' (dispersion, specific, acid–base, etc.) when contact angle data of several liquids on the solid and the surface tensions of liquids are available. In the case of the Fowkes equation, one contact angle measurement is sufficient, for Owens–Wendt two data points (i.e. measurements for two liquids on the same surface) are needed, while the Hansen and van Oss et al. theories require contact angle data for three liquids on the same solid. In the case of van Oss *et al.*, the resulting equation for solid–liquid interfaces (W_{adh} is the work of adhesion) is:

$$W_{adh} = \gamma_l(1+\cos\vartheta) = 2\sqrt{\gamma_s^{LW}\gamma_l^{LW}} + 2\sqrt{\gamma_s^{+}\gamma_l^{-}} + 2\sqrt{\gamma_s^{-}\gamma_l^{+}} \tag{18.13}$$

First, the *LW* part of a solid will be estimated from Equation (18.13) using a non-polar liquid:

$$\gamma_s^{LW} = \frac{\gamma_l(1+\cos\vartheta)^2}{4} \tag{18.14}$$

There are many ways of estimating the solid surface tension of components and sometimes different triplets of liquids can give somewhat different values. Water is typically used as one of the three fluids and is considered to have equal LA–LB (acid–base) components (= 25.5 mN/m) and an *LW* (van der Waals) value of 21.8 mN/m (remember, the dispersion value obtained from the Fowkes equation!). The base (−) values obtained for most solids with the van Oss–Good theory are typically rather high, much higher than the acid (+) values, and this observation should be considered when evaluating the results. Of course this is also a limitation of the method. Balkenende *et al.*[11] discuss the limitations of various recent interfacial theories including van Oss *et al.*'s.

An alternative way of studying wetting phenomena is the Zisman plot and the associated concept of the critical surface tension.[5,7] The Zisman plot is an empirical industrially used technique, based on an observation of Fox and Zisman, that a plot of surface tension against liquid–gas surface tension is often linear:

Zisman plot;

$$\cos\vartheta = 1 \text{ when } \gamma_l = \gamma^{crit} \tag{18.15}$$

An example of the Zisman plot for PET is presented in Figure 18.3. The critical surface tension read from the plot is 43 mN/m. Zisman plots can be used for comparing the wetting of various surfaces against different liquids. Wetting occurs if $\gamma_l \leq \gamma^{crit}$. For example, for PET only liquids with a surface tension of less than 43 mN/m will completely wet the polymer, as shown in Figure 18.3. The lower the critical surface tension, the more difficult the wetting of the solid. Polymers and fabrics have typically much lower critical surface tensions than *clean* metals and ceramics. Among polymers the hydrocarbon plastics and especially the fluorocarbons have the lowest critical surface tensions, e.g. 31 mN/m for polyethylene (PE) and 18 mN/m for Teflon. For this reason, various fluoro-functionalities have been tried when super-hydrophobic ('non-sticky' or 'self-cleaning') surfaces are designed. Finally, it should be noted that solid surfaces are most often rather complex (heterogeneous, contaminated, change with time, rough), which complicates their characterization. Such phenomena result in the so-called 'contact angle hysteresis' and thus use of the Young equation together with experimental contact angle data is not always straightforward.

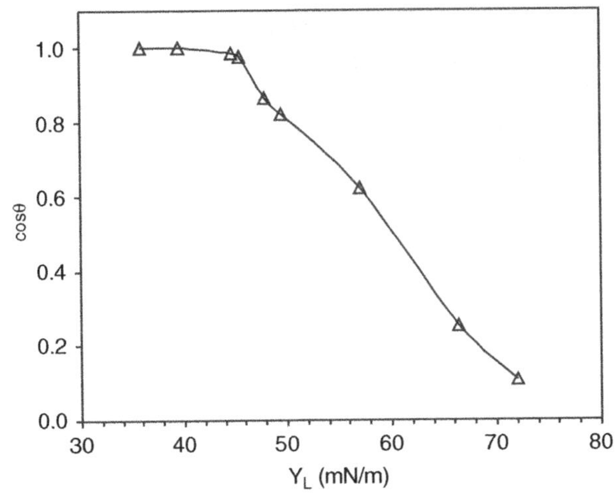

Figure 18.3 *Zisman plot (cosine of contact angle against liquid surface tension) for PET (polyethylene terephthalate). A critical surface tension value equal to 43 mN/m can be read from this plot*

18.2.3 Spreading

Spreading of liquids in immiscible liquids or solids can be expressed via the Harkins spreading coefficient, defined as follows for liquid–liquid and liquid–solid interfaces:

Harkins spreading coefficient (for an oil (O) in water (W); A = air):

$$S = \gamma_{WA} - (\gamma_{OA} + \gamma_{OW}) = W_{adh,OW} - W_{coh,O} \tag{18.16}$$

Harkins spreading coefficient for a liquid on a solid:

$$S = W_{adh,sl} - W_{coh,l} = \gamma_l(\cos\vartheta - 1) \tag{18.17}$$

The Harkins spreading coefficient is a method for determining whether one liquid will spread on another, e.g. 'oil' on water. Positive values of the spreading coefficient indicate spreading, e.g. for polar compounds in water ($S = 36.8$ mN/m for octanol in water). Spreading increases with solubility, but mutual saturation can affect the results, e.g. for benzene–water, $S = 8.9$ mN/m at the start of and -1.4 mN/m (non-spreading) after mutual saturation. Notice that the Harkins coefficient is related to the work of adhesion and cohesion: a positive spreading coefficient and thus spreading are obtained only when the work of adhesion between oil and water is higher than the work of cohesion of the oil or, in other words, when oil 'likes' (adheres to) water more than it likes itself (coheres to itself). Similar concepts are true for solid–liquid interfaces.

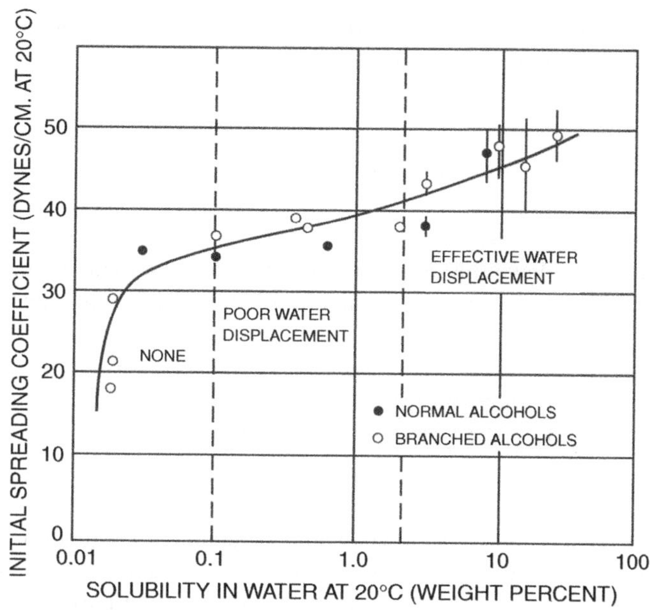

Figure 18.4 *Relation of water solubility to the initial spreading coefficient and water's displacing ability of simple alcohols. The vertical lines through points in the region of strong displacement are proportional in length to the areas of a 2 mm water film displaced by the respective alcohols. Reprinted with permission from Industrial and Engineering Chemistry, Cleaning by Surface Displacement of Water and Oils by H. R. Baker, P. B. Leach, C. R. Singleterry and W. A. Zisman, 59, 6, 29–40 Copyright (1967) American Chemical Society*

Spreading of a liquid on water is a measure of its affinity to water. Thus, it is expected that the spreading coefficient is, as illustrated in Figure 18.4, linked to the solubility in water[12]. The higher the water solubility of a compound, the higher its spreading on water.

18.3 Interparticle forces in colloids and interfaces

18.3.1 Interparticle forces and colloids

While polar and acid–base interactions are often very important in many applications related to interfaces, most studies in colloids related to colloid stability involve a balance between the van der Waals forces (typically attractive) and the (typically repulsive) electric (double-layer) forces. The van der Waals force between spherical colloidal particles of radius a (see also Chapter 2) is often expressed via the potential energy:

$$V_A = -\frac{A_{eff}a}{12H} \tag{18.18}$$

where H is the interparticle or interface distance and a is the radius of the particle. The effective Hamaker constant of particles (1) in a medium (2) is given as:

$$A_{eff} = \left(\sqrt{A_{11}} - \sqrt{A_{22}}\right)^2 \tag{18.19}$$

Equation (18.19) is true when we have the same types of particles in a medium, e.g. water. However, one simple equation for the effective Hamaker constant in the case of two different types of particles (1 and 3) in a medium (2) is:

$$A_{123} = \left(\sqrt{A_{11}} - \sqrt{A_{22}}\right)\left(\sqrt{A_{33}} - \sqrt{A_{22}}\right) \tag{18.20}$$

Thus, in the case of particles of the same type or when air or a vacuum is the medium, Equations (18.19) and (18.20) always lead to a positive Hamaker constant and attractive van der Waals forces ($V_A < 0$ in Equation (18.18)), which is the most usual case. However, in the case of unlike particles, the effective Hamaker constant can be negative, which implies repulsive van der Waals forces ($V_A > 0$ in Equation (18.18)). These repulsive vdW forces find many applications and can be used for screening solubility in blend–solvent mixtures as shown for some systems in Table 18.2. The agreement is satisfactory.

The Hamaker constant can be calculated directly from London's theory (Chapter 2):

$$A = \pi^2 C \rho^2 \tag{18.21}$$

where ρ is the number density (molecules/volume), C is the London coefficient for separation (see Chapter 2) and $\pi = 3.14$. All Hamaker constant values range typically between 0.4 and 4×10^{-19} J. The relatively

Table 18.2 *Negative Hamaker constants and blend–solvent miscibility. After van Oss et al.*[13]

Polymer(1)–solvent(2)–polymer(3)	Hamaker constant, A_{123}	Visual observation
PIB–THF–cellulose acetate	−3.76	*S*
PIB–benzene–PMMA	−3.49	*S*
PIB–cyclohexanone–PS	−1.61	*S*
PIB–benzene–PS	−1.12	*S*
PS–chlorobenzene–PMMA	−0.09	*M*
Cellulose acetate–MEK–PS	+0.04	*M*
PVC–THF–cellulose acetate	+2.80	*M*
PMMA–MEK–cellulose acetate	+7.51	*M*

S = separation, *M* = miscible. PIB = polyisobutylene, PS = polystyrene, PVC = poly(vinyl chloride), PMMA = poly(methyl methacrylate), MEK = methyl ethyl ketone, THF = tetrahydrofurane.

constant values for different compounds are due to the fact that the parameter C is roughly proportional to the square of the polarizability, i.e. the square of the volume, as discussed in Chapter 2.

A more rigorous estimation of the Hamaker constant for mixtures avoids simplified combining rules and uses the Lifshitz theory (based on dielectric constants and refractive indices; see Israelachvili[14] and Chapter 2, Section 2.1.2).

18.3.2 Forces and colloid stability

Colloidal dispersions are inherently unstable systems and in the long run the attractive forces will dominate and the colloidal system will destabilize. The stability of colloidal dispersions is often described via the DLVO theory, which accounts for both the (usually) attractive vdW forces and the (typically) repulsive electrostatic or double-layer forces:

$$V = V_R + V_A \tag{18.22}$$

The vdW forces decrease with the first (sometimes the second) power of the interparticle distance, while the electrostatic (double-layer) forces decrease exponentially with the distance. Both types of interparticle forces are much more long range compared to the intermolecular forces. The electrostatic forces are due to the fact that most particles are charged inside a medium, especially a polar one like water, which has a high dielectric constant. The origin of the surface charge can be complex and there are many mechanisms explaining its creation, e.g. adsorption of ions from the solution or dissociation of surface groups. Other repulsive forces, which help stability, exist especially at low interparticle distances (hydration, steric).

Both the attractive vdW and electrostatic forces:

- are long range (vdW forces dominate at high and small distances, electrostatic forces may dominate at intermediate distances);
- depend on the characteristics of the particle and the medium;
- depend on the shape of the particles/surfaces (sphere, plates, cylinders, etc.), their size and proximity.

Many expressions for the total potential energy V exist, but in the simple case of equal-sized spherical particles of equal surface potentials and small electric double-layer overlap, the DLVO theory takes the form:

$$V = V_R + V_A = 2\pi\varepsilon_0 D a\psi_0^2 \exp(-\kappa H) - \frac{aA_{eff}}{12H} \tag{18.23}$$

The two key properties of this equation are the surface (or zeta) potential $\psi_0 \approx \zeta$, typically obtained from electrophoretic measurements, and the Debye length, κ^{-1}, which is defined (in nm) as:

$$\kappa^{-1} = \sqrt{\frac{\varepsilon_0 DkT}{q^2 N_A \sum_i C_{i(B)} Z_i^2}} \tag{18.24}$$

where:

q = electronic or unit charge
C_i = concentration of ion i (in mol/l)
Z_i = valency of ion i

Performing the summation in Equation (18.24) implies knowledge of the type of electrolyte. The equation can be simplified in specific cases, e.g. for aqueous solution at 25 °C (in nm):

$$\kappa^{-1} = \frac{0.429}{\sqrt{CZ^2}} \tag{18.25}$$

where the concentration of the electrolyte, C, is given in mol/l.

More specifically, we obtain the following expressions if the electrolyte type is known (C in mol/l and the Debye length κ^{-1} in nm):

$$\kappa^{-1} = \frac{0.304}{\sqrt{C}} \quad \text{for 1 : 1 electrolytes, e.g. NaCl} \tag{18.26}$$

$$\kappa^{-1} = \frac{0.176}{\sqrt{C}} \quad \text{for both 1 : 2 and 2 : 1 electrolytes like } CaCl_2 \text{ and } Na_2SO_4 \tag{18.27}$$

$$\kappa^{-1} = \frac{0.152}{\sqrt{C}} \quad \text{for 2 : 2 electrolytes like } MgSO_4 \tag{18.28}$$

$$\kappa^{-1} = \frac{0.124}{\sqrt{C}} \quad \text{for 1 : 3 electrolytes like } AlCl_3 \tag{18.29}$$

Notice that the electrostatic forces are typically repulsive due to the overlap of the double layer between particles. These electrostatic forces are always repulsive between particles of the same kind. They are proportional to the second power of the surface potential (in most cases, see Equation (18.23)) and decrease with decreasing Debye (double-layer) thickness. The Debye thickness is a key parameter, crucial for controlling the double-layer forces. It decreases with increasing electrolyte concentration, even more pronounced for the high-valency electrolytes. The surface potential is estimated, sometimes indirectly, via electrokinetic experiments. Via these experiments we can measure the electrophoretic mobility, which for very small or very large particles, can be related to the so-called zeta potential (Huckel and Smoluckoswki equations). The zeta potential is approximately equal to the surface potential and is used in the DLVO theory, e.g. in Equation (18.23).

When suitable expressions are available for the attractive and repulsive potential energies, e.g. Equation (18.23), then the total potential energy (DLVO theory) can be calculated and plotted as a function of the interparticle distance. High V values, above 15–25kT ($kT = 4.12 \times 10^{-21}$ J at 25 °C), of the repulsive potential barrier, indicate stable (actually metastable) colloidal systems. One example is shown in Figure 18.5.

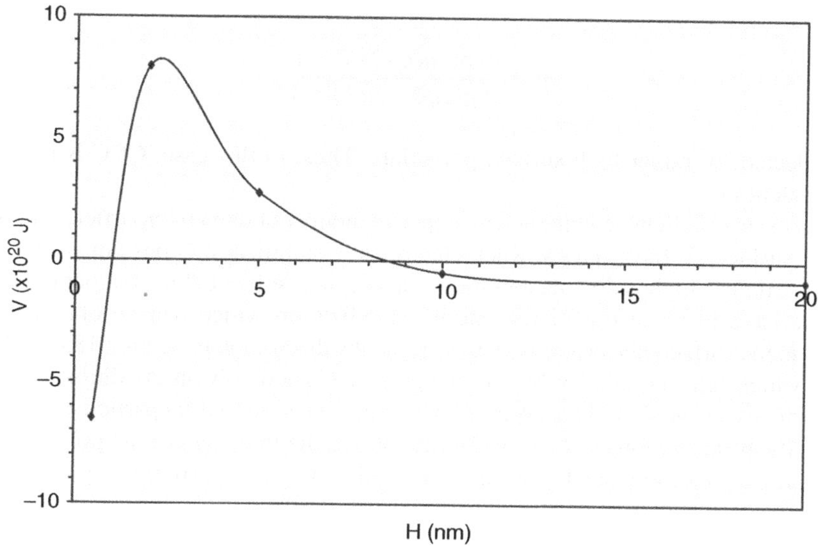

Figure 18.5 *The total interaction energy–distance plot (DLVO theory) for spherical colloidal particles of diameter 10^{-7} m dispersed in 10^{-2} mol/l aqueous 1 : 1 electrolyte solutions at 25 °C. The Hamaker constants of the particles and of the dispersion medium are 1.6×10^{-19} J and 0.4×10^{-19} J, respectively, and the zeta potential is −40 mV. The system is stable since it can be seen that $V_{max}/kT \cong 19.5$*

According to the DLVO theory, stable colloidal dispersions are expected if one or more of the following conditions are fulfilled:

- particles with high surface (zeta) potentials;
- low electrolyte concentration (and especially not having electrolytes with high valency);
- presence of polar media like water;
- low values of the Hamaker constant;
- the rather unusual situation of negative Hamaker constant (leading to repulsive vdW forces) – this can happen in cases of particles of two different kinds in a third medium and if one of the particles attracts the medium more than the two particles attract themselves (see Table 18.2).

The DLVO theory is in good agreement with many empirical rules based on large volumes of experimental data such as the Schulze–Hardy rule, which were actually developed prior to the DLVO theory. This rule indicates that the critical coagulation concentration (CCC) depends largely (decreases) on the sixth power of the valency of the counter-ion (ion of opposite sign to that of the surface). CCC is the concentration of the electrolyte which is just sufficient to coagulate a lyophobic sol (to an arbitrarily defined extent in an arbitrarily chosen time). With some degree of approximation, it could be obtained, e.g. under the conditions where $V=0$ and $dV/dH=0$.

From the DLVO theory and for aqueous dispersions at 25 °C, it can be shown that:

$$\mathrm{CCC} = \frac{3.84 \times 10^{-39}\gamma^4}{A^2(Z)^6}$$

where:

$$\gamma = \frac{\exp(zq\psi_0/2kT)-1}{\exp(zq\psi_0/2kT)+1}$$

which can be neglected for rather high surface potentials. Thus, in this case, CCC is proportional to $1/Z^6$ (Z = counter-ion valency).

The DLVO theory is in excellent agreement with experimentally obtained interparticle forces in many cases, e.g. between mica surfaces.[14] In the cases, when electrostatic stabilization does not work well (uncharged/weakly charged particles, high salt concentration, in organic solvents/low temperatures, high particle concentration), we have to make use of the 'steric' stabilization which can be obtained by coating the particles with polymers, surfactants, proteins, etc. (see also the discussion in Section 18.8.2). It is important to choose polymers which have an affinity for the medium (i.e. good solvent conditions), provide complete coverage of the particles and attain full adsorption. In the opposite case (partial particle coverage), we may get bridging and actually attractive forces. The configurations and the thickness of the polymer layer are crucial factors. The thickness of the polymer layer can be considered to be approximately equal to the radius of gyration.

18.3.3 Interparticle forces and adhesion

The vdW forces between macroscopic surfaces can lead to strong adhesive forces (in the absence of 'cracks', 'voids' and other defects which can often be present), as shown in Table 18.3. This is because such forces between macroscopic particles or surfaces are much longer range compared to the intermolecular forces. In the case of adhesion between plates, the adhesive pressure is expressed as:

$$P = \frac{A}{6\pi H^3} \tag{18.30}$$

The results of Table 18.3 are obtained using three different expressions for the Hamaker constant:

$$\text{'ideal':} \quad \text{using } A = 24\pi H^2\gamma \Rightarrow P = \frac{4\gamma}{H} \tag{18.31}$$

$$\text{'Myers':} \quad A = 12.4\pi H^2\gamma \Rightarrow P = \frac{2.06\gamma}{H} \tag{18.32}$$

Table 18.3 *Real and ideal (using Equations (18.31)–(18.33)) cohesive strengths of some common materials (obtained at contact, distance H = 0.4 nm). All values are expressed in MPa. Real values from Myers*[4]

Material	Ideal – Myers (MPa), Equation (18.32)	Real (MPa)	Ideal (MPa), Equation (18.31)	'New' (MPa), Equation (18.33)
PE (molded)	180	38	349	48
PS (molded)	210	69	407	56
Aramid yarn	7900	2760	15 340	2109
Drawn steel	9800	1960	19 029	2617
Graphite whisker	100 000	24 000	194 174	26 700

$$\text{'new':} \qquad A = \frac{4\pi}{1.2} H^2 \gamma^d \Rightarrow P = \frac{0.55\gamma}{H} \tag{18.33}$$

where H is the distance, a is the sphere radius and γ is the surface tension.

Notice the better agreement achieved with the 'new' more realistic expression of the Hamaker constant in comparison to the 'ideal' equation. All values indicate, however, that very high adhesive (cohesive) pressures are obtained because of the vdW forces. Similar expressions illustrating the long-range character of the vdW forces between colloid particles can be written for other geometries as well:

Force between two spheres:

$$F = \frac{Aa}{12H^2} = 2\pi a\gamma \tag{18.34}$$

Force between sphere and plate:

$$F = \frac{Aa}{6H^2} = 4\pi a\gamma \tag{18.35}$$

where H is the distance, a is the sphere radius and γ is the surface tension. The last right-hand-side form of Equations (18.34) and (18.35) is based on one of the 'ideal' expressions for the Hamaker constant A (given by Equation (18.31)). However, even though the agreement in Table 18.3 is satisfactory, these strong adhesive forces are ideal values assuming no defects in the materials. Polar and acid–base interactions can significantly increase adhesion, especially when the dominant adhesive mechanism is based on 'thermodynamic' considerations ('adsorption' theory). The importance of acid–base concepts in adhesion and wetting phenomena is discussed in the next section.

18.4 Acid–base concepts in adhesion studies

18.4.1 Adhesion measurements and interfacial forces

An impressive result of 'practical' adhesion and its interpretation via the van Oss–Good parameters is shown in Figure 18.6.[15] The peel energy and thus the adhesion for the acidic pressure-sensitive (commercially available) adhesive is highest for the basic oxide film of aluminum. This means that there is greater affinity for this basic oxide film to donate electrons to the acidic polymer. On the other hand, the very basic PMMA polymer is attracted most by the acid oxide Si film. These increased acid–base interactions lead to increased practical adhesion. Although the measured pull-off force for PMMA includes both interfacial and viscoelastic contributions, since the polymer is the same for each metal surface, the viscoelastic contributions may be taken to be the same in all cases. Thus, differences in the measured pull-off forces may be ascribed to differences in interfacial adhesion. We can conclude that results like the ones shown in Figure 18.6 provide direct experimental evidence of the importance of LA–LB effects in practical adhesion of polymers to oxide-covered metals, and possibly other surfaces as well.

There are many examples in the literature where good correlation between adhesion properties and surface properties like surface tension is found. However, in practice, mechanical properties are also important. For example, in their study of various coatings for fouling release systems for ships, Brady and Singer[16] showed

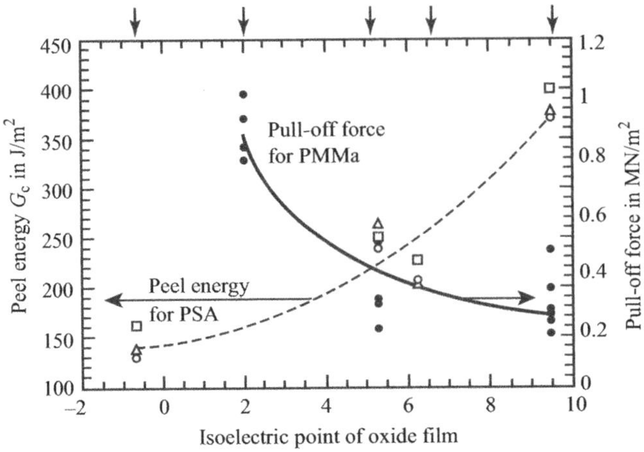

Figure 18.6 *Measured peel energy for a pressure-sensitive adhesive (PSA) and pull-off force for PMMA vs. the isoelectric point of the oxide film. Reprinted with permission from the Journal of Adhesion Science Technology, Acid-base effects in polymer adhesion at metal surfaces by E. McCafferty, 16, 3, 239–255 Copyright (2002) Koninlijke Brill NV*

that there is a strong correlation between surface energy and resistance to bioadhesion (both in ship coatings and other environments like dental implants, artificial joints, etc.). Polymers with low surface energies include Teflon and other highly fluorinated materials (polyvinylidene fluoride and polychlorotrifluoroethylene). Adhesion was found to decrease with decreasing surface energy until a critical value of surface energy around 22 mN/m was reached. Then, adhesion starts to increase. Clearly, surface energy is important, but it is not the only significant factor influencing adhesion. The coating with the lowest elastic modulus shows the lowest bioadhesion even though this does not correspond to the coating with the lowest surface energy. As Figure 18.7

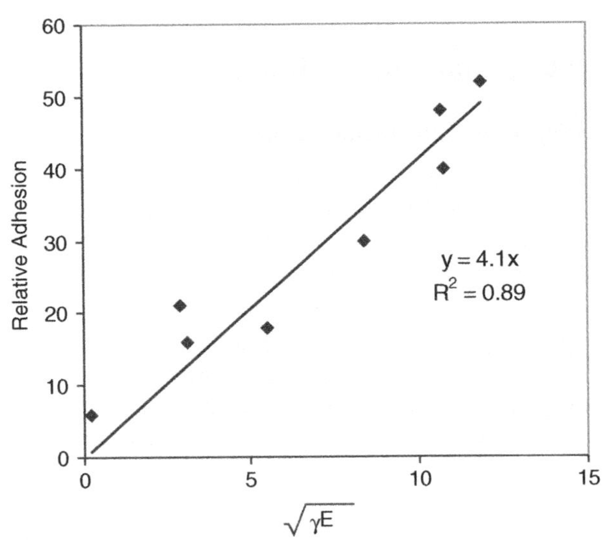

Figure 18.7 *Relative adhesion (dimensionless) for coatings used in fouling release systems for ships as a function of the square root of the product of critical surface tension and elastic modulus. After Brady and Singer*[16]

illustrates, very good correlation is obtained between the adhesion as a function of the square root of surface tension; in fact, better than when the adhesion is plotted against the modulus of elasticity or the surface tension alone. Similar results are obtained from other studies as well.

The relationship between adhesive forces and $\sqrt{\gamma E}$ illustrated in fig. 18.7 has been shown by Kendall[41] and has been used by several researchers, see also Kohl and Singer.[42]

18.4.2 Industrial examples

Two examples from industrial projects are presented below, which make use of the acid–base theory for understanding adhesion phenomena or for characterizing solid surfaces.

Industrial example 1

In the first example, a project in collaboration with a Danish paint company, the adhesion problems observed with several epoxy–silicon interfaces have been studied. The experimental work involved: (1) contact angle measurements of several liquids on each of the six epoxies; (2) AFM (Atomic Force Microscopy) for studying the topography of the surfaces; and (3) experimental determination of adhesion via pull-off tests. On the theoretical side, critical surface tensions were estimated using Zisman plots and the solid surfaces were analyzed using the van Oss *et al.*[10] (acid–base) theory. Table 18.4 summarizes the basic theoretical results which agree well with the experimental observations. The 'most acidic' epoxy (E5) gives the best adhesion with the basic silicone surface. More information is given by Svendsen *et. al.* [40]

Industrial example 2

The purpose of the second project was to synthesize and characterize PEG derivatives of carboxylic acids and their reaction with a PUF binder (= urea-modified phenol–formaldehyde). The final outcome is to have an optimum hydrophilic binder which can be compared favorably to some of the existing industrial products (indicated as industry_1 and industry_2 in Table 18.5). Various materials have been synthesized (listed as 1–5 in

Table 18.4 *Characterization of the epoxy surfaces using the van Oss–Good approach. The total surface tensions and their vdW and acid–base components are given (in mN/m). The values are estimated based on contact angles of water, ethylene glycol and benzaldehyde. Given in parentheses are the acid (+) and base (−) components of the acid–base contribution. After Svendsen et al.*[40]

Epoxy	Total surface tension (mN/m)	Critical surface tension (mN/m)	γ^{LW} (mN/m)	γ^{AB} (mN/m) (acid–base values, Equation (18.10))
E1	43.80	33	37.4	6.38 (0.342/29.80)
E2	42.09	33	36.89	5.21 (0.2578/26.28)
E3	39.66	33	36.26	3.39 (0.1416/20.38)
E4	40.58	33	36.34	4.24 (0.2399/18.75)
E5	44.16	37.4	37.4	6.76 (0.568/20.10)
E6	37.90	30	36.34	1.56 (0.0213/28.64)

Table 18.5 *Total surface tension and surface tension 'components' using the Owens–Wendt and van Oss et al.*[10] *theories for eight different 'polymeric binders'. All surface energies are in mN/m. The dispersive and 'specific' components for the Owens–Wendt and van Oss et al.*[10] *theories are given in parentheses*

Polymer binder	Lab-measured surface energy	Critical surface tension	Owens–Wendt	van Oss *et al.*
1	48–52	48.5	53.3 (45/8.2)	51.6 (48/3.6)
2	48–52	48.8	58.6 (44.4/14.2)	54.8 (48.2/6.6)
3	48–52	48.8	58.3 (44.5/13.8)	48.9 (47.8/1.1)
4	48–52	48.3	55.6 (44.5/11.1)	53.1 (47.7/5.4)
5	48–52	49.2	56.6 (45.2/11.4)	54.4 (48.1/6.3)
Industry_1	52–56	—	67.0 (41.6/25.4)	56.8 (47.4/9.4)
PUF	38–40	46.2	49.8 (42.8/7.0)	48.2 (46.1/2.1)
$\text{PUF} + \text{industry}_2$	34–38	—	68.0 (40/28)	56.0 (46.2/9.8)

Table 18.5) by adding PEG chains in PUF. The various samples have different ratios of phthalic acetate/PEG and various PEGs (i.e. having different molecular weights). Then, contact angles of three liquids on these materials were measured and analyzed first using the Owens–Wendt theory (for estimating dispersive/polar components, Equation (18.5)) and with the Zisman plot to estimate the critical surface tension. From this analysis, it appeared that none of the newly developed samples has surface tension as high as the two existing industrial binders (see Table 18.5). However, this surprising result (not in agreement with the surface tensions obtained with an in-house method and other experimental indications) was essentially reversed when the same experimental data were analyzed with the van Oss *et al.*[10] (acid–base) method, as shown in Table 18.5. We expect that, in most cases, the van Oss *et al.*[10] theory, which properly accounts for the different intermolecular forces, will yield more accurate results compared to the older 'component' theories such as the Owens–Wendt.

18.5 Surface and interfacial tensions from thermodynamic models

Table 18.6 summarizes several recent investigations on using thermodynamic models, especially association equations of state like CPA and SAFT, for calculating surface and interfacial tensions. An additional framework is needed and this is offered, for example, via the gradient theory or the density functional theory. The gradient theory is briefly discussed here.

18.5.1 The gradient theory

One case where thermodynamics is directly related to surface chemistry is in the calculation of surface and interfacial tensions. There are a number of techniques to model the interfacial tension, which range from simple empirical correlations to those based on statistical thermodynamics. The simpler methods, such as the Parachor method, suffer in their accuracy to model for example hydrocarbon–water systems. The techniques derived from statistical thermodynamics and the theory of inhomogeneous fluids have been shown to be theoretically more sound, accurate and robust calculation methods.

Table 18.6 Thermodynamic models used for calculating surface and interfacial tensions

Equation of state	Framework	Application	Reference
SAFT, PR	Gradient theory	Water, hydrocarbons	Kahl and Enders[17]
Sanchez–Lacombe		Alcohol–hydrocarbons	Kahl and Enders[18]
SAFT–VR	DFT	Water	Paricaud *et al.*[19] Gloor *et al.*[20]
QCHB	Gradient theory	Water, alcohol, non-polars, polymers, etc. Alcohol–hydrocarbons	Panayiotou[21]
CPA	Gradient theory	Water, alcohols	Queimada *et al.*[22]
CPA	Gradient theory	Alkanes, alcohols, fluoroalkanes Alcohols + alkanes Alcohols + alcohols Alkanes + alkanes	Oliveira *et al.*[23]
SRK	Linear gradient theory	Water–methane interfacial tension	Schmidt *et al.*[2]

DFT = density functional theory.

The gradient theory of fluid interfaces is maybe the most widely used framework for interfacial tension calculations, providing a way of relating an EoS to interfacial properties. There are numerous investigations on the use of the gradient theory model combined with an EoS to determine the interfacial tension of pure component and mixture systems (Table 18.6). For example, Carey[24] and Miqueu[25] used PR EoS for interfacial tension calculations of several binary and ternary mixtures, and Cornelisse[26] used PR and the associated perturbed anisotropic chain theory (APACT) for surface and interfacial tension calculations of pure fluids and mixtures (varying from non-polar hydrocarbons to very polar and associating mixtures). Kahl and Enders[17] used PR, Sanchez–Lacombe (SL) and SAFT for surface tension calculations of water and hydrocarbons, and Kahl and Enders[18] also used PR and SAFT for interfacial tension calculations of alcohol–hydrocarbons. Finally, Queimada *et al.*[22] used CPA for surface tensions of water, alcohols and hydrocarbons, and Mejía *et al.*[27] used PR with modified Huron–Vidal mixing rules for interfacial tensions of various mixtures.

The gradient theory of fluid interfaces originated from the work of van der Waals[28], but only after Cahn and Hilliard[29] did it find widespread use. The Cahn–Hilliard[29] theory describes the thermodynamic properties of a system where an interface exists between two fluid phases. In contrast to systems of pure components, in binary mixtures the density and composition change across the interface. Before any interfacial properties can be computed, it is necessary to find the thermodynamic equilibrium densities of the two phases, between which the interface being considered is formed, at a given temperature and pressure. According to the gradient theory, the interfacial tension can be determined with:

$$\gamma = \int_{n_1^I}^{n_1^{II}} \sqrt{2c(\Phi(n) - \Phi_B)}\,dn_1 \tag{18.36}$$

where c, $\Phi(n)$, Φ_B are the mixture influence parameter, grand thermodynamic potential energy density and negative pressure ($-P$), while n_i is the number density (mol/m^3) of the component i, and phases I and II are the equilibrium bulk phases. The influence parameter c contains information on the intermolecular geometry of the interface and it relates the deviation of the Helmholtz energy density to the chemical potential in the interface due to the density gradient. This energy deviation ultimately manifests itself in a contribution of the density gradient to the interfacial tension. The grand thermodynamic potential energy density is determined from an EoS via the Helmholtz free energy, $f^o(n)$, and chemical potentials of each species, μ_i. The grand thermodynamic potential energy density is defined by:

$$\Phi(n) = f^o(n) - \sum_{i=1}^{Nc} n_i \mu_i \tag{18.37}$$

The mixture Helmholtz energy density, $f^o(n)$, is obtained from:

$$f^o(n) = A^*(T,V,n) + A^{RES}(T,V,n) \tag{18.38}$$

Using standard thermodynamics, the ideal contribution to the Helmholtz energy density is obtained from:

$$A^*(T,V,n) = \sum_i^{Nc} n_i \left[-RT + \mu_i^*(T,P^{REF}) + RT \ln\left(\frac{n_i RT}{P^{REF} V}\right)\right] \tag{18.39}$$

With the thermodynamic definitions that can be found in Michelsen and Mollerup[30], the chosen EoS can be transformed into an expression for the Helmholtz energy and the chemical potential. For PR or SRK EoS, those expressions can be found in Michelsen and Mollerup[30], while for CPA or SAFT models the derived residual properties (Helmholtz energy and chemical potential) are presented in the appendices of Chapters 8 and 9 (Appendices 8.A and 9.C).

The chemical potential is obtained from the derivative of the Helmholtz energy density with respect to the molar density of component i:

$$\mu_i(T,V,n) = \left(\frac{\partial A(T,V,n)}{\partial n_i}\right)_{T,V,n_j} = \left(\frac{\partial A^*(T,V,n)}{\partial n_i}\right)_{T,V,n_j} + \left(\frac{\partial A^{RES}(T,V,n)}{\partial n_i}\right)_{T,V,n_j} \tag{18.40}$$

$$\left(\frac{\partial A^*(T,V,n)}{\partial n_i}\right)_{T,V,n_j} = \mu_i^*(T,P^{REF}) + RT \ln\left(\frac{n_i RT}{P^{REF} V}\right)$$

It is evident from Equation (18.37) that the calculation of $\Phi(n)$ requires knowledge of the number density (n_i) of component i for each point within the interface. This dependence conforms to the density profiles across the interface. Consider a planar interface between two bulk phases (liquid and vapor) of a mixture and denote by z the direction perpendicular to the surface. Then the densities $n_i(z)$ at the position z obey the following equilibrium conditions:

$$\sum_j^{Nc} \frac{d}{dz}\left(c_{ij}\frac{dn_j}{dz}\right) - \frac{1}{2}\sum_j^{Nc}\sum_k^{Nc} \frac{\partial c_{jk}}{\partial n_i}\frac{dn_j}{dz}\frac{dn_k}{dz} = \frac{\partial \Phi(n)}{\partial n_i} \tag{18.41}$$

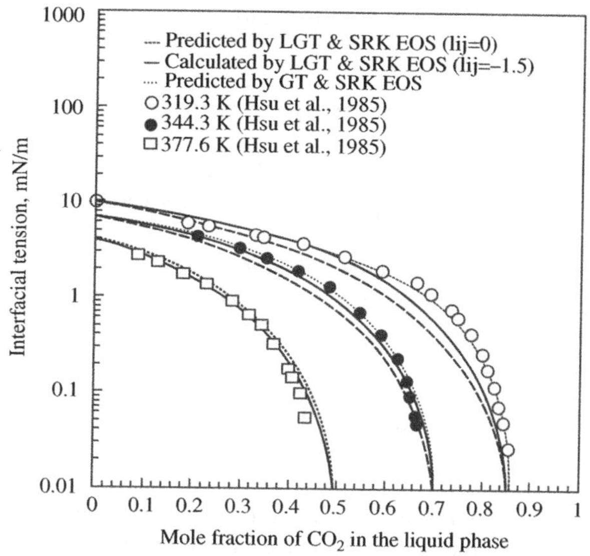

Figure 18.8 *Interfacial tension for carbon dioxide–n-butane mixture at various temperatures, using GT and LGT with PR or SRK EoS. Reprinted with permission from Journal of Colloid and Interface Science, A Linear Gradient Theory Model for Calculating Interfacial Tensions of Mixtures by You-Xiang Zuo and Erling H. Stenby, 182, 1, 126–132 Copyright (1996) Elsevier*

Hence, in order to compute interfacial tension with the gradient theory, the density profiles in the interface, which are the solution of Equation (18.41), should first be determined. The numerical effort in resolving the density profiles has been described in detail by Carey[24], Cornelisse[26] and Miqueu.[25] To ease the burden of computational effort, Zuo and Stenby[31–33] developed the linear gradient theory (LGT), a practical version of the gradient theory, which eliminates the need to solve the set of time-consuming density profile equations that are inherent with the gradient theory approach. In the LGT the number density (n_i) of each component in the mixture is assumed to be distributed linearly between the coexisting equilibrium bulk phases (e.g. vapor and liquid) through a planar interface of height h. The number density of component i at position z in the interface can be represented by:

$$\frac{dn_i(z)}{dz} = D_i \tag{18.42}$$

$$D_i = \frac{\Delta n_i}{h} = \frac{n_i^{II} - n_i^{I}}{h} \tag{18.43}$$

Once the density profile is known, similar to the gradient theory, the interfacial tension is determined by solving the set of Equations (18.36)–(18.41). The LGT has been applied by Zuo and Stenby[31–33] for interfacial tension calculations to a wide range of binary systems and by Schmidt *et al.*[2] over a wide range of temperatures and pressures using the PR and SRK EoS. A typical example is presented in Figure 18.8.

18.6 Hydrophilicity

In thermodynamics hydrophilicity or hydrophobicity is typically expressed via the so-called octanol–water partition coefficient (as discussed in Chapter 17):

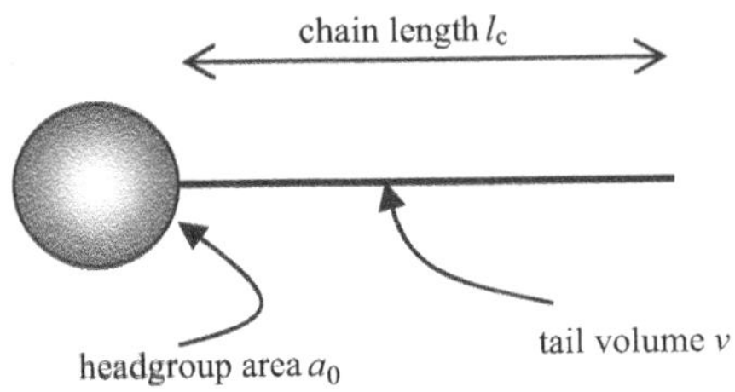

Figure 18.9 *The CPP of a surfactant molecule, where v is the tail chain (or chains) volume (V_{surf} in Equation (18.45)), l_c is the critical tail chain length and a_0 is the head group area at the head–tail interface*

$$K_{OW_i} = \lim_{x_i \to 0}\left(\frac{C_i^O}{C_i^W}\right) = 0.151\frac{\gamma_i^{W,\infty}}{\gamma_i^{O,\infty}} \quad (18.44)$$

where C_i is the concentration of compound i in the octanol (O) or water (W) phase and γ_i^∞ is the infinite dilution activity coefficient of compound i in the octanol (O) or water (W) phase. However, in many applications in colloid and surface science, e.g. related to surfactants and emulsions, we use the so-called CPP and HLB (critical packing parameter, hydrophilic–lipophilic balance).

18.6.1 The CPP parameter

This is defined as follows (see Figure 18.9):

$$CPP = \frac{V_{surf}}{a_0 l_c} \quad (18.45)$$

The CPP is related to the micellar structure in surfactants, as given in Table 18.7.

18.6.2 The HLB parameter

While the CPP is a geometric factor, the HLB is an empirical parameter for which there are many estimation methods, e.g. that of Davies and Rideal (Equation (18.46)), which is used together with the group values shown in Table 18.8. HLB values below 8 indicate hydrophobic emulsifiers suitable for stabilizing water-in-oil (w/o) emulsions, while HLB values above 8 or 10 indicate hydrophilic emulsifiers suitable for use in oil-in-water (o/w) emulsions.

Table 18.7 *Relationship between CPP values and micelle structure*

CPP value	Micelle structure	Example*
<1/3	Spherical micelles	SDS in low salt
1/3–1/2	Cylindrical micelles	SDS and CTAB in high salt
1/2–1	Flexible bilayers	Phosphatidyl choline
~1	Planar bilayers	Phosphatidyl ethanolamine
>1	Reversed micelle	Cardiolipin + Ca^{2+}

*SDS = sodium dodecyl sulfate, CTAB = hexadecyl trimethylammonium bromide.

Table 18.8 Group contribution values for estimating the HLB according to the Davies–Rideal method (Equation (18.46))

Group	HLB	Group	HLB
$-SO_4Na$	38.7	–OH (free)	1.9
–COOK	21.1	–OH (sorbitan)	0.5
–COONa	19.1	Sulfonate	11
–*N*	9.4	CH,CH_2,CH_3	−0.475
Ester (sorbitan)	6.8	$-CH_2CH_2O$	0.33
Ester (free)	2.4	$-CH_2CH_2CH_2O$	−0.15
COOH	2.1	$-CF_2$, $-CF_3$	−0.87
–O–	1.3		

HLB as estimated via the Davies and Rideal method (Table 18.8):

$$HLB = 7 + \sum_i n_i HLB_i \qquad (18.46)$$

The HLB concept can be used in emulsion design according to the following steps:

1. Calculate HLB, either from Equation (18.46) or from specific equations. For example:
   Special method for nonyl phenyl ethoxylates:

$$HLB = 20(17 + 44X_{EO})/(220 + 44X_{EO}) \qquad (18.47)$$

   where X_{EO} is the number of ethylene oxide groups.
   Based on oil–water partition coefficient, $K_{oil/w}$:

$$HLB = 7 + 0.36 \ln K_{oil/w} \qquad (18.48)$$

2. Check whether the HLB value is higher or lower than 8 and then apply the Bancrofft rule for determining the emulsion type: 'The phase in which the emulsifier is more soluble tends to be the dispersion medium.' For example, emulsifiers which are preferably soluble in water are suitable for o/w emulsions.
3. For mixed emulsifiers use the following rule:

$$HLB = \sum_{i=1}^{n} w_i HLB_i \qquad (18.49)$$

   where n is the number of emulsifiers and w_i is the weight percentage fraction of an emulsifier. The latter equation can be used in case one or more emulsifiers must be replaced in an emulsion. The following rule approximately applies: 'Different emulsions have similar characteristics if the emulsifiers have similar HLB numbers.'

The CPP is mostly used in surfactant science and the HLB is often used in the study of emulsions. These parameters are, as expected, correlated, as shown for example in Figure 18.10 for surfactants, which are discussed further in the next section.

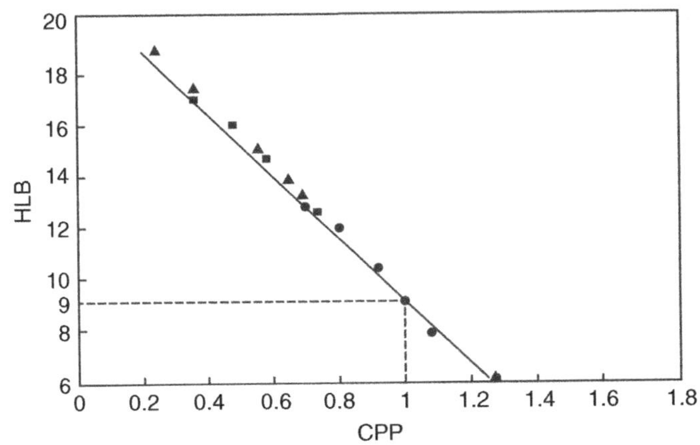

Figure 18.10 *HLB–CPP correlation for surfactants*

18.7 Micellization and surfactant solutions

18.7.1 General

Surfactants are amphiphilic compounds which adsorb on liquid–air and oil–water interfaces and decrease surface and interfacial tensions and the corresponding works of adhesion. The decrease of water surface tension is observed up to a certain concentration, called CMC (Critical Micelle Concentration). After CMC, surfactants create new liquid-like near-spherical structures, called micelles (Figure 18.11).

At even higher concentration more complex (liquid crystalline) structures, e.g. hexagonal or lamellar crystals, may appear, some of which resemble the structure in biological membranes. CMC can be measured by observing the change of several properties, e.g. surface tension with concentration (Figure 18.12). CMC is not, strictly speaking, a thermodynamic transition state; its value roughly depends on the estimation method.

Surfactants are used extensively as part of commercial detergents and assist in cleaning via a combination of various mechanisms:

- improving the wetting of surfaces;
- reducing the interfacial tensions between dirt(d)–water(w) and solid substrate(s)–water(w), thus reducing the work of adhesion ($W_{adh,ds} = \gamma_{sw} + \gamma_{dw} - \gamma_{sd}$);
- solubilization inside micelles, thus avoiding redispersion of dirt.

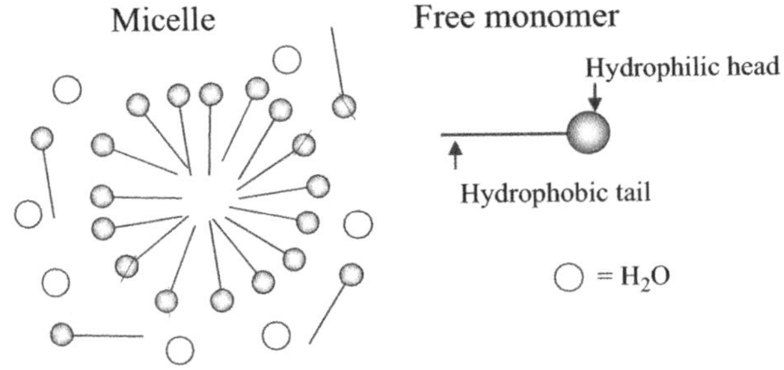

Figure 18.11 *Micelle formation and surfactant molecules*

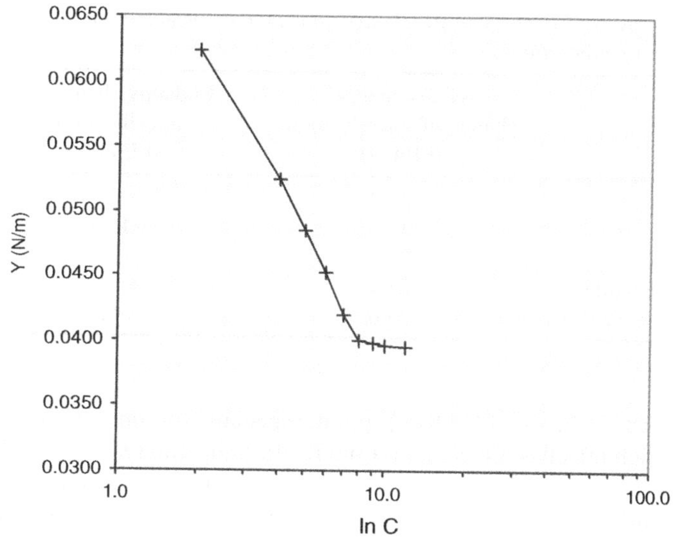

Figure 18.12 *Surface tension against logarithm of concentration for an aqueous solution of sodium dodecyl sulfate (SDS) at 20°C*

Foam formation is not directly related to detergency. The properties (e.g. cleaning and stabilizing capabilities) of surfactants depend on both solution properties (temperature, time, presence of salts and co-surfactants) and their own characteristics, especially CMC, the Krafft point and their chemistry. The surfactant chemistry and especially the balance between hydrophobic and hydrophilic parts are quantified using tools like the CPP or HLB (see section 18.6). For example, it is often observed that detergency increases with surfactant concentration, especially up to CMC, and is best often at CPP values around 1.

18.7.2 CMC, Krafft point and micellization

CMC decreases with increasing chain length of the hydrophobic part, slightly decreases with decreasing hydrophilic part, and upon adding salt (for ionics) or co-surfactants. On the other hand, the temperature dependency of CMC is often rather complex. In general:

- CMC does not depend very much on temperature (due to a balance between enthalpic and entropic effects).
- CMC increases with increasing temperature for ionic surfactants because micellization is an exothermic process, though there are exceptions to this rule. For example, SDS presents a minimum in the CMC–temperature plot, thus at low temperatures the micellization must be endothermic.
- CMC decreases with increasing temperature for non-ionic surfactants, e.g. those belonging to the polyoxyethylene family.

In conclusion, as shown in Table 18.9, the micellization is either endothermic (positive enthalpy change), e.g. for SDS and non-ionic polyoxyethylenes, or exothermic, but most importantly it is an entropic phenomenon with high positive values of the entropy change, which can be attributed to the changes in the structure of water in the presence of surfactants and micelles, i.e. the so-called hydrophobic effect (discussed in Chapter 2). The role of water seems to be much more crucial than that of structure creation when the surfactants are assembled into micelles.

***Table 18.9** Enthalpy, entropy and Gibbs energy change during the micellization process for several ionic and non-ionic surfactants at room temperature*

Surfactant	Gibbs energy change of micellization (kJ/mol)	Enthalpy change of micellization (kJ/mol)	Entropy change of micellization (J/K mol)
SDS	−21.9	+2.51	+81.9
$C_{12}E_6$	−33.0	+16.3	+49.3
Dodecyl pyridinium bromide	−21.0	−4.06	+56.9
N,N-dimethyl dodecyl amine oxide	−25.4	+7.11	+109.0
N-dodecyl-*N,N*- dimethyl glycine	−25.6	−5.86	+64.9

Equally important to CMC is the so-called Krafft point, especially for ionic surfactants. The Krafft point is the temperature above which micelles are created. Low Krafft temperatures are linked to surfactants having high CMC values, i.e. low chain lengths, as Figure 18.13 shows for two surfactant families. Thus, a balance is needed in the choice of surfactants in order to ensure suitable values for the design parameters (economy, effectiveness and time) in addition to good cleaning. This is because long-chain ionic surfactants create micelles at low concentrations, but this happens at rather high temperatures. A chain length of about 12–14 is often a good compromise.

18.7.3 CMC estimation from thermodynamic models

Thermodynamic models like UNIFAC can be used to predict CMC as shown, for example, by Flores *et al.*[34] and Cheng *et al.*[35] Two typical results are shown in Figures 18.14 and 18.15. The approach is based on the work of Chen[36] who proposed a methodology that enables the use of standard thermodynamic (activity coefficient) models for modeling CMC of both non-ionic and ionic surfactants. He assumed the existence of two 'distinct' liquid phases, a monomer surfactant phase (isolated surfactants dispersed in water) and a micelle phase (without any water present). He then assumed a physical equilibrium for the surfactant molecules in the

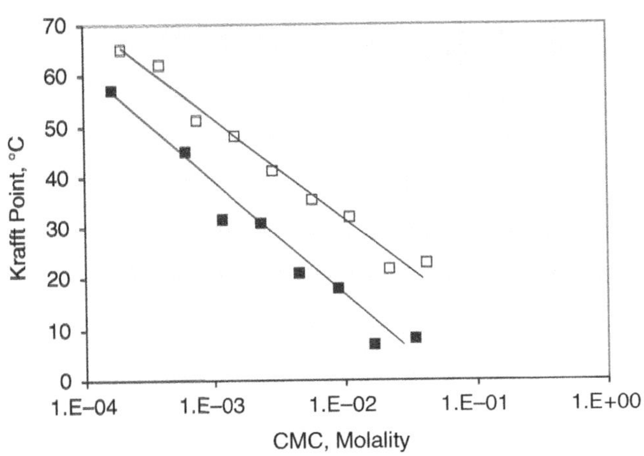

***Figure 18.13** Krafft–CMC relationship for two families of surfactants (filled squares, sodium alkyl sulfates; open squares, sodium alkyl sulfonates)*

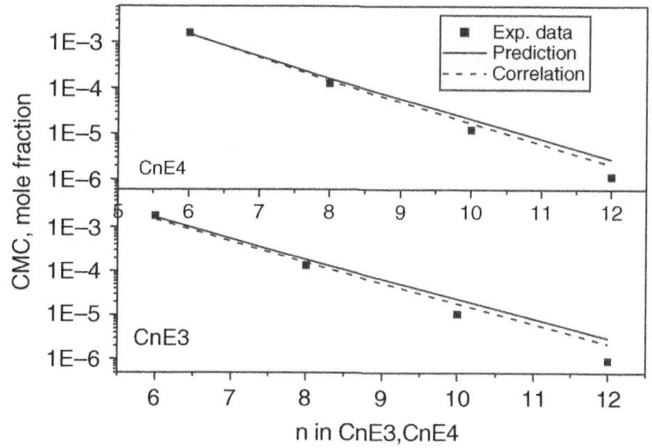

Figure 18.14 Predicted and correlated CMC values with the UNIFAC VLE model at 25 °C for two non-ionic polyoxyethylene (C_iE_j) surfactants (C_nE_3 and C_nE_4) with different hydrophobic alkyl chains. Reprinted with permission from Ind. Eng. Chem. Res., Prediction of Micelle Formation for Aqueous Polyoxyethylene Alcohol Solutions with the UNIFAC Model by Hongyuan Cheng, Georgios M. Kontogeorgis, and Erling H. Stenby, 41, 5, 892–898 Copyright (2002) American Chemical Society

monomeric and micellar states. The activity coefficient can be estimated from a thermodynamic model and, in the work of Cheng *et al.*[35], UNIFAC is used. The basic equation in the calculations is:

$$\begin{aligned} \mu^0_{mon} + RT \ln x_{mon}\gamma_{mon} &= \mu^0_{micelle} + RT \ln x_{micelle}\gamma_{micelle} \\ x_{mon}\gamma_{mon} &= 1.0 \end{aligned} \tag{18.50}$$

It is assumed that the micelles contain only surfactant molecules and that the activity coefficients can be calculated from standard activity coefficient models, e.g. UNIFAC. It has been shown by Cheng *et al.*[35] that

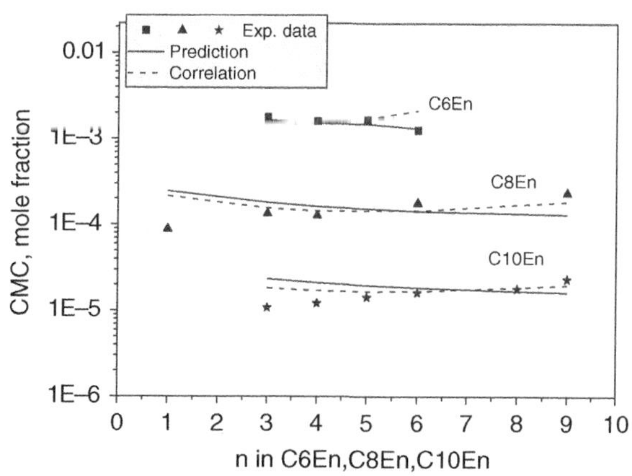

Figure 18.15 Predicted and correlated CMC values with the UNIFAC VLE model at 25 °C for three non-ionic polyoxylethylene (C_iE_j) surfactants (C_6E_n, C_8E_n and $C_{10}E_n$) with different hydrophilic groups. Reprinted with permission from Ind. Eng. Chem. Res., Prediction of Micelle Formation for Aqueous Polyoxyethylene Alcohol Solutions with the UNIFAC Model by Hongyuan Cheng, Georgios M. Kontogeorgis, and Erling H. Stenby, 41, 5, 892–898 Copyright (2002) American Chemical Society

existing UNIFAC models with published parameter tables do not perform very well for aqueous non-ionic surfactants of the polyoxyethylene type. When, however, the oxyethylene group (OCH_2CH_2) is defined as a separate group and the parameters are fitted to phase equilibrium data for low-molecular-weight (non-surfactant) solutions, very good results are obtained, as can be seen in Figures 18.14 and 18.15 for the CMC of two surfactants as a function of the chain length. The results are as good as when the parameter values are estimated directly using the CMC data. Equally satisfactory results are obtained for the CMC of non-ionic surfactants as a function of the size of hydrophilic group. In this case the CMC increases slightly; for all practical purposes it can be considered to be almost constant for a surfactant with the same hydrophobic group. It is clear that the change in the length of the hydrophobic groups affects the CMC much more than the change in the size of the hydrophilic groups.

18.8 Adsorption

18.8.1 General

A topic which is undoubtedly at the junction of thermodynamics and colloids and interfaces is adsorption. Adsorption is a universal phenomenon in colloid and surface science, manifested via adsorption of high-molecular-weight amphiphilic compounds for monolayer creation, adsorption of gases on solids, of surfactants, polymers or proteins (biomolecules), adsorption on solid surfaces and/or from solutions. There are numerous applications of adsorption: stabilization, cleaning, heterogeneous catalysis, separations like drying, surface modification and change of properties like foaming or rheological ones. A fundamental equation in adsorption studies is the Gibbs equation for the relative adsorption discussed next.

The Gibbs adsorption equation

The Gibbs adsorption equation can, for dilute solutions, be written as:

$$\Gamma_i = \frac{n_i^\sigma}{A} = -\frac{C_i}{RT}\frac{d\gamma}{dC_i} = -\frac{1}{RT}\frac{d\gamma}{d\ln C_i} \tag{18.51}$$

where n_i^σ is the number of moles at the surface, γ is the surface tension, A is the surface area and C_i is the concentration.

The area occupied by an adsorbed molecule, e.g. a surfactant molecule, is given by the equation:

$$A = \frac{10^{18}}{N_A \Gamma} \tag{18.52}$$

(in nm^2 when the adsorption Γ is given in mol/m^2).

It can be shown in Table 18.10 that adsorption isotherms, surface tension–concentration equations and the so-called two-dimensional EoS (surface pressure vs. area relationships) are interrelated. The two-dimensional EoS are the analogues in two dimensions of the well-known from thermodynamics 'classical' three-dimensional EoS (P–V).

Table 18.10 Surface tension vs. concentration equations, adsorption isotherms and two-dimensional EoS from three well-known theories (Henry, Langmuir and van der Waals). Typical C and Γ units are mol/l and mol/g. The surface pressure $\pi = \gamma_0 - \gamma$ is the difference between the surface tension of water and the surface tension of the solution

Surface tension–concentration, $\gamma(C)$	Adsorption isotherm equation, Γ	Two-dimensional EoS, π–A
$\gamma = \gamma_0 - (RTk\Gamma_{\max})C$	$\Gamma = \Gamma_{\max}kC$ (Henry)	$\pi A = n_i^\sigma RT = kT$
$\gamma = \gamma_0 - RT\Gamma_{\max}\ln(1+kC)$	$\Gamma = \frac{\Gamma_{\max}kC}{1+kC}$ (Langmuir)	$\pi = \frac{kT}{A-b}$
	$kC = \frac{\Theta}{1-\Theta}\exp\left(\frac{\Theta}{1-\Theta} - \frac{2a\Theta}{bkT}\right)$ $\Theta = \Gamma/\Gamma_{\max}$ (van der Waals)	$\pi = \frac{kT}{A-b} - \frac{a}{A^2}$

The area of an adsorbed molecule at the interface, A, can be calculated using the Langmuir equation from the monolayer coverage if the specific surface area of the solid is known:

$$A = \frac{A_{spec}}{\Gamma_{\max}N_A} \tag{18.53}$$

A is in m^2, $\Gamma_{\max}$ in mol/g and A_{spec} in m^2/g. The specific solid surface area is typically obtained from BET adsorption measurements (of N_2 or other gases on the solid):

$$A_{spec} = \frac{V_m N_A}{V_g} A_o \tag{18.54}$$

where V_m is the monolayer volume, V_g in this equation is the gas volume at the standard temperature and pressure, i.e. 1 atm and 273 K (= 22 414 cm^3/mol), A_o is the area occupied by one gas molecule (0.162 nm^2 for the typically used N_2 at 77 K, 0.138 nm^2 for Ar and 0.195 nm^2 for Kr).

Except for the localized adsorption models such as Langmuir (shown in Table 18.10), approaches particularly suitable for multicomponent adsorption are the thermodynamic adsorption theory, especially in the ideal form presented by Myers and Prausnitz[43] and the potential theory of adsorption presented by Shapiro and Stenby[44] and recently applied to many systems (including adsorption from solutions) by Monsalvo and Shapiro.[45,46] In the potential adsorption theory, a thermodynamic model, e.g. the SRK EoS, is used to describe the volumetric behavior at high pressures.

Both approaches (thermodynamic and potential adsorption theories), despite significant differences in their background and derivation, yield similar and equally satisfactory results and are predictive for adsorption of multicomponent gas systems, i.e. the parameters are adjusted to adsorption of pure gases on the same solid.

18.8.2 Some applications of adsorption

Even the simple ideal gas film equation (the equivalent of the ideal gas equation in three dimensions; first row in Table 18.10, the Henry equation) finds important applications. For example, it can be used to estimate the molecular weight of proteins from surface pressure data (assuming ideal gas film behavior at low

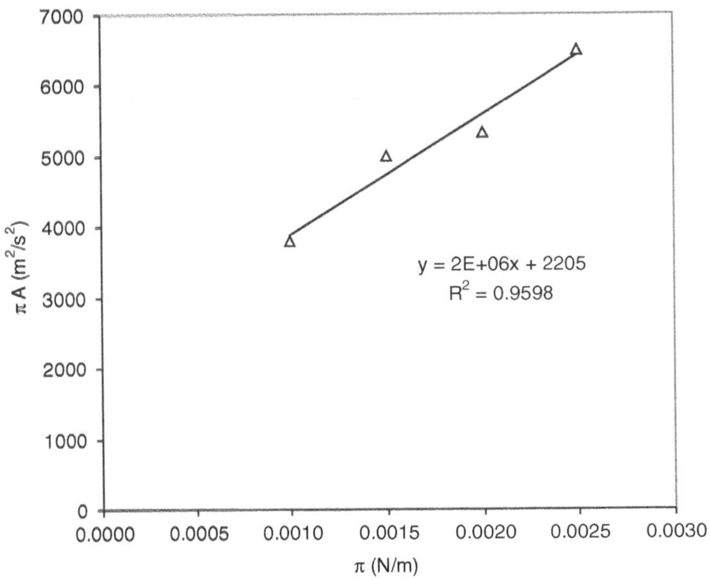

Figure 18.16 *Surface pressure multiplied by area against the surface pressure for valinomycin. The molecular weight is estimated via Equation (18.55) to be 1152 g/mol*

pressures):

$$\lim_{\pi \to 0}(\pi A) = \frac{RT}{M} \tag{18.55}$$

One example for valinomycin is shown in Figure 18.16.

In particular, the Langmuir adsorption isotherm is useful as it has been successfully applied to the adsorption of surfactants, polymers and occasionally even adsorption of proteins from solutions. For hydrophobic surfaces, the adsorption of surfactants increases with increasing hydrophobicity, e.g. with increasing CPP of the surfactant. The Langmuir isotherm is also widely used for the description of chemical adsorption of gases on solids.

Adsorption has several more 'thermodynamic links':

- role of solvent and surface in adsorption from solution (Figure 18.17);
- use of LA–LB concepts for understanding and predicting adsorption (Figure 18.18);
- link with steric stabilization of polymers and theta conditions.

Adsorption of solutes (from solution) on solids is, in some respects, more complex than the adsorption of gases. The effects of solvent and competition of solute and solvent for sites on the solid surface are very important, as Figure 18.17 illustrates. In this example, we observe the adsorption of different polar compounds on a polar surface (silica gel) from solutions in a non-polar solvent (carbon tetrachloride). The longer the chain of acids, the stronger the interaction (affinity) with the non-polar solvent, the more the acid wants to stay in the solution and thus the lower the adsorption. The adsorption of surfactants is simpler than that of polymers and proteins and depends on their CPP: the higher the CPP, the higher the adsorption (on hydrophobic surfaces). The nature of the surface is less important than that of the surfactant. All factors that increase the CPP (increase of hydrophobic chain or temperature, decrease of hydrophilic length, addition of salt, use of surfactants with

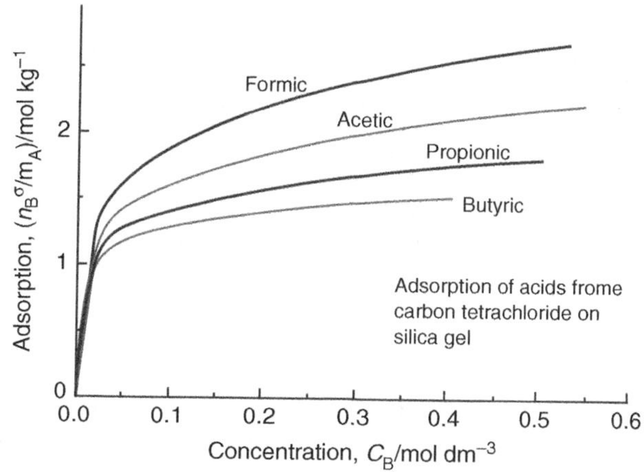

Figure 18.17 *Adsorption of carboxylic acids from carbon tetrachloride solution onto a polar solid silica gel illustrating the effect of acetyl chain length. Reprinted with permission from Interfacial Science-An Introduction by G. T. Barnes and I. R. Gentle, Oxford University Press, Oxford, UK Copyright (2005) Oxford University Press*

two hydrocarbon chains, addition of surfactants with opposite charge, adding long-chain alcohols or amines, etc.) result in increased adsorption.

The adsorption of polymers is very important for understanding and controlling the steric stabilization of colloids. The best stabilization is obtained when the coverage of particles is complete (maximum adsorption, the plateau of Langmuir isotherms). The adsorption of polymers typically increases with polymer molecular

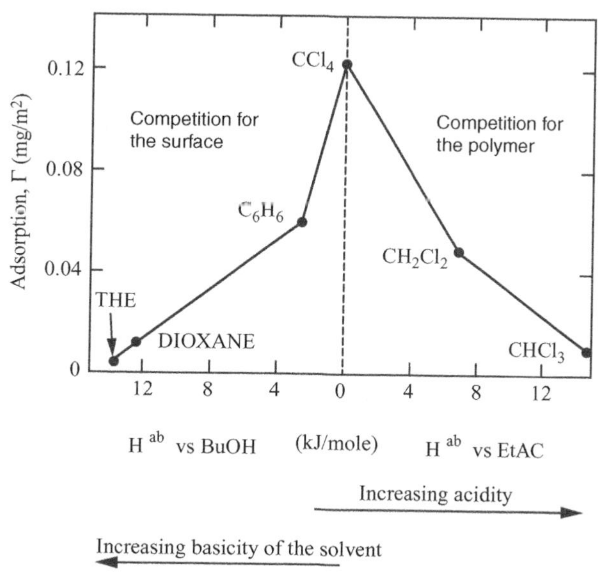

Figure 18.18 *Adsorption of PMMA on silica from a variety of solvents that differ with respect to their acid–base properties. Reprinted with permission from Surfactants and Polymers in Aqueous Solutions by B. Jonsson, B. Lindman, K. Holmberg and B. Kronberg, John Wiley & Sons, Chichester, UK Copyright (2001) John Wiley & Sons, Ltd*

Table 18.11 Comparison of critical flocculation temperatures (CFT) and theta temperatures (both in K) obtained for some sterically stabilized dispersions. After Shaw[5] and Hiemenz and Rajagopalan[6]

Stabilizer	Dispersion medium	CFT (K)	Theta temperature (K)
Polyethylene oxide	0.39 mol/l $MgSO_4$	316–318	315
Polyacrylic acid	0.20 mol/l HCl (aq)	281–287	287
Polyisobutylene	2-methylbutane	325	325
Polyvinyl alcohol	2 mol/l NaCl	301–312	300

weight and is higher and optimum (e.g. thicker, denser adsorbed layers) when it occurs from a solution containing a poor solvent. For these reasons, block co-polymers are excellent steric stabilizers. The LA–LB concept is a useful tool for evaluating the relative adsorption of polymers vs. solvents on the same solid surface, as illustrated in Figure 18.18[38]. Due to competition from solvent and surface, in many cases solvents of 'balanced' acidity–basicity can be best choices for achieving maximum adsorption.

Moreover, solubility concepts are very important in understanding adsorption and steric stabilization due to polymers. For this reason, the best steric stabilization of colloidal dispersions is obtained with co-polymers, especially block co-polymers, which have one part (polymer) with low solubility in the medium (thus adsorbing well on the particles) and another one having a high solubility in the medium (thus expanding well in the medium). Thick, firmly sticking to the surface, dense and fully adsorbing layers are also very important for good steric stabilization. As Table 18.11 shows, of special relevance in the Gibbs free energy of the interpenetrating chains is the critical flocculation temperature (CFT) which is closely related to the theta temperature (upper or lower) of the polymer in the solvent.

18.8.3 Multicomponent Langmuir adsorption and the vdW–Platteeuw solid solution theory

As already mentioned in Chapter 10, the difference in the chemical potential of water in the hydrate phase and water in the empty hydrate phase, which is derived from the solid solution theory of van der Waals and Platteeuw[39] is essentially a combination of the Gibbs–Duhem equation with the Langmuir multicomponent adsorption equation. The derivation is shown bellow, adapted from Hendriks and Meijer[47], and Michael L. Michelsen (2006, private communication).

The site occupancy of cavity m by a component i, Θ_{im}, is calculated from the Langmuir multicomponent expression (based on fugacities):

$$\Theta_{i1} = \frac{C_{i1} f_i}{S_1}$$
$$\Theta_{i2} = \frac{C_{i2} f_i}{S_2} \qquad (18.56)$$
$$S_k = 1 + \sum_j C_{jk} f_j$$

This gives the component contents:

$$\frac{x_i}{x_w} = \frac{n_i}{n_w} = \nu_1 \Theta_{i1} + \nu_2 \Theta_{i2} = \left(\frac{\nu_1 C_{i1}}{S_1} + \frac{\nu_2 C_{i2}}{S_2} \right) f_i \qquad (18.57)$$

Applying the Gibbs–Duhem equation:

$$\sum_i n_i d \ln f_i = 0 \Rightarrow d \ln f_w = -\sum_{i \neq w} \frac{n_i}{n_w} d \ln f_i \tag{18.58}$$

Combining Equations (18.58) and (18.57):

$$d \ln f_w = -\sum_{i \neq w} \left(\frac{\nu_1 C_{i1}}{S_1} + \frac{\nu_2 C_{i2}}{S_2} \right) df_i = -\nu_1 d \ln S_1 - \nu_2 d \ln S_2 \tag{18.59}$$

Finally, integrating from the empty lattice ($f_i = 0, S_1 = S_2 = 1$):

$$\ln f_w^H - \ln f_w^{EH} = -\nu_1 \ln S_1 - \nu_2 \ln S_2 \tag{18.60}$$

This equation is essentially the one obtained using the solid solution theory of van der Waals and Platteeuw[39], see equation 10.8. (Chapter 10).

18.9 Conclusions

There are many applications at the interface between thermodynamics and colloid and surface science. Some of these applications have been illustrated in this chapter.

Interfacial theories were discussed first. The most widely used theory today is that of van Oss and co-workers which accounts for the van der Waals and acid–base interactions. Via this theory and experimental measurements of contact angles, solid surfaces can be characterized, e.g. we can determine the solid surface tension (and its 'distinct terms' such as those due to dispersion and polar character), wetting and adhesion characteristics.

The stability of colloidal dispersions depends on the balance between attractive van der Waals and repulsive electrostatic (double-layer) forces, according to the DLVO theory. The van der Waals forces between colloid particles or surfaces are of much longer range than those between molecules and often dominate the behavior of colloidal systems. The van der Waals forces in combination with expressions for the Hamaker constant can be used to determine miscibility in ternary systems, e.g. the effect of solvents in polymer blends. They can also be used as a direct measure of adhesive forces, although adhesion in practice is a more complex phenomenon and both surface and bulk (e.g. mechanical) properties of the materials involved are important.

The hydrophilic character of surfactants as used, for instance, in cleaning formulations or emulsions is quantified via the hydrophilic–lipophilic balance (HLB) or the critical packing parameter (CPP). These parameters also help to determine the type of emulsions and the structure of surfactants in solutions. They are related to, but are not identical to, the octanol–water partition coefficient discussed in Chapter 17. The micellization is a pseudo-phase transition, largely due to entropic effects, and it can be approximately predicted by thermodynamic models, e.g. UNIFAC, although different parameters (than those estimated from low-molecular-weight compounds) may be needed for the groups involved in typical surfactants. The adsorption of surfactants, polymers and other molecules on solid surfaces is of paramount importance in many applications and is often described by the Langmuir adsorption theory, but other theories have also been developed. The Langmuir multicomponent adsorption theory is also the basis of the derivation of the van der Waals–Platteeuw model used in gas hydrate studies.

References

1. H.Y. Jennings, *J. Colloid Interface Sci.*, **1967**, *24*, 323.
2. K.A.G. Schmidt, G.K. Folas, B. Kvamme, *Fluid Phase Equilib.*, **2007**, *261(1)*, 230.
3. B. Kvamme, T. Kuznetsova, K.A.G. Schmidt, *WSEAS Trans. Biol. Biomed.*, **2007**, *3*, 517.
4. D. Myers, *Surfaces, Interfaces, and Colloids: Principles and Applications*, VCH, 1991.
5. D. Shaw, *Introduction to Colloid & Surface Chemistry* (4th edition). Butterworth–Heinemann, 1992.
6. P.C. Hiemenz, R. Rajagopalan, *Principles of Colloid and Surface Science* (3rd edition). Marcel Dekker, 1997.
7. J. Goodwin, *Colloids and Interfaces with Surfactants and Polymers: An Introduction.* John Wiley & Sons, Ltd, 2004.
8. D.W. van Krevelen, P.J. Hoftyzer, *Properties of Polymers: Correlations with chemical structure.* Elsevier, 1972.
9. C.M. Hansen, *Hansen Solubility Parameters: A User's Handbook.* CRC Press, 2000.
10. C.J. van Oss, M.K. Chaudhurym, R.J. Good, *Adv. Colloid Interface Sci.*, **1987**, *28*, 35.
11. A.R. Balkenende, H.J.A.P. van de Boogaard, M. Scholten, N.P. Willard, *Langmuir*, **1988**, *14*, 5907.
12. H.R. Baker, P.B. Leach, C.R. Singleterry, W.A. Zisman, *Ind. Eng. Chem.*, **1967**, *59(6)*, 29.
13. C.J. van Oss, D.R. Absolom, A.W. Neumann, *Colloids Interfaces*, **1980**, *1*, 45.
14. J. Israelachvilli, *Intermolecular and Surface Forces.* Academic Press, **1985**.
15. E. McCafferty, *J. Adhesion Sci. Technol.*, **2002**, *16(3)*, 239.
16. R.F. Brady, I.L. Singer, *Biofouling*, **2000**, *15(1–3)*, 73.
17. H. Kahl, S. Enders, *Fluid Phase Equilib.*, **2000**, *172*, 27.
18. H. Kahl, S. Enders, *Phys. Chem. Chem. Phys.*, **2002**, *4*, 931.
19. P. Paricaud, A. Galindo, G. Jackson, *Fluid Phase Equilib.*, **2002**, *194–197*, 87.
20. G.J. Gloor, F.J. Blas, E.M. Rio, E. Miguel, G. Jackson, *Fluid Phase Equilib.*, **2002**, *194–197*, 521.
21. C. Panayiotou, *J. Colloid Interface Sci.*, **2003**, *267*, 418.
22. A.J. Queimada, Ch. Miqueu, I.M. Marrucho, G.M. Kontogeorgis, J.A.P. Coutinho, *Fluid Phase Equilib.*, **2005**, *228–229*, 479.
23. M.B. Oliveira, I.M. Marrucho, J.A.P. Coutinho, A.J. Queimada, *Fluid Phase Equilib.*, **2008**, *267*, 83.
24. B.S. Carey,The gradient theory of fluid interfaces. PhD Dissertation, University of Minnesota, 1979.
25. C. Miqueu, Modélisation á température et pression elevées de la tension superficielle de composants des fluides pétroliers et de leurs mélanges synthétiques ou réels. PhD Dissertation, Université de Pau, 2001.
26. P.M.W. Cornelisse, The square gradient theory applied: simultaneous modelling of interfacial tension and phase behaviour. PhD Dissertation, Technische Universiteit Delft, 1997.
27. A. Mejía, H. Segura, L.F. Vega, J. Wisniak, *Fluid Phase Equilib.*, **2005**, *227*, 225.
28. J.D. Van der Waals, *Z. Phys. Chem. (Leipzig)*, **1894**, 657.
29. J.W. Cahn, J.E. Hilliard, *J. Chem. Phys.*, **1958**, *28*, 258.
30. M.L. Michelsen, J.M. Mollerup, *Thermodynamic Models, Fundamentals and Computational Aspects.* Tie-Line Publications, 2004.
31. Y.X. Zuo, E.H. Stenby, *J. Colloid Interface Sci.*, **1996**, *182*, 126.
32. Y.X. Zuo, E.H. Stenby, *J. Chem. Eng. Jpn*, **1996**, *29*, 159.
33. Y.X. Zuo, E.H. Stenby, *SPE J.*, **1998**, *3*, 134.
34. M.V. Flores, E.C. Voutsas, N. Spiliotis, G.M. Eccleston, G. Bell, D.P. Tassios, *J. Colloid Interface Sci.*, **2001**, *240*, 277.
35. H. Cheng, G.M. Kontogeorgis, E.H. Stenby, *Ind. Eng. Chem. Res.*, **2002**, *41(5)*, 892.
36. C.-C. Chen, *AIChE J.*, **1996**, *42*, 3231.
37. G.T. Barnes, I.R. Gentle, *Interfacial Science – An introduction.* Oxford University Press, 2005.
38. B. Jonsson, B. Lindman, K. Holmberg, B. Kronberg, *Surfactants and Polymers in Aqueous Solution.* John Wiley & Sons, Ltd, 2001.
39. J.H. van der Waals, J.C. Platteeuw, *Adv. Chem. Phys.*, **1959**, *2*, 1.
40. J.R. Svendsen, G.M. Kontogeorgis, S. Kiil, C. Weinell, M. Grønlund, *J. Colloid Interface Sci.*, **2007**, *316*, 678.
41. K. Kendall, *J. Phys. D: Appl. Phys.*, **1971**, *4*, 1186.
42. J.G. Kohl, I.L. Singer, *Prog. Org. Coatings*, **1999**, *36*, 15.

43. A.L. Myers, J.M. Prausnitz, *AIChE J.*, **1965**, *11*, 121.
44. A.A. Shapiro, E.H. Stenby, *J. Colloid Interface Sci.*, **1998**, *201*, 146.
45. M.A. Monsalvo, A.A. Shapiro, *Fluid Phase Equilib.*, **2007**, *254*, 91.
46. M.A. Monsalvo, A.A. Shapiro, *Fluid Phase Equilib.*, **2007**, *261*, 292.
47. E. Hendriks, H. Meijer, An introduction to modeling of gas hydrates. In: G.M. Kontogeorgis, R. Gani, *Computer-Aided Property Estimation for Process and Product Design*. Elsevier, 2004.

19

Thermodynamics for Biotechnology

19.1 Introduction

Many applications in biotechnology, especially those related to bioseparations such as crystallization and extraction using aqueous two-phase systems, require thermodynamic data in order to achieve an improved understanding of the process and optimum design. The molecules involved are more complex than those encountered in the petroleum and chemical industries. They are multifunctional molecules (amino acids), oligomeric complex molecules (pharmaceuticals), charged biopolymers (proteins), etc. An overview of the most typical biomolecules is given in Figure 19.1.

Some applications of 'classical' models to some simple biomolecules have been discussed in Chapter 4:

Section 4.5.3: regular solution theory applied to solvent selection for pharmaceuticals (see Figure 4.4 and Problem 10 in Chapter 4 on the companion website at www.wiley.com/go/Kontogeorgis)

Section 4.5.5.4: combination of the Flory–Huggins model and solubility parameters together with transport properties for predicting the controlled release of steroids from polymers

and in Chapter 5, where Table 5.8 gives some references on UNIFAC applications to pharmaceuticals. Finally, the concluding Chapter 20 will summarize challenges in current biothermodynamics research. In this chapter, we will briefly outline: (1) the current status of pharmaceutical thermodynamics highlighting two promising recent approaches; (2) modeling attempts for amino acids and peptides; and finally (3) simple semi-predictive models for estimating thermodynamic properties (solubility, partition coefficients) in protein-containing systems. The models which will be presented for proteins are tailor-made for some important specific applications. We will also discuss protein adsorption as applied to chromatographic processes.

19.2 Models for pharmaceuticals

19.2.1 General

Screening for solvents and calculating the solubility of pharmaceuticals in the selected solvents are important tasks in the pharmaceutical industry. The number of functional groups involved in the various pharmaceuticals is very large and pharmaceutical molecules often contain more than one functional group. This is illustrated in

Thermodynamic Models for Industrial Applications: From Classical and Advanced Mixing Rules to Association Theories
Georgios M. Kontogeorgis and Georgios K. Folas

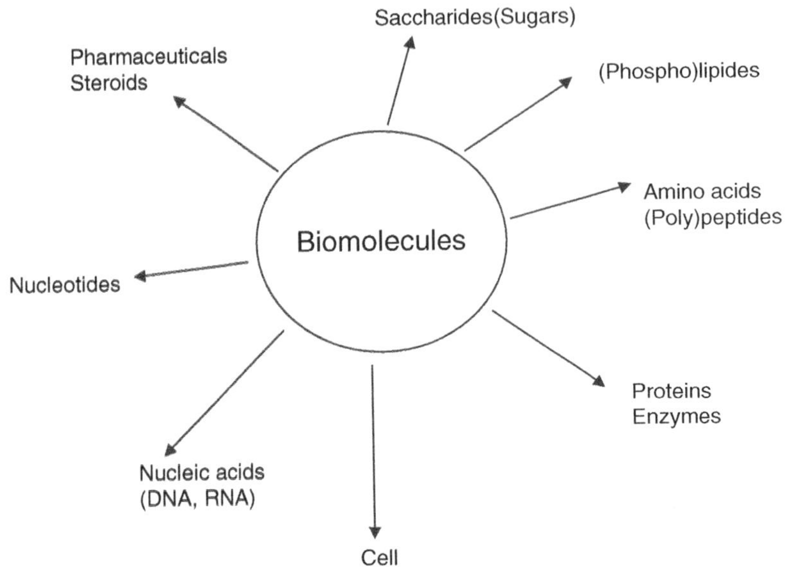

Figure 19.1 *The different types of biomolecules. The upper ones are the smaller biomolecules, the building blocks, with molecular weights between 200 and 350 g/mol. The biological macromolecules such as nucleic acids, polysaccharides and proteins/enzymes have molecular weights which can exceed 1000 and they can reach 10^9 g/mol*

Figure 19.2 for some anti-inflammatory drugs. Extensive data are available in a recent DECHEMA publication.[1] Other complications arise from the frequent use of mixed solvents as well as the existence of many pharmaceuticals in different polymorphs.

Despite these difficulties, attention has been given to the thermodynamics of pharmaceuticals and both data and models have appeared in the literature. Recently, several industrial studies have also been reported, e.g. from Mitsubishi, Dow, Merck, ASPEN Technology and Astra Zeneca[2–6]. Many of the classical modeling attempts include QSAR, Flory–Huggins/solubility parameters and UNIFAC, as shown in Table 19.1, which provides an overview of some important approaches. Some of the most recent methods include the use of the quantum chemical COSMO–RS model (see the monograph by Klamt for an extensive discussion[7] as well as Chapter 16), a segment-type version of NRTL developed by ASPEN Technology researchers (called NRTL–SAC) and a recent extension of the NRHB equation of state to pharmaceuticals.

NRTL–SAC and NRHB are discussed in the next sections. Some general comments on the applicability of various thermodynamic models for pharmaceuticals are:

1. With the exception of NRHB, the other approaches in Table 19.1 can only be used at low pressures, which is not a serious problem for many pharmaceutical applications.
2. Even the simplest semi-empirical models (first three modeling types in Table 19.1) are of practical use. They are semi-predictive in the sense that certain solubility data are needed to estimate some model parameters and then the models can be used for other solvents (and sometimes also at other temperatures).
3. The simple approaches are based on semi-theoretical principles (solubility parameter concept, group contributions, local compositions), but they do not explicitly account for complex interactions which are due to strong intermolecular forces, especially polarity and hydrogen bonding.
4. 'Hybrid' models, i.e. combination of models based on semi-predictive schemes (group contributions, solubility parameters, etc.), with some experimental data, are often more useful for pharmaceutical

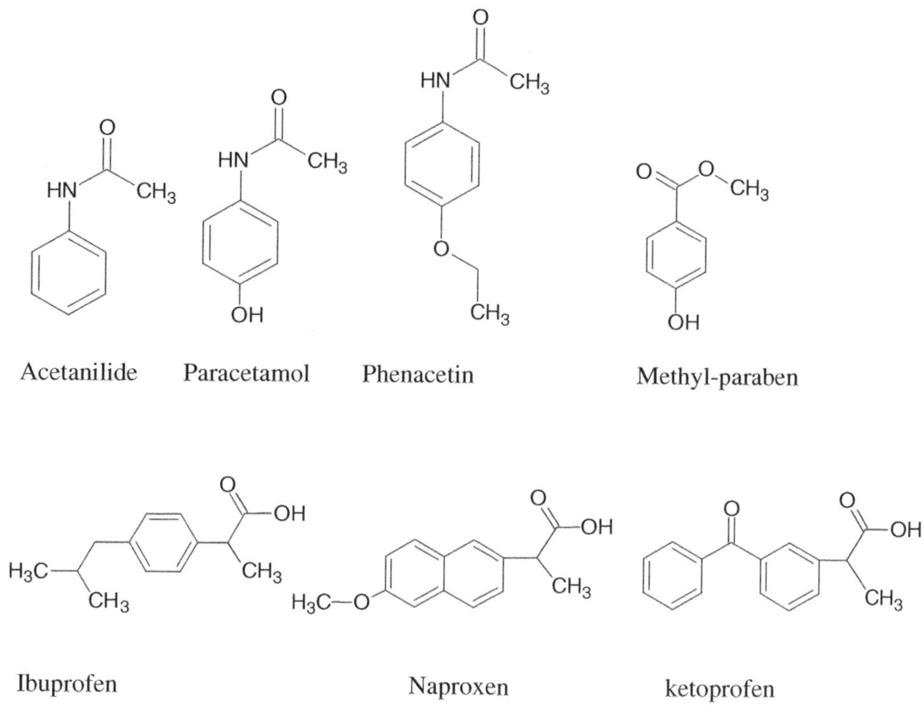

Figure 19.2 *Examples of pharmaceutical compounds and intermediates. The molecules shown here have been studied with the NRHB equation of state (Section 19.2.3)*

applications compared to 'purely predictive' models such as UNIFAC with parameters obtained from low-molecular-weight compounds data. For example, von Stockar and van der Wielen[8,9] have shown that UNIFAC cannot satisfactorily predict the partition of Penicillin G over organic and aqueous phases (neither quantitatively nor qualitatively correct) and similar conclusions have been reached by other researchers. Thus, most successful UNIFAC applications[10–13] are essentially local UNIFAC models where some parameters are obtained from pharmaceutical–solvent solubility data. In this case satisfactory results have been obtained even for mixed solvents.[14] A recent approach has been reported by Modarresi *et al.*[15] who used the Lindvig *et al.*[16] Flory–Huggins/Hansen solubility parameter (HSP) model (see Chapter 4) to calculate solubilities of felodipine in various solvents. Very good results are obtained, while a model based on the Hansen solubility parameter (HSP) alone is not equally successful. The HSPs are estimated either from a third-order group contribution method or using an approach based on connectivity indices.

19.2.2 The NRTL–SAC model

NRTL–SAC[23,24] is a relatively new model developed by ASPEN researchers[23] and applied to pharmaceutical–solvent phase behavior. The model has already received some attention and case studies and successful applications have been reported both from industry (Merck pharmaceuticals by Mathias[6], cimetidine from Astra Zeneca[5,24]) and from academia[25]. NRTL–SAC is based on polymer NRTL by Chen[28] and the equations

Table 19.1 *Some modeling approaches for pharmaceuticals*

Modeling type	Reference	Pharmaceuticals and related compounds to which the methods have been applied
QSAR incl. chemoinformatics	Bergstrom *et al.*[17] Jorgensen and Duffy[70] Hansen *et al.*[19] Kouskoumvekaki *et al.*[73]	Many different molecules
Flory–Huggins and solubility parameters incl. HSP	Franck *et al.*[2] Kolar *et al.*[4] Abildskov and O'Connell[12–14] Lin and Nash[71] Modarresi *et al.*[15]	Aspirin, Morphine Ephedrine, Hydrocortisone Diuron, Monouron, Paracetamol, Phenobarbital, Vinbarbital, Estriol, Ibuprofen Felodipine
UNIFAC incl. local UNIFAC models	Franck *et al.*[2] Kolar *et al.*[4] Abildskov and O'Connell[12–14] Von Stockar and van der Wielen[8,9] Prausnitz *et al.*[18] Gani *et al.*[10] Gracin *et al.*[72] Modarresi *et al.*[15]	Aspirin, Penicillin G Aromatic acids, ibuprofen Paracetamol Carbomycin-A Ephedrine, Hydrocortisone Diuron, Monouron, Phenobarbital, Vinbarbital, Estriol, Felodipine
COSMO–RS	Klamt *et al.*[20–21]	Etoricoxib, Rofecoxib, Lovastatin, Simvastatin
COSMO–SAC	Tung *et al.*[22]	
NRTL–SAC	Chen and Song[23] Crafts[5] Tung *et al.*[22] Chen and Crafts[24] Mota *et al.*[25]	Paracetamol, Budesonide, Furosemide, Allopurinol, Aspirin, Cimetidine, Acetaminophen, Sulfadiazine, Sulfamerazine, aromatic acids, steroids, morphine, hydrocortisone, piroxicam, camphor, theophylline, haloperidol Etoricoxib, Rofecoxib, Lovastatin, Simvastatin
NRHB	Tsivintzelis *et al.*[26,27]	Acetanilide, Ibuprofen, Phenacetin, Ketoprofen, Naproxen, Methyl paraben, Paracetamol, Benzoic acid, ethyl paraben

are presented in Appendix 19.A. The novelty compared to polymer NRTL is the modeling of non-ideality in terms of interactions between three different types of conceptual segments:

- polar (P), Y^- and Y^+ (electrostatic solvation and electrostatic polar);
- hydrophobic (HPB), X;
- hydrophilic (HPL), Z.

Each solute and solvent molecule is described by the above four types of conceptual segments which characterize the effective surface interactions. X and Z are related to hydrogen bonding (adversity to formation or creation), while all other surface interactions are lumped within the 'polar' segment and, possibly for this reason, two parameters are included, Y^- and Y^+. The individual segments are assumed to interact with each other through binary interactions alone, estimated via binary interaction parameters, which are constant in the model. The electrostatic solvation segment (Y^-) is attractive to the hydrophilic segment (Z), while the electrostatic polar segment (Y^+) is repulsive to the hydrophilic segment (Z). Not all compounds will be characterized with all four parameters. For example, hydrocarbons and ethers are mostly hydrophobic, while ketones, esters and amides are both hydrophobic and polar. Alcohols, glycols and amines may have both hydrophobic and hydrophilic character, whereas certain complex compounds such as organic acids may require all four parameters.

Chen and Song[23] and Chen and Crafts[24] present P/HPB/HPL parameters for 62 common solvents which are determined by fitting VLE and LLE data of these solvents with three reference molecules (hexane, acetonitrile and water) that are considered to cover a wide range of interactions. When the solvent parameters are known, then for every new pharmaceutical, X/Y/Z parameters are determined from some selected (five or six) drug–solvent SLE data. It is important to select mixtures which cover all range of interactions.[5] When the pharmaceutical segment parameters have been determined, then the NRTL–SAC model can be applied to other solvents, temperatures and mixed solvents.[24,5] The results are frequently very successful and two examples for some Merck drugs are shown in Figure 19.3. The results are comparable and often better than other models, e.g. COSMO–SAC.[22]

Crafts[5] showed that NRTL–SAC can represent reasonably well large amounts of cimetidine–solvent data, including solvent mixtures (ethanol–water), using solely two fitting parameters for the solute (X and Z). Successful modeling results for the solubility of pharmaceuticals in mixed solvents have been presented by Chen and Crafts.[24] It is shown that NRTL–SAC can capture the maximum of the solubility at a specific concentration in common mixed solvents such as alcohols–water and ethers–water, including the shape of the curves.

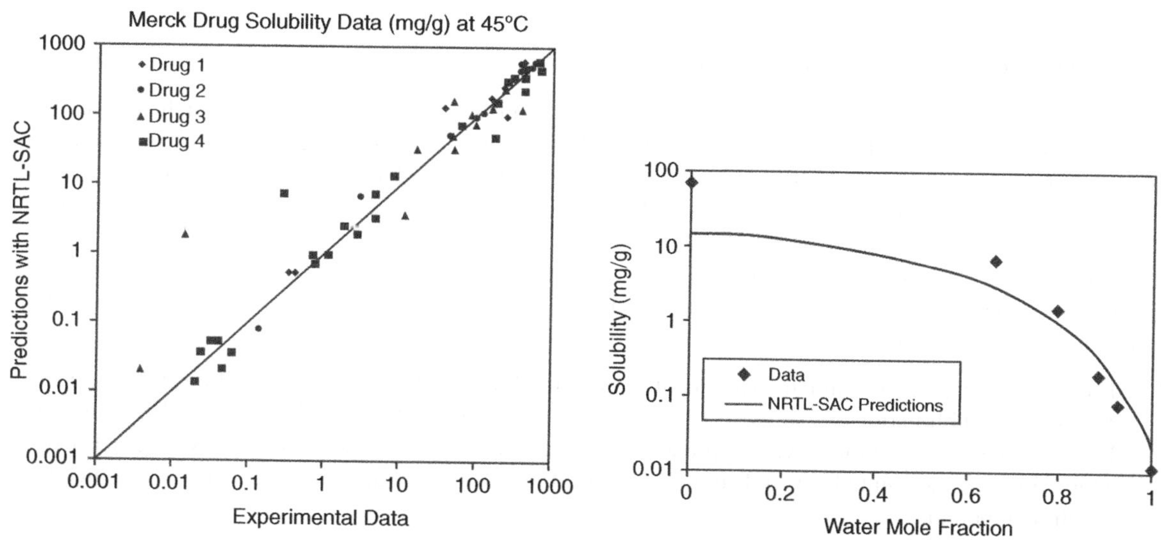

Figure 19.3 Prediction of drug solubility with NRTL–SAC. Left: Prediction of a Merck drug solubility in various solvents. Right: Prediction of a confidential 'Drug 2' solubility in acetone–water mixture at 22.5 °C. Reproduced with permission from Fluid Phase Equilibria, Applied thermodynamics in chemical technology: current practice and future challenges by Paul M. Mathias, PPEPPD 2004 Proceedings 228–229, 49–57 Copyright (2005) Elsevier. More results on Merck drugs are presented by Tung et al.[22]

It is thus understood that, like 'local' UNIFAC approaches, NRTL–SAC is a 'hybrid' model. This means that it is a semi-predictive model using plausible assumptions and few selected data for pharmaceutical–solvent mixtures in order to estimate the drug parameters. Thus the model can be used for other conditions, having ensured that 'own' data can be used in parameter estimation.
Certain limitations of the model can be identified:

1. All calculations are based on the simplified SLE equation:

$$\ln K_{sp} = \ln(x_i \gamma_i) = \frac{\Delta S^{fus}}{R}\left(1 - \frac{T_m}{T}\right) = A + \frac{B}{T} \tag{19.1}$$

2. This implies that the ΔC_p term (see Chapter 1, Table 1.3) is ignored, whereas recent investigations have shown that this term may be important in pharmaceutical calculations.[26] Moreover, in several NRTL–SAC publications, due to the uncertainties in the values of entropy of fusion and melting points of the pharmaceutical involved, these two parameters, or in other words the A, B parameters in Equation (19.1) or the K_{sp}, have been fitted together with the adjustable parameters of the activity coefficient model.
3. The only temperature dependency in the SLE calculations with NRTL–SAC is via Equation (19.1) as all interaction parameters are temperature independent (see Appendix 19.A).
4. The results for some pharmaceuticals, e.g. sulfadiazine, are not very satisfactory.[24] The model in this case systematically predicts higher solubilities in many solvents compared to the experimental data. Although the reasons are not clear, it is important to identify the role of polymorphism in phase equilibrium calculations involving pharmaceuticals.

19.2.3 The NRHB model for pharmaceuticals

It is somewhat surprising that association models, e.g. those in the SAFT family such as the ones discussed in Chapters 8–14, have not been systematically applied to pharmaceuticals. Certain applications involving drugs with supercritical fluids such as CO_2 can best be handled with equations of state due to the high pressures involved. The PC–SAFT equation of state is currently under development for pharmaceuticals[75] and the first results are promising.

A novel approach based on the NRHB equation of state has been recently applied to pharmaceuticals–solvent phase behavior (SLE as well as supercritical fluids).[26,27] The equation of state is described in Appendix 19.B. The novelty of the approach is two-fold:

1. The efficient use of the limited experimental data (total and hydrogen bonding solubility parameters, densities and sublimation pressures) available for pharmaceuticals for estimating the model's parameters.
2. The adoption of a segment-type approach which ensures that all the involved hydrogen bonding interactions are accounted for and, moreover, that the parameters are obtained from low-molecular-weight compounds.

See, for example, Figure 19.2 and Table 19.2, where the COOH···COOH parameters are obtained from propionic acid, the OH···OH interaction in Paracetamatol is obtained from alcohol mixtures and the interaction of NH···NH from *N*-methyl aniline. Similarly for the other interaction parameters, which are thus obtained from low-molecular-weight compounds for which large amounts of data are available. The same parameters are used for the interactions between pharmaceuticals and solvents as well as for the interactions

Table 19.2 *Types of self associating interactions for the pharmaceuticals studied with the NRHB equation of state. The molecules are shown in Figure 19.2. After Tsivintzelis et al.*[26,27]

Hydrogen bonding groups	Acetanilide	Ibuprofen	Phenacetin	Ketoprofen	Naproxen	Methyl-paraben	Paracetamol
–OH· · ·OH–	—	—	—	—	—	✓	✓
>NH· · ·NH <	✓	—	✓	—	—	—	✓
>NH· · ·OH–	—	—	—	—	—	—	✓
–OH· · ·O=C<	—	—	—	—	—	—	✓
>NH· · ·O=C<	✓	—	✓	—	—	—	✓
–OH· · ·O=C–O–	—	—	—	—	—	✓	—
>NH· · ·–O–	—	—	✓	—	—	—	—
–COOH· · ·HOOC–	—	✓	—	✓	✓	—	—
–COOH· · ·O=C–	—	—	—	✓	—	—	—
–COOH· · ·–O–	—	—	—	—	✓	—	—

within the pharmaceutical molecule itself. Finally, cross-interactions (when no data are available) can be taken from combining rules: that is, for the cross-association between two self-associating groups:

$$E_{ij}^{H} = \frac{E_i^H + E_j^H}{2}, \quad S_{ij}^{H} = \left[\frac{S_i^{H^{1/3}} + S_j^{H^{1/3}}}{2}\right]^3 \tag{19.2}$$

and for the cross-association between one self- and one non-associating group:

$$E_{ij}^{H} = \frac{E_i^H}{2}, \quad S_{ij}^{H} = \frac{S_i^H}{2} \tag{19.3}$$

Nine pharmaceuticals and intermediates have so far been considered (Figure 19.2), which contain a variety of hydrogen bonding forming groups (Table 19.2).

Excellent results are obtained, as shown for a few ketoprofen and naproxen systems in Figures 19.4 and 19.5 and in the overall comparison shown in Figure 19.6 (against another novel approach, the COSMO–RS model). Similar results are obtained for the other pharmaceuticals as well (Figure 19.7). It can thus be concluded that NRHB together with the proposed approach can become a promising tool in the pharmaceutical industry. Unlike the approach used for NRTL–SAC, it has been found that the ΔC_p term (see Chapter 1, Table 1.3) is important and should explicitly be accounted for when performing SLE calculations for pharmaceuticals.

19.3 Models for amino acids and polypeptides

19.3.1 Chemistry and basic relationships

The structural units of proteins (and many enzymes) are the amino acids, multifunctional compounds containing carboxyl and amino groups. There are 20 different amino acids in nature, which are parts of proteins. Condensation of amino acids gives rise to dipeptide and tripeptide bonds, thus at first oligopeptides

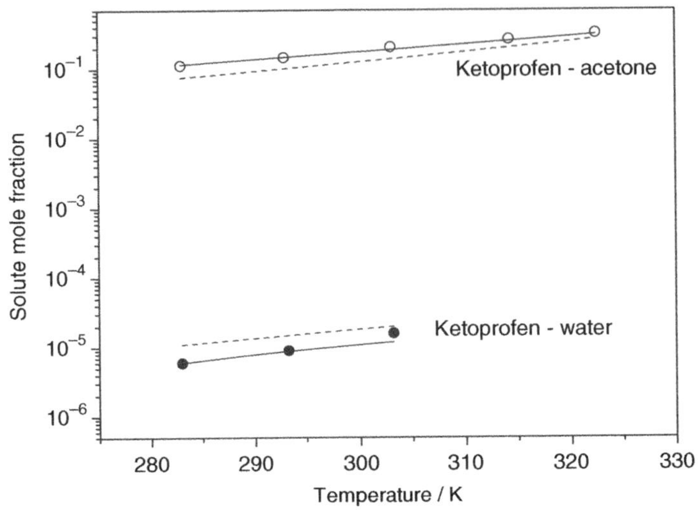

Figure 19.4 *Solubility of ketoprofen in acetone and water with NRHB. Experimental data (points), NRHB predictions (dashed lines) and correlations (solid lines), using a single interaction parameter per solvent. The predictions imply $k_{ij} = 0$, while correlations are performed using a single temperature-independent interaction parameter k_{ij}. Reprinted with permission from Journal of Physical Chemistry B, Modeling the Phase Behavior in Mixtures of Pharmaceuticals with Liquid or Supercritical Solvents by Ioannis Tsivintzelis, Ioannis G. Economou and Georgios M. Kontogeorgis, 113, 18, 6446–6458 Copyright (2009) American Chemical Society*

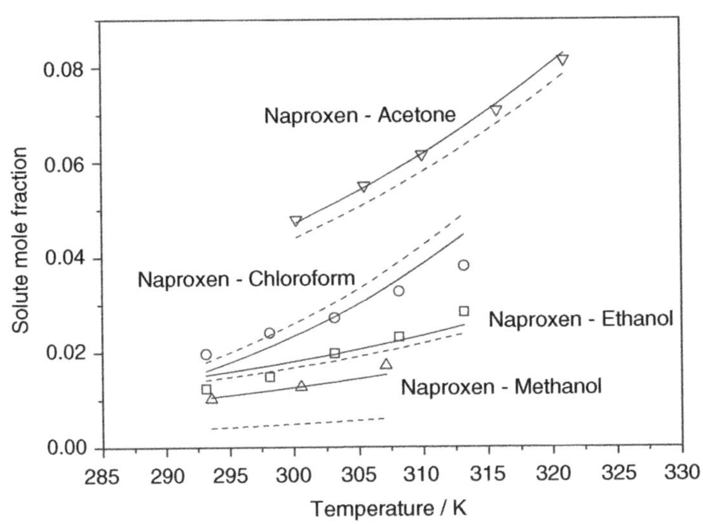

Figure 19.5 *Solubility of naproxen in various solvents with NRHB. Experimental data (points), NRHB predictions (dashed lines) and correlations (solid lines) using a single interaction parameter per solvent. The predictions imply $k_{ij} = 0$, while correlations are performed using a single temperature-independent interaction parameter k_{ij}. Reprinted with permission from Journal of Physical Chemistry B, Modeling the Phase Behavior in Mixtures of Pharmaceuticals with Liquid or Supercritical Solvents by Ioannis Tsivintzelis, Ioannis G. Economou and Georgios M. Kontogeorgis, 113, 18, 6446–6458 Copyright (2009) American Chemical Society*

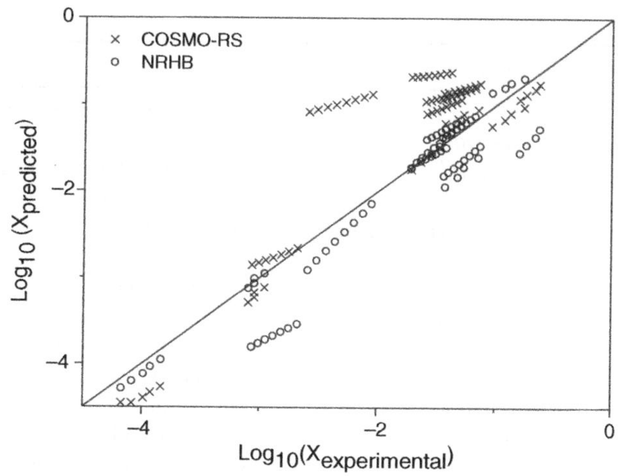

Figure 19.6 *Summarized prediction results for investigated binary pharmaceutical–solvent systems with NRHB (circles) and COSMO–RS (crosses). No interaction parameters are used. Reprinted with permission from AICHE, Modeling the Solid-Liquid Equilibrium in Pharmaceutical-Solvent Mixtures: Systems with Complex Hydrogen Bonding Behavior by Ioannis Tsivintzelis, Ioannis G. Economou and Georgios M. Kontogeorgis, 55, 3, 756–770 Copyright (2009) Wiley-VCH*

and then polypeptides, and finally proteins. The process is illustrated in Figure 19.8. Amino acids exist in various forms: a positive form at low pH < pI, and a negative form at high pH > pI, where pI is the isoelectric point. There is also a pH value where they do not appear to have any charge and are in the form of zwitterions (neutral dipolar species). This happens at pH = pI. They have the lowest solubility in water at pI. Zwitterions are strong dipolar ions and they have very high dipole moments. For example, the dipole moment of glycine is 15 Debye compared to 1.84 Debye for water. In principle, for accurate representation of the electrostatic

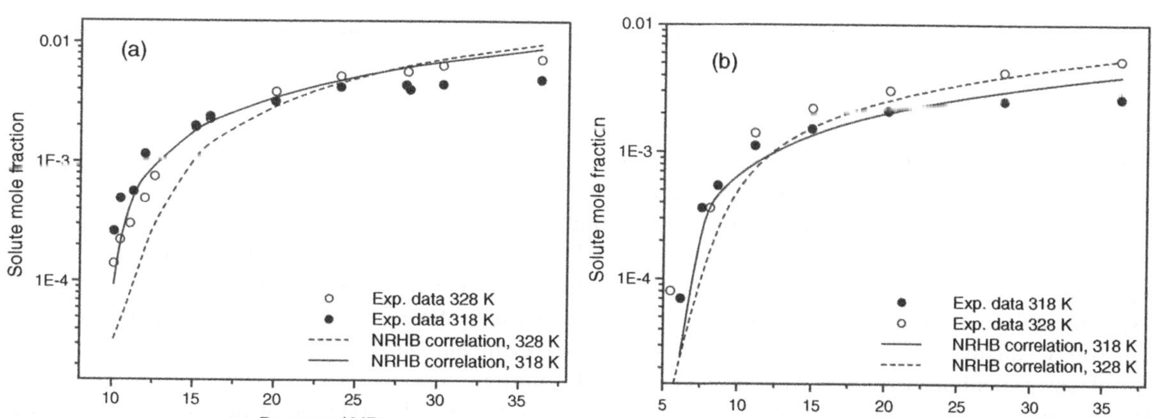

Figure 19.7 *Solubility of Benzoic acid in (a) supercritical CO_2 and (b) supercritical ethane. Experimental data (points) and NRHB correlations (lines) with $k_{ij} = -0.04417$ for CO_2 and -0.06974 for ethane*

Table 19.3 *Models for amino acids (in approximate chronological order)*

Model	Reference	Comments
NRTL or Wilson + extended Debye–Hückel + Born	Nass[34]	Alanine, glycine in water–ethanol mixed solvent
	Chen *et al.*[35]	
Electrolyte-NRTL	Chen *et al.*[35]	16 amino acid–water systems and water + salts, eight amino acid pairs, mixed water–ethanol solvents
Modified UNIFAC	Gupta and Heidemann[11]	Glycine, serine, alanine, proline, hydroxyl–proline, valine, threonine, amino–butyric acid + water
Wilson, three-suffix Margules, NRTL	Orella and Kirwan[30,31]	Glycine, L-alanine, L-isoleucine, L-phenylalanine, L-aspargine in alcohol–water solvents (miscible solutions)
Electrolyte-UNIQUAC	Peres and Macedo[36]	Nine amino acid–water, five peptides–water
UNIFAC + Debye–Hückel	Pinho *et al.*[37]	15 amino acid–water, five peptides–water
Flory–Huggins + Margules	Gude *et al.*[32,33]	Seven amino acids (serine, glycine, alanine, tyrosine, isoleucine, phenylalanine, tryptophan) in water–ethanol, water–propanol, water–butanol and water-octanol (partially miscible solutions)
Polymer-type modified Wilson	Xu *et al.*[38]	14 amino acids and peptides + water, two mixed amino acids + water
Pitzer–Simonsen–Clegg	Ferreira *et al.*[39]	Glycine, DL-alanine + water–KCl and water–Na_2SO_4
NRTL	Ferreira *et al.*[40]	12 amino acids and diglycine + water–alcohols
Electrolyte-UNIQUAC SRK + Kirkwood	Breil and Mollerup[41,42]	Alanine, glycine, serine, threonine, valine, glycylglycine, glycyl-L-alanine in water and water–NaCl (KCl, $NaNO_3$)
PC–SAFT, e-PC–SAFT	Fuchs *et al.*[43]	DL-methionine, Glycine, Alanine + water–alcohols, incl. mixed solvents
	Cameretti and Sadowski[29]	L-arginine, DL-serine in aqueous salt solutions
	Held *et al.*[44]	Proline, Valine
		Oligopeptides based on glycine and alanine–glycine

contributions in mixtures containing amino acids (or proteins, peptides) and salts, both ion–ion and zwitterion–ion interactions should be considered.

As amino acids and polypeptides are building blocks of proteins, understanding their thermodynamic properties may be the first step in developing models for protein solutions. Moreover, the recovery and separation of amino acids often involve crystallization for which the solubility is the key property. Experimental data for the solubility of many amino acids in water over an extensive temperature range (25–100 °C) are available. During the past two decades experimental data for the solubility of amino acids in mixed water–alcohol solvents as well as in aqueous salt solutions have also become available. There are also data for some mixed amino acid–water systems, some dipeptide– and oligopeptide–water mixtures, as well as a few solubility data at high pressures. Important topics are the study of the effect of temperature, salt type, salt concentration and pH on the amino acid solubility. The addition of ethanol or other alcohols in water has a significant effect on amino acid solubility and the study of this dependency is also of interest (the solubility in pure alcohol can be 1000 times lower than in pure water).

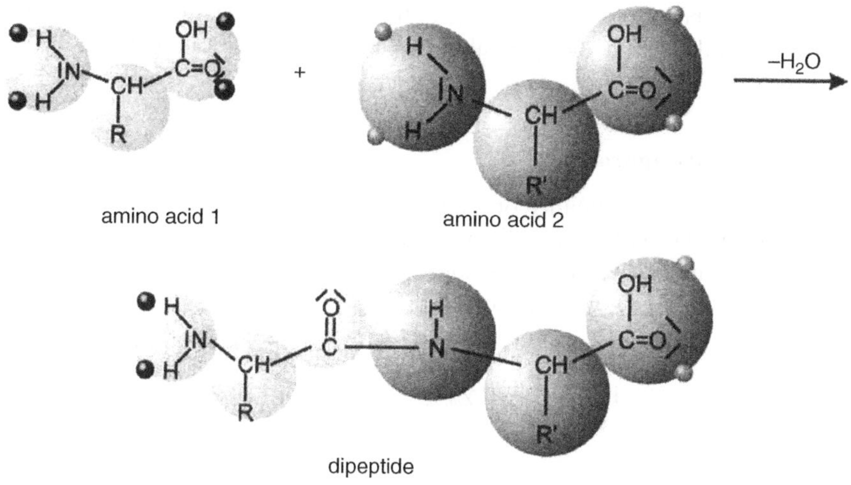

Figure 19.8 *The formation of a dipeptide out of two amino acids. The figure can be used also for estimating the association sites in models such as PC–SAFT. Due to the condensation reaction four association sites are 'lost'. Thus, the dipeptide has four (instead of eight) association sites of two different types, exactly like the amino acids, which constitute its building blocks. Reprinted with permission from Chemical Engineering and Processing, Modeling of aqueous amino acid and polypeptide solutions with PC-SAFT by Luca F. Camaretti and Gabriele Sadowski, 47, 6, 1018–1025 Copyright (2008) Elsevier*

When the amino acids are dissolved in water, they are converted to the zwitterion form and/or to one of the ions with a net charge. The dissociation of amino acids can be described by the following reactions:

Formation of neutral zwitterion:

$$NH_2RCOOH \leftrightarrow NH_3^+RCOO^- \qquad K_0 \tag{19.4}$$

Formation of positively charged species (amino acid cation) – protonation of carboxyl group:

$$NH_3^+RCOOH \leftrightarrow H^+ + NH_3^+RCOO^- \qquad K_{A1} \tag{19.5}$$

Formation of negatively charged species (amino acid anion) – deprotonation of the amino group:

$$NH_3^+RCOO^- \leftrightarrow H^+ + NH_2RCOO^- \qquad K_{A2} \tag{19.6}$$

K_0 values are of the order of 10^5 or 10^6, which means that the amount of uncharged amino acid in aqueous solutions is negligible. It can be shown[11] that the effect of pH on solubility can be expressed for divalent amino acids (only COOH and NH_2 groups) by the equation:

$$x_{HA,total}^L = x_{HA\pm}^L\left(1 + \frac{10^{-pH}}{K_{A1}} + \frac{K_{A2}}{10^{-pH}}\right) \tag{19.7}$$

According to this equation, based on the knowledge of the solubility of an uncharged amino acid, e.g. at its isoelectric point, the solubility at another pH value can be calculated, assuming that the two equilibrium constants are known.

19.3.2 The excess solubility approach

A useful approach to express the solubility of amino acids in mixed solvents is via the excess solubility approach introduced by Orella and Kirwan[30,31] and Gude *et al.*[32,33] The excess solubility of a solute (amino acid) in a mixed solvent can, under certain plausible assumptions (pure solid phase, low mole fractions of amino acids on saturated solutions), be given as:

$$\ln x_{aa}^{E} = -\ln \gamma_{aa,mix}^{\infty} + \sum_{i=1}^{N} x'_i \ln \gamma_{aa,i}^{\infty} \tag{19.8}$$

where x'_i is the mole fraction of the solvent in solute-free basis and *aa* indicates the amino acid. The activity coefficients could be obtained by a local composition or other suitable activity coefficient models, see Table 19.3. The partition coefficient of the amino acid over a mixed water–organic compound mixed solvent, e.g. water–butanol or water–octanol, can then be estimated as:

$$\ln K_{aa} = \ln\left(\frac{x_{aa}^{o}}{x_{aa}^{w}}\right) = \ln x_{aai}^{E,o} + \sum_{j}\left[x_j^{\prime,a} - x_j^{\prime,w}\right] \ln x_{j,aa}^{sat} \tag{19.9}$$

where *aa* indicates the amino acid, *w* the aqueous phase, *o* the alcohol(organic) phase, and *N* solvent components *j*.

19.3.3 Classical modeling approaches

There are many modeling attempts for describing the phase behavior of amino acids in water and other solvents. Even classical local composition models such as NRTL, Wilson, UNIQUAC and UNIFAC have been extensively used. Important models presented over the past 30 years are summarized in Table 19.3. They can be roughly divided into:

1. Local composition models which do not consider the long-range interactions (Wilson, NRTL, UNIQUAC and UNIFAC), since very often the neutral dipolar species are the predominant species present.
2. Local composition and other models which account explicitly for electrostatic long-range interactions, i.e. the dissociation of amino acids (electrolyte versions of NRTL and UNIQUAC or combinations of Margules with electrolyte terms).
3. Perturbation models, e.g. recent applications of PC–SAFT.

Macedo[45] and Khoshbarchi and Vera[46] provided excellent reviews of the most important models which have been applied to calculate activity coefficients and solubilities in amino acid solutions. Figures 19.9–19.13 show some results which are characteristic of the capabilities of the various models, especially for correlation purposes. Important conclusions from the literature studies are:

- Local composition models such as Wilson and UNIQUAC, even without accounting for electrostatic interactions, can describe well amino acid–water solubilities, in combination with a temperature-dependent thermodynamic solubility constant (Equation (19.4)), which is also adjusted to experimental data.

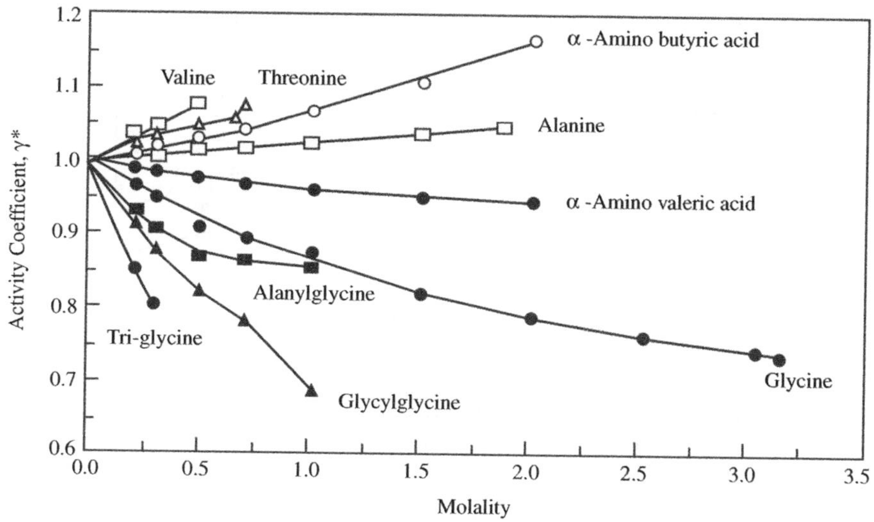

Figure 19.9 *Experimental (points) and calculated (lines) values with the electrolyte UNIQUAC for the activity coefficients of some amino acids and some peptides in water at 298.15 K. Reprinted with permission from Chemical Engineering Science, Representation of solubilities of amino acids using the uniquac model for electrolytes by António M. Peres and Eugénia A. Macedo, 49, 22, 3803–3812 Copyright (1994) Elsevier*

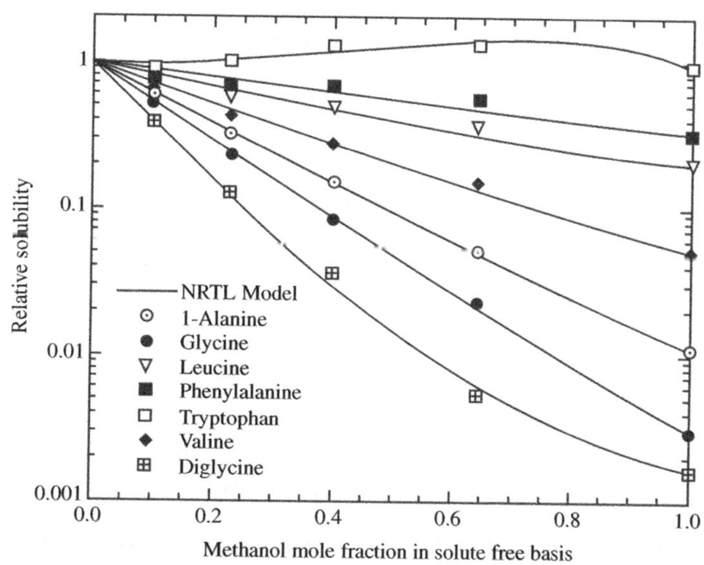

Figure 19.10 *Relative solubilities of amino acids and diglycine in water–methanol solutions at 25 °C, calculated from the NRTL equation. Reprinted with permission from Chemical Engineering Science, Solubility of amino acids and diglycine in aqueous–alkanol solutions by L. A. Ferreira, E. A. Macedo and S. P. Pinho, 59, 15, 3117–3124 Copyright (2004) Elsevier*

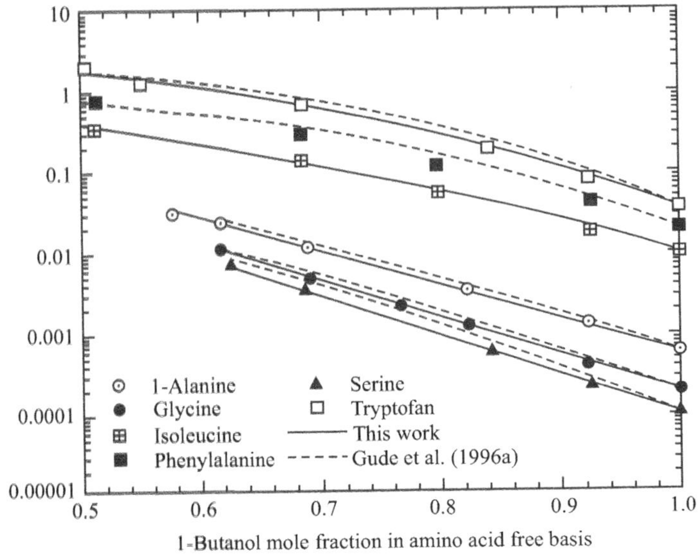

Figure 19.11 Relative solubilities of amino acids in water–butanol solutions at 25 °C, calculated from the NRTL equation and the model of Gude et al.[32,33] Reprinted with permission from Chemical Engineering Science, Solubility of amino acids and diglycine in aqueous–alkanol solutions by L. A. Ferreira, E. A. Macedo and S. P. Pinho, 59, 15, 3117–3124 Copyright (2004) Elsevier

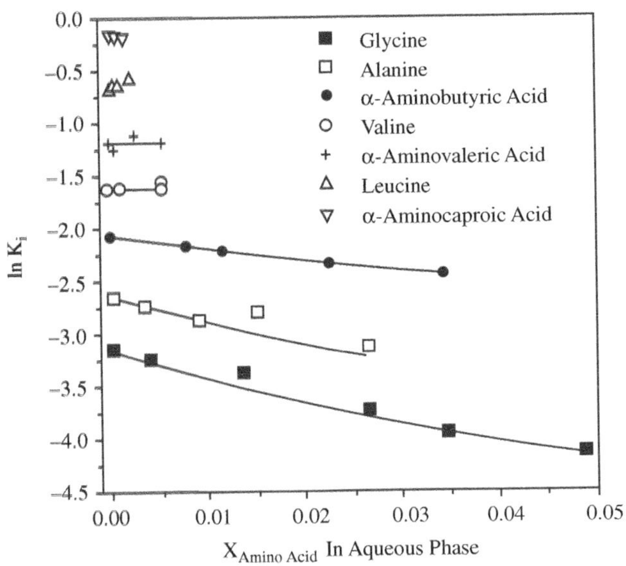

Figure 19.12 Thermodynamic partition coefficients of seven amino acids in water + 1-butanol solutions against the amino acid mole fraction in the aqueous phase. The points are the experimental data while the solid lines are the correlations using Equation (19.9) and an excess Gibbs energy activity coefficient. The model used has a Flory–Huggins combinatorial term and a Margules residual term, having a single adjustable parameter per ternary system. Reprinted with permission from Fluid Phase Equilibria, Phase behavior of α-amino acids in multicomponent aqueous alkanol solutions by Michael T. Gude, Luuk A. M. van der Wielen and Karel Ch. A. M. Luyben, 116, 1–2, 110–117 Copyright (1996) Elsevier

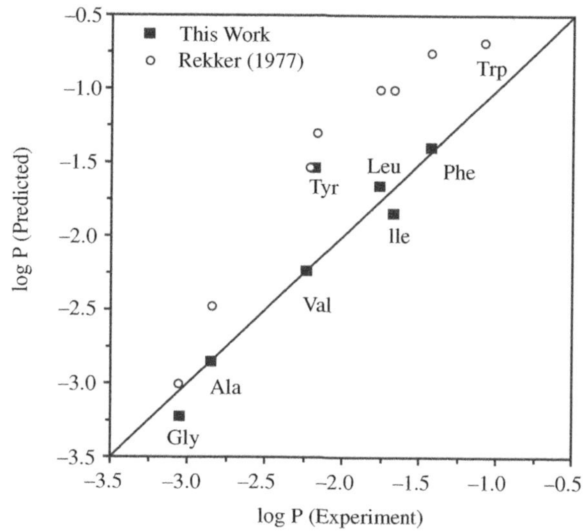

Figure 19.13 *Comparison between experimental and predicted 1-octanol–water partition coefficients for seven amino acids. The calculations are made using Equation (19.9) and an excess Gibbs energy activity coefficient. The model used has a Flory–Huggins combinatorial term and a Margules residual term, having a single adjustable parameter per ternary system. Reprinted with permission from Fluid Phase Equilibria, Phase behavior of α-amino acids in multicomponent aqueous alkanol solutions by Michael T. Gude, Luuk A. M. van der Wielen and Karel Ch. A. M. Luyben, 116, 1–2, 110–117 Copyright (1996) Elsevier*

- The models are useful in predicting the effects of temperature, pH and – when an electrolyte term is used – also the ionic strength.
- The excess solubility approach (introduced by Orella and Kirwan[30,31] and Gude *et al.*[32,33]) combined with NRTL can be used to describe very well amino acid solubilities in mixed solvents, e.g. water–ethanol at different temperatures,[40] better than earlier approaches using the same number of adjustable parameters.

The 'correlative' local composition models, e.g. UNIQUAC and NRTL, are to be preferred over UNIFAC for describing amino acid solutions. This conclusion is verified when considering the results obtained with such UNIFAC models presented in two systematic investigations for developing UNIFAC for amino acid solutions.[11,37] While there are differences in the two methods (the Gupta UNIFAC has no electrolyte term and uses larger groups), both investigations illustrate several of the problems of UNIFAC related to amino acid systems:

1. Temperature extrapolations should be done with caution if data used in parameter estimation are limited to 25 °C.
2. UNIFAC cannot differentiate between optical isomers.
3. (Most importantly) predictions for amino acids not used in parameter estimation are often both quantitatively and qualitatively incorrect, indicating that the interaction parameters estimated are not generally applicable.

19.3.4 Modern approaches

PC–SAFT

Recently, the PC–SAFT equation of state has been used for correlating and predicting amino acid (and peptide) solubilities in water, mixed solvent (water–alcohol) and mixed solvent–salt solutions. For the first

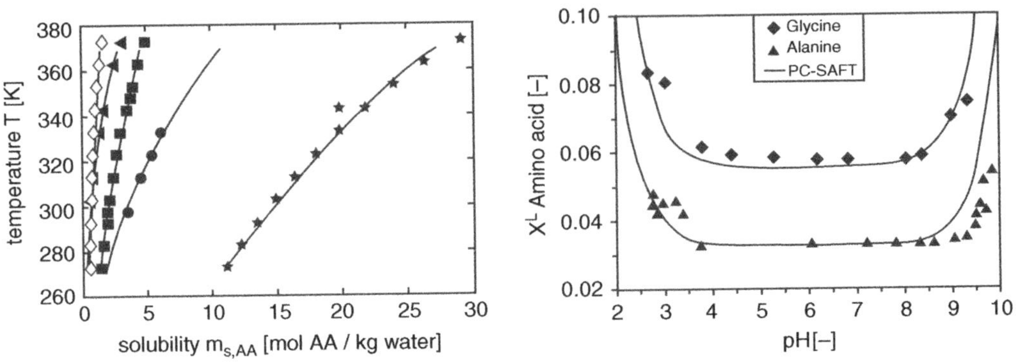

Figure 19.14 *Calculations (lines) with the PC–SAFT equation of state for amino acid–water systems. Left: Solubilities (curves from left to right) of valine, serine, alanine, glycine and praline. Reprinted with permission from Chemical Engineering and Processing, Modeling of aqueous amino acid and polypeptide solutions with PC-SAFT by Luca F. Camaretti and Gabriele Sadowski, 47, 6, 1018–1025 Copyright (2008) Elsevier Right: The solubility of glycine and* DL-*alanine in aqueous electrolyte solutions at 298.15 K as a function of pH. Reprinted with permission from Ind. Eng.Chem. Res., Solubility of Amino Acids: Influence of the pH value and the Addition of Alcoholic Cosolvents on Aqueous Solubility by Dominik Fuchs, Jan Fischer, Feelly Tumakaka, and Gabriele Sadowski, 45, 19, 6578–6584 Copyright (2006) American Chemical Society*

two applications, the original PC–SAFT was used, while for the latter the e-PC–SAFT was used as presented in Chapter 15 (i.e. using the Debye–Hückel term for the electrostatic interactions). Parameters for six amino acids have already been published[29,43] but more parameters are on the way.[44] In all these applications, water has been modeled as a 2B-site molecule using a five-parameter temperature dependency of the segment diameter in order to represent the experimental densities (and vapor pressures) over the whole temperature range. Amino acids are modeled as 4C molecules (all sites being equal). The pH dependency is estimated from Equation (19.7), and the enthalpy and temperature of melting of the amino acids (which are not available) are fitted to experimental solubility data. These regressed thermophysical properties of amino acids should be regarded as adjustable parameters with little physical meaning. Cross-interactions are expressed by Equations (8.50) (Chapter 8) and one k_{ij} is used for each water–alcohol and water–amino acid system.

Even with this limited number of experimental data and T-independent parameters, it has been shown that PC–SAFT can satisfactorily describe the solubility of the various amino acids in water and alcohols, and it can predict the solubilities in mixed water–alcohols without additional parameters and over extensive ranges of T. Equally satisfactory results are obtained for the densities and vapor pressures of amino acid solutions. The pH dependency is also well represented, as shown in Figure 19.14. It should be emphasized that, for each amino acid, in total seven pure compound parameters are estimated (the five PC–SAFT parameters as well as the enthalpy and temperature of melting). In addition a k_{ij} value is used for the water–amino acid interactions (values around -0.07 for different amino acids). These eight parameters were fitted simultaneously to densities, vapor pressures and solubility data for aqueous amino acid solutions. The solubility data are the most sensitive ones, while k_{ij} values are not needed for representing densities and vapor pressure of amino acid solutions alone.

Equally good results are obtained for the density of oligopeptides (glycine units and alanine–glycine) which are modeled with the co-polymer extension of PC–SAFT described in Section 14.2.4. Thus, the oligopeptides are treated as co-polymers built up by the respective amino acids, and, like their building blocks, they also have four (equal) association sites, as explained in Figure 19.8. Essentially, the advantage of using an approach such as PC–SAFT is demonstrated in this application as the properties of polypeptide solutions can be estimated from knowledge of the pure component parameters of the respective amino acids. Following the extension of PC–SAFT to co-polymers (Section 14.2.4) the additional information needed is the segment-type fraction (i.e. the fraction of segments of each type in the molecule) and the bonding fraction (i.e. the amount of bonds between the two segment types in the molecule). Both parameters are exactly given for a polypeptide because its amino acid sequence is known a priori. Thus, (almost) no additional parameters are needed for the peptides except for the segment number of the peptide, although to a first approximation even this parameter could be obtained from the following equations:

$$\left(\frac{m_{seg}}{M}\right)_{PP} = \frac{1}{M_{PP}} \sum_{AA}^{all\ types} N_{AA} M_{AA} \left(\frac{m_{seg}}{M}\right)_{AA} \tag{19.10}$$

$$M_{PP} = \sum_{AA}^{all\ types} N_{AA} M_{AA} - N \times 18.015 \tag{19.11}$$

where N is the number of peptide bonds, PP refers to peptide and AA to amino acid, and $(m_{seg}/M)_{AA}$ is the segment number of the respective amino acid AA divided by its molar mass M_{AA}.

Although Equations (19.10) and (19.11) provide a first approximation of the polypeptide's segment number, when they are used in connection with PC–SAFT the density in aqueous solution is underestimated, see Figure 19.15. This might be expected since the segment of peptides should depend on the sequence length and

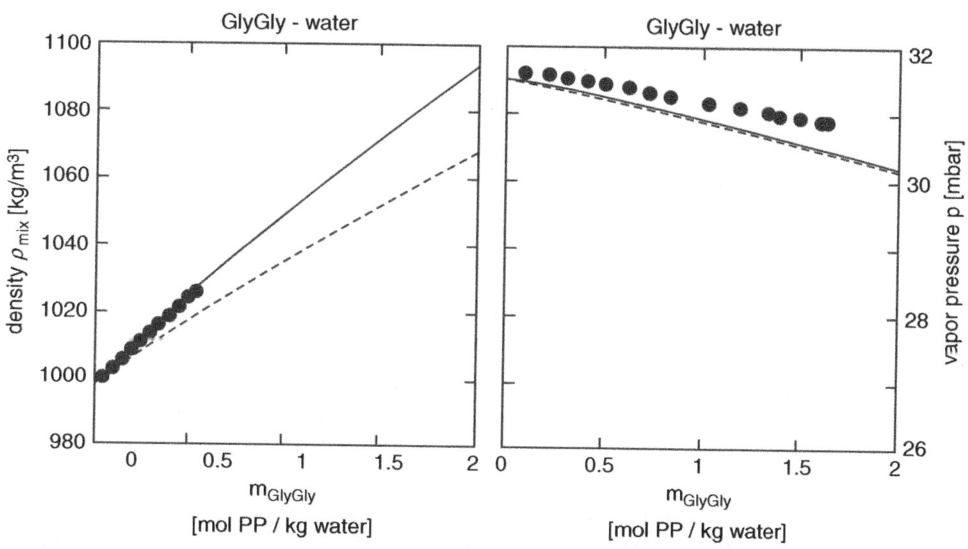

Figure 19.15 Densities and vapor pressures of aqueous diglycine (GlyGly) solutions at 298.15 K. The symbols are the experimental data, while the solid curves are calculations with PC–SAFT using a fitted value for the peptide's segment number. The dashed lines are PC–SAFT calculations with a predicted value for the peptide's segment number (based entirely on the segment numbers of the constituent amino acids; Equations (19.10) and (19.11)). Reprinted with permission from Chemical Engineering and Processing, Modeling of aqueous amino acid and polypeptide solutions with PC-SAFT by Luca F. Camaretti and Gabriele Sadowski, 47, 6, 1018–1025 Copyright (2008) Elsevier

entanglements and steric effects in the polypeptide chain should have an influence as well (secondary and tertiary formations at high molecular weights). The fitted segment numbers have values about 1–5 less than the values calculated from Equations (19.10) and (19.11) and the difference increases approximately linearly with the number of peptide bonds.

The Kirkwood approach

One theoretical shortcoming of the previously mentioned approaches is the use of only the Debye–Hückel equation to account for the electrostatic interactions for mixtures containing amino acids. This is indeed a limitation because the Debye–Hückel equation accounts solely for ion–ion interactions and not for the zwitterion–ion interaction present in mixtures containing amino acids and salts. A model derived by Kirkwood[47,48] accounts also for the zwitterion–ion contribution to the Helmholtz energy. Following Breil and Mollerup[41,42], the final expression for the excess Helmholtz energy is, after some simplifying assumptions:

$$\frac{A^E}{RT} = -\frac{N_A}{4\pi\varepsilon RT}\sum_i \frac{x_i q_i^2}{(\kappa\alpha_i)^3}\kappa\left[\frac{3}{2} + \ln(1+\kappa\alpha_i) - 2(1+\kappa\alpha_i) + \frac{1}{2}(1+\kappa\alpha_i)^2\right]$$
$$-\frac{3}{8}\frac{N_A}{4\pi\varepsilon RT}\sum_i \frac{x_i \mu_i^2}{\kappa^2\tau_i^2\alpha_i}\left[\kappa^2\tau_i - \ln(1+\kappa^2\tau_i)\right] \qquad (19.12)$$

where:

α_i = distance of closest approach (in general different from the sum of the two radii of ions due to the presence of water molecules); it is close to the hydrated ion diameter

$$\tau = b^3/6\alpha$$

b = diameter of the complex ion

$$q_i = Z_i e$$
$$\varepsilon = \varepsilon_0 D$$

The key property, the inverse Debye screening length κ, is defined as:

$$\kappa^2 = \frac{N_A}{kT\varepsilon}\sum_i c_i q_i^2 = \frac{e^2 N_A^2}{RTV\varepsilon}\sum_{ions} n_i Z_i^2 = \frac{e^2 N_A}{kTV\varepsilon}\sum_{ions} n_i Z_i^2 = \frac{F^2}{\varepsilon RT}\sum_{ions} c_i Z_i^2 \qquad (19.13)$$

The first term in Equation (19.12) is the Debye–Hückel term but the second one is due to zwitterion–ion interactions. Breil and Mollerup[41,42] have combined Equation (19.12) with the Redlich–Kwong equation of state and applied it to correlate salt activities in aqueous amino acid solutions. Conventional mixing rules and the Berthelot–Lorentz rules were used in the cubic part of the equation of state. The correlation performance of the Redlich–Kwong and Kirkwood model and its predictions of amino acid solubility in salt solutions have been compared to the results from extended UNIQUAC. The most important conclusions are:

- The performance of the two approaches is comparable using the same types of data in the parameter estimation (salt activity coefficients in amino acid solutions and osmotic coefficients of water in aqueous amino acid solutions) and subsequently predicting the solubilities (activity coefficients) of amino acids in salt solutions. Both models have the necessary flexibility to describe the experimental data but there are differences in their predictive capabilities.

- Extended UNIQUAC has problems in correlating the activity coefficients in $NaNO_3$ amino acid solutions (and irrespective of the amino acid type considered), while the performance in other salts (NaCl and KCl) is satisfactory.
- Satisfactory results are obtained with extended UNIQUAC and the Redlich–Kwong and Kirkwood model for the activity coefficients of NaCl in the dipeptides glycylglycine and glycyl-L-alanine (except for very low molalities, at and below 0.1 molal).
- The models do not predict very satisfactorily the solubilities of the amino acids and dipeptides (when such data are not included in the parameter estimation), with the extended UNIQUAC being slightly better. The solubilities in NaCl are much better predicted than in KCl with both approaches.

19.4 Adsorption of proteins and chromatography

19.4.1 Introduction

Downstream processes in biotechnology involve a variety of separations, of which – beyond the mechanical separations – crystallization and chromatography are two of the most widely used separation processes. Other novel processes, e.g. use of supercritical fluids, using reverse micelles or aqueous two-phase systems, are emerging but have not as yet received practical attention by the biotech industry (see also Section 19.5). Crystallization is very important for medium-sized biomolecules, e.g. pharmaceuticals (see Sections 19.2 and 19.3), but it is not always optimal for proteins, e.g. insulin can only be purified up to 90%.[49] For such high-molecular-weight biomolecules, chromatography is often the preferred separation process. In this section we will discuss in some detail the adsorption phenomena connected to chromatographic separations, which are of significant importance for obtaining highly pure biological products (e.g. proteins). The discussion, following Mollerup and co-workers,[49–53] will illustrate the thermodynamic background as well as similarities and differences between the adsorption phenomena with two common chromatographic techniques, the ion-exchange chromatography (IEC) and hydrophobic interaction chromatography (HIC). Similar ideas could be applied to certain other processes such as reversed-phase chromatography. A simple model developed by Mollerup and co-workers[49–53] will be presented, which can represent adsorption phenomena with a small number of physically meaningful adjustable parameters that are rather easy to determine.

19.4.2 Fundamentals of adsorption related to two chromatographic separations

In chromatographic separations, adsorption is a process where proteins bind to ligands, but there are differences between ion-exchange chromatography (IEC) and hydrophobic chromatography (HIC). The fundamentals of the two methods are discussed below, and we will generally use the symbol q to denote the concentration of the adsorbed protein and C the concentrations in the fluids phase. The subscripts P, L, s will indicate protein, ligand and salt (counter-ion) respectively. K will be the equilibrium constants of the involved reactions (complexations), which is generally given, as a function of the Gibbs free energy of adsorption (ΔG^0), by the equation:

$$RT \ln K = -\Delta G^0 \tag{19.14}$$

We start with the HIC. When the proteins bind to the hydrophobic ligand, the protein undergoes conformational changes (partial unfolding). The binding interaction between the hydrophobic moieties in the protein and the hydrophobic ligands is modeled like a reaction scheme where the 'product' is the ligand bounded protein. Thus, in HIC, the adsorption implies a reversible association of the protein (P) with the immobilized hydrophobic ligand (L), forming a weak-bonded complex (PL_n) described as:

HIC:

$$P + nL \leftrightarrow PL_n \quad K = \frac{a_{PL_n}}{a_P a_L^n} = \frac{q_{PL_n} C^n}{C_P C_L^n} \frac{\gamma_{PL_n}}{\gamma_P \gamma_L^n} = \exp\left(-\frac{\Delta G^0}{RT}\right) \tag{19.15}$$

where a and γ are, respectively, the activities and activity coefficients of protein, ligand and the complex.

In the case of IEC, the interaction is electrostatic and the counter-ions are released when the protein adsorbs. This process can be described by the equation:

IEC:

$$\text{Protein}_{solution} + \nu\text{Ion}_{adsorbed} \leftrightarrow \text{Protein}_{adsorbed} + \nu\text{Ion}_{solution}$$

$$K = \frac{a_P^{adsorbed}}{a_P^{solution}}\left(\frac{a_S^{solution}}{a_S^{adsorbed}}\right)^{\nu} = \exp\left(-\frac{\Delta G_P^0}{RT} + \nu\frac{\Delta G_S^0}{RT}\right) \cong \frac{q_P}{C_P}\left(\frac{C_s}{q_s}\right)^{\nu} \tag{19.16}$$

where $\nu = Z_P/Z_s$, Z is the binding charge, q_s and C_s are the salt (counter-ion) concentrations in the adsorbed and fluid phases, respectively.

In both the chromatographic cases (IEC and HIC), Mollerup and co-workers[49–53] have derived from fundamental thermodynamic principles simple equations that describe the adsorption of proteins. The equations are similar for the two cases of adsorption but *not* identical, and in the case of multicomponent adsorption (various proteins adsorbing), they can be expressed by the equations (where K_i is given by Equations (19.15) and (19.16)):

HIC:

$$\frac{q_{P_i}}{C_{P_i}} = K_i\left(\frac{\Lambda}{C}\right)^{n_i}\left(1 - \sum_j \frac{q_{P_j}}{q_{P_j}^{\max}}\right)^{n_i} \gamma_{P_i}, \quad q_{P_j}^{\max} = \frac{\Lambda}{\zeta_j + n_j} \tag{19.17}$$

where Λ is the ligand concentration in the adsorbent, ζ is the so-called surface coverage or steric hindrance factor, and $q_p^{\max}$ is the maximum concentration (capacity) of the protein.

IEC:

$$\frac{q_{P_i}}{C_{P_i}} = K_i\left(\frac{\Lambda}{C_s Z_s}\right)^{\nu_i}\left(1 - \sum_j \frac{q_{P_j}}{q_{P_j}^{\max}}\right)^{\nu_i} \gamma_{P_i}, \quad q_{P_j}^{\max} = \frac{\Lambda}{\zeta_j + Z_{P_j}} \tag{19.18}$$

In the case of IEC, the activity coefficient is often ignored. Equations (19.17) and (19.18) are simplified if a single protein adsorbs. *A* indicates the initial slope of the isotherms, in other words the partition coefficient between the protein in the adsorbed and fluid phases:

HIC:

$$\frac{q_P}{C_P} = K\left(\frac{\Lambda}{C}\right)^n \left(1 - \frac{q_P}{q_P^{\max}}\right)^n \gamma_P = A\left(1 - \frac{q_P}{q_P^{\max}}\right)^n \gamma_P, \quad A = K\left(\frac{\Lambda}{C}\right)^n \gamma_P \tag{19.19}$$

IEC:

$$\frac{q_P}{C_P} = K\left(\frac{\Lambda}{C_s Z_s}\right)^\nu \left(1 - \frac{q_P}{q_P^{\max}}\right)^\nu \gamma_P = A\left(1 - \frac{q_P}{q_P^{\max}}\right)^\nu \gamma_P, \quad A = K\left(\frac{\Lambda}{C_s Z_s}\right)^\nu \gamma_P \tag{19.20}$$

The parameters in the models have a well-defined physical significance and can be determined from measurements of the isocratic retention volumes and a few capacity measurements. The approach is discussed next.

19.4.3 A simple adsorption model (low protein concentrations)

Equations (19.17) and (19.18) can be further simplified and combined with Equations (19.15) and (19.16), resulting in a model for correlating adsorption data, in the case of very low protein concentration. Then, a linear form of Equations (19.19) and (19.20) is obtained (note that we have dropped the subscripts i and p for protein in all cases except for the activity coefficient):

HIC:

$$\ln A = \ln\left(\frac{q}{C}\right) = \left[-\frac{\Delta G^0}{RT} + n \ln \frac{\Lambda}{C} + \ln \gamma_{P,W}^{\infty}\right] + \ln \tilde{\gamma}_{C_{P=0}} \Rightarrow \ln A = \ln A_0 + \ln \tilde{\gamma}_{C_{P=0}} \Rightarrow$$
$$\ln A = \ln A_0 + K_s C_s \tag{19.21}$$

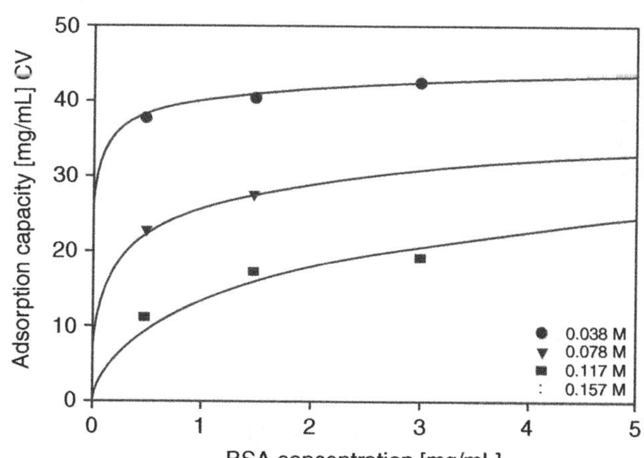

Figure 19.16 *IEC: experimental and correlated adsorption isotherm of BSA on Source 30Q media (a polymer) at pH = 8 and at four concentrations of sodium chloride. Reprinted with permission from Fluid Phase Equilibria, Applied thermodynamics: A new frontier for biotechnology by Jørgen M. Mollerup, 241, 1–2, 205–215 Copyright (2006) Elsevier*

where the asymmetric activity coefficient of protein is introduced:

$$\ln \tilde{\gamma}_P = \ln \gamma_P - \ln \gamma_{P,W}^{\infty}$$

Here $\gamma_{P,W}^{\infty}$ is the activity coefficient of the protein at infinite dilution in pure water (which depends on only the protein–water interaction).

IEC (activity coefficients ignored):

$$\ln A = \ln\left(\frac{q}{C}\right) = \left[-\frac{\Delta G_P}{RT}\right] + \nu\left[\frac{\Delta G_s}{RT} + \ln\frac{\Lambda}{C_s}\right] - \nu \ln C_s \Rightarrow$$
$$\ln A = A + \nu B - \nu \ln C_s \qquad (19.22)$$

Equations (19.21) and (19.22) are functionally very similar, despite the different underlying mechanisms in HIC and IEC.

According to Equations (19.21) and (19.22), the distribution ratio A (at low concentrations) is proportional to the salt concentration. This is indeed the case as shown in Figure 19.17 below for the distribution ratio of lysozyme in ammonium sulfate (HIC) at pH = 7 and various ligands. Similar results are obtained at pH = 6 and 8 as well. The slopes of these straight lines are dependent on the pH, while the intersection with the ordinate characterizes the HIC media used. Despite their similarities, there are some important differences, which are discussed later (Section 19.4.4). Equations (19.21) and (19.22) permit the interpretation of chromatographic

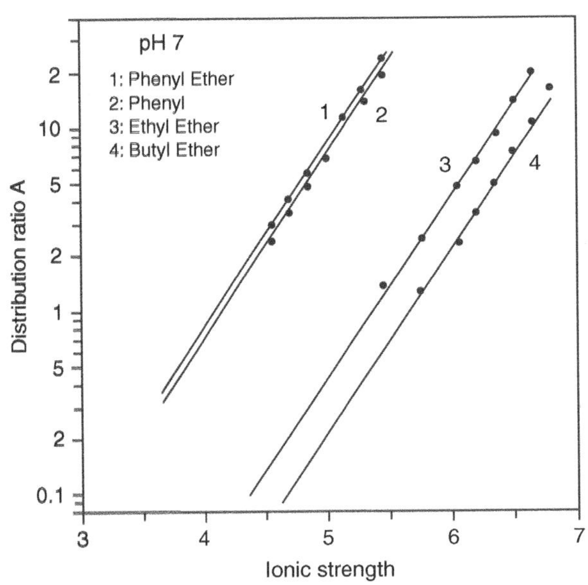

Figure 19.17 *HIC: semi-logarithmic plot of the experimental and calculated distribution ratio A of lysozyme on four HIC media (resins) against the molar salt ionic strength of ammonium sulfate at pH = 7. The ionic strength of ammonium sulfate is equal to three times the molarity. The straight lines with identical slopes are correlations using Equation (19.21). Reprinted with permission from Fluid Phase Equilibria, Applied thermodynamics: A new frontier for biotechnology by Jørgen M. Mollerup, 241, 1–2, 205–215 Copyright (2006) Elsevier*

data (for inter- and extrapolations), as well as estimations of the Gibbs free energy of adsorption and the activity coefficient of proteins in solution (see Section 19.4.4). However, the distribution factor A should be known first.

As mentioned, the parameter A is determined by retention volume experiments, specifically by measuring the isocratic elution volume at various salt concentrations at low protein concentrations using Equations (19.23) and (19.24), where the reduced retention volume is defined as:

$$\tilde{V}_R = \frac{V_R}{V_{column}}$$

HIC:

$$\begin{aligned}\ln(\tilde{V}_R - \tilde{V}_{NA}) &= \ln(1-\varepsilon)\varepsilon_P k_d + \ln A_i = \ln(1-\varepsilon)\varepsilon_P k_d + \ln A_{0,i} + \ln\tilde{\gamma}_i \Rightarrow \\ \ln(\tilde{V}_R - \tilde{V}_{NA}) &= \ln(1-\varepsilon)\varepsilon_P k_d + \ln A_0 + K_s C_s\end{aligned} \tag{19.23}$$

IEC (activity coefficients ignored):

$$\begin{aligned}\ln(\tilde{V}_R - \tilde{V}_{NA}) &= \ln(1-\varepsilon)\varepsilon_P k_d - \frac{\Delta G_P^0}{RT} + \nu\left(\frac{\Delta G_s^0}{RT} + \ln\frac{\Lambda}{Z_s}\right) - \nu\ln C_s \Rightarrow \\ \ln(\tilde{V}_R - \tilde{V}_{NA}) &= A_0 + \nu B_0 - \nu\ln C_s\end{aligned} \tag{19.24}$$

where:

$\tilde{V}_{NA}$ = reduced retention volume due to size exclusion (at high salt concentrations where the protein does not bind to ligands)

k_d = the exclusion factor (unity for common ions)

$\varepsilon, \varepsilon_P$ = the porosities (interstitial and particle)

$\ln\tilde{\gamma}_i = \ln\gamma_i - \ln\gamma_i^\infty$, the asymmetric activity coefficient

The parameters k_d and $\varepsilon, \varepsilon_P$ are determined from experiments under conditions where no adsorption takes place, e.g. buffer solutions without salt (HIC) or containing 0.5 1 M NaCl (IEC).[52]

It is thus clear that a plot of $\ln(\tilde{V}_R - \tilde{V}_{NA})$ against C_s (HIC) or $-\nu\ln C_s$ (IEC) is linear and then the values of the Gibbs free energy change of the salt and the protein can be determined. Linear plots in the case of IEC shown by Mollerup[52] verify that the activity coefficients are of minor importance. When the A parameter is determined, then q^{max} can be obtained from a few capacity measurements.

19.4.4 Discussion

Mollerup's work has shown that it is possible to model two important types of adsorption phenomena (HIC and IEC) based on different principles using simple and similar equations for the adsorption isotherms (see Figures 19.16 and 19.17). There is a minimum number of parameters, one pH-dependent and one protein/media-dependent parameter, but still there is a very good fit to the chromatographic data tested. It should be mentioned that the models have been validated over a limited data set and further validation is required. Two important aspects of the Mollerup model, its link with the Langmuir equation and the role of the activity coefficient are discussed next.

Beyond the well-known forms of the Langmuir equation

Mollerup's expressions can be used to derive a form of the multicomponent Langmuir equation which differs from the ones often presented in the literature. The Langmuir equation has been discussed in Chapter 18. In its classical form, written here with the symbols used in this section, it can be expressed as:

$$q_P = \frac{q_P^{\max} K C_p}{1 + K C_p} \tag{19.25}$$

or alternatively as:

$$\frac{q_P}{C_p} = K q_P^{\max}\left(1 - \frac{q_P}{q_P^{\max}}\right) \tag{19.26}$$

The latter form resembles the equations used for HIC and IEC and can indeed be derived from Equations (19.19) and (19.20) on setting all exponents and activity coefficients equal to unity. This resemblance can be used to derive new expressions for the multicomponent Langmuir equation. The classical form of the multicomponent Langmuir equation typically presented in the literature is:

$$q_{P_i} = \frac{q_{P_i}^{\max} K_i C_{P_i}}{1 + \sum_j K_j C_{P_j}} \tag{19.27}$$

However, based on the resemblance between the Langmuir and HIC/IEC equations for single protein adsorption, a more exact multicomponent Langmuir equation could be derived from Equations (19.17) and (19.18) again on setting all exponents and activity coefficients equal to unity:

$$\frac{q_{P_i}}{C_{P_i}} = K_i\left(\frac{\Lambda}{C}\right)\left(1 - \sum_j \frac{q_{P_j}}{q_{P_j}^{\max}}\right) \tag{19.28}$$

The role of the activity coefficients in the adsorption models

In the applications so far, Mollerup has used activity coefficients in the HIC but not in the IEC models. Mollerup[49,52] justifies the need for activity coefficients in HIC over IEC based on the salts which are used in the two cases. While sulfates and phosphates are used in HIC, chlorides are used in IEC. Thus, although the IEC model appears simpler (no activity coefficient used) than the HIC model, this simplicity may be apparent, reflecting the choice of salts used. This choice remains to be further validated by future experiments. In the case of HIC, the development of expressions of A as a function of salt concentration (Equation (19.21)) requires use of a model for the protein activity coefficient. Mollerup[49] used the van der Waals equation of state, ignoring the repulsive/entropic term, and derived the following expression for the protein activity coefficient (where 1 refers to water, 2 to protein, 3 to salt, and a_{ij} are the vdW EoS parameters):

$$\ln \tilde{\gamma}_P = 2\left(\frac{a_{12} - a_{22}}{RT}\right) C_P + 2\left(\frac{a_{12} - a_{32}}{RT}\right) C_s = K_P C_P + K_s C_s \tag{19.29}$$

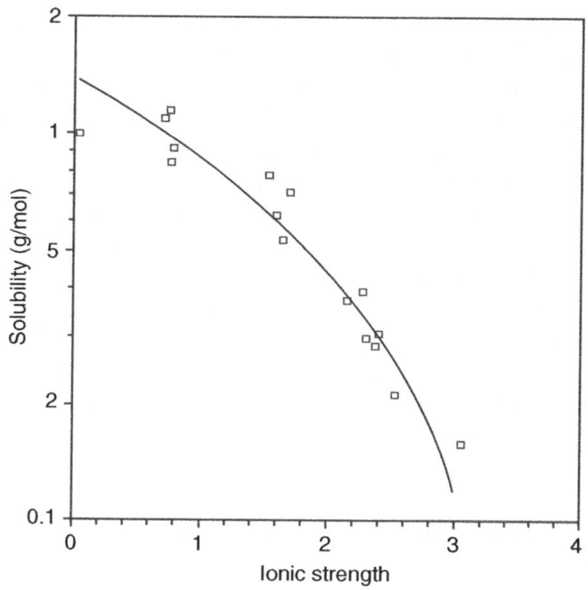

Figure 19.18 *Correlation of the solubility of lysozyme in ammonium sulfate at pH = 8. Reprinted with permission from Fluid Phase Equilibria, Applied thermodynamics: A new frontier for biotechnology by Jørgen M. Mollerup, 241, 1–2, 205–215 Copyright (2006) Elsevier*

It is thus clear that at low protein concentrations the protein activity coefficient is proportional to the salt concentration, as indicated in Equation (19.29). Moreover, this equation provides an insight on the significance of the K parameters as a function of the various interactions and it can be used to estimate the solubility of proteins. In the case of lysozyme at pH = 8, we have equilibria with crystal protein, and using standard thermodynamics and Equation (19.29), the solubility of lysozyme (S_P) can be calculated as:[49]

$$\ln S_P + K_P S_P = \left(\frac{\Delta\mu}{RT} + \ln M_P\right) + K_s C_s \tag{19.30}$$

The model has three parameters, K_P, K_s and $(\Delta\mu/RT) + \ln M_P$, which fit the existing data very well, as shown in Figure 19.18. From the values of the parameters determined, it can be concluded that the size of interactions can be ranked as: protein–water > protein–salt > protein–protein.

19.5 Semi-predictive models for protein systems

Chen and Mathias[3] presented an overview with recommended thermodynamic models in various process industries, e.g. petrochemicals, chemicals, polymers, etc. Under both the so-called 'primary' and 'secondary' choice models, they found none which could be suitable for pharmaceutical and biological applications. Since Chen and Mathias's work the situation has improved for pharmaceuticals and many options are now available (discussed in Section 19.2). Biotechnology and bioprocesses represent an area where applied thermodynamics has so far had little impact. This is largely because most biomolecules (e.g. proteins, enzymes, DNA, etc.) are polar and often charged polymers. Significant efforts have been made to describe the units present in these macromolecules, e.g. sugar molecules and especially the amino acids (see Section 19.3), but proteins contain often over 100 amino acids. Even one of the most well-studied enzymes, lysozyme, is composed of 129 amino acids. An additional complexity is related to the fact that proteins have a multiplicity of structures beyond the primary ones (which is defined by the sequence of amino acids). Secondary (helix and sheet) and tertiary structures owe their existence to bonds much weaker than the primary structure bonds, i.e. they are due to the vdW, hydrogen bonding, hydrophobic bonds. Many properties of proteins and enzymes depend on these secondary/tertiary structures as well, which are much affected by temperature, pH and other parameters. It may take several years to develop practical engineering models for handling polyelectrolytes and thus to predict the phase behavior of proteins/enzymes and the like as a function of all the parameters which are of significance in applications, e.g. temperature, pH, ionic strength, etc. In the meantime, simple engineering tools have been developed for predicting protein solubility or partition coefficients in specific applications. These models are semi-predictive and based heavily on the available experimental data, but they can be used for extrapolations and understanding similar systems, at least to some extent. Three of the successful approaches are discussed in this section.

19.5.1 The osmotic second virial coefficient and protein solubility: a tool for modeling protein precipitation

The osmotic second virial coefficient B has long been considered a valuable tool in studies of solutions containing biomolecules.[54,55] It is defined via the osmotic virial equation:

$$\frac{\Pi}{C_2} = RT\left[\frac{1}{M_2} + BC_2 + CC_2^2 + \cdots\right] \tag{19.31}$$

where Π is the osmotic pressure, C_2 the solute concentration (g/l), M_2 the molecular weight (g/mol) of the solute and B and C, respectively, the second and third osmotic virial coefficients. Like the 'ordinary' virial coefficient (see Section 2.4.4, Equations (2.21) and (2.22)), the osmotic second virial coefficient reflects the intermolecular forces between molecules.

Equation (19.31) can be used, when truncated at its second term, to estimate the molecular weight of macromolecules, e.g. polymers and proteins, from osmotic pressure measurements at low concentrations. In addition, the osmotic second virial coefficient can be estimated from Equation (19.31). B values can be alternatively obtained from low-angle laser light scattering (LALLS) measurements, and this method sometimes provides more accurate values of the virial coefficients than osmotic pressure measurements. Using Π–C_2 data, Prausnitz *et al.*[18] presented molecular weights and osmotic second virial coefficient values for three proteins: a-chymotrypsin, lysozyme and ovalbumin. While for the last two proteins B values are positive, the osmotic second virial coefficient of the first protein is negative, which indicates that there is an attraction in dilute solutions between a-chymotrypsin molecules.

Osmotic second virial coefficients are used in biothermodynamics in different respects, e.g. understanding protein–protein interactions and in models for aqueous two-phase systems as discussed later. An alternative, somewhat more direct, use of osmotic second virial coefficients has been presented by Ruppert *et al.*[55] who showed that there is a direct relationship between B and the protein solubility (C_P). Under certain simplifying assumptions, this relationship can be written as:

$$B = \frac{1}{2M_P C_P}\left[A_c - (1-K)\ln\left(\frac{V_w}{M_P}C_P\right)\right] - \frac{\bar{V}_P^\infty}{M_P^2} - \frac{dn/dC_P}{n_0 M_P} \tag{19.32}$$

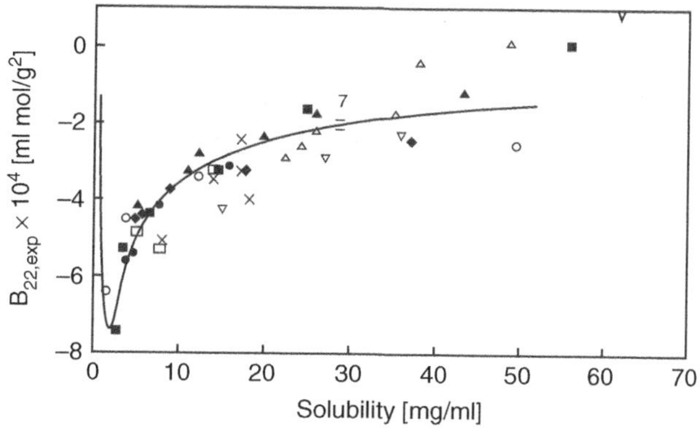

Figure 19.19 *Relationship between the osmotic second virial coefficient (B_{22}; denoted as B in the text and Equation (19.32)) and the solubility of lysozyme in water for different temperatures, salts, salt concentrations and pH. The line is the prediction from Equation (19.32). Reprinted with permission from Biotechnol. Prog., Correlation between the Osmotic Second Virial Coefficient and the Solubility of Proteins by S. Ruppert, S. I. Sandler and A. M. Lenhoff, 17, 1, 182–187 Copyright (2001) Wiley-VCH*

where: B is expressed in ml mol/g^2 and the protein solubility C_P in mg/ml; V_w is the solvent molar volume (ml/mol); the properties with subscript P refer to protein; M is molecular weight (g/mol); $\bar{V}_P^{\infty}$ is the partial molar volume (ml/mol); n_0 is the refractive index and dn/dC_P the refractive index increment (ml/mg). A_c and K are the only two parameters that must be fitted to experimental data.

Equation (19.32) has been derived from thermodynamic principles independent of any specific thermodynamic model or statistical mechanics theory. Using only two parameters (A_c and K) fitted to all available data, Equation (19.32) can capture well the variation of the osmotic second virial coefficient with the protein solubility, as shown in Figure 19.19 in the case of lysozyme. Similarly good results are obtained for ovalbumin. The agreement is excellent for protein solubilities up to 30 ml/ml, but at higher values the model underestimates the experimental data. The simple model of Equation (19.32) does not account for changes in the nature of salt and the pH. Nevertheless, it illustrates the feasibility of using the osmotic second virial coefficient for the modeling of protein precipitation. It may be possible for example[54] to define ranges of the osmotic second virial coefficients of protein molecules where they may precipitate as crystals from aqueous solutions.

19.5.2 Partition coefficients in protein–micelle systems

Reverse micelles can be used as mini-reactors for enzyme-catalyzed reactions (Figure 19.20). These reactions require an aqueous environment but often the reactants are only sparingly soluble in water. It has therefore been suggested to form reverse micelles in an organic solvent. The reverse micelle contains the enzyme in a tiny pool of water. Reactants dissolved in the organic solvent diffuse into the reverse micelle and products diffuse out. For the design of such reactors, it is necessary to know the distribution coefficient K for the enzyme between the reverse micelle and a continuous aqueous phase.[56,57] This coefficient depends on numerous variables, especially the electric charge of the enzyme, which depends on pH and the surfactant concentration.

Woll and Hatton[57] have developed a simple model assuming that protein solubilization can be interpreted as the interaction between a protein molecule P and n empty micelles M to form a protein–micelle complex, PM. Using this 'pseudo-chemical equilibrium' principle and a number of simplifying assumptions, especially that

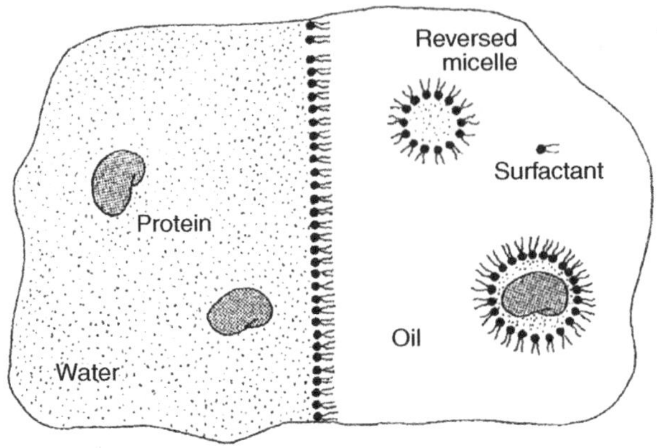

Figure 19.20 Representation of the protein partitioning in reversed micellar extraction processes. Reprinted with permission from Bioprocess and Biosystems Engineering, A simple phenomenological thermodynamic model for protein partitioning in reversed micellar systems by J. M. Woll, 4, 5, 193–199 Copyright (1989) Springer Science + Business Media

the aggregation number is independent of surfactant concentration and that the protein charge and the size of the complex n are linear functions of the pH and pI difference (pI = isoelectric point), they arrived at the following expression for the partition coefficient K against the surfactant concentration S and pH:

$$\ln K = A + B(\text{pH}) + [C + D(\text{pH})]\ln S \tag{19.33}$$

According to the theory, the four parameters A–D are expressed as:

$$\begin{aligned} A &= -[\Delta G^0 - aF\Delta\psi(\text{pI})]/RT - [n_0 - \varepsilon a(\text{pI})]\ln N_{agg} \\ B &= -(aF\Delta\psi)/RT - \varepsilon a \ln N_{agg} \\ C &= n_0 - \varepsilon a(pI) \\ D &= \varepsilon a \end{aligned} \tag{19.34}$$

where: F is Faraday's constant; pI is the protein isoelectric point; N is the aggregation number of the micelle; a is the proportionality factor relating protein charge to pH (from protein titration data); ε is the change in micelle and bulk solution; n_0 is the number of empty micelles required at pH = pI; $\Delta\psi$ is the electrostatic potential difference between the micelle and bulk solution; and ΔG^0 is the standard-state Gibbs free energy for protein transfer in the absence of electric charge effects.

A key assumption in the model development was that the surfactant associated with the protein–micelle complex was small relative to the total surfactant present in the system. This is confirmed by the fact that K is shown to be rather independent of the total protein concentration.

Figure 19.21 shows two K–pH plots for the two proteins studied by Woll and Hatton.[57] There is very good agreement between the experimental data (points) and the correlations (lines). The model is not predictive; it contains four parameters which are fitted to experimental data. But the parameters *do* have a physical significance and this is verified via numerical calculations based on Equations (19.34). When using the fitted A–D parameters for the two proteins and experimental values for some of the properties (pI, aggregation

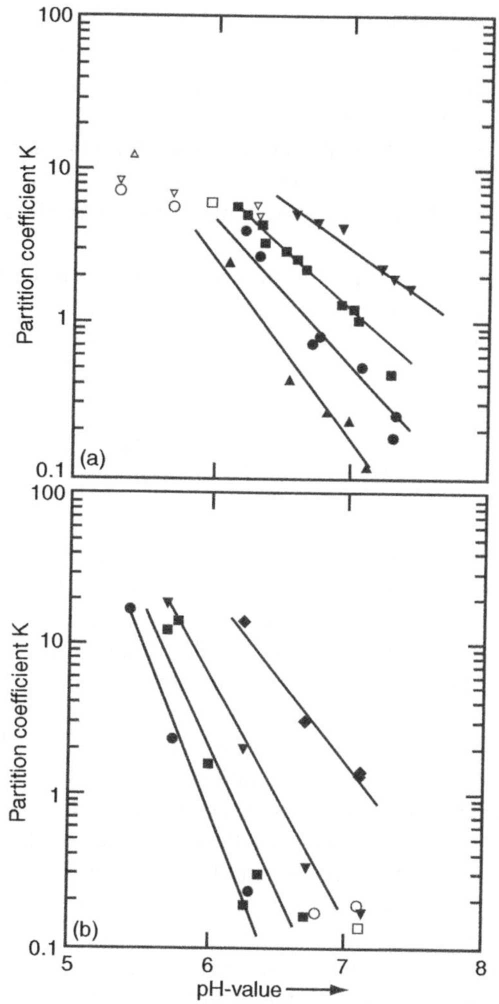

Figure 19.21 *Partition coefficient calculated with Equation (19.33) for Ribonuclease (upper figure) and Concanavalin (lower figure) in reversed micelles against pH at various salt concentrations. The micelles are formed by the ionic surfactants Aerosol–OT in iso-octane. Reprinted with permission from Bioprocess and Biosystems Engineering, A simple phenomenological thermodynamic model for protein partitioning in reversed micellar systems by J. M. Woll, 4, 5, 193–199 Copyright (1989) Springer Science + Business Media*

number), the remaining parameters have physically reasonable values, e.g. a potential difference around 70 mV and n_0 around 2–3. The values are in the ranges anticipated and thus the model can be used for interpreting data and extrapolation to other systems where only few or no data exist.

19.5.3 Partition coefficients in aqueous two-phase systems for protein separation

Liquid–liquid extraction using aqueous two-phase systems (ATPS) formed with two water-soluble polymers can be used for the selective separation of proteins. Typical polymers for this are PEG and

Dextran and the separation can be further facilitated with the use of salts (also for pH control). Since Albertsson[58] introduced ATPS in 1956, they have received great interest, especially in academia, and their applicability in separating a variety of biomolecules has been demonstrated. They have not yet received great acceptance for industrial separations, but this trend may change when improved understanding is obtained on how the protein solubility depends on numerous factors such as temperature, pH, surface charge of proteins, ionic strength, electrolyte type and hydrophobicity. For a mixture of proteins, their different solubilities, depending on the factors above, may result in different partitioning in ATPS thus facilitating separation.

While a detailed precise model for protein solubility may seem out of reach at this time, a useful model for predicting the protein partition coefficient (ratio of protein concentration in the two phases) has been developed by Prausnitz and co-workers.[59–62] In the case of dilute protein systems, the partition coefficient can be given by the equation:

$$\ln K_P = \ln\left(\frac{\gamma_P^\beta}{\gamma_P^\alpha}\right) + \left[\frac{Z_P F \Delta\varphi}{RT}\right] \tag{19.35}$$

where: K_P is the partition coefficient of the protein (concentration of protein in phase a over concentration of protein in phase β); $\ln(\gamma_P^\beta/\gamma_P^\alpha)$ is the ratio of chemical activity coefficients, which expresses the partition coefficient of the protein when the potential generated from the salt is zero or when the protein charge is zero; R and F are the ideal gas and Faraday constants; T is the absolute temperature; Z_P is the protein charge; and $\Delta\varphi$ is the electrical potential between the two liquid phases generated due to the different affinity (partition) of the salts for the two phases.

For many ATPS, Equation (19.35) captures the basic effects of added electrolytes (at low concentrations) on protein partitioning. This is possibly due to the special importance in Equation (19.35) of the $\Delta\varphi$ factor, which is the electric potential difference between the two phases caused by the different salt concentrations in the two phases. Clearly, the larger the affinity difference for the two phases, the larger the potential difference that will be generated and the larger the partition coefficient of the protein. The partitioning of biomolecules will thus depend significantly on $\Delta\varphi$, associated factors such as pH and the charge of proteins, but also on the properties of phase-forming polymers.

The two terms of Equation (19.35) need to be estimated for quantitative calculations. For very dilute protein solutions, the ratio of the activity coefficients of proteins can be estimated using the osmotic virial expansion:[59]

$$\ln\left(\frac{\gamma_P^\beta}{\gamma_P^\alpha}\right) = a_{2P}\left(m_2^\beta - m_2^\alpha\right) + a_{3P}\left(m_3^\beta - m_3^\alpha\right) \tag{19.36}$$

where the subscripts 2, 3 indicate the two water-soluble polymers (often PEG and Dextran), m are the molalities of solutes in the water–salt pseudo-solvent (if salts are also present, as often is the case), and a_{ij} are constants characterizing the interaction between polymer molecules in aqueous solvents.

The interaction parameters are related to the osmotic second virial coefficients B_{ij} and the molar masses (M) via:

$$a_{ij} = \frac{2M_i M_j B_{ij}}{1000^2} \tag{19.37}$$

The osmotic second virial coefficients are typically obtained from low-angle laser light scattering (LALLS) or osmotic measurements. The electric potential difference can be estimated using the quasi-electrostatic potential theory and depends on the charges of anions and cations of the salt:[62]

$$\Delta\varphi = \frac{RT}{[Z_+ - Z_-]F} \ln\left[\frac{(\gamma_-^\beta/\gamma_-^\alpha)}{(\gamma_+^\beta/\gamma_+^\alpha)^{Z_+/Z_-}}\right] \tag{19.38}$$

where Z_+, Z_- are the charges of the salt.

Equation (19.38) can be simplified using mean ionic activity coefficients and molalities if the electrolyte type is known. For example, for a 1 : 1 electrolyte (e.g. NaCl), Equation (19.38) can be simplified to:

$$\Delta\varphi = \frac{RT}{F} \ln\left[\frac{\gamma_\pm^\beta}{\gamma_\pm^\alpha}\right] = \frac{RT}{F} \ln\left[\frac{m_{salt}^\alpha}{m_{salt}^\beta}\right] = \frac{RT}{F} \ln K_s \tag{19.39}$$

where m is the molality of the salt, K_s is the salt partition coefficient and $\gamma_\pm$ is the mean ionic activity coefficient of the salt.

Some results and discussion

The simplified model illustrated above ignores essentially all 'electrolyte' effects except for the electrolyte potential difference generated by the salts. Even though the $\Delta\varphi$ term is small (about 5–10 millivolts), it can have a significant effect on the protein partitioning, especially as it is multiplied in Equation (19.35) by the protein charge. Thus, the term may in some cases overshadow the activity coefficient ratio. Still, improvements in the activity coefficient models may be of importance and better models than the osmotic virial equation could result in better results. Such models have been proposed and rely upon adding electrolyte terms, e.g. via the Debye–Hückel[61] or the MSA[60] theories (Chapter 15). The latter is possibly the preferred approach, as shown by Haynes *et al.*,[60] and one typical result is shown in Figure 19.22. Good agreement between theory

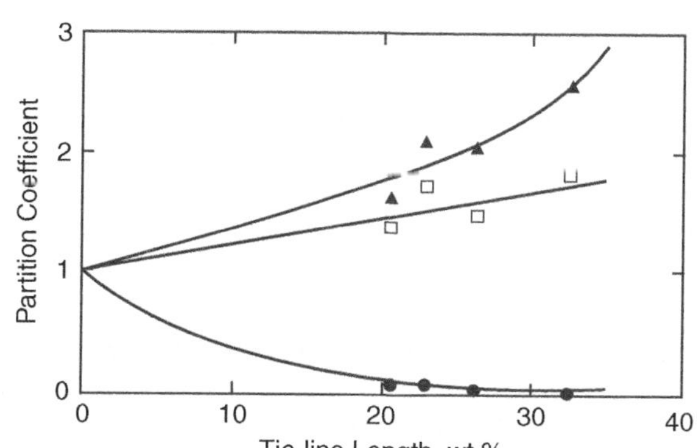

Figure 19.22 *Predicted (lines) with the Haynes et al.[60] model and experimental (points) partition coefficients for a dilute protein mixture in ATPS containing PEG-3350, Dextran T-70 and 50 mM KCl at pH = 7.3 and 25 °C. The circles are albumin (Z = −8), squares are chymotrypsin (Z = 2) and triangles are lysozyme (Z = 7). The tie-line length is zero when the two phases are identical. Reprinted with permission from AICHE, Application of Integral-Equation Theory to Aqueous Two-Phase Partitioning Systems by C. A. Haynes, F. J. Benitez, H. W. Blanch and J. M. Prausnitz, 39, 9, 1539–1557 Copyright (1993) John Wiley & Sons, Inc*

and experiment is observed. Moreover, the analysis of Haynes *et al.*[62] illustrates the great importance of salts in protein partitioning, which can significantly assist in designing ATPS processes. The salt partitioning is a strong function of anion size and charge; partitioning is strongest for salts with small, highly charged anions, a trend which is in agreement with the Hofmeister series.

19.6 Concluding remarks

Biomolecules are complex molecules ranging from medium-molecular-weight compounds such as pharmaceuticals and amino acids up to polymers such as proteins and enzymes. Even the simplest biomolecules are multifunctional molecules and their thermodynamics is often complex as a result of a variety of interactions present. We have presented in this chapter models which can handle different types of biomolecules. We can conclude that:

- There are a few successful models for pharmaceuticals, typically extensions of the NRTL and the lattice–fluid equation of state extended to account for hydrogen bonding. Even simpler models such as variations of the Flory–Huggins/Hansen solubility parameters or extensions of UNIFAC may provide qualitatively successful results in some cases.
- Rigorous thermodynamic models have been developed for amino acids (and to a lesser degree also for peptides). Essentially none of them accounts for the strong zwitterion–ion interactions which are expected to be present due to the dipolar ion nature of amino acids. Nevertheless, correlation results are successful with variations of the Wilson or NRTL equations, while the group contribution concept is less successful e.g. for amino acids. Phase behavior of aqueous mixtures of amino acids (without salts) can often be well correlated even by models which do not explicitly account for electrostatic interactions. This is because amino acids exist as zwitterions around their isoelectric point. Recently, PC–SAFT has been extended to amino acids and peptides and this approach is a promising alternative compared to the more empirical local composition models.
- Simple successful models have been developed for describing the adsorption of proteins related to two key separation methods, the ion-exchange and the hydrophobic interaction chromatography. The adsorption theories developed differ from the 'classical' Langmuir approach and can describe adsorption data. Moreover, the solubility of proteins can be interpreted with these models.
- Rigorous thermodynamic models have not been developed for routine applications in protein-containing mixtures. However, some simple models, especially for estimating partition coefficients, have been developed. In many models for protein mixtures, osmotic second virial coefficients play a key role and can also be linked to protein solubility. These semi-predictive models have had some success in applications related to aqueous two-phase systems for protein separation.

Appendix 19.A The NRTL–SAC activity coefficient model

The NRTL–SAC (segment activity coefficient model) is an activity coefficient model which builds on the polymer NRTL by Chen.[28] Both the polymer NRTL and NRTL–SAC have significant differences from the 'classical' NRTL, described in Chapter 5, as will be apparent from the equations below. The most important differences are:

- the introduction of a combinatorial term;
- the use of temperature-independent parameters (no in-built temperature dependency);
- the use of a segment approach and interactions.

In NRTL–SAC, the activity coefficient is given, similar to UNIQUAC or UNIFAC models, as a sum of a combinatorial (C) and a residual (R) term:

$$\ln \gamma_I = \ln \gamma_I^C + \ln \gamma_I^R \tag{19.40}$$

I (and J) denote components, while the symbols i, j, k, m, m' below are the segment-based species indices, i.e. X, Y^-, Y^+ and Z. The combinatorial part is calculated from the Flory–Huggins term:

$$\ln \gamma_I^C = \ln\left(\frac{\phi_I}{x_I}\right) + 1 - r_I \sum_J \frac{\phi_J}{r_J} \tag{19.41}$$

$$r_I = \sum_i r_{i,I}$$
$$\phi_I = \frac{x_I r_I}{\sum_J x_J r_J} \tag{19.42}$$

where r_I, φ_I are the total segment number and segment mole fraction of component I, respectively.

The residual part is essentially identical to the local composition interaction from the polymer NRTL model:

$$\ln \gamma_I^R = \ln \gamma_I^{lc} = \sum_m r_{m,I}\left[\ln \Gamma_m^{lc} - \ln \Gamma_m^{lc,I}\right] \tag{19.43}$$

The segment activity coefficients (of species m in a mixture, Γ_m^{lc}, and in the pure component I, $\Gamma_m^{lc,I}$) are computed as:

$$\ln \Gamma_m^{lc} = \frac{\sum_j x_j G_{jm}\tau_{jm}}{\sum_k x_k G_{km}} + \sum_{m'} \frac{x_{m'} G_{mm'}}{\sum_k x_k G_{km'}}\left(\tau_{mm'} - \frac{\sum_j x_j G_{jm'}\tau_{jm'}}{\sum_k x_k G_{km'}}\right) \tag{19.44}$$

$$\ln \Gamma_m^{lc,I} = \frac{\sum_j x_{j,I} G_{jm}\tau_{jm}}{\sum_k x_{k,I} G_{km}} + \sum_{m'} \frac{x_{m',I} G_{mm'}}{\sum_k x_{k,I} G_{km'}}\left(\tau_{mm'} - \frac{\sum_j x_{j,I} G_{jm'}\tau_{jm'}}{\sum_k x_{k,I} G_{km'}}\right) \tag{19.45}$$

$$x_j = \frac{\sum_J x_J r_{j,J}}{\sum_I \sum_i x_I r_{i,I}} \tag{19.46}$$

$$x_{j,I} = \frac{r_{j,I}}{\sum_i r_{i,I}} \tag{19.47}$$

$$G_{jm} = \exp(-a\tau_{jm}) \tag{19.48}$$

Here: x_j is the segment-based mole fraction of segment species j; x_J is the mole fraction of component J; $r_{m,I}$ is the number of segment species m in component i; G_{jm}, τ_{jm} and a are the well-known NRTL parameters and the non-randomness factor.

For monosegment solvent components, the residual term of the activity coefficient reduces to the classical NRTL model (except for the temperature dependency of the interaction parameters). The NRTL binary parameters (τ_{12}, τ_{21}, a) for the conceptual segments in NRTL–SAC for all interaction possibilities (X/Y^-, X/Z, Y^-/Z, Y^+/Z and X/Y^+) have been provided by Chen and Song[23] and have not been changed in subsequent publications. The X, Y^-, Y^+ and Z for 62 solvents were presented first by Chen and Song[23] and then revised by Chen and Crafts.[24] The X, Y^-, Y^+ and Z for various pharmaceuticals have been presented by Chen and co-workers[23,24] and are always estimated from SLE pharmaceutical data in at least five–six different solvents.

Appendix 19.B The NRHB equation of state

The NRHB theory is a compressible lattice model, where holes are used to account for density variation as a result of temperature and pressure changes. NRHB accounts explicitly for the non-random distribution of molecular sites while Veytsman's statistics are used to calculate the contribution of hydrogen bonding to the thermodynamics of the system.[63] Thus, the model is suitable for property calculations of highly non-ideal fluids. According to this, N molecules are assumed to be arranged on a quasi-lattice of N_r sites, N_0 of which are empty, with a lattice coordination number, z. Each molecule of type i in the system occupies r_i sites of the quasi-lattice. It is characterized by three scaling parameters and one geometric, or surface-to-volume-ratio factor, s. The first two scaling parameters, ε_h^* and ε_s^*, are used for the calculation of the mean interaction energy per molecular segment, ε^*, according to the following equation:

$$\varepsilon^* = \varepsilon_h^* + (T-298.15)\varepsilon_s^* \tag{19.49}$$

while the third scaling parameter, $v_{sp,0}^*$, is used for calculation of the close-packed density, $\rho^* = 1/v_{sp}^*$, as described by the following equation:

$$v_{sp}^* = v_{sp,0}^* + (T-298.15)v_{sp,1}^* \tag{19.50}$$

The parameter $v_{sp,1}^*$ in Equation (19.50) is treated as a constant for a given homologous series.[64,65] Furthermore, the hard-core volume per segment, v^*, is constant and equal to 9.75 cm^3/mol for all fluids. The following relation holds:

$$r = \frac{MWv_{sp}^*}{v^*} \tag{19.51}$$

Finally, the shape factor is defined as the ratio of molecular surface to molecular volume, $s = q/r$, and is calculated from the UNIFAC group contribution method.[63] The equation of state for a fluid mixture assumes the following form:[64]

$$\tilde{P} + \tilde{T}\left[\ln(1-\tilde{\rho}) - \tilde{\rho}\left(\sum_i \phi_i \frac{l_i}{r_i} - v_{hb}\right) - \frac{z}{2}\ln\left(1-\tilde{\rho}+\frac{q}{r}\tilde{\rho}\right) + \frac{z}{2}\ln\Gamma_{00}\right] = 0 \tag{19.52}$$

while the chemical potential for the component i is given by:

$$\frac{\mu_i}{RT} = \ln\frac{\varphi_i}{\omega_i r_i} - r_i\sum_j \frac{\varphi_j l_j}{r_j} + \ln\tilde{\rho} + r_i(\tilde{v}-1)\ln(1-\tilde{\rho}) - \frac{z}{2}r_i\left[\tilde{v}-1+\frac{q_i}{r_i}\right]$$
$$\ln\left[1-\tilde{\rho}+\frac{q}{r}\tilde{\rho}\right] + \frac{zq_i}{2}\left[\ln\Gamma_{ii} + \frac{r_i}{q_i}(\tilde{v}-1)\ln\Gamma_{00}\right] + r_i\frac{\tilde{P}\tilde{v}}{\tilde{T}} - \frac{q_i}{\tilde{T}_i} + \frac{\mu_{i,hb}}{RT} \tag{19.53}$$

where φ_i is the site fraction of component i, while l_i and ω_i are characteristic quantities for each fluid. Parameters Γ_{00} and Γ_{ii} are non-random factors for the distribution of empty sites around an empty site and of molecular segments of component i around a molecular segment of component i, respectively. Here $\mu_{i,hb}/RT$ is the hydrogen bonding contribution to the chemical potential of component i. Finally, parameters $\tilde{T} = T/T^*$, $\tilde{P} = P/P^*$ and $\tilde{v}\ (= 1/\tilde{\rho} = \rho^*/\rho)$ are the reduced temperature, pressure and specific volume respectively. The characteristic temperature, T^*, and pressure, P^*, are related to the mean intersegmental energy by:

$$\varepsilon^* = RT^* = P^* v^* \tag{19.54}$$

Detailed expressions for the calculation of all these parameters can be found elsewhere.[64] The expression for the chemical potential of a pure component can be obtained from Equation (19.53) by setting $\varphi_i = \theta_i = 1$ and the number of components in the summations equal to one.

19.B.1 Modeling of hydrogen bonding fluids

NRHB can model properties of hydrogen bonding fluids of any number of donor and acceptor groups. In the following equations, m is the type of proton donors and n the type of proton acceptors that exist in the mixture, while d_α^k is the number of donor groups of type α in each molecule of type k and a_β^k the number of acceptor groups of type β in each molecule of type k. $N_{\alpha\beta}^{hb}$ is the total number of hydrogen bonds between a donor of type α and an acceptor of type β in the system. In Equation (19.55) the parameter ν_{hb} is the average number of hydrogen bonds per molecular segment in the system and is given by:

$$\nu_{hb} = \sum_\alpha^m \sum_\beta^n \nu_{\alpha\beta} = \sum_\alpha^m \sum_\beta^n \frac{N_{\alpha\beta}^{hb}}{rN} \tag{19.55}$$

The hydrogen bonding contribution in the chemical potential of component i is given by the expression:

$$\frac{\mu_{i,hb}}{RT} = r_i\nu_{hb} - \sum_{\alpha=1}^m d_\alpha^i \ln\frac{\nu_d^\alpha}{\nu_{\alpha 0}} - \sum_{\beta=1}^n a_\beta^i \ln\frac{\nu_a^\beta}{\nu_{0\beta}} \tag{19.56}$$

where the number of donors of type α per molecular segment, ν_d^α, is:

$$\nu_d^\alpha = \frac{N_d^\alpha}{rN} = \frac{\sum_k d_\alpha^k N_k}{rN} \tag{19.57}$$

and the number of acceptors of type β per molecular segment, ν_a^β, is:

$$\nu_a^\beta = \frac{N_a^\beta}{rN} = \frac{\sum_k a_\beta^k N_k}{rN} \tag{19.58}$$

while the number of non-bonded donors of type a per molecular segment, ν_{a0}, is:

$$\nu_{\alpha 0} = \nu_d^\alpha - \sum_{\beta=1}^{n} \nu_{\alpha\beta} \tag{19.59}$$

Similarly the number of unbonded acceptors of type β per molecular segment, $\nu_{\beta 0}$, is:

$$\nu_{0\beta} = \nu_a^\beta - \sum_{\alpha=1}^{m} \nu_{\alpha\beta} \tag{19.60}$$

According to the model, the $\nu_{\alpha\beta}$ satisfy the minimization conditions:

$$\frac{\nu_{\alpha\beta}}{\nu_{\alpha 0}\nu_{0\beta}} = \tilde{\rho}\exp\left(\frac{-G_{\alpha\beta}^{hb}}{RT}\right) \quad \text{for all}(\alpha, \beta) \tag{19.61}$$

In Equation (19.61), $G_{\alpha\beta}^{hb}$ is the free enthalpy of formation of the hydrogen bond of type α–β and is given in terms of the energy (E), volume (V) and entropy (S) of hydrogen bond formation by the equation:

$$G_{\alpha\beta}^{hb} = E_{\alpha\beta}^{hb} + PV_{\alpha\beta}^{hb} - TS_{\alpha\beta}^{hb} \tag{19.62}$$

The formalism is general and sufficient for solving phase equilibrium problems in systems of hydrogen bonding fluids of any number of donor and acceptor groups. Based on the above, for associating fluids, NRHB has three more parameters, namely the energy, $E_{\alpha\beta}^{hb}$, the volume, $V_{\alpha\beta}^{hb}$, and the entropy change, $S_{\alpha\beta}^{hb}$, for the formation of hydrogen bonds between proton donors of type α and proton acceptors of type β in different molecules. However, usually the volume change for the formation of a hydrogen bond, $V_{\alpha\beta}^{hb}$, is set equal to zero, so the number of hydrogen bonding parameters is reduced to two without compromising the performance of the model.[64] In this version of the NRHB model polar and dispersive interactions are not treated separately and are characterized as physical interactions, in contrast to the hydrogen bonding or chemical interactions. Subsequently, it is not possible to estimate the polar or the dispersive partial solubility parameters. However, the partial hydrogen bonding and the total solubility parameter can be calculated directly. The corresponding equation for the cohesive energy is:

$$E_{coh} = E_{ph} + E_{hb} \tag{19.63}$$

where E_{ph} is the potential energy due to physical (dispersive and polar interactions) and E_{hb} is the contribution due to hydrogen bonding. For pure fluids, these quantities are calculated from the following equations:

$$E_{ph} = \sum_i N_{ii}\varepsilon_{ii} = \Gamma_{11} q N \Theta_r \varepsilon^* \quad (19.64)$$

$$E_{hb} = -\sum_a \sum_\beta N_{a\beta}^{hb} E_{a\beta}^{hb} = -rN \sum_a \sum_\beta \nu_{a\beta} E_{a\beta}^{hb} \quad (19.65)$$

where: N_{ii} is the number of all physical intersegmental interactions; ε^* denotes the average intersegmental interaction energy; $N_{\alpha\beta}^{hb}$ is the number of hydrogen bonds of type α–β in the system, while $\nu_{\alpha\beta}$ is the number of hydrogen bonds of type α–β per molecular segment; and $E_{\alpha\beta}^{hb}$ is the energy change on the formation of one hydrogen bond of the same type. The volume of the system is given by:

$$V = rN\tilde{v}v^* + \sum_a \sum_\beta N_{a\beta}^{hb} V_{a\beta}^{hb} \quad (19.66)$$

As mentioned earlier, usually the volume change for the formation of a hydrogen bond, V_{ab}^{hb}, is set equal to zero. With the above definitions the total and partial hydrogen bonding solubility parameters are:

$$\delta = \sqrt{\frac{E_{coh}}{V}} = \sqrt{\frac{\Gamma_{11} q \Theta_r \varepsilon^* - r \sum_a \sum_\beta \nu_{a\beta} E_{a\beta}^{hb}}{r\tilde{v}v^*}} \quad (19.67)$$

and:

$$\delta_{hb} = \sqrt{\frac{E_{hb}}{V}} = \sqrt{\frac{-r \sum_a \sum_\beta \nu_{a\beta} E_{a\beta}^{hb}}{r\tilde{v}v^*}} \quad (19.68)$$

19.B.2 Other applications of NRHB

A special recent development of NRHB for pharmaceuticals has been presented in this chapter. However, NRHB is a general model, an equation of state belonging to the family of lattice fluid theories which accounts, moreover, explicitly for hydrogen bonding. It can – indeed has – been applied to a variety of mixtures including aqueous systems and other associating fluids as well as polymers. More specifically:

- Panayiotou *et al.*[63] have presented NRHB parameters for numerous fluids including 17 polymers. They have shown that NRHB can describe a variety of properties of pure fluids (vapor pressures, densities also at supercritical state, polymer densities over extensive pressure–temperature ranges as well as surface tensions of low-molecular-weight liquids and polymers as a function of temperature).
- Stefanis *et al.*[66] have developed a group contribution method for estimating the NRHB parameters (for non-hydrogen bonding compounds) using both first- and second-order groups in a way similar to that described in Section 8.5 for the sPC–SAFT EoS. A group contribution method for estimating the influence parameter needed in surface tension calculations has been also presented.

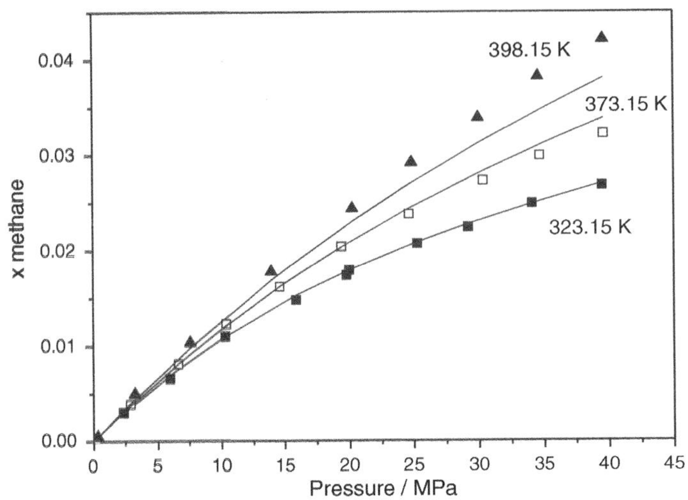

Figure 19.23 *CH_4–TEG VLE. Experimental data (points) and NRHB calculations with $k_{ij} = 0.05482$*

- Panayiotou *et al.*[64] presented VLE calculations with NRHB for a variety of mixtures containing CO_2, hydrocarbons and alcohols. A few polymer–solvent systems were considered as well and the model was shown to correlate successfully and even predict the solubility of methane in water.
- Missopolinou *et al.*[67] illustrated how intramolecular association can be incorporated in NRHB and in this way satisfactorily predicted excess enthalpies in four 2-ethoxyethanol–alkane mixtures in the 298–318 K range.

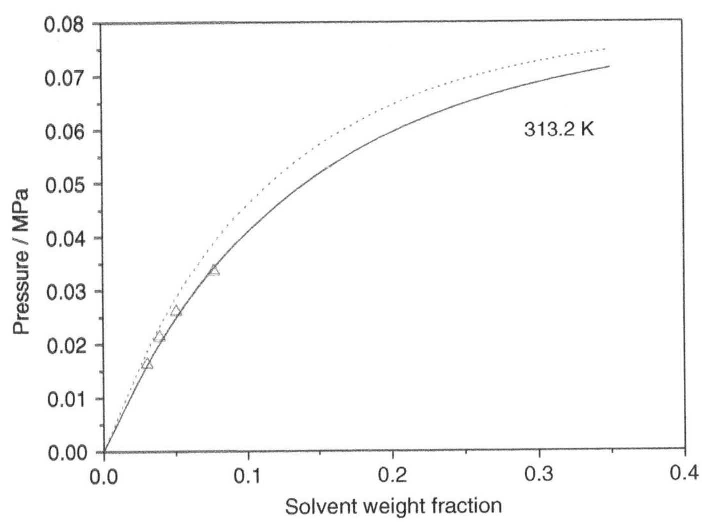

Figure 19.24 *1-Propylamine–PVAC VLE. Experimental data (points), NRHB predictions (dotted line) and correlations (solid line) with $k_{ij} = -0.006919$. The cross-association between the amine (NH_2) and acetate group (COO) is modeled using Equations (19.3). Reprinted with permission from Fluid Phase Equilibria, Modeling the vapor–liquid equilibria of polymer–solvent mixtures: Systems with complex hydrogen bonding behavior by Ioannis Tsivintzelis and Georgios M. Kontogeorgis, 280, 1–2, 100–109 Copyright (2009) Elsevier*

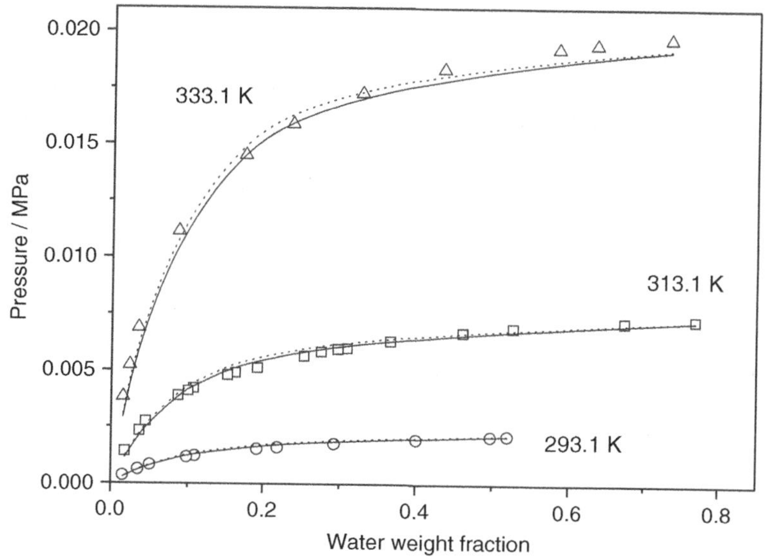

Figure 19.25 VLE of binary PEG ($M_w = 600$ g/mol) and water systems. Experimental data (points), NRHB predictions ($k_{ij} = 0$, dotted lines) and NRHB correlations ($k_{ij} \neq 0$, solid lines). Reprinted with permission from Fluid Phase Equilibria, Modeling the vapor–liquid equilibria of polymer–solvent mixtures: Systems with complex hydrogen bonding behavior by Ioannis Tsivintzelis and Georgios M. Kontogeorgis, 280, 1–2, 100–109 Copyright (2009) Elsevier

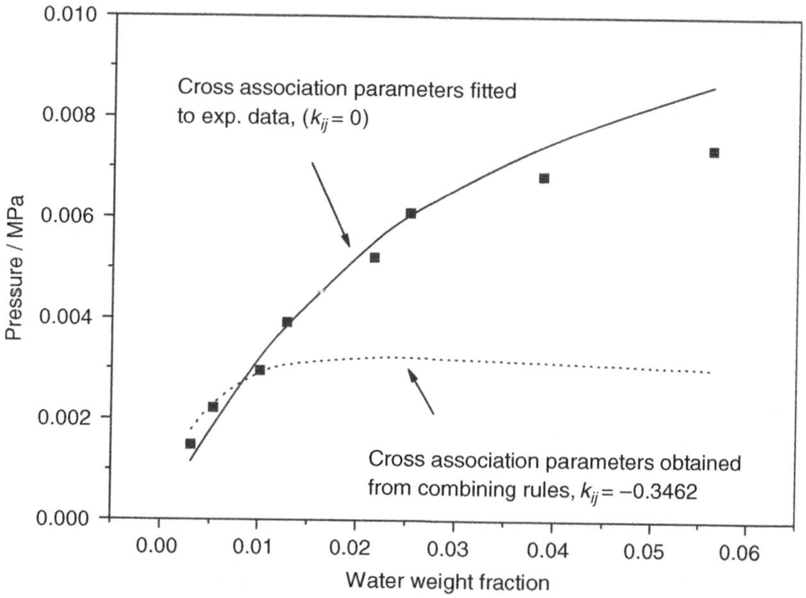

Figure 19.26 Water–PVAC VLE. Experimental data (points) and NRHB calculations using parameters for the cross-association obtained from the combining rules of Equation (19.3) (dotted line) and by fitting the predictions of the theory to the experimental data (solid line). Reprinted with permission from Fluid Phase Equilibria, Modeling the vapor–liquid equilibria of polymer–solvent mixtures: Systems with complex hydrogen bonding behavior by Ioannis Tsivintzelis and Georgios M. Kontogeorgis, 280, 1–2, 100–109 Copyright (2009) Elsevier

- Recently,[65,68,69] a systematic comparison between NRHB and SAFT models (CK–SAFT and sPC–SAFT) has been carried out. This comparison considered pure compound properties, monomer fractions for associating fluids, as well as numerous binary VLE and LLE systems covering all ranges of complexity and polarity, including aqueous mixtures with hydrocarbons and alcohols. Overall, NRHB compares favorably with SAFT, and while the models each reveal their strengths and weaknesses, on average they perform similarly (see also Figures 13.9 and 13.13). NRHB performs similarly to PC–SAFT also in other respects, e.g. the necessity of explicitly accounting for induced association (solvation) in polar-associating mixtures including water–aromatic hydrocarbons. Figure 19.23 shows a typical result with NRHB for TEG–methane.
- NRHB has not as yet been systematically applied to polymer mixtures (unlike other lattice–fluid models) but the first results are promising, as can be seen for three systems in Figures 19.24–19.26. Equally successful correlations are obtained for LLE, but the performance depends on the choice of the interaction parameters.

References

1. J. Marrero, J. Abildskov, Solubility and related properties of large complex chemicals. Part 1: Organic solutes ranging from C4 to C40. *Chemistry Data Series XV*, DECHEMA, Frankfurt/Main, 2003.
2. T.C. Frank, J.R. Downey, S.K. Gupta, *Chem. Eng. Prog.*, **1999**, *95*(*12*), 41.
3. C.C. Chen, P.M. Mathias, *AIChE J.*, **2002**, *48*(2), 194.
4. P. Kolar, J.-W. Shen, A. Tsuboi, T. Ishikawa, *Fluid Phase Equilib.*, **2002**, *194–197*, 771.
5. P. Crafts, The role of solubility modeling and crystallization in the design of active pharmaceutical ingredients. In: N.M. Ng, R. Gani, K. Dam-Johansen, Eds, *Chemical Product Design: Toward a perspective through case studies.* Elsevier, 2007, Chapter 2.
6. P.M. Mathias, *Fluid Phase Equilib.*, **2005**, *228*, 49.
7. A. Klamt, *COSMO-RS: From Quantum Chemistry to Fluid Phase Thermodynamics and Drug Design*. Elsevier, 2005.
8. U. von Stockar, L.A.M. van der Wielen, *J. Biotechnol.*, **1997**, *59*, 25.
9. U. von Stockar, L.A.M. van der Wielen, *Adv. Biochem. Eng./Biotechnol.*, **2003**, *80*, 1.
10. R. Gani, J. Abildskov, G.M. Kontogeorgis, Application of property models in chemical product design. In: G.M. Kontogeorgis, R. Gani, *Computer-Aided Property Estimation for Process and Product Design*. Elsevier, 2004.
11. R.M. Gupta, R.A. Heidemann, *AIChE J.*, **1990**, *36*, 333.
12. J. Abildskov, J.P. O'Connell, *Ind. Eng. Chem. Res.*, **2003**, *42*, 5622.
13. J. Abildskov, J.P. O'Connell, *Mol. Simulation*, **2004**, *30*(*6*), 367.
14. J. Abildskov, J.P. O'Connell, *Fluid Phase Equilib.*, **2005**, *228–229*, 395.
15. H. Modarresi, E. Conte, J. Abildskov, R. Gani, P. Crafts, *Ind. Eng. Chem. Res.*, **2008**, *47*, 5234.
16. Th. Lindvig, M.L. Michelsen, G.M. Kontogeorgis, *Fluid Phase Equilib.*, **2002**, *203*(*1–2*), 247.
17. C.A.S. Bergstrom, U. Norinder, K. Luthman, P. Artursson, *Pharm. Res.*, **2002**, *19*(2), 182.
18. J.M. Prausnitz, R.N. Lichtenthaler, E.G. de Azevedo, *Molecular Thermodynamics of Fluid-Phase Equilibria* (3rd edition). Prentice Hall International, 1999.
19. N.T. Hansen, I. Kouskoumvekaki, F.S. Jorgensen, S. Brunak, S.O. Jonsdottir, *J. Chem. Inf. Model.*, **2006**, *46*(*6*), 2601.
20. A. Klamt, F. Eckert, M. Hornig, M.E. Beck, Th. Burger, *J. Comput. Chem.*, **2002**, *23*(2), 275.
21. A. Klamt, F. Eckert, M. Hornig, *J. Comput.-Aided Mol. Des.*, **2001**, *15*, 355.
22. H.-H. Tung, J. Tabora, N. Variankaval, D. Bakken, C.-C. Chen, *J. Pharm. Sci.*, **2008**, *97*(*5*), 1813.
23. C.C. Chen, Y. Song, *Ind. Eng. Chem. Res.*, **2004**, *43*, 8354.
24. C.C. Chen, P.A. Crafts, *Ind. Eng. Chem. Res.*, **2006**, *45*, 4816.
25. F.L. Mota, A.J. Queimade, S.P. Pinho, E.A. Macedo, Solubilities of some pharmaceuticals compounds in water. Proceedings of the ESAT Conference, Cannes, France, 2008, pp. 106–110.

26. I. Tsivintzelis, I.G. Economou, G.M. Kontogeorgis, *AIChE J.*, **2009**, *55(3)*, 756.
27. I. Tsivintzelis, I.G. Economou, G.M. Kontogeorgis, Modeling the phase behavior in mixtures of pharmaceuticals with liquid or supercritical solvents *J. Phys. Chem. B*, **2009**, *113(18)*, 6446.
28. C.C. Chen, *Fluid Phase Equilib.*, **1993**, *83*, 301.
29. L.F. Cameretti, G. Sadowski, *Chem. Eng. Process.*, **2008**, *47*, 1018.
30. C.J. Orella, D.J. Kirwan, *Biotechnol. Prog.*, **1989**, *5*, 89.
31. C.J. Orella, D.J. Kirwan, *Ind. Eng. Chem. Res.*, **1991**, *30*, 1040.
32. M.T. Gude, H.H.J. Meuwissen, L.A.M. van der Wielsen, K.Ch.A.M. Luyben, *Ind. Eng. Chem. Res.*, **1996**, *35*, 4700.
33. M.T. Gude, L.A.M. van der Wielen, K.Ch.A.M. Luyben, *Fluid Phase Equilib.*, **1996**, *116*, 110.
34. K.K. Nass, *AIChE J.*, **1988**, *34*, 1257.
35. C.C. Chen, Y. Zhu, L.B. Evans, *Biotechnol. Prog.*, **1989**, *5*, 111.
36. A.M. Peres, E.A. Macedo, *Chem. Eng. Sci.*, **1994**, *49*, 3803.
37. S.P. Pinho, C.M. Silva, E.A. Macedo, *Ind. Eng. Chem. Res.*, **1994**, *33*, 1341.
38. X. Xu, S.P. Pinho, E.A. Macedo, *Ind. Eng. Chem. Res.*, **2004**, *43*, 3200.
39. L.A. Ferreira, E.A. Macedo, S.P. Pinho, *Chem. Eng. Sci.*, **2004**, *59*, 3117.
40. L.A. Ferreira, E.A. Macedo, S.P. Pinho, *Ind. Eng. Chem. Res.*, **2005**, *44*, 8892.
41. M.P. Breil, J. Mollerup, Modelling of salt activities in aqueous amino acid solutions: I. The UNIQUAC model., SEP 0437, Internal Report, Department of Chemical and Biochemical Engineering, Technical University of Denmark, 2004.
42. M.P. Breil, J. Mollerup, Modelling of salt activities in aqueous amino acid solutions: II. The Kirkwood theory. SEP 0438, Internal Report, Department of Chemical and Biochemical Engineering, Technical University of Denmark, 2004.
43. D. Fuchs, J. Fischer, F. Tumakaka, G. Sadowski, *Ind. Eng. Chem. Res.*, **2006**, *45*, 6578.
44. C. Held, L.F. Cameretti, G. Sadowski, Thermodynamic properties of aqueous electrolyte/amino acid-solutions. Presented at the ESAT Conference, Cannes, France, 2008.
45. E.A. Macedo, *Pure Appl. Chem.*, **2005**, *77*, 559.
46. M.K. Khoshkbarchi, J.H. Vera, *Ind. Eng. Chem. Res.*, **1996**, *35*, 4319.
47. G. Scatchard, J. Kirkwood, *Phys. Z.*, **1932**, *33*, 297.
48. J. Kirkwood, *J. Chem. Phys.*, **1934**, *2*, 351.
49. J.M. Mollerup, *Fluid Phase Equilib.*, **2006**, *241*, 205.
50. J.M. Mollerup, *J. Biotechnol.*, **2007**, *132*, 187.
51. J.M. Mollerup, *Fluid Phase Equilib.*, **2007**, *261*, 133.
52. J.M. Mollerup, *Chem. Eng. Technol.*, **2008**, *31*, 864.
53. J.M. Mollerup, T.B. Hansen, S. Kidal, A. Staby, *J. Chromatogr. A*, **2008**, *1177*, 200.
54. S. Gupta, J.D. Olson, *Ind. Eng. Chem. Res.*, **2003**, *42*, 6359.
55. S. Ruppert, S.I. Sandler, A.M. Lenhoff, *Biotechnol. Prog.*, **2001**, *17*, 182.
56. J.M. Prausnitz, *Fluid Phase Equilib.*, **1996**, *116*, 12.
57. J.M. Woll, T.A. Hatton, *Bioprocess Eng.*, **1989**, *4*, 193.
58. P.A. Albertsson, *Partition of Cell Particles and Macromolecules* (3rd edition). Wiley Interscience, 1986.
59. R.S. King, H.W. Blanch, J.M. Prausnitz, *AIChE J.*, **1988**, *34(10)*, 1585.
60. C.A. Haynes, F.J. Benitez, H.W. Blanch, J.M. Prausnitz, *AIChE J.*, **1993**, *39(9)*, 1539.
61. C.A. Haynes, H.W. Blanch, J.M. Prausnitz, *Fluid Phase Equilib.*, **1989**, *53*, 463.
62. C.A. Haynes, J. Carson, H.W. Blanch, J.M. Prausnitz, *AIChE J.*, **1991**, *37(9)*, 1401.
63. C. Panayiotou, M. Pantoula, E. Stefanis, I. Tsivintzelis, I.G. Economou, *Ind. Eng. Chem. Res.*, **2004**, *43*, 6592.
64. C. Panayiotou, I. Tsivintzelis, I.G. Economou, *Ind. Eng. Chem. Res.*, **2007**, *46*, 2628.
65. I. Tsivintzelis, T. Spyriouni, I.G. Economou, *Fluid Phase Equilib.*, **2007**, *253*, 19.
66. E. Stefanis, L. Constantinou, I. Tsivintzelis, C. Panayiotou, *Int. J. Thermophys.*, **2005**, *26*, 1369.
67. D. Missopolinou, I. Tsivintzelis, C. Panayiotou, *Fluid Phase Equilib.*, **2006**, *245*, 89.
68. I. Tsivintzelis, A. Grenner, G.M. Kontogeorgis, I.G. Economou, C. Panayiotou, *Ind. Eng. Chem. Res.*, **2008**, *47(15)*, 5651.

69. A. Grenner, I. Tsivintzelis, G.M. Kontogeorgis, I.G. Economou, C. Panayiotou, *Ind. Eng. Chem. Res.*, **2008**, *47*(*15*), 5636.
70. W.L. Jorgensen, E.M. Duffy, *Adv. Drug Delivery Rev.*, **2002**, *54*, 355.
71. H.-M. Lin, R.A. Nash, *J. Pharm. Sci.*, **1993**, *82*(*10*), 1018.
72. S. Gracin, T. Brinck, A.C. Rasmuson, *Ind. Eng. Chem. Res.*, **2002**, *41*, 5114.
73. I. Kouskoumvekaki, N.T. Hansen, F. Bjorkling, S.M. Vadlamudi, S.O. Jonsdottir, *SAR QSAR Environ Res.*, **2008**, *19*(*1–2*), 167.
74. I. Tsivintzelis, G.M. Kontogeorgis, *Fluid Phase Equilib.*, **2009**, *280*, 100.
75. F. Ruether, G. Sadowski, *J. Pharmaceutical Sciences*, **2009**, *98*, 4205.

20

Epilogue: Thermodynamic Challenges in the Twenty-First Century

20.1 In brief

The chemical and other engineering disciplines have undergone dramatic changes during the last few decades and this trend is expected to continue. In answer to societal and industrial needs, we can anticipate significant challenges in thermodynamics in order to meet the needs emerging due to these changes. Future developments are needed or expected especially in the following sectors:

- Oil and gas industry, especially towards environmentally oriented and safe operations and productions.
- Novel environmentally friendly separations without hazardous solvents, using solvents such as ionic liquids and supercritical fluids as well as processes for capturing CO_2 after the combustion of coal and other fuels.
- Green engineering in broader terms.
- Life sciences and biotechnology.
- Applied nanotechnology, materials science and complex product design towards knowledge-based materials.

The references at the end of this chapter (from academia and industry) discuss some of these trends as well as some of the new modeling approaches that may be required. Table 20.1 summarizes some of the anticipated challenges for various industrial sectors, while Table 20.2 presents an overview of the articles cited in the references, divided according to industrial sector. The discussion which follows in the remainder of the chapter is categorized according to the various industrial sectors where major changes and thermodynamic needs are expected.

Thermodynamic Models for Industrial Applications: From Classical and Advanced Mixing Rules to Association Theories
Georgios M. Kontogeorgis and Georgios K. Folas

Table 20.1 Some thermodynamic challenges according to industrial sector

Industrial sector	Challenges
Oil and gas	Multicomponent, multiphase equilibria (VLLE, SLE) when associating and/or polar molecules are present Solids in a unified way (gas hydrates–asphaltenes–wax) Environmental distribution of chemicals Hydrate kinetics
Chemicals including polymers	Reacting systems for CO_2 removal (e.g. CO_2–water–alkanolamines) New polymer structures, e.g. dendrimers, hydrogels, etc. Simultaneous description of electrolyte and association phenomena Water-based paints, polymer membranes Supercritical fluid extraction and supercritical water oxidation including co-solvent choice
Biotechnology including pharmaceuticals	Controlled drug release – biomedical engineering Combined description of electrolyte and association effects together with biomolecules, e.g. aqueous two-phase systems Protein aggregation and precipitation
Food	Communication with colloid chemists and understanding of surface phenomena Metastability and emulsions Microemulsions Surfactants–micelles Thermodynamics and nanotechnology

20.2 Petroleum and chemical industries

Almost 20 years ago Zeck[4] wrote:

> *There is still considerable potential for improvement for phase equilibrium thermodynamics even in long-established areas of the chemical industry. With all the enthusiasm about the possibilities of thermodynamics in new areas, it is necessary first to concentrate on performing the basic tasks.*

Challenges related to complex distillations, adsorption or extraction are what Zeck refers to as 'basic tasks'. Today, we feel that, despite great progress, there are still thermodynamic challenges in these 'most mature' areas of chemical engineering:

- The prediction of multicomponent, multiphase equilibria, based exclusively on binary parameters, e.g. for partially immiscible mixtures such as water–alcohols (glycols)–hydrocarbons.
- The simultaneous prediction of different types of phase equilibria (vapor–liquid, liquid–liquid and solid–liquid equilibria) over extended temperature ranges with a common set of parameters.
- The need for highly predictive models for a wide range of applications.
- The prediction of thermal properties such as heats of mixing and heat capacities at different conditions based on parameters estimated from phase equilibrium data.
- The prediction of a variety of properties from equations of state, e.g. densities, entropies, speed of sound, heat capacities, etc., important for fluid flow calculations.

Table 20.2 *Reviews on thermodynamic challenges per (industrial) sector*

(Industrial) sector	References from academia	References from industry
Oil and gas		Tsonopoulos and Heidman[1]
Chemical industry including polymers	Prausnitz[2,3] Assael[5] Agarwal *et al.*[56] Harvey and Laesecke[6] Prausnitz and Tavares[7] Arlt *et al.*[8]	Zeck[4] Zeck and Wolf[9] Carlson[10] Rhodes[11] Daubert[12] Bokis *et al.*[13] (polymers) Chen and Mathias[14] Dohrn and Pfohl[15] Gupta and Olson[16] Mathias[17]
Pharmaceuticals and agrochemicals	Teja and Eckert[18] Abildskov and O'Connell[19–21]	Larson and King[22] Franck *et al.*[23] Sangster[24] Perrut[25] Kolar *et al.*[26]
(Complex) product design	Villadsen[27] Favre and Kind[28] Prausnitz[29] Abildskov and Kontogeorgis[30]	Bruin[31] (food)
Biotechnology	Prausnitz[32,3,33] Stockar and van der Wielen[34,35] Blanch *et al.*[36] Mollerup[37]	
Environmental applications	Sandler[38,39] Sandler and Orbey[40] Prausnitz[41]	
Quantum chemistry	Sandler[42] Prausnitz and Tavares[7] Sandler and Castier[43]	Klamt *et al.*[44,45]
Several industrial sectors	Gubbins[46] Sandler[47] Whiting[48] Marsh[49] Rainwater *et al.*[50] NIST 975[51]	Wilson[52] Thomson and Larsen[53] Mathias and Klotz[54] Moorwod[55] Chen and Mathias[14] Gupta and Olson[16]

- The simultaneous calculation of physical and chemical equilibria, e.g. the absorption of CO_2 or H_2S in aqueous alkanolamine solutions, required for CO_2 and H_2S removal using a minimum amount of data or interaction parameters.
- Calculations in the presence of multiple solid phases such as waxes, asphaltenes and gas hydrates.

Tsonopoulos and Heidman[1] stated nearly 25 years ago that cubic equations of state are largely adequate for the needs of the oil and gas industry. Cubic EoS are indeed reliable high-pressure models for mixtures containing gases and hydrocarbons, but even when advanced mixing rules are used they often cannot provide good descriptions of multicomponent LLE of water–alcohol–hydrocarbon systems or other liquid phase properties such as enthalpies and heat capacities, just to mention a few examples. Difficult separations, such as those involving azeotropic mixtures or close boiling mixtures, require additional accuracy, which may not be possible with simple cubic EoS. We feel that many classical models (cubic EoS, local composition/group contribution methods such as UNIFAC) may have reached their limit of application, unless process-specific parameters are used, and much promise lies in the advanced association EoS (CPA, SAFT) discussed in the third part of the book (Chapters 7–14).

20.3 Chemicals including polymers and complex product design

In the development of 'new materials and chemical products', we foresee substantial challenges for thermodynamics – especially in combination with other fields, e.g. transport phenomena, mathematical modeling and kinetics. A highly multidisciplinary effort is required here. Advanced chemical products often include specialty chemicals, microstructured materials and colloidal formulations and are typically multicomponent systems containing polymers, surfactants, solid particles, as well as water or other solvents. Multiple solid–liquid phases exist, often in (non-equilibrium) metastable states (emulsions). The new products range from pharmaceuticals and paints to food, detergents and chemical devices, to mention just a few. Physical chemistry (thermodynamics, stability, interfacial science) is crucial for the design of various non-traditional manufacturing/separation processes, e.g. emulsification, foaming, gelation, granulation and crystallization. Future theories and models should meet these challenges towards an understanding of:

- The behavior of structured polymers such as dendrimers, hyperbranched polymers, inorganic polymers used in antifouling paints, new complex blends and co-polymers. Thermodynamics and transport phenomena are often equally important (swelling, diffusivity and permeability of gases), e.g. in the design of hydrogels and other polymers used for controlled drug release of pharmaceuticals.
- The multiple simultaneous interactions in systems containing polymers, electrolytes and mixed solvents. The existing models have been developed for each subsystem independently, e.g. only for polymeric or electrolyte mixtures, and such independent developments should be abandoned if future models must have wide applicability.[14]
- Surface phenomena that are crucial in colloidal systems require joint developments of advanced models like SAFT (Statistical Association Fluid Theory) together with separate frameworks such as the DFT (Density Functional Theory) or the gradient theory.
- Multicomponent, multiphase systems used in many consumer products, e.g. shampoos or paints and coatings. These applications range from the description of water–oil–surfactant-*co*-surfactant mixtures of importance to microemulsions up to the complex paint formulations containing mixed solvents. Multiple liquid and/or solid phases are present.
- Non-equilibrium phenomena (metastability) of emulsions.
- Complex interactions, polymorphism and isomeric effects in pharmaceuticals and agrochemicals, of importance in solvent selection and for environmental purposes – these problems account for up to 30% of the thermodynamic work in these fields.

20.4 Biotechnology including pharmaceuticals

The role of thermodynamics in 'life sciences and biotechnology' is under continuous discussion and special courses in biothermodynamics have been organized.[34,35] A short introduction to current models with applications in biotechnology was presented in Chapter 19. The challenges are numerous and just a few of them are summarized here:

- Estimation of the properties of pure complex molecules (e.g. penicillin) for use in thermodynamic calculations.
- Solvent selection for pharmaceuticals, which are compounds with numerous heteroatoms and complex functional groups, thus often more complex than most 'ordinary' chemicals.
- Models to describe charged macromolecules and systems containing biomolecules (e.g. water–polymers–electrolytes-proteins) are important in the separation of proteins (e.g. aqueous two-phase systems).
- Phase separation/agglomeration phenomena in protein systems are of interest in understanding certain diseases such as cataract and Alzheimer's[33].
- Computer-aided drug design using molecular modeling techniques and quantum mechanics as well as QSAR-type property (activity) models.

Table 20.3 provides an overview of some major applications and challenges for thermodynamics in biotechnology illustrated in several references, while the thermodynamics needed and some comments are included in the last two columns.

Table 20.3 *Major challenges and applications of thermodynamics in biotechnology*

Application	References	Thermodynamics needed	Comments/status
Controlled drug release – drug delivery via polymeric devices	Prausnitz[29]	Solubility of pharmaceuticals in polymers (SLE)	Michaels *et al.*[58] model (FH + diffusion) with limited success for a few pharmaceuticals
Solvent screening for pharmaceuticals (separation and recovery of amino acids and antibiotics)	Franck *et al.*[23] Kolar *et al.*[26] Abildskov *and* O'Connell[19–21] Stockar *et al.*[34] Prausnitz *et al.*[57]	SLE (drug solubility in liquids)	QSAR of various types[59,60] Classical models (HSP, UNIFAC) – capabilities but also limitations Hybrid models: NRTL–SAC from ASPEN Quantum chemistry: COSMO–RS New theoretically oriented EoS, e.g. PC–SAFT
Novel separations for pharmaceuticals and other biomolecules – SCFE	Larson and King[22] Prausnitz *et al.*[57]	Solubility of biomolecules in CO_2–co-solvent (SGE)	Limited modeling success for complex molecules and co-solvents
Novel separations for proteins – reverse micelles	Prausnitz[32]		

(*continued*)

Table 20.3 *(Continued)*

Application	References	Thermodynamics needed	Comments/status
Understanding of protein–protein interactions in solutions	Prausnitz *et al.*[57] Gupta and Olson[16] Sandler[38]	Osmotic second virial coefficient – as a function of solubility and link to intermolecular potential	Obtained via osmometry or LALS
Understanding of the mechanisms of certain diseases (cataract, Alzheimer's, sickle-cell anemia)	Prausnitz[33] Galkin *et al.*[61] Pande *et al.*[62]	Protein aggregation and complexing (SLE, LLE)	Thermodynamics and colloid theory (DLVO)
Polyelectrolyte hydrogels for controlled release devices, contact lenses, etc.	Prausnitz[32]	Swelling	
'Classical' separation methods for proteins, e.g. protein precipitation from aqueous solutions using salts and selective separation	Prausnitz[2,29] Sandler[38] Chen and Mathias[14]	SLE, LLE of water–protein–salt as a function of numerous parameters (temperature, salt–protein concentration, ionic strength, pH, etc.)	Various models available of unknown predictive value
ATPS for protein separation	Prausnitz[32] Prausnitz and Tavares[7] Prausnitz *et al.*[57] Albertsson *et al.*[63] King *et al.*[64]	LLE of water–protein–salt (s)–polymer (s), e.g. PEG or Dextran	

SLE = solid–liquid equilibria, SGE = solid–gas equilibria, LLE = liquid–liquid equilibria, ATPS = aqueous two-phase systems, LALS = low-angle light scattering, PC–SAFT = perturbed chain statistical association fluid equation of state, SCFE = supercritical fluid extraction, FH = Flory–Huggins, HSP = Hansen solubility parameters, QSAR = quantitative structure activity relationships, PEG = (poly)ethylene glycol, UNIFAC = universal quasi-chemical functional group activity coefficient.

20.5 How future needs will be addressed

Traditional models have reached their limits in most cases, although a number of specially designed models from the traditional domain may find applications even in some of the most advanced fields. Among these models can be mentioned:

- Specially designed extensions of group contribution methods, e.g. UNIFAC, with parameters obtained from specific system data or molecular connectivities[65] and other predictive approaches.
- Solubility parameter techniques and theoretically based QSAR, e.g. for estimating the distribution coefficients of biomolecules in aqueous two-phase systems with polymers and salts.
- Semi-empirical tools in the biotechnology industry, e.g. the use of osmotic second virial coefficients.

For more demanding applications much is expected from the existing models/theories based on statistical mechanics (SAFT, CPA), especially those based on the Wertheim framework, as well as their anticipated future developments and extensions. This further development should include understanding and incorporation of 'delicate' aspects of hydrogen bonding (cooperativity, intramolecular association, multisite interactions, complex solvation, charged effects) including coupling of these theories with spectroscopy or quantum chemistry in order to obtain the necessary parameters. Additional measurements of phase equilibria are still required for many industrially important compounds such as glycolethers and alkanolamines.[51] It must be emphasized that the traditional and advanced (based on statistical mechanics) thermodynamic models discussed in this book are not the only tools which may make a difference in future thermodynamics. Two alternative approaches are offered by molecular simulation and quantum chemistry.

Molecular simulation techniques, especially as computer power increases and knowledge of intermolecular forces is enhanced, are expected to make a considerable impact in meeting future needs in thermodynamics. Such methods may not always be of the same accuracy as actual measurements (or even some of the models). They can, however, help in understanding complex phenomena (e.g. the nature of the critical or the azeotropic state) and they are also useful in cases where measurements are impossible or difficult because of regulations (sulfur compounds) or other reasons. The molecular simulation methods can also be used to distinguish between conflicting data, as was the case with the chain length dependency of the critical density of *n*-alkanes.[51,66] In some fields such as chemical product design it is important to define the product needs and functions through a set of properties, which are a combination of microscopic (atomic-structure-based) and macroscopic (molecular-structure-based) properties. Traditional models do not always satisfactorily capture molecular structural differences, e.g. between isomers, and molecular simulation techniques offer a possible solution in this case. The methods of quantum mechanics or the tools derived from it[42,44,45] are also expected to find widespread use in the future. As discussed briefly in Chapter 16, a number of variations of these approaches are already available. Many researchers believe that in the future these quantum-chemistry-based models may result in reliable tools for phase equilibria calculations of complex fluids.

References

1. C. Tsonopoulos, J. Heidman, *Fluid Phase Equilib.*, **1986**, *29*, 391.
2. J.M. Prausnitz, *Fluid Phase Equilib.*, **1995**, *104*, 1
3. J.M. Prausnitz, *Fluid Phase Equilib.*, **1996**, *116*, 12.
4. S. Zeck, *Fluid Phase Equilib.*, **1991**, *70(2–3)*, 125.
5. M.J. Assael, *High Temp. –High Pressures*, **2000**, *32*, 159.
6. A.H. Harvey, A. Laesecke, *Chem. Eng. Prog.*, **2002**, 34.
7. J.M. Prausnitz, F.W. Tavares, *AIChE J.*, **2004**, *50(4)*, 739.
8. W. Arlt, O. Spuhl, A. Klamt, *Chem. Eng. Process.*, **2004**, *43*, 221.
9. S. Zeck, D. Wolf, *Fluid Phase Equilib.*, **1993**, *82*, 27.
10. E.C. Carlson, *Chem. Eng. Progr.*, **1996**, 35.
11. C.L. Rhodes, *J. Chem. Eng. Data*, **1996**, *41*, 947.
12. T.E. Daubert, *J. Chem. Eng. Data*, **1996**, *41*, 942.
13. C.P. Bokis, H. Orbey, C.C. Chen, *Chem. Eng. Prog.*, **1999**, 39.
14. C.C. Chen, P.M. Mathias, *AIChE J.*, **2002**, *48*(2), 194.
15. R. Dohrn, O. Pfohl, *Fluid Phase Equilib.*, **2002**, *194–197*, 15.
16. S. Gupta, J.D. Olson, *Ind. Eng. Chem. Res.*, **2003**, *42*, 6359.
17. P.M. Mathias, *Fluid Phase Equilib.*, **2005**, *228*, 49.

18. A. Teja, C.A. Eckert, *Ind. Eng. Chem. Res.*, **2000**, *39*, 4442.
19. J. Abildskov, J.P. O'Connell, *Ind. Eng. Chem. Res.*, **2003**, *42*, 5622.
20. J. Abildskov, J.P. O'Connell, *Mol. Simulation*, **2004**, *30(6)*, 367.
21. J. Abildskov, J.P. O'Connell, *Fluid Phase Equilib.*, **2005**, *228–229*, 395.
22. K.A. Larson, M.L. King, *Biotechnol. Prog.*, **1986**, *2*(2), 73.
23. T.C. Frank, J.R. Downey, S.K. Gupta, *Chem. Eng. Progr.*, **1999**, 41.
24. J. Sangster, *J. Phys. Chem. Ref. Data*, **1999**, *28(4)*, 889.
25. M. Perrut, *Ind. Eng. Chem. Res.*, **2000**, *39*, 4531.
26. P. Kolar, J.-W. Shen, A. Tsuboi, T. Ishikawa, *Fluid Phase Equilib.*, **2002**, *194–197*, 771.
27. J. Villadsen, *Chem. Eng. Sci.*, **1997**, *52*, 2857.
28. E. Favre, M. Kind, Formulation engineering: towards a multidisciplinary and integrated approach of the training of chemical engineers. Paper presented at the 2nd European Congress of Chemical Engineering, Montpellier, France, 1999.
29. J.M. Prausnitz, *Fluid Phase Equilib.*, **1999**, *158–160*, 95.
30. J. Abildskov, G.M. Kontogeorgis, *Chem. Eng. Res. Des.*, **2004**, *82(11)*, 1505.
31. S. Bruin, *Fluid Phase Equilib.*, **1999**, *158–160*, 657.
32. J.M. Prausnitz, *Fluid Phase Equilib.*, **1989**, *53*, 439.
33. J.M. Prausnitz, *J. Chem. Thermodyn.*, **2003**, *35*, 21.
34. U. von Stockar, L.A.M. van der Wielen, *J. Biotechnol.*, **1997**, *59*, 25.
35. U. von Stockar, L.A.M. van der Wielen, *Adv. Biochem. Eng./Biotechnol.*, **2003**, *80*, 1.
36. H.W. Blanch, J.M. Prausnitz, R.A. Curtis, D. Bratko, *Fluid Phase Equilib.*, **2002**, *194–197*, 31.
37. J.M. Mollerup, *Fluid Phase Equilib.*, **2006**, *241*, 205.
38. S.I. Sandler, *Fluid Phase Equilib.*, **1996**, *116*, 343.
39. S.I. Sandler, *J. Chem. Thermodyn.*, **1999**, *31*, 3.
40. S.I. Sandler, H. Orbey, *Fluid Phase Equilib.*, **1993**, *82*, 63.
41. J.M. Prausnitz, *Chem. Eng. Sci.*, **2001**, *56*, 3627.
42. S.I. Sandler, *Fluid Phase Equilib.*, **2003**, *210*, 147.
43. S.I. Sandler, M. Castier, *Pure Appl. Chem.*, **2007**, *79(8)*, 1345.
44. A. Klamt, F. Eckert, M. Hornig, *J. Comput.-Aided Mol. Des.*, **2001**, *15*, 355.
45. A. Klamt, F. Eckert, M. Hornig, M.E. Beck, T. Burger, *J. Comput. Chem.*, **2002**, *23*(2), 275.
46. K.E. Gubbins, *Chem. Eng. Prog.*, **1989**, *85*(2), 38.
47. S.I. Sandler, *Int. J. Thermophys.*, **1994**, *15*, 1013.
48. W.B. Whiting, *J. Chem. Eng. Data*, **1996**, *41*, 935.
49. K.N. Marsh, *J. Chem. Eng. Data*, **2001**, *46*, 1.
50. J.C. Rainwater, D.G. Friend, H.J.M. Hanley, A.H. Harvey, C.D. Holcomb, A. Laesecke, J.W. Magee, C. Muzny, *J. Chem. Eng. Data*, **2001**, *46*, 1002.
51. NIST Special Publication 975 (Rainwater *et al.*, 2001), Report on Forum 2000: Fluid properties for new technologies – connecting virtual design with physical reality (http://Forum2000.Boulder.NIST.Gov/NISTSP975.pdf).
52. G.M. Wilson, *Fluid Phase Equilib.*, **1996**, *116*, 1.
53. G.H. Thomson, A.H. Larsen, *J. Chem. Eng. Data*, **1996**, *41*, 930.
54. P.M. Mathias, H.C. Klotz, *Chem. Eng. Progr.*, **1994**, *90(6)*, 67.
55. T. Moorwood, CAPE tools and techniques for the 21st century. Properties of Materials and Mixtures – Where do we need to be 10 years from now? Report for Eureka Project 3211, **2001**.
56. R. Agarwal, Y.-K. Li, O. Santollani, M.A. Satyro, A. Vieler, *Chem. Eng. Prog.*, **2001**, *97*(5), 42; **2001**, *97(6)*, 64.
57. J.M. Prausnitz, R.N. Lichtenthaler, E.G. de Azevedo, *Molecular Thermodynamics of Fluid-Phase Equilibria* (3rd edition). Prentice Hall International, 1999.
58. A.S. Michaels, P.S.L. Wong, R. Prather, R.M. Gale, *AIChE J.*, **1975**, *21(6)*, 1073.
59. C.A.S. Bergstrom, U. Norinder, K. Luthman, P. Artursson, *Pharm. Res.*, **2002**, *19*(2), 182.
60. W.L. Jorgensen, E.M. Duffy, *Adv. Drug Delivery Rev.*, **2002**, *54*, 355.

61. O. Galkin, K. Chen, R.L. Nagel, R.E. Hirsch, P.G. Vekilov, *Proc. Natl Acad. Sci. USA*, **2002**, *99(13)*, 8479.
62. A. Pande, J. Pande, N. Asherie, A. Lomakin, O. Ogun, J. King, G.B. Benedek, *Proc. Natl Acad. Sci. USA*, **2001**, *98(11)*, 6116.
63. P.A. Albertsson, G. Johansson, F. Tjerneld, Aqueous two-phase separations. In: J.A. Asenjo,Ed., *Separation Processes in Biotechnology*. Marcel Decker, **1990**, Chapter 10.
64. R.S. King, H.W. Blanch, J.M. Prausnitz, *AIChE J.*, **1988**, *34(10)*, 1585.
65. H.E. Gonzalez, J. Abildskov, R. Gani, P. Rousseaux, B. Le Bert, *AIChE J.*, **2007**, *53(6)*, 1620.
66. I.J. Siepmann, S. Karaborni, B. Smit, *Nature*, **1993**, *365(6444)*, 330.

Index

Note: Page numbers in *italics* refer to figures; those in **bold** to tables.

Thermodynamic Models for Industrial Applications: From Classical and Advanced Mixing Rules to Association Theories
Georgios M. Kontogeorgis and Georgios K. Folas

Printed and bound in the UK by
CPI Antony Rowe, Eastbourne

Printed and bound by CPI Group (UK) Ltd, Croydon, CR0 4YY
16/09/2026
14963481-0001